Lecture Notes in Computer Science 7794

Advanced Research in Computing and Software Science

Subline of Lectures Notes in Computer Science

Frank Pfenning (Ed.)

Foundations of Software Science and Computation Structures

16th International Conference, FOSSACS 2013
Held as Part of the European Joint Conferences
on Theory and Practice of Software, ETAPS 2013
Rome, Italy, March 16-24, 2013
Proceedings

Springer

Volume Editor

Frank Pfenning
Carnegie Mellon University
School of Computer Science
Pittsburgh, PA 15213-3891, USA
E-mail: fp@cs.cmu.edu

ISSN 0302-9743
e-ISSN 1611-3349
ISBN 978-3-642-37074-8
e-ISBN 978-3-642-37075-5
DOI 10.1007/978-3-642-37075-5
Springer Heidelberg Dordrecht London New York

Library of Congress Control Number: 2013932563

CR Subject Classification (1998): F.4.1, F.4.3, F.3.1-3, D.3.1, D.2.4, F.1.1-3, I.2.3

LNCS Sublibrary: SL 1 – Theoretical Computer Science and General Issues

Typesetting: Camera-ready by author, data conversion by Scientific Publishing Services, Chennai, India

Printed on acid-free paper

Springer is part of Springer Science+Business Media (www.springer.com)

Foreword

ETAPS 2013 is the sixteenth instance of the European Joint Conferences on Theory and Practice of Software. ETAPS is an annual federated conference that was established in 1998 by combining a number of existing and new conferences. This year it comprised six sister conferences (CC, ESOP, FASE, FOSSACS, POST, TACAS), 20 satellite workshops (ACCAT, AiSOS, BX, BYTECODE, CerCo, DICE, FESCA, GRAPHITE, GT-VMT, HAS, Hot-Spot, FSS, MBT, MEALS, MLQA, PLACES, QAPL, SR, TERMGRAPH and VSSE), three invited tutorials (*e-education*, by John Mitchell; *cyber-physical systems*, by Martin Fränzle; and *e-voting* by Rolf Küsters) and eight invited lectures (excluding those specific to the satellite events).

The six main conferences received this year 627 submissions (including 18 tool demonstration papers), 153 of which were accepted (6 tool demos), giving an overall acceptance rate just above 24%. (ETAPS 2013 also received 11 submissions to the software competition, and 10 of them resulted in short papers in the TACAS proceedings). Congratulations therefore to all the authors who made it to the final programme! I hope that most of the other authors will still have found a way to participate in this exciting event, and that you will all continue to submit to ETAPS and contribute to making it the best conference on software science and engineering.

The events that comprise ETAPS address various aspects of the system development process, including specification, design, implementation, analysis, security and improvement. The languages, methodologies and tools that support these activities are all well within its scope. Different blends of theory and practice are represented, with an inclination towards theory with a practical motivation on the one hand and soundly based practice on the other. Many of the issues involved in software design apply to systems in general, including hardware systems, and the emphasis on software is not intended to be exclusive.

ETAPS is a confederation in which each event retains its own identity, with a separate Programme Committee and proceedings. Its format is open-ended, allowing it to grow and evolve as time goes by. Contributed talks and system demonstrations are in synchronised parallel sessions, with invited lectures in plenary sessions. Two of the invited lectures are reserved for 'unifying' talks on topics of interest to the whole range of ETAPS attendees. The aim of cramming all this activity into a single one-week meeting is to create a strong magnet for academic and industrial researchers working on topics within its scope, giving them the opportunity to learn about research in related areas, and thereby to foster new and existing links between work in areas that were formerly addressed in separate meetings.

ETAPS 2013 was organised by the *Department of Computer Science of 'Sapienza' University of Rome*, in cooperation with

- ▷ European Association for Theoretical Computer Science (EATCS)
- ▷ European Association for Programming Languages and Systems (EAPLS)
- ▷ European Association of Software Science and Technology (EASST).

The organising team comprised:

General Chair: *Daniele Gorla;*

Conferences: *Francesco Parisi Presicce;*

Satellite Events: *Paolo Bottoni* and *Pietro Cenciarelli;*

Web Master: *Igor Melatti;*

Publicity: *Ivano Salvo;*

Treasurers: *Federico Mari* and *Enrico Tronci.*

Overall planning for ETAPS conferences is the responsibility of its Steering Committee, whose current membership is:

Vladimiro Sassone (Southampton, chair), Martín Abadi (Santa Cruz), Erika Ábrahám (Aachen), Roberto Amadio (Paris 7), Gilles Barthe (IMDEA-Software), David Basin (Zürich), Saddek Bensalem (Grenoble), Michael O'Boyle (Edinburgh), Giuseppe Castagna (CNRS Paris), Albert Cohen (Paris), Vittorio Cortellessa (L'Aquila), Koen De Bosschere (Gent), Ranjit Jhala (San Diego), Matthias Felleisen (Boston), Philippa Gardner (Imperial College London), Stefania Gnesi (Pisa), Andrew D. Gordon (MSR Cambridge and Edinburgh), Daniele Gorla (Rome), Klaus Havelund (JLP NASA Pasadena), Reiko Heckel (Leicester), Holger Hermanns (Saarbrücken), Joost-Pieter Katoen (Aachen), Paul Klint (Amsterdam), Jens Knoop (Vienna), Steve Kremer (Nancy), Gerald Lüttgen (Bamberg), Tiziana Margaria (Potsdam), Fabio Martinelli (Pisa), John Mitchell (Stanford), Anca Muscholl (Bordeaux), Catuscia Palamidessi (INRIA Paris), Frank Pfenning (Pittsburgh), Nir Piterman (Leicester), Arend Rensink (Twente), Don Sannella (Edinburgh), Zhong Shao (Yale), Scott A. Smolka (Stony Brook), Gabriele Taentzer (Marburg), Tarmo Uustalu (Tallinn), Dániel Varró (Budapest) and Lenore Zuck (Chicago).

The ordinary running of ETAPS is handled by its management group comprising: Vladimiro Sassone (chair), Joost-Pieter Katoen (deputy chair and publicity chair), Gerald Lüttgen (treasurer), Giuseppe Castagna (satellite events chair), Holger Hermanns (liaison with local organiser) and Gilles Barthe (industry liaison).

I would like to express here my sincere gratitude to all the people and organisations that contributed to ETAPS 2013, the Programme Committee chairs and members of the ETAPS conferences, the organisers of the satellite events, the speakers themselves, the many reviewers, all the participants, and Springer-Verlag for agreeing to publish the ETAPS proceedings in the ARCoSS subline.

Last but not least, I would like to thank the organising chair of ETAPS 2013, Daniele Gorla, and his Organising Committee, for arranging for us to have ETAPS in the most beautiful and historic city of Rome.

My thoughts today are with two special people, profoundly different for style and personality, yet profoundly similar for the love and dedication to our discipline, for the way they shaped their respective research fields, and for the admiration and respect that their work commands. Both are role-model computer scientists for us all.

ETAPS in Rome celebrates *Corrado Böhm*. Corrado turns 90 this year, and we are just so lucky to have the chance to celebrate the event in Rome, where he has worked since 1974 and established a world-renowned school of computer scientists. Corrado has been a pioneer in research on programming languages and their semantics. Back in 1951, years before FORTRAN and LISP, he defined and implemented a *metacircular compiler* for a programming language of his invention. The compiler consisted of just 114 instructions, and anticipated some modern list-processing techniques.

Yet, Corrado's claim to fame is asserted through the breakthroughs expressed by the *Böhm-Jacopini Theorem* (CACM 1966) and by the invention of *Böhm-trees*. The former states that any algorithm can be implemented using only sequencing, conditionals, and while-loops over elementary instructions. Böhm trees arose as a convenient data structure in Corrado's milestone proof of the decidability inside the λ-calculus of the equivalence of terms in β-η-normal form.

Throughout his career, Corrado showed exceptional commitment to his roles of researcher and educator, fascinating his students with his creativity, passion and curiosity in research. Everybody who has worked with him or studied under his supervision agrees that he combines an outstanding technical ability and originality of thought with great personal charm, sweetness and kindness. This is an unusual combination in problem-solvers of such a high calibre, yet another reason why we are ecstatic to celebrate him. *Happy birthday from ETAPS, Corrado!*

ETAPS in Rome also celebrates the life and work of *Kohei Honda*. Kohei passed away suddenly and prematurely on December 4th, 2012, leaving the saddest gap in our community. He was a dedicated, passionate, enthusiastic scientist and –more than that!– his enthusiasm was contagious. Kohei was one of the few theoreticians I met who really succeeded in building bridges between theoreticians and practitioners. He worked with W3C on the standardisation of web services choreography description languages (WS-CDL) and with several companies on *Savara* and *Scribble*, his own language for the description of application-level protocols among communicating systems.

Among Kohei's milestone research, I would like to mention his 1991 epoch-making paper at ECOOP (with M. Tokoro) on the treatment of asynchrony in message passing calculi, which has influenced all process calculi research since. At ETAPS 1998 he introduced (with V. Vasconcelos and M. Kubo) a new concept in type theories for communicating processes: it came to be known as '*session types*,' and has since spawned an entire research area, with practical and multi-disciplinary applications that Kohei was just starting to explore.

Kohei leaves behind him enormous impact, and a lasting legacy. He is irreplaceable, and I for one am proud to have been his colleague and glad for the opportunity to arrange for his commemoration at ETAPS 2013.

My final ETAPS '*Foreword*' seems like a good place for a short reflection on ETAPS, what it has achieved in the past few years, and what the future might have in store for it.

On April 1st, 2011 in Saarbrücken, we took a significant step towards the consolidation of ETAPS: the establishment of *ETAPS e.V.* This is a *non-profit association* founded under German law with the immediate purpose of supporting the conference and the related activities. ETAPS e.V. was required for practical reasons, e.g., the conference needed (to be represented by) a legal body to better support authors, organisers and attendees by, e.g., signing contracts with service providers such as publishers and professional meeting organisers. Our ambition is however to make of '*ETAPS the association*' more than just the organisers of '*ETAPS the conference*'. We are working towards finding a voice and developing a range of activities to support our scientific community, in cooperation with the relevant existing associations, learned societies and interest groups. The process of defining the structure, scope and strategy of ETAPS e.V. is underway, as is its first ever membership campaign. For the time being, ETAPS e.V. has started to support community-driven initiatives such as open access publications (LMCS and EPTCS) and conference management systems (Easychair), and to cooperate with cognate associations (European Forum for ICT).

After two successful runs, we continue to support POST, *Principles of Security and Trust*, as a candidate to become a permanent ETAPS conference. POST was the first addition to our main programme since 1998, when the original five conferences met together in Lisbon for the first ETAPS. POST resulted from several smaller workshops and informal gatherings, supported by IFIP WG 1.7, and combines the practically important subject of security and trust with strong technical connections to traditional ETAPS areas. POST is now attracting interest and support from prominent scientists who have accepted to serve as PC chairs, invited speakers and tutorialists. I am very happy about the decision we made to create and promote POST, and to invite it to be a part of ETAPS.

Considerable attention was recently devoted to our *internal processes* in order to streamline our procedures for appointing Programme Committees, choosing invited speakers, awarding prizes and selecting papers; to strengthen each member conference's own Steering Group, and, at the same time, to strike a balance between these and the ETAPS Steering Committee. A lot was done and a lot remains to be done.

We produced a *handbook* for local organisers and one for PC chairs. The latter sets out a code of conduct that all the people involved in the selection of papers, from PC chairs to referees, are expected to adhere to. From the point of view of the authors, we adopted a *two-phase submission* protocol, with fixed deadlines in the first week of October. We published a *confidentiality policy* to set high standards for the handling of submissions, and a *republication policy* to clarify what kind of material remains eligible for submission to ETAPS after presentation at a workshop. We started an *author rebuttal phase*, adopted by most of the conferences, to improve the author experience. It is important to acknowledge that – regardless of our best intentions and efforts – the quality of reviews is not always what we would like it to be. To remain true to our commitment to the authors who elect to submit to ETAPS, we must endeavour to improve our standards of refereeing. The rebuttal phase is a step in that direction and, according to our experience, it seems to work remarkably well at little cost, provided both authors and PC members use it for what it is. ETAPS has now reached a healthy paper acceptance rate around the 25% mark, essentially uniformly across the six conferences. This seems to me to strike an excellent balance between being selective and being inclusive, and I hope it will be possible to maintain it even if the number of submissions increases.

ETAPS signed a favourable three-year publication contract with Springer for publication in the ARCoSS subline of LNCS. This was the result of lengthy negotiations, and I consider it a good achievement for ETAPS. Yet, publication of its proceedings is possibly the hardest challenge that ETAPS – and indeed most computing conferences – currently face. I was invited to represent ETAPS at a most interesting Dagstuhl Perspective Workshop on the '*Publication Culture in Computing Research*' (seminar 12452). The paper I gave there is available online from the workshop proceedings, and illustrates three of the views I formed also thanks to my experience as chair of ETAPS, respectively on open access, bibliometrics, and the roles and relative merits of conferences versus journal publications. Open access is a key issue for a conference like ETAPS. Yet, in my view it does not follow that we can altogether dispense with publishers – be they commercial, academic, or learned societies – and with their costs. A promising way forward may be based on the '*author-pays*' model, where publications fees are kept low by resorting to learned-societies as publishers. Also, I believe it is ultimately in the interest of our community to de-emphasise the perceived value of conference publications as viable – if not altogether superior – alternatives to journals. A large and ambitious conference like ETAPS ought to be able to rely on quality open-access journals to cover its entire spectrum of interests, even if that means promoting the creation of a new journal.

Due to its size and the complexity of its programme, hosting ETAPS is an increasingly challenging task. Even though excellent candidate *locations* keep being volunteered, in the longer run it seems advisable for ETAPS to provide more support to local organisers, starting e.g., by taking direct control of the organisation of satellite events. Also, after sixteen splendid years, this may be a good time to start thinking about exporting ETAPS to other continents. The US East Coast would appear to be the obvious destination for a first ETAPS outside Europe.

The strength and success of ETAPS comes also from presenting – regardless of the natural internal differences – a homogeneous interface to authors and participants, i.e., to look like one large, coherent, well-integrated conference rather than a mere co-location of events. I therefore feel it is vital for ETAPS to regulate the centrifugal forces that arise naturally in a 'union' like ours, as well as the legitimate aspiration of individual PC chairs to run things their way. In this respect, we have large and solid foundations, alongside a few relevant issues on which ETAPS has not yet found agreement. They include, e.g., submission by PC members, rotation of PC memberships, and the adoption of a rebuttal phase. More work is required on these and similar matters.

January 2013

Vladimiro Sassone
ETAPS SC Chair
ETAPS e.V. President

Preface

FoSSaCS is an annual conference presenting papers on foundational research with a clear significance for software science. It covers research on theories and methods to support the analysis, integration, synthesis, transformation, and verification of programs and software systems. This volume contains contributions to FoSSaCS 2013, which took place March 18-20, 2013, as part of ETAPS 2013.

We received 109 submissions. Of these, 28 were selected for presentation at the conference and inclusion in the proceedings. There were a number of additional strong submissions that we could not accept because of space constraints. Also included in the proceedings is an abstract for the invited talk by Martin Hofmann on "Ten Years of Amortized Resource Analysis." I would like to thank the Program Committee and the additional reviewers for their excellent work. Throughout the submission, selection, and proceedings production process we relied on EasyChair, and we are grateful we were able to use this again. I would also like to thank the ETAPS 2013 General Chair, Daniele Gorla, and the Chair of the ETAPS Steering Committee, Vladimiro Sassone, for their leadership and guidance in the process.

January 2013 Frank Pfenning

Organization

Program Committee

Andreas Abel	LMU Munich, Germany
Umut Acar	MPI for Software Systems, Germany
Rajeev Alur	University of Pennsylvania, USA
Eugene Asarin	Université Paris Diderot, France
Nick Benton	Microsoft Research Cambridge, UK
Krishnendu Chatterjee	Institute of Science and Technology (IST), Austria
Swarat Chaudhuri	Rice University, USA
Adam Chlipala	MIT, USA
Ugo Dal Lago	University of Bologna, Italy
Giorgio Delzanno	University of Genova, Italy
Dan Ghica	University of Birmingham, UK
Jean Goubault-Larrecq	ENS Cachan, France
Bart Jacobs	Radboud University Nijmegen, The Netherlands
Radha Jagadeesan	DePaul University, USA
Patricia Johann	University of Strathclyde, UK
Naoki Kobayashi	University of Tokyo, Japan
Stephane Lengrand	Ecole Polytechnique, France
Aart Middeldorp	University of Innsbruck, Austria
Catuscia Palamidessi	INRIA Saclay, France
Frank Pfenning	Carnegie Mellon University, USA
Ashish Tiwari	SRI International, USA
Pawel Urzyczyn	University of Warsaw, Poland
Kwangkeun Yi	Seoul National University, Korea

Additional Reviewers

Accattoli, Beniamino
Adams, Robin
Aehlig, Klaus
Allais, Guillaume
Ancona, Davide
Atig, Mohamed Faouzi
Atkey, Robert
Avanzini, Martin
Avni, Guy
Baelde, David
Baldan, Paolo
Barth, Stephan
Benes, Nikola
Benzmueller, Christoph
Berger, Martin
Bernadet, Alexis
Bertrand, Nathalie
Birkedal, Lars
Bjorklund, Henrik
Bloem, Roderick

Blute, Richard
Boespflug, Mathieu
Boigelot, Bernard
Boker, Udi
Bollig, Benedikt
Bonchi, Filippo
Boreale, Michele
Bouissou, Olivier
Broadbent, Christopher
Brotherston, James
Bruyère, Véronique
Cacciagrano, Diletta Romana
Cerny, Pavol
Chang, Stephen
Cheval, Vincent
Chong, Stephen
Cimini, Matteo
Cirstea, Corina
Clairambault, Pierre
D'Antoni, Loris
Dagand, Pierre-Evariste
Danos, Vincent
de Carvalho, Daniel
Degorre, Aldric
Deng, Yuxin
Denielou, Pierre-Malo
Dezani-Ciancaglini, Mariangiola
Doyen, Laurent
Durand-Gasselin, Antoine
Duret-Lutz, Alexandre
Dziubiński, Marcin
Ehrig, Hartmut
Emmi, Michael
Esmaeilsabzali, Shahram
Faella, Marco
Faggian, Claudia
Felgenhauer, Bertram
Fernandez, Maribel
Figueira, Diego
Fijalkow, Nathanael
Fischer, Felix
Fitting, Melvin
Fokkink, Wan
Forejt, Vojtech
Francalanza, Adrian
Froeschle, Sibylle
Gabbay, Jamie
Gaboardi, Marco
Gadducci, Fabio
Galmiche, Didier
Ganty, Pierre
Gascon, Adria
Gimenez, Stéphane
Goubault, Eric
Gupta, Vineet
Haase, Christoph
Habermehl, Peter
Haddad, Axel
Haddad, Serge
Hansen, Helle Hvid
Harmer, Russ
Hasuo, Ichiro
Heijltjes, Willem
Hermanns, Holger
Hernich, André
Hildebrandt, Thomas
Hirschkoff, Daniel
Hirschowitz, Tom
Horn, Florian
Hur, Chung-Kil
Hüttel, Hans
Jacquemard, Florent
Jeż, Artur
Jonsson, Bengt
Kaliszyk, Cezary
Kennedy, Andrew
King, Tim
Klin, Bartek
König, Barbara
Komendantskaya, Katya
Komuravelli, Anvesh
Kong, Soonho
Krishnaswami, Neelakantan
Kupke, Clemens
Kurz, Alexander
Kwiatkowska, Marta
Křetínský, Jan
La Torre, Salvatore
Laird, Jim
Lasota, Sławomir

Lazic, Ranko
Lee, Oukseh
Lee, Wonchan
Levy, Paul
Lin, Anthony Widjaja
Loader, Ralph
Loreti, Michele
Madhusudan, P.
Majumdar, Rupak
Malecha, Gregory
Markey, Nicolas
Martí Oliet, Narciso
Mayr, Richard
McCusker, Guy
Melliès, Paul-André
Mendler, Michael
Meyer, Roland
Mezzina, Claudio Antares
Milius, Stefan
Millstein, Todd
Mimram, Samuel
Min, Zhang
Minamide, Yasuhiko
Mio, Matteo
Moggi, Eugenio
Monin, Jean-Francois
Moser, Georg
Muscholl, Anca
Müller-Olm, Markus
Niwiński, Damian
Norman, Gethin
Novotny, Petr
Oh, Hakjoo
Panangaden, Prakash
Pang, Jun
Paolini, Michela
Park, Daejun
Park, Sungwoo
Parrow, Joachim
Parys, Paweł
Perera, Roly
Phillips, Iain
Pientka, Brigitte
Pitcher, Corin
Platzer, André
Popescu, Andrei
Power, John
Prabhu, Vinayak
Praveen, M.
Puppis, Gabriele
Pérez, Jorge A.
Raskin, Jean-Francois
Rathke, Julian
Reddy, Uday
Regnier, Laurent
Reynier, Pierre-Alain
Rezine, Ahmed
Ridder, Bram
Riely, James
Rothenberg, Robert
Rubin, Sasha
Ryu, Sukyoung
Sacerdoti Coen, Claudio
Saha, Indranil
Samanta, Roopsha
Sangiorgi, Davide
Sangnier, Arnaud
Scerri, Guillaume
Schmidt, David
Schoepp, Ulrich
Schröder, Lutz
Schubert, Aleksy
Schöpp, Ulrich
Segala, Roberto
Senjak, Christoph-Simon
Serwe, Wendelin
Setzer, Anton
Sevegnani, Michele
Silva, Alexandra
Simpson, Alex
Skou, Arne
Sobocinski, Pawel
Sokolova, Ana
Sproston, Jeremy
Stampoulis, Antonis
Staton, Sam
Stenman, Jari
Sternagel, Christian
Strejcek, Jan
Suenaga, Kohei

Tabareau, Nicolas
Thiemann, René
Thrane, Claus
Toninho, Bernardo
Tripakis, Stavros
Trivedi, Ashutosh
Troina, Angelo
Tsukada, Takeshi
Turon, Aaron
Turrini, Andrea
Tzevelekos, Nikos
Unno, Hiroshi
Vafeiadis, Viktor
Valencia, Frank
van Breugel, Franck
Van Den Bussche, Jan
Van Glabbeek, Rob
Vaux, Lionel
Velner, Yaron
Wachter, Björn
Walukiewicz, Igor
Wiedijk, Freek
Wies, Thomas
Winkler, Sarah
Zankl, Harald
Zdanowski, Konrad
Ziliani, Beta

Ten Years of Amortized Resource Analysis (Invited Talk)

Martin Hofmann

www.ifi.lmu.de

Amortized resource analysis is a new method for deriving concrete bounds on resource usage, i.e., heap space, stack space, worst-case execution time, of functional and imperative programs that use recursion and complex data structures (lists, trees, graphs) for intermediate results. Of course, it can also be used with straight-line programs or programs without data structures, but in those cases does not offer any benefits over alternative approaches.

Amortized resource analysis started in the early 2000s when we tried to develop an automatic inference for a type system with explicit resource types [1, 2] that had arisen from implicit computational complexity, i.e., the attempt to characterise complexity classes, e.g. polynomial time, by logical or type-theoretic means. The link with classical amortized analysis as pioneered by Tarjan [3] was then realized and exploited.

Amortized resource analyses were subsequently developed for first-order functional programs [4], object-oriented programs [5], higher-order functional programs [6], and recently for lazy functional programming [7]. Initially, only linear resource bounds could be inferred, but recently we found ways for deriving univariate [8] and multivariate [9] polynomial resource bounds. This allows in particular for a fairly precise automatic analysis of various dynamic programming algorithms. Most recently [10] we developed an automatic inference for the object-oriented system from 2006. Amortized analysis has also been employed in three larger application-oriented research projects: [11–13].

In a nutshell, amortized analysis works as follows. Data structures are assigned non-negative numbers, called *potential*, in an *a priori* arbitrary fashion. If done cleverly, it then becomes possible to obtain *constant* bounds on the "amortised cost" of an individual operation, that is, its actual resource usage plus the difference in potential of the data structure before and after performing the operation. This makes it possible to take into account the effect that an operation might have on the resource usage of subsequent operations and also to merely add up amortised costs without having to explicitly track size and shape of intermediate data structures. In traditional amortised analysis [3] where the emphasis lies on the manual analysis of algorithms, the potentials were ascribed to particular data structures such as union-find trees or Fibonacci heaps by some formula that must be manually provided. When amortised analysis is used for automatic resource analysis one uses refined types to define the potentials — typing rules then ensure that potential and actual resource usage is accounted for correctly. Combined with type inference and numerical constraint solving, it then allows for an automatic inference of the potential functions.

The talk will survey these works and some of the related approaches [14–18] and discuss ongoing work and directions for future research.

References

1. Hofmann, M.: Linear types and non-size-increasing polynomial time computation. In: LICS, pp. 464–473. IEEE Computer Society (1999)
2. Hofmann, M.: A type system for bounded space and functional in-place update. Nord. J. Comput. 7(4), 258–289 (2000)
3. Tarjan, R.E.: Amortized computational complexity. SIAM Journal on Algebraic and Discrete Methods 6(2), 306–318 (1985)
4. Hofmann, M., Jost, S.: Static prediction of heap space usage for first-order functional programs. In: Proceedings of the 30th ACM SIGPLAN-SIGACT Symposium on Principles of Programming Languages (POPL), pp. 185–197. ACM (2003)
5. Hofmann, M.O., Jost, S.: Type-Based Amortised Heap-Space Analysis. In: Sestoft, P. (ed.) ESOP 2006. LNCS, vol. 3924, pp. 22–37. Springer, Heidelberg (2006)
6. Jost, S., Hammond, K., Loidl, H.W., Hofmann, M.: Static Determination of Quantitative Resource Usage for Higher-Order Programs. In: 37th ACM Symposium on Principles of Programming Languages (POPL 2010), pp. 223–236. ACM, New York (2010)
7. Simões, H.R., Vasconcelos, P.B., Florido, M., Jost, S., Hammond, K.: Automatic amortised analysis of dynamic memory allocation for lazy functional programs. In: Thiemann, P., Findler, R.B. (eds.) ICFP, pp. 165–176. ACM (2012)
8. Hoffmann, J., Hofmann, M.: Amortized Resource Analysis with Polynomial Potential. In: Gordon, A.D. (ed.) ESOP 2010. LNCS, vol. 6012, pp. 287–306. Springer, Heidelberg (2010)
9. Hoffmann, J., Aehlig, K., Hofmann, M.: Multivariate amortized resource analysis. ACM Trans. Program. Lang. Syst. 34(3), 14 (2012)
10. Hofmann, M., Rodriguez, D.: Automatic type inference for amortised heap-space analysis. In: Gardner, P., Felleisen, M. (eds.) European Symposium on Programming (ESOP). LNCS. Springer (2013)
11. Sannella, D., Hofmann, M., Aspinall, D., Gilmore, S., Stark, I., Beringer, L., Loidl, H.W., MacKenzie, K., Momigliano, A., Shkaravska, O.: Mobile resource guarantees (project evaluation paper). In: [19], pp. 211–226
12. Hammond, K., Dyckhoff, R., Ferdinand, C., Heckmann, R., Hofmann, M., Jost, S., Loidl, H.W., Michaelson, G., Pointon, R.F., Scaife, N., Sérot, J., Wallace, A.: The embounded project (project start paper). In: [19], pp. 195–210
13. Barthe, G., Beringer, L., Crégut, P., Grégoire, B., Hofmann, M.O., Müller, P., Poll, E., Puebla, G., Stark, I., Vétillard, E.: MOBIUS: Mobility, Ubiquity, Security. In: Montanari, U., Sannella, D., Bruni, R. (eds.) TGC 2006. LNCS, vol. 4661, pp. 10–29. Springer, Heidelberg (2007)
14. Albert, E., Arenas, P., Genaim, S., Puebla, G., Zanardini, D.: COSTA: Design and Implementation of a Cost and Termination Analyzer for Java Bytecode. In: de Boer, F.S., Bonsangue, M.M., Graf, S., de Roever, W.-P. (eds.) FMCO 2007. LNCS, vol. 5382, pp. 113–132. Springer, Heidelberg (2008)
15. Albert, E., Arenas, P., Genaim, S., Puebla, G.: Closed-Form Upper Bounds in Static Cost Analysis. Journal of Automated Reasoning (2010)
16. Chin, W.N., David, C., Gherghina, C.: A hip and sleek verification system. In: Lopes, C.V., Fisher, K. (eds.) OOPSLA Companion, pp. 9–10. ACM (2011)

17. Wilhelm, R., et al.: The Worst-Case Execution-Time Problem – Overview of Methods and Survey of Tools. ACM Trans. Embedded Comput. Syst. 7(3), 36:1–36:53 (2008)
18. Hughes, J., Pareto, L., Sabry, A.: Proving the Correctness of Reactive Systems Using Sized Types. In: 23rd ACM Symposium on Principles of Programming Languages (POPL 1996), pp. 410–423. ACM, New York (1996)
19. van Eekelen, M.C.J.D. (ed.): Revised Selected Papers from the Sixth Symposium on Trends in Functional Programming, TFP 2005, Tallinn, Estonia, September 23-24 (2005); In: van Eekelen, M.C.J.D. (ed.) Trends in Functional Programming. Trends in Functional Programming, vol. 6. Intellect (2007)

Table of Contents

Invited Talk

Models of Computation

Reasoning about Processes

Bisimulation

Modal and Higher-Order Logics

Reasoning about Programs

Computational Complexity

Quantitative Models

Categorical Models

Pattern Graphs and Rule-Based Models: The Semantics of Kappa

Jonathan Hayman[1,3,⋆] and Tobias Heindel[2,⋆⋆]

[1] DIENS (INRIA/ÉNS/CNRS), Paris, France
[2] CEA, LIST, Gif sur Yvette, France
[3] Computer Laboratory, University of Cambridge, UK

Abstract. Domain-specific rule-based languages to represent the systems of reactions that occur inside cells, such as Kappa and BioNetGen, have attracted significant recent interest. For these models, powerful simulation and static analysis techniques have been developed to understand the behaviour of the systems that they represent, and these techniques can be transferred to other fields. The languages can be understood intuitively as transforming graph-like structures, but due to their expressivity these are difficult to model in 'traditional' graph rewriting frameworks. In this paper, we introduce pattern graphs and closed morphisms as a more abstract graph-like model and show how Kappa can be encoded in them by connecting its single-pushout semantics to that for Kappa. This level of abstraction elucidates the earlier single-pushout result for Kappa, teasing apart the proof and guiding the way to richer languages, for example the introduction of compartments within cells.

1 Introduction

Rule-based models such as Kappa [6] and BioNetGen [2] have attracted significant recent attention as languages for modelling the systems of reactions that occur inside cells. Supported by powerful simulation and static analysis tools, the rule-based approach to modelling in biochemistry offers powerful new techniques for understanding these complex systems [1].

Many of the ideas emerging from rule-based modelling have the potential to be applied much more widely. Towards this goal, in this paper we frame the semantics of Kappa developed in [4] in a more general setting. In [4], an SPO semantics [10] is described by showing that specific pushouts in categories of partial maps between structures specifically defined for Kappa called Σ-graphs correspond to rewriting as performed by Kappa. The richness of the Kappa language means that this construction is highly subtle: it cannot be understood in the more widely studied DPO approach to graph rewriting.

⋆ JH gratefully acknowledges the support of the ANR *AbstractCell* Chair of Excellence and the ERC Advanced Grant *ECSYM*.
⋆⋆ TH is thankful for the financial support of the ANR project PANDA ANR-09-BLAN-0169.

F. Pfenning (Ed.): FOSSACS 2013, LNCS 7794, pp. 1–16, 2013.

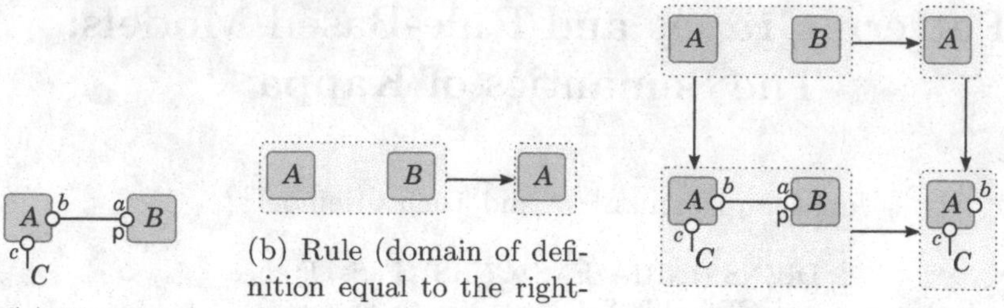

(a) Σ-graph

(b) Rule (domain of definition equal to the right-hand side)

(c) Rule application

Fig. 1. Example Σ-graph, rule and rule application

In this paper, we study pushouts in categories of simpler, more general structures called pattern graphs. Via an encoding of Σ-graphs into pattern graphs, we determine precisely when pushouts exist and what they are. The original motivation for this work was to tease-apart and generalise the pushout construction in [4]; the more abstract structures certainly provide new insight here by revealing the subtlety of the previous categories, for example in their not admitting all pushouts. But additionally, the study of pattern graphs both exports the fundamentals of Kappa rewriting to a more general setting and provides a uniform target for encodings of other rule-based models. The intention is to use this framework to obtain directly a categorical semantics for BioNetGen and for the enhancement of Kappa with regions.

Overview: In Section 2, we give an overview of Kappa and implicit deletion (called side-effects in [4]) using *closed partial maps*. In Section 3, we introduce *pattern graphs* as an expressive form of graph into which the encoding of Kappa proceeds in Section 4. In Section 5, we isolate the role of *coherent* graphs and determine when they have pushouts. Finally, we show in Section 6 that pushouts in the category of Σ-graphs and pushouts in the category of coherent pattern graphs correspond: the pushout of the encoding is the encoding of the pushout.

2 Kappa and Implicit Deletion

We shall give a formal account of the semantics of Kappa in Section 4, but essentially we wish to characterise Kappa rewriting as a pushout in a category of special forms of graph called *Σ-graphs.* A rule $\alpha : L \to R$ can be applied to a Σ-graph S if there is a matching of the *pattern* L in S and there is a pushout as follows, generating a rule application $\beta : S \to T$:

$$\begin{array}{ccc} L & \xrightarrow{\alpha} & R \\ {\scriptstyle m}\downarrow & & \downarrow{\scriptstyle m'} \\ S & \xrightarrow{\beta} & T \end{array}$$

An example Σ-graph is drawn in Figure 1(a). Squares represent entities called *agents* which have circles attached called *sites*. The *signature Σ* describes the labels that can occur on agents and sites. Links can be drawn between sites, and additionally, for use in patterns which will be used to describe transformation rules, we allow *anonymous* links. The anonymous link drawn at the site c on the A-agent, for example, will represent the existence of a link to some site on a C-agent. Finally, sites can have internal properties attached to them; in this case, the site a on the B-agent has an internal property p, perhaps to represent that the site is *phosphorylated*.

Homomorphisms express how the structure of one Σ-graph embeds into that of another. They are functions on the components of the graph sending agents to agents, sites to sites and links to links, that preserve structure in the sense of preserving the source and target of links and preserving the labels on agents and sites. They allow anonymous links to be sent either to proper links that satisfy any requirements on the target, such as that it is on a C-agent, or to other anonymous links that are at least as specific in their description of the target, for example allowing the link drawn to be sent to another anonymous link that specifies a connection to a site with a particular label on an agent labelled C.

To allow rules to represent the deletion of structure, we consider pushouts of *partial maps*. Partial maps generalise homomorphisms by allowing undefinedness. Formally, we view a partial map $f : L \rightharpoonup R$ to be a span consisting of an inclusion $\mathrm{def}(f) \hookrightarrow L$, where $\mathrm{def}(f)$ is the *domain of definition* of the partial map, and a homomorphism $f_0 : \mathrm{def}(f) \to R$. The interpretation of partial maps as describing transformations is that the rule can be applied if the pattern L is matched; if so, the elements of L that are in the domain of definition indicate what is preserved and the elements of L outside the domain of definition indicate what is to be deleted. Anything in R outside the image of the domain of definition is created by application of the rule.

An example rule is presented in Figure 1(b), showing the deletion of an agent labelled B in the presence of an agent labelled A. Note that the rule does not include any sites, so the state of the sites on agents matched by those in the left-hand side does not determine whether the rule can be applied. This is an instance of the "don't care; don't write" principle in Kappa. Consequently, the rule can be applied to give the application drawn in Figure 1(c). Importantly for capturing the semantics of Kappa, the fact that the B-agent cannot be in the pushout forces the sites on B and the link to A also to be absent from the pushout: the link cannot be in the domain of definition of the rule application since, otherwise, the domain of definition and the produced Σ-graph would not be well-formed. We say that the sites and links are *implicitly deleted* as a side-effect of the deletion of B.

Rewriting is abstractly characterised as taking a pushout in the category Σ-graphs with partial maps between them. In the construction of the pushout described in [4], as in the pushout for containment structures in [8], there is a close relationship to the construction in the category of sets and partial functions. For example, in generating the pushout in Figure 1c, we cannot have the B-labelled agent preserved by the rule application due to the morphism α being undefined on the B-agent matching it. On top of this, however, we have implicit deletion: as remarked, we are additionally forced to remove the link connecting the A-agent and the site a on B since the domain of definition has to be a well-formed Σ-graph. A natural abstraction that captures both the set-like features of the pushout and implicit deletion is to encode the structures as labelled graphs, the nodes of which are the agents, sites and links and internal properties in the Σ-graph. In this way, we simplify the analysis by treating all elements of the Σ-graph uniformly. We represent using links the dependencies of the Σ-graph: sites depend on agents, links on the sites that they connect, and so on. Links are labelled to indicate the role of the dependency, such as a site being the source of a link. The construction is described more fully in Section 4.1.

Before proceeding into detail, we demonstrate in the simplest possible setting how *closed* morphisms allow implicit deletion as described above to be captured.

Definition 1. *Let Λ be a fixed set of labels. A* basic graph *is a tuple $G = (V, E)$ where V is the set of* nodes *(or* vertices*) and $E \subseteq V \times \Lambda \times V$ is the set of* edges.

We adopt the convention of adding subscripts to indicate the components of a structure. For example, we write V_G for the vertices of a graph G.

Definition 2. *A graph* homomorphism *$f : G \to H$ is a function on nodes $f : V_G \to V_H$ such that if $(v, \lambda, v') \in E_G$ then $(f(v), \lambda, f(v')) \in V_H$.*

We write $\mathbb{bG}$ for the category of basic graphs with homomorphisms between them. When considering partial maps, a critical feature will be the following condition called *closure*, which ensures that any partial map will be defined on any node reachable from any defined node. For example, if a partial map is defined on a link of a Σ-graph, which will be encoded as a node, it will be defined on both the sites and agents that the link connects since they shall be reachable from the node representing the link.

Definition 3. *A partial map $f : G \rightharpoonup H$ is a span $G \hookleftarrow \mathrm{def}(f) \xrightarrow{f_0} H$ of homomorphisms such that $\mathrm{def}(f)$ is a subgraph of G, i.e. $V_{\mathrm{def}(f)} \subseteq V_G$ and $E_{\mathrm{def}(f)} \subseteq E_G$, and the following* closure property *holds:*

$$\text{if } v \in V_{\mathrm{def}(f)} \text{ and } (v, \lambda, v') \in E_G \text{ then } v' \in V_{\mathrm{def}(f)} \text{ and } (v, \lambda, v') \in E_{\mathrm{def}(f)}.$$

Write $\mathbb{bG}_*$ for the category of basic graphs with partial maps between them. Partial maps $f : G \rightharpoonup H$ and $g : H \rightharpoonup K$ compose as partial functions, with the domain of definition of $g \circ f$ obtained as the *inverse image* of $\mathrm{def}(g)$ along f_0. It consists of vertices v of G such that $f(v)$ is in the domain of definition of g and edges to satisfy the closure condition, and can be shown to be a pullback of f_0 against the inclusion $\mathrm{def}(g) \hookrightarrow H$ in $\mathbb{bG}$.

Using the category-theoretical account of existence of pushouts of partial maps mentioned in the conclusion or directly, it can be shown that $\mathbb{bG}_*$ has pushouts of all spans of partial maps; we do not present the details here since they shall be subsumed by those in the following section for pattern graphs. In Figure 2, we give an example of a pushout in $\mathbb{bG}_*$. The pushout shows in a simplified way how pushing out against a partial map representing deletion of an agent (the node w_2) requires the implicit deletion of a link (the node ℓ). The key is that, following the argument for pushouts in the category of sets and partial functions,

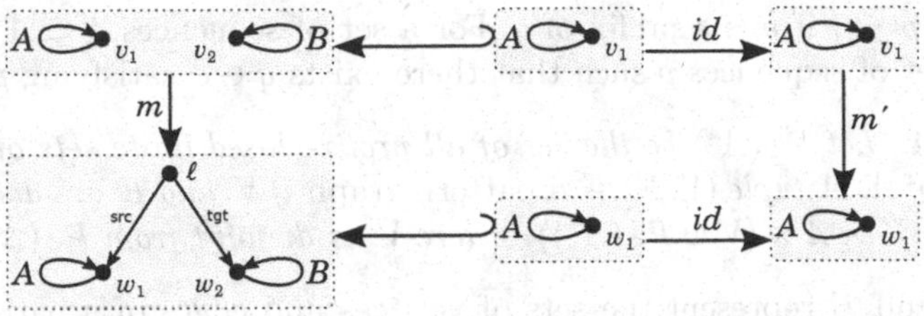

Fig. 2. An example pushout in $\mathbb{bG}_*$. Horizontal spans represent partial maps. Vertical maps m, m' are total and send v_i to w_i for $i \in \{1, 2\}$.

we cannot have the node w_2 in the domain of definition of the lower partial map and hence, by the closure condition, we therefore cannot have ℓ in the domain of definition of the rule application represented by the lower span.

We wish to use the fact that the category $\mathbb{bG}_*$ has pushouts of all spans to consider, via the encoding described above and to be formalised in Section 4.1, pushouts in the category of Σ-graphs. Before we can do so, there are two important details to consider.

The first is that we must ensure that maps between the encoded structures correspond to maps in the category of Σ-graphs: that they send agents to agents, sites to sites and links to links. The standard trick of encoding the types of elements as labelled loops can deal with these aspects, but the presence in Kappa of anonymous links that can be mapped either to proper links or to more specific anonymous links will necessitate a richer structure than basic graphs. There are alternatives, for example recording the type of nodes, but in this paper an elegant treatment is provided by the use of *pattern graphs*.

The second detail is that we will wish the pushout in the category of pattern graphs to correspond to the encoding of some Σ-graph. Certain aspects will follow from the nature of the pushout, but it turns out that a pivotal issue will be *coherence*. The graphs that we form as encodings will be coherent in the sense that any node will have at most one edge with any given label from it. For example, the encoding of a Kappa link will have at most one source and at most one target; Kappa does not have hyper-edges. However, as we shall see, the graphs that we form by taking pushouts in the category of pattern graphs might fail to be coherent, so we desire an operation that forms a coherent pushout from the non-coherent one. This operation can fail: it is not always possible to form a pushout of an arbitrary span of morphisms in the category of coherent pattern graphs with partial morphisms between them, and critically this will inform us that there is no pushout of the given span in the category of Σ-graphs.

3 Pattern Graphs

Pattern graphs add to basic graphs the ability to express, via homomorphisms, the existence of an labelled edge from a node to some node satisfying a specification. Specifications are just prefix-closed sets of sequences of labels to indicate paths that must exist from the specified node. Assuming a set of link labels Λ, let Λ^* denote the set of all finite sequences of elements of Λ. For sequences p

and q, write $p \leq q$ if p is a prefix of q. For a set of sequences $\phi \subseteq \Lambda^*$, we write $\downarrow \phi$ for the set of sequences p such that there exists $q \in \phi$ satisfying $p \leq q$.

Definition 4. *Let $\mathcal{P}_{\leq}(\Lambda^*)$ be the set of all prefix-closed finite sets of sequences of elements of Λ. A tuple (V, E) is a* pattern graph *if V and E are disjoint finite sets and $E \subseteq V \times \Lambda \times (V \cup \mathcal{P}_{\leq}(\Lambda^*))$, where V is disjoint from $\mathcal{P}_{\leq}(\Lambda^*)$.*

The sets V and E represent the sets of vertices and edges of a pattern graph. Edges can either be normal (*i.e.* between vertices) or be specifications, being of the form (v, λ, ϕ): the intention is that specifications are used in patterns when we wish to specify, via homomorphisms, the structure that some other perhaps more refined graph possesses.

Definition 5. *A vertex $v \in V$ in a pattern graph G satisfies $p \in \Lambda^*$, written $v \models_G p$, if either p is empty or $p = \lambda.p_0$ and either:*

- *there exists v' such that $(v, \lambda, v') \in E_G$ and $v' \models_G p_0$, or*
- *there exists $\psi \in \mathcal{P}_{\leq}(\Lambda^*)$ such that $(v, \lambda, \psi) \in E_G$ and $p_0 \in \psi$.*

A vertex $v \in V_G$ satisfies $\phi \in \mathcal{P}_{\leq}(\Lambda^*)$, written $v \models_G \phi$, if $v \models_G p$ for all $p \in \phi$.

Homomorphisms embed the structure of one pattern graph into that of another: they preserve the presence of normal links between vertices and, if a vertex has a specification, the image of the vertex satisfies the specification. Importantly, they do not record exactly *how* the specification is satisfied.

Definition 6. *A* homomorphism *of pattern graphs $f : G \to H$ is a function on vertices $f : V_G \to V_H$ such that, for all $v, v' \in V_G$, $\lambda \in \Lambda$ and $\phi \in \mathcal{P}_{\leq}(\Lambda^*)$:*

- *if $(v, \lambda, v') \in E_G$ then $(f(v), \lambda, f(v')) \in E_H$, and*
- *if $(v, \lambda, \phi) \in E_G$ then there exists $x \in V_H \cup \mathcal{P}_{\leq}(\Lambda^*)$ such that $(f(v), \lambda, x) \in E_H$ and $x \models_G \phi$ if $x \in V_H$ and $\phi \subseteq x$ if $x \in \mathcal{P}_{\leq}(\Lambda^*)$.*

We write $\mathbb{PG}$ for the category of pattern graphs connected by homomorphisms. For any pattern graph τ, denote by $\mathbb{PG}/\tau$ the slice category above τ. The objects of $\mathbb{PG}/\tau$ are pairs (G, γ) where G is a pattern graph and $\gamma : G \to \tau$ is a homomorphism, and a morphism $h : (G, \gamma) \to (G', \gamma')$ in $\mathbb{PG}/\tau$ is a homomorphism such that $\gamma = \gamma' \circ h$. We can regard τ as representing the structure that the pattern graphs being considered are allowed to possess. Where no ambiguity arises, we shall simply write G for the pair (G, γ) and τ_G for γ.

Partial maps extend homomorphisms by allowing them to be undefined on vertices and edges. Again, we require the closure condition of Section 2.

Definition 7. *Let G and H be objects of $\mathbb{PG}/\tau$. A* partial map *$f : G \rightharpoonup H$ consists of a pattern graph* $\mathrm{def}(f) = (V_0, E_0)$ *and a homomorphism $f_0 : \mathrm{def}(f) \to H$ in $\mathbb{PG}/\tau$ where:*

- $\mathrm{def}(f)$ *is a pattern graph satisfying $V_0 \subseteq V_G$ and $E_0 \subseteq E_G$;*
- $\tau_{\mathrm{def}(f)} : \mathrm{def}(f) \to \tau$ *is the restriction of τ_G to* $\mathrm{def}(f)$*; and*
- $\mathrm{def}(f)$ *is closed: for all $(v, \lambda, x) \in E_G$, if $v \in V_0$ then $(v, \lambda, x) \in E_0$ (and hence $x \in V_0$ if $x \in V_G$).*

We write $(\mathbb{PG}/\tau)_*$ for the category of pattern graphs with partial maps between them. As it was for basic graphs, composition is obtained using the inverse image construction and can be shown to be a pullback in $\mathbb{PG}/\tau$.

The category $(\mathbb{PG}/\tau)_*$ can be shown to have pushouts of all spans. However, as we shall see in Section 5, the existence conditions for pushouts of coherent typed pattern graphs will be much more subtle and involve considerable additional work.

Theorem 1. *The category* $(\mathbb{PG}/\tau)_*$ *has pushouts.*

The above result can be proved either by using the category-theoretical conditions for existence of pushouts of partial maps mentioned in the conclusion or directly by showing that the following construction yields a pushout of any span.

Given a span $S \xleftarrow{g} L \xrightarrow{f} R$, we define a cospan $S \xrightarrow{p} T \xleftarrow{q} R$ that forms a pushout in $(\mathbb{PG}/\tau)_*$. Let $V = V_L \cup V_R \cup V_S$. We define $\sim$ to be the least equivalence relation on V such that $v_0 \sim f(v_0)$ for all $v_0 \in V_{\mathsf{def}(f)}$ and $w_0 \sim g(w_0)$ for all $w_0 \in V_{\mathsf{def}(g)}$. For any $v \in V$, we denote by $[v]$ its $\sim$-equivalence class; these equivalence classes will be used to form the vertices of the pushout object T.

Write $[v] \xrightarrow{\lambda} [v']$ iff there exist $v_0 \in [v]$ and $v_0' \in [v']$ such that $(v_0, \lambda, v_0') \in E_L \cup E_R \cup E_S$. The unlabelled transitive closure of this relation, written $[v_1] \rightarrow^* [v_n]$, relates $[v_1]$ to $[v_n]$ if there exist $\lambda_1, \ldots, \lambda_{n-1}$ such that $[v_1] \xrightarrow{\lambda_1} \ldots \xrightarrow{\lambda_{n-1}} [v_n]$. Now define $\mathsf{del}_0([v])$ iff there exists $v_0 \in [v] \cap V_L$ such that $v_0 \notin V_{\mathsf{def}(f)} \cap V_{\mathsf{def}(g)}$ and define $\mathsf{del}([v])$ iff there exists $v' \in V$ such that $[v] \rightarrow^* [v']$ and $\mathsf{del}_0([v'])$. The vertices and edges of the pushout object are given as:

$$\begin{aligned} V_T &= \{[v] \mid v \in V \;\&\; \neg\mathsf{del}([v])\} \\ E_T &= \{([v], \lambda, [v']) \mid [v], [v'] \in V_T \;\&\; [v] \xrightarrow{\lambda} [v']\} \\ &\quad \cup \{([v], \lambda, \phi) \mid [v] \in V_T \;\&\; \exists v_0 \in [v].(v_0, \lambda, \phi) \in E_L \cup E_R \cup E_S\} \end{aligned}$$

The domain of definition of the pushout morphism p is the closed subgraph of S containing all vertices $v \in V_S$ such that $\neg\mathsf{del}([v])$. Where defined, p sends vertices of S to their $\sim$-equivalence classes. The partial map q is defined similarly. The type map $\tau_T : T \rightarrow \tau$ sends an equivalence class $[v]$ to $\tau_L(v)$ if $v \in V_L$, to $\tau_R(v)$ if $v \in V_R$ and $\tau_S(v)$ if $v \in V_S$; well-definedness of this follows from the maps f and g being type-preserving.

4 Kappa and Σ-graphs

We begin this section by briefly describing the semantics given to Kappa in [4], where a fuller explanation and examples can be found. The semantics of Kappa is given over graphs with a given signature Σ, specifying the labels that can occur on agents Σ_{ag}, sites Σ_{st} and internal properties Σ_{prop}, and, for any agent label A, the set of sites $\Sigma_{\mathsf{ag-st}}(A)$ that are permitted to occur on an agent labelled A.

Definition 8. *A* signature *is a 4-tuple* $\Sigma = (\Sigma_{\mathsf{ag}}, \Sigma_{\mathsf{st}}, \Sigma_{\mathsf{ag-st}}, \Sigma_{\mathsf{prop}})$*, where* Σ_{ag} *is a finite set of* agent types, Σ_{st} *is a finite set of* site identifiers, $\Sigma_{\mathsf{ag-st}} : \Sigma_{\mathsf{ag}} \to \mathcal{P}_{\mathsf{fin}}(\Sigma_{\mathsf{st}})$ *is a* site map*, and* Σ_{prop} *is a finite set of* internal property identifiers.

As described in Section 2, Σ-graphs consist of sites on agents; sites can have internal properties indicated and be linked to each other. Sites can also have anonymous links attached to them, typically used when the Σ-graph represents a pattern, so the anonymous link represents that a link is required to exist at the image of the site under a homomorphism. There are three types of anonymous link; the first is represented by a dash '$-$' and indicates that the link connects to any site on any agent, the second by A for $A \in \Sigma_{\mathsf{ag}}$ which indicates that the link connects to some site on an agent of type A, and the final kind of anonymous link is (A, i) which indicates that the link connects to site i on some agent labelled A. These form the set $\mathsf{Anon} = \{-\} \cup \Sigma_{\mathsf{ag}} \cup \{(A, i) \mid A \in \Sigma_{\mathsf{ag}} \ \&\ i \in \Sigma_{\mathsf{ag-st}}(A)\}$.

Definition 9. *A Σ-graph comprises a finite set $\mathcal{A}$ of* agents, *an* agent type *assignment* $\mathsf{type} : \mathcal{A} \to \Sigma_{\mathsf{ag}}$*, a set $\mathcal{S}$ of* link sites *satisfying* $\mathcal{S} \subseteq \{(n, i) \ :\ n \in \mathcal{A} \ \&\ i \in \Sigma_{\mathsf{ag-st}}(\mathsf{type}(n))\}$*, a symmetric* link *relation* $\mathcal{L} \subseteq (\mathcal{S} \cup \mathsf{Anon})^2 \setminus \mathsf{Anon}^2$*, and a* property *set* $\mathcal{P} \subseteq \{(n, i, k) \mid n \in \mathcal{A} \ \&\ i \in \Sigma_{\mathsf{ag-st}}(\mathsf{type}(n)) \ \&\ k \in \Sigma_{\mathsf{prop}}\}$.

We shall conventionally assume that the sets described above are pairwise-disjoint. A normal link is a pair of sites $((n, i), (m, j))$ and an anonymous link is of the form $((n, i), x)$ where (n, i) is a site and $x \in \mathsf{Anon}$. We use x to range over both sites and Anon. Note that $(n, i, k) \in \mathcal{P}$ does not imply $(n, i) \in \mathcal{S}$: as explained in [4], this is to allow Σ-graphs to represent patterns where we represent a property holding at some site but do not specify anything about its linkage.

Homomorphisms between Σ-graphs are structure-preserving functions from the agents, sites, links and internal properties of one Σ-graph to those of another. They preserve structure by preserving the presence of sites on agents, preserving properties held on sites, preserving the source and target of links and ensuring that the source and target of the image of any link is at least as high in the *link information order* as those of the original link. Given a typing function type, this order is the least reflexive, transitive relation $\leq_{\mathsf{type}}$ s.t. for all $A \in \Sigma_{\mathsf{ag}}$ and $i \in \Sigma_{\mathsf{ag-st}}(A)$ and n s.t. $\mathsf{type}_G(n) = A$: $\ - \leq_{\mathsf{type}} A \leq_{\mathsf{type}} (A, i) \leq_{\mathsf{type}} (n, i)$.

Definition 10. *A* homomorphism *of Σ-graphs $h : G \to H$ consists of a function on agents $h_{\mathsf{ag}} : \mathcal{A}_G \to \mathcal{A}_H$, a function on sites $h_{\mathsf{st}} : \mathcal{S}_G \to \mathcal{S}_H$, a function on links $h_{\mathsf{ln}} : \mathcal{L}_G \to \mathcal{L}_H$ and a function on internal properties $h_{\mathsf{prop}} : \mathcal{P}_G \to \mathcal{P}_H$, satisfying:*

- $\mathsf{type}_G(n) = \mathsf{type}_H(h_{\mathsf{ag}}(n))$ *for all* $n \in \mathcal{A}_G$
- $h_{\mathsf{st}}(n, i) = (h_{\mathsf{ag}}(n), i)$ *and* $h_{\mathsf{prop}}(n, i, k) = (h_{\mathsf{ag}}(n), i, k)$
- $h_{\mathsf{ln}}((n, i), x) = (h_{\mathsf{st}}(n, i), y)$ *for some y such that* $\hat{h}(x) \leq_{\mathsf{type}_H} y$*, where we take* $\hat{h}(m, j) = h_{\mathsf{st}}(m, j)$ *for any* $(m, j) \in \mathcal{S}_G$ *and* $\hat{h}(x) = x$ *for any* $x \in \mathsf{Anon}$.

We write $\Sigma\mathbb{G}$ for the category of Σ-graphs with homomorphisms between them. Note that in [4], attention was restricted to graphs with only one link to or from any site, allowing a less general form of homomorphism to be used.

Definition 11. *A partial map $f : G \rightharpoonup H$ between Σ-graphs G and H is a span $G \hookleftarrow \mathrm{def}(f) \xrightarrow{f_0} H$ where f_0 is a homomorphism and* $\mathrm{def}(f)$ *is a Σ-graph that is a subgraph of G, i.e.: $\mathcal{A}_{\mathrm{def}(f)} \subseteq \mathcal{A}_G$ and $\mathcal{S}_{\mathrm{def}(f)} \subseteq \mathcal{S}_G$ and $\mathcal{L}_{\mathrm{def}(f)} \subseteq \mathcal{L}_G$ and $\mathcal{P}_{\mathrm{def}(f)} \subseteq \mathcal{P}_G$.*

Partial maps between Σ-graphs form a category denoted $\Sigma\mathbb{G}_*$, where partial maps $f : G \rightharpoonup H$ and $g : H \rightharpoonup K$ compose in the usual way, with the domain of definition of their composition $\mathrm{def}(g \circ f)$ containing elements of $\mathrm{def}(f)$ such that their image under f is in $\mathrm{def}(g)$. This corresponds to taking a pullback of the homomorphism $f_0 : \mathrm{def}(f) \to H$ against $\mathrm{def}(g) \hookrightarrow H$ in $\Sigma\mathbb{G}$.

4.1 Encoding Σ-graphs as Pattern Graphs

We now show how Σ-graphs can be interpreted as pattern graphs. As stated before, the idea behind the encoding of a Σ-graph G is to build a pattern graph $[\![G]\!]$ with vertices that are the agents, sites, links and properties of G. Labelled edges in $[\![G]\!]$ indicate the dependencies between elements of the graph, so for example that deletion of an agent causes the deletion of any edge connecting to that agent. There will be edges from links to their source and target, labelled src or tgt respectively, and edges labelled ag from sites to agents and internal properties to agents (not to sites since, as mentioned, we use the set of sites specifically to represent link state). There will also be an edge labelled symm between every link and its symmetric counterpart to ensure that morphisms preserve the symmetry of the link relation. Specifications are used in the representation of anonymous links.

The starting point is to define a pattern graph $\hat{\Sigma}$ to represent the structure of encodings, so that the encoding of a Σ-graph will be an object of $(\mathbb{PG}/\hat{\Sigma})_*$.

Definition 12. *With respect to signature Σ, the pattern graph $\hat{\Sigma}$ is*

$$
\begin{aligned}
V_{\hat{\Sigma}} = \;& \{\mathsf{link}\} \;\cup\; \Sigma_{\mathsf{ag}} \;\cup\; \{(A,i) \mid A \in \Sigma_{\mathsf{ag}} \;\&\; i \in \Sigma_{\mathsf{ag-st}}(A)\} \\
& \cup \{(A,i,p) \mid A \in \Sigma_{\mathsf{ag}} \;\&\; i \in \Sigma_{\mathsf{ag-st}}(A) \;\&\; p \in \Sigma_{\mathsf{prop}}\} \\
E_{\hat{\Sigma}} = \;& \{(\mathsf{link}, \mathsf{symm}, \mathsf{link})\} \\
& \cup \{(\mathsf{link}, \mathsf{src}, (A,i)),\;\; (\mathsf{link}, \mathsf{tgt}, (A,i)) \;\mid\; A \in \Sigma_{\mathsf{ag}} \;\&\; i \in \Sigma_{\mathsf{ag-st}}(A)\} \\
& \cup \{((A,i), \mathsf{ag}, A) \mid A \in \Sigma_{\mathsf{ag}} \;\&\; i \in \Sigma_{\mathsf{ag-st}}(A)\} \\
& \cup \{((A,i,p), \mathsf{ag}, A) \mid A \in \Sigma_{\mathsf{ag}} \;\&\; i \in \Sigma_{\mathsf{ag-st}}(A) \;\&\; p \in \Sigma_{\mathsf{prop}}\} \\
& \cup \{((A,i,p), (i,p), (A,i,p)) \mid A \in \Sigma_{\mathsf{ag}} \;\&\; i \in \Sigma_{\mathsf{ag-st}}(A) \;\&\; p \in \Sigma_{\mathsf{prop}}\} \\
& \cup \{(A,A,A) \mid A \in \Sigma_{\mathsf{ag}}\} \cup \{(A,i), i, (A,i) \mid A \in \Sigma_{\mathsf{ag}} \;\&\; i \in \Sigma_{\mathsf{ag-st}}(A)\}
\end{aligned}
$$

Example 1. Given Σ as follows, the graph $\hat{\Sigma}$ is:

$$
\begin{aligned}
\Sigma_{\mathsf{ag}} &= \{A, B\} \\
\Sigma_{\mathsf{st}} &= \{i, j\} \\
\Sigma_{\mathsf{prop}} &= \emptyset \\
\Sigma_{\mathsf{ag-st}} &= \{A \mapsto \{i,j\}, B \mapsto \{i\}\}
\end{aligned}
$$

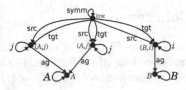

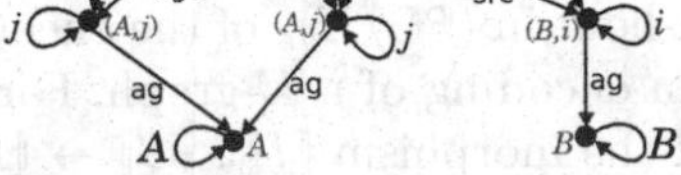

The loops on the vertices for agents and sites will be used in patterns for anonymous links. We now define a functor $[\![\cdot]\!] : \Sigma\mathbb{G}_* \to (\mathbb{PG}/\hat{\Sigma})_*$ that embeds the category of Σ-graphs into the category of pattern graphs over $\hat{\Sigma}$.

Definition 13. *For a Σ-graph G, the pattern graph $[\![G]\!]$ is:*

$$\begin{aligned}
V_{[\![G]\!]} = \;& \mathcal{A}_G \cup \mathcal{S}_G \cup \mathcal{L}_G \cup \mathcal{P}_G \\
E_{[\![G]\!]} = \;& \{(n, \mathsf{type}_G(n), n) \mid n \in \mathcal{A}\} \cup \{((n,i), i, (n,i)) \mid (n,i) \in \mathcal{S}\} \\
& \cup \{((n,i,p), (i,p), (n,i,p)) \mid (n,i,p) \in \mathcal{P}_G\} \\
& \cup \{((n,i), \mathsf{ag}, n) \mid (n,i) \in \mathcal{S}_G\} \cup \{((n,i,p), \mathsf{ag}, n) \mid (n,i,p) \in \mathcal{P}_G\} \\
& \cup \{(((n,i),x), \mathsf{src}, (n,i)) \mid ((n,i),x) \in \mathcal{L}_G\} \\
& \cup \{((x,(n,i)), \mathsf{tgt}, (n,i)) \mid (x,(n,i)) \in \mathcal{L}_G\} \\
& \cup \{((x,(n,i)), \mathsf{src}, anon(x) \mid x \in \mathsf{Anon} \;\&\; (x,(n,i)) \in \mathcal{L}_G\} \\
& \cup \{(((n,i),x), \mathsf{tgt}, anon(x) \mid x \in \mathsf{Anon} \;\&\; ((n,i),x) \in \mathcal{L}_G\}
\end{aligned}$$

where anon gives a specification for anonymous links:

$$anon(-) = \emptyset \qquad anon(A) =\downarrow \{\mathsf{ag}.A\} \qquad anon(A,i) =\downarrow \{i.\mathsf{ag}.A\}$$

The function $\tau_{[\![G]\!]} : [\![G]\!] \to \hat{\Sigma}$ sends links to link, *agents n to* $\mathsf{type}_G(n)$, *sites (n,i) to* $(\mathsf{type}_G(n), i)$ *and properties (n,i,p) to* $(\mathsf{type}_G(n), i, p)$.

The encoding $[\![g]\!] : [\![G]\!] \rightharpoonup [\![H]\!]$ of a partial map $g : G \rightharpoonup H$ in $\Sigma\mathbb{G}_*$ has domain of definition $[\![\mathrm{def}(g)]\!]$ and sends a vertex $v \in V_{[\![\mathrm{def}(g)]\!]}$ to $g_{\mathsf{ag}}(v)$ if $v \in \mathcal{A}_G$, or $g_{\mathsf{st}}(v)$ if $v \in \mathcal{S}_G$, or $g_{\mathsf{lnk}}(v)$ if $v \in \mathcal{L}_G$, or $g_{\mathsf{prop}}(v)$ if $v \in \mathcal{P}_G$.

It is straightforward to check that the encoding defines a functor. The key is that the partial map $[\![f]\!]$ satisfies the closure condition due to the domain of definition of f being a well-formed Σ-graph.

Example 2. The encoding of the unique homomorphism from G to H as drawn is the unique homomorphism from $[\![G]\!]$ to $[\![H]\!]$. Specifications are drawn as kites.

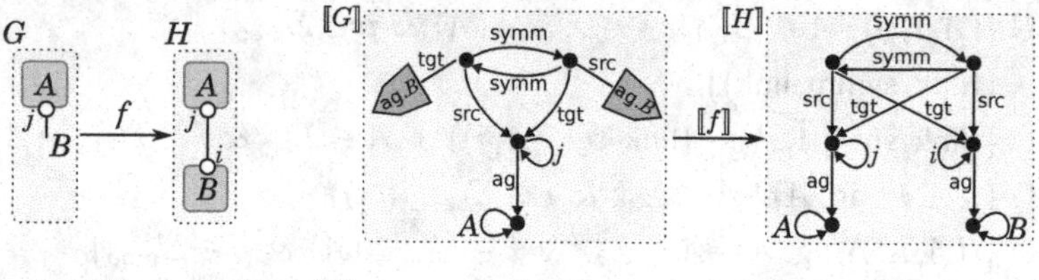

Lemma 1. *The encoding $[\![\cdot]\!]$ is an embedding: it is a full and faithful functor and is injective on objects.*

As such, the image of $[\![\cdot]\!]$ is a full subcategory of $(\mathbb{PG}/\hat{\Sigma})_*$ isomorphic to $\Sigma\mathbb{G}_*$.

5 Coherence

The pushout in $(\mathbb{PG}/\hat{\Sigma})_*$ of the encoding of a span of morphisms in $\Sigma\mathbb{G}_*$ can fail to be an encoding of a Σ-graph. For example, in $(\mathbb{PG}/\hat{\Sigma})_*$ the pushout against itself of the morphism $[\![f]\!] : [\![G]\!] \to [\![H]\!]$ drawn in Example 2 is:

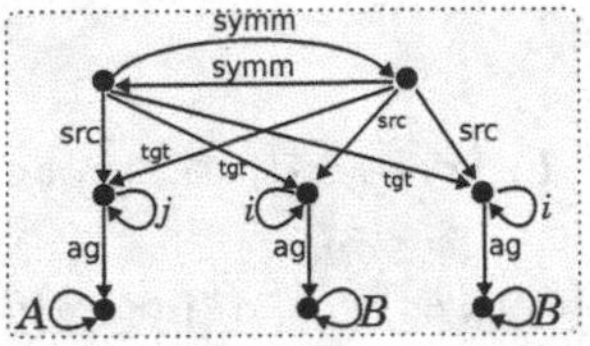

The encodings of Σ-graphs have links with at most one source and at most one target: we say that they are *coherent*.

Definition 14. *A pattern graph G is* coherent *if $(v, \lambda, x_1) \in E_G$ and $(v, \lambda, x_2) \in E_G$ implies $x_1 = x_2$ for all $v \in V_G$, all $\lambda \in \Lambda$ and all $x_1, x_2 \in V_G \cup \mathcal{P}_{\leq}(\Lambda^*)$.*

Lemma 2. *For any Σ-graph G, the encoding $[\![G]\!]$ is coherent.*

We now characterise pushouts in the category $(\mathbb{PG}/\tau)^{\mathrm{c}}_*$, the full subcategory of $(\mathbb{PG}/\tau)_*$ with coherent pattern graphs over τ as objects. By restricting ourselves to coherent graphs, remarkably we lose the property that all spans of partial (and even total) maps have pushouts. It is not hard, for example, to verify that there is no pushout in $(\mathbb{PG}/\hat{\Sigma})^{\mathrm{c}}_*$ of the morphism $[\![f]\!] : [\![G]\!] \to [\![H]\!]$ in Example 2 against itself. The significance of this is that exactly the same phenomenon occurs in the category of Σ-graphs, which fails to have pushouts of all spans for the same reason. For example, analogously there is no pushout of the morphism of Σ-graphs $f : G \to H$ drawn in Example 2 against itself in $\Sigma\mathbb{G}_*$ (noting that the form of homomorphism in this paper includes a component to give the target of links, unlike in [4] where the less general form of homomorphism could be used).

We shall see in Section 6 that pushouts of encodings in $(\mathbb{PG}/\hat{\Sigma})^{\mathrm{c}}_*$ will correspond to those in $\Sigma\mathbb{G}_*$. We now study when pushouts in $(\mathbb{PG}/\tau)^{\mathrm{c}}_*$ exist. We do so by characterising the largest full subcategory $\mathcal{C}$ of $(\mathbb{PG}/\tau)_*$ for which $(\mathbb{PG}/\tau)^{\mathrm{c}}_*$ is a reflective subcategory of $\mathcal{C}$. When the pushout in $(\mathbb{PG}/\tau)_*$ lies in the category $\mathcal{C}$, the pushout in coherent graphs will be obtained by applying the left adjoint of the reflection. Otherwise, if the pushout in $(\mathbb{PG}/\tau)_*$ is outside $\mathcal{C}$, there is no pushout of the span in $(\mathbb{PG}/\tau)^{\mathrm{c}}_*$.

The process for determining if a pattern graph G in $(\mathbb{PG}/\tau)_*$ lies in $\mathcal{C}$ is somewhat intricate. We begin by removing from G vertices on which any partial map to any coherent graph in $(\mathbb{PG}/\tau)_*$ must be undefined due to the type constraint τ; we go through this in more detail below. We then successively merge *joinable* edges, continuing until there is no joinable pair of edges. If the result is a coherent graph, then G lies in $\mathcal{C}$ and the left adjoint applied to G is the constructed graph; otherwise, G is not in $\mathcal{C}$. It is convenient to begin the formalisation of this with the merging operation.

Definition 15. *Given a pattern graph G, a distinct pair of edges $e_1 = (v_1, \lambda_1, x_1)$ and $e_2 = (v_2, \lambda_2, x_2) \in E_G$ is* engaged *if $v_1 = v_2$ and $\lambda_1 = \lambda_2$.*

Definition 16. *Let $e_1 = (v, \lambda, x_1)$ and $e_2 = (v, \lambda, x_2)$ be an engaged pair of edges in a graph G. The graph obtained by merging them, denoted $G[e_1 \bowtie e_2]$, and homomorphism $(e_1 \bowtie e_2) : G \to G[e_1 \bowtie e_2]$ is defined as follows:*

– *if* x_1 *and* x_2 *are vertices,*

$$G[e_1 \bowtie e_2] = (V_G \setminus \{x_2\}, E_G[x_2 \mapsto x_1]) \qquad (e_1 \bowtie e_2)(w) = \begin{cases} w & \text{if } w \neq x_2 \\ x_1 & \text{if } w = x_2 \end{cases}$$

where $E_G[x_2 \mapsto x_1] = \{((e_1 \bowtie e_2)w, \lambda', (e_1 \bowtie e_2)w') \mid (w, \lambda', w') \in E_G\}$

– *if* x_1 *is a specification and* x_2 *is a vertex,*

$$G[e_1 \bowtie e_2] = (V, E \setminus \{e_1\} \cup \{(x_2, \lambda_1, \{p_1\}) | \lambda_1.p_1 \in x_1\}) \qquad (e_1 \bowtie e_2)(w) = w$$

– *if* x_2 *is a specification and* x_1 *is a vertex,*

$$G[e_1 \bowtie e_2] = (V, E \setminus \{e_2\} \cup \{(x_1, \lambda_2, \{p_2\}) | \lambda_2.p_2 \in x_2\}) \qquad (e_1 \bowtie e_2)(w) = w$$

– *if* x_1 *and* x_2 *are specifications,*

$$G[e_1 \bowtie e_2] = (V_G, E_G \setminus \{e_1, e_2\} \cup \{(v, \lambda, x_1 \cup x_2)\}) \qquad (e_1 \bowtie e_2)(w) = w$$

The process of merging engaged pairs of edges is locally confluent up to isomorphism.

Lemma 3. *Let* G *be a graph containing engaged pairs of edges* e_1, e_2 *and* f_1, f_2*. Either* $G[e_1 \bowtie e_2] \cong G[f_1 \bowtie f_2]$ *or there exist engaged pairs of edges* e'_1, e'_2 *and* f'_1, f'_2 *such that* $G[e_1 \bowtie e_2][f'_1 \bowtie f'_2] \cong G[f_1 \bowtie f_2][e'_1 \bowtie e'_2]$*.*

Repeatedly joining engaged nodes, we obtain (up to isomorphism) a coherent graph denoted $\mathsf{collapse}(G)$ and a homomorphism $\mathsf{collapse}_G : G \to \mathsf{collapse}(G)$.

For a vertex v of G, let $\lceil v \rceil_G$ denote the pattern graph obtained by restricting G to vertices and edges reachable from v.

Definition 17. *Let* $e_1 = (v_1, \lambda_1, x_1)$ *and* $e_2 = (v_2, \lambda_2, x_2)$ *be an engaged pair of edges. They are* joinable *if for all* $i, j \in \{1, 2\}$ *such that* $i \neq j$*, if* x_i *is a vertex then either* $x_i \to^* v$ *or* x_j *is a specification and* $\mathsf{collapse}_{\lceil x_i \rceil_G}(x_i) \models x_j$ *in* $\mathsf{collapse}(\lceil x_i \rceil_G)$*, where* $\to^*$ *denotes reachability in* G*.*

We now return to the initial stage, deleting vertices on grounds of the 'type' τ. As an example, let $G = \tau$ be the non-coherent graph with $V_G = \{v, w_1, w_2\}$ and $E_G = \{(v, a, w_1), (v, a, w_2), (w_1, b, w_1), (w_2, c, w_2)\}$. Let C be any coherent graph with a homomorphism $\tau_C : C \to \tau$ and $f : G \to C$ be a morphism in $(\mathbb{PG}/\tau)_*$. We cannot have $v \in \mathrm{def}(f)$ since, if it were, by closure and coherence there would have to exist a vertex $f(w_1) = f(w_2)$ with outgoing edges labelled b and c, contradicting the assumption of a homomomorphism from C to τ. The graph remaining after removal of vertices that cannot be in the domain of definition of any partial map to any coherent graph is defined as follows. Let $\tau_{\lceil v \rceil_G} : \lceil v \rceil_G \to \tau$ be the restriction of the homomorphism $\tau_G : G \to \tau$ to $\lceil v \rceil_G$.

Definition 18. *For a pattern graph* G *and homomorphism* $\tau_G : G \to \tau$*, define the graph* $\mathsf{td}(G, \tau_G) = (V_0, E_0)$ *to have vertices* $v \in V_G$ *such that there exists a homomorphism* $\tau' : \mathsf{collapse}(\lceil v \rceil_G) \to \tau$ *such that* $\tau_{\lceil v \rceil_G} = \tau' \circ \mathsf{collapse}_{\lceil v \rceil_G}$ *and* $E_0 = \{(v, \lambda, x) \in E_G \mid v \in V_0\}$*.*

Write $v \notin \mathsf{td}(G, \tau_G)$ *if* $v \in V_G \setminus V_{\mathsf{td}(G,\tau_G)}$*; it is the predicate that determines if the vertex* v *is deleted on grounds of type.*

Note that for each v, the morphism τ', if it exists, must be the unique such morphism since $\mathsf{collapse}_{\lceil v \rceil_G}$ is an epimorphism (it is surjective on vertices).

Lemma 4. *Let G be a pattern graph and $\tau_G : G \to \tau$ and $\mathsf{td}(G, \tau_G) = (V_0, E_0)$. Let $\tau_{\mathsf{td}(G,\tau_G)} : \mathsf{td}(G, \tau_G) \to \tau$ be the restriction of τ_G to V_0. The span $\mathsf{td}_G = (G \hookleftarrow \mathsf{td}(G, \tau_G) \xrightarrow{\mathsf{id}_{\mathsf{td}(G,\tau_G)}} \mathsf{td}(G, \tau_G))$ is a partial map in $(\mathbb{PG}/\tau)_*$. Furthermore, for any coherent pattern graph C and morphism $f : G \to C$ in $(\mathbb{PG}/\tau)_*$, there is a unique morphism $f^\sharp : \mathsf{td}(G, \tau_G) \to C$ such that $f = f^\sharp \circ \mathsf{td}_G$.*

The process for determining if a pattern graph lies in the category $\mathcal{C}$ begins by forming the graph $\mathsf{td}(G, \tau_G)$. We then repeatedly merge *joinable* pairs of vertices. If the resulting graph is coherent, the graph is in $\mathcal{C}$. Otherwise, if we obtain a graph with no joinable pair of edges but some engaged pair of edges, the graph cannot be in the category $\mathcal{C}$. Importantly, as we merge vertices we never re-introduce grounds for removal of other vertices due to type incompatibility.

Lemma 5. *Let G be in $(\mathbb{PG}/\tau)_*$ and there be no $v \in V_G$ such that $v \notin \mathsf{td}(G, \tau_G)$. For any joinable pair of edges e, e' in E_G, there is a unique homomorphism $\tau_{G[e \bowtie e']}$ such that the following diagram commutes:*

$$\begin{array}{ccc} G & \xrightarrow{e \bowtie e'} & G[e \bowtie e'] \\ {\scriptstyle \tau_G}\big\downarrow & \swarrow{\scriptstyle \tau_{G[e\bowtie e']}} & \\ \tau & & \end{array}$$

Furthermore, there is no $v \in V_{G[e \bowtie e']}$ such that $v \notin \mathsf{td}(G[e \bowtie e'], \tau_{G[e\bowtie e']})$.

The following lemma represents one step of proving that the constructed graph lies in the subcategory in reflection with coherent graphs.

Lemma 6. *Let G in $(\mathbb{PG}/\tau)_*$ contain no vertex v such that $v \notin \mathsf{td}(G, \tau_G)$. Let e, e' be any joinable pair of edges in G. For any C in $(\mathbb{PG}/\tau)_*$ such that C is coherent and morphism $f : G \to C$ in $(\mathbb{PG}/\tau)_*$, there is a unique morphism $f' : G[e \bowtie e'] \to C$ such that $f = f' \circ (e \bowtie e')$.*

Conversely, the following lemma is used to show that if the process of merging joinable nodes from $\mathsf{td}(G, \tau_G)$ fails, leaving a non-coherent graph with no joinable pair of edges, the graph G lies outside the category $\mathcal{C}$.

Lemma 7. *Let G in $(\mathbb{PG}/\tau)_*$ contain no vertex v such that $v \notin \mathsf{td}(G, \tau_G)$ and no pair of joinable edges. If G is not coherent, there is no P in $(\mathbb{PG}/\tau)_*$ and morphism $\phi : G \to P$ in $(\mathbb{PG}/\tau)_*$ such that P is coherent and any morphism $f : G \to C$ in $(\mathbb{PG}/\tau)_*$ to a coherent graph C factors uniquely through ϕ.*

For a sequence of pairs of edges $\mathcal{E}$ and pattern graph G, let $G\lfloor e \bowtie e' \rfloor_{\mathcal{E}}$ denote G with all pairs up to but not including (e, e') in $\mathcal{E}$ merged as in Definition 16 in sequence. Let $G[\mathcal{E}]$ denote G with all pairs in $\mathcal{E}$ merged in sequence.

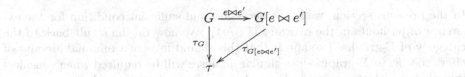

Theorem 2. *The largest full subcategory $\mathcal{C}$ of $(\mathbb{PG}/\tau)_*$ for which there is a reflection*

$$(\mathbb{PG}/\tau)^{\mathrm{c}}_* \underset{F}{\overset{\hookleftarrow}{\underset{\top}{\rightleftarrows}}} \mathcal{C}$$

consists of pattern graphs G for which there exists a sequence of pairs of edges $\mathcal{E}$ such that $\mathsf{collapse}(G_0) = G_0[\mathcal{E}]$, where $G_0 = \mathsf{td}(G, \tau_G)$, and e and e' are joinable in $G_0\lfloor e \bowtie e' \rfloor_{\mathcal{E}}$ for all $(e, e') \in \mathcal{E}$. The functor F sends G to $\mathsf{collapse}(G_0)$.

It follows categorically that this is sufficient to show the key required characterisation of pushouts:

Theorem 3. *Let $R \xleftarrow{f} L \xrightarrow{g} S$ be a span in $(\mathbb{PG}/\tau)^{\mathrm{c}}_*$ and let the cospan $R \xrightarrow{g'} T \xleftarrow{f'} S$ be its pushout in $(\mathbb{PG}/\tau)_*$.*

- *If T is not in $\mathcal{C}$ then the span $R \xleftarrow{f} L \xrightarrow{g} S$ has no pushout in $(\mathbb{PG}/\tau)^{\mathrm{c}}_*$.*
- *If T is in $\mathcal{C}$ then the cospan $R \xrightarrow{\mathsf{collapse}_{T_0} \circ \mathsf{td}_T \circ g'} \mathsf{collapse}(T_0) \xleftarrow{\mathsf{collapse}_{T_0} \circ \mathsf{td}_T \circ f'} S$ is a pushout of the span $R \xleftarrow{f} L \xrightarrow{g} S$ in $(\mathbb{PG}/\tau)^{\mathrm{c}}_*$, where $T_0 = \mathsf{td}(T, \tau_T)$.*

6 Pushouts of Σ-graphs

In the previous section, we saw a necessary and sufficient condition for the existence of pushouts in the category $(\mathbb{PG}/\tau)^{\mathrm{c}}_*$. We now tie the result back to the category of Σ-graphs. The aim is that this should involve a minimal amount of effort specific to Σ-graphs since similar analyses will be required when considering other models, for example Σ-graphs equipped with regions. In fact, the only requirement that has to be proved specifically for Kappa is Lemma 10, which establishes that if there is a regular epi from the encoding of a Σ-graph to a coherent pattern graph C then C is isomorphic to the encoding of some Σ-graph.

Lemma 8. *Let $\mathcal{C}$ be a full subcategory of $\mathcal{D}$ and suppose that every morphism in $\mathcal{D}$ is equal to a regular epi followed by a mono, $\mathcal{C}$ has finite coproducts preserved by the inclusion and any regular epi $e : C \to D$ in $\mathcal{D}$ from some C in $\mathcal{C}$ implies that $D \cong C'$ for some C' in $\mathcal{C}$. For a span of morphisms in $\mathcal{C}$, any pushout in $\mathcal{D}$ is also a pushout in $\mathcal{C}$ and any pushout in $\mathcal{C}$ is also a pushout in $\mathcal{D}$.*

Following the remark after Lemma 1, we regard the category $\Sigma\mathbb{G}_*$ as a full subcategory of $(\mathbb{PG}/\hat{\Sigma})^{\mathrm{c}}_*$. Firstly, note that $\Sigma\mathbb{G}_*$ has coproducts obtained by taking the disjoint union of Σ-graphs and that these are preserved by the inclusion. For any τ, the regular epis of $(\mathbb{PG}/\tau)^{\mathrm{c}}_*$ are characterised as follows:

Lemma 9. *A morphism $f : G \to H$ in $(\mathbb{PG}/\tau)^{\mathrm{c}}_*$ is a regular epi if, and only if:*

- *for all $w \in V_H$ there exists $v \in V_G$ such that $f(v) = w$,*
- *if $(w, \lambda, w') \in E_H$ then there exist $v, v' \in V_G$ such that $(v, \lambda, v') \in E_G$ and $f(v) = w$ (and hence $f(v') = w'$), and*
- *if $(w, \lambda, \phi) \in E_H$ for $\phi \in \mathcal{P}_{\leq}(\Lambda^*)$ then $\mathcal{S} = \{\psi \mid (v, \lambda, \psi) \in E_G \;\&\; f(v) = w\}$ is non-empty and $\phi = \bigcup \mathcal{S}$.*

Monos in $(\mathbb{PG}/\tau)^{\mathsf{c}}_{*}$ are total, injective functions on vertices. It is easy to see that any morphism in $(\mathbb{PG}/\tau)^{\mathsf{c}}_{*}$ factors as a regular epi followed by a mono. All that remains before we can apply Lemma 8 to obtain the required result about pushouts in $\Sigma\mathbb{G}_{*}$ is the following straightforward result:

Lemma 10. *For any Σ-graph S and regular epi $e : [\![S]\!] \to G$ in $(\mathbb{PG}/\hat{\Sigma})^{\mathsf{c}}_{*}$, there exists a Σ-graph T such that $G \cong [\![T]\!]$.*

We conclude by applying Lemmas 8 and 10 to characterise pushouts in $\Sigma\mathbb{G}_{*}$.

Theorem 4. *If there is no pushout in $\Sigma\mathbb{G}_{*}$ of a span $S \xleftarrow{g} L \xrightarrow{f} R$ then there is no pushout in $(\mathbb{PG}/\hat{\Sigma})^{\mathsf{c}}_{*}$ of $[\![S]\!] \xleftarrow{[\![g]\!]} [\![L]\!] \xrightarrow{[\![f]\!]} [\![R]\!]$. If there is a pushout*

$$\begin{array}{ccc} L & \xrightarrow{f} & R \\ {\scriptstyle g}\downarrow & \ulcorner & \downarrow{\scriptstyle g'} \\ S & \xrightarrow[f']{} & T \end{array} \quad \text{in } \Sigma\mathbb{G}_{*} \text{ then there is a pushout} \quad \begin{array}{ccc} [\![L]\!] & \xrightarrow{[\![f]\!]} & [\![R]\!] \\ {\scriptstyle [\![g]\!]}\downarrow & \ulcorner & \downarrow{\scriptstyle [\![g']\!]} \\ [\![S]\!] & \xrightarrow[{[\![f']\!]}]{} & [\![T]\!] \end{array} \quad \text{in } (\mathbb{PG}/\hat{\Sigma})^{\mathsf{c}}_{*}.$$

In summary, we have the following chain of functors:

$$\Sigma\mathbb{G}_{*} \xrightarrow{[\![\cdot]\!]} (\mathbb{PG}/\hat{\Sigma})^{\mathsf{c}}_{*} \underset{\text{collapse}}{\overset{\top}{\rightleftarrows}} \mathcal{C} \hookrightarrow (\mathbb{PG}/\hat{\Sigma})_{*}$$

To determine the pushout of a span in $\Sigma\mathbb{G}_{*}$, we take the pushout in $(\mathbb{PG}/\hat{\Sigma})_{*}$ of its encoding. If this is outside $\mathcal{C}$ as characterised in Theorem 2, there is no pushout of the span in $\Sigma\mathbb{G}_{*}$. Otherwise, the pushout is the cospan in $\Sigma\mathbb{G}_{*}$ that is isomorphic under the encoding to the collapse of the pushout taken in $(\mathbb{PG}/\hat{\Sigma})_{*}$.

7 Conclusion

This paper has begun the work of placing Kappa in a more general graph rewriting setting, abstracting away features of Σ-graphs that are tailored to the efficient representation of biochemical signalling pathways to arrive at rewriting based on pattern graphs. The central features are the capture of implicit deletion through the use of closed partial maps and the characterisation of pushouts for categories of pattern graphs.

Alongside the work presented in this paper, we have studied categorical conditions for the existence of pushouts in categories of partial maps, that for example can show that $(\mathbb{PG}/\tau)_{*}$ has pushouts. In this paper, we have focused on the encoding of Σ-graphs, showing how pushouts in $\Sigma\mathbb{G}_{*}$ can be obtained in $(\mathbb{PG}/\hat{\Sigma})_{*}$. At the core of this was the intricate consideration of pushouts in the subcategory of coherent pattern graphs.

Though there has not been space to present it here, we expect that this work applies without complications to a wide range of biochemical models. For example, categories for Kappa with regions [8] can be encoded in pattern graphs, resulting in a simpler characterisation of their pushouts. We also intend to give the BioNetGen Language [2] its first categorical interpretation using the framework developed. Another area for further research is to translate the work on dynamic restriction by type in Kappa presented in [5] to the current setting.

On rewriting, there are areas where further generalisation would be of interest. A more expressive logical formalism for patterns could be adopted, connected to the work on application conditions for rules in graph transformation [7]. It would also be interesting to consider the role of negative application conditions, perhaps specified as open maps [9], in this setting; in [4], these were used to constrain matchings. Finally, and more speculatively, by studying bi-pushouts in bi-categories of spans, a new perspective on span-based rewriting approaches (see, e.g. [11,3]) that allows duplication of entities might be obtained.

More abstractly, the restriction to finite structures in this paper can be lifted fairly straightforwardly; we intend to present the details in a journal version of this paper. However, for modelling biochemical pathways, the restriction to finite structures is no limitation.

References

1. Bachman, J.A., Sorger, P.: New approaches to modeling complex biochemistry. Nature Methods 8(2), 130–131 (2011)
2. Blinov, M.L., Yang, J., Faeder, J.R., Hlavacek, W.S.: Graph Theory for Rule-Based Modeling of Biochemical Networks. In: Priami, C., Ingólfsdóttir, A., Mishra, B., Riis Nielson, H. (eds.) Transactions on Computational Systems Biology VII. LNCS (LNBI), vol. 4230, pp. 89–106. Springer, Heidelberg (2006)
3. Corradini, A., Heindel, T., Hermann, F., König, B.: Sesqui-Pushout Rewriting. In: Corradini, A., Ehrig, H., Montanari, U., Ribeiro, L., Rozenberg, G. (eds.) ICGT 2006. LNCS, vol. 4178, pp. 30–45. Springer, Heidelberg (2006)
4. Danos, V., Feret, J., Fontana, W., Harmer, R., Hayman, J., Krivine, J., Thompson-Walsh, C., Winskel, G.: Graphs, rewriting and pathway reconstruction for rule-based models. In: Proc. FSTTCS 2012. LIPICs (2012)
5. Danos, V., Harmer, R., Winskel, G.: Constraining rule-based dynamics with types. MSCS (2012)
6. Danos, V., Laneve, C.: Formal molecular biology. TCS 325 (2004)
7. Habel, A., Pennemann, K.-H.: Nested Constraints and Application Conditions for High-Level Structures. In: Kreowski, H.-J., Montanari, U., Yu, Y., Rozenberg, G., Taentzer, G. (eds.) Formal Methods in Software and Systems Modeling. LNCS, vol. 3393, pp. 293–308. Springer, Heidelberg (2005)
8. Hayman, J., Thompson-Walsh, C., Winskel, G.: Simple containment structures in rule-based modelling of biochemical systems. In: Proc. SASB (2011)
9. Heckel, R.: DPO Transformation with Open Maps. In: Ehrig, H., Engels, G., Kreowski, H.-J., Rozenberg, G. (eds.) ICGT 2012. LNCS, vol. 7562, pp. 203–217. Springer, Heidelberg (2012)
10. Löwe, M.: Algebraic approach to single-pushout graph transformation. TCS 109 (1993)
11. Löwe, M.: Refined Graph Rewriting in Span-Categories. In: Ehrig, H., Engels, G., Kreowski, H.-J., Rozenberg, G. (eds.) ICGT 2012. LNCS, vol. 7562, pp. 111–125. Springer, Heidelberg (2012)

History-Register Automata

Nikos Tzevelekos and Radu Grigore

Queen Mary, University of London

Abstract. Programs with dynamic allocation are able to create and use an unbounded number of fresh resources, such as references, objects, files, etc. We propose History-Register Automata (HRA), a new automata-theoretic formalism for modelling and analysing such programs. HRAs extend the expressiveness of previous approaches and bring us to the limits of decidability for reachability checks. The distinctive feature of our machines is their use of unbounded memory sets (histories) where input symbols can be selectively stored and compared with symbols to follow. In addition, stored symbols can be consumed or deleted by reset. We show that the combination of consumption and reset capabilities renders the automata powerful enough to imitate counter machines (Petri nets with reset arcs), and yields closure under all regular operations apart from complementation. We moreover examine weaker notions of HRAs which strike different balances between expressiveness and effectiveness.

1 Introduction

Program analysis faces substantial challenges due to its aim to devise finitary methods and machines which are required to operate on potentially infinite program computations. A specific such challenge stems from dynamic generative behaviours such as, for example, object or thread creation in Java, or reference creation in ML. A program engaging in such behaviours is expected to generate a possibly unbounded amount of distinct resources, each of which is assigned a unique identifier, a *name*. Hence, any machine designed for analysing such programs is expected to operate on an infinite alphabet of names. The latter need has brought about the introduction of automata over infinite alphabets in program analysis, starting from prototypical machines for mobile calculi [23] and variable programs [18], and recently developing towards automata for verification tasks such as equivalence checks of ML programs [24,25], context-bounded analysis of concurrent programs [7,3] and runtime program monitoring [14].

The literature on automata over infinite alphabets is rich in formalisms each based on a different approach for tackling the infiniteness of the alphabet in a finitary manner (see e.g. [31] for an overview). A particularly intuitive such model is that of *Register Automata (RA)* [18,26], which are machines built around the concept of an ordinary finite-state automaton attached with a fixed finite amount of registers. The automaton can store in its registers names coming from the input, and make control decisions by comparing new input names with those already stored. Thus, by talking about addresses of its memory registers rather than actual names, a so finitely-described automaton can tackle the infinite alphabet of names. Driven by program analysis considerations, register automata have been recently extended with the feature of name-freshness recognition [33], that is, the capability of the automaton to accept specific inputs just if they

F. Pfenning (Ed.): FOSSACS 2013, LNCS 7794, pp. 17–33, 2013.

are *fresh* – they have not appeared before during computation. Those automata, called *Fresh-Register Automata (FRA)*, can account for languages like the following,

$$\mathcal{L}_0 = \{a_1 \cdots a_n \in \mathcal{N}^* \mid \forall i \neq j.\ a_i \neq a_j\}$$

which captures the output of a fresh-name generator ($\mathcal{N}$ is an infinite set of names). FRAs are expressive enough to model, for example, finitary fragments of languages like the π-calculus [33] or ML [24].

The freshness oracle of FRAs administers the automata with perhaps too restricted an access to the full history of the computation: it allows them to detect name freshness, but not non-freshness. Consider, for instance, the following simple language,

$$\mathcal{L}' = \{w \in (\{O,P\} \times \mathcal{N})^* \mid \text{each element of } w \text{ appears exactly once in it} \wedge \text{ each } (O,a) \text{ in } w \text{ is preceded by some } (P,a)\}$$

where the alphabet is made of pairs containing an element from the set $\{O,P\}$ and a name (O and P can be seen as different processes, or agents, exchanging names). The language $\mathcal{L}'$ represents a paradigmatic scenario of a name generator P coupled with a name consumer O: each consumed name must have been created first, and no name can be consumed twice. It can capture e.g. the interaction of a process which creates new files with one that opens them, where no file can be opened twice. The inability of FRAs to detect non-freshness, as well as the fact that names in their history cannot be removed from it, do not allow them to express $\mathcal{L}'$. More generally, the notion of *re-usage* or *consumption* of names is beyond the reach of those machines. Another limitation of FRAs is the failure of closure under concatenation, interleaving and Kleene star.

Aiming at providing a stronger theoretical tool for analysing computation with names, in this work we further capitalise on the use of histories by effectively upgrading them to the status of registers. That is, in addition to registers, we equip our automata with a fixed number of unbounded sets of names (*histories*) where input names can be stored and compared with names to follow. As histories are internally unordered, the kind of name comparison we allow for is name belonging (*does the input name belong to the i-th history?*). Moreover, names can be selected and removed from histories, and individual histories can be emptied/reset. We call the resulting machines *History-Register Automata (HRA)*. For example, $\mathcal{L}'$ is accepted by the HRA with 2 histories

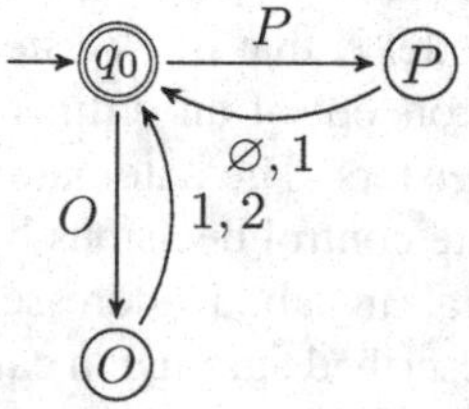

The automaton starts at state q_0 with empty history and non-deterministically makes a transition to state P or Q, accepting the respective symbol. From state P, it accepts any input name a which does not appear in any of its histories (this is what $\varnothing$ stands for), puts it in history number 1, and moves back to q_0. From state O, it accepts any input name a which appears in history number 1, puts it in history number 2, and moves back to q_0.

Fig. 1. History-register automaton accepting $\mathcal{L}'$

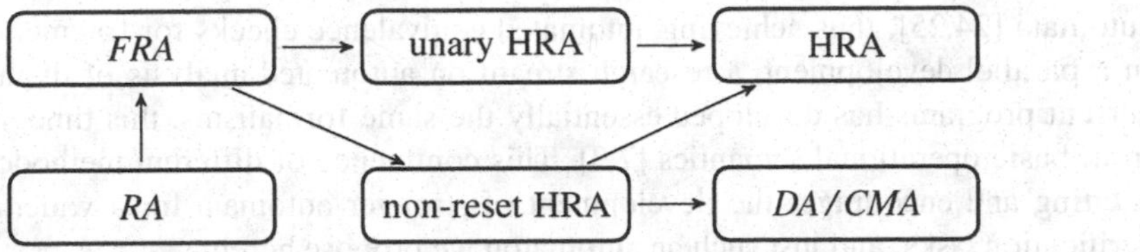

Fig. 2. Expressiveness of history-register automata compared to previous models (in italics). The inclusion $\mathcal{M} \longrightarrow \mathcal{M}'$ means that for each $\mathcal{A} \in \mathcal{M}$ we can effectively construct an $\mathcal{A}' \in \mathcal{M}'$ accepting the same language as $\mathcal{A}$. All inclusions are strict.

depicted in Figure 1, where by convention we model pairs of symbols by sequences of two symbols.[1]

The strengthening of the role of histories substantially increases the expressive power of our machines. More specifically, we identify three distinctive features of HRAs: (1) the capability to reset histories; (2) the use of multiple histories; (3) the capability to select and remove individual names from histories. Each feature allows us to express one of the paradigmatic languages below, none of which are FRA-recognisable.

$$\mathcal{L}_1 = \{a_0 w_1 \cdots a_0 w_n \in \mathcal{N}^* \mid \forall i.\, w_i \in \mathcal{N}^* \wedge a_0 w_i \in \mathcal{L}_0\} \quad \text{for given } a_0$$
$$\mathcal{L}_2 = \{a_1 a_1' \cdots a_n a_n' \in \mathcal{N}^* \mid a_1 \cdots a_n, a_1' \cdots a_n' \in \mathcal{L}_0\}$$
$$\mathcal{L}_3 = \{a_1 \cdots a_n a_1' \cdots a_{n'}' \in \mathcal{N}^* \mid a_1 \cdots a_n, a_1' \cdots a_{n'}' \in \mathcal{L}_0 \wedge \forall i. \exists j.\, a_i' = a_j\}$$

Apart from the gains in expressive power, the passage to HRAs yields a more well-rounded automata-theoretic formalism for generative behaviours as these machines enjoy closure under all regular operations apart from complementation. On the other hand, the combination of features (1-3) above enable us to use histories as counters and simulate counter machines, and in particular Petri nets with reset arcs [2]. We therefore obtain non-primitive recursive bounds for checking language emptiness. Given that language containment and universality are undecidable already for register automata [26], HRAs are fairly close to the decidability boundary for properties of languages over infinite alphabets. Nonetheless, starting from HRAs and weakening them in each of the first two factors (1,2) we obtain automata models which are still highly expressive but computationally more tractable. Overall, the expressiveness hierarchy of the machines we examine is depicted in Figure 1 (weakening in (2) and (1) respectively occurs in the second column of the figure).

Motivation and Related Work. The motivation for this work stems from semantics and verification. In semantics, the use of names to model resource generation originates in the work of Pitts and Stark on the ν-calculus [27] and Stark's PhD [32]. Names have subsequently been incorporated in the semantics literature (see e.g. [16,4,1,19]), especially after the advent of *Nominal Sets* [13], which provided formal foundations for doing mathematics with names. Moreover, recent work in game semantics has produced algorithmic representations of game models using extensions of fresh-register

[1] Although, technically speaking, the machines we define below do not handle constants (as e.g. O, P), the latter are encoded as names appearing in initial registers, in standard fashion.

automata [24,25], thus achieving automated equivalence checks for fragments of ML. In a parallel development, a research stream on automated analysis of dynamic concurrent programs has developed essentially the same formalisms, this time stemming from basic operational semantics [7,3]. This confluence of different methodologies is exciting and encourages the development of stronger automata for a wider range of verification tasks, and just such an automaton we propose herein.

Although our work is driven by program analysis, the closest existing automata models to ours come from XML database theory and model checking. Research in the latter area has made great strides in the last years on automata over infinite alphabets and related logics (e.g. see [31] for an overview from 2006). As we show in this paper, history-register automata fit very well inside the big picture of automata over infinite alphabets (cf. Figure 1) and in fact can be seen as a variant of *Data Automata (DA)* [6] or, equivalently, *Class Memory Automata (CMA)* [5]. This fit leaves space for transfer of technologies and, more specifically, of the associated logics of data automata.

2 Definitions and First Properties

We start by fixing some notation. Let $\mathcal{N}$ be an infinite alphabet of ***names*** (or *data values*, in terminology of [31]), which we range over by a, b, c, etc. For any pair of natural numbers $i \leq j$, we write $[i,j]$ for the set $\{i, i+1, \cdots, j\}$, and for each i we let $[i]$ be the set $\{1, \cdots, i\}$. For any set S, we write $|S|$ for the cardinality of S, $\mathcal{P}(S)$ for the powerset of S, $\mathcal{P}_{\mathsf{fn}}(S)$ for the set of finite subsets of S, and $\mathcal{P}_{\neq\emptyset}(S)$ for the set of non-empty subsets of S. We write $\mathsf{id} : S \to S$ for the identity function on S, and $\mathsf{img}(f)$ for the image of $f \colon S \to T$.

We define automata which are equipped with a fixed number of ***registers*** and ***histories*** where they can store names. Each register is a memory cell where one name can be stored at a time; each history can hold an unbounded set of names. We use the term ***place*** to refer to both histories and registers. Transitions are of two kinds: name-accepting transitions and reset transitions. Those of the former kind have labels of the form (X, X'), for sets of places X and X'; and those of the latter carry labels with single sets of places X. A transition labelled (X, X') means:

- accept name a if it is contained precisely in places X, and
- update places in X and X' so that a be contained precisely in places X' after the transition (without touching other names).

By a being contained precisely in places X we mean that it appears in every place in X, and in no other place. In particular, the label $(\emptyset, X')$ signifies accepting a fresh name (one which does not appear in any place) and inserting it in places X'. On the other hand, a transition labelled by X resets all the places in X, that is, it updates each of them to the empty set. Reset transitions do not accept names; they are ϵ-transitions from the outside. Note then that the label $(X, \emptyset)$ has different semantics from the label X: the former stipulates that a name appearing precisely in X be accepted and then removed from X; whereas the latter clears all the contents of places in X, without accepting anything.

Formally, let us fix positive integers m and n which will stand for the default number of histories and registers respectively in the machines we define below. The set Asn of ***assignments*** and the set Lab of ***labels*** are:

$$\mathsf{Asn} = \{H : [m{+}n] \to \mathcal{P}_{\mathsf{fn}}(\mathcal{N}) \mid \forall i > m.\, |H(i)| \leq 1\}$$
$$\mathsf{Lab} = \mathcal{P}([m{+}n])^2 \cup \mathcal{P}([m{+}n])$$

For example, $\{(i,\emptyset) \mid i \in [m{+}n]\}$ is the empty assignment. We range over elements of Asn by H and variants, and over elements of Lab by ℓ and variants. Moreover, it will be handy to introduce the following notation for assignments. For any assignment H and any $a \in \mathcal{N}$, $S \subseteq \mathcal{N}$ and $X \subseteq [m{+}n]$:

- We set $H@X$ to be the set of names which *appear precisely* in places X in H, that is, $H@X = \bigcap_{i \in X} H(i) \setminus \bigcup_{i \notin X} H(i)$.
  In particular, $H@\emptyset = \mathcal{N} \setminus \bigcup_i H(i)$ is the set of names which do not appear in H.
- $H[X \mapsto S]$ is the update H' of H so that all places in X are mapped to S, that is, $H' = \{(i, H(i)) \mid i \notin X\} \cup \{(i,S) \mid i \in X\}$. E.g. $H[X \mapsto \emptyset]$ resets all places in X.
- $H[a \text{ in } X]$ is the update of H which removes name a from all places and inserts it back in X, that is, $H[a \text{ in } X]$ is the assignment:
  $\{(i, H(i) \cup \{a\}) \mid i \in X \cap [m]\} \cup \{(i, \{a\}) \mid i \in X \setminus [m]\} \cup \{(i, H(i) \setminus \{a\}) \mid i \notin X\}$

Note above that operation $H[a \text{ in } X]$ acts differently in the case of histories ($i \leq m$) and registers ($i > m$) in X: in the former case, the name a is added to the history $H(i)$, while in the latter the register $H(i)$ is set to $\{a\}$ and its previous content is cleared.

We can now define our automata.

Definition 1. *A* **history-register automaton (HRA)** *of type* (m,n) *is a tuple* $\mathcal{A} = \langle Q, q_0, H_0, \delta, F \rangle$ *where:*
- *Q is a finite set of states, q_0 is the initial state, $F \subseteq Q$ are the final ones,*
- *$H_0 \in \mathsf{Asn}$ is the initial assignment, and*
- *$\delta \subseteq Q \times \mathsf{Lab} \times Q$ is the transition relation.*

For brevity, we shall call $\mathcal{A}$ an (m,n)-HRA.

We write transitions in the forms $q \xrightarrow{X,X'} q'$ and $q \xrightarrow{X} q'$, for each kind of transition label. In diagrams, we may unify different transitions with common source and target, for example $q \xrightarrow{X,X'} q'$ and $q \xrightarrow{Y,Y'} q'$ may be written $q \xrightarrow{X,X'\,/\,Y,Y'} q'$; moreover, we shall lighten notation and write i for the singleton $\{i\}$, and ij for $\{i,j\}$.

We already gave an overview of the semantics of HRAs. This is formally defined by means of configurations representing the current computation state of the automaton. A ***configuration*** of $\mathcal{A}$ is a pair $(q, H) \in \hat{Q}$, where:

$$\hat{Q} = Q \times \mathsf{Asn}$$

From the transition relation δ we obtain the configuration graph of $\mathcal{A}$ as follows.

Definition 2. *Let $\mathcal{A}$ be an (m,n)-HRA as above. Its* **configuration graph** $(\hat{Q}, \longrightarrow)$, *where* $\longrightarrow \subseteq \hat{Q} \times (\mathcal{N} \cup \{\epsilon\}) \times \hat{Q}$, *is constructed by setting* $(q,H) \xrightarrow{x} (q',H')$ *just if one of the following conditions is satisfied.*

– $x = a \in \mathcal{N}$ *and there is* $q \xrightarrow{X,X'} q' \in \delta$ *such that* $a \in H@X$ *and* $H' = H[a \,\text{in}\, X']$.
– $x = \epsilon$ *and there is* $q \xrightarrow{X} q' \in \delta$ *such that* $H' = H[X \mapsto \emptyset]$.

The language accepted by $\mathcal{A}$ *is* $\mathcal{L}(\mathcal{A}) = \{ w \in \mathcal{N}^* \mid (q_0, H_0) \xrightarrow{w}\!\!\!\!\rightarrow (q, H)$ *and* $q \in F \}$ *where* $\longrightarrow\!\!\!\!\rightarrow$ *is the reflexive transitive closure of* $\longrightarrow$ *(i.e.* $\hat{q} \xrightarrow{x_1 \cdots x_n}\!\!\!\!\rightarrow \hat{q}'$ *if* $\hat{q} \xrightarrow{x_1} \cdots \xrightarrow{x_n} \hat{q}'$*).*

Note that we use ϵ both for the empty sequence and the empty transition so, in particular, when writing sequences of the form $x_1 \cdots x_n$ we may implicitly consume ϵ's.

Example 3. The language $\mathcal{L}_1$ of the Introduction is recognised by the following $(1,1)$-HRA (leftmost below), with initial assignment $\{(1,\emptyset),(2,a_0)\}$. The automaton starts by accepting a_0, leaving it in register 2, and moving to state q_1. There, it loops accepting fresh names (appearing in no place) which it stores in history 1. From q_1 it goes back to q_1 by resetting its history.

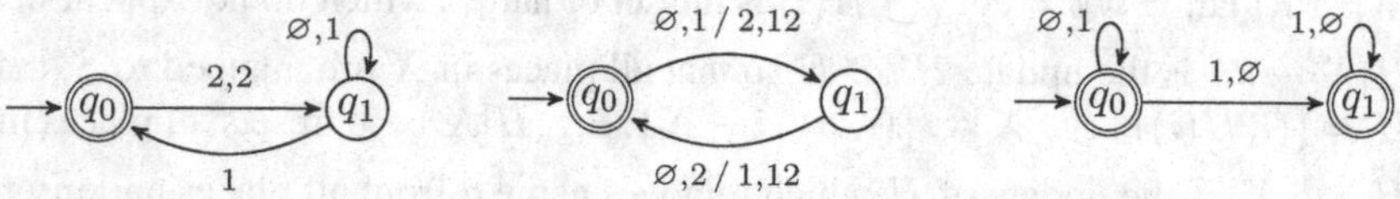

We can also see that the other two HRAs, of type $(2,0)$ and $(1,0)$, accept the languages $\mathcal{L}_2$ and $\mathcal{L}_3$ respectively. Both automata start with empty assignments.
Finally, the automaton we drew in Figure 1 is, in fact, a (2,2)-HRA where its two registers initially contain the names O and P respectively. The transition label O corresponds to $(3,3)$, and P to $(4,4)$.

As mentioned in the introductory section, HRAs build upon *(Fresh) Register Automata* [18,26,33]. The latter can be defined within the HRA framework as follows.[2]

Definition 4. *A* Register Automaton (RA) *of* n *registers is a* $(0,n)$*-HRA with no reset transitions. A* Fresh-Register Automaton (FRA) *of* n *registers is a* $(1,n)$*-HRA* $\mathcal{A} = \langle Q, q_0, H_0, \delta, F\rangle$ *such that* $H_0(1) = \bigcup_i H_0(i)$ *and:*

– *for all* $(q,\ell,q') \in \delta$*, there are* X, X' *such that* $\ell = (X,X')$ *and* $1 \in X'$*;*
– *for all* $(q,\{1\},X',q') \in \delta$*, there is also* $(q,\emptyset,X',q') \in \delta$*.*

Thus, in an FRA all the initial names must appear in its history, and the same holds for all the names the automaton accepts during computation ($1 \in X'$). As, in addition, no reset transitions are allowed, the history effectively contains all names of a run. On the other hand, the automaton cannot recognise *non-freshness*: if a name appearing only in the history is to be accepted at any point then a totally fresh name can be also be accepted in the same way. Now, from [33] we have the following.[3]

Lemma 5. *The languages* $\mathcal{L}_1, \mathcal{L}_2$ *and* $\mathcal{L}_3$ *are not FRA-recognisable.*

Bisimulation Bisimulation equivalence, also called *bisimilarity*, is a useful tool for relating automata, even from different paradigms. It implies language equivalence and is generally easier to reason about than the latter. We will be using it avidly in the sequel.

[2] The definitions given in [18,26,33] are slightly different but can routinely be shown equivalent.
[3] $\mathcal{L}_1$ was explicitly examined in [33]. For $\mathcal{L}_2$ and $\mathcal{L}_3$ we use a similar argument as the one for showing that $\mathcal{L}_0 * \mathcal{L}_0$ is not FRA-recognisable [33].

Definition 6. *Let $\mathcal{A}_i = \langle Q_i, q_{0i}, H_{0i}, \delta_i, F_i \rangle$ be (m,n)-HRAs, for $i = 1,2$. A relation $R \subseteq \hat{Q}_1 \times \hat{Q}_2$ is called a* simulation *on $\mathcal{A}_1$ and $\mathcal{A}_2$ if, for all $(\hat{q}_1, \hat{q}_2) \in R$,*

- *if $\hat{q}_1 \xrightarrow{\epsilon}\!\!\!\!\rightarrow \hat{q}_1'$ and $\pi_1(\hat{q}_1') \in F_1$ then $\hat{q}_2 \xrightarrow{\epsilon}\!\!\!\!\rightarrow \hat{q}_2'$ for some $\pi_1(\hat{q}_2') \in F_2$, where π_1 is the first projection function;*
- *if $\hat{q}_1 \xrightarrow{\epsilon}\!\!\!\!\rightarrow \cdot \xrightarrow{a} \hat{q}_1'$ then $\hat{q}_2 \xrightarrow{\epsilon}\!\!\!\!\rightarrow \cdot \xrightarrow{a} \hat{q}_2'$ for some $(\hat{q}_1', \hat{q}_2') \in R$.*

R is called a **bisimulation** *if both R and R^{-1} are simulations. We say that $\mathcal{A}_1$ and $\mathcal{A}_2$ are* **bisimilar**, *written $\mathcal{A}_1 \sim \mathcal{A}_2$, if $((q_{01}, H_{01}), (q_{02}, H_{02})) \in R$ for some bisimulation R.*

The following is a standard result.

Lemma 7. *If $\mathcal{A}_1 \sim \mathcal{A}_2$ then $\mathcal{L}(\mathcal{A}_1) = \mathcal{L}(\mathcal{A}_2)$.*

As a first taste of HRA reasoning, we sketch a technique for simulating registers by histories in HRAs. The idea is to represent a register by a history whose size is always kept at most 1. To ensure that histories are effectively kept in size ≤ 1 they must be cleared before inserting names, which in turn complicates deciding when a transition can be taken as it may depend on the deleted names. To resolve this, we keep two copies of each register so that, for each transition with label (X, X'), we use one set of copies for the name comparisons needed for the X part of the label, and the other set for the assignments dictated by X'. Resets are used so that one set of copies is always empty.

Proposition 8. *Let $\mathcal{A} = \langle Q, q_0, H_0, \delta, F \rangle$ be an (m,n)-HRA. There is an $(m+2n, 0)$-HRA $\mathcal{A}'$ such that $\mathcal{A} \sim \mathcal{A}'$.*

In Proposition 20 we show that registers can be simulated also without using resets. Both that and the above reductions, though, come at the cost of an increased number of histories and, more importantly, the simulation technique obscures the intuition of registers and produces automata which need close examination even for simple languages like the one which contains all words $a_1 \cdots a_n$ such that $a_i \neq a_{i+1}$ for all i (see Example 21). As, in addition, it is not applicable to the weaker unary HRAs we examine in Section 4, we preferred to explicitly include registers in HRAs. Another design choice regards the use of sets of places in transitions instead e.g. of single places. Although the latter description would lead to an equivalent and probably conciser formalism, it would be inconvenient for combining HRAs e.g. in order to produce the intersection of their accepted languages. In fact, our formulation follows *M-automata* [18], an equivalent presentation of RAs susceptible to closure constructions.

Determinism. We close our presentation here by describing the deterministic class of HRAs. We defined HRAs in such a way that, at any given configuration (q, H) and for any input symbol a, there is at most one set of places X that can match a, i.e. such that $a \in H@X$. As a result, the notion of determinism in HRAs can be ensured by purely syntactic means. Below we write $q \xrightarrow{X}\!\!\!\!\rightarrow q' \in \delta$ if there is a sequence of transitions $q \xrightarrow{X_1} \cdots \xrightarrow{X_n} q'$ in δ such that $X = \bigcup_{i=1}^{n} X_i$. In particular, $q \xrightarrow{\emptyset}\!\!\!\!\rightarrow q \in \delta$.

Definition 9. *We say that an HRA $\mathcal{A}$ is* **deterministic** *if, for any reachable configuration $\hat{q}$ and any name a, if $\hat{q} \xrightarrow{\epsilon}\!\!\!\!\rightarrow \cdot \xrightarrow{a} \hat{q}_1$ and $\hat{q} \xrightarrow{\epsilon}\!\!\!\!\rightarrow \cdot \xrightarrow{a} \hat{q}_2$ then $\hat{q}_1 = \hat{q}_2$.*

$\mathcal{A}$ is **strongly deterministic** *if $q \xrightarrow{Y_1}\!\!\!\!\rightarrow \cdot \xrightarrow{X \setminus Y_1, X_1} q_1 \in \delta$ and $q \xrightarrow{Y_2}\!\!\!\!\rightarrow \cdot \xrightarrow{X \setminus Y_2, X_2} q_2 \in \delta$ imply $q_1 = q_2$, $Y_1 = Y_2$ and $X_1 = X_2$.*

Lemma 10. *If $\mathcal{A}$ is strongly deterministic then it is deterministic.*

3 Closure Properties, Emptiness and Universality

History-register automata enjoy good closure properties with respect to regular language operations. In particular, they are closed under union, intersection, concatenation and Kleene star, but not closed under complementation.

In fact, the design of HRAs is such that the automata for union and intersection come almost for free through a straightforward product construction which is essentially an ordinary product for finite-state automata, modulo reindexing of places to account for duplicate labels (cf. [18]). The constructions for Kleene star and concatenation are slightly more involved as we make use of the following technical gadget. Given an (m, n)-HRA $\mathcal{A}$ and a sequence w of k distinct names, we construct a bisimilar $(m, n{+}k)$-HRA, denoted $\mathcal{A}\,\mathsf{fix}\,w$, in which the names of w appear exclusively in the additional k registers, which, moreover, remain unchanged during computation. The construction allows us, for instance, to create loops such that after each loop transition the same initial configuration occurs (in this case, w would enlist all initial names).

Proposition 11. *Languages recognised by HRAs are closed under union, intersection, concatenation and Kleene star.*

As we shall next see, while universality is undecidable for HRAs, their emptiness problem can be decided by reduction to coverability for transfer-reset vector addition systems with states. In combination, these results imply that HRAs cannot be effectively complemented. In fact, there are HRA-languages whose complements are not recognisable by HRAs. This can be shown via the following example, adapted from [22].

Lemma 12. *HRAs are not closed under complementation.*

Example 13. Consider $\mathcal{L}_4 = \{w \in \mathcal{N}^* \mid \text{not all names of } w \text{ occur exactly twice in it}\}$, which is accepted by the $(2, 0)$-HRA below, where "–" can be any of $\varnothing, 1, 2$.

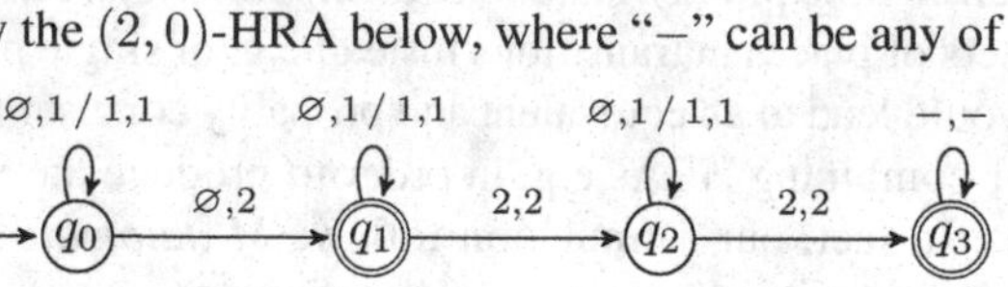

The automaton non-deterministically selects an input name which either appears only once in the input or at least three times.

We claim that $\overline{\mathcal{L}_4}$, the language of all words whose names occur exactly twice in them, is not HRA-recognisable. For suppose it were recognisable (wlog) by an $(m, 0)$-HRA $\mathcal{A}$ with k states. Then, $\mathcal{A}$ would accept the word $w = a_1 \cdots a_k\, a_1 \cdots a_k$ where all a_i's are distinct and do not appear in the initial assignment of $\mathcal{A}$. Let $p = p_1 p_2$ be the path in $\mathcal{A}$ through which w is accepted, with each p_i corresponding to one of the two halves of w. Since all a_is are fresh for $\mathcal{A}$, the non-reset transitions of p_1 must carry labels of the form $(\emptyset, X)$, for some sets X. Let q be a state appearing twice in p_1, say $p_1 = p_{11}(q)p_{12}(q)p_{13}$. Consider now the path $p' = p_1' p_2$ where p_1' is the extension of p_1 which repeats p_{12}, that is, $p_1' = p_{11}(q)p_{12}(q)p_{12}(q)p_{13}$. We claim that p' is an accepting path in $\mathcal{A}$. Indeed, by our previous observation on the labels of p_1, the path

p_1' does not block, i.e. it cannot reach a transition $q_1 \xrightarrow{X,Y} q_2$, with $X \neq \emptyset$, in some configuration (q_1, H_1) such that $H_1@X = \emptyset$. We need to show that p_2 does not block either (in p'). Let us denote (q, H_1) and (q, H_2) the configurations in each of the two visits of q in the run of p on w; and let us write (q, H_3) for the third visit in the run of p_1', given that for the other two visits we assume the same configurations as in p. Now observe that, for each non-empty $X \subseteq [m]$, repeating p_{12} cannot reduce the number of names appearing precisely in X, therefore $|H_2@X| \leq |H_3@X|$. The latter implies that, since p does not block, p' does not block either. Now observe that any word accepted by w' is not in $\overline{\mathcal{L}_4}$, as p_1' accepts more than k distinct names, a contradiction.

We now turn to the question of checking emptiness. The use of unbounded histories effectively renders our machines into counter automata: where a counter automaton would increase (or decrease) a counter, an HRA would add (remove) a name from one of its histories, or set of histories. Nonetheless, HRAs cannot decide their histories for emptiness, which leaves space for decidability.[4] The capability for resetting histories, on the other hand, leads us to consider Transfer-Reset Vector Addition Systems with States [8,2] (i.e. Petri nets with reset and transfer arcs) as appropriate formalisms for this question.

A ***Transfer-Reset Vector Addition System with States (TR-VASS)*** of m dimensions is a tuple $\mathcal{A} = \langle Q, \delta \rangle$, with Q a set of states and $\delta \subseteq Q \times (\{-1,0,1\}^m \cup [m]^2 \cup [m]) \times Q$ a transition relation. Each dimension of $\mathcal{A}$ corresponds to an unbounded counter. Thus, a transition of $\mathcal{A}$ can either update its counters by addition of a vector $\vec{v} \in \{-1,0,1\}^m$, or transfer the value of one counter to another, or reset some counter.

Formally, a configuration of $\mathcal{A}$ is a pair $(q, \vec{v}) \in Q \times \mathbb{N}^m$ consisting of a state and a vector of values stored in the counters. The configuration graph of $\mathcal{A}$ is constructed by including an edge $(q, \vec{v}) \to (q', \vec{v}')$ if:

- there is some $(q, \vec{v}'', q') \in \delta$ such that $\vec{v}' = \vec{v} + \vec{v}''$, or
- there is $(q, i, j, q') \in \delta$ such that $\vec{v}' = (\vec{v}[j \mapsto v_i + v_j])[i \mapsto 0]$,
- or there is some $(q, i, q') \in \delta$ such that $\vec{v}' = \vec{v}[i \mapsto 0]$;

where we write v_i for the ith dimension of $\vec{v}$, and $\vec{v}[i \mapsto v']$ for the update of $\vec{v}$ where the i-th counter is set to v'. An ***R-VASS*** is a TR-VASS without transfer transitions.

The ***control-state reachability*** problem for TR-VASSs is defined as follows. Given a TR-VASS $\mathcal{A}$ of m dimensions, a configuration $(q_0, \vec{v}_0)$ and a state q, is there some $\vec{v} \in \mathbb{N}^m$ such that $(q_0, \vec{v}_0) \twoheadrightarrow (q, \vec{v})$? In such a case, we write $(\mathcal{A}, q_0, \vec{v}_0, q) \in \mathsf{Reach}$.

Fact 14 ([10,30,11]). *Control-state reachability for TR-VASSs and R-VASSs is decidable and has non-primitive recursive complexity.*

We next reduce HRA nonemptiness to TR-VASS control-state reachability. Starting w.l.o.g. from an $(m, 0)$-HRA $\mathcal{A}$, we construct a TR-VASS $\mathcal{A}'$ with 2^m dimensions: one dimension $\widetilde{X}$ for each $X \subseteq [m]$. The dimension $\widetilde{\emptyset}$ is used for garbage collecting. We assign to each state of $\mathcal{A}$ a corresponding state in $\mathcal{A}'$ (and also include a stock of dummy states for intermediate transitions) and translate the transitions of $\mathcal{A}$ into transitions of $\mathcal{A}'$ as follows.

[4] Recall that 2-counter machines with increase, decrease and check for zero are Turing complete.

– Each transition with label (X, X') is mapped into a pair of transitions which first decrease counter $\widetilde{X}$ and then increase $\widetilde{X'}$.
– Each reset transition with label X causes a series of transfers: for each counter $\widetilde{Y}$, we do a transfer from $\widetilde{Y}$ to $\widetilde{Y \setminus X}$.

Thus, during computation in each of $\mathcal{A}'$ and $\mathcal{A}$, the value of counter $\widetilde{X}$ matches the number of names which precisely appear in histories X. Since, for checking emptiness, the specific names inside the histories of $\mathcal{A}$ are of no relevance, the above correspondence extends to matching nonemptiness for $\mathcal{A}$ to (final) control-state reachability for $\mathcal{A}'$.

Proposition 15. *Emptiness is decidable for HRAs.*

Doing the opposite reduction we can show that emptiness of even strongly deterministic HRAs is non-primitive recursive. In this direction, each R-VASS $\mathcal{A}$ of m dimensions is simulated by an $(m, 0)$-HRA $\mathcal{A}'$ so that the value of each counter i of the former is the same as the number of names appearing precisely in history i of the latter. Using a non-trivial encoding of resets we can ensure that if $\mathcal{A}$ adheres to a particular kind of determinacy conditions (which the machines used in [30] for proving non-primitive recursive complexity do adhere to) then $\mathcal{A}'$ is strongly deterministic.

Proposition 16. *Emptiness for strongly deterministic HRAs is non-primitive recursive.*

We finally consider universality and language containment. Note first that our machines inherit undecidability of these properties from register automata [26]. However, these properties are decidable in the deterministic case, as deterministic HRAs are closed under complementation. In particular, given a deterministic HRA $\mathcal{A}$, the automaton $\mathcal{A}'$ accepting the language $\mathcal{N} \setminus \mathcal{L}(\mathcal{A})$ can be constructed in an analogous way as for deterministic finite-state automata, namely by obfuscating the automaton with all "missing" transitions and swapping final with non-final states (modulo ϵ-transitions). We add the missing transitions as follows. For each state q and each set X such that there are no transitions of the form $q \xrightarrow{Y}\!\!\twoheadrightarrow \cdot \xrightarrow{X\setminus Y, X'} q'$ in $\mathcal{A}$, we add a transition $q \xrightarrow{X,\emptyset} q_S$ to some sink non-final state q_S.

Proposition 17. *Language containment and universality are undecidable for HRAs. They are decidable for deterministic HRAs, with non-primitive recursive complexity.*

4 Weakening HRAs

Since the complexity of HRAs is substantially high, e.g. for deciding emptiness, it is useful to seek for restrictions thereof which allow us to express meaningful properties and, at the same time, remain at feasible complexity. As the encountered complexity stems from the fact that HRAs can simulate computations of R-VASSs, our strategy for producing weakenings is to restrict the functionalities of the corresponding R-VASSs. We follow two directions:

(a) We remove reset transitions. This corresponds to removing counter transfers and resets and drops the complexity of control-state reachability to exponential space.

(b) We restrict the number of histories to just one. We thus obtain polynomial space complexity as the corresponding counter machines are simply one-counter automata. This kind of restriction is also a natural extension of FRAs with history resets.

Observe that each of the aspects of HRAs targeted above corresponds to features (1,2) we identified in the Introduction, witnessed by the languages $\mathcal{L}_1$ and $\mathcal{L}_2$ respectively. We shall see that each restriction leads to losing the corresponding language.

Our analysis on emptiness for general HRAs from Section 3 is not applicable to these weaker machines as we now need to take registers into account: the simulation of registers by histories is either not possible or not practical for deriving satisfactory complexity bounds. Additionally, a direct analysis will allow us to reduce instances of counter machine problems to our setting decreasing the complexity size by an exponential, compared to our previous reduction. Solving emptiness for each of the weaker versions of HRAs will involve reduction to a name-free counter machine. In both cases, the reduction shall follow the same concept as in Section 3, namely of simulating computations with names *symbolically*.

4.1 Non-reset HRAs

We first weaken our automata by disallowing resets. We show that the new machines retain all their closure properties apart from Kleene-star closure. The latter is concretely manifested in the fact that language $\mathcal{L}_1$ of the Introduction is lost. On the other hand, the emptiness problem reduces in complexity to exponential space.

Definition 18. *A* **non-reset HRA** *of type* (m,n) *is an* (m,n)*-HRA* $\mathcal{A} = \langle Q, q_0, H_0, \delta, F\rangle$ *such that there is no* $q \xrightarrow{X} q' \in \delta$.

We call such a machine a non-reset (m,n)-HRA. In an analogous fashion, a ***VASS*** of m dimensions (an m*-VASS*) is an R-VASS with no reset transitions. For these machines, control-state reachability is significantly less complex.

Fact 19. *Control-state reachability for VASSs is* EXPSPACE*-complete [21,28], and can be decided in space* $O((M+\log|Q|)\cdot 2^{\kappa m \log m})$*, where* Q *the set of states of the examined instance,* m *the vector size,* M *the maximum initial value and* κ *a constant [29].*

Closure Properties. Of the closure constructions of Section 3, those for union and intersection readily apply to non-reset HRAs, while the construction for concatenation needs some minor amendments. On the other hand, using an argument similar to that of [5, Proposition 7.2], we can show that the language $\mathcal{L}_1$ is not recognised by non-reset HRAs and, hence, the latter are not closed under Kleene star. Finally, note that the HRA constructed for the language $\mathcal{L}_4$ in Example 13 is a non-reset HRA, which implies that non-reset HRAs are not closed under complementation.

Emptiness. We next reduce nonemptiness for non-reset HRAs to control-state reachability for VASSs. Starting from a non-reset (m,n)-HRA $\mathcal{A} = \langle Q, q_0, H_0, \delta, F\rangle$, the reduction maps each non-empty subset of $[m]$ *which appears in* δ to a VASS counter ($Y \subseteq [m]$ appears in δ if there is $(q, X, X', q') \in \delta$ such that $Y \in \{X \cap [m], X' \cap [m]\}$). Thus, the resulting VASS $\mathcal{A}'$ has m' counters, where $m' \leq 2|\delta|$. Although the number of states of $\mathcal{A}'$ is exponential, as the status of the registers needs to be embedded in states, the dominating factor for state-reachability is m', which is linear in the size of $\mathcal{A}$.

For the converse direction, we reduce reachability for a VASS of 2^m-1 counters to nonemptiness for an $(m,0)$-automaton: we map each counter to a non-empty subset of $[m]$. Note that such a $(2^m$–$1)$-to-m reduction would not work for R-VASSs, hence the different reduction in Proposition 16. This is because resets in HRAs cannot fully capture the behaviour of resets in R-VASSs. In HRAs a reset of a set of histories $\{1,2\}$, say, cannot occur without also resetting histories 1 and 2. In addition, resets necessarily cause virtual transfers of names (e.g. resetting history 1 makes all names appearing precisely in $\{1,2\}$ to appear precisely in history 2).

Proposition 20. *Emptiness checking for non-reset HRAs is* EXPSPACE*-complete.*

Non-reset HRAs without registers We now show that non-reset HRAs with only histories are as expressive as general non-reset HRAs. The equivalence we prove is weaker than the one we proved for general HRAs: we obtain language equivalence rather than bisimilarity. Our proof below is based on the *colouring technique* of [5]. Before we proceed with the actual result, let us first demonstrate the technique through an example.

Example 21. It is easy to see that the following language

$$\mathcal{L}_5 = \{a_1 \cdots a_n \in \mathcal{N}^* \mid \forall i.\, a_i \neq a_{i+1}\}$$

is recognised by the $(0,1)$-HRA on its right. What is perhaps not as clear is that the $(2,0)$-HRA on the right below, call it $\mathcal{A}$, accepts the same language.

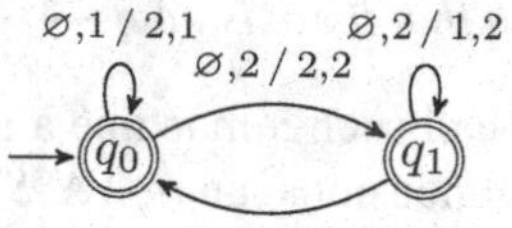

Note first that, by construction, it is not possible for $\mathcal{A}$ to accept the same name in two successive transitions: if we write (X,X') for the labels of incoming transitions to q_0 and (Y,Y') for the outgoing, we cannot match any X' with some Y, and similarly for q_1. This shows $\mathcal{L}(\mathcal{A}) \subseteq \mathcal{L}_5$. To prove the other inclusion, we need to show that for every word $w = a_1 \cdots a_n \in \mathcal{L}_5$ there is an accepting run in $\mathcal{A}$. For this, it suffices to find a sequence $\ell_1, \cdots, \ell_n$ of labels from the set $\{(\emptyset,1),(\emptyset,2),(1,1),(1,2),(2,1),(2,2)\}$, say $(\ell_i = (X_i, X'_i)$, satisfying:

1. For any i, $X'_i \neq X_{i+1}$.
2. If $a_i = a_j$, $i < j$, and for no $i < k < j$ do we have $a_i = a_k$ then $X'_i = X_j$.
3. For any i, if $a_i \neq a_j$ for all $j < i$ then $X_i = \emptyset$.

The first condition ensures that the sequence corresponds to a valid transition sequence in $\mathcal{A}$, and the other two that the sequence accepts the word $w = a_1 \cdots a_n$. Conditions 1 and 2 determine dependencies between the choices of left and right components in ℓ_is. Let us attach to w dependency pointers as follows: attach a pointer of type 1 (dependency right-to-left) from each a_i to its next occurrence in w, say a_j; from each a_{i+1} attach a type 2 pointer (dependency left-to-right) to a_i. Now note that, as there is no cycle in w which alternates between type 1 and type 2 pointers, it is always possible to produce a valid sequence $\ell_1, \cdots, \ell_n$.

We now state the general result. The proof follows the rationale described above, and is omitted for space limitations. We assume automata with their registers initially empty – the general case can be captured by first applying a construction like $\mathcal{A}\,\mathsf{fix}\,w$ of Section 3 (the construction introduces new registers where we would store the initial names, but we can as well use new histories for the same purpose).

Proposition 22. *For each (m,n)-non-reset HRA $\mathcal{A}$ with initially empty registers there is an $(m{+}3n,0)$-non-reset HRA $\mathcal{A}'$ such that $\mathcal{L}(\mathcal{A}) = \mathcal{L}(\mathcal{A}')$.*

4.2 Unary HRAs

Our second restriction concerns allowing resets but bounding the number of histories to just one. Thus, these automata are closer to the spirit of FRAs and, in fact, extend them by rounding up their history capabilities. We show that these automata require polynomial space complexity for emptiness and retain all their closure properties apart from intersection. The latter is witnessed by failing to recognise $\mathcal{L}_2$ from the Introduction. Extending this example to multiple interleavings, one can show that intersection is in general incompatible with bounding the number of histories.

Definition 23. *A $(1,n)$-HRA is called a* **unary HRA** *of n registers.*

In other words, unary HRAs are extensions of FRAs where names can be selectively inserted or removed from the history and, additionally, the history can be reset. These capabilities give us a strict extension.

Example 24. The automata used in Example 3 for $\mathcal{L}_1$ and $\mathcal{L}_3$ were unary HRAs. Note that neither of those languages is FRA-recognisable. On the other hand, in order to recognise $\mathcal{L}_2$, an HRA would need to use at least two histories: one history for the odd positions of the input and another for the even ones. Following this intuition we can show that $\mathcal{L}_2$ is not recognisable by unary HRAs.

Closure properties. The closure constructions of Section 3 readily apply to unary HRAs, with one exception: intersection. For the latter, we can observe that $\mathcal{L}_2 = \mathcal{L}(\mathcal{A}_1) \cap \mathcal{L}(\mathcal{A}_2)$, where $\mathcal{L}(\mathcal{A}_1) = \{a_1a_1' \cdots a_na_n' \in \mathcal{N}^* \mid a_1 \cdots a_n \in \mathcal{L}_0\}$ and $\mathcal{L}(\mathcal{A}_2) = \{a_1a_1' \cdots a_na_n' \in \mathcal{N}^* \mid a_1' \cdots a_n' \in \mathcal{L}_0\}$, and $\mathcal{A}_1$ and $\mathcal{A}_2$ are the unary $(1,0)$-HRAs on the side, with empty initial assignments. On the other hand, unary HRAs are not closed under complementation as well, as one can construct unary HRAs accepting $\overline{\mathcal{L}(\mathcal{A}_1)}$ and $\overline{\mathcal{L}(\mathcal{A}_2)}$, and then take their union to obtain a unary HRA for $\overline{\mathcal{L}_2}$.

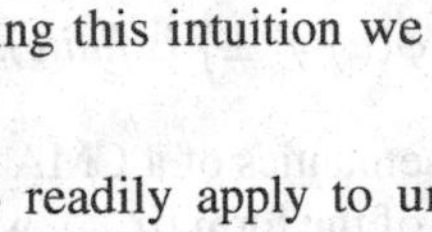

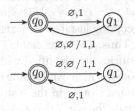

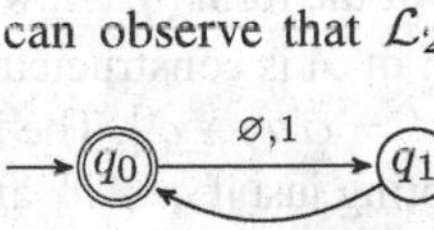

Emptiness In the case of just one history, the results on TR-VASS reachability [30,11] from Section 3 provide rather rough bounds. It is therefore useful to do a direct analysis. We reduce nonemptiness for unary HRAs to control-state reachability for R-VASSs of 1 dimension. Although these machines can be seen as close relatives to several other formalisms, like one-counter automata or pushdown automata on a one-letter alphabet, to the best of our knowledge there has been no direct attack of state reachability for them. Our analysis below, which follows standard techniques, yields square minimal-path length, and hence a polynomial complexity for emptiness (N is the size of the input).

Lemma 25. *Control-state reachability for R-VASSs of dimension 1 can be decided in* $\text{SPACE}(\log^2 N)$.

Proposition 26. *Emptiness for unary HRAs can be decided in* $\text{SPACE}((N \log N)^2)$.

5 Connections with Existing Formalisms

We have already seen that HRAs strictly extend FRAs. In this section we shall draw connections between HRAs and an automata model over infinite alphabets at the limits of decidability, called *Data Automata (DA)*, introduced in [6] in the context of XML theory. DAs operate on *data words*, i.e. over finite sequences of elements from $\mathcal{S} \times \mathcal{N}$, where $\mathcal{S}$ is a finite set of *data tags* and $\mathcal{N}$ is an infinite set of *data values* (but we shall call them *names*). A DA operates in two stages which involve a transducer automaton and a finite-state automaton respectively. Both automata operate on the tag projection of the input, with the second automaton focussing on tags paired with the same name.

For the rest of our discussion we shall abuse data words and treat them simply as strings of names, neglecting data tags. This is innocuous since there are straightforward translations between the two settings.[5] An equivalent formulation of DAs which is closer to our framework is the following [5].

Definition 27. *A* **Class Memory Automaton (CMA)** *is a tuple* $\mathcal{A} = \langle Q, q_0, \phi_0, \delta, F_1, F_2 \rangle$ *where* Q *is a finite set of states,* $q_0 \in Q$ *is initial,* $F_1 \subseteq F_2 \subseteq Q$ *are sets of final states and the transition relation is of type* $\delta \subseteq Q \times (Q \cup \{\bot\}) \times Q$. *Moreover,* ϕ_0 *is an initial* class memory function, *that is, a function* $\phi : \mathcal{N} \to Q \cup \{\bot\}$ *with finite domain* ($\{a \mid \phi(a) \neq \bot\}$ *is finite).*

The semantics of a CMA $\mathcal{A}$ like the above is given as follows. Configurations of $\mathcal{A}$ are pairs of the form (q, ϕ), where $q \in Q$ and ϕ a class memory function. The configuration graph of $\mathcal{A}$ is constructed by setting $(q, \phi) \xrightarrow{a} (q', \phi')$ just if there is $(q, \phi(a), q') \in \delta$ and $\phi' = \phi[a \mapsto q']$. The initial configuration is (q_0, ϕ_0), while a configuration (q, ϕ) is accepting just if $q \in F_1$ and, for all $a \in \mathcal{N}$, $\phi(a) \in F_2 \cup \{\bot\}$.

Thus, CMAs resemble HRAs in that they store input names in "histories", only that histories are identified with states: for each state q there is a corresponding history q (note notation overloading), and a transition which accepts a name a and leads to a state q must store a in the history q. Moreover, each name appears in at most one history (hence the type of ϕ) and, moreover, the finality conditions for configurations allow us to impose that all names appear in specific histories, if they appear in any. For example, here is a CMA (left below, with $F_1 = F_2 = \{q_0\}$) which recognises $\overline{\mathcal{L}_4}$ of Example 13.

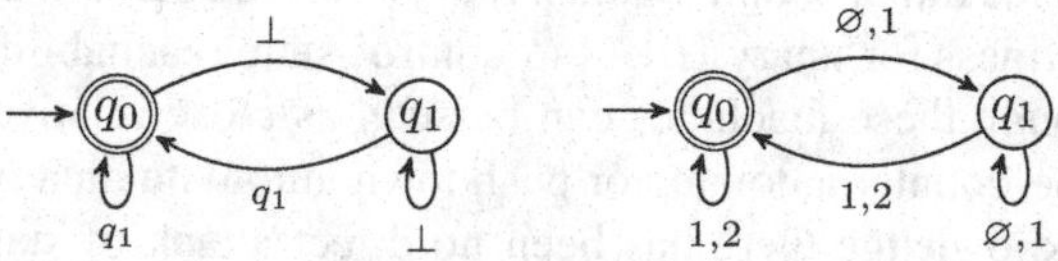

[5] A string of names is the same as a data word over a singleton set of data tags; while data tags can be simulated by names in registers of the initial configuration which do not get moved nor copied during the computation.

Each name is put in history q_1 when seen for the first time, and to q_0 when seen for the second time. The automaton accepts if all its names are in q_0. This latter condition is what makes the essential difference to HRAs, namely the capability to check where the names reside for acceptance. For example, the HRA on the right above would accept the same language were we able to impose the condition that accepting configurations (q, H) satisfy $a \in H@\{2\}$ for all names $a \in \bigcup_i H(i)$.

The above example proves that HRAs cannot express the same languages as CMAs. Conversely, as shown in [5, Proposition 7.2], the fact that CMAs lack resets does not allow them to express languages like, for example, $\mathcal{L}_1$. In the latter sections of [5] several extensions of CMAs are considered, one of which does involve resets. However, the resets considered there do not seem directly comparable to the reset capability of HRAs.

On the other hand, a direct comparison can be made with non-reset HRAs. We already saw in Proposition 22 that, in the latter idiom, histories can be used for simulating register behaviour. In the absence of registers, CMAs differ from non-reset HRAs solely in their constraint of relating histories to states (and their termination behaviour, which is more expressive). As the latter can be easily counterbalanced by obfuscating the set of states, we obtain the following.

Proposition 28. *For each non-reset HRA $\mathcal{A}$ there is a CMA $\mathcal{A}'$ such that $\mathcal{L}(\mathcal{A}) = \mathcal{L}(\mathcal{A}')$.*

6 Further Directions and Acknowledgements

Our goal is to apply automata with histories in static and runtime verification. While FRAs have been successful in modelling programs which, at each point during computation, can have access to a bounded memory fragment [33,24], HRAs allow us to express access to unbounded memory, provided that memory locations can be grouped in a bounded number of equivalence classes. Moreover, with HRAs we can express a significantly wider range of properties, closed under complementation-free regular operations, and in particular we can write properties where the history is used in meaningful ways (cf. the scenario of Figure 1). Although the complexity results derived in this paper may seem discouraging at first, they are based on quite specific representations of hard problems; in practice, we expect programs to yield automata of low complexities. Experience with tools based on TR-VASS coverability, like e.g. BFC [17], positively testify in that respect. On the other hand, an extension we envisage to consider is one with restricted emptiness tests, in analogy to e.g. [12].

A connection we would like to investigate is that between our automata and register automata which use alternation. Such machines with one register express behaviours related to HRAs [9] and enjoy some common properties, such as non-primitive recursive complexity for emptiness. Another interesting connection is with *Data Nets* [20], a class of machines which combine Petri nets with infinite alphabets but are not formalised as language acceptors over them. In terms of complexity, data nets seem substantially more involved than reset Petri nets and our machines. Finally, a problem left open in this work is decidability and complexity of bisimilarity. Although it is known that bisimilarity is undecidable for Petri nets [15], the version which seems of relevance towards an undecidability argument for HRAs is that of *visibly* counter automata with labels, i.e. automata which accept labels at each transition, and the action of each transition is determined by its label. The latter problem is not known to be decidable.

We would like to thank Dino Distefano, Petr Jancar, Ranko Lazic, Philippe Schnoebelen, Sylvain Schmitz and anonymous reviewers for fruitful discussions, suggestions and explanations. This work was supported by EPSRC grant H011749 (Grigore) and a Royal Academy of Engineering research fellowship (Tzevelekos).

References

1. Abramsky, S., Ghica, D.R., Murawski, A.S., Ong, C.-H.L., Stark, I.D.B.: Nominal games and full abstraction for the nu-calculus. In: LICS, pp. 150–159 (2004)
2. Araki, T., Kasami, T.: Some decision problems related to the reachability problem for Petri nets. Theor. Comput. Sci. 3(1), 85–104 (1977)
3. Atig, M.F., Bouajjani, A., Qadeer, S.: Context-bounded analysis for concurrent programs with dynamic creation of threads. Log. Meth. Comput. Sci. 7(4) (2011)
4. Benton, N., Leperchey, B.: Relational Reasoning in a Nominal Semantics for Storage. In: Urzyczyn, P. (ed.) TLCA 2005. LNCS, vol. 3461, pp. 86–101. Springer, Heidelberg (2005)
5. Björklund, H., Schwentick, T.: On notions of regularity for data languages. TCS 411 (2010)
6. Bojanczyk, M., Muscholl, A., Schwentick, T., Segoufin, L., David, C.: Two-variable logic on words with data. In: LICS, pp. 7–16 (2006)
7. Bouajjani, A., Fratani, S., Qadeer, S.: Context-Bounded Analysis of Multithreaded Programs with Dynamic Linked Structures. In: Damm, W., Hermanns, H. (eds.) CAV 2007. LNCS, vol. 4590, pp. 207–220. Springer, Heidelberg (2007)
8. Ciardo, G.: Petri Nets with Marking-Dependent Arc Cardinality. In: Valette, R. (ed.) ICATPN 1994. LNCS, vol. 815, pp. 179–198. Springer, Heidelberg (1994)
9. Demri, S., Lazic, R.: LTL with the freeze quantifier and register automata. TOCL 10(3) (2009)
10. Dufourd, C., Finkel, A., Schnoebelen, P.: Reset Nets Between Decidability and Undecidability. In: Larsen, K.G., Skyum, S., Winskel, G. (eds.) ICALP 1998. LNCS, vol. 1443, pp. 103–115. Springer, Heidelberg (1998)
11. Figueira, D., Figueira, S., Schmitz, S., Schnoebelen, P.: Ackermannian and primitive-recursive bounds with Dickson's lemma. In: LICS, pp. 269–278 (2011)
12. Finkel, A., Sangnier, A.: Mixing Coverability and Reachability to Analyze VASS with One Zero-Test. In: van Leeuwen, J., Muscholl, A., Peleg, D., Pokorný, J., Rumpe, B. (eds.) SOFSEM 2010. LNCS, vol. 5901, pp. 394–406. Springer, Heidelberg (2010)
13. Gabbay, M., Pitts, A.M.: A new approach to abstract syntax with variable binding. Formal Asp. Comput. 13(3-5), 341–363 (2002)
14. Grigore, R., Distefano, D., Petersen, R.L., Tzevelekos, N.: Runtime verification based on register automata. In: TACAS (to appear 2013)
15. Jancar, P.: Undecidability of bisimilarity for Petri nets and some related problems. Theor. Comput. Sci. 148(2), 281–301 (1995)
16. Jeffrey, A., Rathke, J.: Towards a theory of bisimulation for local names. In: LICS (1999)
17. Kaiser, A., Kroening, D., Wahl, T.: Efficient Coverability Analysis by Proof Minimization. In: Koutny, M., Ulidowski, I. (eds.) CONCUR 2012. LNCS, vol. 7454, pp. 500–515. Springer, Heidelberg (2012), `http://www.cprover.org/bfc/`
18. Kaminski, M., Francez, N.: Finite-memory automata. Theor. Comput. Sci. 134(2) (1994)
19. Laird, J.: A Fully Abstract Trace Semantics for General References. In: Arge, L., Cachin, C., Jurdziński, T., Tarlecki, A. (eds.) ICALP 2007. LNCS, vol. 4596, pp. 667–679. Springer, Heidelberg (2007)

20. Lazic, R., Newcomb, T., Ouaknine, J., Roscoe, A.W., Worrell, J.: Nets with tokens which carry data. Fundam. Inform. 88(3), 251–274 (2008)
21. Lipton, R.: The reachability problem requires exponential space. Tech. Rep. 62, Yale (1976)
22. Manuel, A., Ramanujam, R.: Class counting automata on datawords. Int. J. Found. Comput. Sci. 22(4), 863–882 (2011)
23. Montanari, U., Pistore, M.: An introduction to History Dependent Automata. Electr. Notes Theor. Comput. Sci. 10 (1997)
24. Murawski, A.S., Tzevelekos, N.: Algorithmic Nominal Game Semantics. In: Barthe, G. (ed.) ESOP 2011. LNCS, vol. 6602, pp. 419–438. Springer, Heidelberg (2011)
25. Murawski, A.S., Tzevelekos, N.: Algorithmic Games for Full Ground References. In: Czumaj, A., Mehlhorn, K., Pitts, A., Wattenhofer, R. (eds.) ICALP 2012, Part II. LNCS, vol. 7392, pp. 312–324. Springer, Heidelberg (2012)
26. Neven, F., Schwentick, T., Vianu, V.: Finite state machines for strings over infinite alphabets. ACM Trans. Comput. Logic 5(3), 403–435 (2004)
27. Pitts, A.M., Stark, I.: Observable Properties of Higher Order Functions that Dynamically Create Local Names, or: What's new? In: Borzyszkowski, A.M., Sokolowski, S. (eds.) MFCS 1993. LNCS, vol. 711, pp. 122–141. Springer, Heidelberg (1993)
28. Rackoff, C.: The covering and boundedness problems for vector addition systems. TCS (1978)
29. Rosier, L.E., Yen, H.-C.: A multiparameter analysis of the boundedness problem for vector addition systems. J. Comput. Syst. Sci. 32(1), 105–135 (1986)
30. Schnoebelen, P.: Revisiting Ackermann-Hardness for Lossy Counter Machines and Reset Petri Nets. In: Hliněný, P., Kučera, A. (eds.) MFCS 2010. LNCS, vol. 6281, pp. 616–628. Springer, Heidelberg (2010)
31. Segoufin, L.: Automata and Logics for Words and Trees over an Infinite Alphabet. In: Ésik, Z. (ed.) CSL 2006. LNCS, vol. 4207, pp. 41–57. Springer, Heidelberg (2006)
32. Stark, I.: Names and higher-order functions. PhD thesis, University of Cambridge (1994)
33. Tzevelekos, N.: Fresh-register automata. In: POPL, pp. 295–306 (2011)

Fatal Attractors in Parity Games

Michael Huth[1], Jim Huan-Pu Kuo[1], and Nir Piterman[2]

[1] Department of Computing, Imperial College London
London, SW7 2AZ, United Kingdom
{m.huth,jimhkuo}@imperial.ac.uk

[2] Department of Computer Science, University of Leicester
Leicester, LE1 7RH, United Kingdom
nir.piterman@leicester.ac.uk

Abstract. We study a new form of attractor in parity games and use it to define solvers that run in PTIME and are *partial* in that they do not solve all games completely. Technically, for color c this new attractor determines whether player $c\%2$ can reach a set of nodes X of color c whilst avoiding any nodes of color less than c. Such an attractor is *fatal* if player $c\%2$ can attract all nodes in X back to X in this manner. Our partial solvers detect fixed-points of nodes based on fatal attractors and correctly classify such nodes as won by player $c\%2$. Experimental results show that our partial solvers completely solve benchmarks that were constructed to challenge existing full solvers. Our partial solvers also have encouraging run times in practice. For one partial solver we prove that its runtime is in $O(|V|^3)$, that its output game is independent of the order in which attractors are computed, and that it solves all Büchi games.[1]

1 Introduction

Parity games are an important foundational structure in formal verification (see e.g. [11]). Mathematically, they can be seen as a representation of the model checking problem for the modal mu-calculus [4], and its exact computational complexity has been an open problem for over twenty years now.

Parity games are infinite, 2-person, 0-sum, graph-based games that are hard to solve. Their nodes, controlled by different players, are colored with natural numbers and the winning condition of plays depends on the minimal color occurring in cycles. The condition for winning a node, therefore, is an alternation of existential and universal quantification. In practice, this means that the maximal color of its coloring function is the only exponential source for the worst-case complexity of most parity game solvers, e.g. for those in [11,8,10].

Research on solving parity games may be loosely grouped into the following approaches: design of algorithms that solve all parity games by construction and that so far all have exponential or sub-exponential worst-case complexity

[1] A preliminary version of the results reported in this paper was presented at the GAMES 2012 workshop in Naples, Italy, on 11 September 2012.

F. Pfenning (Ed.): FOSSACS 2013, LNCS 7794, pp. 34–49, 2013.

(e.g. [11,8,10,9]), restriction of parity games to classes for which polynomial-time algorithms can be devised as complete solvers (e.g. [1,3]), and practical improvements to solvers so that they perform well across benchmarks (e.g. [5]).

We here propose a new approach that relates to, and potentially impacts, all of these aforementioned activities. We want to design and evaluate a new form of "partial" parity game solver. These are solvers that are well defined for all parity games but that may not solve all games completely, i.e. for some parity games they may not decide the winning status of some nodes. For us, a partial solver has an arbitrary parity game as input and returns two things: a sub-game of the input game, and a classification of the winning status of all nodes of the input game that are not in that sub-game. In particular, the returned sub-game is empty if, and only if, the partial solver classified the winners for all input nodes.

The input/output type of our partial solvers clearly relates them to so called preprocessors that may decide the winner of nodes whose structure makes such a decision an easy static criterion (e.g. in the elimination of self-loops or dead ends [5]). But we here search for dynamic criteria that allow partial solvers to completely solve a range of benchmarks of parity games. This ambition sets our work apart from research on preprocessors but is consistent with it as one can always run a partial solver as preprocessor.

The motivation for the study reported in this paper is that we want to investigate what theoretical building blocks one may create and use for designing partial solvers that run in polynomial time and work well on many games, whether partial solvers can be components of more efficient complete solvers, and whether there are interesting subclasses of parity games for which partial solvers completely solve all games. In particular, one may study the class of output games of a PTIME partial solver in lieu of studying the aforementioned open problem for all parity games.

We summarize the main contributions made in this paper:

- We present a new form of attractor that can be used in fixed-point computations to detect winning nodes for a given player in parity games.
- We propose several designs of partial solvers for parity games by using this new attractor within fixed-point computations.
- We analyze these partial solvers and show, e.g., that they work in PTIME and that one of them is independent of the order of attractor computation.
- And we evaluate these partial solvers against known benchmarks and report that these experiments have very encouraging results.

Outline of paper. Section 2 contains needed formal background and fixes notation. Section 3 introduces the building block of our partial solvers, a new form of attractor. Some partial solvers based on this attractor are presented in Section 4, theoretical results about these partial solvers are proved in Section 5, and experimental results for these partial solvers run on benchmarks are reported and discussed in Section 6. We summarize and conclude the paper in Section 7.

2 Preliminaries

We write $\mathbb{N}$ for the set $\{0, 1, \ldots\}$ of natural numbers. A parity game G is a tuple (V, V_0, V_1, E, c), where V is a set of nodes partitioned into possibly empty node sets V_0 and V_1, with an edge relation $E \subseteq V \times V$ (where for all v in V there is a w in V with (v, w) in E), and a coloring function $c\colon V \to \mathbb{N}$. In figures, $c(v)$ is written within nodes v, nodes in V_0 are depicted as circles and nodes in V_1 as squares. For v in V, we write $v.E$ for node set $\{w \in V \mid (v, w) \in E\}$ of successors of v. By abuse of language, we call a subset U of V a *sub-game* of G if the game graph $(U, E \cap (U \times U))$ is such that all nodes in U have some successor. We write $\mathcal{PG}$ for the class of all finite parity games G, which includes the parity game with empty node set for our convenience. We only consider games in $\mathcal{PG}$.

Throughout, we write p for one of 0 or 1 and $1 - p$ for the other player. In a parity game, player p owns the nodes in V_p. A play from some node v_0 results in an infinite play $r = v_0 v_1 \ldots$ in (V, E) where the player who owns v_i chooses the successor v_{i+1} such that (v_i, v_{i+1}) is in E. Let $\mathsf{Inf}(r)$ be the set of colors that occur in r infinitely often: $\mathsf{Inf}(r) = \{k \in \mathbb{N} \mid \forall j \in \mathbb{N}\colon \exists i \in \mathbb{N}\colon i > j \text{ and } k = c(v_i)\}$. Player 0 wins play r iff $\min \mathsf{Inf}(P)$ is even; otherwise player 1 wins play r.

A strategy for player p is a total function $\tau\colon V_p \to V$ such that $(v, \tau(v))$ is in E for all $v \in V_p$. A play r is consistent with τ if each node v_i in r owned by player p satisfies $v_{i+1} = \tau(v_i)$. It is well known that each parity game is determined: node set V is the disjoint union of two, possibly empty, sets W_0 and W_1, the winning regions of players 0 and 1 (respectively). Moreover, strategies $\sigma\colon V_0 \to V$ and $\pi\colon V_1 \to V$ can be computed such that

- all plays beginning in W_0 and consistent with σ are won by player 0; and
- all plays beginning in W_1 and consistent with π are won by player 1.

Solving a parity game means computing such data (W_0, W_1, σ, π).

Example 1. In the parity game G depicted in Figure 1, the winning regions are $W_1 = \{v_3, v_5, v_7\}$ and $W_0 = \{v_0, v_1, v_2, v_4, v_6, v_8, v_9, v_{10}, v_{11}\}$. Let σ move from v_2 to v_4, from v_6 to v_8, from v_9 to v_8, and from v_{10} to v_9. Then σ is a winning strategy for player 0 on W_0. And every strategy π is winning for player 1 on W_1.

3 Fatal Attractors

In this section we define a special type of attractor that is used for our partial solvers in the next section. We start by recalling the normal definition of attractor, and that of a trap, and then generalize the former to our purposes.

Definition 1. *Let X be a node set in parity game G. For player p in $\{0, 1\}$, set*

$$\mathsf{cpre}_p(X) = \{v \in V_p \mid v.E \cap X \neq \emptyset\} \cup \{v \in V_{1-p} \mid v.E \subseteq X\} \quad (1)$$

$$\mathsf{Attr}_p[G, X] = \mu Z.(X \cup \mathsf{cpre}_p(Z)) \quad (2)$$

where $\mu Z.F(Z)$ denotes the least fixed point of a monotone function $F\colon 2^V \to 2^V$.

The control predecessor of a node set X for p in (1) is the set of nodes from which player p can force to get to X in exactly one move. The attractor for player p to a set X in (2) is computed via a least fixed-point as the set of nodes from which player p can force the game in zero or more moves to get to the set X. Dually, a *trap* for player p is a region from which player p cannot escape.

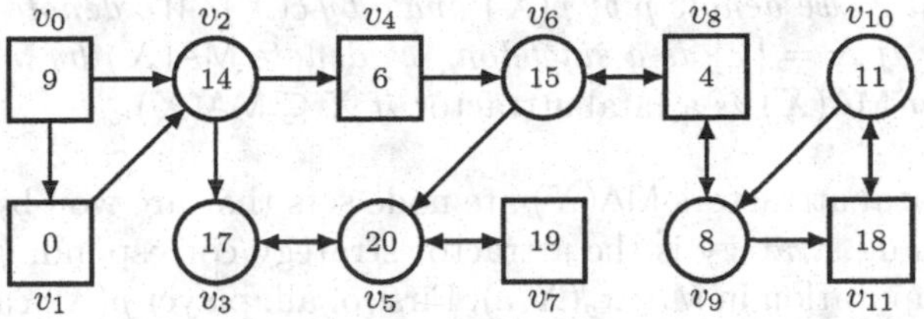

Fig. 1. A parity game: circles denote nodes in V_0, squares denote nodes in V_1

Definition 2. *Node set X in parity game G is a trap for player p (p-trap) if for all $v \in V_p \cap X$ we have $v.E \subseteq X$ and for all $v \in V_{1-p} \cap X$ we have $v.E \cap X \neq \emptyset$.*

It is well known that the complement of an attractor for player p is a p-trap and that it is a sub-game. We state this here formally as a reference:

Theorem 1. *Given a node set X in a parity game G, the set $V \setminus \mathsf{Attr}_p[G, X]$ is a p-trap and a sub-game of G.*

We now define a new type of attractor, which will be a crucial ingredient in the definition of all our partial solvers.

Definition 3. *Let A and X be node sets in parity game G, let p in $\{0, 1\}$ be a player, and c a color in G. We set*

$$\begin{aligned} \mathsf{mpre}_p(A, X, c) &= \{v \in V_p \mid c(v) \geq c \wedge v.E \cap (A \cup X) \neq \emptyset\} \cup \\ &\qquad \{v \in V_{1-p} \mid c(v) \geq c \wedge v.E \subseteq A \cup X\} \\ \mathsf{MAttr}_p(X, c) &= \mu Z.\mathsf{mpre}_p(Z, X, c) \end{aligned} \tag{3}$$

The *monotone* control predecessor $\mathsf{mpre}_p(A, X, c)$ of node set A for p with target X is the set of nodes of color at least c from which player p can force to get to either A or X in one move. The *monotone* attractor $\mathsf{MAttr}_p(X, c)$ for p with target X is the set of nodes from which player p can force the game in one or more moves to X by only meeting nodes whose color is at least c. Notice that the target set X is kept external to the attractor. Thus, if some node x in X is included in $\mathsf{MAttr}_p(X, c)$ it is so as it is attracted to X in at least one step.

Our control predecessor and attractor are different from the "normal" ones in a few ways. First, ours take into account the color c as a formal parameter. They add only nodes that have color at least c. Second, as discussed above, the target set X itself is not included in the computation by default. For example, $\mathsf{MAttr}_p(X, c)$ includes states from X only if they can be attracted to X.

We now show the main usage of this new operator by studying how specific instantiations thereof can compute so called *fatal attractors.*

Definition 4. *Let X be a set of nodes of color c, where $p = c\%2$.*

1. *For such an X we denote p by $p(X)$ and c by $c(X)$. We denote $\mathsf{MAttr}_p(X, c)$ by $\mathsf{MA}(X)$. If $X = \{x\}$ is a singleton, we denote $\mathsf{MA}(X)$ by $\mathsf{MA}(x)$.*
2. *We say that $\mathsf{MA}(X)$ is* a fatal attractor *if $X \subseteq \mathsf{MA}(X)$.*

We note that fatal attractors $\mathsf{MA}(X)$ are node sets that are won by player $p(X)$ in G. The winning strategy is the attractor strategy corresponding to the least fixed-point computation in $\mathsf{MAttr}_p(X, c)$. First of all, player $p(X)$ can force, from all nodes in $\mathsf{MA}(X)$, to reach some node in X in at least one move. Then, player $p(X)$ can do this again from this node in X as X is a subset of $\mathsf{MA}(X)$. At the same time, by definition of $\mathsf{MAttr}_p(X, c)$ and $\mathsf{mpre}_p(A, X, c)$, the attraction ensures that only colors of value at least c are encountered. So in plays starting in $\mathsf{MA}(X)$ and consistent with that strategy, every visit to a node of parity $1-p(X)$ is followed later by a visit to a node of color $c(X)$. It follows that in an infinite play consistent with this strategy and starting in $\mathsf{MA}(X)$, the minimal color to be visited infinitely often is c – which is of p's parity.

Theorem 2. *Let $\mathsf{MA}(X)$ be fatal in parity game G. Then the attractor strategy for player $p(X)$ on $\mathsf{MA}(X)$ is winning for $p(X)$ on $\mathsf{MA}(X)$ in G.*

Let us consider the case when X is a singleton $\{k\}$ and $\mathsf{MA}(k)$ is not fatal. Suppose that there is an edge (k, w) in E with w in $\mathsf{MA}(k)$. We show that this edge cannot be part of a winning strategy (of either player) in G. Since $\mathsf{MA}(k)$ is not fatal, k must be in $V_{1-p(k)}$ and so is controlled by player $1 - p(k)$. But if that player were to move from k to w in a memoryless strategy, player $p(k)$ could then attract the play from w back to k without visiting colors of parity $1-p(k)$ and smaller than $c(k)$, since w is in $\mathsf{MA}(k)$. And, by the existence of memoryless winning strategies [4], this would ensure that the play is won by player $p(k)$ as the minimal infinitely occurring color would have parity $p(k)$. We summarize:

Lemma 1. *Let $\mathsf{MA}(k)$ be not fatal for node k. Then we may remove edge (k, w) in E if w is in $\mathsf{MA}(k)$, without changing winning regions of parity game G.*

Example 2. For G in Figure 1, the only colors k for which $\mathsf{MA}(k)$ is fatal are 4 and 8: $\mathsf{MA}(4)$ equals $\{v_2, v_4, v_6, v_8, v_9, v_{10}, v_{11}\}$ and $\mathsf{MA}(8)$ equals $\{v_9, v_{10}, v_{11}\}$. In particular, $\mathsf{MA}(8)$ is contained in $\mathsf{MA}(4)$ and nodes v_1 and v_0 are attracted to $\mathsf{MA}(4)$ in G by player 0. And v_{11} is in $\mathsf{MA}(11)$ (but the node of color 11, v_{10}, is not), so edge (v_{10}, v_{11}) may be removed.

4 Partial Solvers

We can use the above definitions and results to define partial solvers next. Their soundness will be shown in Section 5.

```
psol(G = (V, V0, V1, E, c)) {
  for (k ∈ V in descending color ordering c(k)) {
    if (k ∈ MA(k)) { return psol(G \ Attr_p(k)[G, MA(k)]) }
    if (∃(k, w) ∈ E: w ∈ MA(k))
    { G  =  G \ {(k, w) ∈ E |  w ∈ MA(k)} }
  }
  return G
}
```

Fig. 2. Partial solver `psol` based on detection of fatal attractors $\mathsf{MA}(k)$ and fatal moves

4.1 Partial Solver psol

Figure 2 shows the pseudocode of a partial solver, named `psol`, based on $\mathsf{MA}(X)$ for singleton sets X. Solver `psol` explores the parity game G in descending color ordering. For each node k, it constructs $\mathsf{MA}(k)$, and aims to do one of two things:

- If node k is in $\mathsf{MA}(k)$, then $\mathsf{MA}(k)$ is fatal for player $1 - p(k)$, thus node set $\mathsf{Attr}_{p(k)}[G, \mathsf{MA}(k)]$ is a winning region of player $p(k)$, and removed from G.
- If node k is not in $\mathsf{MA}(k)$, and there is a (k, w) in E where w is in $\mathsf{MA}(k)$, all such edges (k, w) are removed from E and the iteration continues.

If for no k in V attractor $\mathsf{MA}(k)$ is fatal, game G is returned as is – empty if `psol` solves G completely. The accumulation of winning regions and computation of winning strategies are omitted from the pseudocode for improved readability.

Example 3. In a run of `psol` on G from Figure 1, there is no effect for colors larger than 11. For $c = 11$, `psol` removes edge (v_{10}, v_{11}) as v_{11} is in $\mathsf{MA}(11)$. The next effect is for $c = 8$, when the fatal attractor $\mathsf{MA}(8) = \{v_9, v_{10}, v_{11}\}$ is detected and removed from G (the previous edge removal did not cause the attractor to be fatal). On the remaining game, the next effect occurs when $c = 4$, and when the fatal attractor $\mathsf{MA}(4)$ is $\{v_2, v_4, v_6, v_8\}$ in that remaining game. As player 0 can attract v_0 and v_1 to this as well, all these nodes are removed and the remaining game has node set $\{v_3, v_5, v_7\}$. As there is no more effect of `psol` on that remaining game, it is returned as the output of `psol`'s run.

4.2 Partial Solver psolB

Figure 3 shows the pseudocode of another partial solver, named `psolB` (the "B" suggests a relationship to "Büchi"), based on $\mathsf{MA}(X)$, where X is a set of nodes of the same color. This time, the operator $\mathsf{MA}(X)$ is used within a greatest fixed-point in order to discover the largest set of nodes of a certain color that can be (fatally) attracted to itself. Accordingly, the greatest fixed-point starts from all the nodes of a certain color and gradually removes those that cannot be attracted to the same color. When the fixed-point stabilizes, it includes the set of nodes of the given color that can be (fatally) attracted to itself. This node set can be removed (as a winning region for player $d\%2$) and the residual game analyzed recursively. As before, the colors are explored in descending order.

```
psolB(G = (V, V0, V1,E, c)) {
  for (colors d in descending ordering) {
    X  =  { v in V |  c(v) =  d };
    cache = {};
    while (X ≠  {} && X ≠  cache) {
      cache = X;
      if (X ⊆  MA(X)) { return psolB(G  \  Attr_{d%2}[G, MA(X)])
      } else { X = X ∩  MA(X); }
    }
  }
  return G
}
```

Fig. 3. Partial solver psolB

We make two observations. First, if we were to replace the recursive calls in `psolB` with the removal of the winning region from G and a continuation of the iteration, we would get an implementation that discovers less fatal attractors. Second, edge removal in `psol` relies on the set X being a singleton. A similar removal could be achieved in `psolB` when the size of X is reduced by one (in the operation $X = X \cap \mathsf{MA}(X)$). Indeed, in such a case the removed node would not be removed and the current value of X be realized as fatal. We have not tested this edge removal approach experimentally for this variant of `psolB`.

Example 4. A run of `psolB` on G from Figure 1 has the same effect as the one for `psol`, except that `psolB` does not remove edge (v_{10}, v_{11}) when $c = 11$.

A way of comparing partial solvers P_1 and P_2 is to say that $P_1 \leq P_2$ if, and only if, for all parity games G the set of nodes in the output sub-game $P_1(G)$ is a subset of the set of nodes of the output sub-game $P_2(G)$. We note that `psol` and `psolB` are incomparable for this intensional pre-order over partial solvers.

4.3 Partial Solver psolQ

It seems that `psolB` is more general than `psol` in that if there is a singleton X with $X \subseteq \mathsf{MA}(X)$ then `psolB` will discover this as well. However, the requirement to attract to a single node seems too strong. Solver `psolB` removes this restriction and allows to attract to more than one node, albeit of the same color. Now we design a partial solver `psolQ` that can attract to a set of nodes of more than one color (the "Q" is our code name for this "Q"uantified layer of colors of the same parity). Solver `psolQ` allows to combine attraction to multiple colors by adding them gradually and taking care to "fix" visits to nodes of opposite parity.

We extend the definition of mpre and MAttr to allow inclusion of more (safe) nodes when collecting nodes in the attractor.

```
layeredAttr(G,p,X) { // PRE-CONDITION: all nodes in X have parity p
  A = {};
  b = max{c(v) | v ∈ X};
  for (d = p up to b in increments of 2) {
    Y = {v ∈ X | c(v) ≤ d};
    A = PMAttr_p(A ∪ Y,d);
  }
  return A;
}

psolQ(G = (V,V_0,V_1,E,c)) {
  for (colors b in ascending order) {
    X = {v ∈ V | c(v) ≤ b ∧ c(v)%2 = b%2};
    cache = {};
    while (X ≠ {} && X ≠ cache) {
      cache = X;
      W = layeredAttr(G,b%2,X);
      if (X ⊆ W) { return psolQ(G \ Attr_{b%2}[G,W]);
      } else { X = X ∩ W; }
    }
  }
  return G;
}
```

Fig. 4. Operator `layeredAttr`(G, p, X) and partial solver `psolQ`

Definition 5. *Let A and X be node sets in parity game G, let p in $\{0, 1\}$ be a player, and c a color in G. We set*

$$\begin{aligned}\mathsf{pmpre}_p(A, X, c) &= \{v \in V_p \mid (c(v) \geq c \vee v \in X) \wedge v.E \cap (A \cup X) \neq \emptyset\} \cup \\ &\quad \{v \in V_{1-p} \mid (c(v) \geq c \vee v \in X) \wedge v.E \subseteq A \cup X\} \quad (4)\\ \mathsf{PMAttr}_p(X, c) &= \mu Z.\mathsf{pmpre}_p(Z, X, c) \quad (5)\end{aligned}$$

The *permissive monotone* predecessor in (4) adds to the monotone predecessor also nodes that are in X itself even if their color is lower than c, i.e., they violate the monotonicity requirement. The *permissive monotone* attractor in (5) then uses the permissive predecessor instead of the simpler predecessor. This is used for two purposes. First, when the set X includes nodes of multiple colors – some of them lower than c. Then, inclusion of nodes from X does not destroy the properties of fatal attraction. Second, increasing the set X of target nodes allows us to include the previous target and the attractor to it as set of "permissible" nodes. This creates a layered structure of attractors.

We use the permissive attractor to define `psolQ`. Figure 4 presents the pseudo code of operator `layeredAttr`(G, p, X). It is an attractor that combines attraction to nodes of multiple color. It takes a set X of colors of the same parity p. It considers increasing subsets of X with more and more colors and tries to attract fatally to them. It starts from a set Y_p of nodes of parity p with color p and computes $\mathsf{MA}(Y_p)$. At this stage, the difference between pmpre and mpre does not apply as Y_p contains nodes of only one color and A is empty. Then, instead of stopping as before, it continues to accumulate more nodes. It creates the set Y_{p+2} of the nodes of parity p with color p or $p+2$. Then, $\mathsf{PMAttr}_p(A \cup Y_{p+2}, p+2)$ includes all the previous nodes in A (as all nodes in A are now permissible) and all nodes that can be attracted to them or to Y_{p+2} through nodes of color at least $p+2$. This way, even if nodes of a color lower than $p+2$ are included they will be ensured to be either in the previous attractor or of the right parity. Then Y is increased again to include some more nodes of p's parity. This process continues until it includes all nodes in X.

This layered attractor may also be fatal:

Definition 6. *We say that* `layeredAttr`*(G, p, X) is fatal if X is a subset of* `layeredAttr`*(G, p, X).*

As before, fatal layered attractors are won by player p in G. The winning strategy is more complicated as it has to take into account the number of iteration in the for loop in which a node was first discovered. Every node in `layeredAttr`(G, p, X) belongs to a layer corresponding to a maximal color d. From a node in layer d, player p can force to reach some node in $Y_d \subseteq X$ or some node in a lower layer d'. As the number of layers is finite, eventually some node in X is reached. When reaching X, player p can attract to X in the same layered fashion again as X is a subset of `layeredAttr`(G, p, X). Along the way, while attracting through layer d we are ensured that only colors at least d or of a lower layer are encountered. So in plays starting in `layeredAttr`(G, p, X) and consistent with that strategy, every visit to a node of parity $1-p$ is followed later by a visit to a node of parity p of lower color.

Theorem 3. *Let* `layeredAttr`*(G, p, X) be fatal in parity game G. Then the layered attractor strategy for player p on* `layeredAttr`*(G, p, X) is winning for p on* `layeredAttr`*(G, p, X) in G.*

Pseudo code of solver `psolQ` is also shown in Figure 4: `psolQ` prepares increasing sets of nodes X of the same color and calls `layeredAttr` within a greatest fixed-point. For a set X, the greatest fixed-point attempts to discover the largest set of nodes within X that can be fatally attracted to itself (in a layered fashion). Accordingly, the greatest fixed-point starts from all the nodes in X and gradually removes those that cannot be attracted to X. When the fixed-point stabilizes, it includes a set of nodes of the same parity that can be attracted to itself. These are removed (along with the normal attractor to them) and the residual game is analyzed recursively.

We note that the first two iterations of `psolQ` are equivalent to calling `psolB` on colors 0 and 1. Then, every iteration of `psolQ` extends the number of colors considered. In particular, in the last two iterations of `psolQ` the value of b is the maximal possible value of the appropriate parity. It follows that the sets X defined in these last two iterations include all nodes of the given parity. These last two computations of greatest fixed-points are the most general and subsume all previous greatest fixed-point computations. We discuss in Section 6 why we increase the bound b gradually and do not consider these two iterations alone.

Example 5. The run of psolQ on G from Figure 1 finds a fatal attractor for bound $b = 4$, which removes all nodes except v_3, v_5, and v_7. For $b = 19$, it realizes that these nodes are won by player 1, and outputs the empty game. That psolQ is a partial solver can be seen in Figure 5, which depicts a game that is not modified at all by psolQ and so is returned as is.

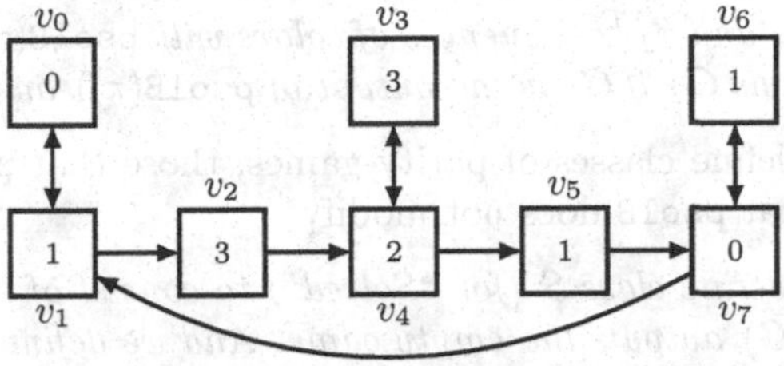

Fig. 5. A 1-player parity game modified by neither psol, psolB nor psolQ

5 Properties of Our Partial Solvers

We now discuss the properties of our partial solvers, looking first at their soundness and computational complexity.

5.1 Soundness and Computational Complexity

Theorem 4. *1. The partial solvers* psol, psolB, *and* psolQ *are sound.*
2. The running time for psol *and* psolB *is in* $O(|V|^2 \cdot |E|)$.
3. And psol *and* psolB *can be implemented to run in time* $O(|V|^3)$.
4. And psolQ *runs in time* $O(|V|^2 \cdot |E| \cdot |c|)$ *with* $|c|$ *the number of colors in* G.

If psolQ were to restrict attention to the last two iterations of the for loop, i.e., those that compute the greatest fixed-point with the maximal even color and the maximal odd color, the run time of psolQ would be bounded by $O(|V|^2 \cdot |E|)$. For such a version of psolQ we also ran experiments on our benchmarks and do not report these results, except to say that this version performs considerably worse than psolQ in practice. We believe that this is so since psolQ more quickly discovers small winning regions that "destabilize" the rest of the games.

5.2 Robustness of psolB

Our pseudo-code for psolB iterates through colors in descending order. A natural question is whether the computed output game depends on the order in which these colors are iterated. Below, we formally state that the outcome of psolB is indeed independent of the iteration order. This suggests that these solvers are a form of polynomial-time projection of parity games onto sub-games.

Let us formalize this. Let π be some sequence of colors in G, that may omit or repeat some colors from G. Let psolB(π) be a version of psolB that checks for (and removes) fatal attractors according to the order in π (including any color repetitions in π). We say that psolB(π) is *stable* if for every color c_1, the input/output behavior of psolB(π) and psolB($\pi \cdot c_1$) are the same. That is, the sequence π leads psolB to stabilization in the sense that every extension of the version psolB(π) with one color does not change the input/output behavior.

Theorem 5. *Let π_1 and π_2 be sequences of colors with* psolB(π_1) *and* psolB(π_2) *stable. Then G_1 equals G_2 if G_i is the output of* psolB(π_i) *on G, for $1 \leq i \leq 2$.*

Next, we formally define classes of parity games, those that psolB solves completely and those that psolB does not modify.

Definition 7. *We define class $\mathcal{S}$ (for "Solved") to consist of those parity games G for which* psolB(G) *outputs the empty game. And we define $\mathcal{K}$ (for "Kernel") as the class of those parity games G for which* psolB(G) *outputs G again.*

The meaning of psolB is therefore a total, idempotent function of type $\mathcal{PG} \to \mathcal{K}$ that has $\mathcal{S}$ as inverse image of the empty parity game. By virtue of Theorem 5, classes $\mathcal{S}$ and $\mathcal{K}$ are *semantic* in nature.

We now show that $\mathcal{S}$ contains the class of Büchi games, which we identify with parity games G with color 0 and 1 and where nodes with color 0 are those that player 0 wants to reach infinitely often.

Theorem 6. *Let G be a parity game whose colors are only* 0 *and* 1. *Then G is in $\mathcal{S}$, i.e.* psolB *completely solves G.*

We point out that $\mathcal{S}$ does not contain some game types for which polynomial-time solvers are known. For example, not all 1-player parity games are in $\mathcal{S}$ (see Figure 5). Class $\mathcal{S}$ is also not closed under sub-games.

6 Experimental Results

6.1 Experimental Setup

We wrote Scala implementations of psol, psolB, and psolQ, and of Zielonka's solver (zlka) that rely on the same data structures and do not compute winning strategies – which has routine administrative overhead. The (parity) *Game* object has a map of *Nodes* (objects) with node identifiers (integers) as the keys. Apart from colors and owner type (0 or 1), each *Node* has two lists of identifiers, one for successors and one for predecessors in the game graph (V, E). For attractor computation, the predecessor list is used to perform "backward" attraction.

This uniform use of data types allows for a first informed comparison. We chose zlka as a reference implementation since it seems to work well in practice on many games [5]. We then compared the performance of these implementations on all eight non-random, structured game types produced by the PGSolver tool [6]. Here is a list of brief descriptions of these game types.

– `Clique`: fully connected games with alternating colors and no self-loops.
– `Ladder`: layers of node pairs with connections between adjacent layers.
– `Recursive Ladder`: layers of 5-node blocks with loops.
– `Strategy Impr`: worst cases for strategy improvement solvers.
– `Model Checker Ladder`: layers of 4-node blocks.
– `Tower Of Hanoi`: captures well-known puzzle.
– `Elevator Verification`: a verification problem for an elevator model.
– `Jurdzinski`: worst cases for small progress measure solvers.

The first seven types take as game parameter a natural number n as input, whereas `Jurdzinski` takes a pair of such numbers n, m as game parameter.

For regression testing, we verified for all tested games that the winning regions of `psol`, `psolB`, `psolQ` and `zlka` are consistent with those computed by PGSolver. Runs of these algorithms that took longer than 20 minutes (i.e. 1200K milliseconds) or for which the machine exhausted the available memory during solver computation are recorded as aborts ("`abo`") – the most frequent reason for `abo` was that the used machine ran out of memory. All experiments were conducted on the same machine with an Intel® Core^{TM} i5 (four cores) CPU at 3.20GHz and 8G of RAM, running on a Ubuntu 11.04 Linux operating system.

For most game types, we used *unbounded binary search* starting with 2 and then iteratively doubling that value, in order to determine the `abo` boundary value for parameter n within an accuracy of plus/minus 10. As the game type `Jurdzinski`$[n, m]$ has two parameters, we conducted three unbounded binary searches here: one where n is fixed at 10, another where m is fixed at 10, and a third one where n equals m. We used a larger parameter configuration ($10 \times$ power of two) for `Jurdzinski` games.

We report here only the last two powers of two for which one of the partial solvers didn't timeout, as well as the boundary values for each solver. For game types whose boundary value was less than 10 (`Tower Of Hanoi` and `Elevator Verification`), we didn't use binary search but incremented n by 1. Finally, if a partial solver didn't solve its input game completely, we ran `zlka` on the remaining game and added the observed running times for `zlka` to that of the partial solver. (This occurred for `Elevator Verification` for `psol` and `psolB`.)

6.2 Experiments on Structured Games

Our experimental results are depicted in Figures 6 and 7, colored green (respectively red) for the partial solver with best (respectively worst) result. Running times are reported in milliseconds. The most important outcome is that partial solvers `psol` and `psolB` solved seven of the eight game types *completely* for all runs that did not time out, the exception being `Elevator Verification`; and that `psolQ` solved all eight game types completely. This suggests that partial solvers can actually be used as solvers on a range of structured game types.

We now compare the performance of these partial solvers and of `zlka`. There were ten experiments, three for `Jurdzinski` and one for each of the remaining seven game types. For seven out of these ten experiments, `psolB` had the largest boundary value of the parameter and so seems to perform best overall. The solver

Clique[n]

n	psol	psolB	psolQ	zlka
2**11	6016.68	48691.72	3281.57	12862.92
2**12	abo	164126.06	28122.96	76427.44
20min	$n = 3680$	$n = 5232$	$n = 4608$	$n = 5104$

Ladder[n]

n	psol	psolB	psolQ	zlka
2**19	abo	22440.57	26759.85	24406.79
2**20	abo	47139.96	59238.77	75270.74
20min	$n = 14712$	$n = 1596624$	$n = 1415776$	$n = 1242376$

Model Checker Ladder[n]

n	psol	psolB	psolQ	zlka
2**12	119291.99	90366.80	117006.17	79284.72
2**13	560002.68	457049.22	644225.37	398592.74
20min	$n = 11528$	$n = 12288$	$n = 10928$	$n = 13248$

Recursive Ladder[n]

n	psol	psolB	psolQ	zlka
2**12	abo	abo	138956.08	abo
2**13	abo	abo	606868.31	abo
20min	$n = 1560$	$n = 2064$	$n = 11352$	$n = 32$

Strategy Impr[n]

n	psol	psolB	psolQ	zlka
2**10	174913.85	134795.46	abo	abo
2**11	909401.03	631963.68	abo	abo
20min	$n = 2368$	$n = 2672$	$n = 40$	$n = 24$

Tower Of Hanoi[n]

n	psol	psolB	psolQ	zlka
9	272095.32	54543.31	610264.18	56780.41
10	abo	397728.33	abo	390407.41
20min	$n = 9$	$n = 10$	$n = 9$	$n = 10$

Elevator Verification[n]

n	psol	psolB	psolQ	zlka
1	171.63	120.59	147.32	125.41
2	646.18	248.56	385.56	237.51
3	2707.09	584.83	806.28	512.72
4	223829.69	1389.10	2882.14	1116.85
5	abo	11681.02	22532.75	3671.04
6	abo	168217.65	373568.85	41344.03
7	abo	abo	abo	458938.13
20min	$n = 4$	$n = 6$	$n = 6$	$n = 7$

Fig. 6. First experimental results for partial solvers run over benchmarks

Jurdzinski[10, m]

m	psol	psolB	psolQ	zlka
10*2**7	abo	179097.35	abo	abo
10*2**8	abo	833509.48	abo	abo
20min	$n = 560$	$n = 2890$	$n = 1120$	$n = 480$

Jurdzinski[n, 10]

n	psol	psolB	psolQ	zlka
10*2**7	308033.94	106453.86	abo	abo
10*2**8	abo	406621.65	abo	abo
20min	$n = 2420$	$n = 4380$	$n = 1240$	$n = 140$

Jurdzinski[n, n]

n	psol	psolB	psolQ	zlka
10*2**3	215118.70	23045.37	310665.53	abo
10*2**4	abo	403844.56	abo	abo
20min	$n = 110$	$n = 200$	$n = 100$	$n = 50$

Fig. 7. Second experimental results run over `Jurdzinski` benchmarks

`zlka` was best for `Model Checker Ladder` and `Elevator Verification`, and about as good as `psolB` for `Tower Of Hanoi`. And `psolQ` was best for `Recursive Ladder`. Thus `psol` appears to perform worst across these benchmarks.

Solvers `psolB` and `zlka` seem to do about equally well for game types `Clique`, `Ladder`, `Model Checker Ladder`, and `Tower Of Hanoi`. But solver `psolB` appears to outperform `zlka` dramatically for game types `Recursive Ladder`, and `Strategy Impr` and is considerably better than `zlka` for `Jurdzinski`.

We think these results are encouraging and corroborate that partial solvers based on fatal attractors may be components of faster solvers for parity games.

6.3 Number of Detected Fatal Attractors

We also recorded the number of fatal attractors that were detected in runs of our partial solvers. One reason for doing this is to see whether game types have a typical number of dynamically detected fatal attractors that result in the complete solving of these games.

We report these findings for `psol` and `psolB` first: for `Clique`, `Ladder`, and `Strategy Impr` these games are solved by detecting two fatal attractors only; `Model Checker Ladder` was solved by detecting one fatal attractor. For the other game types `psol` and `psolB` behaved differently. For `Recursive Ladder`[n], `psolB` requires $n = 2^k$ fatal attractors whereas `psolQ` needs only 2^{k-2} fatal attractors. For `Jurdzinski`[n, m], `psolB` detects $mn + 1$ many fatal attractors, and `psol` removes x edges where x is about $nm/2 \leq x \leq nm$, and detects slightly more than these x fatal attractors. Finally, for `Tower Of Hanoi`[n], `psol` requires the detection of 3^n fatal attractors whereas `psolB` solves these games with detecting two fatal attractors only.

We also counted the number of recursive calls for `psolQ`: it equals the number of fatal attractors detected by `psolB` for all game types except `Recursive Ladder`, where it is 2^{k-1} when n equals 2^k.

6.4 Experiments on Variants of Partial Solvers

We performed additional experiments on variants of these partial solvers. Here, we report results and insights on two such variants. The first variant is one that modifies the definition of the monotone control predecessor to

$$\begin{aligned}\mathsf{mpre}_p(A, X, c) = \{v \in V_p \mid ((c(v)\%2 = p) \vee c(v) \geq c) \wedge v.E \cap (A \cup X) \neq \emptyset\} \cup \\ \{v \in V_{1-p} \mid ((c(v)\%2 = p) \vee c(v) \geq c) \wedge v.E \subseteq A \cup X\}\end{aligned}$$

The change is that the constraint $c(v) \geq c$ is weakened to a disjunction $(c(v)\%2 = p) \vee (c(v) \geq c)$ so that it suffices if the color at node v has parity p even though it may be smaller than c. This implicitly changes the definition of the monotone attractor and so of all partial solvers that make use of this attractor; and it also impacts the computation of A within `psolQ`. Yet, this change did not have a dramatic effect on our partial solvers. On our benchmarks, the change improved things slightly for `psol` and made it slightly worse for `psolB` and `psolQ`.

A second variant we studied was a version of `psol` that removes at most one edge in each iteration (as opposed to all edges as stated in Fig. 2). For games of type `Ladder`, e.g., this variant did much worse. But for game types `Model Checker Ladder` and `Strategy Impr`, this variant did much better. The partial solvers based on such variants and their combination are such that `psolB` (as defined in Figure 3) is still better across all benchmarks.

6.5 Experiments on Random Games

It is our belief that comparing the behavior of parity game solvers on random games does not give an impression of how these solvers perform on parity games in practice. However, evaluating our partial solvers over random games gives an indication of how often partial solvers completely solve random games, and of whether partial solvers can speed up complete solvers as preprocessors. So we generated $130,000$ random games with the `randomgame` command of PGSolver.

Each game had between 10 and 500 nodes (average of 255). Each node v had out-degree (i.e. the size of $v.E$) at least 1, and at most 2, 3, 4, or 5 – where this number was determined at random. These games contained no self-loops and no bound on the number of different colors. Then `psolB` solved 82% of these $130,000$ random games completely. The average run-time over these $130,000$ games was 319ms for `psolB` (which includes run-time of `zlka` on the residual game where applicable), whereas the full solver `zlka` took 505ms on average. And only about $22,000$ of these games (less than 17%) were such that `zlka` solved them faster than the variant of `zlka` that used `psolB` as preprocessor.

7 Conclusions

We proposed a new approach to studying the problem of solving parity games: partial solvers as polynomial algorithms that correctly decide the winning status of some nodes and return a sub-game of nodes for which such status cannot be decided. We demonstrated the feasibility of this approach both in theory and in practice. Theoretically, we developed a new form of attractor that naturally lends itself to the design of such partial solvers; and we proved results about the computational complexity and semantic properties of these partial solvers. Practically, we showed through extensive experiments that these partial solvers can compete with extant solvers on benchmarks – both in terms of practical running times and in terms of precision in that our partial solvers completely solve such benchmark games.

In future work, we mean to study the descriptive complexity of the class of output games of a partial solver, for example of `psolQ`. We also want to research whether such output classes can be solved by algorithms that exploit invariants satisfied by these output classes. Furthermore, we mean to investigate whether classes of games characterized by structural properties of their game graphs can be solved completely by partial solvers. Such insights may connect our work to that of [3], where it is shown that certain classes of parity games that can be solved in PTIME are closed under operations such as the join of game graphs. Finally, we want to investigate whether and how partial solvers can be integrated into solver design patterns such as the one proposed in [5].

A technical report [7] accompanies this paper and contains – amongst other things – selected proofs, the pseudo-code of our version of Zielonka's algorithm, and further details on experimental results and their discussion.

References

1. Berwanger, D., Dawar, A., Hunter, P., Kreutzer, S.: DAG-Width and Parity Games. In: Durand, B., Thomas, W. (eds.) STACS 2006. LNCS, vol. 3884, pp. 524–536. Springer, Heidelberg (2006)
2. Chatterjee, K., Henzinger, M.: An $O(n^2)$ time algorithm for alternating Büchi games. In: SODA, pp. 1386–1399 (2012)
3. Dittmann, C., Kreutzer, S., Tomescu, A.I.: Graph operations on parity games and polynomial-time algorithms. arXiv:1208.1640 (2012)
4. Emerson, E.A., Jutla, C.: Tree automata, μ-calculus and determinacy. In: FOCS, pp. 368–377 (1991)
5. Friedmann, O., Lange, M.: Solving Parity Games in Practice. In: Liu, Z., Ravn, A.P. (eds.) ATVA 2009. LNCS, vol. 5799, pp. 182–196. Springer, Heidelberg (2009)
6. Friedmann, O., Lange, M.: The PGSolver Collection of Parity Game Solvers. Tech. report, Institut für Informatik, LMU Munich, Version 3 (February 2010)
7. Huth, M., Kuo, J.H., Piterman, N.: Fatal attractors in parity games. Tech. report no. 2013/1, Dep. of Computing, Imperial College London (January 2013) `http://www.doc.ic.ac.uk/research/technicalreports/2013/DTR13-1.pdf`
8. Jurdziński, M.: Small Progress Measures for Solving Parity Games. In: Reichel, H., Tison, S. (eds.) STACS 2000. LNCS, vol. 1770, pp. 290–301. Springer, Heidelberg (2000)
9. Jurdziński, M., Paterson, M., Zwick, U.: A deterministic subexponential algorithm for solving parity games. In: SODA, pp. 117–123. ACM/SIAM (2006)
10. Vöge, J., Jurdziński, M.: A Discrete Strategy Improvement Algorithm for Solving Parity Games. In: Emerson, E.A., Sistla, A.P. (eds.) CAV 2000. LNCS, vol. 1855, pp. 202–215. Springer, Heidelberg (2000)
11. Zielonka, W.: Infinite games on finitely coloured graphs with applications to automata on infinite trees. Theoretical Computer Science 200(1–2), 135–183 (1998)

On Unique Decomposition of Processes in the Applied π-Calculus

Jannik Dreier, Cristian Ene, Pascal Lafourcade, and Yassine Lakhnech

Université Grenoble 1, CNRS, Verimag, France
firstname.lastname@imag.fr

Abstract. Unique decomposition has been a subject of interest in process algebra for a long time (for example in BPP [2] or CCS [11,13]), as it provides a normal form with useful cancellation properties. We provide two parallel decomposition results for subsets of the Applied π-Calculus: we show that any closed normed (i.e. with a finite shortest complete trace) process P can be decomposed uniquely into prime factors P_i with respect to strong labeled bisimilarity, i.e. such that $P \sim_l P_1|\ldots|P_n$. We also prove that closed finite processes can be decomposed uniquely with respect to weak labeled bisimilarity.

Keywords: Applied π-Calculus, Unique Decomposition, Normal Form, Weak Bisimilarity, Strong Bisimilarity, Factorization, Cancellation.

1 Introduction

Process Algebras or Calculi allow one to formally model and analyze distributed systems. Famous examples include the Calculus of Communicating Systems (CCS) due to Milner [10], or Basic Parallel Processes (BPP) [2]. These calculi contain basic operations such as emission and reception of messages as well as parallel composition or interleaving. In an extension to CCS, Milner, Parrow and Walker developed the π-Calculus [12], which also features channel passing and scope extrusion. Abadi and Fournet [1] subsequently proposed the Applied π-Calculus, a variant of the π-Calculus designed for the verification of cryptographic protocols. It additionally features equational theories and active substitutions.

In all of these process algebras the question of unique process decomposition naturally arises. Can we rewrite a process P as $P =^1 P_1|P_2|\ldots|P_n$, where $|$ is the parallel composition operator, and each P_i is prime in the sense that it cannot be rewritten as the parallel composition of two non-zero processes?

Such a decomposition provides a maximally parallelized version of a given program P. Additionally, it is useful as it provides a normal form, and a cancellation result in the sense that $P|Q = P|R$ implies $Q = R$. This is convenient in proofs, for example when proving the equivalence of different security notions

[1] Here $=$ does not designate syntactical identity, but rather some behavioral equivalence or bisimilarity relation.

F. Pfenning (Ed.): FOSSACS 2013, LNCS 7794, pp. 50–64, 2013.

in electronic voting [3]: one can show that coercion of one voter and coercion of multiple voters are (under some realistic hypotheses) equivalent. This simplifies the analysis of e-voting protocols, in particular some proofs of observational equivalence. If there is an efficient procedure to transform a process into its normal form, such a decomposition can also be used to verify the equivalence of two processes [5]: once the processes are in normal form, one only has to verify if the factors on both sides are identical.

However, existing results [2,6,11,13] on the unique decomposition focus on "pure" calculi such as CCS or BPP or variants thereof. The Applied π-Calculus, as an "impure" variant of the π-Calculus designed for the verification of cryptographic protocols, has a more complex structure and semantics. For example, it features an equational theory to model cryptographic primitives, and active substitutions, i.e. substitutions that apply to all processes. This creates an element that is not zero, but still exhibits no transitions.

Additionally, the Applied π-Calculus inherits the expressive power of the π-Calculus including *channel* or *link passing* (sometimes also called *mobility*) and *scope extrusion*. Consider three parallel processes P, Q and R, where P and Q synchronize using an internal reduction τ_c, i.e. $P|Q|R \xrightarrow{\tau_c} P'|Q'|R$ (see Figure 1). Channel passing allows a process P to send a channel y he shares with R to process Q (Figure 1a). Scope extrusion arises for example when P sends a restricted channel y he shares with R to Q, since the scope after the transition includes Q' (Figure 1b). This is of particular importance for unique decomposition since two parallel processes sharing a restricted channel might not be decomposable and hence a simple reduction might "fuse" two prime factors.

1.1 Our Contributions

We provide two decomposition results for subsets of the Applied π-Calculus. In a first step, we prove that closed normed (i.e. with a finite shortest complete trace) processes can be uniquely decomposed with respect to strong labeled bisimilarity. In the second step we show that any closed finite (i.e. with a finite longest complete trace) process can be uniquely decomposed with respect to (weak) labeled bisimilarity, the standard bisimilarity notion in the Applied π-Calculus. Note that although we require the processes to be finite or normed, no further hypothesis is needed, i.e. they may use the full power of the calculus including channel passing and scope extrusion. As a direct consequence of the uniqueness of the decomposition, we also obtain cancellation results for both cases.

1.2 Outline of the Paper

In the next section, we recall the Applied π-Calculus, and establish different subclasses of processes. In Section 3 we provide our first unique decomposition result with respect to strong bisimilarity. In the next Section we show the second result w.r.t. weak bisimilarity. Then we discuss related work in Section 5 and conclude in Section 6. The full proofs can be found in our technical report [4].

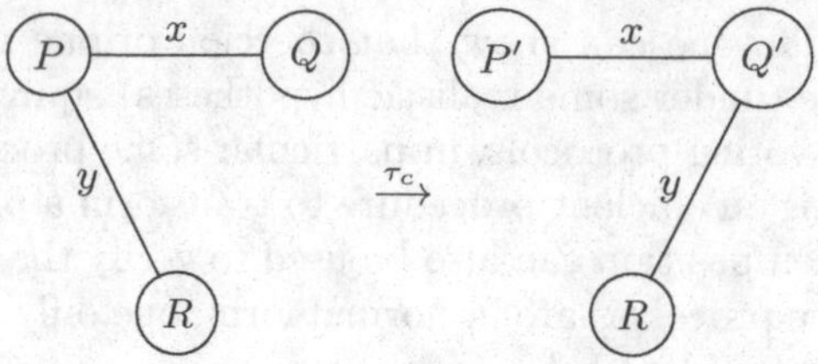

(a) Channel/Link Passing

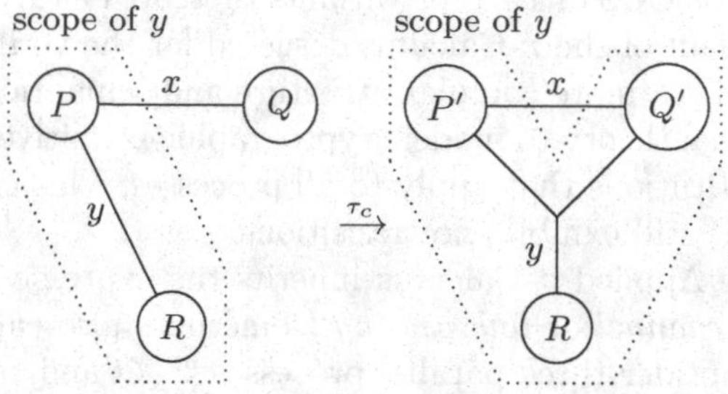

(b) Scope extrusion

Fig. 1. Features of the Applied π-Calculus

2 Preliminaries

In this section we recall briefly the Applied π-Calculus proposed by Abadi and Fournet [1] as an extension of the π-Calculus [12].

2.1 Applied π-Calculus

The Applied π-Calculus is a formal language for describing concurrent processes. The calculus consists of *names* (which typically correspond to data or channels), *variables*, and a *signature* Σ of *function symbols* which can be used to build *terms*. Functions typically include encryption and decryption (for example $\mathtt{enc}(message, key)$, $\mathtt{dec}(message, key)$), hashing, signing etc. Terms are correct combinations of names and functions, i.e. respecting arity and sorts. We distinguish the type "channel" from other *base* types. Equalities are modeled using an equational theory E which defines a relation $=_E$. A classical example, which describes the correctness of symmetric encryption, is $\mathtt{dec}(\mathtt{enc}(message, key), key) =_E message$.

Plain processes are constructed using the grammar depicted in Figure 2a. *Active* or *extended processes* are plain processes or active substitutions as shown in Figure 2b. Note that we do not include the "+"-operator which implements a nondeterministic choice, yet we can implement something similar using a restricted channel (see Example 4). For more details on encoding the operator with respect to different semantics, see [14,15].

$P, Q :=$	plain processes
0	null process
$P\|Q$	parallel composition
$!P$	replication
$\nu n.P$	name restriction ("new")
$\texttt{if}\ M = N\ \texttt{then}\ P\ \texttt{else}\ Q$	conditional (M, N terms)
$\texttt{in}(u, x).P$	message input
$\texttt{out}(u, M).P$	message output

(a) Plain Processes

$A, B, P, Q :=$	active processes
P	plain process
$A\|B$	parallel composition
$\nu n.A$	name restriction
$\nu x.A$	variable restriction
$\{^M/_x\}$	active substitution

(b) Active/Extended Processes

Fig. 2. Process Grammars

The substitution $\{^M/_x\}$ replaces the variable x with a term M. Note that we do not allow two active substitutions to define the same variable, as this might lead to situations with unclear semantics. We denote by $fv(A)$, $bv(A)$, $fn(A)$, $bn(A)$ the free variables, bound variables, free names or bound names respectively.

As an additional notation we write $\nu S.A$ for $\nu s_1.\nu s_2 \ldots \nu s_n.A$ where $s_1, \ldots s_n$ are the elements of a set of variables and names S. By abuse of notation we sometimes leave out ".0" at the end of a process. We will also write A^k for $A| \ldots |A$ (k times), in particular $A^0 = 0$ as 0 is the neutral element of parallel composition.

The *frame* $\Phi(A)$ of an active process A is obtained by replacing all plain processes in A by 0. This frame can be seen as a representation of what is statically known to the environment about a process. The domain $dom(\Phi)$ of a frame Φ is the set of variables for which Φ defines a substitution. By abuse of notation, we also write $dom(A)$ to denote the domain of the frame $\Phi(A)$ of an active process A. Note that $dom(A) \subseteq fv(A)$, and that – as we cannot have two active substitutions for the same variable – $P = Q|R$ implies $dom(P) = dom(Q) \cup dom(R)$ and $dom(Q) \cap dom(R) = \emptyset$. A frame or process is *closed* if all variables are bound or defined by an active substitution. An evaluation context $C[\_]$ denotes an active process with a hole for an active process that is not under replication, a conditional, an input or an output.

The semantics of the calculus presupposes a notion of *Structural Equivalence* ($\equiv$), which is defined as the smallest equivalence relation on extended processes that is closed under application of evaluation contexts, α-conversion on bound names and bound variables such that:

PAR-0	$A\|0 \equiv A$	
PAR-A	$A\|(B\|C) \equiv (A\|B)\|C$	
PAR-C	$A\|B \equiv B\|A$	
NEW-0	$\nu n.0 \equiv 0$	
NEW-C	$\nu u.\nu v.A \equiv \nu v.\nu u.A$	
NEW-PAR	$A\|\nu u.B \equiv \nu u.(A\|B)$	if $u \notin fn(A) \cup fv(A)$
REPL	$!P \equiv P\|!P$	
REWRITE	$\{^M/_x\} \equiv \{^N/_x\}$	if $M =_E N$
ALIAS	$\nu x.\{^M/_x\} \equiv 0$	
SUBST	$\{^M/_x\}\|A \equiv \{^M/_x\}\|A\{^M/_x\}$	

Note the contagious nature of active substitutions: by rule SUBST they apply to any parallel process.

Example 1. Consider the following running example, where x and y are variables, and c, d, k, l, m and n are names:

$$P_{ex} = \nu k.\nu l.\nu m.\nu d.\,(\{^l/_y\}\,|\mathtt{out}(c, enc(n,k))|\mathtt{out}(d,m)|\mathtt{in}(d,x).\mathtt{out}(c,x))$$

We have $dom(P_{ex}) = \{y\}$, $fv(P_{ex}) = \{y\}$, $bv(P_{ex}) = \{x\}$, $fn(P_{ex}) = \{n, c\}$, $bn(P_{ex}) = \{k, l, m, d\}$ and

$$\Phi(P_{ex}) = \nu k.\nu l.\nu m.\nu d.\,(\{^l/_y\}\,|0|0|0) \equiv \nu k.\nu l.\nu m.\nu d.\,(\{^l/_y\})$$

Internal Reduction ($\xrightarrow{\tau}$) is the smallest relation on extended processes closed by structural equivalence and application of evaluation contexts such that:

COMM	$\mathtt{out}(a,x).P \mid \mathtt{in}(a,x).Q \xrightarrow{\tau_c} P \mid Q$
THEN	$\mathtt{if}\ M = M\ \mathtt{then}\ P\ \mathtt{else}\ Q \xrightarrow{\tau_t} P$
ELSE	$\mathtt{if}\ M = N\ \mathtt{then}\ P\ \mathtt{else}\ Q \xrightarrow{\tau_e} Q$
	for any ground terms such that $M \neq_E N$

Note that in accordance with the original notations [1], we sometimes omit the labels τ_c, τ_t and τ_e, and write $P \rightarrow P'$ for $P \xrightarrow{\gamma} P'$ with $\gamma \in \{\tau_c, \tau_t, \tau_e\}$.

Interactions of extended processes are described using labeled operational semantics ($\xrightarrow{\alpha}$), where α can be an input or an output of a channel name or variable of base type, e.g. $\mathtt{out}(a, u)$ where u is a variable or a name.

IN	$\mathtt{in}(a,x).P \xrightarrow{\mathtt{in}(a,M)} P\{^M/_x\}$
OUT-ATOM	$\mathtt{out}(a,u).P \xrightarrow{\mathtt{out}(a,u)} P$
OPEN-ATOM	$\dfrac{A \xrightarrow{\mathtt{out}(a,u)} A' \quad u \neq a}{\nu u.A \xrightarrow{\nu u.\mathtt{out}(a,u)} A'}$
SCOPE	$\dfrac{A \xrightarrow{\alpha} A' \quad u \text{ does not occur in } \alpha}{\nu u.A \xrightarrow{\alpha} \nu u.A'}$
PAR	$\dfrac{A \xrightarrow{\alpha} A' \quad bv(\alpha) \cap fv(B) = bn(\alpha) \cap fn(B) = \emptyset}{A \mid B \xrightarrow{\alpha} A' \mid B}$
STRUCT	$\dfrac{A \equiv B \quad B \xrightarrow{\alpha} B' \quad B' \equiv A'}{A \xrightarrow{\alpha} A'}$

Labeled *external transitions* are not closed under evaluation contexts. Note that a term M (except for channel names and variables of base type) cannot be output directly. Instead, we have to assign M to a variable, which can then be output. This is to model that the output of $\mathtt{enc}(m,k)$ (message m encrypted with key k) does not give the environment access to m.

Example 2. Consider our running example process P_{ex}. Using an internal reduction, we can execute the following transition:

$$\begin{aligned}
P_{ex} &= \nu k.\nu l.\nu m.\nu d.\ (\{^l/_y\}\,|\mathtt{out}(c, enc(n,k)).0|\mathtt{out}(d,m).0|\mathtt{in}(d,x).\mathtt{out}(c,x).0)\\
&\equiv \nu k.\nu l.\nu m.\nu d.(\{^l/_y\}\,|\mathtt{out}(c, enc(n,k))|\boldsymbol{\nu x.(\{^m/_x\})}|\mathtt{out}(d,m)|\\
&\quad \mathtt{in}(d,x).\mathtt{out}(c,x)) \qquad\qquad \text{by PAR-0, ALIAS}\\
&\equiv \nu k.\nu l.\nu m.\nu d.(\{^l/_y\}\,|\mathtt{out}(c, enc(n,k))|\nu x.(\,\{^m/_x\}\,|\mathtt{out}(d,\boldsymbol{x})|\\
&\quad \mathtt{in}(d,x).\mathtt{out}(c,x))) \qquad\qquad \text{by SUBST, NEW-PAR}\\
&\xrightarrow{\tau_c} \nu k.\nu l.\nu m.\nu d.\ (\{^l/_y\}\,|\mathtt{out}(c, enc(n,k))|\nu x.\ (\{^m/_x\}\,|\mathtt{out}(c,x)))\\
&\equiv \nu k.\nu l.\nu m.\nu d.\ (\{^l/_y\}\,|\mathtt{out}(c, enc(n,k))|\mathtt{out}(c,\boldsymbol{m}))\\
&\qquad\qquad \text{by SUBST, ALIAS, NEW-PAR, PAR-0}
\end{aligned}$$

Similarly, we can also execute an external transition:

$$\begin{aligned}
P_{ex} &\equiv \nu k.\nu l.\nu m.\nu d.(\{^l/_y\}\,|\nu z.\,\{^{enc(n,k)}/_z\}\,|\mathtt{out}(c,z)|\mathtt{out}(d,m)|\\
&\quad \mathtt{in}(d,x).\mathtt{out}(c,x))\\
&\xrightarrow{\nu z.\mathtt{out}(c,z)} \nu k.\nu l.\nu m.\nu d.\ (\{^l/_y\}\,|\,\{^{enc(n,k)}/_z\}\,|\mathtt{out}(m,d)|\mathtt{in}(x,d).\mathtt{out}(x,c))
\end{aligned}$$

The following two definitions allow us to reason about the messages exchanged with the environment.

Definition 1 (Equivalence in a Frame [1]). *Two terms M and N are equal in the frame $\phi \equiv \nu\tilde{n}.\sigma$, written $(M = N)\phi$, if and only if $M\sigma =_E N\sigma$, and $\{\tilde{n}\} \cap (fn(M) \cup fn(N)) = \emptyset$.*

Note that any frame ϕ can be written as $\nu\tilde{n}.\sigma$ modulo structural equivalence, i.e. using rule NEW-PAR.

Definition 2 (Static Equivalence ($\approx_s$) [1]). *Two closed frames ϕ and ψ are statically equivalent, written $\phi \approx_s \psi$, when $dom(\phi) = dom(\psi)$ and when for all* *terms M and N we have $(M = N)\phi$ if and only if $(M = N)\psi$. Two extended processes A and B are statically equivalent ($A \approx_s B$) if their frames are statically equivalent.*

The intuition behind this definition is that two processes are statically equivalent if the messages exchanged previously with the environment cannot be distinguished, i.e. all operations on both sides gave the same results.

2.2 Depth and Norm of Processes

We prove unique decomposition for different subsets of processes, namely finite and normed processes. This requires to formally define the length of process traces. Let $\mathbf{Int} = \{\tau_c, \tau_t, \tau_e\}$ denote the set of labels corresponding to internal reductions or *silent* transitions, and $\mathbf{Act} = \{\mathtt{in}(a, M), \mathtt{out}(a, u), \nu u.\mathtt{out}(a, u)\}$ for any channel name n, term M and variable or name u, denote the set of labels of possible external or *visible* transitions. By construction $\mathbf{Act} \cap \mathbf{Int} = \emptyset$.

The *visible depth* is defined as the length of the longest complete trace of visible actions, i.e. labeled transitions, excluding internal reductions. Note that this may be infinite for processes including replication. We write $P \not\rightarrow$ if P cannot execute any transition, and $P \xrightarrow{\mu_1\mu_2\ldots\mu_n} P'$ for $P \xrightarrow{\mu_1} P_1 \xrightarrow{\mu_2} P_2 \xrightarrow{\mu_3} \ldots \xrightarrow{\mu_n} P'$.

Definition 3 (Visible Depth). *Let $length_v : (\mathbf{Act} \cup \mathbf{Int})^* \mapsto \mathbb{N}$ be a function where $length_v(\epsilon) = 0$ and $length_v(\mu w) = \begin{cases} 1 + length_v(w) & \text{if } \mu \in \mathbf{Act} \\ length_v(w) & \text{otherwise} \end{cases}$
Then the* visible depth *$|P|_v \in (\mathbb{N} \cup \{\infty\})$ of a closed process P is defined as follows:*

$$|P|_v = \sup\left\{ length_v(w) : P \xrightarrow{w} P' \not\rightarrow, w \in (\mathbf{Act} \cup \mathbf{Int})^* \right\}$$

The *total depth* is defined as the length of the longest complete trace of actions (including internal reductions).

Definition 4 (Total Depth). *Let $length_t : (\mathbf{Act} \cup \mathbf{Int})^* \mapsto \mathbb{N}$ be a function where $length_t(\epsilon) = 0$ and $length_t(\mu w) = 1 + length_t(w)$. The* total depth *$|P|_t \in (\mathbb{N} \cup \{\infty\})$ of a closed process P is defined as follows:*

$$|P|_t = \sup\left\{ length_t(w) : P \xrightarrow{w} P' \not\rightarrow, w \in (\mathbf{Act} \cup \mathbf{Int})^* \right\}$$

The norm of a process is defined as the length of the shortest complete trace, including internal reductions, where communications are counted as two. This is necessary to ensure that the norm of $P|Q$ is the sum of the norm of P and the norm of Q.

Definition 5 (Norm of a Process). *Let $length_n : (\mathbf{Act} \cup \mathbf{Int})^* \mapsto \mathbb{N}$ be a function where $length_n(\epsilon) = 0$ and $length_n(\mu w) = \begin{cases} 1 + length_n(w) & \text{if } \mu \neq \tau_c \\ 2 + length_n(w) & \text{if } \mu = \tau_c \end{cases}$
The norm $\mathcal{N}(P) \in (\mathbb{N} \cup \{\infty\})$ of a closed process P is defined as follows:*

$$\mathcal{N}(P) = \inf\left\{ length_n(w) : P \xrightarrow{w} P' \not\rightarrow, w \in (\mathbf{Act} \cup \mathbf{Int})^* \right\}$$

Example 3. We have $|P_{ex}|_v = 2$, $|P_{ex}|_t = 3$ and $\mathcal{N}(P_{ex}) = 4$.

The above definitions admit some simple properties.

Property 1. For any closed extended processes P, Q and R we have

- $P = Q|R$ implies $|P|_v = |Q|_v + |R|_v$
- $P = Q|R$ implies $|P|_t = |Q|_t + |R|_t$
- $P = Q|R$ implies $\mathcal{N}(P) = \mathcal{N}(Q) + \mathcal{N}(R)$
- $|P|_v \leq |P|_t$

Now we can define the two important subclass of processes: finite processes, i.e. processes with a finite longest complete trace, and normed processes, i.e. processes with a finite shortest complete trace.

Definition 6 (Finite and normed processes). *A closed process P is called* finite, *if $|P|_t$ is finite (which implies $|P|_v$ is finite). A closed process P is called* normed, *if $\mathcal{N}(P)$ is finite.*

It is easy to see that any finite process is normed, but not all normed processes are finite, as the following example illustrates.

Example 4. Consider $P = \nu a.(out(a,m)|(in(a,x).(!in(b,y)))|in(a,x))$. Then we have $P \rightarrow P' \sim_l 0$, hence P is normed. However we also have $P \rightarrow P'' \sim_l !in(b,y)$, which has infinite traces. Hence P is not finite.

It is also clear that not all processes are normed. Consider the following example.

Example 5. Consider $P = !(\nu x.out(c,x))$. It is easy to see that for no sequence of transitions s we have $P \xrightarrow{s} P' \not\rightarrow$, i.e. P has no finite traces.

3 Decomposition w.r.t. Strong Labeled Bisimilarity

We begin with the simpler case of strong labeled bisimilarity, defined as follows.

Definition 7 (Strong Labeled Bisimilarity ($\sim_l$)). *Strong labeled bisimilarity is the largest symmetric relation $\mathcal{R}$ on closed active processes, such that $A \mathcal{R} B$ implies:*

1. $A \approx_s B$,
2. *if $A \rightarrow A'$, then $B \rightarrow B'$ and $A' \mathcal{R} B'$ for some B'*
3. *if $A \xrightarrow{\alpha} A'$ and $fv(\alpha) \subseteq dom(A)$ and $bn(\alpha) \cap fn(B) = \emptyset$, then $B \xrightarrow{\alpha} B'$ and $A' \mathcal{R} B'$ for some B'.*

Note that $P \sim_l Q$ implies $|P|_t = |Q|_t$ and $\mathcal{N}(P) = \mathcal{N}(Q)$ for any closed processes P and Q. To ensure that labeled bisimilarity is a congruence w.r.t to parallel composition ("|") and closed under the application of contexts, we will require that active substitutions are only defined on variables of base type [7].

We define strong parallel primeness as follows: A process is prime if it cannot be decomposed into non-trivial subprocesses (w.r.t. strong labeled bisimilarity). We require the processes to be closed, which is necessary as our bisimulation relation is only defined on closed processes.

Definition 8 (Strongly Parallel Prime). *A closed process P is* strongly parallel prime, *if*

- *$P \not\sim_l 0$ and*
- *for any two closed processes Q and R such that $P \sim_l Q|R$, we have $Q \sim_l 0$ or $R \sim_l 0$.*

Example 6. Consider our running example:

$$P_{ex} = \nu k.\nu l.\nu m.\nu d.\ (\{^l/_y\}\,|\mathtt{out}(c, enc(n, k))|\mathtt{out}(d, m)|\mathtt{in}(d, x).\mathtt{out}(c, x))$$

We can decompose P_{ex} as follows:

$$P_{ex} \sim_l (\nu l.\ \{^l/_y\})|(\nu k.\mathtt{out}(c, enc(n, k)))|(\nu d.(\nu m.\mathtt{out}(d, m)|\mathtt{in}(d, x).\mathtt{out}(c, x)))$$

The first factor $S_1 = \nu l.\ \{^l/_y\}$ is prime since we cannot have two substitutions defining the same variable. It is easy to see that the second factor $S_2 = \nu k.\mathtt{out}(c, enc(n, k))$ is prime, as it can only perform one external transition. The third factor $S_3 = \nu d.(\nu m.\mathtt{out}(d, m)|\mathtt{in}(d, x).\mathtt{out}(c, x))$ is prime because its two parts can synchronize using a shared restricted channel and then perform a visible external transition. Since $dom(S_3) = \emptyset$ and $\mathcal{N}(S_3) = 2$, the only possible decomposition would be into two factors of norm 1 each, i.e. such that $S_3 \sim_l S_3'|S_3''$. This would however mean that both transitions of $S_3'|S_3''$ can be executed in any order, whereas in S_3 we have to start with the internal reduction. Hence no such decomposition exists.

With respect to applications in protocol analysis, this illustrates that shared restricted names, for example private channels or shared keys, can prohibit decomposition. This is unavoidable, since a decomposition should not change the behavior of the processes (up to $\sim_l$), yet it might appear to hinder the usefulness of the decomposition on first view. However, a decomposition that does not preserve the behavioral equivalence is probably not useful either, and note that – since our definition is solely based on the semantics and the bisimilarity notion – it allows to decompose as far as possible without changing the observed behavior, and thus any further decomposition will change the behavior. As a side-effect, the decomposition will show where shared restricted names (modeling for example keys) are actually used in a noticeable (w.r.t. to $\sim_l$) way, and where they can be ignored and processes can be further decomposed.

Note also that within a prime factor we can recursively apply the decomposition as our bisimilarity notion is closed under the application of contexts. For example if we have a prime factor $P = \nu a.P'$, we can bring P' into normal form, i.e. $P' \sim_l P_1'|...|P_n'$, and rewrite $P = \nu a.P'$ as $P \sim_l \nu a.(P_1'|...|P_n')$.

It is clear that not all processes can be written as a unique decomposition of parallel primes according to our definition.

Example 7. Consider $!P$ for a process $P \not\sim_l 0$. By definition we have $!P = P|!P$, hence $!P$ is not prime. At the same time any such decomposition contains again $!P$, a non-prime factor, which needs to be decomposed again. Thus there is no decomposition into prime factors.

However we can show that any closed normed process has a unique decomposition with respect to strong labeled bisimilarity. In a first step, we prove the existence of a decomposition.

Theorem 1 (Existence of Factorization). *Any closed normed process P can be expressed as the parallel product of strong parallel primes, i.e.* $P \sim_l P_1 | \dots | P_n$ *where for all* $1 \leq i \leq n$ P_i *is strongly parallel prime.*

Proof. Sketch: The proof proceeds by induction on the norm of P, and inside each case by induction on the size of the domain. The second induction is necessary to deal with active substitutions, which cannot perform transitions. The main idea is simple: If a process is not prime, by definition it can be decomposed into two "smaller" processes, where we can apply the induction hypothesis. □

To show the uniqueness of the decomposition, we need some preliminary lemmas about transitions and the domain of processes. The first lemma captures the fact that intuitively any process which cannot perform any transition and has an empty domain, is bisimilar to 0 (the empty process).

Lemma 1. *For any closed process A with $dom(A) = \emptyset$ and $\mathcal{N}(A) = 0$, we have* $A \sim_l 0$.

We also need to show that if a normed process can execute a transition, it can also execute a norm-reducing transition.

Lemma 2. *Let A be a closed normed process with* $A \xrightarrow{\mu} A'$ *where μ is an internal reduction or visible transition. Then* $A \xrightarrow{\mu'} A''$ *with* $\mathcal{N}(A'') < \mathcal{N}(A)$.

These lemmas allow us to show the uniqueness of the decomposition.

Theorem 2 (Uniqueness of Factorization). *The strong parallel factorization of a closed normed process P is unique (up to $\sim_l$).*

Proof. Sketch: In the proof we have to deal with numerous cases due to the complex semantics of the calculus. Here we focus on the main differences compared to existing proofs for simpler calculi (e.g. [13]).

The proof proceeds by induction on the norm of P, and inside each case by induction on the size of the domain. By Lemma 1, each prime factor can either perform a transition, or has a non-empty domain. A transition may not always be norm-reducing since processes can be infinite, but in this case Lemma 2 gives us that if a normed process can execute a transition, it can also execute a norm-reducing one - which we can then consider. We suppose the existence of two different factorizations, and show that this leads to a contradiction. Consider the following four cases:

- If we have a process that cannot do any transition and has an empty domain, by Lemma 1 we have the unique factorization 0.
- If the process cannot perform a transition but has a non-empty domain, we can apply a restriction on part of the domain to hide all factors but one (since we cannot have two substitutions defining the same variable). We can

then use the fact that labeled bisimilarity is closed under the application of contexts to exploit the induction hypothesis, which eventually leads to a contradiction to the primality of the factors.
- In the case of a process with empty domain, but that can perform a transition, we can execute a transition and then apply the induction hypothesis. However, we have to be careful since in case of an internal reduction factors could fuse using scope extrusion (see Figure 1b). Hence, whenever possible, we choose a visible transition. If no such transition exists, processes cannot fuse using an internal reduction either, since this would mean they synchronized on a public channel, which implies the existence of visible transitions. Thus we can safely execute the invisible transition.
- In the last case (non-empty domain and visible transitions) we have to combine the above two techniques.

□

As a direct consequence, we have the following cancellation result.

Lemma 3 (Cancellation Lemma). *For any closed normed processes A, B and C, we have*

$$A|C \sim_l B|C \Rightarrow A \sim_l B$$

Proof. Sketch: All processes have a unique factorization and can be rewritten accordingly. As both sides are bisimilar, they have the same unique factorization, hence A and B must be composed of the same factors, thus they are bisimilar.

□

4 Decomposition w.r.t. Weak Labeled Bisimilarity

In this part, we discuss unique decomposition with respect to (weak) labeled bisimilarity. This is the standard bisimilarity notion in the Applied π-Calculus as defined by Abadi and Fournet in their original paper [1].

Definition 9 ((Weak) Labeled Bisimilarity ($\approx_l$) [1]). *(Weak) Labeled Bisimilarity is the largest symmetric relation $\mathcal{R}$ on closed active processes, such that $A \mathcal{R} B$ implies:*

1. $A \approx_s B$,
2. *if* $A \rightarrow A'$, *then* $B \rightarrow^* B'$ *and* $A' \mathcal{R} B'$ *for some* B',
3. *if* $A \xrightarrow{\alpha} A'$ *and* $fv(\alpha) \subseteq dom(A)$ *and* $bn(\alpha) \cap fn(B) = \emptyset$, *then* $B \rightarrow^* \xrightarrow{\alpha} \rightarrow^* B'$ *and* $A' \mathcal{R} B'$ *for some* B'.

The resulting bisimilarity notion is weak in the sense that only visible external transitions have to be matched exactly, and there may be a number of silent internal reductions in the background which are not taken into account. Note that $P \approx_l Q$ implies $|P|_v = |Q|_v$ for any closed processes P and Q.

Again we will assume that active substitutions can only be defined on variables of base type to ensure that labeled bisimilarity is a congruence w.r.t. to parallel composition ("|") and closed under the application of contexts. Under this condition, it also coincides with observational equivalence [7]. This was claimed in the original paper [1] without requiring the additional condition, but turned out to be untrue when a counterexample was found (see [7] for more details).

To obtain our unique decomposition result for weak labeled bisimilarity, we need to define parallel prime with respect to weak labeled bisimilarity.

Definition 10 (Weakly Parallel Prime). *A closed extended process P is* weakly parallel prime, *if*

- $P \not\approx_l 0$ *and*
- *for any two closed processes Q and R such that $P \approx_l Q|R$, we have $Q \approx_l 0$ or $R \approx_l 0$.*

This definition is analogous to strongly parallel prime. However, as the following example shows, in contrast to strong bisimilarity, not all normed processes have a unique decomposition w.r.t. to weak bisimilarity.

Example 8. Consider $P = \nu a.(out(a,m)|(in(a,x).(!in(b,y)))|in(a,x))$. Then we have $P \approx_l P|P$, hence we have no unique decomposition. Note that this example does not contradict our previous result, as we have $P \not\sim_l P|P$, as $P \to P' \sim_l 0$, but $P|P \to P'' \sim_l P$ and $P|P \not\to P'''$ for any $P''' \sim_l 0$. Hence, w.r.t. strong labeled bisimilarity, P is prime.

If however we consider normed processes that contain neither restriction ("ν") nor conditionals, we have that any normed process is finite (and hence has a unique decomposition, as we show below).

Lemma 4. *For any process P that does not contain restriction ("ν") or conditionals ("if then else"), we have that P is finite if and only if P is normed.*

Similarly any process that does not contain replication is finite.

In the following we show that all finite processes have a unique decomposition w.r.t. to (weak) labeled bisimilarity. Again, in a first step, we show that a decomposition into prime factors exists.

Theorem 3 (Existence of Factorization). *Any closed finite active process P can be expressed as the parallel product of parallel primes, i.e. $P \approx_l P_1|\ldots|P_n$ where for all $1 \le i \le n$ P_i is weakly parallel prime.*

The proof is analogous to the proof of Theorem 1, but we have to proceed by induction on the visible depth instead of the norm, as two weakly bisimilar processes may have a different norm.

To prove uniqueness, we again need some preliminary lemmas about transitions and the domain of processes. This first lemma captures the fact that intuitively any process that cannot perform any visible transition and has an empty domain, is weakly bisimilar to 0 (the empty process).

Table 1. Summary of unique factorization results for the Applied π-Calculus

Type of Process	Strong Bisimilarity ($\sim_l$)	Weak Bisimilarity ($\approx_l$)
finite	Theorem 1	Theorem 3
normed	Theorem 1	Counterexample 4
general	Counterexample 7	Counterexample 7

Lemma 5. *If for a closed process A with $dom(A) = \emptyset$ there does not exist a sequence of transitions $A \rightarrow^* \xrightarrow{\alpha} A'$, then we have $A \approx_l 0$.*

Now we can show the uniqueness of the decomposition.

Theorem 4 (Uniqueness of Factorization). *The parallel factorization of a closed finite process P is unique (up to $\approx_l$).*

Proof. Sketch: In the proof we show the following statement: Any closed finite processes P and Q with $P \approx_l Q$ have the same factorization (up to $\approx_l$). The proof proceeds by induction on the sum of the total depth of both factorizations, and in each case on the size of the domain. We show that if we suppose the existence of two different factorizations, this leads to a contradiction.

The proof follows the same structure as the one for strong bisimilarity. In the case of processes with non-empty domain and no visible transition, we use the same idea and apply restrictions to use the induction hypothesis. In the other cases, when executing a transition to apply the induction hypothesis, we have to be more careful since each transition can be simulated using additionally several internal reductions. This can affect several factors, and prime factors could fuse using an internal reduction and scope extrusion (see Figure 1b). We can circumvent this problem by choosing transitions that decrease the visible depth by exactly one (such a transition must always exist). A synchronization of two factors in the other factorization would use at least two visible actions and the resulting processes cannot be bisimilar any more, since bisimilar processes have the same depth. Using Lemma 5 we know that each prime factor has either a non-empty domain or can execute a visible transition, which allows us to conclude. □

Again we have a cancellation result using the same proof as above.

Lemma 6 (Cancellation Lemma). *For any closed finite processes A, B and C, we have*

$$A|C \approx_l B|C \Rightarrow A \approx_l B$$

5 Related Work

Unique decomposition (or factorization) has been a field of interest in process algebra for a long time. The first results for a subset of CCS were published by Moller and Milner [11,13]. They showed that finite processes with interleaving can be uniquely decomposed with respect to strong bisimilarity. The same is true for finite processes with parallel composition, where – in contrast to interleaving – the parallel processes can synchronize. They also proved that finite processes with parallel composition can be uniquely decomposed w.r.t. weak bisimilarity. Compared to the Applied π-Calculus, BPP and CCS do not feature channel passing, scope extrusion and active substitutions.

Later on Christensen [2] proved a unique decomposition result for normed processes (i.e. processes with a finite shortest complete trace) in BPP with interleaving or parallel composition w.r.t. strong bisimilarity.

Luttik and van Oostrom [9] provided a generalization of the unique decomposition results for ordered monoids. They show that if the calculus satisfies certain properties, the unique decomposition result follows directly. Recently Luttik also extended this technique for weak bisimilarity [8]. Unfortunately this result cannot be employed in the Applied π-Calculus as active substitutions are minimal elements (with respect to the transition relation) different from 0.

6 Conclusion and Future Work

We presented two unique decomposition results for subsets of the Applied π-Calculus. We showed that any closed finite process can be decomposed uniquely with respect to weak labeled bisimilarity, and that any normed process can be decomposed uniquely with respect to strong labeled bisimilarity. Table 1 sums up our results.

As the concept of parallel prime decomposition has its inherent limitations with respect to replication ("!", see Example 7), a natural question is to find an extension to provide a normal form even in cases with infinite behavior. A first result in this direction has been obtained by Hirschkoff and Pous [6] for a subset of CCS with top-level replication. They define the *seed* of a process P as the process Q, Q bisimilar to P, of least size (in terms of prefixes) whose number of replicated components is maximal (among the processes of least size), and show that this representation is unique. They also provide a result for the Restriction-Free-π-Calculus (i.e. no "ν"). It remains however open if a similar result can be obtained for the full calculus.

Another interesting question is to find an efficient algorithm that converts a process into its unique decomposition. It is unclear if such an algorithm exists and can be efficient, as simply deciding if a process is finite can be non-trivial.

References

1. Abadi, M., Fournet, C.: Mobile values, new names, and secure communication. In: Proceedings of the 28th ACM SIGPLAN-SIGACT Symposium on Principles of Programming Languages, POPL 2001, pp. 104–115. ACM, New York (2001)
2. Christensen, S.: Decidability and Decompostion in Process Algebras. PhD thesis, School of Computer Science, University of Edinburgh (1993)
3. Dreier, J., Lafourcade, P., Lakhnech, Y.: Defining Privacy for Weighted Votes, Single and Multi-voter Coercion. In: Foresti, S., Yung, M., Martinelli, F. (eds.) ESORICS 2012. LNCS, vol. 7459, pp. 451–468. Springer, Heidelberg (2012)

4. Dreier, J., Lafourcade, P., Lakhnech, Y.: On parallel factorization of processes in the applied pi calculus. Technical Report TR-2012-3, Verimag Research Report (March 2012), `http://www-verimag.imag.fr/TR/TR-2012-3.pdf`
5. Groote, J.F., Moller, F.: Verification of Parallel Systems via Decomposition. In: Cleaveland, W.R. (ed.) CONCUR 1992. LNCS, vol. 630, pp. 62–76. Springer, Heidelberg (1992)
6. Hirschkoff, D., Pous, D.: On Bisimilarity and Substitution in Presence of Replication. In: Abramsky, S., Gavoille, C., Kirchner, C., Meyer auf der Heide, F., Spirakis, P.G. (eds.) ICALP 2010. LNCS, vol. 6199, pp. 454–465. Springer, Heidelberg (2010)
7. Liu, J.: A proof of coincidence of labeled bisimilarity and observational equivalence in applied pi calculus. Technical Report ISCAS-SKLCS-11-05 (2011), `http://lcs.ios.ac.cn/~jliu/`
8. Luttik, B.: Unique parallel decomposition in branching and weak bisimulation semantics. Technical report (2012), `http://arxiv.org/abs/1205.2117v1`
9. Luttik, B., van Oostrom, V.: Decomposition orders – another generalisation of the fundamental theorem of arithmetic. Theoretical Computer Science 335(2-3), 147–186 (2005)
10. Milner, R.: Communication and Concurrency. International Series in Computer Science. Prentice Hall (1989)
11. Milner, R., Moller, F.: Unique decomposition of processes. Theoretical Computer Science 107(2), 357–363 (1993)
12. Milner, R., Parrow, J., Walker, D.: A calculus of mobile processes. Information and Computation 100(1), 1–40 (1992)
13. Moller, F.: Axioms for Concurrency. PhD thesis, School of Computer Science, University of Edinburgh (1989)
14. Nestmann, U., Pierce, B.C.: Decoding choice encodings. Information and Computation 163(1), 1–59 (2000)
15. Palamidessi, C., Herescu, O.M.: A randomized encoding of the pi-calculus with mixed choice. Theoretical Computer Science 335(23), 373–404 (2005)

Bounded Context-Switching and Reentrant Locking

Rémi Bonnet[1] and Rohit Chadha[2]

[1] LSV, ENS Cachan & CNRS
[2] University of Missouri

Abstract. Reentrant locking is a *recursive locking* mechanism which allows a thread in a multi-threaded program to acquire the reentrant lock multiple times. The thread must release this lock an equal number of times before another thread can acquire this lock. We consider the control state reachability problem for recursive multi-threaded programs synchronizing via a finite number of reentrant locks. Such programs can be abstracted as multi-pushdown systems with a finite number of counters. The pushdown stacks model the call stacks of the threads and the counters model the reentrant locks. The control state reachability problem is already undecidable for non-reentrant locks. As a consequence, for non-reentrant locks, under-approximation techniques which restrict the search space have gained traction. One popular technique is to limit the number of context switches. Our main result is that the problem of checking whether a control state is reachable within a bounded number of context switches is decidable for recursive multi-threaded programs synchronizing via a finite number of reentrant locks if we restrict the lock-usage to contextual locking: a release of an instance of reentrant lock can only occur if the instance was acquired before in the same procedure and each instance of a reentrant lock acquired in a procedure call must be released before the procedure returns. The decidability is obtained by a reduction to the reachability problem of Vector Addition Systems with States (VASS).

1 Introduction

A mutex lock is a synchronization primitive used in multi-threaded programs to enable communication amongst threads and guide their computations. A lock is either *free* or is *held (owned)* by a thread. If the lock is free then any thread can *acquire it* and in that case the lock is said to be held (owned) by that thread. The lock becomes free when the owning thread *releases* it. If the lock is held by some thread then any attempt to acquire it by any thread (including the owning thread) fails and the requesting thread blocks. However, some programming languages such as Java support non-blocking *reentrant locks*. In a reentrant locking mechanism, if a thread attempts to acquire a reentrant lock it already holds then the thread succeeds. The lock becomes free only when the owning thread releases the lock as many times as it has acquired the lock.

F. Pfenning (Ed.): FOSSACS 2013, LNCS 7794, pp. 65–80, 2013.

Verification of multi-threaded programs is an important challenge as they often suffer from subtle programming errors. One approach to tackle this challenge is static analysis, and this paper investigates this approach for multi-threaded recursive programs using reentrant locks. In static analysis of sequential recursive programs, a program is often abstracted into a pushdown system that captures the control flow of the program and where the stack models recursion [19]. Several static analysis questions are then formulated as reachability questions on the abstracted pushdown system. In a similar fashion, multi-threaded programs can be abstracted as multi-pushdown systems that synchronize using the synchronization primitives supported by the programming language. Important safety verification questions, such as data-race detection and non-interference, can then be formulated as control state reachability problem: given a global state $\boldsymbol{q}$ of a concurrent program, is $\boldsymbol{q}$ reachable?

The control state reachability problem for multi-pushdown systems is undecidable. As a consequence, under-approximation techniques which restrict the search space have become popular. One such restriction is to *bound the number of context switches* [18]: a context is a contiguous sequence of actions in a computation belonging to the same thread. The *bounded context-switching reachability problem* asks if given a global state $\boldsymbol{q}$, is $\boldsymbol{q}$ reachable within a bounded number of context switches. This was shown to be decidable for multi-threaded programs [18]. Such analyses are used to detect errors in programs.

Our Contributions. In this paper, we study the bounded context-switching reachability problem for multi-threaded recursive programs using a finite set of reentrant locks. Such programs can be abstracted by using standard abstraction techniques as multi-pushdown systems with a finite number of counters. Each counter corresponds to a lock and is used to model the number of times the corresponding lock has been acquired by its owning thread. Acquisition of the corresponding lock increments the counter and a release of the corresponding lock decrements the counter. There is, however, no explicit zero-test on the counters: when a thread P successfully acquires a lock l, it happens either because nobody held l before or P itself held l before. A successful acquisition does not explicitly distinguish these cases. An "explicit" zero-test can, however, be simulated by communication amongst threads.

Furthermore, we restrict our attention to *contextual* reentrant locking: we assume that a release of an instance of a reentrant lock can only occur if this instance was acquired before in the same procedure and that each instance of a reentrant lock acquired in a procedure call is released before the procedure returns. Not only is this restriction natural, several higher-level programming constructs automatically ensure contextual locking. For example, the `synchronized(o) { ...}` statement in Java enforces contextual locking.[1]

Our main result is that the bounded context-switching reachability problem of multi-threaded recursive programs using contextual reentrant locks is decidable. The proof of this fact is carried out in two steps.

[1] Please note that not all uses of reentrant locks in Java are contextual.

First, we associate to each computation a *switching vector*. Switching vectors were introduced in [21,13] for multi-threaded programs. A switching vector is a "snapshot" of the computation at the positions where context switches happen. A switching vector is a sequence; if a computation has r context switches then its switching vector has $r+1$ elements. The i-th element of the switching vector records the global state at the beginning of the i-th context and the active thread in the i-th context. For multi-threaded programs with reentrant locks, the i-th element also records the *lock ownership status*, i.e., which locks are owned by each thread at the start of the i-th context. Observe that the number of switching vectors $\leq r+1$ is finite. Thus, in order to decide whether a global state $\boldsymbol{q}$ is reachable within a bounded number of context switches, it suffices to check whether given a switching vector sig, is there a computation whose switching vector is sig and which leads to $\boldsymbol{q}$. This check is done iteratively: for each prefix sig' of sig, we check if there is a computation whose switching vector is sig'.

The iterative step above is reduced to checking whether a control state of a pushdown counter system is reachable by computations in which at most a bounded number of zero-tests are performed. The status of the latter problem, i.e., whether it is decidable or not is open. However, in our case, we exploit the fact that the constructed pushdown counter system is also *contextual*: the values of a counter in a procedure call is always greater than the value of the counter before the procedure call and the value of the counter immediately before a procedure return is the same as the value of the counter before the procedure call. We show that the control state reachability problem on a *contextual* pushdown counter system with bounded number of zero-tests is decidable. This is achieved by first showing that we only need to consider stacks of bounded height and thus the problem reduces to the problem of checking control-state reachability on counter systems with bounded number of zero-tests. The latter problem is easily seen to be equivalent to the (configuration) reachability problem of vector addition systems (VASS)[2]. The latter is known to be decidable [12,17,15].

We then show that the problem of bounded context-switching reachability is at least as hard as the configuration reachability problem of VASS even when the context switch bound is taken to be 1. Since the configuration reachability problem of VASS is EXPSPACE-hard [3], we conclude that the bounded context-switching reachability problem for VASS is also EXPSPACE-hard.

The rest of the paper is organized as follows. We give our formal model in Section 2. The result of deciding reachability in contextual pushdown counter systems with a bounded number of zero-tests is given in Section 3 and our main result in Section 4. We conclude and discuss future work in Section 5.

Related Work. For multi-threaded programs (without reentrant locks), bounded context-switching reachability problem was first posed and shown to be decidable in [18]. Several different proofs of this fact have been discovered since then (see, for example, [16,13,21]). The technique of switching vectors that we have

[2] For our purposes, VASS are counter systems in which there are no zero-test transitions.

adapted to establish our result for the case of multi-threaded programs with reentrant locks was first introduced in [13,21]. (Please note that switching vectors sometimes go by the name of interfaces).

For non-reentrant locks, it was shown in [10] that if we abstract away the data and assume that threads follow *nested locking* then the control state reachability (even with unbounded context-switching) is decidable. A thread is said to follow nested locking [10] if locks are released in the reverse order in which they are acquired. For contextual (non-reentrant) locking, we showed a similar result for 2-threaded programs in [4].

For reentrant locks, [11,14] observe that if threads follow both contextual and nested locking, then the stack of a thread can be used to keep track of both recursive calls as well as recursive lock acquisitions. Thus, the bounded context-switching reachability problem in this case reduces to the case of bounded context-switching reachability problem in multi-pushdown systems. The restriction of both contextual and nested locking is naturally followed by many programs. In Java, for example, if *only* `synchronized(o) { ...}` blocks are used for synchronization then locking is both contextual and nested. However, the assumption of nested locking in presence of other synchronizing primitives, for example, `wait/notify{ ...}` construct can break nested locking while preserving contextual locking.

A modular approach for verifying concurrent non-recursive Java programs with reentrant locks is proposed in [2]. In this approach, first a "lock interface" s guessed: a "lock interface" characterizes the sequence of lock operations that can happen in an execution of the program. Then, they check if each thread respects the lock interface. Since the number of possible lock interfaces is infinite, termination is not guaranteed. Thus, they check programs against specific lock interfaces and thus this approach is another way of restricting the search space.

The control state reachability problem for pushdown counter systems with no zero-tests has long been an open problem. The only non-trivial cases that we are aware of, for which decidability has been established, is when counters are decremented only the stack contents are empty [20,9,5] (in which case it is EXPSPACE-complete [6]), or when the stack is restricted to *index-bounded* behaviors [1] (equivalent to VASS with hierarchical zero-tests, complexity unknown), or when the number counter reversals are bounded [7,8] (in which case it is NP-complete).

2 Model

The set of natural numbers shall be denoted by $\mathbb{N}$. The set of functions from a set A to B shall be denoted by B^A. Given a function $f \in B^A$, the function $f|_{a \mapsto b}$ shall be the unique function g defined as follows: $g(a) = b$ and for all $a' \neq a$, $g(a') = f(a')$. If $\bar{a} = (a_1, \ldots, a_n) \in A_1 \times \cdots \times A_n$ then $\pi_i(a) = a_i$ for each $1 \leq i \leq n$.

Pushdown Systems. Recursive programs are usually modeled as pushdown systems for static analysis. We are modeling threads in concurrent recursive programs that synchronize via reentrant locks. A single thread can be modeled as follows:

Definition 1. *Given a finite set* Lcks*, a pushdown system (PDS)* $\mathcal{P}$ *using (reentrant locks)* Lcks *is a tuple* (Q, Γ, qs, δ) *where*

- Q *is a finite set of control states.*
- Γ *is a finite stack alphabet.*
- qs *is the initial state.*
- $\delta = \delta_{\mathsf{int}} \cup \delta_{\mathsf{cll}} \cup \delta_{\mathsf{rtn}} \cup \delta_{\mathsf{acq}} \cup \delta_{\mathsf{rel}}$ *is a finite set of transitions where*
  - $\delta_{\mathsf{int}} \subseteq Q \times Q$.
  - $\delta_{\mathsf{cll}} \subseteq Q \times (Q \times \Gamma)$.
  - $\delta_{\mathsf{rtn}} \subseteq (Q \times \Gamma) \times Q$.
  - $\delta_{\mathsf{acq}} \subseteq Q \times (Q \times \mathsf{Lcks})$.
  - $\delta_{\mathsf{rel}} \subseteq (Q \times \mathsf{Lcks}) \times Q$.

  A transition in δ_{int} *is said to be a* state transition, *a transition in* δ_{cll} *is said to be a* push transition, *a transition in* δ_{rtn} *is said to be a* pop transition, *a transition in* δ_{acq} *is said to be an* acquire transition *and a transition in* δ_{rel} *is said to be a* release transition.

The semantics of a PDS $\mathcal{P}$ using Lcks is given as a transition system. The set of configurations of $\mathcal{P}$ using Lcks is $\mathsf{Conf}_{\mathcal{P}} = Q \times \Gamma^* \times \mathbb{N}^{\mathsf{Lcks}}$. Intuitively, the elements of a configuration (q, w, hld) have the following meaning: q is the "current" control state of $\mathcal{P}$, w the contents of the pushdown stack and $\mathsf{hld} : \mathsf{Lcks} \to \mathbb{N}$ is a function that tells the number of times each lock has been acquired by $\mathcal{P}$. The transition relation is not a relation between configurations of $\mathcal{P}$, since a thread *executes* in an *environment*, namely the set of free locks (i.e., locks not being held by any thread). Thus, the transition relation can be seen as a binary relation on $2^{\mathsf{Lcks}} \times \mathsf{Conf}_{\mathcal{P}}$, i.e., a transition takes a pair (fr, c) (set of free locks and the configuration) and gives the resulting pair (fr', c'). In order to emphasize that the set of free locks is an "environment", we shall write $fr : c$ instead of the usual notation of (fr, c). In addition to the usual push, pop and internal actions of a PDS; a thread can acquire or release a lock. The thread can only acquire a lock if it is either free or was held by itself before. The thread can only release a lock if it is held by itself. A lock held by a thread is freed only after the thread releases all instances held by it. Formally,

Definition 2. *A PDS* $\mathcal{P} = (Q, \Gamma, qs, \delta)$ *using* Lcks *gives a labeled transition relation* $\longrightarrow_{\mathcal{P}} \subseteq (2^{\mathsf{Lcks}} \times (Q \times \Gamma^* \times \mathbb{N}^{\mathsf{Lcks}})) \times \mathsf{Labels} \times (2^{\mathsf{Lcks}} \times (Q \times \Gamma^* \times \mathbb{N}^{\mathsf{Lcks}}))$ *where* $\mathsf{Labels} = \{\mathsf{int}, \mathsf{cll}, \mathsf{rtn}\} \cup \{\mathsf{acq}(l), \mathsf{rel}(l) \mid l \in \mathsf{Lcks}\}$ *and* $\longrightarrow_{\mathcal{P}}$ *is defined as follows.*

- $\mathsf{fr} : (q, w, \mathsf{hld}) \xrightarrow{\mathsf{int}}_{\mathcal{P}} \mathsf{fr} : (q', w, \mathsf{hld})$ *if* $(q, q') \in \delta_{\mathsf{int}}$.
- $\mathsf{fr} : (q, w, \mathsf{hld}) \xrightarrow{\mathsf{cll}}_{\mathcal{P}} \mathsf{fr} : (q', wa, \mathsf{hld})$ *if* $(q, (q', a)) \in \delta_{\mathsf{cll}}$.
- $\mathsf{fr} : (q, wa, \mathsf{hld}) \xrightarrow{\mathsf{rtn}}_{\mathcal{P}} \mathsf{fr} : (q', w, \mathsf{hld})$ *if* $((q, a), q') \in \delta_{\mathsf{rtn}}$.
- $\mathsf{fr} : (q, w, \mathsf{hld}) \xrightarrow{\mathsf{acq}(l)}_{\mathcal{P}} \mathsf{fr} \setminus \{l\} : (q', w, \mathsf{hld}|_{l \mapsto \mathsf{hld}(l)+1})$ *if* $(q, (q', l)) \in \delta_{\mathsf{acq}}$ *and either* $l \in \mathsf{fr}$ *or* $\mathsf{hld}(l) > 0$.

– $\mathsf{fr} : (q, w, \mathsf{hld}) \xrightarrow{\mathsf{rel}(l)}_{\mathcal{P}} \mathsf{fr} : (q', w, \mathsf{hld}|_{l \mapsto \mathsf{hld}(l)-1})$ *if* $((q, l), q') \in \delta_{\mathsf{rel}}$ *and* $\mathsf{hld}(l) > 1$.
– $\mathsf{fr} : (q, w, \mathsf{hld}) \xrightarrow{\mathsf{rel}(l)}_{\mathcal{P}} \mathsf{fr} \cup \{l\} : (q', w, \mathsf{hld}|_{l \mapsto 0})$ *if* $((q, l), q') \in \delta_{\mathsf{rel}}$ *and* $\mathsf{hld}(l) = 1$.

2.1 Multi-pushdown Systems

Concurrent programs are usually modeled as multi-pushdown systems. For our paper, we assume that threads in a concurrent program also synchronize through reentrant locks which leads us to the following definition.

Definition 3. *Given a finite set* Lcks*, a n-pushdown system (n-PDS)* $\mathcal{CP}$ *communicating via (reentrant locks)* Lcks *and shared state* Q *is a tuple* $(\mathcal{P}_1, \dots, \mathcal{P}_n)$ *where*

– *Q is a finite set of states.*
– *Each $\mathcal{P}_i$ is a PDS using* Lcks.
– *The set of control states of $\mathcal{P}_i$ is $Q \times Q_i$. Q_i is said to be the set of* local states *of thread i.*
– *There is a $qs \in Q$ s.t. for each i the initial state of $\mathcal{P}_i$ is (qs, qs_i) for some $qs_i \in Q_i$. The state qs is said to be the* initial shared state *and qs_i is said to be the* initial local state *of $\mathcal{P}_i$.*

Given a n-PDS $\mathcal{CP}$, we will assume that the set of local states and the stack symbols of the threads are mutually disjoint.

Definition 4. *The semantics of a n-PDS $\mathcal{CP} = (\mathcal{P}_1, \dots, \mathcal{P}_n)$ communicating via* Lcks *and shared state Q is given as a labeled transition system $T = (S, s_0, \longrightarrow)$ where*

– *S, said to be the set of configurations of $\mathcal{CP}$, is the set $Q \times (Q_1 \times \Gamma_1^* \times \mathbb{N}^{\mathsf{Lcks}}) \times \dots \times (Q_n \times \Gamma_n^* \times \mathbb{N}^{\mathsf{Lcks}})$ where Q_i is the set of local states of $\mathcal{P}_i$ and Γ_i is the stack alphabet of $\mathcal{P}_i$.*
– *s_0, said to be the initial configuration, is $(qs, (qs_1, \epsilon, \overline{0}), \cdots, (qs_m, \epsilon, \overline{0}))$ where qs is the initial shared state, qs_i is the initial local state of $\mathcal{P}_i$ and $\overline{0} \in \mathbb{N}^{\mathsf{Lcks}}$ is the function which takes the value 0 for each $l \in$* Lcks.
– *The set of labels on the transitions is* Labels $\times \{1, \dots, n\}$ *where* Labels $= \{\mathsf{int}, \mathsf{cll}, \mathsf{rtn}\} \cup \{\mathsf{acq}(l), \mathsf{rel}(l) \mid l \in \mathsf{Lcks}\}$. *The labeled transition relation $\xrightarrow{(\lambda,i)}$ is defined as follows*

$$(q, (q_1, w_1, \mathsf{hld}_1), \cdots, (q_n, w_n, \mathsf{hld}_n)) \xrightarrow{(\lambda,i)} (q', (q'_1, w'_1, \mathsf{hld}'_1), \cdots, (q'_n, w'_n, \mathsf{hld}'_n))$$

iff for all $j \neq i$, $q_j = q'_j$, $w_j = w'_j$ and $\mathsf{hld}_j = \mathsf{hld}'_j$ and

$$\mathsf{Lcks} \setminus \{l \mid \cup_{1 \le r \le n} \mathsf{hld}_r(l) > 0\} : ((q, q_i), w_i, \mathsf{hld}_i) \xrightarrow{\lambda}_{\mathcal{P}_i} \mathsf{Lcks} \setminus \{l \mid \cup_{1 \le r \le n} \mathsf{hld}'_r(l) > 0\} : ((q', q'_i), w'_i, \mathsf{hld}'_i).$$

Notation: A *global state* is $(q, q_1, \dots, q_n)$ where $q \in Q$ and $q_i \in Q_i$. Given a configuration $s = (q, (q_1, w_1, \mathsf{hld}_1), \cdots, (q_n, w_n, \mathsf{hld}_n))$ of a n-PDS $\mathcal{CP}$, we say that $\mathsf{ShdSt}(s) = q$, $\mathsf{GlblSt}(s) = (q, q_1, \cdots, q_n)$, $\mathsf{LckHld}(s) = (\mathsf{hld}_1, \cdots, \mathsf{hld}_n)$, $\mathsf{LckOwnd}(s) = (held_1, \cdots, held_n)$ and $\mathsf{LckOwnd}_i(s) = held_i$ where $held_i = \{l \mid \mathsf{hld}_i(l) > 0\}$, $\mathsf{Conf}_i(s) = (q_i, w_i, \mathsf{hld}_i)$, $\mathsf{CntrlSt}_i(s) = q_i$, $\mathsf{Stck}_i(s) = w_i$, $\mathsf{StHt}_i(s) = |w_i|$, the length of w_i and $\mathsf{LckHld}_i(s) = \mathsf{hld}_i$.

Computations. A *computation* of the n-PDS $\mathcal{CP}$, is a sequence $\sigma = s_0 \xrightarrow{(\lambda_1,i_1)} s_1 \cdots \xrightarrow{(\lambda_m,i_m)} s_m$, such that s_0 is the initial configuration of $\mathcal{CP}$. The transition $s_j \xrightarrow{(\mathsf{cll},i)} s_{j+1}$ is said to be a *procedure call by thread* i. Similarly, we can define *procedure return, internal action, acquisition of lock* l and *release of lock* l by thread i. A procedure return $s_j \xrightarrow{(\mathsf{rtn},i)} s_{j+1}$ is said to *match* a procedure call $s_p \xrightarrow{(\mathsf{cll},i)} s_{p+1}$ iff $p < j$, $\mathsf{StHt}_i(s_p) = \mathsf{StHt}_i(s_{j+1})$ and for all $p+1 \leq t \leq j$, $\mathsf{StHt}_i(s_{p+1}) \leq \mathsf{StHt}_i(s_t)$. A release of a lock $s_j \xrightarrow{(\mathsf{rel}(l),i)} s_{j+1}$ is said to *match* an acquisition $s_p \xrightarrow{(\mathsf{acq}(l),i)} s_{p+1}$ iff $p < j$, $\mathsf{LckHld}_i(s_p)(l) = \mathsf{LckHld}_i(s_{j+1})(l)$ and for each $p+1 \leq t \leq j$, $\mathsf{LckHld}_i(s_{p+1})(l) \leq \mathsf{LckHld}_i(s_t)(l)$.

2.2 Contextual Locking

We recall the notion of *contextual locking* [4] and adapt the notion to the reentrant locking mechanism. Informally, contextual locking means that –

- each instance of a lock acquired by a thread in a procedure call must be released before the corresponding return is executed, and
- the instances of locks held by a thread just before a procedure call is executed are not released during the execution of the procedure.

Formally,

Definition 5. *A thread i in a n-PDS $\mathcal{CP} = (\mathcal{P}_1, \dots, \mathcal{P}_n)$ is said to follow contextual locking if whenever $s_\ell \xrightarrow{(\mathsf{cll},i)} s_{\ell+1}$ and $s_j \xrightarrow{(\mathsf{rtn},i)} s_{j+1}$ are matching procedure call and return along a computation $s_0 \xrightarrow{(\lambda_1,i)} s_1 \cdots \xrightarrow{(\lambda_m,i)} s_m$, we have that*

$$\mathsf{LckHld}_i(s_\ell) = \mathsf{LckHld}_i(s_{j+1}) \text{ and for all } \ell \leq r \leq j.\ \mathsf{LckHld}_i(s_\ell) \leq \mathsf{LckHld}_i(s_r).$$

Example 1. Consider the 3-threaded program shown in Figure 1. Threads P0 and P1 follow contextual locking, but thread P2 does not follow contextual locking.

2.3 Bounded Context-Switching

A context [18] is a contiguous sequence of actions in a computation belonging to the same thread:

Definition 6. *Given a computation $\sigma = s_0 \xrightarrow{(\lambda_1,i_1)} s_1 \cdots \xrightarrow{(\lambda_m,i_m)} s_m$ of a n-PDS $\mathcal{CP}$, we say that a context switch happens at position $j \in \{1, \cdots, m\}$ if $i_j \neq i_{j+1}$. The number of context switches in σ is the number of positions at which a context switch happens.*

The bounded context-switching reachability problem [18] is defined formally as:

```
int a(){
   acq l1;
   acq l2;
   if (..) then{
       ...
       rel l2;
       rel l1;
     };
   else{
       ...
      rel l1
      rel l2
     };
    return i;
};

public void P0() {
   n=a();
}
```

```
int b(){
   acq l1;
   ...
   rel l1;
   return j;
};

public void P1() {
  l=a();
}
```

```
int c(){
   rel l2;
   acq l1;
   ...
   return i;
};
public void P2(){
acq l2;
n=c();
rel l1;
}
```

Fig. 1. Threads P0 and P1 follow contextual locking. Thread P2 does not follow contextual locking.

Definition 7. *For $k \in \mathbb{N}$, the k-bounded context-switching reachability problem asks that given a n-PDS $\mathcal{CP} = (\mathcal{P}_1, \dots, \mathcal{P}_n)$ communicating via* Lcks *and shared state Q, and a global state $\boldsymbol{q}$ of $\mathcal{CP}$, if there is a computation $\sigma = s_0 \xrightarrow{(\lambda_1, i_1)} s_1 \cdots \xrightarrow{(\lambda_m, i_m)} s_m$ of $\mathcal{CP}$ such that i) $\mathsf{GlblSt}(s_m) = \boldsymbol{q}$ and ii) there are at most k context switches in σ.*

3 Contextual Pushdown Counter Systems

In order to establish our main result, we shall need an auxiliary result about pushdown counter systems. A pushdown counter system is an automaton which in addition to a pushdown stack also has counters. Formally, a k-counter pushdown system (k-counter PDS), $\mathcal{M}$, is a tuple (Q, Γ, qs, δ) where Q is a finite set of *control* states, Γ is a finite *stack alphabet*, qs is the *initial* state and δ, the set of *transitions* of $\mathcal{M}$, is a tuple $(\delta_{\mathsf{int}}, \delta_{\mathsf{cll}}, \delta_{\mathsf{rtn}}, \{\delta_{inc_i}, \delta_{dec_i}, \delta_{z_i}\}_{1 \le i \le k})$ where $\delta_{\mathsf{int}} \subseteq Q \times Q$ is the set of *state* transitions, $\delta_{\mathsf{cll}} \subseteq Q \times (Q \times \Gamma)$ is the set of *push* transitions, $\delta_{\mathsf{rtn}} \subseteq (Q \times \Gamma) \times Q$ is the set of *pop* transitions, and for each $1 \le i \le k$, $\delta_{inc_i} \subseteq Q \times Q$ is the set of *increment* transitions of the counter i, $\delta_{dec_i} \subseteq Q \times Q$ is the set of *decrement* transitions of the counter i and $\delta_{z_i} \subseteq Q \times Q$ is the set of *zero-tests* of the counter i.

The semantics of the k-counter PDS $\mathcal{M}$ is given in terms of a labeled transition system $\to_{\mathcal{M}}$. The definition of the semantics is as expected; we set out some notations here. The set of configurations of the transition system is the set

$Q \times \Gamma^* \times \mathbb{N}^k$. In a configuration $(q, w, j_1, \cdots, j_k)$, q is the control state, $w \in \Gamma^*$ is the stack contents and $j_i \in \mathbb{N}$ is the value of the i-th counter. The set of transition labels are $\{\mathsf{int}, \mathsf{cll}, \mathsf{rtn}\} \cup \{inc_i, dec_i, z_i \mid 1 \leq i \leq k\}$. The initial configuration is $s_0 = (qs, \epsilon, 0, \cdots, 0)$. The definition of computations and the definition of matching push and pop transitions along a computation are as expected.

Given a k-counter PDS $\mathcal{M} = (Q, \Gamma, qs, \delta)$ and a computation $C = s_0 \xrightarrow{\lambda_1}_{\mathcal{M}} s_1 \cdots \xrightarrow{\lambda_m}_{\mathcal{M}} s_m$ of $\mathcal{M}$, the *number of zero-tests* along the computation is $|\{\lambda_j \mid \lambda_j = z_i \text{ for some } 1 \leq i \leq n\}|$. We are interested in the problem of checking whether a control state is reachable by computations in which the number of zero-tests are bounded. However, we will be only interested in contextual counter PDSs. Contextual counter PDSs are analogous to threads that follow contextual locking; in any computation, a) there are an equal number of increments and decrements in a procedure call and b) counter values during procedure call are at least as large as the counter values before the procedure call. Formally,

Definition 8. *The k-counter PDS $\mathcal{M}$ is said to be* contextual *if whenever $s_\ell \xrightarrow{\mathsf{cll}}_{\mathcal{M}} s_{\ell+1}$ and $s_j \xrightarrow{\mathsf{rtn}}_{\mathcal{M}} s_{j+1}$ are matching push and pop transitions along a computation $s_0 \xrightarrow{\lambda_1}_{\mathcal{M}} s_1 \cdots \xrightarrow{\lambda_m}_{\mathcal{M}} s_m$ then for each $1 \leq i \leq k$, a) $c_i(s_\ell) = c_i(s_{j+1})$ and b) for all each $\ell \leq r \leq j$, $c_i(s_\ell) \leq c_i(s_r)$, where $c_i(s)$ is the value of i-th counter in the configuration s.*

In order to establish our result about contextual counter PDSs, we need one auxiliary lemma. The stack in configuration $s = (q, w, j_1, \cdots, j_k)$ is said to be *strictly larger than the stack* in configuration $s' = (q', w', j'_1, \cdots, j'_k)$ if $w = w'u$ where u is a nonempty word over Γ.

Lemma 1. *Let $\mathcal{M} = (Q, \Gamma, qs, \delta)$ be a k-counter contextual PDS. Consider three computations of $\mathcal{M}$:*

$$C_1 : (q_1, w, j_1, \ldots, j_k) \xrightarrow{cll}_{\mathcal{M}} \cdots (q_1, ww', j'_1, \ldots, j'_k)$$
$$C_2 : (q_1, ww', j'_1, \ldots, j'_k) \xrightarrow{cll}_{\mathcal{M}} \cdots \xrightarrow{rtn}_{\mathcal{M}} (q_2, ww', j'_1, \ldots, j'_k)$$
$$C_3 : (q_1, ww', j'_1, \ldots, j'_k) \cdots \xrightarrow{rtn}_{\mathcal{M}} (q_2, w, j_1, \ldots, j_k)$$

such that the stack stays strictly larger than w in the intermediate states of C_1 and C_3 and stays strictly larger than ww' (with w' non-empty) in the intermediate states of C_2. Then, if $\{i \mid j_i = 0\} = \{i \mid j'_i = 0\}$ then there is a computation from $(q_1, w, j_1, \ldots, j_k)$ leading to $(q_2, w, j_1, \ldots, j_k)$ by using exactly the transitions used in C_2.

Proof. Consider the computation $C_2 = (q_1, ww', j'_1, \ldots, j'_k) \xrightarrow{cll}_{\mathcal{M}} s'_0 \xrightarrow{\lambda_1}_{\mathcal{M}} \cdots \xrightarrow{\lambda_p}_{\mathcal{M}} s'_p \xrightarrow{rtn}_{\mathcal{M}} (q_2, ww', l'_1, \ldots, l'_k)$. As the stack stays strictly larger than ww' during this computation, it means that the initial call matches the final return. Therefore as $\mathcal{M}$ is contextual, the counter values in any intermediate state of C_2 are at least as large as $(j'_1, \ldots, j'_k)$; and by contextuality, $(j'_1, \cdots, j'_k)$ is itself larger than $(j_1, \ldots, j_k)$.

We establish the following invariant by induction on t: if $(q_2, ww', j'_1, \dots, j'_k) \xrightarrow{cll}_{\mathcal{M}} s'_0 \dots \xrightarrow{\lambda_t}_{\mathcal{M}} (q_3, ww'\sigma, l'_{t,1}, \dots, l'_{t,n})$ then $(q_1, w, j_1, \dots, j_k) \dots \xrightarrow{\lambda_t}_{\mathcal{M}} (q_3, w\sigma, l_{t,1}, \dots, l_{t,n})$ with $l'_{t,i} = l_{t,i} + (j'_i - j_i)$. The base case, $t = 0$, is immediate. In the inductive step, we proceed by cases on the label λ_t. The case of push, pop and internal state transitions is immediate because the w' part of the stack is never used in C_2. A transition that increments a counter also fulfills this invariant immediately. We now consider a transition that can decrement a counter i. Then, because $\mathcal{M}$ is contextual, we have $l'_{t+1,i} \geq j'_i$ which means that $l'_t \geq j'_i + 1$, and thus $l_{t,i} = (l'_{t,i} - j'_i) + j_i \geq 1$ and the decrement can be performed. For a zero-test, we have that $l'_{t,i} = 0$, which means by contextuality that $j'_i = j_i = 0$. Thus, by induction hypothesis, $l_{t,i} = 0$ and the zero-test can be performed. This concludes the demonstration of the invariant and the statement of the lemma follows directly. □

We are ready to establish that the control state reachability problem for contextual pushdown counter systems with a bounded number of zero tests is decidable.

Theorem 1. *The following problem is decidable: Given a k-counter contextual PDS $\mathcal{M}$ with initial configuration s_0, a control state q of $\mathcal{M}$ and a number $r \in \mathbb{N}$, check if there is a computation $C \doteq s_0 \xrightarrow{\lambda_1}_{\mathcal{M}} s_1 \cdots \xrightarrow{\lambda_m}_{\mathcal{M}} s_m$ s.t.*

– *there are at most r zero-tests along C, and*
– *$s_m = (q, w, j_1, \cdots j_k)$ for some $w \in \Gamma^*$, $j_1, \cdots, j_k \in \mathbb{N}$.*

Proof. We first turn the problem into one where the final state must have an empty stack as follows.

We first encode in the control states of the PDS counter system the information about whether the stack is empty as follows. When symbol a is to be pushed on an empty stack, we push a marked symbol a^* instead. Popping a marked symbol indicates that the resulting stack is empty.

Now, if a stack symbol is pushed but never popped in a computation leading to q; it means that this stack symbol is never subsequently accessed, and thus can be ignored. Therefore, when the stack is empty and a symbol is ready to be pushed, we allow a non-deterministic choice: either to push the symbol (guessing that it would have been popped later) or to not push it (guessing that it would have never been popped subsequently). In the latter case, we perform the same change of control state. As the discarded symbols were at the bottom of the stack, we don't expose symbols that could be used in the computation.

The resulting system is still contextual (because any transition sequence between matching push and pop in the new system was already present in the old one). Moreover, one can reach a state $(q, \epsilon, j_1, \dots, j_k)$ for some $(j_1, \dots, j_k)$ in the new system if and only if one could reach $(q, w, j_1, \dots, j_k)$ for some $(j_1, \dots, j_k)$ and w in the old one.

For the case with a final empty stack, we show that if such a computation exists, then there exists one such that the stack size in any intermediate state is bounded by $|Q|^2 2^k$. Indeed, if we assume this is not the case, fix a computation C whose length is minimal amongst computations ending in control state q

with empty stack. By assumption, there are at least $|Q|^2 2^k + 1$ nested pairs of matching push and pop in C. But, if to each pair of matching push and pop $(q_1, w, j_1, \dots, j_k) \xrightarrow{cll}_{\mathcal{M}} \dots \xrightarrow{rtn}_{\mathcal{M}} (q_2, w, j_1, \dots, j_k)$ in C we associate the vector $(q_1, q_2, \{i \mid j_i = 0\})$, by the pigeonhole principle, there exist two nested pairs such that:

$$\begin{aligned}(q_1, w, j_1, \dots, j_k) &\xrightarrow{cll}_{\mathcal{M}} \dots (q_1, ww', j'_1, \dots, j'_k) \\ &\xrightarrow{cll}_{\mathcal{M}} \dots \xrightarrow{rtn}_{\mathcal{M}} (q_2, ww', j'_1, \dots, j'_k) \\ &\dots \xrightarrow{rtn}_{\mathcal{M}} (q_2, w, j_1, \dots, j_k)\end{aligned}$$

such that a) (due to the fact that we are consider matching pushes and pops) the stack stays strictly larger than w in the intermediate states of the first and third part and stays strictly larger than ww' (with w' non-empty) during the second part, and b) $\{i \mid j_i = 0\} = \{i \mid j'_i = 0\}$. Thanks to Lemma 1, we get a shorter computation, which contradicts the assumption of minimality of C.

Now, because the stack is bounded, it means we can just encode it in the control state, which gives us a reachability problem in a counter system with restricted zero-tests. We reduce it to reachability in Vector Addition Systems [12,17,15]. As the number of possible zero-tests is known, we encode in the control state the number of zero-tests remaining, and work on t copies of each counter, where t is the number of zero-tests remaining. When a zero-test is performed, we only change the control state, remembering the index of the counter on which the zero-test is supposed to be performed, continue to work on $t-1$ copies of the counters, leaving the remaining counters frozen. At the end of the computation, we test whether all frozen counters which should have been zero when they were frozen are indeed zero. □

4 Bounded Context-Switching Reachability

We shall now establish the decidability of the bounded context-switching reachability problem. A key technique we will use is the technique of *switching vectors* developed for bounded context-switching reachability for multi-pushdown systems [21,13]. Intuitively, a switching vector is "snapshot" of a computation in a multi-pushdown system: it is the sequence of active threads and the global states at the beginning of a context in the computation. We extend this definition to n-PDS communicating via reentrant locks by also taking into account which locks are held by which thread at the positions where context-switches happen.

We start by fixing some definitions. Fix a n-PDS $\mathcal{CP} = (\mathcal{P}_1, \dots, \mathcal{P}_n)$ communicating via Lcks and shared state Q. Let Q_i be the set of local states of $\mathcal{P}_i$. Recall that a *global state* is $(q, q_1, \dots, q_n)$ where $q \in Q$ and $q_i \in Q_i$; given a configuration $s = (q, (q_1, w_1, \mathsf{hld}_1), \dots, (q_n, w_n, \mathsf{hld}_n))$, $\mathsf{GlblSt}(s) = (q, q_1, \cdots, q_n)$, $\mathsf{LckOwnd}(s) = (held_1, \cdots, held_n)$ and $\mathsf{LckOwnd}_i(s) = held_i$ where $held_i = \{l \mid \mathsf{hld}_i(l) > 0\}$. We say that $\mathsf{LckOwnd}_i(s)$ is the set of locks *owned by* $\mathcal{P}_i$ and the tuple $\mathsf{LckOwnd}(s)$ is the *lock ownership status*. We are ready to define switching vectors formally.

Definition 9. *Let $\mathcal{CP} = (\mathcal{P}_1, \ldots, \mathcal{P}_n)$ be a n-PDS communicating via* Lcks *and shared state Q. For each $1 \leq i \leq n$, let Q_i be the set of local states of $\mathcal{P}_i$. A sequence $(gs_0, ls_0, p_0), \cdots, (gs_r, ls_r, p_r)$ is said to be a $\mathcal{CP}$-*switching vector *if the following holds for each $0 \leq t \leq r$:*

- *p_t is an element of the set $\{1, \cdots, n\}$ and for $0 \leq t < r$, $p_t \neq p_{t+1}$.*
- *$gs_t \in Q \times Q_1 \times \cdots \times Q_n$. gs_0 is the global state of the initial configuration, and for all $t > 0$, If $gs_{t-1} = (q, q_1, \cdots, q_n)$ and $gs_t = (q', q'_1, \ldots, q'_n)$ then $q_x = q'_x$ for each $x \neq pr_{t-1}$.*
- *$ls_t \in (2^{\mathsf{Lcks}})^n$, $ls_0 = (\emptyset, \ldots, \emptyset)$, and for all $t > 0$, if $ls_t = (held'_1, \ldots, held'_n)$ then $held'_y \cap held'_z = \emptyset$ for each $y \neq z$; and if $ls_{t-1} = (held_1, \ldots, held_n)$ then $held_x = held'_x$ for each $x \neq pr_{t-1}$.*

Note that the last two conditions are consistency checks: an active thread cannot affect the local states of other threads and the locks owned by them. The following definition captures the intuitive meaning of a switching vector being the "snapshot" of a computation.

Definition 10. *Let $\mathcal{CP} = (\mathcal{P}_1, \ldots, \mathcal{P}_n)$ be a n-PDS and let $sig = (gs_0, ls_0, p_0), \cdots, (gs_r, ls_r, p_r)$ be a $\mathcal{CP}$-switching vector. We say that a computation $C = s_0 \xrightarrow{(\lambda_1, i_1)} s_1 \cdots \xrightarrow{(\lambda_m, i_m)} s_m$ of $\mathcal{CP}$ is* compatible *with sig if*

- *C has r context switches.*
- *$gs_0 = \mathsf{GlblSt}(s_0), ls_0 = (\emptyset, \cdots, \emptyset)$ and $p_0 = i_1$.*
- *If the context switches occur at positions $j_1, \cdots, j_r$ then for each $1 \leq t \leq r$,*
  - *$gs_t = \mathsf{GlblSt}(s_t)$.*
  - *$p_t = i_{j_t+1}$.*
  - *If $ls_t = (held_1, \ldots, held_n)$ then $\mathsf{LckOwnd}_i(s_{j_t}) \subseteq held_i$ for each $1 \leq i \leq n$.*
  - *Let j_{r+1} be m. If $ls_t = (held_1, \ldots, held_n)$ then in the sequence $s_{j_t} \xrightarrow{(\lambda_{j_t+1}, p_t)} \cdots \xrightarrow{(\lambda_{j_{t+1}}, p_t)} s_{j_{t+1}}$, the thread p_t does not do any lock acquisitions and releases of locks in the set $\cup_{i \neq p_t} held_i$.*

It is easy to see that that if $\boldsymbol{q}$ is reachable by a computation C at most r bounded context-switches then C must be compatible with a switching vector sig of length $\leq r+1$ (the compatible switching vector is the sequence of the global state, the identifier of the active thread and lock ownership status at the beginning of each context). Hence, we can decide the bounded context-switching reachability problem if we can give an algorithm that given an a n-PDS $\mathcal{CP}$ communicating via Lcks, a $\mathcal{CP}$-switching vector sig and a global state $\boldsymbol{q}$, checks if there is a computation C compatible with sig leading to $\boldsymbol{q}$. We establish this result next.

Lemma 2. *The following problem is decidable:*

Given a n-PDS $\mathcal{CP} = (\mathcal{P}_1, \ldots, \mathcal{P}_n)$ communicating via Lcks *and shared state Q, s.t. each thread is contextual, a $\mathcal{CP}$-switching vector sig and a global state $\boldsymbol{q}$ of $\mathcal{CP}$, is there a computation that is a) compatible with sig and b) ends in global state $\boldsymbol{q}$?*

Proof. We give an algorithm that decides the above problem. Note if $r = 0$, then we can decide the problem by using Theorem 1. So, we only consider the case $r > 0$. Let num_{locks} be the cardinality of Lcks. Fix an enumeration $l_1, l_2, \cdots, l_{num_{locks}}$ of the elements of Lcks. Let $\mathcal{P}_i = (Q \times Q_i, \Gamma_i, (qs, qs_i), \delta_i)$ where qs_i is the initial local state of $\mathcal{P}_i$. Let $sig = (gs_0, ls_0, p_0), \cdots, (gs_r, ls_r, p_r)$. For each $1 \leq t \leq r$, let sig_t be $(gs_0, ls_0, p_0), \cdots, (gs_t, ls_t, p_t)$. First note that if $\boldsymbol{q} = (q, q_1, \ldots, q_n)$ and $gs_r = (q', q'_1, \ldots, q'_n)$ then for each $i \neq p_r$, q_i must be the same q'_i (since p_r is the last active thread). Therefore the algorithm immediately outputs "NO" if there is some $i \neq p_r$ s.t. $q_i \neq q'_i$.

Otherwise, the algorithm proceeds iteratively and will have at most $r + 1$ iterations. At the end of each iteration $t \leq r$, the algorithm will either output "NO" or move to the next iteration. If the algorithm outputs "NO" at the end of iteration $t < r$, then it would mean that there is no computation of $\mathcal{CP}$ compatible with sig_t. If the algorithm moves to the next iteration then it would mean that there is a computation of $\mathcal{CP}$ compatible with sig_t ending in a configuration s such that $\mathsf{GlblSt}(s) = gs_{t+1}$ and $\mathsf{LckOwnd}_i(s) \subseteq \pi_i(ls_{t+1})$ for each $1 \leq i \leq n$.

In each iteration t, the algorithm constructs n pushdown counter systems $\mathcal{M}_1^t, \ldots, \mathcal{M}_n^t$. Intuitively, the pushdown counter system $\mathcal{M}_i^t$ will "simulate" the actions of the ith thread up-to the t-th context switch. Each $\mathcal{M}_i^t$ has num_{locks} counters: the counter j keeps track of number of times lock l_j has been acquired by thread i. The algorithm proceeds as follows. For the sake of brevity, we only illustrate the first iterative step. The other iterative steps are similar.

- (Iterative step 1.) For each $1 \leq i \leq n$, we pick new states $q_i^\dagger, test_{i,1}, \ldots, test_{i,n}$ and let $Q_{new,i} = \{q_i^\dagger, test_{i,1}, \ldots, test_{i,num_{locks}}\}$.
  In the first iterative step, the active thread is supposed to be p_0. For $i \neq p_0$, let $\mathcal{M}_i^1 = (Q_i^1, \Gamma_i, qs_i^1, \delta_i^1)$ be the num_{locks}-counter PDS where $Q_i^1 = Q_{new,i} \times \{1\}$, $qs_i^1 = (q_i^\dagger, 1)$ and δ_i^1 is $\emptyset$.
  The num_{locks}-counter PDS $\mathcal{M}_{p_0}^1 = (Q_{p_0}^1, \Gamma_i, qs_{p_0}^1, \delta_{p_0}^1)$ is constructed as follows. Intuitively, $\mathcal{M}_{p_0}^1$ simulates the thread p_0.
  - $Q_{p_0}^1 = ((Q \times Q_{p_0}) \cup Q_{new,p_0}) \times \{1\}$. The initial state $qs_{p_0}^1 = ((qs, qs_{p_0}), 1)$.
  - $\delta_{p_0}^1$ is constructed as follows. If (qp, qn) is a state transition of $\mathcal{P}_{p_0}$ then $((qp, 1), (qn, 1))$ is a state transition of $\mathcal{M}_{p_0}^1$. If $(qp, (qn, a))$ $(((qp, a), qn)$ respectively) is a stack push (stack pop respectively) transition of $\mathcal{P}_{p_0}$ then $((qp, 1), ((qn, 1), a))$ $((((qp, 1), a), (qn, 1))$ respectively) is a stack push (stack pop respectively) transition of $\mathcal{M}_{p_0}^1$. If $(qp, (qn, l_j))$ $((qp, l_j), qn)$ respectively) is a lock acquisition (lock release respectively) transition of $\mathcal{P}_{p_0}$ then $((qp, 1), (qn, 1))$ is an increment (decrement respectively) transition of the jth counter.

  In addition there are some zero-test transitions and one additional state transition constructed as follows. These extra transitions are to ensure that just before the first context switch happens, the set of the locks owned by $\mathcal{P}_{p_0}$ is a subset of $\pi_{p_0}(ls_1)$. This is achieved as follows. Let $gs_1 = (q, q_1, \ldots, q_n)$. If $\pi_{p_0}(ls_1) = \mathsf{Lcks}$ then we add a state transition that takes $(q, q_{p_0}, 1)$ to $(q_{p_0}^\dagger, 1)$. Otherwise, let $\ell_1 < \cdots < \ell_m$ be the indices of the elements in $\mathsf{Lcks} \setminus \pi_{\ell_0}(ls_1)$. We add a zero-test of the counter ℓ_1 which takes the state $((q, q_{po}), 1)$ to the state $(test_{po,\ell_1}, 1)$. For each $1 \leq x < m$, we add a zero-test of the counter ℓ_{x+1} which takes the state $(test_{po,\ell_x}, 1)$ to $(test_{po,\ell_{x+1}}, 1)$. From the state $(test_{po,\ell_m}, 1)$ we add a state transition to $(q^{\dagger}_{po}, 1)$.

It is easy to see that the PDS $\mathcal{M}^1_{po}$ is a contextual PDS (since every thread of $\mathcal{CP}$ is contextual) and the state $(q^{\dagger}_{po}, 1)$ is reachable iff it is reachable with $\leq$ num_{locks} zero-test. Thus, after constructing $\mathcal{M}^1_{po}$, we check whether $(q^{\dagger}_{po}, 1)$ is reachable or not (thanks to Theorem 1). If it is not, the algorithm outputs "NO." If it is reachable then we can conclude that there is a computation compatible with sig_1. The next iteration begins.

The details of the other iterative steps are similar. The main difference is that in the iterative step t, the thread p_{t-1} cannot manipulate counters that correspond to the locks in the set $\cup_{i \neq t} held_i$. Furthermore, in the last iterative step, we check for reachability of $\boldsymbol{q}$. □

Hence, we can establish the main result of the paper.

Theorem 2. *Given $k \in \mathbb{N}$, the k-bounded context-switching reachability problem is decidable for n-PDS in which each thread exhibits contextual locking. For any fixed $k > 0$, the problem is at least as hard as the VASS configuration reachability problem.*

Proof. The decidability follows from Lemma 2. The VASS configuration reachability problem is as follows:

Given a n-counter system $\mathcal{M} = (Q, qs, \{\delta_{inc_i}, \delta_{dec_i}\}_{1 \leq i \leq n})$ with no zero-test transitions and a control state $q \in Q$, check if there is a computation starting with $(qs, \bar{0})$ that leads to $(q, \bar{0})$.

Now, given a n-counter system $\mathcal{M}$, we construct a 2-PDS $\mathcal{CP} = (\mathcal{P}_1, \mathcal{P}_2)$ that synchronizes only using reentrant locks as follows:

1. The set of locks, Lcks has $n+1$ elements, $l_0, l_1, \ldots, l_n$.
2. $\mathcal{P}_1$ is non-recursive and simulates $\mathcal{M}$. The initial state of $\mathcal{P}_1$ is qs. For each $i > 0$, the value of the counter c_i is maintained by the number of times l_i is acquired by $\mathcal{P}_i$. The sum of the counters $c_1 + \cdots + c_n$ is maintained by the number of times l_0 is acquired. The simulation is achieved as follows. Whenever $\mathcal{M}$ makes an internal transition, so does $\mathcal{P}_1$. Whenever $\mathcal{M}$ increments (decrements respectively) counter i, $\mathcal{P}_1$ acquires (releases respectively) locks l_i and l_0.
3. $\mathcal{P}_2$ is also non-recursive and has two states $\{qs_2, qf_2\}$. qs_2 is the initial state of $\mathcal{P}_2$. There is only one transition of $\mathcal{P}_2$: $\mathcal{P}_2$ can acquire lock l_0 and transit to qf_2 from qs_2.

It is easy to see that there is a computation of $\mathcal{M}$ starting with $(qs, \bar{0})$ that leads to $(q, \bar{0})$ iff the state (q, qf_2) is reachable in $\mathcal{CP}$ by a computation with at most 1 context switch. □

Remark 1. Since the VASS configuration reachability problem is EXPSPACE-hard [3], Theorem 2 implies that the bounded context-switching reachability problem for n-PDS communicating via contextual reentrant locks is EXPSPACE-hard.

5 Conclusions

We have investigated the bounded context-switching problem for multi-threaded recursive programs synchronizing with contextual reentrant locks, showing it to be decidable. The decidability result is established by proving a novel result on pushdown counter systems: if the pushdown counter system is contextual then the problem of deciding whether a control state is reachable with a bounded number of zero tests is decidable. The result on pushdown counter systems is obtained by a reduction to the configuration reachability problem of VASS (Vector Addition System with States) and may be of independent interest. We also establish that the bounded context-switching reachability problem problem is at least as hard as the configuration reachability problem for VASS.

There are a few open problems. The status of the bounded context-switching reachability problem for the case when the locks are not contextual is open. This appears to be a very difficult problem. Our techniques imply that this problem is equivalent to the problem of checking configuration reachability in pushdown counter systems with a bounded number of zero-tests. The latter has been a longstanding open problem.

Another line of investigation is to explore other under-approximation techniques such as bounded phases [16]. It would also be useful to account for other synchronization primitives such as thread creation and barriers in addition to reentrant locks.

Practical aspects of our decision algorithm is left to future investigation. Since earlier static analysis techniques for analyzing programs with reentrant locks [11] mainly consider locks to be both nested and contextual, our techniques should be useful in analyzing a larger class of problems.

References

1. Atig, M.F., Ganty, P.: Approximating Petri net reachability along context-free traces. In: Foundations of Software Technology and Theoretical Computer Science. LIPIcs, vol. 13, pp. 152–163. Schloss Dagstuhl - Leibniz-Zentrum fuer Informatik (2011)
2. Bultan, T., Yu, F., Betin-Can, A.: Modular verification of synchronization with reentrant locks. In: MEMOCODE, pp. 59–68 (2010)
3. Cardoza, E., Lipton, R., Meyer, A.R.: Exponential space complete problems for Petri nets and commutative semigroups. In: Proceedings of the ACM Symposium on Theory of Computing, pp. 50–54 (1976)
4. Chadha, R., Madhusudan, P., Viswanathan, M.: Reachability under Contextual Locking. In: Flanagan, C., König, B. (eds.) TACAS 2012. LNCS, vol. 7214, pp. 437–450. Springer, Heidelberg (2012)

5. Chadha, R., Viswanathan, M.: Decidability Results for Well-Structured Transition Systems with Auxiliary Storage. In: Caires, L., Vasconcelos, V.T. (eds.) CONCUR 2007. LNCS, vol. 4703, pp. 136–150. Springer, Heidelberg (2007)
6. Ganty, P., Majumdar, R.: Algorithmic verification of asynchronous programs. ACM Transactions on Programming Languages and Systems 34(1), 6 (2012)
7. Hague, M., Lin, A.W.: Model Checking Recursive Programs with Numeric Data Types. In: Gopalakrishnan, G., Qadeer, S. (eds.) CAV 2011. LNCS, vol. 6806, pp. 743–759. Springer, Heidelberg (2011)
8. Hague, M., Lin, A.W.: Synchronisation- and Reversal-Bounded Analysis of Multithreaded Programs with Counters. In: Madhusudan, P., Seshia, S.A. (eds.) CAV 2012. LNCS, vol. 7358, pp. 260–276. Springer, Heidelberg (2012)
9. Jhala, R., Majumdar, R.: Interprocedural analysis of asynchronous programs. In: Proceedings of the ACM Symposium on the Principles of Programming Languages, pp. 339–350 (2007)
10. Kahlon, V., Ivančić, F., Gupta, A.: Reasoning About Threads Communicating via Locks. In: Etessami, K., Rajamani, S.K. (eds.) CAV 2005. LNCS, vol. 3576, pp. 505–518. Springer, Heidelberg (2005)
11. Kidd, N., Lal, A., Reps, T.W.: Language strength reduction. In: Static Analysis, pp. 283–298 (2008)
12. Rao Kosaraju, S.: Decidability of reachability in vector addition systems (preliminary version). In: Proceedings of the ACM Symposium on Theory of Computing, pp. 267–281 (1982)
13. Lal, A., Reps, T.W.: Reducing concurrent analysis under a context bound to sequential analysis. Formal Methods in System Design 35(1), 73–97 (2009)
14. Lammich, P., Müller-Olm, M., Wenner, A.: Predecessor Sets of Dynamic Pushdown Networks with Tree-Regular Constraints. In: Bouajjani, A., Maler, O. (eds.) CAV 2009. LNCS, vol. 5643, pp. 525–539. Springer, Heidelberg (2009)
15. Leroux, J.: Vector addition system reachability problem: a short self-contained proof. In: Proceedings of the ACM Symposium on the Principles of Programming Languages, pp. 307–316. ACM (2011)
16. Madhusudan, P., Parlato, G.: The tree width of auxiliary storage. In: Proceedings of the ACM Symposium on the Principles of Programming Languages, pp. 283–294 (2011)
17. Mayr, E.W.: An algorithm for the general Petri net reachability problem. In: Proceedings of the ACM Symposium on Theory of Computing, pp. 238–246 (1981)
18. Qadeer, S., Rehof, J.: Context-Bounded Model Checking of Concurrent Software. In: Halbwachs, N., Zuck, L.D. (eds.) TACAS 2005. LNCS, vol. 3440, pp. 93–107. Springer, Heidelberg (2005)
19. Reps, T.W., Horwitz, S., Sagiv, S.: Precise interprocedural dataflow analysis via graph reachability. In: Proceedings of the ACM Symposium on the Principles of Programming Languages, pp. 49–61 (1995)
20. Sen, K., Viswanathan, M.: Model Checking Multithreaded Programs with Asynchronous Atomic Methods. In: Ball, T., Jones, R.B. (eds.) CAV 2006. LNCS, vol. 4144, pp. 300–314. Springer, Heidelberg (2006)
21. La Torre, S., Madhusudan, P., Parlato, G.: The Language Theory of Bounded Context-Switching. In: López-Ortiz, A. (ed.) LATIN 2010. LNCS, vol. 6034, pp. 96–107. Springer, Heidelberg (2010)

Reachability of Communicating Timed Processes*

Lorenzo Clemente[1], Frédéric Herbreteau[1], Amelie Stainer[2], and Grégoire Sutre[1]

[1] Univ. Bordeaux, CNRS, LaBRI, UMR 5800, Talence, France
[2] University of Rennes 1, Rennes, France

Abstract. We study the reachability problem for communicating timed processes, both in discrete and dense time. Our model comprises automata with local timing constraints communicating over unbounded FIFO channels. Each automaton can only access its set of local clocks; all clocks evolve at the same rate. Our main contribution is a complete characterization of decidable and undecidable communication topologies, for both discrete and dense time. We also obtain complexity results, by showing that communicating timed processes are at least as hard as Petri nets; in the discrete time, we also show equivalence with Petri nets. Our results follow from mutual topology-preserving reductions between timed automata and (untimed) counter automata. To account for urgency of receptions, we also investigate the case where processes can test emptiness of channels.

1 Introduction

Communicating automata are a fundamental model for studying concurrent processes exchanging messages over unbounded channels [23,12]. However, the model is Turing-powerful, and even basic verification questions, like reachability, are undecidable. To obtain decidability, various restrictions have been considered, including making channels unreliable [3,14] or restricting to half-duplex communication [13] (later generalized to mutex [18]). Decidability can also be obtained when restricting to executions satisfying additional restrictions, such as bounded context-switching [21], or bounded channels. Finally, and this is the restriction that we consider here, decidability is obtained by constraining the communication topology. For communicating finite-state machines (CFSMs), it is well-known that reachability is decidable if, and only if, the topology is a polyforest [23,21]; in this case, considering channels of size one suffices for deciding reachability.

On a parallel line of research, *timed automata* [9] have been extensively studied as a finite-state model of timed behaviours. Recently, there have been several works bringing time into infinite-state models, including *timed Petri nets* [10,4], *timed pushdown automata* [2], and *timed lossy channel systems* [1]. In this paper, we study *communicating timed processes* [20], where a finite number of timed

* This work was partially supported by the ANR project VACSIM (ANR-11-INSE-004).

F. Pfenning (Ed.): FOSSACS 2013, LNCS 7794, pp. 81–96, 2013.

automata synchronize over the elapsing of time and communicate by exchanging messages over unbounded channels. Note that, when processes can synchronize, runs cannot be re-ordered to have uniformly bounded channels (contrary to polyforest CFSMs). For example, consider two communicating processes p and q, where p can send to q unboundedly many messages in the first time unit, and q can receive messages only after the first time unit has elapsed. Clearly, all transmissions of p have to occur before any reception by q, which excludes the possibility of re-ordering the run into another one with bounded channels.

We significantly extend the results of [20], by giving a complete characterization of the decidability border of reachability properties w.r.t. the communication topology. Quite surprisingly, we show that despite synchronization increases the expressive power of CFSMs, the undecidability results of [20] are not due to just synchronous time, but to an additional synchronization facility called *urgency* (cf. below). Our study comprises both dense and discrete time.

Dense Time: Communicating Timed Automata. Our main result is a complete characterization of the decidability frontier for communicating timed automata: We show that reachability is decidable if, and only if, the communication topology is a polyforest. Thus, adding time does not change the decidability frontier w.r.t. CFSMs. However, the complexity worsens: From our results it follows that communicating timed automata are at least as hard as Petri nets.[1]

Our decidability results generalize those of [20] over the standard semantics for communicating automata. In the same work, undecidability results are also presented. However, they rely on an alternative *urgent semantics*, where, if a message can be received, then all internal actions are disabled: This provides an extra means of synchronization, which makes already the very simple topology $p \rightarrow q \rightarrow r$ undecidable [20]. We show that, without urgency, this topology remains decidable.

Here, we do not consider urgency directly, but we rather model it by introducing an additional *emptiness test* operation on channels on the side of the receiver. This allows us to discuss topologies where emptiness tests (i.e., urgency) are restricted to certain components. We show that, with emptiness tests, undecidable topologies include not only the topology $p \rightarrow q \rightarrow r$ (as shown in [20]), but also $p \rightarrow q \leftarrow r$ and $p \leftarrow q \rightarrow r$. Thus, we complete the undecidability picture for dense time.

All our results for dense time follow from a mutual, topology-preserving reduction to a discrete-time model (discussed below). Over polyforest topologies, we reduce from dense to discrete time when no channel can be tested for emptiness. Over arbitrary topologies, we reduce from discrete to dense time, even in the presence of emptiness tests. While the latter is immediate, the former is obtained via a *Rescheduling Lemma* for dense-time timed automata which is interesting on its own, allowing us to schedule processes in fixed time-slots where senders are always executed before receivers.

---

[1] And probably exponentially worse, due to a blow-up when translating from dense to discrete time.

Discrete Time: Communicating Tick Automata. We provide a detailed analysis of communication in the discrete-time model, where actions can only happen at integer time points. As a model of discrete time, we consider *communicating tick automata*, where the flow of time is represented by an explicit tick action: A process evolves from one time unit to the next by performing a tick action, forcing all the other processes to perform a tick as well; all the other actions are asynchronous. This model of discrete-time is called *tick automata* in [17], which is related to the *fictitious-time* model of [9].

We provide a complete characterization of decidable and undecidable topologies for communicating tick automata: We show that reachability is decidable if, and only if, the topology is a polyforest (like for CFSMs), and, additionally, each weakly-connected component can test at most one channel for emptiness. Our results follow from topology-preserving mutual reductions between communicating tick automata and counter automata. As a consequence of the structure of our reductions, we show that channels and counters are mutually expressible, and similarly for emptiness tests and zero tests. This also allows us to obtain complexity results for communicating tick automata: We show that reachability in a system of communicating tick automata over a weakly-connected topology without emptiness tests has the same complexity as reachability in Petri nets.[2]

Related Work. Apart from [20], communication in a dense-time scenario has also been studied in [15,8,6]. In particular, [15] proposes timed message sequence charts as the semantics of communicating timed automata, and studies the scenario matching problem where timing constraints can be specified on local processes, later extended to also include send/receive pairs [8]. Communicating event-clock automata, a strict subclass of timed automata, are studied in [6] where, instead of considering the decidability frontier w.r.t. the communication topology, it is shown, among other results, that reachability is decidable for arbitrary topologies over existentially-bounded channels. A crucial difference w.r.t. our work is that we do not put any restriction on the channels, and we consider full timed automata. In a distributed setting, the model of global time we have chosen is not the only possible. In particular, [7] studies decidability of networks of (non-communicating) timed asynchronous automata in an alternative setting where each automaton has a local drift w.r.t. global time. In the discrete-time setting, we mention the work [19], which generalizes communicating tick automata to a loosely synchronous setting, where local times, though different, can differ at most by a given bound. While [19] shows decidability for a restricted two-processes topology, we characterize decidability for arbitrary topologies. We finally mention [16,5] that address reachability for parametrized ad hoc networks in both discrete and dense time. They consider an infinite number of processes and broadcast handshake communications over arbitrary topologies while our models have a finite number of processes that exchange messages over unbounded (unicast) channels.

[2] The latter problem is known to be ExpSpace-hard [22], and finding an upper bound is a long-standing open problem.

Outline. In Sec. 2 we introduce general notation; in particular, we define communicating timed processes, which allow us to uniformly model communication in both the discrete and dense time. In Sec. 3 we study the decidability and complexity for communicating tick automata (discrete time), while in Sec. 4 we deal with communicating timed automata (dense time). Finally, Sec. 5 ends the paper with conclusions and future work.

2 Communicating Timed Processes

A *labeled transition system* (LTS for short) is a tuple $\mathcal{A} = \langle S, S_I, S_F, A, \rightarrow\rangle$ where S is a set of *states* with *initial states* $S_I \subseteq S$ and *final states* $S_F \subseteq S$, A is a set of *actions*, and $\rightarrow \subseteq S \times A \times S$ is a *labeled transition relation*. For simplicity, we write $s \xrightarrow{a} s'$ in place of $(s, a, s') \in \rightarrow$. A *path* in $\mathcal{A}$ is an alternating sequence $\pi = s_0, a_1, s_1, \ldots, a_n, s_n$ of states $s_i \in S$ and actions $a_i \in A$ such that $s_{i-1} \xrightarrow{a_i} s_i$ for all $i \in \{1, \ldots, n\}$. We abuse notation and shortly denote π by $s_0 \xrightarrow{a_1 \cdots a_n} s_n$. The word $a_1 \cdots a_n \in A^*$ is called the *trace* of π. A *run* is a path starting in an initial state ($s_0 \in S_I$) and ending in a final state ($s_n \in S_F$).

We consider systems that are composed of several processes interacting with each other in two ways. Firstly, they implicitly synchronize over the passing of time. Secondly, they explicitly communicate through the asynchronous exchange of messages. For the first point, we represent delays by actions in a given *delay domain* $\mathbb{D}$. Typically, the delay domain is a set of non-negative numbers when time is modeled quantitatively, or a finite set of abstract delays when time is modeled qualitatively. Formally, a *timed process* over $\mathbb{D}$ is a labeled transition system $\mathcal{A} = \langle S, S_I, S_F, A, \rightarrow\rangle$ such that $A \supseteq \mathbb{D}$. Actions in A are either synchronous *delay actions* in $\mathbb{D}$, or *asynchronous actions* in $A \setminus \mathbb{D}$.

For the second point, we introduce FIFO channels between processes. Formally, a communication *topology* is a triple $\mathcal{T} = \langle P, C, E\rangle$, where $\langle P, C\rangle$ is a directed graph comprising a finite set P of *processes* and a set of communication *channels* $C \subseteq P \times P$. Additionally, the set $E \subseteq C$ specifies those channels that can be tested for emptiness. Thus, a channel $c \in C$ is a pair (p, q), with the intended meaning that process p can send messages to process q. For a process p, let $C[p] = C \cap (\{p\} \times P)$ be its set of outgoing channels, and let $C^{-1}[p] = C \cap (P \times \{p\})$ be its set of incoming channels. Processes may send messages to outgoing channels, receive messages from incoming channels, as well as test emptiness of incoming channels (for testable channels). Formally, given a finite set M of messages, the set of possible *communication actions* for process p is $A^p_{\mathrm{com}} = \{c!m \mid c \in C[p], m \in M\} \cup \{c?m \mid c \in C^{-1}[p], m \in M\} \cup \{c{==}\varepsilon \mid c \in E \cap C^{-1}[p]\}$. The set of all communication actions is $A_{\mathrm{com}} = \bigcup_{p \in P} A^p_{\mathrm{com}}$. While send actions ($c!m$) and receive actions ($c?m$) are customary, we introduce the extra test action ($c{==}\varepsilon$) to model the *urgent semantics* of [20]. Actions not in ($\mathbb{D} \cup A_{\mathrm{com}}$) are called *internal actions*.

Definition 1. *A* system of communicating timed processes *is a tuple* $\mathcal{S} = \langle \mathcal{T}, M, \mathbb{D}, (\mathcal{A}^p)_{p \in P}\rangle$ *where* $\mathcal{T} = \langle P, C, E\rangle$ *is a topology,* M *is a finite set of* messages, $\mathbb{D}$ *is a delay domain, and, for each* $p \in P$, $\mathcal{A}^p = \langle S^p, S_I^p, S_F^p, A^p, \rightarrow^p \rangle$ *is a timed process over* $\mathbb{D}$ *such that* $A^p \cap A_{\text{com}} = A^p_{\text{com}}$.

States $s^p \in S^p$ are called *local states* of p, while a *global state* $\boldsymbol{s} = (s^p)_{p \in P}$ is a tuple of local states in $\prod_{p \in P} S^p$. We give the semantics of a system of communicating timed processes in terms of a global labeled transition system. The contents of each channel is represented as a finite word over the alphabet M. Processes move asynchronously, except for delay actions that occur simultaneously. Formally, the *semantics of a system of communicating timed processes* $\mathcal{S} = \langle \mathcal{T}, M, \mathbb{D}, (\mathcal{A}^p)_{p \in P} \rangle$ is the labeled transition system $[\![\mathcal{S}]\!] = \langle S, S_I, S_F, A, \rightarrow \rangle$ where $S = (\prod_{p \in P} S^p) \times (M^*)^C$, $S_I = (\prod_{p \in P} S_I^p) \times \{\lambda c \,.\, \varepsilon\}$, $S_F = (\prod_{p \in P} S_F^p) \times \{\lambda c \,.\, \varepsilon\}$, $A = \bigcup_{p \in P} A^p$, and there is a transition $(\boldsymbol{s}_1, w_1) \xrightarrow{a} (\boldsymbol{s}_2, w_2)$ under the following restrictions:

- if $a \in \mathbb{D}$, then $s_1^p \xrightarrow{a} s_2^p$ for all $p \in P$,
- if $a \notin \mathbb{D}$, then $s_1^p \xrightarrow{a} s_2^p$ for some $p \in P$, and $s_1^q = s_2^q$ for all $q \in P \setminus \{p\}$
  - if $a = c!m$, then $w_2(c) = w_1(c) \cdot m$ and $w_2(d) = w_1(d)$ for all $d \in C \setminus \{c\}$,
  - if $a = c?m$, then $m \cdot w_2(c) = w_1(c)$ and $w_2(d) = w_1(d)$ for all $d \in C \setminus \{c\}$,
  - if $a = (c \mathtt{==} \varepsilon)$, then $w_1(c) = \varepsilon$ and $w_1 = w_2$, and
  - if $a \notin A_{\text{com}}$, then $w_1 = w_2$.

To avoid confusion, states of $[\![\mathcal{S}]\!]$ will be called *configurations* in the remainder of the paper. Given a path π in $[\![\mathcal{S}]\!]$, its *projection* to process p is the path $\pi|_p$ in $\mathcal{A}^p$ obtained by projecting each transition of π to process p in the natural way.

The *reachability problem* asks, given a system of communicating timed processes $\mathcal{S}$, whether there exists a run in its semantics $[\![\mathcal{S}]\!]$. Note that we require all channels to be empty at the end of a run, which simplifies our constructions later by guaranteeing that every sent message is eventually received. (This is w.l.o.g. since reachability and control-state reachability are easily inter-reducible.) Two systems of communicating timed processes $\mathcal{S}$ and $\mathcal{S}'$ are said to be *equivalent* if $[\![\mathcal{S}]\!]$ has a run if and only if $[\![\mathcal{S}']\!]$ has a run.

Definition 2. *A* system of communicating tick automata *is a system of communicating timed processes* $\mathcal{S} = \langle \mathcal{T}, M, \mathbb{D}, (\mathcal{A}^p)_{p \in P} \rangle$ *such that* $\mathbb{D} = \{\tau\}$ *and each* $\mathcal{A}^p$ *is a* tick automaton, *i.e., a timed process over* $\mathbb{D}$ *with finitely many states and actions.*

Thus, tick automata communicate with actions in A_{com} and, additionally, synchronize over the tick action τ. This global synchronization makes communicating tick automata more expressive than CFSMs, in the sense that ticks can forbid re-orderings of communication actions that are legitimate without ticks. Notice that there is only one tick symbol in $\mathbb{D}$. With two different ticks, reachability is already undecidable for the one channel topology $p \rightarrow q$ without emptiness test.

3 Decidability of Communicating Tick Automata

In this section, we study decidability and complexity of communicating tick automata. Our main technical tool consists of mutual reductions to/from counter automata, showing that, in the presence of tick actions, 1) each channel is equivalent to a counter, and 2) each emptiness test on a channel is equivalent to a zero test on the corresponding counter. This allows us to derive a complete characterization of decidable topologies, and to also obtain complexity results. We begin by defining communicating counter automata.

Communicating Counter Automata. A *counter automaton* is a classical Minsky machine $\mathcal{C} = \langle L, L_I, L_F, A, X, \Delta \rangle$ with finitely many locations L, initial locations $L_I \subseteq L$, final locations $L_F \subseteq L$, alphabet of actions A, finitely many non-negative counters in X, and transition rules $\Delta \subseteq L \times A \times L$. Operations on a counter $x \in X$ are x++ (increment), x-- (decrement) and x==0 (zero test). Let $\mathtt{Op}(X)$ be the set of operations over counters in X. We require that $A \supseteq \mathtt{Op}(X)$. As usual, the semantics is given as a labelled transition system $[\![\mathcal{C}]\!] = \langle S, S_I, S_F, A, \rightarrow \rangle$ where $S = L \times \mathbb{N}^X$, $S_I = L_I \times \{\lambda x.0\}$, $S_F = L_F \times \{\lambda x.0\}$, and the transition relation $\rightarrow$ is defined as usual. Acceptance is with zero counters.

A *system of communicating counter automata* is a system of communicating timed processes $\mathcal{S} = \langle \mathcal{T}, M, \mathbb{D}, ([\![\mathcal{C}^p]\!])_{p \in P} \rangle$ such that $\mathbb{D} = \emptyset$ and each $\mathcal{C}^p$ is a counter automaton. By Definition 1, this entails that each counter automaton performs communicating actions in A^p_{com}. Notice that, since the delay domain is empty, no synchronization over delay action is possible.

From Tick Automata to Counter Automata. Let $\mathcal{S}$ be a system of communicating tick automata over an arbitrary (i.e., possibly cyclic) weakly-connected[3] topology. We build an equivalent system of communicating counter automata $\mathcal{S}'$ over the same topology. Intuitively, we implement synchronization on the delay action τ in $\mathcal{S}$ by communication in $\mathcal{S}'$ (by definition, no synchronization on delay actions is allowed in $\mathcal{S}'$). We introduce a new type of message, also called τ, which is sent in broadcast by all processes in $\mathcal{S}'$ each time there is a synchronizing tick action in $\mathcal{S}$. Since communication is by its nature asynchronous, we allow the sender and the receiver to be momentarily desynchronized during the computation. However, we restrict the desynchronization to be asymmetric: The receiver is allowed to be "ahead" of the sender (w.r.t. the number of ticks performed), but never the other way around. This ensures causality between transmissions and receptions, by forbidding that a message is received before it is sent.

To keep track of the exact amount of desynchronization between sender and receiver (as the difference in number of ticks), we introduce counters in $\mathcal{S}'$: We endow each process p with a non-negative counter x^p_c for each channel $c \in C^{-1}[p]$ from which p is allowed to receive. The value of counter x^p_c measures the difference in number of ticks τ between p and the corresponding sender along c. Whenever a process p performs a synchronizing tick action τ in $\mathcal{S}$, in $\mathcal{S}'$ it sends a message τ in broadcast onto all outgoing channels; at the same time, all its counters x^p_c are incremented, recording that p, as a receiver process, is one more step ahead of its corresponding senders. When one such τ-message is received by a process

[3] A topology $\mathcal{T}$ is *weakly-connected* if, for every two processes, there is an undirected path between them.

q in $\mathcal{S}'$ along channel c, the corresponding counter $\mathtt{x}_c^q$ is decremented; similarly, this records that the sender process along c is getting one step closer to the receiver process q. The topology needs to be weakly-connected for the correct propagation of τ's.

While proper ordering of receptions and transmissions is ensured by non-negativeness of counters, testing emptiness of the channel is more difficult: In fact, a receiver, which in general is ahead of the sender, might see the channel as empty at one point (thus the test is positive), but then the sender might later (i.e., after performing some tick) send some message, and the earlier test should actually have failed (false positive). We avoid this difficulty by enforcing that the receiver q is synchronized with the corresponding sender along channel c on emptiness tests, by adding to the test action $c{=}{=}\varepsilon$ by q a zero test $\mathtt{x}_c^q{=}{=}0$.

Formally, let $\mathcal{S} = \langle \mathcal{T}, M, \mathbb{D}, (\mathcal{A}^p)_{p \in P} \rangle$ with $\mathbb{D} = \{\tau\}$ be a system of communicating tick automata over topology $\mathcal{T} = \langle P, C, E \rangle$, where, for each $p \in P$, $\mathcal{A}^p = \langle L^p, L_I^p, L_F^p, A^p, \rightarrow^p \rangle$ is a tick automaton, i.e., $\tau \in A^p$. We define the system of communicating counter automata $\mathcal{S}' = \langle \mathcal{T}, M', \mathbb{D}', ([\![\mathcal{C}^p]\!])_{p \in P} \rangle$, *over the same topology* $\mathcal{T}$ *as* $\mathcal{S}$, s.t. $M' = M \cup \{\tau\}$, $\mathbb{D}' = \emptyset$, and, for every process $p \in P$, we have a counter automaton $\mathcal{C}^p$, which is defined as follows: $\mathcal{C}^p = \langle L^p, L_I^p, L_F^p, B^p, \mathtt{X}^p, \Delta^p \rangle$, where control locations L^p, initial locations L_I^p, and final locations L_F^p are the same as in the corresponding tick automaton $\mathcal{A}^p$, and counters are those in $\mathtt{X}^p = \{\mathtt{x}_c^p \mid c \in C^{-1}[p]\}$. For simplifying the definition of transitions, we allow sequences of actions instead of just one action—these can be clearly implemented by introducing more intermediate states. Transitions in Δ^p for $\mathcal{C}^p$ are defined as follows:

- Let $\ell \xrightarrow{\tau} \ell'$ be a transition in $\mathcal{A}^p$, and assume that outgoing channels of p are those in $C[p] = \{c_0, \ldots, c_h\}$, and that counters in $\mathtt{X}^p$ are those in $\{x_0, \ldots, x_k\}$. Then, $\ell \xrightarrow{c_0!\tau;\ldots;c_h!\tau;x_0\text{++};\ldots;x_k\text{++}} \ell'$ is a transition in $\mathcal{C}^p$.
- For every $\ell \in L^p$ and input channel $c \in C^{-1}[p]$, there is a transition $\ell \xrightarrow{c?\tau;\mathtt{x}_c^p\text{--}} \ell$ in $\mathcal{C}^p$.
- If $\ell \xrightarrow{c==\varepsilon} \ell'$ is a transition in $\mathcal{A}^p$, then $\ell \xrightarrow{\mathtt{x}_c^p==0;c==\varepsilon} \ell'$ is a transition in $\mathcal{C}^p$.
- Every other transition $\ell \xrightarrow{a} \ell'$ in $\mathcal{A}^p$ is also a transition in $\mathcal{C}^p$.

The action alphabet of $\mathcal{C}^p$ is thus $B^p = (A^p \setminus \{\tau\}) \cup \{c?\tau \mid c \in C^{-1}[p]\} \cup \{c!\tau \mid c \in C[p]\}$; in particular, τ is no longer an action, but a message that can be sent and received. We show that $\mathcal{S}$ and $\mathcal{S}'$ are equivalent, obtaining the following result.

Proposition 1. *Let $\mathcal{T}$ be a weakly-connected topology with α channels, of which β can be tested for emptiness. For every system of communicating tick automata $\mathcal{S}$ with topology $\mathcal{T}$, we can produce, in linear time, an equivalent system of communicating counter automata $\mathcal{S}'$ with the same topology $\mathcal{T}$, containing α counters, of which β can be tested for zero.*

While the proposition above holds for arbitrary weakly-connected topologies, it yields counter automata *with channels*, which are undecidable in general. To

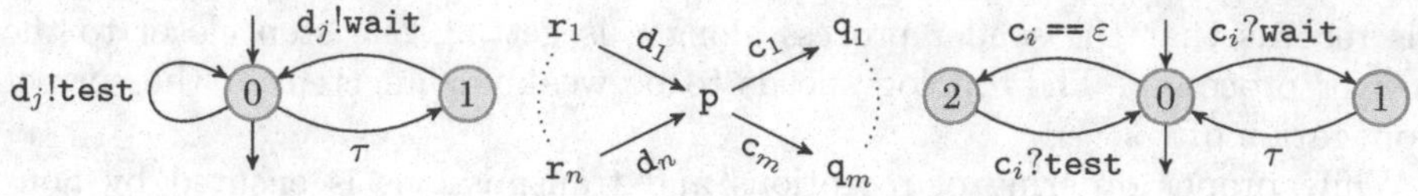

Fig. 1. Simulation of a counter automaton by a system of communicating tick automata: Tick automata for $\mathtt{r}_j$ (left) and $\mathtt{q}_i$ (right), Topology (middle)

avoid undecidability due to communication, we need to forbid cycles (either directed or undirected) in the topology. It has been shown that, on polytrees[4], runs of communicating processes (even infinite-state) can be rescheduled as to satisfy the so-called *eagerness* requirement, where each transmission is *immediately* followed by the matching reception [18]. Their argument holds also in the presence of emptiness tests, since an eager run cannot disable $c == \varepsilon$ transitions (eager runs can only make the channels empty more often). Thus, by restricting to eager runs, communication behaves just as a rendezvous synchronization, and we obtain a global counter automaton by taking the product of all component counter automata.

Theorem 1. *For every polytree topology $\mathcal{T}$ with α channels, of which β can be tested for emptiness, the reachability problem for systems of communicating tick automata with topology $\mathcal{T}$ is reducible, in linear time, to the reachability problem for products of (non-communicating) counter automata, with overall α counters, of which β can be tested for zero.*

From counter automata to tick automata. We reduce the reachability problem for (non-communicating) counter automata to the reachability problem for systems of communicating tick automata with star topology. Formally, a topology $\mathcal{T} = \langle P, C, E\rangle$ is called a *star topology* if there exist two disjoint subsets Q, R of P and a process p in $P \setminus (Q \cup R)$ such that $P = \{p\} \cup Q \cup R$ and $C = (R \times \{p\}) \cup (\{p\} \times Q)$. The idea is to simulate each counter with a separate channel, thus the number of counters fixes the number of channels in $\mathcal{T}$. However, our reduction is uniform in the sense that it works independently of the exact arrangement of channels in $\mathcal{T}$, which we take *not* to be under our control. W.l.o.g., we consider counter automata where all actions are counter operations (i.e., $\Delta \subseteq L \times \mathtt{Op}(X) \times L$).

For the remainder of this section, we consider an arbitrary star topology $\mathcal{T} = \langle P, C, E\rangle$ with set of processes $P = \{\mathtt{p}, \mathtt{q}_1, \ldots, \mathtt{q}_m, \mathtt{r}_1, \ldots, \mathtt{r}_n\}$, where $m, n \in \mathbb{N}$, and set of channels $C = \{\mathtt{p}\} \times \{\mathtt{q}_1, \ldots, \mathtt{q}_m\} \cup \{\mathtt{r}_1, \ldots, \mathtt{r}_n\} \times \{\mathtt{p}\}$ and $E = C$. This topology is depicted in Figure 1 (middle). Note that we allow the limit cases $m = 0$ and $n = 0$. To simplify the presentation, we introduce shorter notations for the channels of this topology: we define $\mathtt{c}_i = (\mathtt{p}, \mathtt{q}_i)$ and $\mathtt{d}_j = (\mathtt{r}_j, \mathtt{p})$ for every $i \in \{1, \ldots, m\}$ and $j \in \{1, \ldots, n\}$.

[4] A *polytree* is a weakly-connected graph with neither directed, nor undirected cycles.

Let $\mathcal{C} = \langle L, L_I, L_F, X \cup Y, \Delta \rangle$ be a counter automaton with $m + n$ counters, namely $X = \{\mathtt{x}_1, \ldots, \mathtt{x}_m\}$ and $Y = \{\mathtt{y}_1, \ldots, \mathtt{y}_n\}$. The counters are split into X and Y to reflect the star topology $\mathcal{T}$, which is given a priori. We build, from $\mathcal{C}$, an equivalent system of communicating tick automata $\mathcal{S}$ with topology $\mathcal{T}$. Basically, the process $\mathtt{p}$ simulates the control-flow graph of the counter automaton, and the counters $\mathtt{x}_i$ and $\mathtt{y}_j$ are simulated by the channels $\mathtt{c}_i$ and $\mathtt{d}_j$, respectively. In order to define $\mathcal{S}$, we need to provide its message alphabet and its tick automata, one for each process p in P. The message alphabet is $M = \{\mathtt{wait}, \mathtt{test}\}$. Actions performed by processes in P are either communication actions or the delay action τ. Processes $\mathtt{r}_j$'s are assigned the tick automaton of Figure 1 (left), and processes $\mathtt{q}_i$'s are assigned the tick automaton of Figure 1 (right). Intuitively, communications on $\mathtt{wait}$ messages are loosely synchronized using the τ actions in $\mathtt{q}_i$ and $\mathtt{r}_j$, so that $\mathtt{p}$ can control the rate of their reception and transmission.

We now present the tick automaton $\mathcal{A}^\mathtt{p}$. As mentioned above, the control-flow graph of $\mathcal{C}$ is preserved by $\mathcal{A}^\mathtt{p}$, so we only need to translate counter operations of $\mathcal{C}$ by communication actions and τ actions. Each counter operation of $\mathcal{C}$ is simulated by a finite sequence of actions in $\Sigma^\mathtt{p}$. To simplify the presentation, we directly label transitions of $\mathcal{A}^\mathtt{p}$ by words in $(\Sigma^\mathtt{p})^*$. The encoding of counter operations is given by the mapping η from $\mathtt{Op}(X \cup Y)$ to $(\Sigma^\mathtt{p})^*$ defined as follows:

$$\eta(\mathtt{x}_i\text{++}) = \mathtt{c}_i!\mathtt{wait} \qquad \eta(\mathtt{x}_i\text{--}) = (\mathtt{c}_h!\mathtt{wait})_{1 \le h \le m, h \ne i} \cdot \tau \cdot (\mathtt{d}_k?\mathtt{wait})_{1 \le k \le n}$$
$$\eta(\mathtt{y}_j\text{--}) = \mathtt{d}_j?\mathtt{wait} \qquad \eta(\mathtt{y}_j\text{++}) = (\mathtt{c}_h!\mathtt{wait})_{1 \le h \le m} \cdot \tau \cdot (\mathtt{d}_k?\mathtt{wait})_{1 \le k \le n, k \ne j}$$
$$\eta(\mathtt{x}_i\text{==}0) = \mathtt{c}_i!\mathtt{test} \qquad \eta(\mathtt{y}_j\text{==}0) = (\mathtt{d}_j \text{==} \varepsilon) \cdot (\mathtt{d}_j?\mathtt{test})$$

where $i \in \{1, \ldots, m\}$ and $j \in \{1, \ldots, n\}$. We obtain $\mathcal{A}^\mathtt{p}$ from $\mathcal{C}$ by replacing each counter operation by its encoding. Observe that these replacements require the addition of a set $S^\mathtt{p}_\diamond$ of fresh intermediate states to implement sequences of actions. Formally, $\mathcal{A}^\mathtt{p}$ is the tick automaton $\mathcal{A}^\mathtt{p} = \langle L \cup S^\mathtt{p}_\diamond, L_I, L_F, \Sigma^\mathtt{p}, \{\ell \xrightarrow{\eta(\mathtt{op})} \ell' \mid (\ell, \mathtt{op}, \ell') \in \Delta\} \rangle$. This completes the definition of the system of communicating tick automata $\mathcal{S} = \langle \mathcal{T}, M, \{\tau\}, (\mathcal{A}^p)_{p \in P} \rangle$.

Let us show that $[\![\mathcal{C}]\!]$ has a run if and only if $[\![\mathcal{S}]\!]$ has a run. We only explain the main ideas behind this simulation of $\mathcal{C}$ by $\mathcal{S}$. The number of $\mathtt{wait}$ messages in channels $\mathtt{c}_i$ and $\mathtt{d}_j$ encodes the value of counters $\mathtt{x}_i$ and $\mathtt{y}_j$, respectively. So, incrementing $\mathtt{x}_i$ amounts to sending $\mathtt{wait}$ in $\mathtt{c}_i$, and decrementing $\mathtt{y}_j$ amounts to receiving $\mathtt{wait}$ from $\mathtt{d}_j$. Both actions can be performed by $\mathtt{p}$. Decrementing $\mathtt{x}_i$ is more involved, since $\mathtt{p}$ cannot receive from the channel $\mathtt{c}_i$. Instead, $\mathtt{p}$ performs a τ action in order to force a τ action in $\mathtt{q}_i$, hence, a receive of $\mathtt{wait}$ by $\mathtt{q}_i$. But all other processes also perform the τ action, so $\mathtt{p}$ compensates (see the definition of $\eta(\mathtt{x}_i\text{--})$) in order to preserve the number of $\mathtt{wait}$ messages in the other channels. The simulation of $\mathtt{y}_j\text{++}$ by $\eta(\mathtt{y}_j\text{++})$ is similar. Let us now look at zero test operations. When $\mathtt{p}$ simulates $\mathtt{x}_i\text{==}0$, it simply sends $\mathtt{test}$ in the channel $\mathtt{c}_i$. This message is eventually received by $\mathtt{q}_i$ since all channels must be empty at the end of the simulation. The construction guarantees that the first receive action of $\mathtt{q}_i$ after the send action $\mathtt{c}_i!\mathtt{test}$ of $\mathtt{p}$ is the matching receive $\mathtt{c}_i?\mathtt{test}$. This means, in particular, that the channel is empty when $\mathtt{p}$ sends $\mathtt{test}$ in $\mathtt{c}_i$. The same device is used to simulate a zero test of $\mathtt{y}_j$, except that the roles of p and its peer (here, r_j) are reversed. Clearly, channels that need to be tested for emptiness are those encoding counters that are tested for zero. We obtain the following theorem.

Theorem 2. *Let $\mathcal{T}$ be an a priori given star topology with α channels, of which β can be tested for emptiness. The reachability problem for (non-communicating) counter automata with α counters, of which β can be tested for zero, is reducible, in linear time, to the reachability problem for systems of communicating tick automata with topology $\mathcal{T}$.*

Decidability and Complexity Results for Communicating Tick Automata. Thanks to the mutual reductions to/from counter automata developed previously, we may now completely characterize which topologies (not necessarily weakly-connected) have a decidable reachability problem, depending on exactly which channels can be tested for emptiness. Intuitively, decidability holds even in the presence of multiple emptiness tests, provided that each test appears in a different weakly-connected component.

Theorem 3 (Decidability). *Given a topology $\mathcal{T}$, the reachability problem for systems of communicating tick automata with topology $\mathcal{T}$ is decidable if and only if $\mathcal{T}$ is a polyforest*[5] *containing at most one testable channel in each weakly-connected component.*

Proof. For one direction, assume that the reachability problem for systems of communicating tick automata with topology $\mathcal{T}$ is decidable. The topology $\mathcal{T}$ is necessarily a polyforest, since the reachability problem is undecidable for non-polyforest topologies even without ticks [23,21]. Suppose that $\mathcal{T}$ contains a weakly-connected component with (at least) two channels that can be tested for emptiness. By an immediate extension of Theorem 2 to account for the undirected path between these two channels, we can reduce the reachability problem for two-counter automata to the reachability problem for systems of communicating tick automata with topology $\mathcal{T}$. Since the former is undecidable, each weakly-connected component in $\mathcal{T}$ contains at most one testable channel.

For the other direction, assume that $\mathcal{T}$ is a polyforest with at most one testable channel in each weakly-connected component, and let $\mathcal{S}$ be a system of communicating tick automata with topology $\mathcal{T}$. Thus, $\mathcal{S}$ can be decomposed into a disjoint union of independent systems $\mathcal{S}_0, \mathcal{S}_1, \ldots, \mathcal{S}_n$, where each $\mathcal{S}_k$ has an undirected tree topology containing exactly one testable channel. But we need to ensure that the $\mathcal{S}_k$'s perform the same number of ticks. By (the construction leading to) Theorem 1, each $\mathcal{S}_k$ can be transformed into an equivalent counter automaton $\mathcal{C}_k$ (by taking the product over all processes in $\mathcal{S}_k$), where exactly one counter, let us call it x_k, can be tested for zero. We may suppose, w.l.o.g., that the counters of $\mathcal{C}_0, \ldots, \mathcal{C}_n$ are disjoint. Moreover, $\mathcal{C}_k$ can maintain, in an extra counter y_k, the number of ticks performed by $\mathcal{S}_k$. We compose the counter machines $\mathcal{C}_0, \ldots, \mathcal{C}_n$ sequentially, and check, at the end, that $y_0 = \cdots = y_n$. Since

[5] A topology $\mathcal{T}$ is a *polyforest* if it is a directed acyclic graph with no undirected cycle.

all counters must be zero in final configurations, this check can be performed by adding, on the final state, a loop decrementing all the $\mathtt{y}_k$'s simultaneously. The construction guarantees that the resulting global counter machine $\mathcal{C}$ is equivalent to $\mathcal{S}$. However, $\mathcal{C}$ contains zero tests on many counters: $\mathtt{x}_0, \ldots, \mathtt{x}_n$. Fortunately, these counters are used one after the other, and they are zero at the beginning and at the end. Therefore, we may reuse $\mathtt{x}_0$ in place of $\mathtt{x}_1, \ldots, \mathtt{x}_n$. We only need to check that $\mathtt{x}_0$ is zero when switching from $\mathcal{C}_k$ to $\mathcal{C}_{k+1}$. Thus, we have reduced the reachability problem for systems of communicating tick automata with topology $\mathcal{T}$ to the reachability problem for counter automata where only one counter can be tested for zero. As the latter is decidable [24,11], the former is decidable, too.

When no test is allowed, we obtain a simple characterization of the complexity for polyforest topologies. A topology $\mathcal{T} = \langle P, C, E\rangle$ is *test-free* if $E = \emptyset$.

Corollary 1 (Complexity). *The reachability problem for systems of communicating tick automata with test-free polyforest topologies has the same complexity as the reachability problem for counter automata without zero tests (equivalently, Petri nets).*

Remark 1. Even though global synchronization makes communicating tick automata more expressive than CFSMs, our characterization shows that they are decidable for exactly the same topologies (polyforest). However, while reachability for CFSMs is PSPACE-complete, systems of communicating tick automata are equivalent to Petri nets, for which reachability is EXPSPACE-hard [22] (the upper bound being a long-standing open problem).

4 Decidability of Communicating Timed Automata

In this section, we consider communicating timed automata, which are communicating timed processes synchronizing over the dense delay domain $\mathbb{D} = \mathbb{R}_{\geq 0}$. We extend the decidability results for tick automata of Section 3 to the case of timed automata. To this end, we present mutual, topology-preserving reductions between communicating tick automata and communicating timed automata. We first introduce the latter model.

Communicating Timed Automata. A *timed automaton* $\mathcal{B} = \langle L, L_I, L_F, X, \Sigma, \Delta\rangle$ is defined by a finite set of locations L with initial locations $L_I \subseteq L$ and final locations $L_F \subseteq L$, a finite set of *clocks* X, a finite alphabet Σ and a finite set Δ of transitions rules $(\ell, \sigma, g, R, \ell')$ where $\ell, \ell' \in L$, $\sigma \in \Sigma$, the guard g is a conjunction of constraints $x \# c$ for $x \in X$, $\# \in \{<, \leq, =, \geq, >\}$ and $c \in \mathbb{N}$, and $R \subseteq X$ is a set of clocks to reset.

The semantics of $\mathcal{B}$ is given by the timed process $[\![\mathcal{B}]\!] = \langle S, S_I, S_F, A, \rightarrow\rangle$, where $S = L \times \mathbb{R}^X_{\geq 0}$, $S_I = L_I \times \{\lambda x.0\}$, $S_F = L_F \times \{\lambda x.0\}$, $A = \Sigma \cup \mathbb{R}_{\geq 0}$ is the set of actions, and there is a transition $(\ell, v) \xrightarrow{d} (\ell, v')$ if $d \in \mathbb{R}_{\geq 0}$ and $v'(x) = v(x) + d$ for every clock x, and $(\ell, v) \xrightarrow{\sigma} (\ell', v')$ if there exists a rule $(\ell, \sigma, g, R, \ell') \in \Delta$ such that g is satisfied by v (defined in the natural way) and $v'(x) = 0$ when $x \in R$, $v'(x) = v(x)$ otherwise. We decorate a path $(\ell_0, u_0) \xrightarrow{a_0, t_0} (\ell_1, u_1) \xrightarrow{a_1, t_1} \cdots (a_n, u_n)$ in $[\![\mathcal{B}]\!]$ with additional *timestamps* $t_i = \sum\{a_j \mid j = 0, \ldots, i-1 \text{ and } a_j \in \mathbb{R}_{\geq 0}\}$. Note that we require clocks to be zero in final configurations, as this simplifies the forthcoming construction from tick automata to timed automata. It can be implemented by duplicating final locations, and by resetting all clocks when entering the new final locations. Without loss of generality, we do not consider location invariants as they can be encoded in the guards. To ensure that the invariant of the last state in a run is satisfied, we duplicate the final locations and we add an edge guarded by the invariant, to the new accepting copy. Combining the two constructions, we get that the invariant in the final configuration is satisfied as the clocks have value zero in accepting configurations.

A *system of communicating timed automata* is a system of communicating timed processes $\mathcal{S} = \langle \mathcal{T}, M, \mathbb{R}_{\geq 0}, ([\![\mathcal{B}^p]\!])_{p \in P} \rangle$ where each $\mathcal{B}^p$ is a timed automaton. Note that each timed automaton has access only to its local clocks. By Definition 1, each timed automaton performs communicating actions in A^p_{com} and synchronizes with all the other processes over delay actions in $\mathbb{R}_{\geq 0}$.

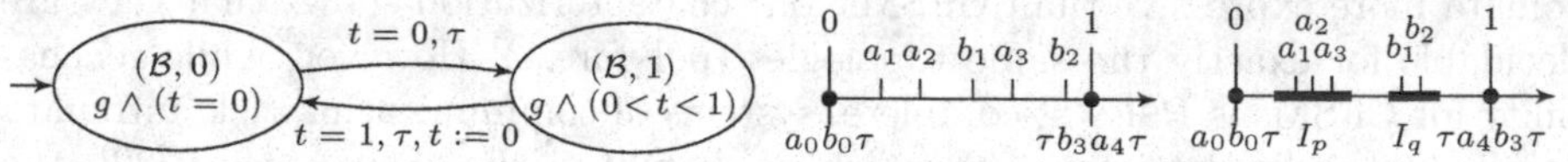

Fig. 2. From timed to tick automata: instrumentation of a timed automaton $\mathcal{B}$ with τ-transitions (left), addition of τ's along a run (middle) and rescheduling of a run (right)

From Timed Automata to Tick Automata. On test-free acyclic topologies, we show a topology-preserving reduction from communicating timed to communicating tick automata. We insist on a reduction that only manipulates processes locally, thus preserving the topology. The absence of emptiness tests on the channels enables such a modular construction.

Naïvely, one would just apply the classical region construction to each process [9]. However, while this preserves local reachability properties, it does not respect the global synchronization between different processes. While quantitative synchronization cannot be obtained by locally removing dense time, a qualitative synchronization suffices in our setting. We require that all processes are either at the same integer date $k \in \mathbb{N}$, or in the same open interval $(k, k+1)$. This suffices because, at integer dates (in fact, at any time-point), any interleaving is allowed, and, in intervals $(k, k+1)$, we can reschedule all processes s.t., for every channel $c = (p, q)$, all actions of p occur before all actions of q (cf. the Rescheduling Lemma below). The latter property ensures the causality between transmissions and receptions.

Qualitative synchronization is achieved by forcing each automaton $\mathcal{B}^p$ to perform a synchronizing tick action τ at each date k and at each interval $(k, k+1)$. See Figure 2 on the left, where $\mathcal{B}^p$ is split into two copies $(\mathcal{B}^p, 0)$ and $(\mathcal{B}^p, 1)$: Actions occurring on integer dates k are performed in $(\mathcal{B}^p, 0)$, and those in $(k, k+1)$ happen in $(\mathcal{B}^p, 1)$. This is ensured by adding a new clock $\mathtt{t}$ and τ-transitions that switch from one mode to the other. Formally, the *τ-instrumentation* of $\mathcal{B} = \langle L, L_I, L_F, X, \Sigma, \Delta\rangle$ is the timed automaton $\mathsf{Instr}(\mathcal{B}, \tau) = \langle L \times \{0,1\}, L_I \times \{1\}, F \times \{0,1\}, X \cup \{\mathtt{t}\}, \Sigma \cup \{\tau\}, \Delta'\rangle$, where $\mathtt{t} \notin X$ and Δ' is defined by: $(\ell, 0) \xrightarrow{a,(g\wedge \mathtt{t}=0),R} (\ell', 0)$ and $(\ell, 1) \xrightarrow{a,(g\wedge 0<\mathtt{t}<1),R} (\ell', 1)$ for all rules $\ell \xrightarrow{a,g,R} \ell'$ in Δ, and $(\ell, 0) \xrightarrow{\tau,\mathtt{t}=0,\emptyset} (\ell, 1)$ and $(\ell, 1) \xrightarrow{\tau,\mathtt{t}=1,\{\mathtt{t}\}} (\ell, 0)$ for all locations $\ell \in L$.

Finally, we obtain an equivalent system of tick automata by applying the exponential region construction to each instrumented process.

Theorem 4. *Let $\mathcal{T}$ be a test-free acyclic topology. For every system of communicating timed automata $\mathcal{S} = \langle \mathcal{T}, M, \mathbb{R}_{\geq 0}, (\llbracket\mathcal{B}^p\rrbracket)_{p\in P}\rangle$ with topology $\mathcal{T}$, we can produce, in exponential time, an equivalent system of communicating tick automata $\mathcal{S}' = \langle \mathcal{T}, M, \{\tau\}, (\mathcal{A}^p)_{p\in P}\rangle$ over the same topology $\mathcal{T}$, where the tick automaton $\mathcal{A}^p$ is obtained by applying the region graph construction to $\mathsf{Instr}(\mathcal{B}^p, \tau)$.*

One direction of the equivalence between $\mathcal{S}$ and $\mathcal{S}'$ is immediate, since every run in $\mathcal{S}$ induces a run in $\mathcal{S}'$ by just inserting τ actions in the right position. For the other direction, let ρ' be a run of $\mathcal{S}'$, and we show how to build a corresponding run ρ of $\mathcal{S}$. We have to schedule all the actions in ρ' on timestamps that are consistent with the guards in $\mathcal{S}$ and that preserve dependencies between transmissions and receptions of messages. Consider a channel $c = (p, q)$ without emptiness test. If p and q are untimed processes, it is always possible to first schedule transmissions of p, and then receptions of q. The Rescheduling Lemma below ensures the same for timed processes. This is depicted in Figure 2 in the middle (before rescheduling) and on the right (after rescheduling) where the a's are emissions of p and the b's are receptions of q.

Lemma 1 (Rescheduling Lemma). *Let $\mathcal{B}$ be a timed automaton, and $I \subseteq (0,1)$ an open interval. Then, every run of $\mathcal{B}$ $(\ell_0, v_0) \xrightarrow{a_0,t_0} \cdots (\ell_n, v_n)$ can be rescheduled such that integral timestamps $t_i \in \mathbb{N}$ are kept the same, and non-integral timestamps $t_i \in (k, k+1)$ belong to $k + I$.*

Intuitively, the lemma above allows us to restrict non-integer timestamps in $(k, k+1)$ to occur in a predefined sub-interval $I + k$. Let us first see how this helps in constructing ρ'. To each process p, we associate an open interval $I_p \subseteq (0,1)$, such that, for every channel (p, q), I_p and I_q are disjoint, and I_p comes before I_q. This is always possible on acyclic topologies. Then, all actions of process p in $(k, k+1)$ are rescheduled to occur in $k + I_p$ (according to the Rescheduling Lemma), which ensures causality between transmissions and receptions. Finally, the τ actions added by instrumentation tell, for each action performed by process p in ρ', whether it should be scheduled at an integer date k, or in $k + I_p$.

Remark 2. Our reduction is incorrect in the presence of emptiness tests. There are essential difficulties in rescheduling senders and receivers in fixed intervals, as emptiness tests introduce a sort of circular dependency and seem to require unboundedly many intervals.

We now comment about the correctness of the Rescheduling Lemma. Resets and guards in a timed automaton allow to enforce minimal and/or maximal delays between timestamps on a path. Since clocks are compared to integers only, it suffices to just distinguish between integral and non-integral dates. While for closed guards like $x \leq 1$ a non-integral time-point $t \in (0, 1)$ would suffice to represent all non-integral dates, to accommodate open guards like $x < 1$ we need a dense interval $I \subseteq (0, 1)$. The following characterization of decidable test-free topologies follows from Theorems 3 and 4.

Theorem 5 (Decidability). *Given a test-free topology $\mathcal{T}$, the reachability problem for systems of communicating timed automata with topology $\mathcal{T}$ is decidable if and only if $\mathcal{T}$ is a polyforest.*

Remark 3. While the reachability problem is known to be decidable for a system of two communicating timed automata with only one channel and emptiness test [20], that proof does not preserve the topology and it looks hardly adaptable to arbitrary polyforest topologies.

From tick automata to timed automata. Given a system of communicating tick automata $\mathcal{S}$, we produce an equivalent system of communicating timed automata $\mathcal{S}'$, over the same topology. The synchronization on τ's is easily simulated using clocks in $\mathcal{S}'$ by ensuring that all the processes elapse 1 time unit exactly when they (synchronously) perform a τ in $\mathcal{S}$. Thus, every run in $\mathcal{S}$ has a corresponding run in $\mathcal{S}'$. For the converse to hold, we have to make sure that for every run of $\mathcal{S}'$, all the processes perform the same number of τ's on the corresponding run of $\mathcal{S}$. This is ensured since we require clocks to be zero at the end of accepting runs, thus preventing time from elapsing on final locations.

The simple topology $p \rightarrow q \rightarrow r$ is known to be undecidable when both channels can be tested for emptiness [20]. Thanks to Theorem 3, we obtain generalized undecidability for every weakly-connected topology containing at least two testable channels.

Theorem 6 (Undecidability). *Given a weakly-connected topology $\mathcal{T}$ with two testable channels, the reachability problem for systems of communicating timed automata with topology $\mathcal{T}$ is undecidable.*

5 Conclusions and Future Work

We have studied the decidability and complexity of communicating timed processes. In discrete time, we give a complete characterization of decidable topologies with emptiness tests, as well as a tight connection with Petri nets in the test-free case. In dense time, we prove decidability for polyforest test-free topologies, and we generalize the undecidability results of [20] to arbitrary weakly-connected topologies containing two testable channels. We leave open whether one can obtain, in the presence of emptiness tests, the same characterization as in discrete time. We conjecture that this is possible, although the techniques used here do not seem to easily extend to deal with emptiness tests. Finally, as another direction for future work one can study richer models where processes are allowed to send timestamps or clocks along channels, in the spirit of [1].

Acknowledgements. The authors wish to thank Jérôme Leroux, Anca Muscholl, and Igor Walukiewicz for helpful discussions. We also thank the anonymous referees for their useful comments and suggestions.

References

1. Abdulla, P.A., Atig, M.F., Cederberg, J.: Timed lossy channel systems. In: FSTTCS. LIPIcs (2012)
2. Abdulla, P.A., Atig, M.F., Stenman, J.: Dense-timed pushdown automata. In: LICS, pp. 35–44 (2012)
3. Abdulla, P.A., Jonsson, B.: Verifying programs with unreliable channels. Information and Computation 127(2), 91–101 (1996)
4. Abdulla, P.A., Nylén, A.: Timed Petri Nets and BQOs. In: Colom, J.-M., Koutny, M. (eds.) ICATPN 2001. LNCS, vol. 2075, pp. 53–70. Springer, Heidelberg (2001)
5. Abdulla, P.A., Delzanno, G., Rezine, O., Sangnier, A., Traverso, R.: On the Verification of Timed Ad Hoc Networks. In: Fahrenberg, U., Tripakis, S. (eds.) FORMATS 2011. LNCS, vol. 6919, pp. 256–270. Springer, Heidelberg (2011)
6. Akshay, S., Bollig, B., Gastin, P.: Automata and Logics for Timed Message Sequence Charts. In: Arvind, V., Prasad, S. (eds.) FSTTCS 2007. LNCS, vol. 4855, pp. 290–302. Springer, Heidelberg (2007)
7. Akshay, S., Bollig, B., Gastin, P., Mukund, M., Narayan Kumar, K.: Distributed Timed Automata with Independently Evolving Clocks. In: van Breugel, F., Chechik, M. (eds.) CONCUR 2008. LNCS, vol. 5201, pp. 82–97. Springer, Heidelberg (2008)
8. Akshay, S., Gastin, P., Mukund, M., Kumar, K.N.: Model checking time-constrained scenario-based specifications. In: FSTTCS. LIPIcs, vol. 8, pp. 204–215 (2010)
9. Alur, R., Dill, D.: A theory of timed automata. TCS 126(2), 183–235 (1994)
10. Bérard, B., Cassez, F., Haddad, S., Lime, D., Roux, O.H.: Comparison of Different Semantics for Time Petri Nets. In: Peled, D.A., Tsay, Y.-K. (eds.) ATVA 2005. LNCS, vol. 3707, pp. 293–307. Springer, Heidelberg (2005), `http://dx.doi.org/10.1007/11562948_/23`
11. Bonnet, R.: The Reachability Problem for Vector Addition System with One Zero-Test. In: Murlak, F., Sankowski, P. (eds.) MFCS 2011. LNCS, vol. 6907, pp. 145–157. Springer, Heidelberg (2011)
12. Brand, D., Zafiropulo, P.: On communicating finite-state machines. J. ACM 30(2), 323–342 (1983)
13. Cécé, G., Finkel, A.: Verification of programs with half-duplex communication. Information and Computation 202(2), 166–190 (2005)

14. Chambart, P., Schnoebelen, P.: Mixing Lossy and Perfect Fifo Channels. In: van Breugel, F., Chechik, M. (eds.) CONCUR 2008. LNCS, vol. 5201, pp. 340–355. Springer, Heidelberg (2008)
15. Chandrasekaran, P., Mukund, M.: Matching Scenarios with Timing Constraints. In: Asarin, E., Bouyer, P. (eds.) FORMATS 2006. LNCS, vol. 4202, pp. 98–112. Springer, Heidelberg (2006)
16. Delzanno, G., Sangnier, A., Zavattaro, G.: Parameterized Verification of Ad Hoc Networks. In: Gastin, P., Laroussinie, F. (eds.) CONCUR 2010. LNCS, vol. 6269, pp. 313–327. Springer, Heidelberg (2010)
17. Gruber, H., Holzer, M., Kiehn, A., König, B.: On Timed Automata with Discrete Time – Structural and Language Theoretical Characterization. In: De Felice, C., Restivo, A. (eds.) DLT 2005. LNCS, vol. 3572, pp. 272–283. Springer, Heidelberg (2005)
18. Heußner, A., Leroux, J., Muscholl, A., Sutre, G.: Reachability analysis of communicating pushdown systems. Logical Methods in Comp. Sci. 8(3:23), 1–20 (2012)
19. Ibarra, O.H., Dang, Z., Pietro, P.S.: Verification in loosely synchronous queue-connected discrete timed automata. Theor. Comput. Sci. 290(3), 1713–1735 (2003)
20. Krcal, P., Yi, W.: Communicating Timed Automata: The More Synchronous, the More Difficult to Verify. In: Ball, T., Jones, R.B. (eds.) CAV 2006. LNCS, vol. 4144, pp. 249–262. Springer, Heidelberg (2006)
21. La Torre, S., Madhusudan, P., Parlato, G.: Context-Bounded Analysis of Concurrent Queue Systems. In: Ramakrishnan, C.R., Rehof, J. (eds.) TACAS 2008. LNCS, vol. 4963, pp. 299–314. Springer, Heidelberg (2008)
22. Lipton, R.J.: The Reachability Problem Requires Exponential Space. Department of Computer Science, Yale University (1976)
23. Pachl, J.K.: Reachability problems for communicating finite state machines. Research Report CS-82-12, University of Waterloo (May 1982)
24. Reinhardt, K.: Reachability in petri nets with inhibitor arcs. ENTCS 223, 239–264 (2008)

Modular Bisimulation Theory for Computations and Values

Martin Churchill and Peter D. Mosses

Department of Computer Science, Swansea University, Swansea, UK
{m.d.churchill,p.d.mosses}@swansea.ac.uk

Abstract. For structural operational semantics (SOS) of process algebras, various notions of bisimulation have been studied, together with rule formats ensuring that bisimilarity is a congruence. For programming languages, however, SOS generally involves auxiliary entities (e.g. stores) and computed values, and the standard bisimulation and rule formats are not directly applicable.

Here, we first introduce a notion of bisimulation based on the distinction between computations and values, with a corresponding liberal congruence format. We then provide metatheory for a modular variant of SOS (MSOS) which provides a systematic treatment of auxiliary entities. This is based on a higher order form of bisimulation, and we formulate an appropriate congruence format. Finally, we show how algebraic laws can be proved sound for bisimulation with reference only to the (M)SOS rules defining the programming constructs involved in them. Such laws remain sound for languages that involve further constructs.

Keywords: structural operational semantics, programming languages, congruence formats, Modular SOS, higher-order bisimulation.

1 Introduction

Background. Structural operational semantics (SOS) [16] is a well-established framework for specifying computational behaviour, where the behaviour of programs is modelled by labelled transition systems, defined inductively by axioms and inference rules. The metatheory of SOS provides various notions of bisimulation [7,15] for proving behavioural equivalence. Bisimilarity is guaranteed to be a congruence when the rules used to define transition relations are restricted to particular formats, e.g. *tyft/tyxt* [3].

SOS is particularly suitable for specifying process calculi such as CCS: the states of the transition system are simply (closed) process terms, and the labels on transitions represent actions corresponding to steps of process execution. For programming languages, however, transition relations often involve auxiliary entities as arguments, e.g. stores (recording the values of imperative variables before and after transitions) and environments (determining the bindings of currently visible identifiers); they also use terminal states to represent computed values. These extra features entail that rules do not conform to the usual congruence formats.

F. Pfenning (Ed.): FOSSACS 2013, LNCS 7794, pp. 97–112, 2013.

The need to specify auxiliary entities in SOS rules also undermines their modularity, which can be a significant pragmatic problem for larger languages. Modular SOS (MSOS) [8] is a simple variant of SOS where auxiliary entities are incorporated in labels. The notation used for label terms in MSOS eliminates references to stores and environments (etc.) in most rules. MSOS provides foundations for the component-based approach to semantics [9] currently being developed by the PLanCompS project (`www.plancomps.org`).

Contribution. In this paper, we introduce a notion of bisimulation for MSOS and a corresponding congruence format supporting auxiliary entities and computations (processes) which compute values. This work stands on the shoulders of previous work on congruence formats used for the SOS of process calculi, but develops it in a direction more suitable for use with programming languages.

Our notion of bisimulation is higher-order, and tailored for use with MSOS; in particular, so-called writeable label components (e.g. the resulting store, or a thrown exception) may vary up to the bisimulation relation in the 'step'. We provide an appropriate rule format – an enhancement of the well-studied *tyft* format [3] – which ensures that bisimilarity is a congruence.

In our setting, there is a strict dichotomy between value terms (which may be inspected) and more general computational terms (which may be run, and their behaviour observed) following e.g. [6]. In particular, the treatment of values is disciplined with fixed built-in rules, and our definition of bisimilarity ensures that bisimilar values have bisimilar subterms. We use silent rewrites to deal with operations on values. This mechanism also allows unit laws to hold for our notion of bisimulation, avoiding the need for weak versions of bisimulation.

MSOS rules for a particular construct need only mention the auxiliary entities relevant to that construct, allowing modular specifications. We describe a complementary notion of *modular bisimulation*, allowing laws to be shown just with respect to the rules relevant to the particular constructs mentioned in that law. This notion of bisimulation identifies when a law is *robust* and cannot be broken by the presence or absence of other constructs that might be in the language. Thus, we can specify the operational semantics of programming languages incrementally, proving laws which remain valid (and a congruence) as the language is extended.

Related Work. There has been a variety of work on notions of bisimulation and congruence formats, reviewed in [14]. No formats in this review, or any other we are aware of, allow value terms as arguments in the source of the conclusion.

Our notion of higher-order bisimulation for MSOS generalises stateless bisimulation from [13] (but we allow information flow between the data and process components). Like the higher-order PANTH format of [12], labels may vary up to bisimulation in the step. However, the distinction between readable and writeable MSOS label components admits a much simpler rule format.

We can also compare it to applicative bisimulation [4], in which bisimilar abstractions must yield bisimilar outputs for bisimilar inputs. In our framework, bindings and abstractions can be dealt with using an environment, which is 'just another auxiliary entity' and treated as such. Our notion of bisimulation does not explicitly require that bisimilar environments yield bisimilar computation results, but in fact this is a consequence of our rule format.

Outline. The rest of this paper is arranged as follows: In Sect. 2, we introduce a notion of bisimulation and congruence format focusing on the distinction between values and computations. In Sects. 3 and 4, we lift our notion of bisimulation to the higher-order setting of MSOS, and define a liberal congruence format for bisimulation in MSOS. In Sect. 5, we discuss how we can formulate and prove robust bisimulations, which continue to hold as additional constructs are added to the language. In Sect. 6, we consider further directions. Full proofs of results in this paper are available at `http://www.plancomps.org/churchill2013a/`.

2 Value-Computation Bisimulation

Structural operational semantics uses terms over a first-order algebraic signature as the states and labels of a transition system. A key distinction at the heart of our notions of equivalence is between *computational* terms and *value* terms. Values consist of structure that can be interrogated, while computational terms generally have behaviour. An appropriate slogan from [6] is that values *are*, while computations *do*. Examples of values include Booleans, integers, and closed function abstractions. Computations model the potential behaviour of expressions, statements, declarations, processes and entire programs. This distinction is important in programming languages, c.f. call-by-name vs. call-by-value in Algol or Scala, corresponding to whether functions take computations or values as parameters.

From the point of view of program (term) equivalence, the distinction is also important. In particular, equivalences must be sound with respect to observational tests (contexts). Contexts, as with programs in general, must be able to interrogate values – for example, true must be distinguishable from false. On the other hand, it must not be possible for a program to be able to interrogate the structure of a computational term such as 'if true then C else D', for then it could distinguish it from C and equivalence would reduce to syntactic identity. Thus, only values may be interrogated. But note that values may also contain computational terms as subexpressions. For example, a function value may also contain a body (a computational term) and a closing environment (a value term, which may include computations as substructure).

In this section we formalise this distinction and the corresponding notion of equivalence, and define a simple bisimulation congruence format.

2.1 Value-Computation Transition Systems

The terms we consider are freely generated from an algebraic signature. Rather than quotienting by an equational congruence, we equip our systems with a rewriting relation $\Rrightarrow$. This represents internal silent functional transitions. Unlike a transition under a distinguished silent label τ, $\Rrightarrow$ is context insensitive (a precongruence). This can be used to avoid polluting traces with silent steps – a goal shared by Plotkin in [16] – and it allows unit laws to hold up to strong bisimulation. Rules can also be kept simple – for example, we can define sequencing using $\frac{s \xrightarrow{l} s'}{\mathsf{seq}(s,t) \xrightarrow{l} \mathsf{seq}(s',t)}$ and $\mathsf{seq}(\mathsf{skip}, s) \Rrightarrow s$ where skip represents successful termination of a command. The relation $\Rrightarrow$ is asymmetric, and we intend that the RHS is simpler than the LHS – this keeps the search space small in bisimulation proofs and animation.

Our notion of value is derived from that of a value constructor, as in [4].

Definition 1 (value-computation signature). *A* value-computation signature *Σ consists of a set of constructors C_Σ (function symbols – f,g,...), each with an arity* $\mathsf{ar}_\Sigma : C_\Sigma \to \mathbb{N}$, *and a set of value constructors $VC_\Sigma \subseteq C_\Sigma$. We let T_Σ denote the set of (closed) terms (s,t,...), and $V_\Sigma \subseteq T_\Sigma$ the set of* value terms *whose outermost constructor is in VC_Σ.*

A *precongruence* with respect to Σ is a reflexive transitive relation R such that if $f \in C_\Sigma$ with $\mathsf{ar}(f) = n$ and $s_i \; R \; t_i$ for $1 \leq i \leq n$ then $f(s_1, \ldots, s_n) \; R \; f(t_1, \ldots, t_n)$. For symmetric relations, we may also call such an R a *congruence.*

Definition 2 (value-computation transition system). *A* value-computation transition system *is a tuple $(\Sigma, L, \rightarrow, \Rrightarrow)$ where Σ is a value-computation signature, L a set of* labels, $\rightarrow \subseteq T_\Sigma \times L \times T_\Sigma$ *a* transition relation *and* $\Rrightarrow \subseteq T_\Sigma \times T_\Sigma$ *a* rewriting relation *such that:*

- *$\Rrightarrow$ is a precongruence*
- *$s \xrightarrow{l} s'$ implies $s \notin V_\Sigma$ (value terms have no computational behaviour)*
- *$s \Rrightarrow s'$ with $s = v(s_1, \ldots, s_n)$ for $v \in VC_\Sigma$ implies $s' = v(s'_1, \ldots, s'_n)$ with $s_i \Rrightarrow s'_i$ for $1 \leq i \leq n$ (rewriting preserves value constructors)*
- *If $s \Rrightarrow s_1$, $s_1 \xrightarrow{l} s_2$ and $s_2 \Rrightarrow s'$ then $s \xrightarrow{l} s'$ (saturation).*

A term made entirely out of value constructors is a *ground value.* Ground values are just as they appear: pure syntactic values, which can be constructed and inspected. The meaning of computational terms in $T_\Sigma - V_\Sigma$ is determined by the $\rightarrow$ and $\Rrightarrow$ relations, representing their behaviour. Non-ground values can be deconstructed to yield computational terms, which may have behaviour.

A value-computation transition system will typically be specified by a set of inductive rules. If Σ is a value-computation signature, let OT_Σ denote the set of *open Σ-terms*, constructed inductively from *term variables* $(x, y, \ldots)$, *value variables* $(v_1, v_2, \ldots)$ and constructors in C_Σ.

Definition 3 (value-computation specification). *A* value-computation specification *consists of a tuple* (Σ, L, D) *where* Σ *is a value-computation signature,* L *a label set, and* D *a set of rules over formulas* $f(s_1, \ldots, s_n) \xrightarrow{l} s'$ *or* $f(s_1, \ldots, s_n) \Rightarrow s'$ *with* $s_i, s' \in OT_\Sigma$ *and* $f \notin VC$*. These rules generate a transition system over* T_Σ *and* L *inductively, after being extended with rules for reflexivity, precongruence, transitivity and saturation (below), where value variables range over value terms.*

$$x \Rightarrow x \qquad \frac{x_1 \Rightarrow y_1 \quad \cdots \quad x_n \Rightarrow y_n}{f(x_1, \ldots, y_n) \Rightarrow f(x_1, \ldots, y_n)} f \in C_\Sigma, \mathsf{ar}_\Sigma(f) = n$$

$$\frac{x \Rightarrow y \qquad y \Rightarrow z}{x \Rightarrow z} \qquad \frac{x \Rightarrow x_1 \qquad x_1 \xrightarrow{l} y_1 \qquad y_1 \Rightarrow y}{x \xrightarrow{l} y}$$

Each such specification generates a value-computation transition system.

Example 4. We consider a value-computation system of basic constructs. The signature Σ contains binary sequencing seq; a ternary conditional cond; nullary constants true, false and skip; operations print_l for $l \in \{a, b\}$; and unary operations thunk for wrapping computations as values, and force for forcing evaluation of a thunk. The value constructors are true, false, skip, and thunk. For labels, $L = \{a, b\}$. The rules are given in Fig. 1.

$$\frac{x \xrightarrow{l} x'}{\mathsf{cond}(x, y_1, y_2) \xrightarrow{l} \mathsf{cond}(x', y_1, y_2)} \quad (1) \qquad \frac{x \xrightarrow{l} x'}{\mathsf{seq}(x, y) \xrightarrow{l} \mathsf{seq}(x', y)} \quad (6)$$

$$\mathsf{cond}(\mathsf{true}, y_1, y_2) \Rightarrow y_1 \quad (2) \qquad \mathsf{seq}(\mathsf{skip}, y) \Rightarrow y \quad (7)$$

$$\mathsf{cond}(\mathsf{false}, y_1, y_2) \Rightarrow y_2 \quad (3)$$

$$\mathsf{print}_a \xrightarrow{a} \mathsf{skip} \quad (4) \qquad \frac{x \xrightarrow{l} x'}{\mathsf{force}(x) \xrightarrow{l} \mathsf{force}(x')} \quad (8)$$

$$\mathsf{print}_b \xrightarrow{b} \mathsf{skip} \quad (5) \qquad \mathsf{force}(\mathsf{thunk}(x)) \Rightarrow x \quad (9)$$

Fig. 1. Operational rules for Example 4

We next introduce our notion of equivalence for value-computation transition systems. This consists of extending the usual bisimulation step condition with two further cases dealing with rewriting and values. For example, if two values are bisimilar, the outermost value constructor must be the same (up to rewriting), and the arguments pointwise bisimilar.

Definition 5 (value-computation bisimulation). *A* value-computation bisimulation *(or* vc-bisimulation*) over a given value-computation transition system* $(\Sigma, L, \rightarrow, \Rightarrow)$ *is a symmetric relation* $R \subseteq T_\Sigma \times T_\Sigma$ *such that*

1. *If* $s\,R\,t$ *and* $s \xrightarrow{l} s'$ *then* $\exists t'$ *with* $s'\,R\,t'$ *and* $t \xrightarrow{l} t'$.
2. *If* $s\,R\,t$ *and* $s \Rightarrow s'$ *then* $\exists t'$ *with* $s'\,R\,t'$ *and* $t \Rightarrow t'$.
3. *If* $v(s_1, \ldots, s_n)\ R\ t$ *with* $v \in VC$, *then* $t \Rightarrow v(t_1, \ldots, t_n)$ *with* $s_i\ R\ t_i$ *for* $1 \leq i \leq n$.

Two terms s *and* t *are* value-computation bisimilar, *written* $s \approx_{vc} t$, *if there exists a value-computation bisimulation* R *with* $s\,R\,t$.

In Example 4, for any terms s, t, r we have $\mathsf{seq}(\mathsf{seq}(s,t),r) \approx_{vc} \mathsf{seq}(s,\mathsf{seq}(t,r))$, and also $\mathsf{thunk}(\mathsf{seq}(\mathsf{seq}(s,t),r)) \approx_{vc} \mathsf{thunk}(\mathsf{seq}(s,\mathsf{seq}(t,r)))$. The use of rewrites $\Rightarrow$ also allows us to prove unit laws up to bisimulation, which usually only hold up to weak bisimulation: for example, $\mathsf{seq}(\mathsf{skip}, s) \approx_{vc} s$.

2.2 Congruence Format

We next define a format guaranteeing that value-computation bisimilarity is a congruence.

Definition 6 (pattern). *A* pattern *is a term constructed inductively from variables and value constructors such that each variable appears at most once.*

Definition 7 (value-added tyft). *A rule is in the* value-added tyft *format if it is of the following shape, where each* $\rightsquigarrow$, $\rightsquigarrow_i$ *may be* $\Rightarrow$ *or* $\xrightarrow{a}$ *for some* a.

$$\frac{\{s_i \rightsquigarrow_i u_i : i \in I\}}{f(w_1, \ldots, w_n) \rightsquigarrow t}$$

Here, t, s_i *range over arbitrary open terms; and* u_i, w_j *over patterns. Further, each variable may occur in at most one of* u_i *or* w_j. *A value-computation specification is in the* value-added tyft *format if all of its rules are.*

By inspecting Fig. 1, we see that Example 4 is in the value-added tyft format. The restriction of certain subterms to patterns ensures that only value constructors may be inspected. To see that this is necessary for congruence, consider any instance of $\mathsf{seq}(\mathsf{seq}(\mathsf{print}_a, t), r) \approx_{vc} \mathsf{seq}(\mathsf{print}_a, \mathsf{seq}(t, r))$. Then w_i must be a pattern, as otherwise f defined by $\mathsf{f}(\mathsf{seq}(\mathsf{seq}(x,y),z)) \Rightarrow \mathsf{true}$ provides a distinguishing context. Each u_i must be a pattern, as otherwise f defined by $\frac{x \xrightarrow{l} \mathsf{seq}(\mathsf{seq}(x,y),z)}{\mathsf{f}(x) \xrightarrow{l} \mathsf{true}}$ provides a distinguishing context. We require uniqueness of variables as otherwise $\mathsf{g}(x,x) \Rightarrow \mathsf{true}$, $\mathsf{f}(x) \Rightarrow \mathsf{g}(x, \mathsf{seq}(\mathsf{seq}(\mathsf{print}_a, t), r))$ provides a distinguishing context.

The above format is built on the *tyft* format of [3], generalised so that u_i, w_j may range over patterns rather than just variables. In *loc. cit.*, *tyxt* rules are also allowed, where the source of the conclusion is just a variable. We have excluded this here just to ensure that values do not perform computational steps; if this is otherwise guaranteed then *tyxt* rules may be added with congruence intact.

We next show that the value-added tyft format ensures that bisimilarity is a congruence. To show our congruence result, we assume that rules are well-founded: that is, the premises of each rule can be ordered such that variables in the conclusion of a premise appear in no earlier premise. This restriction was also required for the *tyft/tyxt* congruence proof in [3]. It was later shown unnecessary via a translation in [2]; such a translation should be possible for this result also.

If σ is a partial mapping from variables to terms, we write $s[\sigma]$ for the substitution replacing each σ-defined variable x in s by $\sigma(x)$. If s is an open term, we write $\mathsf{vars}(s)$ for the variables occurring in s. The *reflexive congruence closure* of a relation R is the least reflexive relation containing R such that $s_1 \; R \; t_1, \ldots, s_n \; R \; t_n$ implies $f(s_1, \ldots, s_n) \; R \; f(t_1, \ldots, t_n)$.

Lemma 8. *Let r be a pattern and σ a substitution with $\mathsf{dom}(\sigma) = \mathsf{vars}(r)$. Let R' denote the reflexive congruence closure of a vc-bisimulation R and let $r[\sigma] \; R' \; t$. Then exists τ with $\mathsf{dom}(\tau) = \mathsf{vars}(r)$ such that $t \Rightarrow r[\tau]$ and for each x, $\sigma(x) \; R' \; \tau(x)$.*

Theorem 9. *If all rules in a value-computation transition system are defined in the value-added tyft format and well-founded, then vc-bisimilarity in that system is a congruence.*

Proof. Let R be a vc-bisimulation, and let R' denote the reflexive congruence closure of R. We will show that R' is also a vc-bisimulation, and since R' contains R we can conclude that vc-bisimilarity is a congruence. To show that R' is a vc-bisimulation, we show the three conditions in Definition 5. Conditions 1 and 2 are shown simultaneously, showing that $s \rightsquigarrow s'$ and $s \; R' \; t$ implies there exists t' with $t \rightsquigarrow t'$ with $s' \; R' \; t'$ for any $\rightsquigarrow$ of the form $\Rightarrow$ or $\xrightarrow{a}$. The proof proceeds by induction on the derivation of $s \rightsquigarrow s'$ making use of the known rule shape, together with Lemma 8 for patterns in the targets of premises and source of conclusion. Condition 3 follows immediately from Lemma 8: If $s = v(s_1, \ldots, s_n) \; R' \; t$ then $s = r[\sigma]$ where $r = v(x_1, \ldots, x_n)$ and $\sigma = \{x_i \mapsto s_i\}$. By the lemma, $t \Rightarrow r[\tau]$ with $\tau = \{x_i \mapsto t_i\}$ and $s_i \; R' \; t_i$. Then $t = v(t_1, \ldots, t_n)$ as required. □

3 Modular SOS

In this section, we first recall the differences between Modular SOS (MSOS) [8] and the original SOS framework [16], explaining how MSOS incorporates auxiliary entities in labels. We then enrich the MSOS specifications of [8] with the notion of value, and illustrate our framework by specifying rules for various constructs.

3.1 MSOS Labels

In SOS (and in the value-computation specifications introduced in Sect. 2) the set of labels can be chosen arbitrarily. In practice, however, when specifying the semantics of concurrent or reactive processes, labels usually represent emitted signals or events; and when specifying sequential programming languages, they are often not used at all. Any auxiliary entities, such as environments (ρ) and stores (σ), are incorporated as sub-terms of states, together with the usual process terms. For example, a state might be a triple (s, ρ, σ).[1]

MSOS differs from SOS by incorporating auxiliary entities in labels, instead of in states. Thus states are simply process terms (including computed values). Moreover, the set of labels forms a *category* (with the labels as the morphisms) and the labels on successive transitions have to be *composable*. The constraint of composability is crucial: for instance, it ensures that the environment in the label does not change between adjacent transitions, and that changes to the store are single-threaded. There is also a notion of *unobservable* label, corresponding to identity morphisms.

Since the various auxiliary entities are (in principle) independent, the label category is obtained as a product of a component category for each auxiliary entity. Following [8] we use indexed products, and write a label using ML-style record value notation as '$\{i_1 = t_1, \ldots, i_n = t_n\}$' (the order in which the components are listed is insignificant).

Three simple kinds of label component category, identified in [8], are sufficient to ensure that MSOS is at least as expressive as SOS:

- *Read-only (RO):* label components are composable only when identical, and always unobservable.
- *Write-only (WO):* label components are always composable, and there is a unique unobservable entity corresponding to an identity morphism.
- *Read-write (RW):* label components are pairs of entities, (x, x') is composable with (y, y') iff x' and y are identical, and (x, x') is unobservable iff x and x' are identical.

For notational convenience we write labels using an unprimed index for each read-only component (e.g. **env**=ρ), a primed index for each write-only component (e.g. **output**′=o), and both an unprimed and a primed index for the two entities of each read-write component (e.g. **store**=σ_0, **store**′=σ_1). Formally:

Definition 10 (MSOS labels). *A* label profile *is a triple of disjoint sets* $\mathcal{L} = (\mathcal{L}_{RO}, \mathcal{L}_{RW}, \mathcal{L}_{WO})$*. The set* $\mathsf{reads}(\mathcal{L})$ *consists of the* unprimed elements $\mathbf{x} \in \mathcal{L}_{RO} \uplus \mathcal{L}_{RW}$*. The set* $\mathsf{writes}(\mathcal{L})$ *consists of the* primed elements $\{\mathbf{x}' : \mathbf{x} \in \mathcal{L}_{WO} \uplus \mathcal{L}_{RW}\}$*. For any set* T*, the* label set $\mathcal{L}(T)$ *is the set of maps* $\mathsf{reads}(\mathcal{L}) \uplus \mathsf{writes}(\mathcal{L}) \to T$*. For a label* $L \in \mathcal{L}(T)$*, we write* $\mathsf{reads}(L)$ *and* $\mathsf{writes}(L)$ *for the restriction of* L *to* $\mathsf{reads}(\mathcal{L})$ *and* $\mathsf{writes}(\mathcal{L})$ *respectively.*

We intend to instantiate T with a set of terms – for example, we can represent stores and environments as terms by using applicative lists. Accordingly, we will use σ, ρ, ... as additional term variables.

3.2 MSOS Specifications

An MSOS specification with respect to a label profile $\mathcal{L}$ and set of terms T generates a transition system specification with states in T and labels in $\mathcal{L}(T)$. Such

[1] Transitions which do not change ρ are usually written $\rho \vdash (s, \sigma) \to (s', \sigma')$.

specifications typically only mention a relevant subset of the label components, treating ellipses '...' as variables ranging over the remaining components, which may be propagated between premises and conclusion. A dash '–' indicates that the rest of the label L is unobservable: concretely, if $\mathbf{x} \in \mathcal{L}_{RW}$ then $L(\mathbf{x}') = L(\mathbf{x})$ and if $\mathbf{x} \in \mathcal{L}_{WO}$ then $L(\mathbf{x}') = \iota_{\mathbf{x}}$ where $\iota_{\mathbf{x}}$ is a distinguished nullary constant associated to $\mathbf{x}$.

Rules may combine labels using *composition* ($\circ$). A pair of labels (L_1, L_2) is *composable* if for $\mathbf{x} \in \mathcal{L}_{RO}$, $L_1(\mathbf{x}) = L_2(\mathbf{x})$ and for $\mathbf{x} \in \mathcal{L}_{RW}$, $L_1(\mathbf{x}') = L_2(\mathbf{x})$. Given a composable pair (L_1, L_2) the *composition* $L_2 \circ L_1$ is defined to be:

- For $\mathbf{x} \in \mathcal{L}_{RO}$, $(L_2 \circ L_1)(\mathbf{x}) = L_1(\mathbf{x}) = L_2(\mathbf{x})$.
- For $\mathbf{x} \in \mathcal{L}_{WO}$, $(L_2 \circ L_1)(\mathbf{x}') = \odot_{\mathbf{x}}(L_1(\mathbf{x}'), L_2(\mathbf{x}'))$ where $\odot_{\mathbf{x}}$ is a distinguished binary constructor associated to $\mathbf{x}$. Typically, $(\odot_{\mathbf{x}}, \iota_{\mathbf{x}})$ will form a monoid on a subset of the terms.
- For $\mathbf{x} \in \mathcal{L}_{RW}$, $(L_2 \circ L_1)(\mathbf{x}) = L_1(\mathbf{x})$ and $(L_2 \circ L_1)(\mathbf{x}') = L_2(\mathbf{x}')$.

In rules, we use label variables (l, '...'). In a given rule, each label must have the same set of explicitly mentioned label components E. Labels in that rule then consist of a map $E \to OT_{\Sigma}$ denoted by a list of equations, followed by a composition (sequence) of label variables (the empty sequence is denoted '–', representing an unobservable label).

Definition 11 (MSOS specification). *An* MSOS specification *consists of a tuple* $(\mathcal{L}, \Sigma, D, M)$ *where* $\mathcal{L}$ *is a label profile,* Σ *a value-computation signature, and* D *is a set of rules over formulas* $f(s_1, \ldots, s_n) \xrightarrow{l} s'$ *or* $f(s_1, \ldots, s_n) \Rightarrow s'$ *with* $s_i, s' \in OT_{\Sigma}$, $f \notin VC_{\Sigma}$ *with labels as immediately above. Finally,* M *specifies for each* $\mathbf{x} \in \mathcal{L}_{WO}$ *a nullary* $\iota_{\mathbf{x}} \in C_{\Sigma}$ *and binary* $\odot_{\mathbf{x}} \in C_{\Sigma}$*. There are* built-in *rules for reflexivity, precongruence, transitivity and saturation, consisting of those in Definition 3 in addition to:*

$$\frac{x \Rightarrow z \qquad y \xrightarrow{\{\mathbf{x}=z,\ldots\}} t'}{y \xrightarrow{\{\mathbf{x}=x,\ldots\}} t'} \qquad \frac{y \xrightarrow{\{\mathbf{x}'=x,\ldots\}} t' \qquad x \Rightarrow z}{y \xrightarrow{\{\mathbf{x}'=z,\ldots\}} t'}$$

An MSOS specification generates a value-computation transition system over $\mathcal{L}(T_{\Sigma})$ *inductively after being extended with the built-in rules, where value variables range over value terms. Unobservability, composability and composition are interpreted as described above.*

Example 12. In Fig. 2, we give some example constructors and their MSOS rules. The label profile includes read-only **env**, write-only **exc** and **output**, and read-writeable **store**. We include all constructors and rules from Example 4, except for print, which has been generalised.

We add value constructors for maps (ternary update, nullary empty), lists (binary cons, nullary nil) and function values (ternary abs). The value abs(x, s, ρ) denotes a closed function, with formal parameter x, body s, and closing environment ρ. We include a set of nullary values $I = \{\mathsf{x}, \mathsf{y}, \ldots\}$ for identifiers and imperative variables.

$$\mathsf{bound}(x) \xrightarrow{\{\mathbf{env}=\rho,\ldots\}} \mathsf{lookup}(\rho, x) \quad (10)$$

$$\frac{y \xrightarrow{\{\ldots\}} y'}{\mathsf{let}(x,y,t) \xrightarrow{\{\ldots\}} \mathsf{let}(x,y',t)} \quad (11)$$

$$\frac{y \xrightarrow{\{\mathbf{env}=\mathsf{update}(\rho,x,v),\ldots\}} y'}{\mathsf{let}(x,v,y) \xrightarrow{\{\mathbf{env}=\rho,\ldots\}} \mathsf{let}(x,v,y')} \quad (12)$$

$$\mathsf{let}(x, v_1, v_2) \Rightarrow v_2 \quad (13)$$

$$\mathsf{throw}(x) \xrightarrow{\{\mathbf{exc}'=\mathsf{cons}(x,\mathsf{nil}),-\}} \mathsf{stuck} \quad (14)$$

$$\frac{x \xrightarrow{\{\mathbf{exc}'=\mathsf{nil},\ldots\}} x'}{\mathsf{catch}(x,z) \xrightarrow{\{\mathbf{exc}'=\mathsf{nil},\ldots\}} \mathsf{catch}(x',z)} \quad (15)$$

$$\frac{x \xrightarrow{\{\mathbf{exc}'=\mathsf{cons}(y,\mathsf{nil}),\ldots\}} x'}{\mathsf{catch}(x,z) \xrightarrow{\{\mathbf{exc}'=\mathsf{nil},\ldots\}} \mathsf{apply}(z,y)} \quad (16)$$

$$\mathsf{catch}(v,z) \Rightarrow v \quad (17)$$

$$\mathsf{assign}(x,v) \xrightarrow{\{\mathbf{store}=\sigma,\ \mathbf{store}'=\mathsf{update}(\sigma,x,v),-\}} \mathsf{skip} \quad (18)$$

$$\frac{y \xrightarrow{\{\ldots\}} y'}{\mathsf{assign}(x,y) \xrightarrow{\{\ldots\}} \mathsf{assign}(x,y')} \quad (19)$$

$$\mathsf{deref}(x) \xrightarrow{\{\mathbf{store}=\sigma,-\}} \mathsf{lookup}(\sigma,x) \quad (20)$$

$$\mathsf{print}(x) \xrightarrow{\{\mathbf{output}'=\mathsf{cons}(x,\mathsf{nil}),-\}} \mathsf{skip} \quad (21)$$

$$\frac{x \xrightarrow{\{\ldots\}} x'}{\mathsf{apply}(x,y) \xrightarrow{\{\ldots\}} \mathsf{apply}(x',y)} \quad (22)$$

$$\frac{y \xrightarrow{\{\ldots\}} y'}{\mathsf{apply}(v,y) \xrightarrow{\{\ldots\}} \mathsf{apply}(v,y')} \quad (23)$$

$$\frac{y \xrightarrow{\{\mathbf{env}=\mathsf{update}(\rho,x,v),\ldots\}} y'}{\mathsf{apply}(\mathsf{abs}(x,y,\rho),v) \xrightarrow{\{\mathbf{env}=\rho_1,\ldots\}} \mathsf{apply}(\mathsf{abs}(x,y',\rho),v)} \quad (24)$$

$$\mathsf{apply}(\mathsf{abs}(x,v_1,\rho),v_2) \Rightarrow v_1 \quad (25)$$

$$\mathsf{lambda}(x,y) \xrightarrow{\{\mathbf{env}=\rho,\ldots\}} \mathsf{abs}(x,y,\rho) \quad (26)$$

$$\frac{x \xrightarrow{l_1} x' \qquad \mathsf{atomic}(x') \xrightarrow{l_2} v}{\mathsf{atomic}(x) \xrightarrow{l_2 \circ l_1} v} \quad (27)$$

$$\mathsf{atomic}(v) \xrightarrow{\{-\}} v \quad (28)$$

$$\mathsf{append}(\mathsf{cons}(x,y),z) \Rightarrow \mathsf{cons}(x,\mathsf{append}(y,z)) \quad (29)$$

$$\mathsf{append}(\mathsf{nil},x) \Rightarrow x \quad (30)$$

$$\frac{\mathsf{lookup}(\mu,j) \Rightarrow v_1}{\mathsf{lookup}(\mathsf{update}(\mu,i,v),j) \Rightarrow v_1}\ i \neq j \in I \quad (31)$$

$$\frac{}{\mathsf{lookup}(\mathsf{update}(\mu,i,v),i) \Rightarrow v}\ i \in I \quad (32)$$

Fig. 2. Operational rules for Example 12

Additional computational constructors include static bindings (let and bound), volatile store (assign and deref), functions (lambda and apply) and exceptions (throw and catch). The term $\mathsf{let}(x, s, t)$ binds x to s in t.

We include binary operations lookup and append for maps and lists respectively. Note that lookup(empty, x) is a stuck computational term – this is an example of how undefinedness can be handled in our setting.

We also include an atomic constructor. The computation $\mathsf{atomic}(s)$ runs s and combines the trace into a single transition. This can be used to block the context interrupting the trace, e.g. by catching a thrown exception. Note that the first rule for atomic only applies when (l_1, l_2) is a composable pair.

We set $\iota_{\mathbf{output}} = \iota_{\mathbf{exc}} = \mathsf{nil}$ and $\odot_{\mathbf{output}} = \odot_{\mathbf{exc}} = \mathsf{append}$.

The notational burden of heavily loaded arrows can be avoided by writing the MSOS rules using conventional SOS notation, following techniques in [11].

4 Bisimulation Metatheory for MSOS

We next revisit our goal of ensuring that bisimilarity is a congruence, this time in the MSOS setting. Even though we still generate value-computation transition systems, the value-added tyft format is of limited use, since it does not allow information flow between labels and other computational terms in rules (this is needed in Example 12 for e.g. Rule (12) for let). Note that we cannot allow such flow arbitrarily: if $s_1 \approx s_2$ is to imply $\mathsf{let}(\mathsf{x}, s_1, t) \approx \mathsf{let}(\mathsf{x}, s_2, t)$ then t can only test the **env** label component up to pointwise bisimilarity.

4.1 Bisimulation in MSOS

We next generalise vc-bisimulation to a higher-order version for the MSOS setting. In particular, in the step writeable label components may themselves vary up to bisimulation. This is required, for example, so that $s \approx t$ implies $\mathsf{assign}(\mathsf{x}, s) \approx \mathsf{assign}(\mathsf{x}, t)$. Given a relation R, for maps σ and τ we write $\sigma\ R\ \tau$ just if $\mathsf{dom}(\sigma) = \mathsf{dom}(\tau)$ and for each $x \in \mathsf{dom}(\sigma)$, $\sigma(x)\ R\ \tau(x)$.

Definition 13 (MSOS bisimulation). *Given a value-computation transition system $(\Sigma, \mathcal{L}(T_\Sigma), \rightarrow, \Rightarrow)$ generated from an MSOS specification, an* MSOS bisimulation *is a symmetric relation $R \subseteq T_\Sigma \times T_\Sigma$ such that:*

1. *If $s\ R\ t$ and $s \xrightarrow{L} s'$ then $\exists t', L'$ with $s'\ R\ t'$, $t \xrightarrow{L'} t'$, $\mathsf{reads}(L') = \mathsf{reads}(L)$ and $\mathsf{writes}(L)\ R\ \mathsf{writes}(L')$.*
2. *If $s\ R\ t$ and $s \Rightarrow s'$ then $\exists t'$ with $s'\ R\ t'$ and $t \Rightarrow t'$.*
3. *If $v(s_1, \ldots, s_n)\ R\ t$ with $v \in VC_\Sigma$, then $t \Rightarrow v(t_1, \ldots, t_n)$ with $s_i\ R\ t_i$ for $1 \leq i \leq n$.*

Two terms s and t are MSOS bisimilar, *written $s \approx_{msos} t$, if there exists an MSOS bisimulation R with $s\ R\ t$.*

In MSOS rules, usually only a few label components are mentioned explicitly, while in the above definition all label components are mentioned. However, in any particular bisimulation proof, one can set $L'(i) = L(i)$ for unmentioned i.

Since vc-bisimulations are also MSOS bisimulations, the associativity and unit laws for seq hold up to MSOS bisimilarity in Example 12. We can also show $\mathsf{catch}(\mathsf{print}(v), x) \approx_{msos} \mathsf{print}(v)$, for example. We may seek to prove laws for state such as $\mathsf{seq}(\mathsf{assign}(\mathsf{x}, v), \mathsf{deref}(\mathsf{x})) \approx_{msos} \mathsf{seq}(\mathsf{assign}(\mathsf{x}, v), v)$. However, this law is not sound with respect to arbitrary contexts. In particular, $C[s]$ may run one step of computation of s and then roll back the store before continuing. Instead, we may prove a modified law which blocks interruption of the trace:

$$\mathsf{atomic}(\mathsf{seq}(\mathsf{assign}(\mathsf{x}, v), \mathsf{deref}(\mathsf{x}))) \approx_{msos} \mathsf{atomic}(\mathsf{seq}(\mathsf{assign}(\mathsf{x}, v), v)).$$

4.2 Congruence Format

We now present a rule format which ensures that MSOS bisimilarity is a congruence. We will need to consider the substructure of labels in rules.

Definition 14 (well-founded MSOS tyft). *A rule is in the* well-founded MSOS tyft *format if it has the following form:*

$$\frac{\{s_i \rightsquigarrow_i u_i : i \in I\}}{f(w_1, \ldots, w_n) \rightsquigarrow t}$$

where premises are ordered and:

- *t, s_i range over arbitrary open terms and u_i, w_j over patterns.*
- *$\rightsquigarrow_i$ is either $\Rightarrow$ or $\xrightarrow{L_i}$ where L_i consists of a sequence of equations $\{l = t_{l,i}\}$ possibly followed by a label variable. Further, if l is primed then $t_{l,i}$ must be a pattern.*
- *$\rightsquigarrow$ is either $\Rightarrow$ or $\xrightarrow{L}$, where L consists of a sequence of equations $\{l = t_l\}$ possibly followed by a composition of label variables. Each such label variable must occur in the premise. If label variable X is to the left of label variable Y in the composition, Y must occur in an earlier premise than X. Further, if l is unprimed then t_l is a pattern.*
- *The set of variables must be disjoint for u_i, w_j, $t_{l,i}$ for primed l, t_l for unprimed l. Variables in u_i or $t_{l,i}$ for primed l must not appear in an earlier premise.*

An MSOS specification is in the well-founded MSOS tyft format just if all its rules are.

This follows a discipline of information flow from readable components in the conclusion to readable components in the premise, to writeable components in the premise to writeable components in the conclusion. Writeable components ($t_{l,i}$ and t_l for primed l) are treated like additional targets, and readable components ($t_{l,i}$ and t_l for unprimed l) like additional sources. To see why each t_l

must be a pattern for unprimed l, consider $\mathsf{g} \xrightarrow{\{\text{env}=\mathsf{update}(y,\mathsf{x},\mathsf{thunk}(\mathsf{seq}(z,w)))\}} \mathsf{true}$, $\mathsf{f}(y) \Rightarrow \mathsf{let}(\mathsf{x}, \mathsf{thunk}(y), \mathsf{g})$ and note that f provides a distinguishing context for $\mathsf{seq}(\mathsf{skip}, \mathsf{print}(\mathsf{true})) \approx_{msos} \mathsf{print}(\mathsf{true})$. To see why each $t_{l,i}$ must be a pattern for primed l, note that $\dfrac{\mathsf{throw}(x) \xrightarrow{\{\mathbf{exn}'=\mathsf{cons}(\mathsf{seq}(z,w),\mathsf{nil}),\ldots\}} y}{\mathsf{f}(x) \xrightarrow{\{\mathbf{exn}'=\mathsf{nil},\ldots\}} \mathsf{true}}$ provides a distinguishing context for the same equation. The same examples given in Sect. 2.2 show why the u_i, w_j must be patterns and why variables may not be shared.

Note that in this format composition expressions and unobservable labels may only occur in the conclusion of a rule. The restriction on ordering of label variables in the conclusion ensures that when composition is made explicit, the pattern restrictions above are satisfied.

The distinction between readable and writeable label components is related to label arguments in [1] and the notion of *volatility* from [12]. In each, certain terms in the label are restricted to be a generalised notion of fresh variable and replacement of bisimilar terms in this component will lead to bisimilar outputs.

By inspection of Figs. 1 and 2 we see that Example 12 is in MSOS tyft format (we view the rules for lookup as a family of rules indexed over I). In the rest of this section we will show that for systems with rules in the well-founded MSOS tyft format, MSOS bisimilarity is a congruence.

Definition 15 (explicit MSOS tyft). *An MSOS specification is in* explicit MSOS tyft format *if it is in the well-founded MSOS tyft format and contains no label variables.*

Proposition 16. *Each well-founded MSOS tyft system is equivalent to one in the explicit MSOS tyft format.*

Given an MSOS transition system T over label profile $\mathcal{L}$ we produce an equivalent set of rules removing all uses of label variables, exhibiting all information flow in labels explicitly following the definitions in Sect. 3. We only give an example. The rules for the atomic constructor are translated as follows:

$$\frac{s \xrightarrow{\lambda_1} s' \qquad \mathsf{atomic}(s') \xrightarrow{\lambda_2} v}{\mathsf{atomic}(s) \xrightarrow{\lambda} v} \qquad \mathsf{atomic}(v) \xrightarrow{\mu} v$$

where

$$\begin{aligned}
\lambda_1 &= \{\mathbf{env} = \rho, \mathbf{store} = \sigma_1, \mathbf{store}' = \sigma_2, \mathbf{output}' = \alpha_1, \mathbf{exc}' = \eta_1\}\\
\lambda_2 &= \{\mathbf{env} = \rho, \mathbf{store} = \sigma_2, \mathbf{store}' = \sigma_3, \mathbf{output}' = \alpha_2, \mathbf{exc}' = \eta_2\}\\
\lambda &= \{\mathbf{env} = \rho, \mathbf{store} = \sigma_1, \mathbf{store}' = \sigma_3, \mathbf{output}' = \mathsf{append}(\alpha_1, \alpha_2),\\
&\qquad\qquad\qquad\qquad\qquad\qquad\qquad\qquad \mathbf{exc}' = \mathsf{append}(\eta_1, \eta_2)\}\\
\mu &= \{\mathbf{env} = \rho, \mathbf{store} = \sigma,\ \mathbf{store}' = \sigma,\ \mathbf{output}' = \mathsf{nil}, \mathbf{exc}' = \mathsf{nil}\}
\end{aligned}$$

Proposition 17. *Consider an MSOS specification in explicit MSOS tyft format. Let R be an MSOS bisimulation over the generated transition system and let R' denote the reflexive transitive congruence closure of R. Suppose $s\ R'\ t$. Then:*

1. *If* $s = r[\sigma]$ *with* $\mathsf{dom}(\sigma) = \mathsf{vars}(r)$ *and* r *is a pattern then there exists* τ *with* $\mathsf{dom}(\tau) = \mathsf{vars}(r)$ *such that* $t \Rightarrow r[\tau]$ *with* $\sigma(x)\ R'\ \tau(x)$ *for each* $x \in \mathsf{vars}(r)$.
2. *If* $s \Rightarrow s'$ *and* $s\ R'\ t$ *then there exists* t' *with* $t \Rightarrow t'$ *and* $s'\ R'\ t'$.
3. *If* $s \xrightarrow{L} s'$ *and* $\mathsf{reads}(L)\ R'\ trs$ *then there exists* t', tws *such that* $s'\ R'\ t'$, $\mathsf{writes}(L)\ R'\ tws$ *and* $t \xrightarrow{L'} t'$ *for* $\mathsf{reads}(L') = trs$ *and* $\mathsf{writes}(L') = tws$.

Proof. The proof proceeds by simultaneous induction on R'. Condition 1 corresponds to Lemma 8 but must be proved simultaneously due to the fact that we also close our relation up to transitivity. For Conditions 2 and 3, we perform an inner induction on the proof of the transition, exploiting the rule format. □

Theorem 18. *Consider an MSOS specification* T *in the well-founded MSOS tyft format. Let* R *be an MSOS bisimulation and* R' *denote the reflexive transitive congruence closure of* R*. Then* R' *is an MSOS bisimulation.*

Proof. We first convert T into an equivalent system in explicit MSOS tyft format following Proposition 16. We then show that R' is an MSOS bisimulation by considering the three conditions in turn, each of which follows straightforwardly from Proposition 17. □

Corollary 19. *MSOS-bisimilarity is a congruence for specifications in the well-founded MSOS tyft format.*

Proposition 17 claim 3 ensures that if $s \approx_{msos} t$ then each composable trace from s can be matched by a corresponding composable trace from t, up to bisimilarity in the subsequent steps and labels. In the case of Example 12, it also has the following consequence: for any term, (pointwise) bisimilar environments yield bisimilar outputs. The fact that bisimilarity is a congruence ensures that bisimilar abstractions yield bisimilar outputs when applied. This is part of the definition of bisimilar abstractions in applicative bisimulation [4].

5 Modular Bisimulations

Bisimulation examples in this paper were given explicitly with respect to our example systems. But in fact the proofs did not make use of the particular closed set of constructors. For example, for $\mathsf{seq}(\mathsf{skip}, s) \approx s$, presence of constructors other than seq and skip had no influence whatsoever on the proof of bisimulation. The proof would work just as well in any system with those rules for the constructors in question; the law and proof are *modular* in nature. On the other hand, if a bisimulation proof performs explicit case analysis on all terms or label components, this is not possible. How can we formalise this distinction?

Given a constructor f, an f*-defining rule* is a rule where the source of the conclusion has f as its outermost symbol. A *disjoint extension* of an MSOS system $(\Sigma, \mathcal{L}, D)$ is an MSOS system $(\Sigma', \mathcal{L}', D')$ with $\Sigma' \supseteq \Sigma$, $\mathcal{L}' \supseteq \mathcal{L}$ and such that each rule in $D' - D$ is f-defining for some f in $\Sigma' - \Sigma$.

Let S be a subset of the constructors of Example 12. We define $\hat{S}$ to be the least set containing S such that for all f in $\hat{S}$, any constructor appearing in an f-defining rule also appears in $\hat{S}$. We define $\mathbf{E}_S$ to be the subsystem of Example 12 restricted to the constructors in $\hat{S}$ and rules that are f-defining for some $f \in \hat{S}$. Given a candidate algebraic law for MSOS, we advocate proving this law with respect to all disjoint extensions of $\mathbf{E}_S$, where S is the set of constructors appearing in that law. We isolate the particular subsystem that makes the law hold, and are guaranteed that any system containing this will validate the law. For associativity of seq, we show: $\mathsf{seq}(\mathsf{seq}(s,t),r) \approx_{msos} \mathsf{seq}(s,\mathsf{seq}(t,r))$ in any disjoint extension of $\mathbf{E}_{\{\mathsf{seq}\}}$.

We call such statements *modular bisimulations.* Since the quantification over extensions is external to the particular notion of bisimulation, meta-results such as congruence can be used directly. All examples of bisimulations in this paper can indeed be formulated and proved as modular bisimulations.

If we wish to internalise this notion, we are led to *fh-bisimulation* [10]. In this setting, the step conditions must hold in the presence of arbitrary *hypotheses*, of the form $x \xrightarrow{a} y$ for variables x, y. More specifically, it is *provable ruloids* that must step – a provable ruloid of $\frac{\Gamma}{s \xrightarrow{a} s'}$ is a proof of $s \xrightarrow{a} s'$ which may have open leaves found in Γ. In [10], it was shown that fh-bisimilarity is preserved under disjoint extensions which preserve the label set for the positive GSOS format. (We have subsequently generalised this result to arbitrary positive source-dependent rules.) In future, we hope to adapt these results to our MSOS bisimulation format.

6 Further Directions

We intend to use our framework to give formal semantics, and prove laws about, real-world programming languages. One reason this has been lacking in the literature is due to the scalability of the usual techniques, and our use of MSOS and modular bisimulations help to address these issues. As a start, we are currently providing dynamic semantics for Caml Light [5] by translating it into the kind of basic constructors found in Example 12, called *funcons*[2] [9]. This includes higher-order functions, pattern matching, records and variants, mutually recursive declarations, exceptions and reference cells. Crucially, all rules for constructors used are in the MSOS tyft format. Thus, if program fragments P and Q have funcon translations P' and Q' respectively, and P' and Q' have been proved equivalent using our techniques, we can conclude that P and Q are equivalent and soundly interchangeable in Caml Light programs.

A possible useful extension of this work would be treatment of multisorted algebras. In particular, the right unit law for seq only holds if the only value that left operand could compute to is skip, i.e. it has type unit, or is a command. We could also consider parametrising bisimulations by the current label components cf. state-based bisimilarity [13], which would increase the number of equivalences one could prove. This could be particularly interesting in the MSOS setting, where labels are both open and higher-order.

Another further direction is to consider rules with negative premises, which we have avoided here by matching on values. We have also avoided special treatment

[2] See `http://www.plancomps.org/churchill2013a/`

for variable binders/names, which are handled by the environment. Fresh name generation is possible using read-write label components.

Acknowledgements. Many thanks to Mohammad Mousavi, Cristian Prisacariu, Paolo Torrini and the anonymous referees for their useful comments. This work was supported by an EPSRC grant (EP/I032495/1) to Swansea University in connection with the *PLanCompS* project (`www.plancomps.org`).

References

1. Bernstein, K.L.: A congruence theorem for structured operational semantics of higher-order languages. In: 13th Annual IEEE Symposium on Logic in Computer Science, pp. 153–164. IEEE (1998)
2. Fokkink, W.: The Tyft/Tyxt Format Reduces to Tree Rules. In: Hagiya, M., Mitchell, J.C. (eds.) TACS 1994. LNCS, vol. 789, pp. 440–453. Springer, Heidelberg (1994)
3. Groote, J.F., Vaandrager, F.: Structured operational semantics and bisimulation as a congruence. Inf. and Comput. 100(2), 202–260 (1992)
4. Howe, D.J.: Equality in lazy computation systems. In: Fourth Annual IEEE Symposium on Logic in Computer Science, pp. 198–203. IEEE (1989)
5. Leroy, X.: The Caml Light system, documentation and user's guide (1997), `http://caml.inria.fr/pub/docs/manual-caml-light/`
6. Levy, P.B.: Call-by-Push-Value: A Subsuming Paradigm. In: Girard, J.-Y. (ed.) TLCA 1999. LNCS, vol. 1581, pp. 228–243. Springer, Heidelberg (1999)
7. Milner, R.: A Calculus of Communicating Systems. LNCS, vol. 92. Springer, Heidelberg (1980)
8. Mosses, P.D.: Modular structural operational semantics. J. Log. Algebr. Program. 60-61, 195–228 (2004)
9. Mosses, P.D.: Component-based semantics. In: Huisman, M. (ed.) Eighth Intl. Workshop on Specification and Verification of Component-Based Systems, pp. 3–10. ACM, New York (2009)
10. Mosses, P.D., Mousavi, M.R., Reniers, M.A.: Robustness of equations under operational extensions. In: Fröschle, S., Valencia, F.D. (eds.) 17th International Workshop on Expressiveness in Concurrency. EPTCS, arXiv, vol. 41, pp. 106–120 (2010)
11. Mosses, P.D., New, M.J.: Implicit propagation in structural operational semantics. In: Hennessy, M., Klin, B. (eds.) Fifth Workshop on Structural Operational Semantics. Electr. Notes Theor. Comput. Sci., vol. 229(4), pp. 49–66. Elsevier, Amsterdam (2009)
12. Mousavi, M.R., Gabbay, M., Reniers, M.: SOS for Higher Order Processes. In: Abadi, M., de Alfaro, L. (eds.) CONCUR 2005. LNCS, vol. 3653, pp. 308–322. Springer, Heidelberg (2005)
13. Mousavi, M.R., Reniers, M.A., Groote, J.F.: Notions of bisimulation and congruence formats for SOS with data. Inf. and Comput. 200(1), 107–147 (2005)
14. Mousavi, M.R., Reniers, M.A., Groote, J.F.: SOS formats and meta-theory: 20 years after. Theor. Comput. Sci. 373(3), 238–272 (2007)
15. Park, D.: Concurrency and Automata on Infinite Sequences. In: Proc. 5th GI-Conference on Theoretical Computer Science, pp. 167–183. Springer, London (1981)
16. Plotkin, G.D.: A structural approach to operational semantics. J. Log. Algebr. Program. 60-61, 17–139 (2004); Originally Tech. Rep. DAIMI FN-19, Dept. of Computer Science, Univ. Aarhus (1981)

Checking Bisimilarity for Attributed Graph Transformation

Fernando Orejas[1,*], Artur Boronat[1,2,**], Ulrike Golas[3], and Nikos Mylonakis[1]

[1] Universitat Politècnica de Catalunya, Spain
{orejas,nicos}@lsi.upc.edu
[2] University of Leicester, UK
aboronat@mcs.le.ac.uk
[3] Konrad-Zuse-Zentrum für Informationstechnik Berlin, Germany
golas@zib.de

Abstract. Borrowed context graph transformation is a technique developed by Ehrig and Koenig to define bisimilarity congruences from reduction semantics defined by graph transformation. This means that, for instance, this technique can be used for defining bisimilarity congruences for process calculi whose operational semantics can be defined by graph transformation. Moreover, given a set of graph transformation rules, the technique can be used for checking bisimilarity of two given graphs. Unfortunately, we can not use this ideas to check if attributed graphs are bisimilar, i.e. graphs whose nodes or edges are labelled with values from some given data algebra and where graph transformation involves computation on that algebra. The problem is that, in the case of attributed graphs, borrowed context transformation may be infinitely branching. In this paper, based on borrowed context transformation of what we call symbolic graphs, we present a sound and relatively complete inference system for checking bisimilarity of attributed graphs. In particular, this means that, if using our inference system we are able to prove that two graphs are bisimilar then they are indeed bisimilar. Conversely, two graphs are not bisimilar if and only if we can find a proof saying so, provided that we are able to prove some formulas over the given data algebra. Moreover, since the proof system is complex to use, we also present a tableau method based on the inference system that is also sound and relatively complete.

Keywords: Attributed graph transformation, symbolic graph transformation, borrowed contexts, bisimilarity.

1 Introduction

Bisimilarity [18] is a core concept in Computer Science and, thus, it has been studied in very different contexts, especially in the framework of process calculi. However, the case where processes include data has received relatively little attention. We think that there are two main reasons for this. On the one hand, abstracting from data allows us to

* This work has been partially supported by the CICYT project (ref. TIN2007-66523) and by the AGAUR grant to the research group ALBCOM (ref. 00516).

** Supported by a Study Leave from University of Leicester.

F. Pfenning (Ed.): FOSSACS 2013, LNCS 7794, pp. 113–128, 2013.

concentrate better on the study of communication and interaction. On the other hand, in general, bisimilarity is already undecidable. Hence, adding values and computation will not only add another source of undecidability, but also of incompleteness, if the data domain is rich enough.

Borrowed context (BC) graph transformation [6] is a technique developed by Ehrig and Koenig to define bisimilarity congruences from reduction semantics defined by graph transformation. This means that, for instance, this technique can be used for defining bisimilarity congruences for process calculi whose operational semantics can be defined by graph transformation (as e.g. CCS [1], the π-calculus [7], or the ambient calculus [2]). As usual in the area of graph transformation [4], the results in [6] apply to all kinds of graphs that form a category that is M-adhesive [13,5], i.e. most classes of graphical structures. In [6] they also show how this technique can be used for checking bisimilarity of two given graphs. Unfortunately, even if attributed graphs (i.e. graphs whose nodes or edges are labelled with values from some given data algebra and where graph transformation involves computation on that algebra) are an M-adhesive category, their techniques can not be used for checking bisimilarity of this kind of graphs, because BC transformation may be infinitely branching.

In this paper, using BC transformation, but applied to a class of *symbolic graphs*, we present an inference system for checking bisimilarity of attributed graphs. The key issue is that, using symbolic graphs, we can decouple the proof of properties about the graph structure of the given graphs from the proof of properties of data and computations, in a similar way that constraint logic programming [12] decouples computation or constraint solving from deduction. The paper builds on [15], where we showed that bisimilarity of attributed graphs is in a way equivalent to a relation, which we call s-bisimilarity, of symbolic graphs. However, in [15] it was unclear how we could use those results to check bisimilarity, since the notion of s-bisimilarity is somewhat involved.

Our inference system is shown to be sound and refutationally complete. This means that, if using our inference system we are able to prove that two graphs are bisimilar, then they are indeed bisimilar. Conversely, two graphs are not bisimilar if and only if we can find a proof saying so, provided that we are able to prove some formulas over the given data algebra. In this sense, it could be better said that our inference system is relatively complete. In addition, since it may be not obvious how to use this inference system, we also present a related tableau method that is also sound and complete.

The paper is organized as follows. In Sections 2 and 3, we introduce borrowed context transformation and attributed and symbolic graphs. In Section 4, we recall the main results from [15]. Sections 5 and 6 are devoted to present the inference system and the tableau method. Finally, in Section 7 we review related work and draw some conclusions.

2 Graph Transformation with Borrowed Contexts

Graph transformation is a powerful approach to describe local computations on systems whose states can be described by graphs. In our context, transformations are specified by rules $p : L \stackrel{l}{\leftarrow} K \stackrel{r}{\rightarrow} R$, which are spans of graph inclusions (or, in general, of some kind of monomorphisms).

A rule p can be applied to a graph G if there is a *match* monomorphism $m : L \to G$ such that pushout (1) on the right exists. The result is the transformation $G \Longrightarrow_{p,m} H$ (or just $G \Longrightarrow H$ if p and m are implicit), where H is defined by the diagram on the right and (2) is also a pushout.

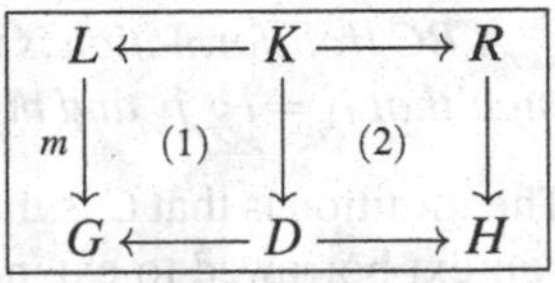

Intuitively, the *pushout complement* D is obtained by deleting from G the images through m of all the elements (nodes and edges) in L which are not in K, and H is obtained by adding to D all the elements in R that are not in K.

Graph transformation with borrowed contexts [6] is a technique that allows us to study the behavior of systems described by graph transformation. In particular, it allows us to analyze how a graph can evolve when embedded in different contexts for a given set of transformation rules.

The first idea behind this technique is that we have to specify explicitly what is the *open* (or visible) part of the given graph G, i.e. what part of G can be extended by a context. This is called the *interface* of the graph and it may be any arbitrary subgraph of G. This means that a graph with interface is an inclusion or, in general, a monomorphism $J \to G$. Then, a context should be a graph with two interfaces $J \to C \leftarrow J'$, so that, when we embed $J \to G$ in the context $J \to C \leftarrow J'$, the result is a graph $J' \to C[G]$, where $C[G]$ is obtained gluing G and C by a pushout, as shown on the diagram on the right.

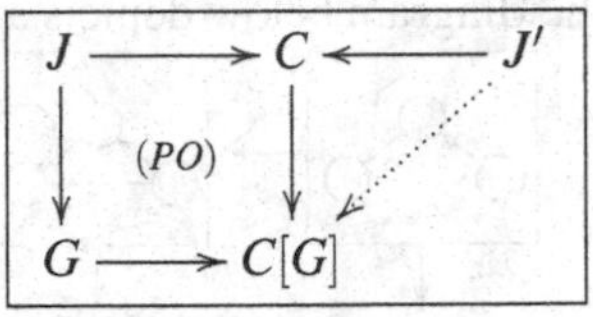

Then, we can model the behavior of a graph G by extending it with minimal contexts allowing the application of the given rules. This means that, to apply a rule $p : L \leftarrow K \to R$, we look for a *partial match* of L in G and add to G the missing part of L, so that we can apply a standard transformation via p. As this context is the part of L that has not been matched with G, we say that *G borrows this context* from the rule. We consider these transformations as transitions labelled by the context borrowed. However, some BC transformations are not useful for studying the behavior of a graph. This is the case of *independent* transformations, where the partial match is included in the part of the interface that remains invariant after the transformation [6].

Definition 1. *Given a graph with interface $J \to G$ and a graph transformation rule $p : L \leftarrow K \to R$, we say that there is a transition from $J \to G$ to $I \to H$ with label $J \to F \leftarrow I$, denoted $(J \to G) \xrightarrow{J \to F \leftarrow I}_{p,m} (I \to H)$ (or just $(J \to G) \xrightarrow{J \to F \leftarrow I} (I \to H)$, if the partial match m and the rule p can remain implicit) if there are graphs C, G^+, D and additional morphisms such that all the squares in the diagram below are pushouts (PO) or pullbacks (PB) and all the morphisms are injective:*

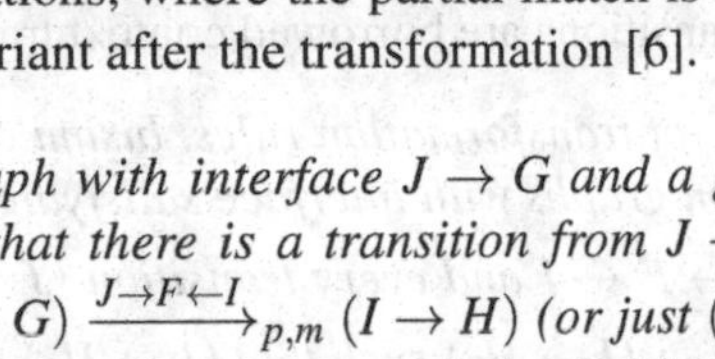

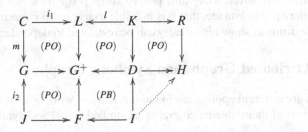

A BC transformation is independent *if there are morphisms* $j_1 : C \to K$ *and* $j_2 : C \to J$ *such that* $i_1 = l \circ j_1$ *and* $m = i_2 \circ j_2$.

The intuition is that C is the subgraph of L that completely matches G; $J \to F \leftarrow I$ is the context borrowed to extend G; G^+ is the graph G enriched with the borrowed context, and H is the result of the transformation. More precisely, F, defined as the pushout complement (if it exists) of the left lower square, extends J with all the elements in G^+ which are not in G. For instance, given the rule below on the left, and the graph with interface $J \to G$ below on the right

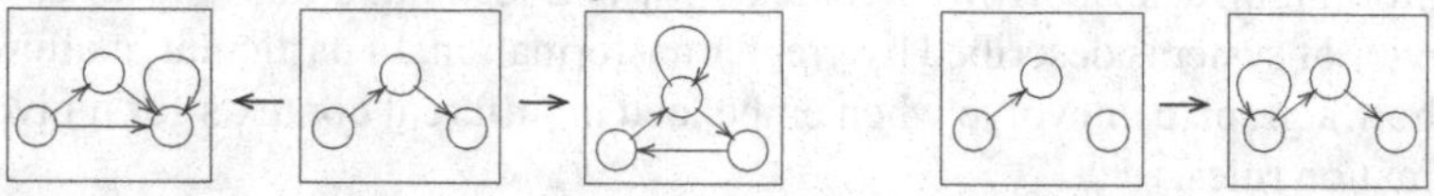

the diagram below depicts a BC transformation of $J \to G$ using that rule.

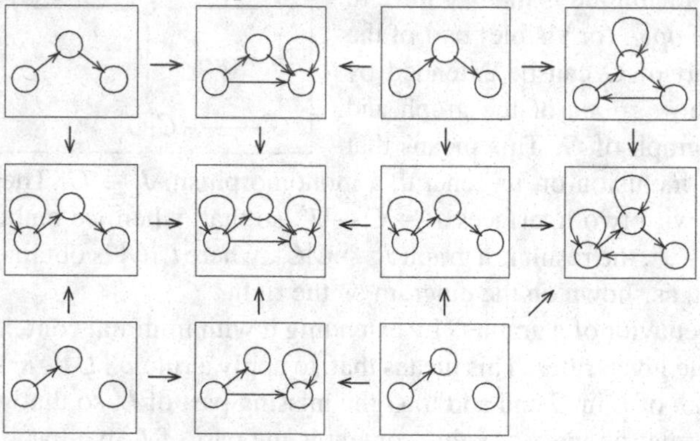

Bisimilarity is the largest symmetric relation between states that is compatible with their observational behaviour. This means that if two states s_1 and s_2 are bisimilar then for every transition from s_1 labelled with ℓ there should be a transition from s_2 with the same label such that the resulting states should again be bisimilar. In our case, states are graphs with interface and transitions are borrowed context transformations.

Definition 2. *Given a set* $\mathcal{T}$ *of transformation rules, bisimilarity, denoted by* $\sim$*, is the largest symmetric relation on graphs with interface satisfying that if* $(J \to G_1) \sim (J \to G_2)$*, for every label* $\ell = J \to F \leftarrow I$ *and every transition* $(J \to G_1) \xrightarrow{\ell} (I \to H_1)$ *there exists a transition* $(J \to G_2) \xrightarrow{\ell} (I \to H_2)$ *such that* $(I \to H_1) \sim (I \to H_2)$.

Ehrig and König [6] proved that bisimilarity is a congruence, providing a relatively simple technique for deriving bisimulation congruences out of a (graph transformation) reduction semantics. They also proved some properties that are useful for checking bisimilarity, for instance, that it is possible to use *up to context* techniques [20] or that the condition to show bisimilarity can be restricted to dependent transformations.

3 Attributed Graphs and Symbolic Graphs

There are different approaches in the literature to work with attributed graphs. We consider two of them: attributed graphs as studied in [4] and symbolic graphs [16]. They are both defined as a special kind of labeled graphs called E-graphs (e.g., see [4]). An attributed graph G, in the sense of [4], consists of two parts: an algebra $\mathcal{A}$, and an E-graph EG, where the labels of EG are the values of $\mathcal{A}$. Similarly, an attributed graph morphism $h : \langle EG, \mathcal{A}\rangle \to \langle EG', \mathcal{A}'\rangle$ consists of two parts: an algebra homomorphism h_{alg} and an E-graph morphism h_{gr}, such that they are compatible, meaning that, for every value v in $\mathcal{A}$, $h_{alg}(v) = h_{gr}(v)$. Attributed graphs and morphisms form the category **AttGraphs**, which is M-adhesive [4].

Attributed graph transformation rules are usually defined as spans $p : L \leftarrow K \rightarrow R$, where L, K and R are attributed graphs over a term algebra $T_\Sigma(X)$. A match morphism $m : L \to G$, where G is an attributed graph over a Σ-algebra $\mathcal{A}$ must bind each term t in $T_\Sigma(X)$ (and, in particular, each variable in X) to some element in $\mathcal{A}$. The fact that m_{alg} must be a homomorphism ensures that $m(t)$ must be the result of the evaluation of t, after replacing every variable x in t by $m(x)$.

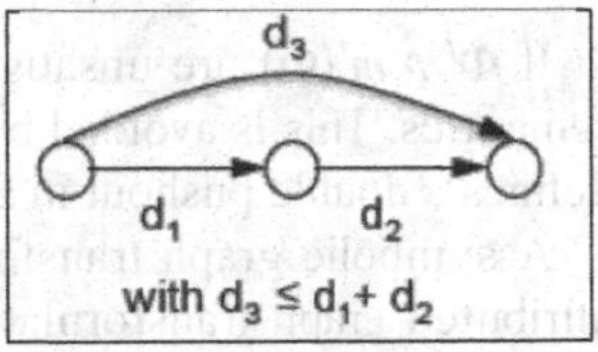

We also work with symbolic graphs because we use them as a tool for checking bisimilarity of attributed graphs. Intuitively, a symbolic graph may be seen as a graph that specifies a class of attributed graphs sharing the same data algebra. In particular, a symbolic graph SG over the algebra $\mathcal{A}$ is an E-graph G, whose labels are variables from a given set X, together with a *condition* (i.e., a first-order formula) Φ over these variables and over the elements in $\mathcal{A}$. For instance, the graph on the right specifies a class of attributed graphs, including distances in the edges, that satisfy the well-known triangle inequality. The intuition is that each substitution $\sigma : X \to \mathcal{A}$, such that $\mathcal{A} \models \sigma(\Phi)$, defines an attributed graph in the semantics of SG, obtained replacing each variable x in G by the corresponding data value $\sigma(x)$. Formally, the semantics of SG is defined:

$$Sem(SG) = \{\langle \sigma(G), \mathcal{A}\rangle \mid \mathcal{A} \models \sigma(\Phi)\}$$

To enhance readability, we refer to the attributed graphs in the semantics of SG just as $\sigma(SG)$, leaving the algebra $\mathcal{A}$ implicit. Moreover, for (technical) simplicity, we assume that in our symbolic graphs no variable is bound to two different elements of the graph. It should be clear that this is not a limitation since it is enough to replace each repeated occurrence of a variable x by a fresh variable y, and to include the equality $x = y$ in the associated formula.

Every attributed graph may be seen as a symbolic graph by just replacing all its values by variables, and by including, for each value v in the graph, an equation $x_v = v$, in the corresponding condition Φ, where x_v is the variable that has replaced the value v. We call this kind of symbolic graphs *grounded symbolic graphs*. In particular, $GSG(G)$ denotes the grounded symbolic graph defined by G.

A morphism $h : \langle G_1, \Phi_1\rangle \to \langle G_2, \Phi_2\rangle$ is a graph morphism $h : G_1 \to G_2$ such that $\mathcal{A} \models \Phi_2 \Rightarrow h(\Phi_1)$, where $h(\Phi_1)$ is the formula obtained when replacing in Φ_1 every variable x_1 in the set of labels of G_1 by $h(x_1)$. Symbolic graphs and morphisms over a given data algebra $\mathcal{A}$ form the category $\mathbf{SymbGraph}_{\mathcal{A}}$, which is M-adhesive [16].

In this paper, a *symbolic graph transformation rule* is a pair $\langle L \hookleftarrow K \hookrightarrow R, \Phi\rangle$, where L, K and R are graphs over a set of variables X and Φ is a condition over X and over the elements in $\mathcal{A}$. We consider that a rule is a span of symbolic graph inclusions $\langle L, \mathbf{true}\rangle \hookleftarrow \langle K, \mathbf{true}\rangle \hookrightarrow \langle R, \Phi\rangle$. Intuitively, Φ defines applicability conditions and relates the attributes in the left and right-hand side of the rule. As usual, we can define the application of a graph transformation rule $\langle L \hookleftarrow K \hookrightarrow R, \Phi\rangle$ by a double pushout in the category of symbolic graphs [17].

Definition 3. *Given a transformation rule* $p = \langle L \hookleftarrow K \hookrightarrow R, \Phi\rangle$ *over a data algebra* $\mathcal{A}$ *and a morphism* $m : L \to G$, $\langle G, \Phi'\rangle \Longrightarrow_{r,m} \langle H, \Phi' \wedge m'(\Phi)\rangle$ *if* (1) *and* (2) *are pushouts and* $\Phi' \wedge m'(\Phi)$ *is satisfiable in* $\mathcal{A}$.

$$\begin{array}{ccccc} L & \hookleftarrow & K & \hookrightarrow & R \\ {\scriptstyle m}\downarrow & (1) & \downarrow & (2) & \downarrow{\scriptstyle m'} \\ G & \hookleftarrow & D & \hookrightarrow & H \end{array}$$

If $\Phi' \wedge m'(\Phi)$ are unsatisfiable, the resulting graph $\langle H, \Phi' \wedge m'(\Phi)\rangle$ has an empty semantics. This is avoided by requiring $\Phi' \wedge m'(\Phi)$ satisfiable. The above construction defines a double pushout in $\mathbf{SymbGraph}_{\mathcal{A}}$ [17].

A symbolic graph transformation rule can be seen as a specification of a class of attributed graph transformation rules. More precisely, we may consider that the rule $p = \langle L \hookleftarrow K \hookrightarrow R, \Phi\rangle$ denotes the class of all rules $\sigma(L) \hookleftarrow \sigma(K) \hookrightarrow \sigma(R)$, where σ is a substitution such that $\mathcal{A} \models \sigma(\Phi)$, i.e.:

$$Sem(p) = \{\sigma(L) \hookleftarrow \sigma(K) \hookrightarrow \sigma(R) \mid \mathcal{A} \models \sigma(\Phi)\}$$

It is not difficult to see [16] that given a rule p and a symbolic graph SG, $SG \Longrightarrow_p SG'$ if for every graph $G \in Sem(SG)$, $G \Longrightarrow_{p'} G'$, with $G' \in Sem(SG')$ and $p' \in Sem(p)$. Vice versa for every $G' \in Sem(SG')$, there is a graph $G \in Sem(SG)$ and a rule $p' \in Sem(p)$ such that $G \Longrightarrow_{p'} G'$.

4 Bisimilarity of Attributed Graphs and S-bisimilarity

Checking bisimilarity of attributed graphs, using directly the notions presented in Section 2, faces a main problem: given an attributed graph with interface $J \to G$ and a finite set of transformation rules, there may exist an infinite number of different transitions $(J \to G) \xrightarrow{\ell} (I \to H)$. For instance, in the example in Section 6, the borrowed context application of any of the given rules to any of the given graphs would require the assignment of a value to the variable x. Hence we would have an infinite number of possible matches, each of them corresponding to each different value.

We may think that we may avoid this infinite branching by using symbolic graph transformation, where we are not forced to substitute every variable in the interface. So that for deciding if two attributed graphs are bisimilar we could check if their associated grounded graphs are bisimilar in the category of symbolic graphs. Unfortunately, in [15] we proved that two attributed graphs may be bisimilar as attributed graphs, while their associated grounded symbolic graphs are not bisimilar as symbolic graphs.

However, in [15] we also proved that the following notion of S-bisimilarity over symbolic graphs could be used for proving bisimilarity of attributed graphs.

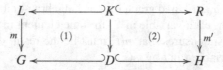

Definition 4. *S-bisimilarity, $\sim_S$, is the largest symmetric relation on symbolic graphs with interface satisfying that if $(J \to SG_1) \sim_S (J \to SG_2)$ then for every dependent transition $(J \to SG_1) \xrightarrow{\ell} (I \to SG'_1)$, with $SG'_1 = \langle G'_1, \Phi'_1 \rangle$ there exists a family of conditions $\{\Psi_i\}_{i \in \mathcal{I}}$ and a family of transitions $\{(J \to SG_2) \xrightarrow{\ell} (I \to SH_i)\}_{i \in \mathcal{I}}$, with $SH_i = \langle H_i, \Pi_i \rangle$ such that:*

1. *For every substitution σ'_1 such that $\mathcal{A} \models \sigma'_1(\Phi'_1)$, there is an index i and a substitution σ_i such that $\mathcal{A} \models \sigma_i(\Psi_i \wedge \Pi_i)$ and $\sigma'_1|_I = \sigma_i|_I$, where $\sigma|_I$ denotes the restriction of σ to the variables in I.*
2. *For every i, $(I \to \langle G'_1, \Phi'_1 \wedge \Psi_i \rangle) \sim_S (I \to \langle H_i, \Pi_i \wedge \Psi_i \rangle)$.*

Moreover, given a label ℓ, we write $(J \to SG_1) \sim_S^{\ell} (J \to SG_2)$ if for every dependent transition $(J \to SG_1) \xrightarrow{\ell} (I \to SG'_1)$ there exists a family of conditions $\{\Psi_i\}_{i \in \mathcal{I}}$ and a family of transitions $\{(J \to SG_2) \xrightarrow{\ell} (I \to SH_i)\}_{i \in \mathcal{I}}$, with $SH_i = \langle H_i, \Pi_i \rangle$ such that conditions 1 *and* 2 *above hold.*

The definition of S-bisimilarity is easy to understand if we think that every symbolic transition $tr = (J \to SG_1) \xrightarrow{\ell} (I \to SG'_1)$ denotes a family of attributed transitions. In particular, every substitution σ'_1 of the variables in SG'_1 such that $\mathcal{A} \models \sigma(\Phi'_1)$ denotes an attributed transition $\sigma'_1(tr) = \sigma'_1(J \to G_1) \xrightarrow{\sigma'_1(\ell)} \sigma'_1(I \to SG'_1)$. Then, each condition Ψ_i should characterize which attributed transitions denoted by tr are simulated by an attributed transition denoted by $tr'_i = (J \to SG_2) \xrightarrow{\ell} (I \to SH_i)$. In this context, conditions 1 and 2 just state that each $\sigma'_1(tr)$ must be simulated by some attributed transition denoted by tr'_i, for some i. Then, as said above, we have:

Theorem 1. *[15] Given transformation rules $\mathcal{T}$, $(J \to G_1) \sim (J \to G_2)$ with respect to $Sem(\mathcal{T})$ if and only if $GSG(J \to G_1) \sim_S GSG(J \to G_2)$ with respect to $\mathcal{T}$.*

In [15] we also proved that S-bisimilarity is a congruence and that up-to-context techniques can also be applied in this setting.

5 An Inference System for Proving Bisimilarity

The results in [15], and in particular Theorem 1, provide a convenient characterization of the bisimilarity relation for attributed graphs that avoids the infinite branching problem associated to the direct application of the results in [6]. However, it is not obvious how this characterization can be actually used for checking bisimilarity. In particular, the main problem is to find the conditions Ψ_i that are needed, according to Def. 4, to play the bisimulation game. Below, we present seven inference rules that describe implicitly how we can compute these conditions.

The judgements that we use in our rules are constrained sequents of the form $\Gamma \vdash (J \to SG_1) \sim_S (J \to SG_2)[\Psi^+, \Psi^-]$ or $\Gamma \vdash (J \to SG_1) \sim_S^{\ell} (J \to SG_2)[\Psi^+, \Psi^-]$, where:

- The antecedent Γ is the context, i.e. a set of facts $(I \to SG) \sim_S (I \to SG')$ that we assume to hold. Contexts are used for *up-to* inference steps.

- The only common variables of SG_1 and SG_2 are the variables in J.
- The succedent $(J \to SG_1)\mathcal{R}(J \to SG_2)[\Psi^+,\Psi^-]$, where $\mathcal{R}$ is either $\sim_S$ or $\sim_S^\ell$ and where Ψ^+ and Ψ^- are formulas including the variables in SG_1 and SG_2, is a statement whose intended meaning is:
  - Ψ^+ is a formula where all its variables not in SG_1 or in SG_2 are (implicitly) quantified universally, such that if it holds then $(J \to SG_1 \wedge \Psi^+)\mathcal{R}(J \to SG_2 \wedge \Psi^+)$ must hold.
  - If Ψ^- is satisfiable then $(J \to SG_1)\mathcal{R}(J \to SG_2)$ does not hold.

  where, if $SG = \langle G, \Phi \rangle$, $SG \wedge \Psi$ denotes the symbolic graph $\langle G, \Phi \wedge \Psi \rangle$.

As a consequence, if we want to check if two attributed graphs, $J \to G$ and $J \to G'$ are bisimilar, and if Φ and Φ' are the conditions of $GSG(G)$ and $GSG(G')$, respectively, we will try to infer judgements of the form $\emptyset \vdash GSG(J \to G) \sim_S GSG(J \to G')[\Psi^+,\Psi^-]$, where $\emptyset$ is the empty context. If Φ and Φ' imply Ψ^+ then we would conclude that $J \to G$ and $J \to G'$ are bisimilar. The reason is that if Φ and Φ' imply Ψ^+, then $GSG(J \to G) = (GSG(J) \to GSG(G) \wedge \Psi^+) \sim_S (GSG(J) \to GSG(G') \wedge \Psi^+) = GSG(J \to G')$ and, by Thm. 1, $(J \to G) \sim (J \to G')$. However, if Ψ^- is satisfiable, also by Thm. 1, we would conclude that $J \to G$ and $J \to G'$ are not bisimilar.

The first rule is just a consequence of how the relation $\sim_S^\ell$ is defined. In particular the rule says that if for each label ℓ, $(J \to SG_1) \sim_S^\ell (J \to SG_2)$ under the condition Ψ_ℓ^+, then $(J \to SG_1) \sim_S (J \to SG_2)$ under the conjunction of all the Ψ_ℓ^+. Conversely, if for each label ℓ, $(J \to SG_1) \not\sim_S^\ell (J \to SG_2)$ under the condition Ψ_ℓ^-, then $(J \to SG_1) \not\sim_S (J \to SG_2)$ under the disjunction of all the Ψ_ℓ^-.

1. Labels

$$\frac{\Gamma \vdash (J \to SG_1) \sim_S^{\ell_1} (J \to SG_2)[\Psi_{\ell_1}^+,\Psi_{\ell_1}^-] \quad \dots \quad \Gamma \vdash (J \to SG_1) \sim_S^{\ell_n} (J \to SG_2)[\Psi_{\ell_n}^+,\Psi_{\ell_n}^-]}{\Gamma \vdash (J \to SG_1) \sim_S (J \to SG_2)[\bigwedge_{i=1}^{n} \Psi_{\ell_i}^+, \bigvee_{i=1}^{n} \Psi_{\ell_i}^-]}$$

If $\{\ell_1,\dots,\ell_n\}$ is the set of all labels ℓ such that there is a dependent transformation $(J \to SG_1) \stackrel{\ell}{\to} (I \to SG_1')$ or $(J \to SG_2) \stackrel{\ell}{\to} (I \to SG_2')$.

If two graphs are equal then they are obviously bisimilar. However, if their underlying E-graphs are equal, but their conditions are different, the rule below tells us that the two graphs are bisimilar under the conjunction of their associated conditions.

2. Equality

$$\Gamma \vdash (J \to \langle G, \Phi \rangle) \sim_S (J \to \langle G, \Phi' \rangle)[\Phi \wedge \Phi', \mathbf{false}]$$

A trivial rule that is needed for technical reasons in the completeness proof:

3. Trivial

$$\Gamma \vdash (I \to SG) \sim_S^\ell (I \to SG')[\mathbf{false}, \mathbf{false}]$$

The fourth rule is also quite simple. Let us assume that $Cond(SG,\ell)$ is the condition that covers all possible transitions of SG with label ℓ, i.e.

$$Cond(SG,\ell) = \bigvee_{p,m} \Phi_{p,m},$$

such that $(J \to SG) \xrightarrow{\ell}_{p,m} (I \to \langle G', \Phi_{p,m}\rangle)$. Then, if $\neg Cond(SG,\ell)$ holds, no transition of SG with label ℓ is possible. Therefore, if $\neg Cond(SG_1,\ell) \wedge \neg Cond(SG_2,\ell)$ holds no transition with label ℓ is possible of neither SG_1 nor SG_2. Thus, under that condition they are ℓ-bisimilar. Conversely, when $(Cond(SG_1,\ell) \setminus Cond(SG_2,\ell)) \vee (Cond(SG_2,\ell) \setminus Cond(SG_1,\ell))$ holds, either there is a transition with label ℓ from SG_1, but not from SG_2, or vice versa, meaning that they are not ℓ-bisimilar.

4. Complement

$$\Gamma \vdash (J \to SG_1) \sim^{\ell}_{S} (J \to SG_2)[\Psi^+, \Psi^-]$$

where

$$\begin{aligned}\Psi^+ &= \neg Cond(SG_1,\ell) \wedge \neg Cond(SG_2,\ell)\\ \Psi^- &= (Cond(SG_1,\ell) \setminus Cond(SG_2,\ell)) \vee\\ &\quad (Cond(SG_2,\ell) \setminus Cond(SG_1,\ell))\end{aligned}$$

The next rule states that if $(J \to SG_1)$ and $(J \to SG_2)$ are bisimilar when Ψ_1^+ holds and, also, when Ψ_2^+ holds, then they are bisimilar when either of them hold. Conversely, if $(J \to SG_1)$ and $(J \to SG_2)$ are not bisimilar when Ψ_1^- is satisfiable and also when Ψ_2^- is satisfiable, then if any of them are satisfiable the two graphs are not bisimilar.

5. Disjunction

$$\frac{\Gamma \vdash (J \to SG_1) \sim^{\ell}_{S} (J \to SG_2)[\Psi_1^+, \Psi_1^-] \quad \Gamma \vdash (J \to SG_1) \sim^{\ell}_{S} (J \to SG_2)[\Psi_2^+, \Psi_2^-]}{\Gamma \vdash (J \to SG_1) \sim^{\ell}_{S} (J \to SG_2)[\Psi_1^+ \vee \Psi_2^+, \Psi_1^- \vee \Psi_2^-]}$$

The following rule is a bit more involved. It essentially follows from the definition of $\sim^{\ell}_{S}$. If $(J \to SG) \xrightarrow{\ell}_{(p,m)} (I \to SH_{(p,m)})$, the disjunction of all the conditions associated with the transformations $(J \to SG') \xrightarrow{\ell}_{(p',m')} (I \to SH'_{(p',m')})$ that are bisimilar to $(I \to SH_{(p,m)})$ should cover $\Phi_{(p,m)}$. But, in general, we cannot ensure this. We can only ensure that, under the condition $\Psi^+_{(p,m)} = \bigvee_{(p',m')} \Psi^+_{(p,m),(p',m')}$, the attributed transitions denoted by $\xrightarrow{\ell}_{(p,m)}$ are simulated by transitions denoted by $\xrightarrow{\ell}_{(p',m')}$. This means that under the condition $\Phi_{(p,m)} \wedge \Psi^+_{(p,m)}$ the transition $(J \to SG) \xrightarrow{\ell}_{(p,m)} (I \to SH_{(p,m)})$ is simulated by transitions from $(J \to SG')$. On the other hand, it may happen that on the condition $\Phi_{(p,m)} \setminus \Psi^+_{(p,m)}$ the transition $(J \to SG) \xrightarrow{\ell}_{(p,m)} (I \to SH_{(p,m)})$ is not simulated by any transition from $(J \to SG')$. Hence, if $\Phi_{(p,m)} \setminus \Psi^+_{(p,m)}$ holds, we cannot ensure that $(J \to SG) \sim^{\ell}_{S} (J \to SG')$. Since this is true for each (p,m), all ℓ- transitions from $(J \to SG)$ are simulated by ℓ'- transitions from $(J \to SG')$ when any of the conditions $\Phi_{(p,m)} \wedge \Psi^+_{(p,m)}$ holds, unless any of the conditions $\Phi_{(p,m)} \setminus \Psi^+_{(p,m)}$ holds, and vice versa for the ℓ- transitions from $(J \to SG')$. Altogether, this means that we can ensure that $(J \to SG) \sim_S^\ell (J \to SG')$ on the condition Ψ^+ as defined in the rule.

Conversely, if $\Psi^-_{(p,m),(p',m')}$ is satisfied then $(J \to SG) \xrightarrow{\ell}_{(p,m)} (I \to SH_{(p,m)})$ is not simulated by $(J \to SG') \xrightarrow{\ell}_{(p',m')} (I \to SH'_{(p',m')})$. So, if the conjunction of conditions $\Psi^-_{(p,m)} = \bigwedge_{(p',m')} \Psi^-_{(p,m),(p',m')}$ is satisfied then $(J \to SG) \xrightarrow{\ell}_{(p,m)} (I \to SH_{(p,m)})$ is not simulated by any ℓ-transition from $(J \to SG')$. But this means that if any of the conditions $\Psi^-_{(p,m)}$ is satisfied then no transition from $(J \to SG)$ can be simulated, and something similar happens with respect to $(J \to SG')$. In short, this means that we can ensure that if Ψ^-, as defined in the rule, is satisfied then $(J \to SG) \nsim_S^\ell (J \to SG')$.

Finally, the rule also states that, when proving $(I \to SH_{(p,m)}) \sim_S (I \to SH'_{(p',m')})$ we may assume that $(J \to SG) \sim_S (J \to SG')$ already holds, so that we can use up-to-context techniques that have been shown valid for S-bisimilarity [15].

6. Bisimulation

$$\frac{\begin{array}{c}\Gamma \cup \{(J \to SG) \sim_S (J \to SG')\} \vdash \\ \bigwedge_{(p,m),(p',m')} (I \to SH_{(p,m)}) \sim_S (I \to SH'_{(p',m')})[\Psi^+_{(p,m),(p',m')}, \Psi^-_{(p,m),(p',m')}]\end{array}}{\Gamma \vdash (J \to SG) \sim_S^\ell (J \to SG')[\Psi^+, \Psi^-]}$$

For all rules p, p' and partial matches m, m' such that $(J \to SG) \xrightarrow{\ell}_{(p,m)} (I \to SH_{(p,m)})$ and $(J \to SG') \xrightarrow{\ell}_{(p',m')} (I \to SH'_{(p',m')})$, and where, if $SH_{(p,m)} = \langle H_{(p,m)}, \Phi_{(p,m)} \rangle$ and $SH'_{(p',m')} = \langle H'_{(p',m')}, \Phi'_{(p',m')} \rangle$, then Ψ^+, Ψ^- are defined:

$$\Psi^+ = \Big(\bigvee(\Phi_{(p,m)} \wedge \Psi^+_{(p,m)}) \setminus \bigvee(\Phi_{(p,m)} \setminus \Psi^+_{(p,m)})\Big) \wedge$$
$$\Big(\bigvee(\Phi_{(p',m')} \wedge \Psi^+_{(p',m')}) \setminus \bigvee(\Phi_{(p',m')} \setminus \Psi^+_{(p',m')})\Big)$$
$$\Psi^- = (\bigvee(\Psi^-_{(p,m)} \wedge \Phi_{(p,m)}) \vee \bigvee(\Psi^-_{(p',m')} \wedge \Phi_{(p',m')}))$$

and where

$$\Psi^+_{(p,m)} = \bigvee_{(p',m')} \Psi^+_{(p,m),(p',m')} \qquad \Psi^+_{(p',m')} = \bigvee_{(p,m)} \Psi^+_{(p,m),(p',m')}$$
$$\Psi^-_{(p,m)} = \bigwedge_{(p',m')} \Psi^-_{(p,m),(p',m')} \qquad \Psi^-_{(p',m')} = \bigwedge_{(p,m)} \Psi^-_{(p,m),(p',m')}$$

The last rule is based on the result from [15] that shows that the up to context technique is sound for proving S-bisimilarity. This means that, when trying to prove $(J \to SG) \sim_S (J \to SG')$, we may assume that for all contexts $J \to F \leftarrow I$: $(I \to F[SG_1]) \sim_S (I \to F[SG_2])$. That is that if $(J \to SG) \sim_S (J \to SG')$ is part of the context, then we could infer $(I \to F[SG_1]) \sim_S (I \to F[SG_2])[\mathbf{true}, \mathbf{false}]$. But this can be generalized to the case where the judgement to infer does not exactly include $F[SG_1]$ and $F[SG_2]$, but $(F[SG_1] \wedge \Phi)$ and $(F[SG_2] \wedge \Phi')$ as the rule shows:

7. Up-to-context

$$\Gamma \cup \{(J \to SG) \sim_S (J \to SG')\} \vdash (I \to SH) \sim_S (I \to SH')[\neg\Phi_1 \wedge \neg\Phi_1', \mathbf{false}]$$

where, $SH = \langle H, \Phi \vee \Phi_1 \rangle$ and $SH' = \langle H', \Phi' \vee \Phi_1' \rangle$, and $\langle H, \Phi \rangle$ and $\langle H', \Phi' \rangle$ are the result of embedding SG and SG', respectively, in a context $J \to F \leftarrow I$.

We can prove that the above rules are sound and complete. More precisely:

Theorem 2 (Soundness of the inference rules). *Given attributed graphs $J \to G_1$ and $J \to G_2$, then:*

- *If we can infer $\emptyset \vdash (J \to GSG(G_1)) \sim_S (J \to GSG(G_2))[\Psi^+, \Psi^-]$, and $\Phi_{GSG(G_1)} \wedge \Phi_{GSG(G_2)}$ implies Ψ^+ in $\mathcal{A}$ then $J \to G_1 \sim J \to G_2$.*
- *If we can infer $\emptyset \vdash (J \to GSG(G_1)) \sim_S (J \to GSG(G_2))[\Psi^+, \Psi^-]$ and Ψ^- is satisfiable in $\mathcal{A}$ then $J \to G_1 \nsim J \to G_2$.*

The proof essentially follows the intuitions of the rules that are given above.

Theorem 3 (Completeness of the inference rules). *Given attributed graphs $J \to G_1$ and $J \to G_2$, if $(J \to G_1) \nsim (J \to G_2)$ then, using the above rules, we can infer $\emptyset \vdash (J \to GSG(G_1)) \sim_S (J \to GSG(G_2))[\Psi^+, \Psi^-]$, where $\emptyset$ is the empty context and Ψ^- is a satisfiable condition.*

The proof is done by induction, using the standard definition of stratified bisimilarity [10]. This is sound, since for each $J \to G$ and each ℓ there is a finite number of transitions $(J \to SG) \xrightarrow{\ell} (I \to SH)$.

6 A Tableau Method for Checking Bisimilarity

In the previous section we have presented a set of rules for proving or disproving bisimilarity of attributed graphs. The problem with these rules is that it may not be obvious how to use them to check whether two given graphs $J \to G_1$ and $J \to G_2$ are bisimilar. In this section, we describe a method with this purpose, based on the construction of a kind of constrained tableau [8], i.e. a tableau whose nodes include constraints, following the inference rules from the previous section.

More precisely, our tableaux are trees whose nodes are labelled by formulas $(J \to SG_1) \sim_S (J \to SG_2)$ or $(J \to SG_1) \sim_S^{\ell} (J \to SG_2)$ and by constraints Ψ^+ and Ψ^-, as our judgements in the proof rules. To construct a tableau for $J \to G_1$ and $J \to G_2$, to check if they are bisimilar, we start creating the root, labelling it with $GSG((J \to G_1)) \sim_S GSG((J \to G_2))[\mathbf{false}, \mathbf{false}]$. Then, we start with an iteration where, at each step, we choose a node in the tableau and we apply to it either an expansion step enlarging the tree (just when the node is a leaf), or a constraint computation step changing the constraints of the node. We stop when the tableau is *closed*, i.e. either when $\Phi_{GSG(G_1)}$ and $\Phi_{GSG(G_2)}$ imply Ψ^+ or when Ψ^- is satisfiable in $\mathcal{A}$, where Ψ^+ and Ψ^- are the constraints in the root. In the former case we would conclude that $J \to G_1$ and $J \to G_2$ are bisimilar, and in the latter case we would conclude that they are not.

As said above, the steps for the construction of the tableau can be either *expansion* steps or *constraint computation* steps. There are two kinds of expansion steps:

1. Label Expansion If a leaf n is labelled with the formula $(J \rightarrow SG_1) \sim_S (J \rightarrow SG_2)$, we create a child of n and we label it with $(J \rightarrow SG_1) \sim_S^{\ell} (J \rightarrow SG_2)[\mathbf{false}, \mathbf{false}]$, for each ℓ such that there is a dependent transition labelled with ℓ from $(J \rightarrow SG_1)$ or from $(J \rightarrow SG_2)$.

2. Bisimulation Expansion If a leaf n is labelled with the formula $(J \rightarrow SG_1) \sim_S^{\ell} (J \rightarrow SG_2)$, for each pair of transitions $(J \rightarrow SG_1) \xrightarrow{\ell} (I \rightarrow SG_1')$ and $(J \rightarrow SG_2) \xrightarrow{\ell} (I \rightarrow SG_2')$, we create a child of n and we label it with $(I \rightarrow SG_1') \sim_S (I \rightarrow SG_2')[\mathbf{false}, \mathbf{false}]$.

There are five kinds of constraint computation steps:

3. Labels Computation If a node n is labelled with $(J \rightarrow SG_1) \sim_S (J \rightarrow SG_2)[\Pi^+, \Pi^-]$, we can compute new constraints $\Psi^+ = \Pi^+ \vee \bigwedge_{i=1}^{n} \Psi_i^+$ and $\Psi^- = \Pi^- \vee \bigvee_{i=1}^{n} \Psi_i^-$, where $\Psi_1^+, \Psi_1^- \ldots, \Psi_n^+, \Psi_n^-$ are the constraints of the descendants of that node.

4. Complement Computation If a node n is labelled with $(J \rightarrow SG_1) \sim_S (J \rightarrow SG_2)[\Psi_1^+, \Psi_1^-]$ then we can compute new constraints Ψ^+ and Ψ^- for n as follows:

$\Psi^+ = \Psi_1^+ \vee (\neg Cond(SG_1, \ell) \wedge \neg Cond(SG_2, \ell))$

$\Psi^- = \Psi_1^- \vee (Cond(SG_1, \ell) \setminus Cond(SG_2, \ell)) \vee (Cond(SG_2, \ell) \setminus Cond(SG_1, \ell))$

5. Equality Computation If a node n is labelled with $(J \rightarrow SG_1) \sim_S (J \rightarrow SG_2)[\Psi_1^+, \Psi_1^-]$, then we can compute a new constraint $\Psi^+ = \Psi_1^+ \vee (\Phi_1 \wedge \Phi_1')$ for n, leaving the negative constraint Ψ_1^- unchanged.

6. Bisimulation Computation If a node n is labelled with $(J \rightarrow SG_1) \sim_S (J \rightarrow SG_2)[\Psi_1^+, \Psi_1^-]$ then we can compute new constraints Ψ^+ and Ψ^- for n as follows:

$$\Psi^+ = \Psi_1^+ \vee \Big(\bigvee (\Phi_{(p,m)} \wedge \Psi^+_{(p,m)}) \setminus \bigvee (\Phi_{(p,m)} \setminus \Psi^+_{(p,m)}) \Big) \wedge \Big(\bigvee (\Phi_{(p',m')} \wedge \Psi^+_{(p',m')}) \setminus \bigvee (\Phi_{(p',m')} \setminus \Psi^+_{(p',m')}) \Big)$$

$$\Psi^- = \Psi_1^- \vee \big(\bigvee (\Psi^-_{(p,m)} \wedge \Phi_{(p,m)}) \vee \bigvee (\Psi^-_{(p',m')} \wedge \Phi_{(p',m')}) \big)$$

where the conditions $\Phi_{(p,m)}$, $\Psi^+_{(p,m)}$, $\Psi^-_{(p,m)}$, $\Phi_{(p',m')}$, $\Psi^+_{(p',m')}$, and $\Psi^-_{(p',m')}$ are as in the Bisimulation inference rule.

7. Up-to-Context Computation If a node n is labelled with $(J \rightarrow SG_1) \sim_S (J \rightarrow SG_2)[\Psi_1^+, \Psi_1^-]$, if there is an ancestor of n labelled with the formula $(I \rightarrow SG_2) \sim_S (I \rightarrow SG_2')$, and if there is a context $I \rightarrow F \leftarrow J$, where $F[SG_2] = \langle G_1, \Pi_1 \rangle$ and $F[SG_2'] = \langle G_1', \Pi_1' \rangle$ then we can compute a new constraint $\Psi^+ = \Psi_1^+ \vee (\neg(\Phi_1 \setminus \Pi_1) \wedge \neg(\Phi_1' \setminus \Pi_1'))$ for n, leaving unchanged the negative constraint Ψ_1^-.

Then, we have:

Theorem 4 (Soundness). *If we can construct a closed tableau for graphs $J \rightarrow G_1$ and $J \rightarrow G_2$ whose root is labelled by the constraints Ψ^+ and Ψ^-, then:*

- *If $\Phi_{GSG(G_1)} \wedge \Phi_{GSG(G_2)}$ implies Ψ^+ in $\mathcal{A}$ then $J \rightarrow G_1 \sim J \rightarrow G_2$.*
- *If Ψ^- is satisfiable in $\mathcal{A}$ then $J \rightarrow G_1 \nsim J \rightarrow G_2$.*

The proof is a direct consequence of the soundness of the inference rules presented in the previous section.

Theorem 5 (Completeness). *If $(J \to G_1) \approx (J \to G_2)$, we can construct a closed tableau for $J \to G_1$ and $J \to G_2$ whose negative constraint at the root Ψ^- is satisfiable in $\mathcal{A}$.*

The proof is very similar to the completeness proof of the inference rules.

Let us now see an example of the construction of a tableau. Suppose that we want to check if the graphs $(J \to SG_1)$ and $(J \to SG_2)$ on the right are bisimilar with respect to the rules depicted below (for simplicity, the rules are presented including only the left and right-hand sides, leaving the intermediate part implicit). Part of the tableau that we would use for this proof is shown in Fig. 1. The interfaces of the graphs are not depicted because, in the transformations considered, J (with the obvious inclusions) would be the interface of all the graphs in the tableau.

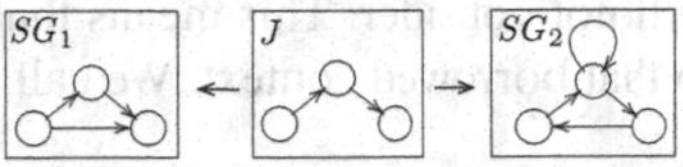

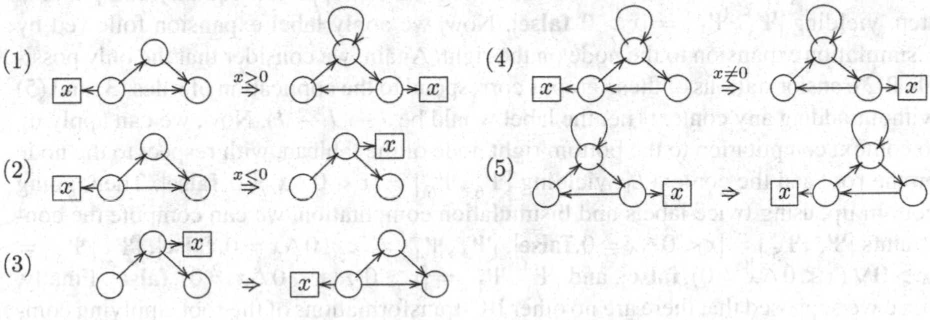

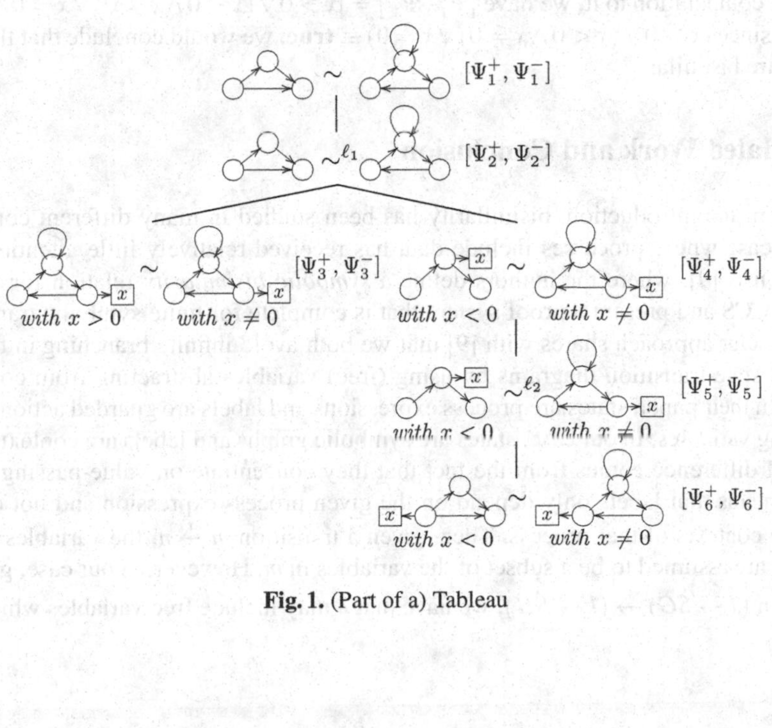

Fig. 1. (Part of a) Tableau

The construction of the tableau starts with the creation of the root and the application of a label expansion step. Due to lack of space, we suppose that we can only transform SG_1 using rules (1) and (2) and SG_2 using rule (4), and using a borrowed context consisting of the square node, together with the attribute x and an edge to the leftmost round node. Actually, there are other transformations with other borrowed contexts that we will not consider. This means that this step would create just one node, corresponding to that borrowed context. We call ℓ_1 this context (and label), which is depicted below.

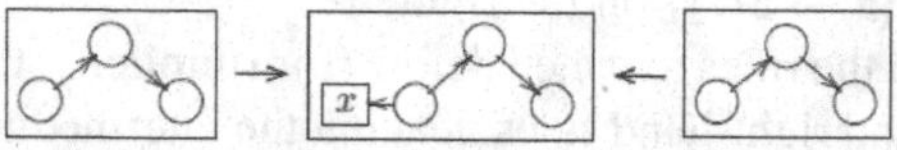

Then, we proceed with bisimulation expansion corresponding to the BC transformations mentioned above. This step creates two nodes. We can see that the graphs in the node on the left are equal (except for the condition), so we can apply an equality computation step, yielding $[\Psi_3^+, \Psi_3^-] = [x > 0, \mathbf{false}]$. Now, we apply label expansion followed by bisimulation expansion to the node on the right. Again, we consider that the only possible BC transformations of these graphs correspond to the application of rules (3) and (5) without adding any context (i.e. the label would be $J \rightarrow J \leftarrow J$). Now, we can apply up to context computation to the bottom right node of the tableau, with respect to the node on the root and the context ℓ_1, yielding $[\Psi_6^+, \Psi_6^-] = [x < 0 \wedge x \neq 0, \mathbf{false}]$. Then, going bottom up, using twice labels and bisimulation computation, we can compute the constraints $[\Psi_5^+, \Psi_5^-] = [x < 0 \wedge x \neq 0, \mathbf{false}]$, $[\Psi_4^+, \Psi_4^-] = [x < 0 \wedge x \neq 0, \mathbf{false}]$, $[\Psi_2^+, \Psi_2^-] = [x > 0 \vee (x < 0 \wedge x \neq 0), \mathbf{false}]$, and $[\Psi_1^+, \Psi_1^-] = [x > 0 \vee (x < 0 \wedge x \neq 0), \mathbf{false}]$. Finally, since we supposed that there are no other BC-transformations of the root, applying complement computation to it, we have $[\Psi_1^+, \Psi_1^-] = [x > 0 \vee (x < 0 \wedge x \neq 0) \vee x = 0, \mathbf{false}]$. To end, since $(x > 0 \vee (x < 0 \wedge x \neq 0) \vee x = 0) \equiv \mathbf{true}$, we would conclude that the two graphs are bisimilar.

7 Related Work and Conclusion

As said in the introduction, bisimilarity has been studied in many different contexts, but the case where processes include data has received relatively little attention. An exception is [9], where the authors define a *symbolic bisimilarity* relation for value-passing CCS and present a proof system that is complete for finite symbolic transition systems. Our approach shares with [9] that we both avoid infinite branching in the associated state-transition diagrams by using (free) variables abstracting from concrete values. In their paper, states are process expressions and labels are guarded actions both including variables. In our case, states are symbolic graphs and labels are contexts. The essential difference comes from the fact that they concentrate on value-passing CCS, which means that labels only depend on the given process expression and not on the possible contexts of that process. Then, given a transition $m \xrightarrow{a} n$, the variables in the action a are assumed to be a subset of the variables in m. However, in our case, given a transition $(J \rightarrow SG) \xrightarrow{\ell} (I \rightarrow SH)$, we have that ℓ may include free variables which are not in SG, representing values from the context needed for the transition. For example, in their case, if p is a ground process expression, its state-transition graph would not include any free variable either. This is not the case in our paper. In particular, the conditions Ψ_i in the definition of S-bisimilarity are needed because of the existence of these context variables. Hence, the inclusion of the constraints Ψ^+ and Ψ^- in our proof rules and in our tableau method, which are needed to compute the conditions Ψ_i, are also a consequence of these context variables.

In our framework, name-passing processes, like the processes in the π-calculus [14], can be seen as a special case of value-passing processes[1]. In that context, open bisimilarity [21] could correspond to bisimilarity of attributed graphs, as defined directly in terms of BC transformations on that category, and its symbolic version would be somewhat related to S-bisimilarity.

With respect to BC graph transformation, in [19] an algorithm for checking bisimilarity of graphs is presented, but this algorithm would not be applicable to the case of attributed graphs. Moreover, no correctness proof is included. On the other hand, in [11], the authors extend BC-transformation to the case of conditional transformation systems. Even if their results mainly apply to the case of non-attributed graphs, their notion of context transition is, in a way, related to our symbolic transitions and so they are the corresponding notions of bisimilarity. The reason is that we both deal with conditional rules and transitions defined by borrowed context transformations. This causes that, in the bisimulation game, a transition on one graph with condition A needs not to be simulated by a single transition on the other graph, but by a set of transitions with associated conditions A_i, such that these conditions cover A. The main difference, which is a substantial one, is based on the different nature of conditions. In [11], conditions are related to the structure of the graphs that we are transforming. In our case, conditions refer to properties of the attributes. Actually, both papers are orthogonal. On the one hand, dealing with application conditions would be an interesting extension of our paper. On the other hand, if the graphs considered in [11] were attributed graphs, in general, their associated state-transition diagrams would be infinitely branching, which is the problem that we address in our paper.

Finally, our tableau method for proving bisimilarity can be seen as an extension of the method presented in [3]. However, in that paper, processes do not include data, which means that their tableaux are considerably simpler. In particular they are basic unconstrained tableaux.

In this paper, we have presented a proof system and a related tableau method for checking bisimilarity of attributed graphs, using the notion of S-bisimilarity presented in [15], proving their soundness and refutational completeness. We think that the main advantages of our approach are, first, its generality, since it could be used to check bisimilarity of any kind of formalism whose semantics is expressed in terms of graph transformation; and, second, the way in which our approach decouples the proofs on the graph structure from the proofs on the given data algebra.

---

1 Obviously, the π-calculus is not a special case of value-passing CCS. However, if we represented the π-calculus in terms of attributed or symbolic graphs (e.g. similarly to [7]), we would consider names as elements of an algebra whose signature includes only the equality predicate.

References

1. Bonchi, F., Gadducci, F., König, B.: Synthesising CCS bisimulation using graph rewriting. Inf. Comput. 207(1), 14–40 (2009)
2. Bonchi, F., Gadducci, F., Monreale, G.V.: Labelled transitions for mobile ambients (as synthesized via a graphical encoding). Electr. Notes Theor. Comput. Sci. 242(1), 73–98 (2009)
3. Christensen, S., Hirshfeld, Y., Moller, F.: Bisimulation Equivalence is Decidable for Basic Parallel Processes. In: Best, E. (ed.) CONCUR 1993. LNCS, vol. 715, pp. 143–157. Springer, Heidelberg (1993)
4. Ehrig, H., Ehrig, K., Prange, U., Taentzer, G.: Fundamentals of Algebraic Graph Transformation. In: EATCS Monographs of Theoretical Comp. Sc. Springer (2006)
5. Ehrig, H., Golas, U., Habel, A., Lambers, L., Orejas, F.: M-adhesive transformation systems with nested application conditions. part 1. Math. Struct. in Com. Sc. (2012) (to appear)
6. Ehrig, H., König, B.: Deriving bisimulation congruences in the DPO approach to graph rewriting with borrowed contexts. Math. Struct. in Com. Sc. 16(6), 1133–1163 (2006)
7. Gadducci, F.: Graph rewriting for the pi-calculus. Math. Struct. in Com. Sc. 17(3), 407–437 (2007)
8. Giese, M., Hähnle, R.: Tableaux + constraints. In: TABLEAUX 2003 position paper (2003)
9. Hennessy, M., Lin, H.: Symbolic bisimulations. Theor. Comput. Sci. 138(2), 353–389 (1995)
10. Hennessy, M., Milner, R.: Algebraic laws for nondeterminism and concurrency. J. ACM 32(1), 137–161 (1985)
11. Hülsbusch, M., König, B.: Deriving Bisimulation Congruences for Conditional Reactive Systems. In: Birkedal, L. (ed.) FOSSACS 2012. LNCS, vol. 7213, pp. 361–375. Springer, Heidelberg (2012)
12. Jaffar, J., Maher, M., Marriot, K., Stuckey, P.: The semantics of constraint logic programs. The Journal of Logic Programming 37, 1–46 (1998)
13. Lack, S., Sobocinski, P.: Adhesive and quasiadhesive categories. Theor. Inf. App. 39, 511–545 (2005)
14. Milner, R., Parrow, J., Walker, D.: A calculus of mobile processes, I and II. Inf. Comput. 100(1), 1–77 (1992)
15. Orejas, F., Boronat, A., Mylonakis, N.: Borrowed Contexts for Attributed Graphs. In: Ehrig, H., Engels, G., Kreowski, H.-J., Rozenberg, G. (eds.) ICGT 2012. LNCS, vol. 7562, pp. 126–140. Springer, Heidelberg (2012)
16. Orejas, F., Lambers, L.: Symbolic attributed graphs for attributed graph transformation. ECEASST 30 (2010)
17. Orejas, F., Lambers, L.: Lazy graph transformation. Fund. Inf. 118, 65–96 (2012)
18. Park, D.: Concurrency and Automata on Infinite Sequences. In: Deussen, P. (ed.) GI-TCS 1981. LNCS, vol. 104, pp. 167–183. Springer, Heidelberg (1981)
19. Rangel, G., König, B., Ehrig, H.: Bisimulation verification for the DPO approach with borrowed contexts. ECEASST 6 (2007)
20. Sangiorgi, D.: On the Proof Method for Bisimulation. In: Hájek, P., Wiedermann, J. (eds.) MFCS 1995. LNCS, vol. 969, pp. 479–488. Springer, Heidelberg (1995)
21. Sangiorgi, D.: A theory of bisimulation for the pi-calculus. Acta Inf. 33(1), 69–97 (1996)

Comodels and Effects in Mathematical Operational Semantics

Faris Abou-Saleh[1] and Dirk Pattinson[2]

[1] Department of Computing, Imperial College London
[2] Research School of Computer Science, Australian National University

Abstract. In the mid-nineties, Turi and Plotkin gave an elegant categorical treatment of denotational and operational semantics for process algebra-like languages, proving compositionality and adequacy by defining operational semantics as a distributive law of syntax over behaviour. However, its applications to stateful or effectful languages, incorporating (co)models of a countable Lawvere theory, have been elusive so far. We make some progress towards a coalgebraic treatment of such languages, proposing a congruence format related to the evaluation-in-context paradigm. We formalise the denotational semantics in suitable Kleisli categories, and prove adequacy and compositionality of the semantic theory under this congruence format.

1 Introduction

Operational models of programming languages and process algebras are often described by a transition system, with transitions given by elementary, atomic evolution steps. For stateful languages, this involves an explicit notion of state, such as the values of program variables at each step of the execution. More abstractly, this state is described by a comodel [19]; computational effects characterise the dependency on state, and other phenomena, in terms of computational branches [5]. We may then understand the denotation of a program as an accumulation of state transformations mapping initial to final states, or an effect-tree describing every possible branch of the computation.

This gives us a powerful tool for reasoning about programs: two programs can be substituted for one another as long as they have the same behaviour, i.e. represent the same mapping from initial to final states, or effect-tree. This reasoning is aided by two key properties: the denotational semantics should be *adequate*, i.e. behaviourally equivalent programs should receive the same denotation, and *compositional*, i.e. the denotation of a program can be expressed in terms of the denotations of its components. These properties must often be proved on a case-by-case basis for different languages. To simplify this task, one often shows these properties are satisfied if the languages are given by operational rules in a particular *congruence format*.

Turi and Plotkin applied this approach in an abstract categorical setting [24], and obtained an elegant proof of adequacy and compositionality for a variety of process algebras. They represented program syntax as an initial algebra, and the semantic domain of denotations as a final coalgebra. Behavioural equivalence was given by coalgebraic bisimilarity, and the congruence format was expressed by a *distributive law* of syntax

F. Pfenning (Ed.): FOSSACS 2013, LNCS 7794, pp. 129–144, 2013.

over behaviour. In concrete instances, this leads to generalisations of the well-known GSOS rule format for a large class of process algebras [9].

However, so far the applications have centered around process algebra; applications of the theory to effectful [15] and comodel-based [14] languages has only been hinted at in the literature. The main stumbling block in applying the theory to these languages is that the final coalgebra is a very fine-grained semantic domain, recording the entire sequence of comodel manipulations or effects, rather than their accumulation.

To solve this problem, we may break the symmetry of the original approach, with syntax as an initial-algebra as before, but expressing behaviour, and the semantic domain of the final coalgebra, in a Kleisli category for a suitable monad. This approach gives a more appropriate characterisation of program behaviour, accumulating state manipulations and/or effects. However, it requires a new treatment of syntactic rule formats, and requires a different approach to proving adequacy and compositionality.

In our previous paper [1], we outlined how the existence of a semantic domain in the Kleisli category requires an enrichment with respect to the category $\mathsf{Cpo}_!$ of ω-complete partial orders with strict, continuous maps. We gave a method for extending operational specifications with effects, and by restricting to a rule format related to evaluation-in-context, we sketched a proof of adequacy and compositionality in terms of syntactic effect trees. However, without a categorical proof, we could not account for effects with equations, or complete the analysis for languages with comodels.

In this paper, we formalise and extend the analysis of our previous paper to incorporate both effects and comodels. We begin by defining transition systems to describe operational models incorporating comodels, effects, or both. We propose congruence formats related to the concept of evaluation-in-context [7], and after characterising behavioural and denotational equivalence in Kleisli categories, we prove adequacy and compositionality of the resulting denotational semantics.

Related Work. Mathematical operational semantics is described in [22,24] and applied to process algebras in [23]. Effects and monads in the semantics of languages are introduced by Moggi in [12] and subsequently by Plotkin, Power et al in [15,16]. Comodels for global state are discussed in [14,19]. Operational rules and adequacy proofs for a purely-effectful, functional language are given in [15]; the conclusion contains the words “one would wish to reconcile this work with the co-algebraic treatment of operational semantics in [33]”.

2 Syntax and Behaviour for Stateful and Effectful Languages

This section introduces three kinds of transition systems to represent effectful and/or stateful programming languages, before a formal coalgebraic treatment. We employ two languages as running examples, called While and NDWhile.

Definition 1. (ND)While Syntax. *We define syntax for a language* While *as follows, in three sorts – numeric and boolean expressions, and commands.*

$$N ::= x \mid n \mid N + N \mid +_n(N) \qquad E ::= b \mid N{==}N \mid \;{==}_n(N) \mid \neg E \mid E \wedge E$$
$$P ::= x{=}N \mid \mathtt{skip} \mid P\,;P \mid \mathtt{while}\,(E)\,\mathtt{do}\,\{P\} \mid \mathtt{if}\,(E)\,\mathtt{then}\,\{P\}\,\mathtt{else}\,\{P\}$$

Here, x is a global variable drawn from locations L, n is a numeral in $\mathbb{N}$, and b is a boolean in $\mathbb{B} = \{true, false\}$. The auxiliary operators $+_n(\cdot)$ and $==_n(\cdot)$ can be read as "add n to ·" and "n equals ·". The language NDWhile *adds binary* choose(·,·) *operators at each type, representing a non-deterministic choice of either expression.*

Although this syntax is multi-sorted, for theoretical simplicity we work in a single-sorted setting ($N = E = P$); badly-typed terms will produce an error return value.

2.1 Transition Systems for Stateful and Effectful Languages

We introduce three kinds of transition system to represent operational models with effects and/or persistent state, to be given by a comodel.

The first kind of transition system consists of pairs $\langle p, s\rangle$ of a program p and a state s (drawn from a collection S). This is typified by While, where the states $S = \mathbb{N}^L$ are assignments to global variables $x \in L$ ('locations') of natural numbers $\mathbb{N}$. Program execution is represented by changes in the program and state $\langle p, s\rangle \to \langle p', s'\rangle$, and it may eventually return a final state s' and a value v drawn from some collection V: $\langle p, s\rangle \to \langle \underline{v}, s'\rangle$. Typical transitions are $\langle x, s\rangle \to \langle \underline{s(x)}, s\rangle$ – looking up and returning the value of a variable x in the store – and variable updates, $\langle x = 5, s\rangle \to \langle \underline{*}, s[x \mapsto 5]\rangle$. (We write $*$ for the 'void' return value.) In general, a notion of state S may be derived canonically from a comodel (Definition 2); with this in mind, we will refer to such a stateful transition system as a *comodel-based transition system*, or CTS.

The second kind of operational model, an *effectful transition system* or ETS, records the paths a program execution may take, in terms of syntactic effects or their semantic equivalence classes (see [7]). For instance, given a global variable x, the first step of evaluating an expression like $3+x$ depends on the value of x. If x is 0, the first step will be $3+0$; and so on for other values of x. We may record this information syntactically by a 'read' effect, as follows: $\mathbf{rd}_x(3+0, 3+1, \ldots)$. Similarly, the first step of evaluating $x = 1; x = 2$ involves setting x to 1, leaving us to evaluate $x = 2$. We record the request to update x by a 'write' effect: $\mathbf{wr}_{x,1}(x = 2)$. A further evaluation step involves another update, giving the result $\mathbf{wr}_{x,1}(\mathbf{wr}_{x,2}(\underline{*}))$ (where $*$ is again the 'void' return value). Here is an example execution combining both effects:

$$x{=}y \to \mathbf{rd}_y(x{=}0, x{=}1, x{=}2, \ldots) \to \mathbf{rd}_y(\mathbf{wr}_{x,0}(\underline{*}), \mathbf{wr}_{x,1}(\underline{*}), \mathbf{wr}_{x,2}(\underline{*}), \ldots).$$

Thus, instead of tracking the state s as in a CTS, one could evaluate While expressions in terms of 'read' and 'write' effects. One must also have a means of identifying syntax trees which we would not want to distinguish semantically, such as $\mathbf{wr}_{x,1}(\mathbf{wr}_{x,2}(\underline{*}))$ and $\mathbf{wr}_{x,2}(\underline{*})$. This amounts to an equational theory for the effects [16].

Another example is given by non-determinism. Given a non-deterministic 'zero or one' function zo := choose(0, 1), evaluating zo + 5 gives either $0+5$ or $1+5$, and we represent this situation using a binary effect operator: $\mathbf{or}(0+5, 1+5)$. Another step produces the final result: $\mathbf{or}(\underline{5}, \underline{6})$. Evaluating zo + zo gives multiple nested **or**'s.

Again, one does not want to distinguish some effect-trees; such equivalences can be enforced by three equations: idempotence $\mathbf{or}(x, x) = x$, symmetry $\mathbf{or}(x, y) = \mathbf{or}(y, x)$, and associativity $\mathbf{or}(x, \mathbf{or}(y, z)) = \mathbf{or}(\mathbf{or}(x, y), z)$.

Note that some execution paths may terminate while others do not: for instance, a 'maybe stop' program could behave as follows: $\mathsf{ms} \to \mathbf{or}(\underline{*}, \mathsf{ms})$. Thus, we express the general form of ETS transitions with notation $p \to \delta((b_i)_{i \in I})$ where δ is an effect-syntax term with I-indexed arguments b_i which may be either terminal values $\underline{v}_i$ or programs p'_i. For convenience, we sometimes use vector notation: $p \to \delta(\tilde{b})$. We will call such a transition system a *syntactic* ETS if we see effects purely as syntax, ignoring the semantic equations on the effects; if we instead quotient the syntactic effect-trees δ by these equations, obtaining transitions involving *equivalence classes* of effect-trees, we call the resulting transition system a *semantic* ETS. For instance, in a syntactic ETS we would distinguish transitions $p \to p'$ and $p \to \mathbf{or}(p', p')$, but not a semantic ETS.

The final kind of transition system combines state S with effects, as needed for languages like NDWhile. Here, keeping track of global variables again require a store; but non-deterministic execution means we must track multiple possible stores and program states, as illustrated by this example execution (again using the 'zero or one' function):

$$\langle x{=}\mathsf{zo}, s\rangle \to \mathbf{or}(\langle x{=}0, s\rangle, \langle x{=}1, s\rangle) \to \mathbf{or}(\langle \underline{*}, s[x \mapsto 0]\rangle, \langle \underline{*}, s[x \mapsto 1]\rangle)$$

Generally, such transitions are of form $\langle p, s\rangle \to \delta(\langle b_i, s_i\rangle_{i \in I})$, where each b_i is either a terminal value or a program term (like an ETS). We call such a transition system a *syntactic comodel and effect-based transition system* (CETS), and again, if we choose to identify semantically equivalent effect-trees, we call the result a *semantic* CETS.

2.2 Transition Systems, Categorically

Now we give an overview of the categorical structure we will use to build a semantic theory for the transition systems described above. The main three constructions are (1) program syntax; (2) effect syntax, or semantic equivalence classes as given by a Lawvere theory; and (3) a semantic domain for programs. Syntax can be constructed by initial algebras for suitable polynomial functors. Effect structure will be described by models of a countable Lawvere theory (Definition 2); we may build effect-trees, as given by *free* models of the theory, if the category is locally countably presentable (l.c.p.) [17]. Moreover, to manipulate effect-trees (Definition 7), we need monadic strength which we will obtain using ($\otimes$-)monoidal-closure of the category, and assume $\otimes$ distributes over coproducts $+$: i.e. that $[\mathsf{inl} \otimes \mathsf{id}, \mathsf{inr} \otimes \mathsf{id}]$ is a natural isomorphism $A{\otimes}C{+}B{\otimes}C \to (A{+}B){\otimes}C$, with an inverse we call $\mathsf{dist}_{A,B,C}$. Finally, the least-fixpoint construction of a semantic domain requires the category to be order-enriched [8], with left and right-strict composition; we assume $\mathsf{Cpo}_!$-enrichment (see below), so that denotations may be canonically assigned to programs by a final coalgebra morphism in a Kleisli category (Proposition 4).

The structure we need is exemplified by the category $\mathsf{Cpo}_!$ of ω-complete pointed partial-orders, with strict ω-continuous maps. It is l.c.p. as it is essentially algebraic (see [3] p.163). Its closed monoidal structure $\otimes$ is the smash product $A \otimes B$, with strict function space $A \to_\perp B$ as exponential B^A; the cartesian product $A \times B$ is pointwise ordered, and coproducts are coalesced sums $A + B$. Lastly, it is enriched over itself, with strict composition. Most of the other requirements are met because we are assured the existence of *initial algebras and final coalgebras* for locally continuous functors

F.[1] These include polynomial functors incorporating constants, $+$ and $\times$, and what we call '$\otimes$-polynomial' functors, where $\otimes$ replaces $\times$.

Syntax. We may represent syntax constructors for a programming language in terms of a $\otimes$-polynomial functor Σ. The functor mapping $X \mapsto X \otimes \cdots \otimes X$ constructs a collection of n-tuples over X; coproducts of such functors combine these collections. For convenience, given a set S, we write $S \cdot A$ for the S-fold coproduct of A.

Example 1. To represent the syntax of the first line of Definition 1, we would take $\Sigma X = L_\perp + \mathbb{N}_\perp + X \otimes X + \mathbb{N} \cdot X + \mathbb{B}_\perp + X \otimes X$. Its elements comprise: constants x and n drawn from the flat cpos $L_\perp$ and $\mathbb{N}_\perp$; pairs (x_1, x_2) in $X \otimes X$ representing $x_1 + x_2$; elements (n, x_1) of $\mathbb{N} \cdot X$ representing $+_n(x_1)$; and so on.

For a syntax functor Σ, we write TX for the *free Σ-algebra over X* (equivalently, the initial $(X + \Sigma)$-algebra), with structure $\psi_X : \Sigma TX \to TX$. Assuming the category $\mathcal{C}$ is suitably concrete, we may consider the 'elements' of TX as individual syntax terms, as we did in Section 2.1. Above, the use of a $\otimes$-polynomial functor Σ is motivated by the fact that it constructs *finite* syntax terms in $\mathsf{Cpo}_!$; ordinary polynomial functors generate countably-deep syntax.

The closed program terms of the language are given by $T0$ where 0 is the initial object. We recall that the *free Σ-algebra functor T* is in fact a monad [22].

Effects and Comodels. An equational theory of effects may be encapsulated by a countable Lawvere theory $\mathcal{L}$ [17], in which the objects n represent n-tuples and n-ary effects become arrows $e : n \to 1$; composing and tupling these gives arrows $n \to m$.

Definition 2. *Let $\aleph_1$ be a skeleton of the category of countable sets. (See [5] Definition 1). A* (countable) Lawvere theory *is a category $\mathcal{L}$ with countable products and a countable strictly product-preserving, identity-on-objects functor $F : \aleph_1^{op} \to \mathcal{L}$. If $\mathcal{C}$ has countable products, the* category of models $\mathsf{Mod}(L, C)$ *of $\mathcal{L}$ in $\mathcal{C}$ has as objects all countable product-preserving functors $L \to C$, and as arrows all natural transformations between them. The category of* comodels *has as objects the countable* coproduct-*preserving functors $L^{op} \to C$, with arrows given by natural transformations.*

A model of a theory is a functor $G : \mathcal{L} \to \mathcal{C}$ with carrier $G1$; an n-ary effect e induces a corresponding function $(G1)^n \cong Gn \overset{G(e)}{\to} G1$, where the isomorphism is by product-preservation. (Following [5], our 'models' are up to such isomorphisms.) Similarly, comodels $C : \mathcal{L}^{op} \to \mathcal{C}$ have carrier $C1$, but the effect e now corresponds to a comodel-transition function $C1 \overset{C(e^{op})}{\to} Cn \cong n \cdot (C1)$ which, given a state in $C1$, 'chooses' a branch $\{1, \ldots, n\}$ of the effect, and returns a new state.

Example 2. The standard notion of state for While programs, $\mathbb{N}^L$, is the carrier $C1$ of a comodel (in fact, the final comodel) for global store; see [19] for details.

Given a set $\mathbb{E}$ of 'effects', arrows $e : n \to 1$, one may define a corresponding polynomial syntax functor Δ. Then the countably-deep syntactic effect-trees over X – which

[1] This is because it is *algebraically ω-compact*, being cocomplete (by local presentability) and $\mathsf{Cpo}_!$-enriched [2].

we have notated $\delta((x_i)_{i\in I})$ or $\delta(\tilde{x})$ – are generated by the *free-Δ-algebra monad*, which we will call T_e.

However, one often wishes to impose equations on this effect syntax T_eX, obtaining equivalence classes. This amounts to seeking the free model of the Lawvere theory over X, which we denote N_eX. It is given by UFX, where F is left adjoint to the forgetful functor $U : \mathsf{Mod}(L, \mathcal{C}) \to \mathcal{C}$. The left adjoint exists as $\mathcal{C}$ is l.c.p. [17].

By giving N_eX a natural Δ-algebra structure (via F), we obtain a Δ-algebra morphism $\mathsf{quot}_X : T_eX \to N_eX$ which performs this quotienting. We may prove:

Proposition 1. *The maps* $\mathsf{quot}_X : T_eX \to N_eX$ *define a monad morphism.*

To ensure existence of our semantic domain for programs (Proposition 4), we must ensure the monads T_e, N_e are $\mathsf{Cpo}_!$-monads. This rules out nullary effects $e : 0 \to 1$ like exceptions, and indirectly enforces effect equations $e(\bot, \ldots, \bot) = \bot$. We may prove directly that T_e is a $\mathsf{Cpo}_!$-monad, as effect-syntax Δ now cannot have constants $(-) + A$. To show N_e is a $\mathsf{Cpo}_!$-monad, one may consider the enriched [5] or discrete [6] $\mathsf{Cpo}_!$-Lawvere theories freely generated by $\mathcal{L}$, and use the results in [6] as follows. By assumption, $\mathcal{C}$ is l.c.p., so by Theorems 14 and 15 of [6], for either freely generated theory the forgetful $\mathsf{Cpo}_!$-functor $\mathsf{Mod}(L', \mathcal{C}) \to \mathcal{C}$ has a left adjoint which induces a $\mathsf{Cpo}_!$-monad N'_e whose underlying, ordinary monad coincides with N_e.

$\mathsf{Cpo}_!$-enrichment also equips a monad M with a *monadic strength* with respect to the monoidal structure $\otimes$ – a natural transformation $\mathsf{st}_{X,Y} : X \otimes MY \to M(X \otimes Y)$ satisfying certain coherence conditions (see [12] Definition 3.2 and Remark 3.3).

Transition Systems. The transition systems of Section 2.1 are equivalent to *coalgebras* for suitable endofunctors. First, we define a primitive transition functor, $BX = V + X$, describing atomic transition steps $x \to x'$ or termination $x \to \underline{v}$.

Given a comodel state-space $C1$ and values V, we may consider a comodel-based transition system (CTS) as a function $(P \otimes C1) \to ((V + P) \otimes C1)$. By $\otimes$-closedness, this is equivalent to a function $P \to (BP \otimes C1)^{C1}$, where $BP = V + P$. By defining a monad $N_cX = (X \otimes C1)^{C1}$ (essentially the side-effect monad), the CTS becomes a function $P \to N_cBP$, i.e. an N_cB-coalgebra.

Using B, given a set of effects E from a Lawvere theory $\mathcal{L}$, we may also express a syntactic effect-based transition system (ETS) as a T_eB-coalgebra, and a semantic ETS as an N_eB-coalgebra. We may quotient a syntactic ETS into a semantic ETS by post-composing with the monad morphism quot.

Finally, we may represent a syntactic or semantic CETS by an arrow $(P \otimes C1) \to M((V + P) \otimes C1)$ where $M = T_e$ or N_e respectively. One may consider a CETS as combining effects from two Lawvere theories via their tensor $\mathcal{L}_1 \otimes \mathcal{L}_2$ [6], where the effects E come from $\mathcal{L}_1$, and the comodel $C1$ is for $\mathcal{L}_2$.Either way, defining monads $T_{ce}X := (T_e(X \otimes C1))^{C1}$ and $N_{ce}X := (N_e(X \otimes C1))^{C1}$, we may express a CETS as a $T_{ce}B$ or $N_{ce}B$-coalgebra.

3 Three Evaluation-in-Context Rule Formats

Having described operational models as coalgebraic transition systems, we give concrete presentations of rule formats for specifying these operational models, which will give rise to compositional and adequate semantics.[2] The formats are based on the Evaluation-In-Context paradigm for sequential languages (c.f. [7]).

Here are some of the (standard) operational rules for While, considered as a CTS.

$$\frac{\langle p,s\rangle \to \langle p',s'\rangle}{\langle p\,;q,s\rangle \to \langle p'\,;q,s'\rangle} \quad \frac{\langle p,s\rangle \to \langle \underline{*},s'\rangle}{\langle p\,;q,s\rangle \to \langle q,s'\rangle} \quad \frac{\langle u,s\rangle \to \langle u',s'\rangle}{\langle x{=}u,s\rangle \to \langle x{=}u',s'\rangle} \quad \frac{\langle u,s\rangle \to \langle \underline{n},s'\rangle}{\langle x{=}u,s\rangle \to \langle \underline{*},s'[x\mapsto n]\rangle}$$

$$\overline{\langle \mathtt{while}\,(e)\,\mathtt{do}\,\{p\},s\rangle \to \langle \mathtt{if}\,(e)\,\mathtt{then}\,\{p;\mathtt{while}\,(e)\,\mathtt{do}\,\{p\}\}\,\mathtt{else}\,\{\mathtt{skip}\},s\rangle}$$

These rules divide the syntax constructors into what we call *context* and *redex* terms. Examples of the former are addition operators $+, +_n$, `if` statements, sequential composition ; and assignments $x{=}u$ (see below). To evaluate them, we must evaluate a distinguished argument; when it terminates with some value, we may have to evaluate another term, or produce another terminal value.

By contrast, a redex term evaluates independently of how its arguments behave. This includes: elementary terms $n \in \mathbb{N}, b \in \mathbb{B}$, `skip` which terminate as $\underline{n}$, $\underline{b}$, and $\underline{*}$; variable lookups $x \in L$, returning the value $\underline{s(x)}$ of the store at x; and `while` statements. This generalises to our first rule-format for specifying operational models as CTS's.

Definition 3. *Given a countable set of syntax variables P, the first kind of* Evaluation-In-Context specification (EIC1) *for values V and comodel C consists of the following, where for each rule below we assume $\tilde{x} = (x_i)_{i\in I}$ and $\tilde{y}_\bullet = (y_j)_{j\in J_\bullet}$ are such that $\{x_i : i\in I\} \subseteq P$ are pairwise distinct and disjoint from $\{x, x'\} \subseteq P$; and $t_\bullet(\tilde{y}_\bullet)$ stands for a terminal value $\underline{v}$ or a syntax term, where $\{y_j : j \in J_\bullet\} \subseteq \{x_i : i \in I\}$.*
– For every context-term constructor σ, the left-hand rule (CTXL) *below, and an instance of the right-hand rule* (CTXR) *for every $v \in V$ and comodel state $c \in C1$, with corresponding terminal value or term $t_{v,c}(\tilde{y}_{v,c})$ and new comodel state $c'_{v,c}$:*

$$\frac{\langle x,s\rangle \to \langle x',s'\rangle}{\langle \sigma(x,\tilde{x}),s\rangle \to \langle \sigma(x',\tilde{x}),s'\rangle}\ (\mathsf{CTXL}) \qquad \frac{\langle x,s\rangle \to \langle \underline{v},c\rangle}{\langle \sigma(x,\tilde{x}),s\rangle \to \langle t_{v,c}(\tilde{y}_{v,c}),c'_{v,c}\rangle}\ (\mathsf{CTXR})$$

– For redex constructors ρ, a rule $\overline{\langle \rho(\tilde{x}),c\rangle \to \langle t_c(\tilde{y}_c),c'_c\rangle}$ (REDX) *for each comodel state c, with terminal value or term $t_c(\tilde{y}_c)$ and new comodel state c'_c.*

Example 3. Consider the four operational rules for While given earlier (with syntax variables $P = \{p, q, u, e\}$). The first rule for sequential composition $\sigma(x, \tilde{x}) = p\,; q$ is merely (CTXL) and the other is an instance of (CTXR) where $v = *$ and $t_{v,c}(\tilde{x}_{v,c}) = q$. The first rule for variable update $x{=}u$ is again (CTXL), and the second is (CTXR) where v is n, $t_{v,c}(x_{j\in J})$ is $\underline{*}$ and $c' = c[x \mapsto n]$ for every $n \in \mathbb{N}$.

This approach also permits us to specify languages combining effects and comodels. Below are the rules for branching `choose` and assignments $x{=}u$ in NDWhile, where

[2] For simplicity, order-theoretic details are omitted; the important point is that divergence $p \to \bot$ is not treated as an ordinary terminal value, but rather propagated in the natural manner.

$\delta(\ldots)$ stands for an arbitrary **or**-tree of pairs $\langle b_k, c_k\rangle$ in which b_k is either a terminal value $\underline{v}$ or a program state u', and c_k a comodel-state; the general format follows.

$$\frac{}{\langle \mathtt{choose}(x,y),s\rangle \to \mathbf{or}(\langle x,s\rangle,\langle y,s\rangle)} \qquad \frac{\langle u, s\rangle \to \delta(\langle b_k, c_k\rangle_{k\in K})}{\langle x{=}u, s\rangle \to \delta\left(\begin{Bmatrix} \langle x{=}u', c_k\rangle & \text{if } b_k{=}u' \\ \langle \underline{*}, c_k[x\mapsto n]\rangle & \text{if } b_k{=}\underline{n}\end{Bmatrix}_{k\in K}\right)}$$

Definition 4. *Analogously to Definition 3, an* Evaluation-In-Context 2 (EIC2) Specification *for values V, comodel C, and a collection $\mathbb{E}$ of syntactic effects, consists of:*

– *For redex constructors ρ, rules $\overline{\langle \rho(\tilde{x}), c\rangle \to \epsilon_c(\langle t_l(\tilde{y}_l), c_l'\rangle_{l\in L_c})}$ (REDX) for all comodel states c, with syntactic effect-trees ϵ_c whose L_c-indexed leaves $\langle t_l(\tilde{y}_l), c_l'\rangle$ are pairs of a terminal value or term, and a new comodel state.*
– *For every context-term constructor σ, a rule (CTXB) for every effect-tree δ with leaves $\{\langle b_k, c_k\rangle : k\in K\}$ given by pairs $\langle b_k, c_k\rangle$ of either a syntax variable x_k or a terminal value $\underline{v}_k$, and a comodel-state c_k. We assume these x_k are all distinct, and do not include x or x_i for $i\in I$. Each rule is given by a $V\otimes C1$-indexed collection of effect-trees $\epsilon_{v,c}$ with $L_{v,c}$-indexed leaves $\langle t_l(\tilde{y}_l), c_l'\rangle$ as above.*

$$\frac{\langle x, s\rangle \to \delta(\langle b_k, c_k\rangle_{k\in K})}{\langle \sigma(x,\tilde{x}), s\rangle \to \delta\left(\begin{Bmatrix} \langle \sigma(x_k,\tilde{x}), c_k\rangle & \text{if } b_k = x_k \\ \epsilon_{v,c_k}(\langle t_l(\tilde{y}_l), c_l'\rangle_{l\in L_{v,c_k}}) & \text{if } b_k = \underline{v}\end{Bmatrix}_{k\in K}\right)} \text{ (CTXB)}$$

Finally, by removing all mention of comodels from the above format, we gain a rule format for specifying ETS's. We could specify an ETS for While with rules such as the following; the general rule format is given below.

$$\frac{}{x \to \mathbf{rd}_x(\underline{0},\underline{1},\underline{2},\ldots)} \qquad \frac{u \to \delta((b_k)_{k\in K})}{x{=}u \to \delta\left(\begin{Bmatrix} x{=}u' & \text{if } b_k = u' \\ \mathbf{wr}_{x,n}(\underline{*}) & \text{if } b_k = \underline{n}\end{Bmatrix}_{k\in K}\right)}$$

Definition 5. Evaluation-In-Context 3 (EIC3) *is analogous to EIC2, but with rules*

$$\frac{}{\rho(\tilde{x}) \to \epsilon((t_l(\tilde{y}_l))_{l\in L})} \qquad \frac{x \to \delta(b_{k\in K})}{\sigma(x,\tilde{x}) \to \delta\left(\begin{Bmatrix} \sigma(x_k,\tilde{x}) & \text{if } b_k = x_k \\ \epsilon_v((t_l(\tilde{y}_l))_{l\in L_v}) & \text{if } b_k = \underline{v}\end{Bmatrix}_{k\in K}\right)}$$

These three kinds of EIC rule format allows us to specify operational models as CTS's or syntactic (C)ETS's. As formalised below, structural recursion then defines transition behaviour for program terms TX over syntax variables X, once we have specified transition behaviour for the variables X – the 'base cases' of the recursion.

3.1 From EIC Specifications to Operational Models

We now formalise the specifications of the previous section as natural transformations, and show how they induce coalgebraic operational models by structural recursion.

There are various ways of expressing operational specifications as distributive laws of syntax over behaviour [11]. For our purposes, the 'abstract GSOS' specifications of [24] suffice: natural transformations $\epsilon : \Sigma(Id \times B) \Rightarrow BT$, where Σ is the program syntax functor, T the free Σ-algebra monad, and B a coalgebraic behaviour functor. These specifications induce operational models $T0 \to BT0$ by structural recursion:

Proposition 2. *[24] Given an arrow $h : \Sigma(TX \times Y) \to Y$ and an arrow $s : X \to Y$, there is a unique arrow $! : TX \to Y$ making the below diagram commute.*

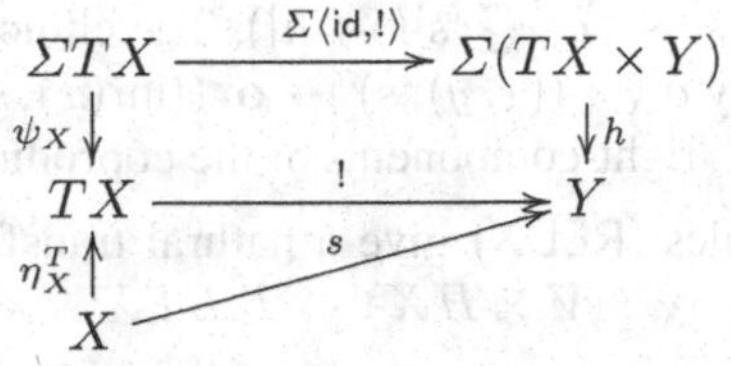

Given an operational specification $\epsilon : \Sigma(Id \times B) \Rightarrow BT$ and a transition structure $\gamma : X \to BX$, we can derive a transition structure $T^{\epsilon}(\gamma) : TX \to BTX$ for terms TX over generators X as follows. Defining $\epsilon' := B\mu^T \circ \epsilon_T : \Sigma(T \times BT) \Rightarrow BT$ [24], we apply the result with $Y = BTX$, $s = B\eta^T \circ \gamma : X \to BTX$, and $h = \epsilon'_X$. The resulting map $! : TX \to BTX$ is the required transition structure on terms TX. In particular, taking $X = 0$ and the unique arrow $\gamma : 0 \to B0$, we obtain an operational model for closed program terms, an arrow $T0 \to BT0$ which we call $T^{\epsilon}(0)$.

The operational models considered in this paper are MB-coalgebras for various monads M, where $BX = V + X$. Replacing B above with MB, operational specifications ϵ of type $\Sigma(X \times MBX) \to MBTX$ induce operational models of form $T^{\epsilon}(0) : T0 \to MBT0$.

It remains to encode EIC specifications as natural transformations ϵ. Example 9 will demonstrate that non-EIC specifications ϵ may result in semantics which are not adequate or compositional; however, EIC-specified languages will be shown to have these properties. First, we express redex term constructors by a syntax functor R, and context terms by $X \otimes HX$, with active argument X and context HX.

Definition 6. *In a symmetric $\otimes$-monoidal category $\mathcal{C}$ with coproducts, an endofunctor Σ is said to be* Redex-Context (R-C) *if $\Sigma X = RX + X \otimes HX$ for some functors R, H.*

Formalising the data of EIC specifications, we represent redex-terms $\rho(\tilde{x})$ over X by RX, context terms $\sigma(x, \tilde{x})$ by $X \otimes HX$, and arbitrary terms by TX. A 'terminal value or term $t(\tilde{x})$' is a basic transition over terms in BTX. Syntactic effect-trees $\delta(\tilde{x})$ over X are given by the free Δ-algebra monad: $T_{\Delta}X$. For EIC 1.0 (i.e. CTS), the rules (REDX) combine into a natural transformation $\alpha_X : (RX \otimes C1) \to (BTX \otimes C1)$, indicating, for each redex RX and initial comodel-state $C1$, the transition behaviour BTX and new comodel-state $C1$. Similarly, (CTXR) gives a natural transformation $\beta_X : (V \otimes HX \otimes C1) \to (BTX \otimes C1)$.

Example 4. Consider the fragment of While given by variable lookups l in L (so $RX = L_{\bot}$) and updates $l = x$ (context terms $l = (-)$ with context-data HX given by a location, so that $HX = L_{\bot}$). These commands are specified by $\alpha_X : (l, s) \mapsto (\mathsf{inl}(\underline{s(l)}), s)$

and $\beta_X : (n, l, s) \mapsto (\mathsf{inl}(\underline{*}), s[l \mapsto n])$ where we underline the 'terminal values', left components of the coproduct $BTX = V + TX$.

For EIC 2.0 (i.e. CETS), the rules (REDX) give a natural transformation $\alpha_X : (RX \otimes C1) \to T_e(BTX \otimes C1)$ – where the codomain describes syntactic effect-trees of transition behaviours. Similarly, (CTXR) gives $\beta_X : (V \otimes HX \otimes C1) \to T_e(BTX \otimes C1)$.

Example 5. Variable lookups and updates in NDWhile are specified using the unit η^{T_e} of the effect-syntax monad T_e (i.e. 'no non-determinism'): we define $\alpha_X : (l, s) \mapsto \eta^{T_e}(\underline{s(l)}, s)$ and $\beta_X : (n, l, s) \mapsto \eta^{T_e}(\underline{*}, s[l \mapsto n])$. The choice operator $\mathtt{choose}(x, y)$ ($RX = X \otimes X$) is given by $\alpha_X : ((x, y), s) \mapsto \mathbf{or}((\mathsf{inr}(x), s), (\mathsf{inr}(y), s))$ where the inr are 'new program terms', right-components of the coproduct $BTX = V + TX$.

Finally, for EIC 3.0, the rules (REDX) give a natural transformation $\alpha_X : RX \to T_eBTX$ and (CTXR) gives $\beta_X : (V \otimes HX) \to T_eBTX$.

Example 6. Considering While as a syntactic ETS, variable lookups would be specified by $\alpha_X : l \mapsto \mathbf{rd}_l(\mathsf{inl}(\underline{0}), \mathsf{inl}(\underline{1}), \ldots)$ and updates by $\beta_X : (n, l) \mapsto \mathbf{wr}_{l,n}(\mathsf{inl}(\underline{*}))$.

We may now use the $\otimes$-monoidal closed structure to show that in all cases (REDX) corresponds to natural transformations $r_X : RX \to MBTX$ where $M = N_c, T_e$ and T_{ce} respectively. Similarly, (CTXR) corresponds to $e_X : X \otimes HX \to MBTX$. This data induces an operational specification $\epsilon_X : \Sigma(X \times MBX) \to MBTX$ as follows:

Definition 7. *Let Σ be an R-C syntax functor, M a $\otimes$-strong monad with costrength* cost*, and B a behaviour functor $BX = V + X$. For given natural transformations $r_X : RX \to MBTX$ and $e_X : V \otimes HX \to MBTX$, the corresponding* abstract EIC specification *$\epsilon_X : \Sigma(TX \times MBTX) \to MBTX$ is given by*

$$\epsilon_X : R(TX \times MBTX) + (H \otimes Id)(TX \times MBTX) \xrightarrow{[\mathsf{aosr}_X, \mathsf{aosc}_X]} MBTX$$

where aosr *and* aosc *are defined below (we have abbreviated* $\mathsf{cost}_{BTX,HTX}$*).*

$$\mathsf{aosr}_X : \quad R(TX \times MBTX) \xrightarrow{R\pi_1} RTX \xrightarrow{r_{TX}} MBTTX \xrightarrow{MB\mu_X} MBTX$$
$$\mathsf{aosc}_X : \quad (Id \times H)(TX \times MBTX) \xrightarrow{\pi_2 \otimes H\pi_1} MBTX \otimes HTX$$
$$\xrightarrow{\mathsf{cost}} M(BTX \otimes HTX) \xrightarrow{M\mathsf{dwc}_X} M^2BTX \xrightarrow{\mu_{BTX}} MBTX$$

Here, dwc *('deal with contexts') is defined as follows, with sub-cases handled by* $\mathsf{dwc}^{(v)}$ *('values') and* $\mathsf{dwc}^{(b)}$ *('non-terminal behaviour'). We abbreviate* $\mathsf{dist}_{V,TX,HTX}$*. Recall that $\psi_X : \Sigma TX \to TX$ is the Σ-algebra structure of TX, the free Σ-algebra over X.*

$$\mathsf{dwc}_X : (V{+}TX){\otimes}HTX \xrightarrow{\mathsf{dist}} V{\otimes}HTX{+}TX{\otimes}HTX \xrightarrow{[\mathsf{dwc}^{(v)}{}_X, \mathsf{dwc}^{(b)}{}_X]} MBTX$$
$$\mathsf{dwc}^{(v)}{}_X : V \otimes HTX \xrightarrow{e_X} MBTTX \xrightarrow{MB\mu_X} MBTX$$
$$\mathsf{dwc}^{(b)}{}_X : TX \otimes HTX \xrightarrow{\mathsf{inr}} \Sigma TX \xrightarrow{\psi_X} TX \xrightarrow{\mathsf{inr}} BTX \xrightarrow{\eta_{BTX}} MBTX$$

The rule (REDX) is described by aosr. For context terms $TX \otimes HTX$, given the behaviour $MBTX$ of the active term TX (isolated by the first line of aosc), the costrength attaches the context HTX to each computation branch. The map dwc decides what to do for each computation branch; if a branch terminates, it is handled by $\mathsf{dwc}^{(v)}$ which corresponds to (CTXR); otherwise, it is handled by $\mathsf{dwc}^{(b)}$, corresponding to (CTXL).

4 Behavioural Equivalence in a Kleisli Category

Following Turi and Plotkin's method, we would take the semantic domain to be the final MB-coalgebra, where $BX = V + X$ and M is the monad for the transition systems under consideration; however, this distinguishes comodel-manipulations and effects at every execution step. For instance, the (CTS or ETS) While programs $x{=}0\,; x{=}1$ and $x{=}2\,; x{=}1$ would be considered to have different behaviour.

A natural solution to this problem is to move to a Kleisli category for the monad M, and construe transition systems as $\overline{B}$-coalgebras, where $\overline{B}$ is a lifting of B [1]. In fact, MB-coalgebras $X \to MBX$ coincide with $\overline{B}$-coalgebras $X \to \overline{B}X$; but $\overline{B}$-coalgebra morphisms, and the final $\overline{B}$-coalgebra, which we write $\langle \overline{D}, \overline{s} : \overline{D} \to \overline{BD} \rangle$, are generally different. For any ($\overline{B}$ or) MB-coalgebra $\gamma : X \to MBX$, there is a unique $\overline{B}$-coalgebra morphism $\beta : X \to \overline{D}$, of underlying type $X \to M\overline{D}$.

Thus, if a lifting $\overline{B}$ and the final $\overline{B}$-coalgebra $\overline{D}$ exist, we obtain a map into $M\overline{D}$, which we will see gives a more appropriate characterisation of program behaviour. The first point is easily addressed: liftings $\overline{B}$ of B are in 1-1 correspondence with distributive laws $\lambda : BM \Rightarrow MB$ (see e.g. [20,4]).

Remark 1. For the functor $BX = V{+}X$ and any monad M, the natural transformation $\lambda_X := [\eta^M \circ \mathsf{inl}, M\mathsf{inr}] : V{+}MX \to M(V{+}X)$ is a distributive law of B over M.

As for existence of the final $\overline{B}$-coalgebra, we draw on the following result (quoted and proved in [4] as Proposition 3.9):

Proposition 3. *Let $\mathcal{D}$ be a $\mathsf{Cpo}_!$-enriched category with left-strict composition, and G a locally-continuous endofunctor on $\mathcal{D}$. An initial G-algebra $\theta : G\overline{D} \cong \overline{D}$, if it exists, yields a final G-coalgebra $\theta^{-1} : \overline{D} \cong G\overline{D}$. Given a G-coalgebra $\gamma : X \to GX$, the corresponding unique G-coalgebra morphism $\beta : X \to \overline{D}$ is the least fixpoint of the operator $\Phi : (X \xrightarrow{f} \overline{D}) \mapsto (X \xrightarrow{\gamma} GX \xrightarrow{Gf} G\overline{D} \xrightarrow{\theta} \overline{D})$ – i.e. β is the join of the approximants $\beta^{(n)} := \Phi^n(\bot_{X,\overline{D}})$ for $n < \omega$.*

We apply this result to the Kleisli category $\mathcal{D} = \mathsf{Kl}(M)$ and $G = \overline{B}$, a lifting of B; this will give the required final $\overline{B}$-coalgebra $\overline{D}$. The $\mathsf{Cpo}_!$-enrichedness of $\mathsf{Kl}(M)$ follows from that of the underlying category $\mathcal{C}$; if B is locally continuous, it is easy to show $\overline{B}$ must also be. If B has an an initial algebra $\alpha : B\overline{D} \to \overline{D}$, we may show $\overline{D}$ is also an initial algebra for $\overline{B}$, with structure $\eta^M \circ \alpha$ ([4] Proposition 3.2). Finally, if M is a $\mathsf{Cpo}_!$-monad, one may show $M\bot_{A,B} = \bot_{MA,MB}$; if composition in $\mathcal{C}$ is both left and right-strict, this implies left-strictness of Kleisli-composition. In summary:

Proposition 4. *Let $\mathcal{C}$ be a $\mathsf{Cpo}_!$-enriched category with strict composition, on which M is a $\mathsf{Cpo}_!$-enriched monad, and B a locally continuous functor with a lifting $\overline{B}$ to $\mathsf{Kl}(M)$. If B has an initial algebra $\alpha : B\overline{D} \cong \overline{D}$, the final $\overline{B}$-coalgebra is given by $\eta^M \circ \alpha^{-1} : \overline{D} \to \overline{BD}$.*

Example 7. We illustrate how the fixpoint construction of Proposition 4 assigns denotations to While programs. Taking $BX = V{+}X$, the initial B-algebra $\overline{D}$ has carrier $\mathbb{N}\cdot V$, whose elements $(n, \underline{v})$ characterise individual computation branches by their length n and return-value v. Its algebra-structure $\alpha : B\overline{D} \to \overline{D}$ is defined by $\alpha(\mathsf{inl}(\underline{v})) = (1, \underline{v})$

and $\alpha(\mathsf{inr}(n,\underline{v})) = (n+1,\underline{v})$. The functor B has a lifting $\overline{B}$ given by the distributive law $\lambda : BM \Rightarrow MB$ of Remark 1.

If we work in $\mathsf{Cpo}_!$ and take M to be a $\mathsf{Cpo}_!$-monad, then Proposition 4 applies. Thus, $\overline{D}$ is also the carrier of the final $\overline{B}$-coalgebra. For any operational model $T^\epsilon(0) \colon T0 \to MBT0$ (construed as a $\overline{B}$-coalgebra), there will be a $\overline{B}$-coalgebra morphism β from $T0$ into the semantic domain $\overline{D}$, of underlying type $T0 \to M\overline{D}$. It is given by the join of the approximants $\beta^{(n)}$ which may be defined as follows: $\beta^{(0)} = \bot_{T0,M\overline{D}} : T0 \to M\overline{D}$ (i.e. the bottom arrow $T0 \to \overline{D}$ in $\mathsf{Kl}(M)$); and then

$$\beta^{(n+1)} : T0 \overset{T^\epsilon(0)}{\longrightarrow} MBT0 \overset{MB\beta^{(n)}}{\longrightarrow} MBM\overline{D} \overset{M\lambda}{\longrightarrow} M^2B\overline{D} \overset{\mu}{\longrightarrow} MB\overline{D} \overset{M\alpha}{\longrightarrow} M\overline{D}$$

We now instantiate these results for While as a CTS, taking $M = N_{\mathrm{c}}$ with the comodel of Example 2. This gives a semantic domain of $M\overline{D} = ((\mathbb{N}\cdot V)\otimes C1)^{C1}$: each program p is assigned a function which, given an initial comodel state s, tells us about the execution of the program p with initial comodel-state s: a pair (n,v) of the number of steps-to-termination n and the return-value v, as well as the final comodel state s'. (Non-terminating pairs (p,s) receive the value $\bot$.)

Suppose we have obtained an (un-curried) operational model $T^\epsilon(0) : T0 \to (BT0 \otimes C1)^{C1}$ for While as a CTS, as in Section 3.1 (see Example 4). We illustrate the action of the approximants $\beta^{(n)}$ on the programs $p_n := (x = n)$ and $q := (x = 5; x = 8)$. (We abbreviate the series of maps $(\overline{B}\beta^{(0)})^\dagger = \mu^{N_{\mathrm{c}}} \circ N_{\mathrm{c}}\lambda \circ N_{\mathrm{c}}B\beta^{(0)}$, which have no effect on the terminated value $\mathsf{inl}(\underline{*})$.)

$$\begin{aligned}\beta^{(1)} : x = n &\overset{T^\epsilon(0)}{\longmapsto} \lambda s.(\mathsf{inl}(\underline{*}), s[x \mapsto n]) \overset{(\overline{B}\beta^{(0)})^\dagger}{\longmapsto} \lambda s.(\mathsf{inl}(\underline{*}), s[x \mapsto n])\\ &\overset{N_{\mathrm{c}}\alpha}{\longmapsto} \lambda s.((1,\underline{*}), s[x \mapsto n])\end{aligned}$$

This yields the desired denotation of the assignment $x = n$. Note that higher approximants $\beta^{(n)}$ will assign the same value. We may now show that $\beta^{(2)}$ assigns the desired denotation to the program $x = 5; x = 8$:

$$\begin{aligned}\beta^{(2)} : (x = 5; x = 8) &\overset{T^\epsilon(0)}{\longmapsto} \lambda s.(\mathsf{inr}(x = 8), s[x \mapsto 5])\\ &\overset{N_{\mathrm{c}}B\beta^{(1)}}{\longmapsto} \lambda s.(\mathsf{inr}(\lambda s'.((1,\underline{*}), s'[x \mapsto 8])), s[x \mapsto 5])\\ &\overset{N_{\mathrm{c}}\lambda}{\longmapsto} \lambda s.((\lambda s'.(\mathsf{inr}(1,\underline{*}), s'[x \mapsto 8])), s[x \mapsto 5])\\ &\overset{\mu^{N_{\mathrm{c}}}}{\longmapsto} \lambda s.(\mathsf{inr}(1,\underline{*}), s[x \mapsto 8]) \overset{N_{\mathrm{c}}\alpha}{\longmapsto} \lambda s.((2,\underline{*}), s[x \mapsto 8])\end{aligned}$$

This illustrates that in a CTS, the map β identifies two programs p, q if and only if: for every initial comodel-state s, $\langle p,s\rangle$ and $\langle q,s\rangle$ both: (a) terminate with the same final comodel-state s' and terminal value $\underline{v}$ in the same number of steps n; or (b) do not terminate. One may check that in a syntactic ETS, the map β identifies two programs p, q if their executions produce the same effect-tree $\delta((n_i, \underline{v}_i)_{i\in I})$ of terminal values $\underline{v}_i$ paired with the number of steps n_i they took to appear; and the situation for CETS's combines features of both CTS's and ETS's. On this basis, we may take the final coalgebra maps β as a characterisation of behavioural equivalence. However, the EIC specifications induce *syntactic* ETS's and CETS's; the corresponding maps $\beta : T0 \to T_{\mathrm{e}}\overline{D}$, $\beta : T0 \to T_{\mathrm{ce}}\overline{D}$ distinguish semantically-equivalent effect-trees. An appropriate behavioural equivalence must quotient by the equations on the effects:

Definition 8. *[Kleisli Behavioural Equivalence] Under the assumptions of Proposition 3, two states p, q of a* CTS *satisfy $p \cong_{\mathrm{c}} q$ if they are identified by the final $\overline{B}$-coalgebra morphism β into the final $\overline{B}$-coalgebra $\overline{D}$ in $\mathsf{Kl}(N_{\mathrm{c}})$. For the appropriate final coalgebra maps β in $\mathsf{Kl}(T_{\mathrm{e}})$ or $\mathsf{Kl}(T_{\mathrm{ce}})$, states p, q of a* syntactic ETS *satisfy $p \cong^{\mathrm{N}}_{\mathrm{e}} q$, if they are identified by $\mathsf{quot}_{\overline{D}} \circ \beta : T0 \to N_{\mathrm{e}}\overline{D}$; and states p, q of a* syntactic CETS *satisfy $p \cong^{\mathrm{N}}_{\mathrm{ce}} q$ if they are identified by $(\mathsf{quot}_{\overline{D}\otimes C1})^{C1} \circ \beta : T0 \to N_{\mathrm{ce}}\overline{D}$.*

Example 8. Suppose we construe While as a CTS, and NDWhile as a CETS. Define

$$p_1 := (x{=}0\,;x{=}1),\quad p_2 := (x{=}1\,;x{=}1),\quad p_3 := (x{=}1),\quad \text{and}\quad p_4 := \mathtt{choose}(p_3, p_3).$$

For While as a CTS, we have $\beta(p_1) = \beta(p_2) = \lambda c.((2, \underline{*}), c[x \mapsto 1])$ and $\beta(p_3) = \lambda c.((1, \underline{*}), c[x \mapsto 1])$; hence $p_1 \cong_{\mathrm{c}} p_2 \not\cong_{\mathrm{c}} p_3$. As programs of NDWhile, we will have $p_2 \not\cong^{\mathrm{T}}_{\mathrm{ce}} p_4$, due to the appearance of a syntactic $\mathbf{or}$-effect when we evaluate p_4; however, the semantic equation $\mathbf{or}(x, x) = x$ will imply that $p_2 \cong^{\mathrm{N}}_{\mathrm{ce}} p_4$.

Example 9. Relaxing the EIC rule formats may give non-compositional syntax constructors. Considering While as a CTS, the 'one-step timeout' $p \rhd q$ executes the first step of p, and continues with q; the interleaver $p \mid q$ alternates steps of p and q.

$$\frac{\langle p, s\rangle \to \langle p', s'\rangle}{\langle p \rhd q, s\rangle \to \langle q, s'\rangle}\quad \frac{\langle p, s\rangle \to \langle \underline{v}, s'\rangle}{\langle p \rhd q, s\rangle \to \langle q, s'\rangle}\quad \frac{\langle p, s\rangle \to \langle p', s'\rangle}{\langle p \mid q, s\rangle \to \langle q \mid p', s'\rangle}\quad \frac{\langle p, s\rangle \to \langle \underline{v}, s'\rangle}{\langle p \mid q, s\rangle \to \langle q, s'\rangle}$$

Letting $p_1 := (x{=}0\,;x{=}2)$, $p_2 := (x{=}1\,;x{=}2)$, and $p_3 := (y{=}x)$, we have $p_1 \cong_{\mathrm{c}} p_2$, but $(p_1 \rhd p_3) \not\cong_{\mathrm{c}} (p_2 \rhd p_3)$ and $(p_1 \mid p_3) \not\cong_{\mathrm{c}} (p_2 \mid p_3)$. Syntax constructors sensitive to individual execution steps can take 'behaviourally equivalent' arguments with different results, breaking compositionality. This cannot occur in EIC specifications.

5 Compositionality and Adequacy

We have defined operational equivalence in terms of mappings $T0 \to M\overline{D}$ into semantic domains $M\overline{D}$ for various monads M. We now define a corresponding denotational model, and prove adequacy and compositionality of the resulting denotational semantics. We do this first for CTS's, and then for syntactic (C)ETS's.

In order to treat the semantic domain $M\overline{D}$ as a denotational model, we must define interpretations $[\![\sigma]\!]$ of syntax constructors σ on denotations. First, we assign transition behaviour to denotations, by giving the semantic domain a natural MB-coalgebra structure: $\eta^M \circ M\alpha^{-1} : M\overline{D} \to MB\overline{D} \to MBM\overline{D}$.

We now take denotations d_i as base cases for structural recursion (Proposition 2), yielding an MB- (i.e. $\overline{B}$-)coalgebra transition structure for *program terms over denotations* $TM\overline{D}$, like $\sigma((d_i)_{i\in I})$. As a result, we obtain a $\overline{B}$-coalgebra morphism ζ, of underlying type $TM\overline{D} \to M\overline{D}$, characterising the behaviour of such terms by mapping back into the semantic domain $M\overline{D}$ (by finality). This gives a Σ-algebra structure χ to the semantic domain $M\overline{D}$, interpreting the syntax constructors $[\![\sigma]\!]$ on denotations, via the following composition (where ψ_X is the free Σ-algebra structure of TX):
$\Sigma M\overline{D} \xrightarrow{\Sigma\eta^T} \Sigma TM\overline{D} \xrightarrow{\psi_{M\overline{D}}} TM\overline{D} \xrightarrow{\zeta} M\overline{D}$.

By repeated application of the functions $[\![\sigma]\!]$ (with base-cases given by nullary symbols), one inductively constructs denotations $[\![t]\!]$ of arbitrary terms t. Formally, this assignment $[\![-]\!] : T0 \to M\overline{D}$ is given by the initial Σ-algebra morphism from $T0$ into $M\overline{D}$, equipped with the above structure χ. This gives a suitable denotational semantics for CTS's, where we take $M = N_c$. That it is a Σ-algebra morphism implies *compositionality* of the denotational semantics; i.e. that the denotation $[\![\sigma((t_i)_{i\in I})]\!]$ of a term $\sigma((t_i)_{i\in I})$ can be constructed from the denotations $[\![t_i]\!]$ of its parts: $[\![\sigma((t_i)_{i\in I})]\!] = [\![\sigma]\!]([\![t_i]\!]_{i\in I})$. This makes it a congruence with respect to the syntax constructors of the language: $[\![s_i]\!] = [\![t_i]\!]$ for all i implies $[\![\sigma((s_i)_{i\in I})]\!] = [\![\sigma((t_i)_{i\in I})]\!]$.

However, it remains to show *adequacy*, i.e. that denotational equivalence implies operational equivalence for CTS's. A convenient method is to show that the denotational map $[\![-]\!]$ coincides with the map β characterising behavioural equivalence for CTS's (Definition 8), by showing that the latter is also a Σ-algebra morphism; by initiality, there can be only one such map. This is achieved through the following theorem.

Theorem 1. *Given an abstract EIC specification ϵ (Definition 7) inducing an operational model $T^{\epsilon}(0) : T0 \to MBT0$, the underlying arrow $\beta : T0 \to M\overline{D}$ of the $\overline{B}$-coalgebra morphism into the final $\overline{B}$-coalgebra $\overline{D}$ is a Σ-algebra morphism.*

The broad strategy is to factor the 'coarse-grained' denotational map β through its 'fine-grained' analogue, the final MB-coalgebra D. The proof involves manipulating colimit diagrams and limit-colimit coincidences in the Kleisli category, and a detailed inspection of the mechanics of the EIC specifications.

Now we consider denotational semantics for ETS's (CETS's are exactly analogous). Recall that we characterised behavioural equivalence not by the semantic domain $T_e\overline{D}$, consisting of syntactic effect-trees, but by the equivalence classes $N_e\overline{D}$ generated by applying the quotienting map $\mathsf{quot}_{\overline{D}} : T_e\overline{D} \to N_e\overline{D}$. To treat the domain $N_e\overline{D}$ as a denotational model, we must give it a Σ-algebra structure.

We may follow an analogous method to the one outlined above, by moving from syntactic ETS's (T_eB-coalgebras) to semantic ETS's (N_eB-coalgebras) and considering rule-formats and structural recursion directly in terms of equivalence-classes of effect-trees. This amounts to seeking an abstract EIC specification (Definition 7) in terms of the monad N_e rather than T_e. Note that a syntactic EIC specification in terms of T_e – given by natural transformations $r : RX \Rightarrow T_eBT$ and $e : V \otimes HX \Rightarrow T_eBT$ – translates into a specification in terms of N_e, by postcomposing with $\mathsf{quot}_{BT} : T_eBT \Rightarrow N_eBT$. Formally, structural recursion induces an operational model $T0 \to NBT0$ directly in terms of equivalence-classes. The semantic analysis of Section 4 may also be transplanted from $\mathsf{Kl}(T_e)$ to $\mathsf{Kl}(N_e)$: again, there is a lifting $\overline{B}^{N_e}$ of B given by Remark 1, and Proposition 4 guarantees the existence of a final $\overline{B}^{N_e}$ coalgebra in $\mathsf{Kl}(N_e)$, whose carrier is the initial B-algebra $\overline{D}$. There is a unique $\overline{B}^{N_e}$-coalgebra morphism β^{N_e} from the operational model into $\overline{D}$, now of underlying type $T0 \to N_e\overline{D}$.

As above, structural recursion over the semantic domain $M\overline{D}$, where M is now N_e, induces a Σ-algebra structure χ^{N_e} on $N_e\overline{D}$; initiality gives a unique Σ-algebra morphism $[\![-]\!] : T0 \to N_e\overline{D}$, i.e. a denotational semantics which is automatically *compositional*. Theorem 1 implies that β^{N_e} is a Σ-algebra morphism, and so $\beta^{N_e} = [\![-]\!]$. To prove *adequacy* of this denotational semantics for ETS's, we may show that the denotational map $[\![-]\!]$ coincides with the map $\mathsf{quot}_{\overline{D}} \circ \beta : T0 \to N_e\overline{D}$ characterising behavioural equivalence. The following result will imply that $\mathsf{quot}_{\overline{D}} \circ \beta = \beta^{N_e}$, and hence $\mathsf{quot}_{\overline{D}} \circ \beta = [\![-]\!]$ as required, by taking the monad morphism $m = \mathsf{quot}$.

Proposition 5. *Let M, N be strong monads on $\mathcal{C}$, $m : M \Rightarrow N$ a strong monad morphism, and B the endofunctor $BX = V + X$ with liftings $\overline{B}^M$ and $\overline{B}^N$ to $\mathsf{Kl}(M)$ and $\mathsf{Kl}(N)$, with final coalgebras of carrier $\overline{D}$ and underlying structure $\eta^M \circ \alpha^{-1}$ and $\eta^N \circ \alpha^{-1}$ for some arrow $\alpha^{-1} : \overline{D} \to B\overline{D}$. Given an abstract EIC specification ϵ in terms of monad M, and its translation via m into a specification in terms of monad N, the corresponding final coalgebra maps β^M, β^N from the induced operational models $T0 \to MBT0, T0 \to NBT0$ into $M\overline{D}$ and $N\overline{D}$ satisfy $\beta^N = m \circ \beta^M$, provided the distributive laws λ^M, λ^N lifting B satisfy this equation (†): $\lambda^N_X \circ Bm_X = m_{BX} \circ \lambda^M_X$.*

Note that if M and N are $\mathsf{Cpo}_!$-monads, then the monad morphism m is strong if it is a $\mathsf{Cpo}_!$-natural transformation (see Remark 1.4 of [10]). As the functor $\mathsf{Cpo}_!(I, -) : \mathsf{Cpo}_! \to \mathsf{Set}$ is faithful, $\mathsf{Cpo}_!$-naturality is essentially equivalent to ordinary naturality ([8] Section 1.3); thus the monad morphism quot may be assumed to be strong. In addition, the condition (†) is easily verified for the liftings $\overline{B}$ we used in Remark 1.

One may give compositional and adequate denotational semantics to CETS's by exactly the same method, where $M = T_{\mathrm{ce}}, N = N_{\mathrm{ce}}$, and the monad morphism is given by $(\mathsf{quot}_{\overline{D} \otimes C1})^{C1} : T_{\mathrm{ce}} \Rightarrow N_{\mathrm{ce}}$. This gives us the main result of our paper:

Corollary 1. *For a language induced by an EIC specification – a (syntactic)* CTS, ETS *or* CETS *transition system – the above assignments of denotations $[\![-]\!]$ to programs are adequate and compositional with respect to the behavioural equivalences $\cong_{\mathrm{c}}, \cong^{\mathrm{N}}_{\mathrm{e}}, \cong^{\mathrm{N}}_{\mathrm{ce}}$. That is, two programs have the same denotation iff they are operationally equivalent; and the assignment of denotations is a congruence.*

6 Conclusion

In this paper, we have given operational and denotational semantics, and syntactic rule formats, for a class of sequential imperative languages with notions of state and/or effects. We have given proofs that under these rule formats, the induced semantics are adequate and compositional. We anticipate applications with various combinations of user input/output, probabilistic non-determinism, and perhaps local state [18,21].

However, there are some limitations on the effect-theories permitted, due to the $\mathsf{Cpo}_!$-enrichment ensuring existence of a final Kleisli-coalgebra. Exceptions are inexpressible in $\mathsf{Cpo}_!$, and the semantics of user I/O is unsatisfactory: divergent programs are identified even if their I/O behaviour is clearly different. In addition, the coalgebraic semantics may still be too fine-grained, in that it records the number of execution steps of computations. One way around these limitations might be to move towards a *weakly-final* semantics in non-strict Cpo, again taking the semantic domain to be $M\overline{D}$ where $\overline{D}$ is the initial B-algebra, $M\overline{D}$, but where the semantic map is now a least fixpoint.

For commutative effect monads M, another way of extending Turi and Plotkin's framework may be to lift both syntax and behaviour functors into the Kleisli category $\mathsf{Kl}(M)$ and use it as the base category (rather than splitting the approach as we have done). An application might be trace semantics for CCS-like languages, as in [13].

References

1. Abou-Saleh, F., Pattinson, D.: Towards effects in mathematical operational semantics. Electr. Notes Theor. Comput. Sci. 276, 81–104 (2011)
2. Adamek, J.: Recursive data types in algebraically w-complete categories. Information and Computation 118(2), 181–190 (1995)
3. Adámek, J., Rosický, J.: Locally Presentable and Accessible Categories. Cambridge University Press (1994)
4. Hasuo, I., Jacobs, B., Sokolova, A.: Generic trace semantics via coinduction. Logical Methods in Computer Science 3(4) (2007)
5. Hyland, M., Plotkin, G., Power, J.: Combining effects: sum and tensor. Theor. Comput. Sci. 357, 70–99 (2006)
6. Hyland, M., Power, J.: Discrete lawvere theories and computational effects. Theor. Comput. Sci. 366(1-2), 144–162 (2006)
7. Johann, P., Simpson, A., Voigtländer, J.: A generic operational metatheory for algebraic effects. In: Proc. LICS 2010, pp. 209–218. IEEE Computer Society (2010)
8. Kelly, G.M.: Basic concepts of enriched category theory. Reprints in Theory and Applications of Categories (10), 1–136 (2005)
9. Klin, B.: Bialgebraic methods in structural operational semantics. Electron. Notes Theor. Comput. Sci. 175(1), 33–43 (2007)
10. Kock, A.: Strong functors and monoidal monads. Archiv der Mathematik 23 (1972)
11. Lenisa, M., Power, J., Watanabe, H.: Category theory for operational semantics. Theor. Comput. Sci. 327(1-2), 135–154 (2004)
12. Moggi, E.: Notions of computation and monads. Inf. Comput. 93(1), 55–92 (1991)
13. Monteiro, L.: A Coalgebraic Characterization of Behaviours in the Linear Time – Branching Time Spectrum. In: Corradini, A., Montanari, U. (eds.) WADT 2008. LNCS, vol. 5486, pp. 251–265. Springer, Heidelberg (2009)
14. Plotkin, G., Power, J.: Tensors of comodels and models for operational semantics. Electron. Notes Theor. Comput. Sci. 218, 295–311 (2008)
15. Plotkin, G.D., Power, J.: Adequacy for Algebraic Effects. In: Honsell, F., Miculan, M. (eds.) FOSSACS 2001. LNCS, vol. 2030, pp. 1–24. Springer, Heidelberg (2001)
16. Plotkin, G., Power, J.: Notions of Computation Determine Monads. In: Nielsen, M., Engberg, U. (eds.) FOSSACS 2002. LNCS, vol. 2303, pp. 342–356. Springer, Heidelberg (2002)
17. Power, J.: Countable lawvere theories and computational effects. Electr. Notes Theor. Comput. Sci. 161, 59–71 (2006)
18. Power, J.: Semantics for local computational effects. Electr. Notes Theor. Comput. Sci. 158, 355–371 (2006)
19. Power, J., Shkaravska, O.: From comodels to coalgebras: State and arrays. Electron. Notes Theor. Comput. Sci. 106 (2004)
20. Power, J., Turi, D.: A coalgebraic foundation for linear time semantics. In: Category Theory and Computer Science. Elsevier (1999)
21. Staton, S.: Completeness for Algebraic Theories of Local State. In: Ong, L. (ed.) FOSSACS 2010. LNCS, vol. 6014, pp. 48–63. Springer, Heidelberg (2010)
22. Turi, D.: Functorial Operational Semantics and its Denotational Dual. PhD thesis, Free University, Amsterdam (June 1996)
23. Turi, D.: Categorical modelling of structural operational rules: Case studies. In: Category Theory and Computer Science, pp. 127–146 (1997)
24. Turi, D., Plotkin, G.D.: Towards a mathematical operational semantics. In: LICS, pp. 280–291 (1997)

Preorders on Monads and Coalgebraic Simulations

Shin-ya Katsumata and Tetsuya Sato

Research Institute for Mathematical Sciences, Kyoto University, Kyoto, 606-8502, Japan
{sinya,satoutet}@kurims.kyoto-u.ac.jp

Abstract. We study the construction of preorders on **Set**-monads by the semantic ⊤⊤-lifting. We show the universal property of this construction, and characterise the class of preorders on a monad as a limit of a $\mathbf{Card}^{op}$-chain. We apply these theoretical results to identifying preorders on some concrete monads, including the powerset monad, maybe monad, and their composite monad. We also relate the construction of preorders and coalgebraic formulation of simulations.

1 Introduction

In the coalgebraic treatment of labelled transition systems and process calculi, several coalgebraic formulations of *bisimulations* are proposed [1,12,18], and their relationships are well-studied [25]. On the other hand, to express the asymmetry of simulations between coalgebras, we need to generalise the framework of bisimulations. One of the earliest works in this direction is [13], where Hesselink and Thijs introduced a class of relational liftings of **Set**-functors called *relational extensions*, with which simulations can be coalgebraically captured. Hughes and Jacobs took *preordered functors* as a basis for constructing relational extensions of endofunctors. This approach was further developed in the subsequent studies on coalgebraic trace semantics [10] and forward and backward simulations of coalgebras [9]. The key assumption in the last two works is that an *order enrichment* is given to the Kleisli category of a monad.

One natural problem arising in this line of research is how to systematically construct preordered functors. In fact, many coalgebra functors of transition systems contain the functor part of *monads* to describe branching types of transition systems, and they are the focal point when considering relational liftings and preorders on endofunctors. Upon this observation, we address the problem of constructing preorders on monads, and study its relationship to the coalgebraic formulation of simulations.

The main technical vehicle to tackle the problem is *semantic* ⊤⊤*-lifting* [16], which originates from the proof of the strong normalisation of Moggi's computational metalanguage by reducibility candidates [21,22]. We apply the semantic ⊤⊤-lifting to construct preorders on monads, and show that this construction satisfies a universal property. We also characterise the class of preorders on a monad as the limit of a large chain of certain preorders. We then apply these theoretical results to identifying preorders on some concrete monads, including the semiring-valued multiset monad, powerset monad and maybe monad. We finally show that the semantic ⊤⊤-lifting satisfies a couple of properties that are relevant to the coalgebraic formulation of simulations.

F. Pfenning (Ed.): FOSSACS 2013, LNCS 7794, pp. 145–160, 2013.

Preliminaries

Throughout this paper we assume the axiom of choice. We write **Pre** (resp. **Pos**) for the cartesian monoidal category of preorders (resp. posets) and monotone functions between them. For sets I, J, by $I \Rightarrow J$ we mean the set of functions from I to J. Each preorder $\leq$ on a set J extends to the pointwise preorder on a function space $I \Rightarrow J$, which we denote by $\dot{\leq}$. In this paper the metavariable $\mathcal{T}$ (and its variants) is reserved for monads over **Set**. Its components are written by (T, η, μ). For a function $f : I \to TJ$, by $f^\#$ we mean the Kleisli lifting of f, that is, the function $\mu_J \circ Tf$. A *preordered functor* [13,15] consists of an endofunctor F : **Set** $\to$ **Set** and an assignment $I \mapsto \sqsubseteq_I$ of a preorder on FI such that for any function $f : I \to J$, Ff is a monotone function from $(FI, \sqsubseteq_I)$ to $(FJ, \sqsubseteq_J)$.

2 Preorders on Monads

Definition 1. *Let I be a set. We call a binary relation $S \subseteq TI \times TI$* substitutive *if for each function $f : I \to TI$ and $(x, y) \in S$, $(f^\#(x), f^\#(y)) \in S$.*

Especially, a preorder $\leq$ on TI is substitutive if and only if for each function $f : I \to TI$, $f^\#$ is a monotone function of type $(TI, \leq) \to (TI, \leq)$.

Definition 2. *Let I be a set. We call a preorder $\leq$ on TI* congruent *if for each set J and functions $f, g : J \to TI$, $f \dot{\leq} g$ implies $f^\# \dot{\leq} g^\#$.*

Under the correspondence between monads and algebraic theories, TI may be viewed as the set of I-many variable polynomials in the algebraic theory corresponding to $\mathcal{T}$. Then a binary relation $S \subseteq TI \times TI$ is substitutive if for each polynomial pair $(t, u) \in S$ and a simultaneous substitution $[i := v_i]_{i \in I}$ of polynomials, we have $(t[i := v_i]_{i \in I}, u[i := v_i]_{i \in I}) \in S$. The congruence of a preorder $\leq$ on TI means that for each polynomial $v \in TJ$ and two simultaneous substitutions $[j := t_j]_{j \in J}$ and $[j := u_j]_{j \in J}$ such that $t_j \leq u_j$, we have $v[j := t_j]_{j \in J} \leq v[j := u_j]_{j \in J}$.

We introduce the main subject of this paper, *preorders on monads*.

Definition 3. *A* preorder $\sqsubseteq$ *on $\mathcal{T}$ is an assignment of a preorder $\sqsubseteq_I$ on TI to each set I such that*

1. *each preorder $\sqsubseteq_I$ is congruent, and*
2. *for each function $f : I \to TJ$, $f^\#$ is a monotone function from $(TI, \sqsubseteq_I)$ to $(TJ, \sqsubseteq_J)$ (we also call this property* substitutivity*).*

From this definition, $\sqsubseteq_I$ is substitutive for each set I, and $(T, \sqsubseteq)$ is a preordered functor. We write **Pre**$(\mathcal{T})$ for the class of preorders on $\mathcal{T}$. We define a pointwise partial order $\preceq$ on **Pre**$(\mathcal{T})$ by: $\sqsubseteq \preceq \sqsubseteq'$ if $\sqsubseteq_I \subseteq \sqsubseteq'_I$ holds for each set I. The class **Pre**$(\mathcal{T})$ admits intersections of arbitrary size: for a subcollection $\mathbf{\sqsubseteq}$ of **Pre**$(\mathcal{T})$, its intersection is the preorder $\bigcap \mathbf{\sqsubseteq}$ on $\mathcal{T}$ defined by: $a\, (\bigcap \mathbf{\sqsubseteq})_I\, b$ if $a \sqsubseteq_I b$ holds for each preorder $\sqsubseteq \in \mathbf{\sqsubseteq}$.

Example 1. We write $\mathcal{T}_p$ for the powerset monad. For each set I, $T_p I$ has a natural preorder given by the set inclusion. This is a preorder on $\mathcal{T}_p$.

Example 2. We write $\mathcal{T}_l$ for the monad whose functor part is given by $T_l I = I + \{*\}$; this is known as the *maybe monad* in Haskell. We assign to each set I the flat partial order on $T_l I$ that makes $\iota_2(*)$ the least element. This is a preorder on $\mathcal{T}_l$.

Example 3. We write $\mathcal{T}_m$ for the free monoid monad. For each set I, we define a preorder $\sqsubseteq_I$ on $T_m I$ by: $x \sqsubseteq_I y$ if the length of x is equal or shorter than y. This is *not* a preorder on T_m because it is not substitutive.

Suppose that the Kleisli category $\mathbf{Set}_{\mathcal{T}}$ of a monad $\mathcal{T}$ is **Pre**-enriched, and moreover the enrichment is *pointwise*, that is, $(\forall x \in \mathbf{Set}_{\mathcal{T}}(1, I)\,.\, f^{\#} \circ x \sqsubseteq_{1,J} g^{\#} \circ x)$ implies $f \sqsubseteq_{I,J} g$ for all $f, g \in \mathbf{Set}_{\mathcal{T}}(I, J)$. Then the assignment $I \mapsto \sqsubseteq_{1,I}$ gives a preorder on $\mathcal{T}$ under the identification $\mathbf{Set}_{\mathcal{T}}(1, I) \simeq TI$. Conversely, given a preorder $\sqsubseteq$ on $\mathcal{T}$, the assignment of the preorder $\dot{\sqsubseteq}_J$ to $\mathbf{Set}_{\mathcal{T}}(I, J)$ gives a pointwise **Pre**-enrichment. This correspondence between pointwise **Pre**-enrichments on $\mathbf{Set}_{\mathcal{T}}$ and preorders on $\mathcal{T}$ is bijective.

3 Relational Liftings and Preorders on Monads

After reviewing a coalgebraic formulation of (bi)simulations in the category **BRel** of binary relations and relation-respecting functions, we introduce a relational lifting of monads, called *preorder* $\top\top$*-lifting*, and show that it gives rise to preorders on monads.

3.1 The Category BRel of Binary Relations

We define the category **BRel** (which is the same as Rel in [15]) by the following data. An object in **BRel** is a triple (X, I_1, I_2) such that $X \subseteq I_1 \times I_2$. A morphism from (X, I_1, I_2) to (Y, J_1, J_2) is a pair (f_1, f_2) of functions $f_1 : I_1 \to J_1$ and $f_2 : I_2 \to J_2$ such that for each $(i_1, i_2) \in X$, $(f_1(i_1), f_2(i_2)) \in Y$. We use bold letters $\mathbf{X}, \mathbf{Y}, \mathbf{Z}$ to range over objects in **BRel**, and refer to each component of $\mathbf{X} \in \mathbf{BRel}$ by $(\mathbf{X}_0, \mathbf{X}_1, \mathbf{X}_2)$. We write $i_{\mathbf{X}} : \mathbf{X}_0 \to \mathbf{X}_1 \times \mathbf{X}_2$ for the inclusion function. We say that $\mathbf{X} \in \mathbf{BRel}$ is *above* $(I_1, I_2) \in \mathbf{Set}^2$ if $\mathbf{X}_1 = I_1$ and $\mathbf{X}_2 = I_2$. Objects above the same $\mathbf{Set}^2$-object are ordered by the inclusion of their relation part. We denote this order by $\subseteq$. For each object $\mathbf{X}, \mathbf{Y}$ in **BRel** and morphism $(f_1, f_2) : (\mathbf{X}_1, \mathbf{X}_2) \to (\mathbf{Y}_1, \mathbf{Y}_2)$ in $\mathbf{Set}^2$, we abbreviate $(f_1, f_2) \in \mathbf{BRel}(\mathbf{X}, \mathbf{Y})$ to $(f_1, f_2) : \mathbf{X} \dot{\to} \mathbf{Y}$. We call a pair $(\mathbf{X}, \mathbf{Y})$ of objects in **BRel** *composable* if $\mathbf{X}_2 = \mathbf{Y}_1$. Their composition $\mathbf{X} * \mathbf{Y}$ is given by the relational composition of $\mathbf{X}_0$ and $\mathbf{Y}_0$:

$$\mathbf{X} * \mathbf{Y} = (\{(x_1, y_2) \mid \exists z \in \mathbf{X}_2\,.\,(x_1, z) \in \mathbf{X}_0, (z, y_2) \in \mathbf{Y}_0\}, \mathbf{X}_1, \mathbf{Y}_2).$$

A preorder $\leq$ on a set I determines a **BRel**-object $(\leq, I, I)$, which we also denote by $\leq$. We write Eq_I for the **BRel**-object of the identity relation on I.

The category **BRel** arises as the vertex of the pullback of the subobject fibration $p : \mathbf{Sub(Set)} \to \mathbf{Set}$ (see [14, Chapter 0]) along the product functor $D : \mathbf{Set}^2 \to \mathbf{Set}$:

$$\begin{array}{ccc} \mathbf{BRel} & \longrightarrow & \mathbf{Sub(Set)} \\ {\scriptstyle \pi}\downarrow & & \downarrow{\scriptstyle p} \\ \mathbf{Set}^2 & \xrightarrow[D]{} & \mathbf{Set} \end{array} \qquad \text{where } \pi \text{ is} \begin{cases} \pi(\mathbf{X}) = (\mathbf{X}_1, \mathbf{X}_2), \\ \pi(f_1, f_2) = (f_1, f_2) \end{cases}$$

The leg $\pi : \mathbf{BRel} \to \mathbf{Set}^2$ of the pullback is a partial order fibration [14]. For an object $\mathbf{X}$ in $\mathbf{BRel}$ and a morphism $(f_1, f_2) : (I_1, I_2) \to (\mathbf{X}_1, \mathbf{X}_2)$ in $\mathbf{Set}^2$, we define the *inverse image object* $(f_1, f_2)^*\mathbf{X}$ by

$$(f_1, f_2)^*\mathbf{X} = (\{(x_1, x_2) \mid (f_1(x_1), f_2(x_2)) \in X\}, I_1, I_2).$$

The category $\mathbf{BRel}$ has a bi-cartesian closed structure that is strictly preserved by π. The object part of this structure is given as follows:

$$\dot{\prod}_{i\in I}\mathbf{X}_i = (\{(x, y) \mid \forall i \in I \,.\, (\pi_i(x), \pi_i(y)) \in (\mathbf{X}_i)_0\}, \textstyle\prod_{i\in I}(\mathbf{X}_i)_1, \prod_{i\in I}(\mathbf{X}_i)_2)$$
$$\dot{\coprod}_{i\in I}\mathbf{X}_i = (\textstyle\bigcup_{i\in I}\{(\iota_i(x), \iota_i(y)) \mid (x, y) \in (\mathbf{X}_i)_0\}, \coprod_{i\in I}(\mathbf{X}_i)_1, \coprod_{i\in I}(\mathbf{X}_i)_2)$$
$$\mathbf{X} \dot{\Rightarrow} \mathbf{Y} = (\{(f, g) \mid \forall (x, y) \in \mathbf{X}_0 \,.\, (f(x), g(y)) \in \mathbf{Y}_0\}, \mathbf{X}_1 \Rightarrow \mathbf{Y}_1, \mathbf{X}_2 \Rightarrow \mathbf{Y}_2).$$

This structure captures the essence of *logical relations* for product, coproduct and arrow types interpreted in type hierarchies [23]. We note that the equality functor $\mathrm{Eq} : \mathbf{Set} \to \mathbf{BRel}$ also preserves the bi-CC structure (*identity extension*).

3.2 Relational Liftings and Coalgebraic Simulations

Definition 4. *A* relational lifting *of an endofunctor F : $\mathbf{Set} \to \mathbf{Set}$ is an assignment $\dot{F} : |\mathbf{BRel}| \to |\mathbf{BRel}|$ such that for each morphism $(f, g) : \mathbf{X} \dot{\to} \mathbf{Y}$, we have $(Ff, Fg) : \dot{F}\mathbf{X} \dot{\to} \dot{F}\mathbf{Y}$. We say that $\dot{F}$ is*

- reflexive *if* $\mathrm{Eq}_{FI} \subseteq \dot{F}\mathrm{Eq}_I$,
- lax compositional *if* $\dot{F}\mathbf{X} * \dot{F}\mathbf{Y} \subseteq \dot{F}(\mathbf{X} * \mathbf{Y})$,
- compositional *if* $\dot{F}\mathbf{X} * \dot{F}\mathbf{Y} = \dot{F}(\mathbf{X} * \mathbf{Y})$, *and*
- *a* relational extension *[13] if it is reflexive and compositional.*

A relational lifting bijectively corresponds to an endofunctor $\dot{F} : \mathbf{BRel} \to \mathbf{BRel}$ such that $\pi \circ \dot{F} = F^2 \circ \pi$. We later see that the lax compositionality guarantees the *composability* of simulations between coalgebras.

Example 4. The bi-cartesian closed structure on $\mathbf{BRel}$ gives canonical relational extensions of functors consisting of Id, C_A (the constant functor for a set A), $+$ and $\times$. For instance, the canonical lifting of $FX = C_A + X \times X$ is $\dot{F}\mathbf{X} = \mathrm{Eq}_A \dot{+} \mathbf{X} \dot{\times} \mathbf{X}$.

Example 5. The following relational lifting $\overline{F}$ is known to capture the concept of bisimulation between F-coalgebras in many cases (see e.g. [12]):

$$\overline{F}\mathbf{X} = (\mathrm{Im}, F\mathbf{X}_1, F\mathbf{X}_2),$$

where Im is the image of $\langle F\pi_1, F\pi_2\rangle \circ Fi_{\mathbf{X}} : F\mathbf{X}_0 \to F\mathbf{X}_1 \times F\mathbf{X}_2$. It is always reflexive, and also compositional if and only if F preserves weak pullbacks [3].

Example 6. In [13, Section 4.1] Hesselink and Thijs give the following construction of a relational lifting $F^{+(\sqsubseteq)}(\mathbf{X})$ from a preordered functor $(F, \sqsubseteq)$:

$$F^{+(\sqsubseteq)}(\mathbf{X}) = \sqsubseteq_{\mathbf{X}_1} * \overline{F}\mathbf{X} * \sqsubseteq_{\mathbf{X}_2}.$$

They show that every relational extension $\dot{F}$ of a $\mathbf{Set}$-functor F gives rise to a preordered functor $(F, \dot{F}(\mathrm{Eq}_-))$, and $\dot{F}$ can be recovered as $\dot{F} = F^{+(\dot{F}(\mathrm{Eq}_-))}$. In [20], it is shown that the preordered functor $(F, \sqsubseteq)$ is *stable* (Definition 10, [20]) if and only if $F^{+(\sqsubseteq)}$ is a relational extension such that $(F^{+(\sqsubseteq)}, F^2)$ is an endomorphism over π.

A natural generalisation of the coalgebraic formulation of (bi)simulations in [12,15] is to make it parametrised by relational liftings of coalgebra functors.

Definition 5. *Let $\dot{F}$ be a relational lifting of an endofunctor F : **Set** $\rightarrow$ **Set**. An $\dot{F}$-*simulation *from an F-coalgebra (I_1, f_1) to another F-coalgebra (I_2, f_2) is an object $\mathbf{X} \in$ **BRel** above (I_1, I_2) such that $(f_1, f_2) : \mathbf{X} \dot{\rightarrow} \dot{F}\mathbf{X}$.*

Example 7. Hermida and Jacobs formulated bisimulations between F-coalgebras as $\overline{F}$-simulations [12]. Later, Hughes and Jacobs employed $F^{+(\sqsubseteq)}$-simulations to capture the concept of simulations between F-coalgebras [15].

Here are some properties of $\dot{F}$-simulations. I) $\dot{F}$-simulations are closed under the union of arbitrary family. II) If $\dot{F}$ is reflexive, $\dot{F}$-simulations are $\overline{F}$-simulations. III) If $\dot{F}$ is lax compositional, $\dot{F}$-simulations are closed under the relational composition $*$.

We extend the concept of relational liftings of endofunctors to monads.

Definition 6. *A* relational lifting *of $\mathcal{T}$ is an assignment $\dot{T}$: |**BRel**| $\rightarrow$ |**BRel**| such that*

- *For each object* $\mathbf{X}$ *in* **BRel**, *we have $(\eta_{\mathbf{X}_1}, \eta_{\mathbf{X}_2}) : \mathbf{X} \dot{\rightarrow} \dot{T}\mathbf{X}$, and*
- *for each morphism $(f_1, f_2) : \mathbf{X} \dot{\rightarrow} \dot{T}\mathbf{Y}$, we have $(f_1^\#, f_2^\#) : \dot{T}\mathbf{X} \dot{\rightarrow} \dot{T}\mathbf{Y}$.*

A relational lifting of $\mathcal{T}$ bijectively corresponds to a monad $\dot{\mathcal{T}} = (\dot{T}, \dot{\eta}, \dot{\mu})$ over **BRel** such that

$$\pi(\dot{T}\mathbf{X}) = (T\mathbf{X}_1, T\mathbf{X}_2), \ \ \pi(\dot{T}(f_1, f_2)) = (Tf_1, Tf_2), \ \ \dot{\eta}_{\mathbf{X}} = (\eta_{\mathbf{X}_1}, \eta_{\mathbf{X}_2}), \ \ \dot{\mu}_{\mathbf{X}} = (\mu_{\mathbf{X}_1}, \mu_{\mathbf{X}_2}).$$

We note that every relational lifting $\dot{T}$ of $\mathcal{T}$ is a strong monad over **BRel**, and its strength $\dot{\theta}$ satisfies $\pi(\dot{\theta}_{\mathbf{X},\mathbf{Y}}) = (\theta_{\mathbf{X}_1,\mathbf{Y}_1}, \theta_{\mathbf{X}_2,\mathbf{Y}_2})$, where θ is the canonical strength of $\mathcal{T}$.

The relational lifting in Example 5 extends to monads:

Proposition 1. *For each monad $\mathcal{T}$, $\overline{T}$ is a relational lifting of $\mathcal{T}$.*

Larrecq, Lasota and Nowak further generalised this fact using subscones and mono factorisation systems [8]. Hesselink and Thijs's construction in Example 6 also yields relational liftings of monads, when preorders on monads are supplied:

Proposition 2. *For each monad $\mathcal{T}$ and preorder $\sqsubseteq$ on $\mathcal{T}$, $T^{+(\sqsubseteq)}$ is a lifting of $\mathcal{T}$.*

3.3 Preorder ⊤⊤-Lifting

Inspired from [22,21,24], in [16] the first author introduced *semantic ⊤⊤-lifting*, a method to lift strong monads on the base category $\mathbb{B}$ of a certain partial order fibration $p : \mathbb{E} \rightarrow \mathbb{B}$ to its total category $\mathbb{E}$. This method takes a pair (R, S) such that $pS = TR$ as a parameter of the lifting, and by varying this parameter we can derive various liftings of $\mathcal{T}$. In this paper, we apply the semantic ⊤⊤-lifting to the strong monad $\mathcal{T}^2$ over $\mathbf{Set}^2$ and the fibration π : **BRel** $\rightarrow \mathbf{Set}^2$, and we supply congruent (and substitutive) preorders to the semantic ⊤⊤-lifting as parameters.

Definition 7. *A* preorder parameter *for $\mathcal{T}$ is a pair $(R, \leq)$ of a set R and a congruent preorder $\leq$ on TR.*

The following is a special case of the semantic $\top\top$-lifting [16, Definition 3.2], where a preorder parameter is supplied.

Definition 8. *Let $(R, \leq)$ be a preorder parameter for $\mathcal{T}$. We write σ_I for the function $\lambda xk \,.\, k^\#(x) : TI \to (I \Rightarrow TR) \Rightarrow TR$.*[1] *We define the assignment $T^{\top\top(R,\leq)} : |\mathbf{BRel}| \to |\mathbf{BRel}|$ by:*

$$T^{\top\top(R,\leq)}\mathbf{X} = (\sigma_{\mathbf{X}_1}, \sigma_{\mathbf{X}_2})^*((\mathbf{X} \dot{\Rightarrow} \leq) \dot{\Rightarrow} \leq). \tag{1}$$

Below we call $T^{\top\top(R,\leq)}$ *preorder* $\top\top$*-lifting* to distinguish it from the general semantic $\top\top$-lifting. When the preorder parameter is obvious from context, we simply write $T^{\top\top}$ instead of $T^{\top\top(R,\leq)}$. An equivalent definition of $T^{\top\top}\mathbf{X}$ using an auxiliary object $\mathbf{X}^\top$ is:

$$\mathbf{X}^\top = \mathbf{X} \dot{\Rightarrow} \leq = (\{(f_1, f_2) \mid \forall (x_1, x_2) \in \mathbf{X}_0 \,.\, f_1(x_1) \leq f_2(x_2)\}, \mathbf{X}_1 \Rightarrow TR, \mathbf{X}_2 \Rightarrow TR),$$
$$T^{\top\top}\mathbf{X} = ((\{(x_1, x_2) \mid \forall (f_1, f_2) \in (\mathbf{X}^\top)_0 \,.\, f_1^\#(x_1) \leq f_2^\#(x_2)\}, T\mathbf{X}_1, T\mathbf{X}_2).$$

Theorem 1 ([16]). *The preorder $\top\top$-lifting $T^{\top\top}$ is a relational lifting of $\mathcal{T}$.*

Example 8 (Example 3.6, [16]). We regard $T_p 1 = \{\emptyset, 1\}$ as the congruent preorder $\emptyset \leq 1$. The preorder $\top\top$-lifting of $\mathcal{T}_p$ with this preorder parameter is

$$T_p^{\top\top}\mathbf{X} = (\{(P_1, P_2) \mid \forall x_1 \in P_1 \,.\, \exists x_2 \in P_2 \,.\, (x_1, x_2) \in \mathbf{X}_0\}, T_p\mathbf{X}_1, T_p\mathbf{X}_2).$$

Every preorder $\top\top$-lifting of a monad $\mathcal{T}$ yields a preorder on $\mathcal{T}$.

Theorem 2. *Let $(R, \leq)$ be a preorder parameter for $\mathcal{T}$.*

1. *For each set I, we have $T^{\top\top}\mathrm{Eq}_I = (\{(x, y) \mid \forall f : I \to TR \,.\, f^\#(x) \leq f^\#(y)\}, TI, TI)$.*
2. *The assignment $I \mapsto T^{\top\top}\mathrm{Eq}_I$ is a preorder on $\mathcal{T}$ (which we denote by $[\leq]^R$).*

Proof. We note that $(T^{\top\top}\mathrm{Eq}_I)_0 = \{(x, y) \mid \forall f, g : I \to TR \,.\, f \dot{\leq} g \implies f^\#(x) \leq g^\#(y)\}$.

1. ($\supseteq$) Immediate. ($\subseteq$) Let $x, y \in TI$ and assume $\forall h : I \to TR \,.\, h^\#(x) \leq h^\#(y)$. For functions $f, g : I \to TR$ such that $f \dot{\leq} g$, we have $f^\#(x) \leq g^\#(x)$ as $\leq$ is congruent, and $g^\#(x) \leq g^\#(y)$ from the assumption. Therefore $f^\#(x) \leq g^\#(y)$ holds by the transitivity of $\leq$.
2. (Transitivity) Let $(x, y), (y, z) \in T^{\top\top}\mathrm{Eq}_I$. From 1, for any function $f : I \to TR$, we have $f^\#(x) \leq f^\#(y)$ and $f^\#(y) \leq f^\#(z)$, hence $f^\#(x) \leq f^\#(z)$. (Reflexivity) Reflexivity is immediate from the congruence of $\leq$. (Congruence) The Kleisli lifting of $(f, g) : \mathrm{Eq}_I \dot{\to} T^{\top\top}\mathrm{Eq}_J$ satisfies $(f^\#, g^\#) : T^{\top\top}\mathrm{Eq}_I \dot{\to} \dot{T}\mathrm{Eq}_J$. From the reflexivity of $T^{\top\top}\mathrm{Eq}_I$, we have $(f^\#, g^\#) : \mathrm{Eq}_{TI} \subseteq T^{\top\top}\mathrm{Eq}_I \dot{\to} T^{\top\top}\mathrm{Eq}_I$. (Substitutivity) Let $f : I \to TJ$ be a function and $x, y \in TI$ such that $(x, y) \in T^{\top\top}\mathrm{Eq}_I$. For each function $g : J \to TR$, we have
$$g^\#(f^\#(x)) = (g^\# \circ f)^\#(x) \leq (g^\# \circ f)^\#(y) = g^\#(f^\#(y)),$$
implying $(f^\#(x), f^\#(x)) \in T^{\top\top}\mathrm{Eq}_J$.

Below we write $\mathbf{CSPre}(\mathcal{T}, I)$ for the set of congruent and substitutive preorders on TI, ordered by inclusion. The mapping $(-)_I : \sqsubseteq \mapsto \sqsubseteq_I$ is a monotone function of type $\mathbf{Pre}(\mathcal{T}) \to \mathbf{CSPre}(\mathcal{T}, I)$. We characterise the assignment $\leq \mapsto [\leq]^R$ as the right adjoint $[-]^I : \mathbf{CSPre}(\mathcal{T}, I) \to \mathbf{Pre}(\mathcal{T})$ to $(-)_I$.

[1] This is called the *unit* of the continuation monad transformer [4].

Theorem 3. *For each set I, we have the following adjunction $(-)_I \dashv [-]^I$ such that* $[-]^I_I = \mathrm{id}$.

$$(\mathbf{CSPre}(\mathcal{T}, I), \subseteq) \underset{(-)_I}{\overset{[-]^I}{\underset{\top}{\rightleftarrows}}} (\mathbf{Pre}(\mathcal{T}), \preceq). \qquad (2)$$

Proof. Monotonicity of $[-]^I$ is easy. We show $\sqsubseteq \preceq [\sqsubseteq_I]^I$. Let J be a set and suppose $x \sqsubseteq_J y$. Then from the substitutivity of $\sqsubseteq$, for each function $f : J \to TI$, we have $f^\#(x) \sqsubseteq_I f^\#(y)$, that is, $x\, [\sqsubseteq_I]^I_J\, y$. Next, we show $[\leq]^I_I = \leq$. We first calculate $[\leq]^I_I$:

$$[\leq]^I_I = \{(x, y) \mid \forall f : I \to TI \,.\, f^\#(x) \leq f^\#(y)\}$$

Then $\leq\, \subseteq [\leq]^I_I$ is equivalent to the substitutivity of $\leq$, which is already assumed. To show $[\leq]^I_I \subseteq \leq$, use the unit $\eta_I : I \to TI$ of $\mathcal{T}$.

Example 9. 1. We define a congruent preorder $\leq$ on $T_m 2 = 2^*$ by: $x \leq y$ if x is a subsequence of y. Then we have $x\, [\leq]^2_I\, y$ if and only if x is a subsequence of y.
2. For $x \in T_m I$ and $i \in I$, by $o(x, i)$ we mean the number of occurrences of i in x. For each congruent preorder $\leq$ on $T_m 1 \simeq \mathbf{N}$, we have $x\, [\leq]^1_I\, y$ if and only if $\forall i \in I \,.\, o(x, i) \leq o(y, i)$.

4 Characterising Pre($\mathcal{T}$) as the Limit of a Large Chain

Using the family of adjunctions (2), for sets I, J we define the monotone function $\varphi_{I,J}$: $\mathbf{CSPre}(\mathcal{T}, I) \to \mathbf{CSPre}(\mathcal{T}, J)$ by $\varphi_{I,J}(\leq) = [\leq]^I_J$. Theorem 3 asserts $\varphi_{I,I} = \mathrm{id}$.

Lemma 1. *For each $\sqsubseteq\, \in \mathbf{Pre}(\mathcal{T})$ and sets I, J such that* $\mathrm{card}(I) \leq \mathrm{card}(J)$, *we have* $\sqsubseteq_I = [\sqsubseteq_J]^J_I$.

Proof. From $\sqsubseteq\, \preceq [\sqsubseteq_J]^J$, we have $\sqsubseteq_I\, \subseteq [\sqsubseteq_J]^J_I$. We show the converse. We take an injection $i : I \rightarrowtail J$ and a surjection $s : J \twoheadrightarrow I$ such that $s \circ i = \mathrm{id}$. Suppose $x[\sqsubseteq_J]^J_I y$. Then for the function $\eta \circ i : I \to TJ$, the following holds:

$$Ti(x) = (\eta \circ i)^\#(x) \sqsubseteq_J (\eta \circ i)^\#(y) = Ti(y).$$

From the substitutivity of $\sqsubseteq$, we obtain $x \sqsubseteq_I y$, because

$$x = Ts \circ Ti(x) = (\eta \circ s)^\#(Ti(x)) \sqsubseteq_I (\eta \circ s)^\#(Ti(y)) = Ts \circ Ti(y) = y.$$

Lemma 2. *For sets I, J, K such that* $\mathrm{card}(I) \leq \mathrm{card}(J)$, *we have* $\varphi_{J,I} \circ \varphi_{K,J} = \varphi_{K,I}$.

Proof. We have $\varphi_{K,I}(\leq) = [\leq]^K_I \overset{*}{=} [[\leq]^K_J]^J_I = \varphi_{J,I} \circ \varphi_{K,J}(\leq)$; here, $\overset{*}{=}$ is by Lemma 1.

This implies that when $\mathrm{card}(I) \leq \mathrm{card}(J)$, we have $\varphi_{J,I} \circ \varphi_{I,J} = \mathrm{id}$, hence $\varphi_{I,J}$ is a split monomorphism in **Pos**.

Lemma 3. *For each $\sqsubseteq\, \in \mathbf{Pre}(\mathcal{T})$ and sets I, J such that* $\mathrm{card}(I) \leq \mathrm{card}(J)$, *we have* $[\sqsubseteq_I]^I \succeq [\sqsubseteq_J]^J$.

Proof. We have $[\sqsubseteq_J]^J \preceq [[\sqsubseteq_J]_I^J]^I = [\sqsubseteq_I]^I$; the last step is by Lemma 1.

Thus each $\sqsubseteq \in \mathbf{Pre}(\mathcal{T})$ determines a descending chain of preorders on $\mathcal{T}$ indexed by cardinals: $[\sqsubseteq_0]^0 \succeq [\sqsubseteq_1]^1 \succeq \cdots$, and $\sqsubseteq$ is a lower bound by Theorem 3. In fact, $\sqsubseteq$ is the *greatest* lower bound:

Theorem 4. *For each* $\sqsubseteq \in \mathbf{Pre}(\mathcal{T})$*, we have* $\sqsubseteq = \bigcap_{\alpha\in\mathbf{Card}} [\sqsubseteq_\alpha]^\alpha$.

Proof. It is sufficient to show $\bigcap_{\alpha\in\mathbf{Card}} [\sqsubseteq_\alpha]^\alpha \preceq \sqsubseteq$. Let I be a set, $x, y \in TI$ and suppose that $x \, [\sqsubseteq_\alpha]_I^\alpha \, y$ holds for any cardinal α; so this especially holds at card(I). Taking a bijection $h : I \to \mathrm{card}(I)$, we obtain $Th(x) \sqsubseteq_{\mathrm{card}(I)} Th(y)$. As $\sqsubseteq$ is substitutive, we have $x = Th^{-1} \circ Th(x) \sqsubseteq_I Th^{-1} \circ Th(y) = y$.

Let us write **Card** for the linear order of cardinals (recall that we assume the axiom of choice). To clarify the relationship between $\mathbf{Pre}(\mathcal{T})$ and $\mathbf{CSPre}(\mathcal{T}, -)$, we extend the assignment $\alpha \in \mathbf{Card} \mapsto \mathbf{CSPre}(\mathcal{T}, \alpha)$ to a functor $\mathbf{CSPre}(\mathcal{T}, -) : \mathbf{Card}^{\mathrm{op}} \to \mathbf{Pre}$; the morphism part is given by φ. We thus obtain a large chain:

$$\mathbf{CSPre}(\mathcal{T}, 0) \xleftarrow{\varphi_{1,0}} \mathbf{CSPre}(\mathcal{T}, 1) \xleftarrow{\varphi_{2,1}} \cdots \longleftarrow \mathbf{CSPre}(\mathcal{T}, \aleph_0) \xleftarrow{\varphi_{\aleph_0,\aleph_1}} \cdots$$

We characterise $\mathbf{Pre}(\mathcal{T})$ as a limit of this large chain.

Theorem 5. *The family* $(-)_\alpha : \mathbf{Pre}(\mathcal{T}) \to \mathbf{CSPre}(\mathcal{T}, \alpha)$ *is a limiting cone.*

Proof. We first show that $(-)_\alpha : \mathbf{Pre}(\mathcal{T}) \to \mathbf{CSPre}(\mathcal{T}, \alpha)$ is a cone over $\mathbf{CSPre}(\mathcal{T}, -)$. Let $\sqsubseteq \in \mathbf{Pre}(\mathcal{T})$ and α, β be cardinals such that $\alpha \le \beta$. Then $\varphi_{\beta,\alpha}(\sqsubseteq_\beta) = [\sqsubseteq_\beta]_\alpha^\beta = \sqsubseteq_\alpha$ by Lemma 1.

Next, let V be a class and $p : V \to \mathbf{CSPre}(\mathcal{T}, -)$ be a cone. We construct the unique mediating mapping $m : V \to \mathbf{Pre}(\mathcal{T})$ such that $(-)_\alpha \circ m = p_\alpha$. For this, we first prove the following lemma:

Lemma 4. *For each class* V*, cone* $p : V \to \mathbf{CSPre}(\mathcal{T}, -)$ *and cardinals* α, β *such that* $\alpha \le \beta$*, we have* $[p_\alpha(v)]^\alpha \succeq [p_\beta(v)]^\beta$.

Proof. As p is a cone, for any cardinal $\alpha \le \beta$, we have $\varphi_{\beta,\alpha}(p_\beta(v)) = [p_\beta(v)]_\alpha^\beta = p_\alpha(v)$. Then $[p_\alpha(v)]^\alpha = [[p_\beta(v)]_\alpha^\beta]^\alpha \succeq [p_\beta(v)]^\beta$; the last step is by Lemma 3.

Therefore every $v \in V$ determines a decreasing sequence of preorders on $\mathcal{T}$: $[p_0(v)]^0 \succeq [p_1(v)]^1 \succeq \cdots$. We then define a mapping $m : V \to \mathbf{Pre}(\mathcal{T})$ by

$$m(v) = \bigcap_{\alpha\in\mathbf{Card}} [p_\alpha(v)]^\alpha.$$

This mapping satisfies $m(v)_\alpha = p_\alpha(v)$ because

$$m(v)_\alpha = \bigcap_{\beta\in\mathbf{Card}} [p_\beta(v)]_\alpha^\beta = \bigcap_{\beta\in\mathbf{Card},\alpha\le\beta} [p_\beta(v)]_\alpha^\beta = \bigcap_{\beta\in\mathbf{Card},\alpha\le\beta} p_\alpha(v) = p_\alpha(v).$$

When another mapping $m' : V \to \mathbf{Pre}(\mathcal{T})$ satisfies $m'(v)_\alpha = p_\alpha(v)$, then $m'(v) = m(v)$ because

$$m'(v) = \bigcap_{\alpha\in\mathbf{Card}} [m'(v)_\alpha]^\alpha = \bigcap_{\alpha\in\mathbf{Card}} [p_\alpha(v)]^\alpha = m(v).$$

Corollary 1. *We have an isomorphism* $\mathbf{CSPre}(\mathcal{T},\alpha) \simeq \mathbf{Pre}(\mathcal{T})$ *if* $\varphi_{\beta,\alpha}$ *is an isomorphism for each cardinal* $\beta \geq \alpha$.

Finding such a cardinal α is not obvious and depends on $\mathcal{T}$. Below we present a convenient condition for finding such α; see Example 11 for a concrete case.

Definition 9. *We say that a cardinal* α *is* large enough for preorder axioms *on* $\mathcal{T}$ *if for each cardinal* $\beta \geq \alpha$ *and* $x, y \in T\beta$, *there exists functions* $g : \beta \to T\alpha$ *and* $f : \alpha \to T\beta$ *(depending on* x, y*) such that* $f^\# \circ g^\#(x) = x$ *and* $f^\# \circ g^\#(y) = y$.

Theorem 6. *If* α *is large enough for preorder axioms on* $\mathcal{T}$, *then* $\mathbf{CSPre}(\mathcal{T},\alpha) \simeq \mathbf{Pre}(\mathcal{T})$.

Proof. We show that $\varphi_{\alpha,\beta}$ is surjective as a function for any cardinal $\beta \geq \alpha$. When this is shown, $\varphi_{\alpha,\beta}$ becomes the inverse of $\varphi_{\beta,\alpha}$ in **Pos** because $\varphi_{\alpha,\beta}$ is a split monomorphism.

Let β be a cardinal such that $\beta \geq \alpha$, and suppose that it is witnessed by an injection $w : \alpha \rightarrowtail \beta$. For each congruent and substitutive preorder $\leq \in \mathbf{CSPre}(T,\beta)$, we define a binary relation $\leq' \subseteq T\alpha \times T\alpha$ by

$$a \leq' b \iff \text{there exists an injection } m : \alpha \rightarrowtail \beta \text{ such that } Tm(a) \leq Tm(b).$$

Lemma 5. $\leq' \in \mathbf{CSPre}(\mathcal{T},\alpha)$.

We omit the proof of this lemma. We next show that $\leq$ is the image of $\leq'$ by $\varphi_{\alpha,\beta}$.

Lemma 6. $\varphi_{\alpha,\beta}(\leq') = \leq$.

Proof. Let $x, y \in T\beta$ such that $x \leq y$. For each function $f : \beta \to T\alpha$, we obtain

$$Tw \circ f^\#(x) = (Tw \circ f)^\#(x) \leq (Tw \circ f)^\#(y) = Tw \circ f^\#(y)$$

from the substitutivity of $\leq$, thus $f^\#(x) \leq' f^\#(y)$. Therefore we obtain $x\,[\leq']^\alpha_\beta\, y$.

Conversely, suppose $x\,[\leq']^\alpha_\beta\, y$. From the assumption, we have $g : \beta \to T\alpha$ and $f : \alpha \to T\beta$ such that $f^\# \circ g^\#(x) = x$ and $f^\# \circ g^\#(y) = y$. We thus have $g^\#(x) \leq' g^\#(y)$, hence there is an injection $m : \alpha \rightarrowtail \beta$ such that $Tm \circ g^\#(x) \leq Tm \circ g^\#(y)$. Now take a surjection $s : \beta \twoheadrightarrow \alpha$ such that $s \circ m = \mathrm{id}_\alpha$. Then we have a function $f \circ s : \beta \to T\beta$, and as the preorder $\leq$ is substitutive, we have

$$x = (f \circ s)^\# \circ Tm \circ g^\#(x) \leq (f \circ s)^\# \circ Tm \circ g^\#(y) = y.$$

Theorem 7. *The rank of a monad* $\mathcal{T}$, *if it exists, is large enough for preorder axioms on* $\mathcal{T}$.

Proof. We write α for the rank of $\mathcal{T}$. Let β be a cardinal such that $\beta \geq \alpha$ and $x_1, x_2 \in T\beta$. There exists a cardinal $0 < \gamma < \alpha$ (witnessed by an injection $i' : \gamma \rightarrowtail \alpha$), $m_1, m_2 \in T\gamma$ and an injection $i : \gamma \rightarrowtail \beta$ such that $T(i)(m_i) = x_i$ $(i = 1, 2)$. We then take surjections $s : \beta \twoheadrightarrow \gamma$ and $s' : \alpha \twoheadrightarrow \gamma$ that are left inverses to i and i', respectively. Then $f = \eta \circ i' \circ s : \beta \to T\alpha$ and $g = \eta \circ i \circ s' : \alpha \to T\beta$ satisfy $g^\# \circ f^\#(x_i) = x_i$ because

$$g^\# \circ f^\#(x_i) = Ti \circ Ts' \circ Ti' \circ Ts \circ Ti(m_i) = Ti(m_i) = x_i \quad (i = 1, 2).$$

5 Enumerating and Identifying Preorders on Monads

The understanding of the categorical status of $\mathbf{Pre}(\mathcal{T})$ allows us to identify its contents in several ways. Below we illustrate some methods with concrete monads.

5.1 Showing the Adjunction (2) being an Isomorphism

Let M be a semiring. We write $\mathcal{T}_c^M$ for the M-valued finite multiset monad, whose functor part is given by $T_c^M I = \{f : I \to M \mid \mathrm{supp}(f) \text{ is finite}\}$; here, $\mathrm{supp}(f) = \{i \in I \mid f(i) \neq 0\}$. Below we show that the adjunction (2) becomes an isomorphism for $I = 1$. The following is the key lemma, which states that each preorder on $\mathcal{T}_c^M$ is pointwise:

Lemma 7. *Each preorder* $\sqsubseteq$ *on* $\mathcal{T}_c^M$ *satisfies:* $d \sqsubseteq_I d' \iff \forall i \in I \,.\, d(i) \sqsubseteq_1 d'(i)$.

This implies $[\sqsubseteq_1]^1 \leq \sqsubseteq$. Therefore from Theorem 3, we obtain:

Theorem 8. *We have* $\mathbf{CSPre}(\mathcal{T}_c^M, 1) \simeq \mathbf{Pre}(\mathcal{T}_c^M)$.

By letting M be the two-point boolean algebra and removing the finiteness restriction, $\mathcal{T}_c^M$ becomes the powerset monad $\mathcal{T}_p$. A similar argument then identifies $\mathbf{Pre}(\mathcal{T}_p)$:

Theorem 9. *We have* $4 \simeq \mathbf{CSPre}(\mathcal{T}_p, 1) \simeq \mathbf{Pre}(\mathcal{T}_p)$. *The preorders on* $\mathcal{T}_p$ *are: I) the discrete order, II) the inclusion order, III) the opposite of II and IV) the trivial preorder (that is,* $\sqsubseteq_I = T_p I \times T_p I$*).*

5.2 Collecting Preorders of the Form $[\leq]^R$

From Theorem 4, every preorder $\sqsubseteq$ on $\mathcal{T}$ is the intersection of preorders of the form $[\leq]^R$. Therefore if the collection $\{[\leq]^R \mid R \in \mathbf{Set},\ \leq\ \in \mathbf{CSPre}(\mathcal{T}, R)\}$ is closed under intersections of arbitrary size, then it is equal to $\mathbf{Pre}(\mathcal{T})$. Below we identify $\mathbf{Pre}(\mathcal{T}_l)$ using this fact. We note that Levy identified $\mathbf{Pre}(\mathcal{T}_l)$ using a different method called *boolean precongruences* [19]; see Section 7.

Example 10. Let $(R, \leq)$ be a preorder parameter for $\mathcal{T}_l$. Then $[\leq]^R$ is either I) the discrete order, II) the flat order with $\iota_2(*)$ being the least element, III) the opposite of II, or IV) the trivial order. For proving this statement, we consider the combinations of two subcases: A) whether $\iota_2(*)$ is the least element in $(R, \leq)$ or not, and B) whether $\iota_2(*)$ is the greatest element in $(R, \leq)$ or not. From this, we conclude that I—IV are the only preorders on $\mathcal{T}_l$.

5.3 Computing $\mathbf{CSPre}(\mathcal{T}, \alpha)$ with a Large Enough α for Preorder Axioms

In the previous method, we have managed to find a good case analysis of preorder parameters. However, when the monad $\mathcal{T}$ becomes more complex, we immediately have no idea what kind of case analysis on preorder parameters is sufficient for classifying all the preorders on the monad. The second method presented in this section circumvents this problem by exploiting Theorem 6. We find a finite cardinal α that is large enough for preorder axioms on $\mathcal{T}$, then compute $\mathbf{CSPre}(\mathcal{T}, \alpha)$. Below we examine the case where this computation is feasible. First, we assume that $T\alpha$ is finite. We introduce the following preorder $\lhd$ on $T\alpha \times T\alpha$:

$$(x_1, y_1) \lhd (x_2, y_2) \iff \exists f : \alpha \to T\alpha\,.\,(f^\# x_1, f^\# y_1) = (x_2, y_2)$$

and the following *congruent closure operator* C:

$$\mathsf{C}(B) = \{(f^\#(w), g^\#(w)) \mid X \in \mathbf{Set}, w \in \mathcal{T}X, (f,g) : \mathrm{Eq}_X \dot{\to} (B, T\alpha, T\alpha)\}.$$

For a finite set D, a subset $A \subseteq D$ and a monotone increasing function f over $T_p D$, the following function `lfp` computes the least fixpoint of f including A:

```
lfp(A, f){
    if(A = f(A)) { return A; } else { return lfp(f(A), f); }
}
```

If f is computable then `lfp` terminates in finite steps.

We construct the following algorithm `Naive` to compute $\mathbf{CSPre}(\mathcal{T}, \alpha)$:

```
 CTU(A) { return lfp( A, C ∘ T ∘ U ); }
  f1(L) { return L ∪ { CTU(B ∪ {(x, y)}) | B ∈ L, (x, y) ∈ Tα × Tα \ B }; }
Naive() { return lfp( {Eq_Tα}, f1 ); }
```

where, U is the upward closure operator on $(T\alpha \times T\alpha, \lhd)$ and T is the transitive closure operator; they are both computable. The function CTU thus computes the congruent transitive upward closure of a given binary relation over $T\alpha$. When C is computable, the above algorithm is also computable.

Proposition 3. `Naive()` $= \mathbf{CSPre}(\mathcal{T}, \alpha)$.

We explain how the algorithm `Naive` runs with the following example.

Example 11. First, the cardinal 3 is large enough for preorder axioms on the nonempty powerset monad $\mathcal{T}_{p^+}$, because for each pair $(x, y) \in T_{p^+}X \times T_{p^+}X$, the following two functions $f : X \to T_{p^+}3$ and $g : 3 \to T_{p^+}X$ satisfy $g^\# \circ f^\# x = x$ and $g^\# \circ f^\# y = y$:

$$f(a) = \begin{cases} \{0\} & a \in x \backslash y \\ \{1\} & a \in y \backslash x \\ \{2\} & \text{otherwise} \end{cases}, \qquad g(b) = \begin{cases} y & b = 1 \\ x \cap y & b = 2 \text{ and } x \cap y \neq \emptyset \\ x & \text{otherwise} \end{cases}$$

Since $T_{p^+}3$ is finite and the multiplication of $\mathcal{T}_{p^+}$ is the set union operation, R is congruent if and only if R satisfies $(x_1, y_1), (x_2, y_2) \in R \implies (x_1 \cup x_2, y_1 \cup y_2) \in R$. Therefore, the following algorithm computes C:

```
C(A){ return lfp( A, f2 ); } where f2(B){ return B ∪ {x ∪ y | x, y ∈ B}; }
```

We have $\mathbf{CSPre}(\mathcal{T}, \alpha) \simeq 4$. The orders on $\mathcal{T}_{p^+}$ remains the same as the one for $\mathcal{T}_p$.

Table 1. All Preorders on $\mathcal{T}_{pl}$ (we omit opposite ones)

Type of preorders	The definition of $x \sqsubseteq_I y$
Trivial preorder	true
Equivalence relations	$x = y$, $(x = y) \vee (\bot \in x \wedge \bot \in y)$, $x \backslash \{\bot\} = y \backslash \{\bot\}$
Partial orders	$x \subseteq y$, $x = y \vee x = y \backslash \{\bot\}$, $x = y \vee (x \subseteq y \wedge \bot \in x)$, $x = y \vee (x \subseteq y \wedge \bot \in y)$, $(x = y) \vee (x \backslash \{\bot\} \subseteq y \backslash \{\bot\} \wedge \bot \in x)$
Proper preorders	$x = y \vee \bot \in x$, $x \subseteq y \vee \bot \in y$, $(x \subseteq y) \vee (x \backslash \{\bot\} \subseteq y \backslash \{\bot\} \wedge \bot \in x)$

We rewrite the naive algorithm to an efficient one. The basic idea to improve the efficiency is to work on the poset $(T\alpha \times T\alpha/\sim, [\lhd])$ rather than the preorder $(T\alpha \times T\alpha, \lhd)$, where $\sim$ is the equivalence relation $\lhd \cap \rhd$ and $[\lhd]$ is the extension of $\lhd$ to the partial order on $\sim$-equivalence classes.

Since $T\alpha$ is finite, the set of all $\sim$-equivalence classes and the order $[\lhd]$ between them are computable. We then rewrite the naive algorithms CTU and `Naive` to,

$$\mathtt{CTU}(A)\ \{\ \mathtt{return\ lfp}(\ A,\ \mathtt{C}' \circ \mathtt{T}' \circ \mathtt{U}'\ );\ \}$$
$$\mathtt{f3}(L)\ \{\ \mathtt{return}\ L \cup \{\ \mathtt{CTU}(B \cup \{d\}) \mid B \in L, d \in (T\alpha \times T\alpha/\sim) \setminus B\};\ \}$$
$$\mathtt{Modified}()\ \{\ \mathtt{return\ lfp}(\ \{\{[(x,y)] \mid (x,y) \in \mathrm{Eq}_{T\alpha}\}\},\ \mathtt{f3}\ );\ \}$$

respectively. Here, $\mathtt{U}'$ is the upward closure operator on $(T\alpha \times T\alpha/\sim, [\lhd])$, $\mathtt{C}'(B) = \{[(x,y)] \mid (x,y) \in \mathtt{C}(\bigcup B)\}$, and $\mathtt{T}'(B) = \{[(x,y)] \mid (x,y) \in \mathtt{T}(\bigcup B)\}$. Since an upward closed subset B of $(T\alpha \times T\alpha, \lhd)$ is the union $\bigcup B'$ of an upward closed subset B' of $(T\alpha \times T\alpha/\sim, [\lhd])$, we have $\{\bigcup B \mid B \in \mathtt{Modified}()\} = \mathbf{CSPre}(\mathcal{T}, \alpha)$.

Algorithm `Modified` is faster than `Naive` because the upward closure operator `U'` and the set comprehension in `f3` works on the smaller poset $(T\alpha \times T\alpha/\sim, [\lhd])$ than $(T\alpha \times T\alpha, \lhd)$. Function `f1` also has a redundant computation: it computes $\mathtt{CTU}(B \cup \{(x,y)\})$ for each pair $(x,y) \in T\alpha \times T\alpha \setminus B$, but the results of this computation are the same when $\sim$-equivalent pairs are supplied. The function `f3` avoids such duplicated computation by working on $\sim$-equivalence classes.

We demonstrate an execution of `Modified`. Below, we write $\mathcal{T}_{pl}$ for the composite monad $\mathcal{T}_p \circ \mathcal{T}_l$ using the canonical distributive law between $\mathcal{T}_p$ and $\mathcal{T}_l$.

Example 12. The cardinal $2 = \{a, b\}$ is large enough for preorder axioms on $\mathcal{T}_{pl}$. First we calculate all $\sim$-equivalence classes and the partial order $[\lhd]$. We have $\mathcal{T}_{pl}2 \times \mathcal{T}_{pl}2/\sim = \{p_1, p_2, \cdots, p_{28}\}$ where,

$p_1 = [(\{a\},\{b\})]$	$p_8 = [(\{a,\bot\},\{b\})]$	$p_{15} = [(\{a\},\{b,\bot\})]$	$p_{22} = [(\{a,\bot\},\{b,\bot\})]$
$p_2 = [(\{a,b\},\{b\})]$	$p_9 = [(\{a,b,\bot\},\{b\})]$	$p_{16} = [(\{a,b\},\{b,\bot\})]$	$p_{23} = [(\{a,b,\bot\},\{b,\bot\})]$
$p_3 = [(\{a\},\{a,b\})]$	$p_{10} = [(\{a,\bot\},\{a,b\})]$	$p_{17} = [(\{a\},\{a,b,\bot\})]$	$p_{24} = [(\{a,\bot\},\{a,b,\bot\})]$
$p_4 = [(\{a\},\{a\})]$	$p_{11} = [(\{a,\bot\},\{a\})]$	$p_{18} = [(\{a\},\{a,\bot\})]$	$p_{25} = [(\{a,\bot\},\{a,\bot\})]$
$p_5 = [(\{a\},\emptyset)]$	$p_{12} = [(\{a,\bot\},\emptyset)]$	$p_{19} = [(\{a\},\{\bot\})]$	$p_{26} = [(\{a,\bot\},\{\bot\})]$
$p_6 = [(\emptyset,\{a\})]$	$p_{13} = [(\{\bot\},\{a\})]$	$p_{20} = [(\emptyset,\{a,\bot\})]$	$p_{27} = [(\{\bot\},\{a,\bot\})]$
$p_7 = [(\emptyset,\emptyset)]$	$p_{14} = [(\{\bot\},\emptyset)]$	$p_{21} = [(\emptyset,\{\bot\})]$	$p_{28} = [(\{\bot\},\{\bot\})]$

We draw the following Hasse diagram of the poset $(\mathcal{T}_{pl}2 \times \mathcal{T}_{pl}2/\!\sim, [\lhd])$.

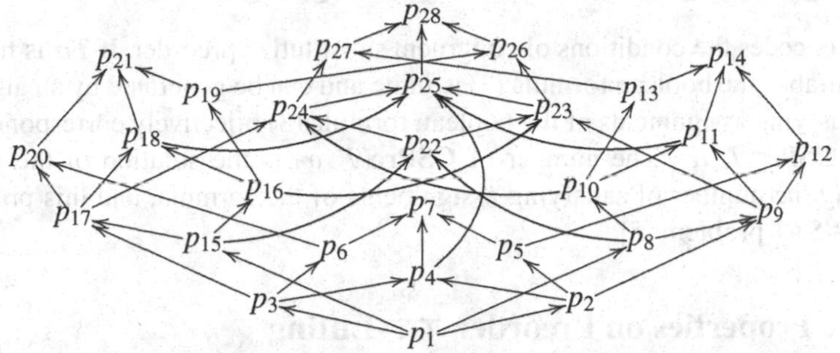

Next, we demonstrate the execution of `Modified`(). It computes the least fixpoint of `f3` containing $\{\{p_4, p_7, p_{25}, p_{28}\}\}$. We now see the first loop of `lfp` in the execution of `Modified`() in detail. The function `f3` picks up an equivalence class other than $\{p_4, p_7, p_{25}, p_{28}\}$, say p_6, then pass $\{p_4, p_7, p_{25}, p_{28}, p_6\}$ to CTU . The function CTU processes its argument by the closure operators `U'`, `T'`, `C'` repeatedly until it gets stationary. The following is the first pass of this process:

- $\mathsf{U}'(\{p_4, p_7, p_{25}, p_{28}, p_6\}) = \{p_4, p_6, p_7, p_{20}, p_{21}, p_{25}, p_{28}\}$
- $\mathsf{T}'(\{p_4, p_6, p_7, p_{20}, p_{21}, p_{25}, p_{28}\}) = \{p_4, p_6, p_7, p_{20}, p_{21}, p_{25}, p_{28}\}$
- $\mathsf{C}'(\{p_4, p_6, p_7, p_{20}, p_{21}, p_{25}, p_{28}\}) = \{p_3, p_4, p_6, p_7, p_{17}, p_{18}, p_{20}, p_{21}, p_{24}, p_{25}, p_{27}, p_{28}\}$

The result of the last calculation by C′, which we call H below, is already closed under U′, T′ and C′. Therefore, $\mathsf{CTU}(\{p_4, p_7, p_{25}, p_{28}, p_6\}) = H$. The function `f3` similarly calculates $\mathsf{CTU}(\{p_4, p_7, p_{25}, p_{28}, p\})$ for each equivalence class p other than $p_4, p_7, p_{25}, p_{28}, p_6$, and returns the union of the results of the calculations of CTUs and `f3`'s argument L. This finishes the first call of `f3`. The function `lfp` in `Modified` repeats calling `f3` until we obtain the least fixpoint of `f3`. The algorithm `Modified`() yields 20 sets of equivalence classes, hence $\mathbf{CSPre}(\mathcal{T}_{pl}, 2) \simeq 20$ (see also Section 7 for Levy's result).

After this computation, we manually extract the definitions of preorders on $\mathcal{T}_{pl}$ from each set of equivalence classes. The 20 preorders are listed in Table 1. For this extraction, we first identify the meaning of the binary relation $\bigcup B$ over $T_{pl}2$ for each set $B \in$ `Modified`() of equivalence classes, then manually characterise $[\bigcup B]_I^2$ for each set I. For instance, $\bigcup H = \subseteq_2$, and from this we obtain $[\subseteq_2]_I^2 = \subseteq_I$.

Another method to enumerate congruent substitutive preorders on $T\alpha$ is to reduce the problem to finding the valuations ρ that satisfy the following boolean formula:

$$\bigwedge_{(Q_1,Q_2)\in V} \Big(\bigwedge_{p\in Q_1} P_p \Longrightarrow \bigwedge_{p\in Q_2} P_q\Big) \tag{3}$$

Here, P_p is the propositional variable assigned to each $p \in T\alpha \times T\alpha/\!\sim$, and V is the set of the following pairs:

- $(\{p\}, \{q\})$ for all $p, q \in T\alpha \times T\alpha/\!\sim$ such that $p \lhd q$
- $(\emptyset, [(x, x)])$ for all $x \in T\alpha$

- $(\{[(x,y)],[(y,z)]\},\{[(x,z)]\})$ for all $x,y,z \in T\alpha$
- $(Q, \mathsf{C}'(Q))$ for all $Q \subseteq T\alpha \times T\alpha/\sim$ such that $\mathsf{C}'(Q) \neq Q$.

The set V encodes the conditions of congruent substitutive preorder. If $T\alpha$ is finite and C is computable, the boolean formula (3) is finite and can be generated by an algorithm.

The satisfying assignments of the boolean formula (3) bijectively correspond to preorders in $\mathbf{CSPre}(\mathcal{T},\alpha)$. The number of $\mathbf{CSPre}(\mathcal{T},\alpha)$ is the solution of the problem of counting the number of satisfying assignments of the formula, and this problem is known as #SAT problem [5].

6 Some Properties on Preorder ⊤⊤-Lifting

We show that preorder ⊤⊤-liftings satisfy a couple of properties that are relevant to the coalgebraic simulations discussed in Section 3.2. The first property relates oplax coalgebra morphisms and simulations. We restrict our attention to TF-coalgebras, where T is the functor part of a monad and F consists of $\mathrm{Id}, C_A, +, \times$ only. Below for each function $g : I \to J$, we define $\mathbf{Gr}(g)$ to be the $\mathbf{BRel}$-object $(\{(i, g(i)) \mid i \in I\}, I, J)$ of the graph of g. We note $\dot{F}(\mathbf{Gr}g) = \mathbf{Gr}(Fg)$.

Theorem 10. *Let $(R,\leq)$ be a preorder parameter, and (I_i, f_i) be TF-coalgebras ($i = 1,2$). For each function $g : I_1 \to I_2$, $\mathbf{Gr}(g)$ is a $T^{\top\top}\dot{F}$-simulation from (I_1, f_1) to (I_2, f_2) if and only if g is an oplax morphism of coalgebras with respect to $[\leq]^R$, that is, $TFg \circ f_1 [\dot{\leq}]^R_{FI_2} f_2 \circ g$.*

In general, preorder ⊤⊤-liftings may not be lax compositional. We here present a condition to guarantee the lax compositionality.

Theorem 11. *Let $(R,\leq)$ be a congruent preorder such that $\leq$ satisfies the following condition for all subsets $X, Y \subseteq TR$:*

$$(\forall x \in X, y \in Y \,.\, x \leq y) \implies \exists z \in TR. \forall x \in X, y \in Y \,.\, x \leq z \wedge z \leq y. \tag{4}$$

Then $T^{\top\top}$ is lax compositional.

For instance, (4) is satisfied when the preorder parameter $(R,\leq)$ is a complete lattice.

7 Conclusion and Related Work

We showed that preorder ⊤⊤-liftings construct preorders on monads, and this construction enjoys a universal property. We gave a characterisation of the collection $\mathbf{Pre}(\mathcal{T})$ of preorders on $\mathcal{T}$ as the limit of the large diagram $\mathbf{CSPre}(\mathcal{T},-) : \mathbf{Card}^{op} \to \mathbf{Set}$. We then applied these theoretical results to identifying preorders on some concrete monads. We also showed the properties of the preorder ⊤⊤-lifting that are relevant to the coalgebraic formulation of simulations.

Besides [13,11,15], we briefly mention some recent works on (bi)simulations and relational liftings. Cîrstea studies modular constructions of relational extensions and modal logics characterising simulations using the categorical structures on **BRel** [7].

Klin studies the least fibred lifting of **Set**-functors across the fibration **ERel** $\to$ **Set**, where **ERel** is the category of equivalence relations [17]. His lifting works for mono-preserving functors, and when they preserve weak pullbacks, his lifting coincides with the one in Example 5. Balan and Kurz give liftings and extensions of finitary **Set**-functors to endofunctors over **Pre** and **Pos** [2]. Their method uses the fact that every finitary **Set**-functor T is presented as $\mathrm{Lan}_I(T \circ I)$, where I : **Finord** $\to$ **Set** is the inclusion functor. Bilkova et al. derive a natural definition of relations between preorders using Sierpinski-space enriched categories, and give relational liftings of endofunctors over **Pre** in this context [6]. Levy extends the characterisation of bisimilarity by final coalgebras to similarity [20].

The novelty of our approach is that we exploit the structure of *monad* to relationally lift functors. The principle of the semantic ⊤⊤-lifting seems fundamentally different from the lifting methods employed in the above works. One distinguishing feature of the semantic ⊤⊤-lifting is its flexibility. By changing the preorder parameter, we can uniformly derive various relational liftings and preorders on monads. The source of this flexibility lies at *continuation monads*, which are a special case of enriched *right* Kan extensions.

Levy introduces the concept called *deterministic / nondeterministic boolean precongruences* (*DBP* and *NDBP* for short) in [19]. They are defined in our language by:

$$\mathrm{DBP}_E = \mathbf{CSPre}(\mathcal{T}_e^E, 2), \quad \mathrm{NDBP}_E = \mathbf{CSPre}(\mathcal{T}_{p^+} \circ \mathcal{T}_e^E, 2);$$

here, $\mathcal{T}_e^E$ is the *error monad*, whose functor part is given by $T_e^E I = I + E$. He shows $\mathbf{CSPre}(\mathcal{T}, 2) \simeq \mathbf{Pre}(\mathcal{T})$ for $\mathcal{T} = \mathcal{T}_e^E$ and $\mathcal{T} = \mathcal{T}_{p^+} \circ \mathcal{T}_e^E$, and enumerates the following boolean precongruences together with their definitions:

$$\mathrm{DBP}_0 \simeq 2, \quad \mathrm{DBP}_1 \simeq 4, \quad \mathrm{DBP}_2 \simeq 13, \quad \mathrm{NDBP}_0 \simeq 4, \quad \mathrm{NDBP}_1 \simeq 20.$$

He also gives modal logics that have Hennesy-Milner property with respect to the concept of simulations derived from boolean precongruences. His results are derived by the method that is specialised to these monads.

Acknowledgement. We are grateful to Naohiko Hoshino, Norihiro Tsumagari and Hasuo Ichiro for valuable discussions. This work was supported by JSPS KAKENHI Grant Number 24700012.

References

1. Aczel, P., Mendler, N.: A Final Coalgebra Theorem. In: Dybjer, P., Pitts, A.M., Pitt, D.H., Poigné, A., Rydeheard, D.E. (eds.) Category Theory and Computer Science. LNCS, vol. 389, pp. 357–365. Springer, Heidelberg (1989)
2. Balan, A., Kurz, A.: Finitary Functors: From Set to Preord and Poset. In: Corradini, A., Klin, B., Cîrstea, C. (eds.) CALCO 2011. LNCS, vol. 6859, pp. 85–99. Springer, Heidelberg (2011)
3. Barr, M.: Relational Algebras. In: MacLane, S., Applegate, H., Barr, M., Day, B., Dubuc, E., Phreilambud, Pultr, A., Street, R., Tierney, M., Swierczkowski, S. (eds.) Reports of the Midwest Category Seminar IV. LNM, vol. 137, pp. 39–55. Springer, Heidelberg (1970)

4. Benton, N., Hughes, J., Moggi, E.: Monads and Effects. In: Barthe, G., Dybjer, P., Pinto, L., Saraiva, J. (eds.) APPSEM 2000. LNCS, vol. 2395, pp. 42–122. Springer, Heidelberg (2002)
5. Biere, A., Heule, M., van Maaren, H., Walsh, T. (eds.): Handbook of Satisfiability. Frontiers in Artificial Intelligence and Applications, vol. 185. IOS Press (February 2009)
6. Bílková, M., Kurz, A., Petrisan, D., Velebil, J.: Relation liftings on preorders and posets. CoRR, abs/1210.1433 (2012)
7. Cîrstea, C.: A modular approach to defining and characterising notions of simulation. Information and Computation 204(4), 469–502 (2006)
8. Goubault-Larrecq, J., Lasota, S., Nowak, D.: Logical Relations for Monadic Types. In: Bradfield, J.C. (ed.) CSL 2002 and EACSL 2002. LNCS, vol. 2471, pp. 553–568. Springer, Heidelberg (2002)
9. Hasuo, I.: Generic Forward and Backward Simulations. In: Baier, C., Hermanns, H. (eds.) CONCUR 2006. LNCS, vol. 4137, pp. 406–420. Springer, Heidelberg (2006)
10. Hasuo, I., Jacobs, B., Sokolova, A.: Generic trace theory. Electr. Notes Theor. Comput. Sci. 164(1), 47–65 (2006)
11. Hermida, C., Jacobs, B.: An Algebraic View of Structural Induction. In: Pacholski, L., Tiuryn, J. (eds.) CSL 1994. LNCS, vol. 933, pp. 412–426. Springer, Heidelberg (1995)
12. Hermida, C., Jacobs, B.: Structural induction and coinduction in a fibrational setting. Inf. Comput. 145(2), 107–152 (1998)
13. Hesselink, W.H., Thijs, A.: Fixpoint semantics and simulation. Theor. Comput. Sci. 238(1-2), 275–311 (2000)
14. Jacobs, B.: Categorical Logic and Type Theory. Elsevier (1999)
15. Jacobs, B., Hughes, J.: Simulations in coalgebra. Electr. Notes Theor. Comput. Sci. 82(1), 128–149 (2003)
16. Katsumata, S.: A Semantic Formulation of ⊤⊤-Lifting and Logical Predicates for Computational Metalanguage. In: Ong, L. (ed.) CSL 2005. LNCS, vol. 3634, pp. 87–102. Springer, Heidelberg (2005)
17. Klin, B.: The Least Fibred Lifting and the Expressivity of Coalgebraic Modal Logic. In: Fiadeiro, J.L., Harman, N.A., Roggenbach, M., Rutten, J. (eds.) CALCO 2005. LNCS, vol. 3629, pp. 247–262. Springer, Heidelberg (2005)
18. Kurz, A.: Logics for Coalgebras and Applications to Computer Science. PhD thesis, Ludwig-Maximilians-Universität, Munchen (2000)
19. Levy, P.: Boolean precongruences. Manuscript (2009)
20. Levy, P.: Similarity Quotients as Final Coalgebras. In: Hofmann, M. (ed.) FOSSACS 2011. LNCS, vol. 6604, pp. 27–41. Springer, Heidelberg (2011)
21. Lindley, S.: Normalisation by Evaluation in the Compilation of Typed Functional Programming Languages. PhD thesis, University of Edinburgh (2004)
22. Lindley, S., Stark, I.: Reducibility and ⊤⊤-Lifting for Computation Types. In: Urzyczyn, P. (ed.) TLCA 2005. LNCS, vol. 3461, pp. 262–277. Springer, Heidelberg (2005)
23. Mitchell, J.: Foundations for Programming Languages. MIT Press (1996)
24. Pitts, A.: Parametric polymorphism and operational equivalence. Mathematical Structures in Computer Science 10(3), 321–359 (2000)
25. Staton, S.: Relating coalgebraic notions of bisimulation. Logical Methods in Computer Science 7(1) (2011)

A Proof System for Compositional Verification of Probabilistic Concurrent Processes

Matteo Mio[1,*] and Alex Simpson[2]

[1] INRIA and LIX, Ecole Polytechnique, France
[2] LFCS, School of Informatics, University of Edinburgh, Scotland

Abstract. We present a formal proof system for compositional verification of probabilistic concurrent processes. Processes are specified using an SOS-style process algebra with probabilistic operators. Properties are expressed using a probabilistic modal μ-calculus. And the proof system is formulated as a sequent calculus in which sequents are given a quantitative interpretation. A key feature is that the probabilistic scenario is handled by introducing the notion of *Markov proof*, according to which proof trees contain probabilistic branches and are required to satisfy a condition formulated by interpreting them as Markov Decision Processes. We present simple but illustrative examples demonstrating the applicability of the approach to the compositional verification of infinite state processes. Our main result is the soundness of the proof system, which is proved by applying the coupling method from probability theory to the game semantics of the probabilistic modal μ-calculus.

1 Introduction

In recent years, *model checking* has established itself as a powerful and widely applicable method for verifying properties of systems, with its techniques adaptable to systems embodying, for example, concurrency, real-time behaviour and probabilistic choice, see [1] for a detailed overview. However, model checking has its limitations. In particular, its applicability is typically restricted to finite-state systems, or to carefully crafted classes of infinite-state systems. Moreover, even in the finite-state case, the applicability of model checking is limited by the *state explosion* problem: the state space of a concurrent system grows exponentially in the number of parallel components.

Many phenomena in computer science give rise to infinite-state systems when modelled at a natural level of abstraction. So it is important to have verification methods that can cope with such systems. Since infinite-state systems are

* This research was partially supported by PhD studentships from LFCS and the IGS at the School of Informatics, University of Edinburgh; by EPSRC research grant EP-F042043-1; by project ANR-09-BLAN-0169-01 PANDA; and by project ANR-11-IS02-0002 LOCALI. It was completed during the tenure of an ERCIM "Alain Bensoussan" Fellowship, supported by the Marie Curie Co-funding of Regional, National and International Programmes (COFUND) of the European Commission.

F. Pfenning (Ed.): FOSSACS 2013, LNCS 7794, pp. 161–176, 2013.

defined using finite descriptions (given using a programming language, process calculus or similar language for system specification), one seeks verification methods that relate descriptions of systems to their properties. Such methods cannot, in general, be fully automatic since, for most interesting cases, the problem of ascertaining whether a description satisfies a property is undecidable.

One general methodology for obtaining such broader verification methods is to develop *formal proof systems* tailored to the goal of establishing that (descriptions of) systems satisfy properties. The verification task then becomes one amenable to the technology of computer-assisted reasoning. An important desideratum for a proof system for verification is that it should support *compositional* reasoning methods, by which the the task of establishing a property of a complex system is broken down into suitable verification goals for the components of the system. As has been extensively discussed in the literature, see, e.g., [9], such compositional methods support methodologies for the modular development and verification of systems. Compositional methods also provide a route to taming the state explosion problem, since the size of a compound system is usually much larger than that of its components.

The purpose of this paper is to present one interesting instantiation of the above general approach to verification. We develop a formal proof system for compositional reasoning about (possibly infinite state) concurrent probabilistic systems. The systems we deal with are ones described by a simple algebra for concurrent probabilistic processes, with SOS-style operational semantics (Section 2). While, in this paper itself, we consider only a few basic process operators, a key feature of our approach is that it is applicable to a wide class of operators (any that can be described in the *probabilistic* GSOS framework of [2]).

Properties are specified using the *probabilistic* (a.k.a. *quantitative*) *modal* μ-*calculus* (pLμ) introduced independently in [10,12] (Section 3). Our reason for not using a standard logic for stating properties of probabilistic systems, such as PCTL [1, §10.2], is that fixed-point logics such as pLμ appear to be better adapted to compositional reasoning. One reason for this is that formulas express properties of states rather than properties of objects of higher complexity, such as paths or Markov chains. Also, powerful proof methods for reasoning about fixed points are available. Nevertheless, as the first author has shown [13], the full expressivity of PCTL (and beyond) can be recovered by extending pLμ with a few additional operators. This makes it plausible that the proof system of the present paper might be similarly extended to provide a system capable of reasoning about arbitrary PCTL-properties.

The main contributions of the paper are the proof system itself and its (nontrivial) soundness theorem (Section 4). Adapting a general methodology for compositional verification, expounded in [19], the proof system is a sequent calculus with sequents of assertions of the form $p : F$. In our quantitative setting, the semantics of $p : F$ is a real number in $[0, 1]$, which roughly (see Section 3 for clarification) expresses the probability that property F holds of process p. The right-hand side of a sequent is a multiset of such assertions, itself given a quantitative meaning as the Łukasiewicz disjunction (see, e.g., [8] and [10]) of the individual assertions. The use of this disjunction from fuzzy logic underpins several features of our proof system, most importantly the soundness of certain crucial proof rules which allow probabilistic choices within different processes to be *coupled* in the reasoning.

To enable the proof system to handle the fixed points in the logic, we allow cyclic derivations and require a combinatorial condition to hold in order for a proof to count as valid. This approach is familiar from proof systems for the ordinary (non-probabilistic) modal μ-calculus [15,21] and other fixed-point logics [3]. However, there is a twist in our probabilistic setting. One of the proof rules in our system introduces probabilistic branching into the proof tree itself. This is addressed by interpreting the proof tree as specifying a Markov Decision Process (MDP), in which the participant, Refuter, is trying to refute the correctness of the proof. Refuter's goal is to try to find an infinite branch through the proof along which all sequences of fixed-point unfoldings illegitimately unfold a least-fixed-point infinitely often. The proof tree is declared valid just in case Refuter almost surely fails in his endeavour; that is when the *value* of the MDP is zero. Due to the critical role played by the probabilistic rule, we call such a valid proof tree a *Markov proof*. An important fact, crucial for the applicability of the approach, is that the property of being a Markov proof is decidable. This follows from known decidability results for one-player stochastic parity games [5].

In Section 5, we present two examples of Markov proofs, illustrating the sort of compositional reasoning possible within our system and establishing nontrivial properties of infinite state systems.

Related Work: Several approaches to compositional verification methods for concurrent probabilistic systems have received attention in the recent literature. In Cardelli *et. al.* [4], new 'spatial' operators are added to a probabilistic modal logic to support compositional reasoning about labelled Markov processes [16] enriched with an algebraic structure defining composition of systems, and a completeness result is obtained for a Hilbert-style axiomatization. However, the logic is limited to expressing *local* properties of systems (that is, it cannot state properties of infinite runs). In Kwiatkowska *et. al.* [11,7], assume-guarantee techniques for compositional verification of (parallel composition of) probabilistic automata [18] are developed. Their approach does handle some global properties of systems, namely safety and liveness properties expressed using automata. However, being based on fully automatizable model-checking techniques, it is restricted to finite models. Similarly, Larsen *et. al.* [6] introduce compositional methodologies for design and verification of finite probabilistic concurrent systems. These are based on *Abstract Probabilistic Automata* which are structures capable of modelling both specifications and implementations, and closed under natural logical operations such as composition and refinement.

Distinguishing features of our approach are: we have a clear separation between the process language and a purely behavioural *endogenous* [17] logic; the nested fixed-points of the logic allow the specification of complex global behaviour; and we are able to establish nontrivial properties of infinite state systems.

2 Probabilistic Concurrent Processes

Ordinary *Labeled Transition Systems* (LTS) allow the description of processes exhibiting nondeterministic behavior. In his PhD thesis, R. Segala introduced a new class of models, nowadays known as *Probabilistic Labeled Transition Systems* (PLTS), for modelling processes exhibiting both nondeterministic and probabilistic behaviours. Since their introduction, PLTS's have been successfully adopted as models for formal languages describing concurrent probabilistic systems, such as the class of PGSOS languages of [2] among others.

Definition 1 (PLTS [18]). *Given a countable set L of labels, a* Probabilistic Labeled Transition System *(PLTS) is a pair $\mathcal{L} = \langle P, \{\xrightarrow{a}\}_{a\in L}\rangle$, where P is a set of states and $\xrightarrow{a} \subseteq P \times \mathcal{D}(P)$, for every $a \in L$, where $\mathcal{D}(P)$ denotes the set of discrete probability distributions over P.*

The intended interpretation of a PLTS $\mathcal{L} = \langle P, \{\xrightarrow{a}\}_{a\in L}\rangle$ is the following: the process states $p \in P$ represent the possible configurations of the system. At a process state p, the system can react to an a-action, for $a \in L$, by changing its state to a process q in accordance with some nondeterministically chosen probability distribution $d \in \mathcal{D}(P)$ such that $p \xrightarrow{a} d$.

$$\frac{}{a.x \xrightarrow{a} \delta(x)}\ \text{prefix}$$

$$\frac{x \xrightarrow{a} \alpha}{x|y \xrightarrow{a} \{\alpha \rightsquigarrow z\}\delta(z|y)}\ |\ \text{left} \qquad \frac{y \xrightarrow{a} \alpha}{x|y \xrightarrow{a} \{\alpha \rightsquigarrow z\}\delta(x|z)}\ |\ \text{right}$$

$$\frac{x \xrightarrow{a} \alpha}{!x \xrightarrow{a} \{\alpha \rightsquigarrow z\}\delta(z|!x)}\ ! \qquad \frac{x \xrightarrow{a} \alpha}{!^{\frac{1}{2}}x \xrightarrow{a} \alpha +_{\frac{1}{2}} \{\alpha \rightsquigarrow y\}\delta(y|!^{\frac{1}{2}}x)}\ !^{\frac{1}{2}}$$

Fig. 1. SOS Rules. The letter a ranges over a fixed set L of labels.

In this paper, we consider PLTS's described by a few very simple process operators chosen to present the examples in Section 5. Our approach, however, adapts straightforwardly to handle arbitrary process algebras described by means of well-behaved operational rules (such as, e.g., the PGSOS format of [2]). The term constructors we consider are the constant 0 denoting the inactive process, the *prefix operation* ($a.\_$) of arity 1, the non-communicating asynchronous parallel composition ($\_ \mid \_$) of arity 2, the *bang* operator ($!\ \_$) and a probabilistic variant of it ($!^{\frac{1}{2}}\ \_$), both of arity 1. Their semantics, specified by the SOS operational rules of Figure 1, allows the derivation of statements of the form $p \xrightarrow{a} d$ where p is a *process term*, the letter a is a label and d is a *process distribution term*. Process distribution terms, denoting probability distributions over processes, are specified by the syntax: $d, e ::= \alpha \mid \delta(p) \mid d +_\lambda e \mid \{d \rightsquigarrow x\}e$, where α is a *process distribution variable* and $\lambda \in [0, 1]$. The term $\delta(p)$ denotes the (Dirac) probability distribution that proceeds deterministically onward to p, and $d +_\lambda e$ denotes the probabilistic choice that chooses d with probability λ and e with probability

$1-\lambda$. The distribution term $\{d \rightsquigarrow x\}e$, first, randomly chooses a process p in accordance with probability distribution d, and then proceeds as $e[p/x]$.

A *SOS-model*, or just a *model*, is a PLTS equipped with sound interpretations for all process constructors under consideration (see, e.g., [19] and [2] for general definitions). In the case of prefix, for example, the PLTS $\langle P, \{\xrightarrow{a}\}_{a\in L}\rangle$ is required to come with a function $f_{a.\_} : P \rightarrow P$ for which $f_{a.\_}(p) \xrightarrow{b} d$ holds if and only if $a=b$ and d is the Dirac distribution with mass at p, for every $p \in P$. In what follows we reserve the letter M to range over models.

Definition 2 (Interpretations). *Given a model $M = \langle P, \{\xrightarrow{a}\}_{a\in L}\rangle$, an* interpretation of the variables *is a function γ mapping process-variables x to states $p \in P$ and process distribution-variables α to probability distributions $d \in \mathcal{D}(P)$. The map γ extends uniquely to a function from process-terms to P and to process distribution terms to $\mathcal{D}(P)$, defined as expected. In particular*

$$\gamma(\{d \rightsquigarrow x\}e)(p) \stackrel{\text{def}}{=} \sum_{q\in P}\Big(\gamma(d)(q)\cdot\gamma[q/x](e)(p)\Big)$$

where $\gamma[q/x](x)=q$ and $\gamma[q/x](y)=\gamma(y)$ for all variables $x \neq y$.

3 Probabilistic Modal μ-Calculus (pLμ)

The *probabilistic* (or *quantitative*) modal μ-calculus (pLμ) [14,12] is a fixed-point logic designed for expressing properties of PLTS's. The syntax of pLμ formulas is the same of the standard modal μ-calculus (Lμ) [20].

Definition 3. *Given a countable set of propositional variables* Var *ranged over by the letters X, Y, Z and a set of labels L ranged over by the letters a, b, c, the formulas of the logic pLμ (in positive form) are defined by the following grammar:*

$$F, G ::= X \mid \langle a\rangle F \mid [a]\, F \mid F \vee G \mid F \wedge G \mid \mu X.F \mid \nu X.F$$

As usual the operators $\nu X.F$ and $\mu X.F$ bind the variable X in F.

Definition 4 (Denotational Semantics [12]). *Given $\mathcal{L} = \langle P, \{\xrightarrow{a}\}_{a\in L}\}$, the denotational semantics of the pLμ formula F under the interpretation ρ:* Var $\rightarrow$ $(P \rightarrow [0,1])$, *is the map $[\![F]\!]_\rho : P \rightarrow [0,1]$ defined by structural induction on F as:*

$$(\!| X |\!)_\rho(p) = \rho(X)(p)$$

$$(\!| G \vee H |\!)_\rho(p) = (\!| G |\!)_\rho(p) \sqcup (\!| H |\!)_\rho(p) \qquad (\!| G \wedge H |\!)_\rho(p) = (\!| G |\!)_\rho(p) \sqcap (\!| H |\!)_\rho(p)$$

$$(\!| \langle a\rangle G |\!)_\rho(p) = \bigsqcup\{(\!| G |\!)_\rho(d) \mid p \xrightarrow{a} d\} \qquad (\!| [a]\, G |\!)_\rho(p) = \bigsqcap\{(\!| G |\!)_\rho(d) \mid p \xrightarrow{a} d\}$$

$$(\!| \mu X.G |\!)_\rho(p) = \mathrm{lfp}\big(\lambda f.((\!| G |\!)_{\rho[f/X]})\big)(p) \qquad (\!| \nu X.G |\!)_\rho(p) = \mathrm{gfp}\big(\lambda f.((\!| G |\!)_{\rho[f/X]})\big)(p)$$

where $\sqcup$, $\sqcap$, lfp *and* gfp *denote the join, meet, least and greatest fixed point operations of the complete lattice $[0,1]$ with its standard order, and $(\!| F |\!)_\rho(d)$ is defined as $(\!| F |\!)_\rho(d) = \sum_{p\in P} d(p)\cdot(\!| F |\!)_\rho(p)$.*

It is easy to verify that the interpretation assigned to every pLμ operator is monotone. Thus, the existence of the least and greatest fixed points is guaranteed by the Knaster-Tarski theorem. Although it is convenient to consider open formulas when defining the semantics of pLμ, logical specifications are generally formulated as closed formulas. For this reason, and for simplifying the presentation of the proof system in Section 4, we shall only consider closed formulas in the rest of this paper. Thus we omit the interpretation ρ from $(\!| F |\!)_\rho$ and just write $(\!| F |\!)$. Given a formula F we denote with $\neg F$ its De Morgan dual obtained by replacing every connective appearing in F with its dual. Note that $F = \neg\neg F$. As customary, we denote with $\top$ the pLμ formula $\nu X.X$ and define $\bot = \neg\top$.

Proposition 5. *For every PLTS* $\mathcal{L} = \langle P, \{\xrightarrow{a}\}_{a \in L}\rangle$*, the equalities* $(\!| \top |\!)(p) = 1$, $(\!| \bot |\!)(p) = 0$ *and* $(\!| \neg F |\!)(p) = 1 - (\!| F |\!)(p)$ *hold, for all* $p \in P$.

It is often suggestive to think of the value of $(\!| F |\!)(p)$ as representing the probability that a property asserted by F holds for p. Technically, this is justified by an alternative semantics for pLμ, which interprets a formula as the *value* of a two-player stochastic parity game [12]. The game in question is obtained by running the usual two-player game for the modal μ-calculus formula over the PLTS. As with ordinary modal μ-calculus games, game configurations $p : \langle a \rangle F$ and $p : [a] F$ are under the control of different players (here called Maximizer and Minimizer respectively) whose move is to choose an a transition out of p. In the case of pLμ, the destination of this transition is a probability distribution, and Nature intercepts in the game to make the probabilistic choice. The winning condition for Maximizer is the usual one that a greatest fixed-point gets unfolded infinitely often. One can then think of the value of the game as the (upper limit) probability with which Maximizer is able to verify the property expressed by the ordinary μ-calculus formula F.

It is a nontrivial fact that the game interpretation of a pLμ formula coincides with the denotational one of Definition 4. This was originally shown just for finite PLTS's in [12], and only recently for general PLTS's in [14].

4 Proof System

We introduce in this section our proof system designed to reason about pLμ-calculus properties of processes given in our process algebra. The system is a *sequent calculus*, in which sequents have the form $\Sigma \vdash \Delta$, where Σ and Δ are multisets of (different kinds of) assertions. We use the letter J to range over *operational assertions* in Σ which are either of the form $d \simeq e$, where d and e are process distribution terms, or of the form $p \xrightarrow{a} d$, where p is a process term, a is an action-label and d is a process distribution term. We use the letters ϕ and ψ to range over *logical assertions* in Δ, which are of the form $p{:}F$ or $d{:}F$, where F is a closed pLμ formula, p a process term and d a distribution term. Given an assertion ϕ of the form $t{:}F$, with $t \in \{p, d\}$, we write $\neg\phi$ for the assertion $t{:}\neg F$.

Definition 6 (Semantics of assertions). *Given a model M and an interpretation of the variables γ, the meaning $[\![\_]\!]^M_\gamma$ of the assertions is defined as:*

1. $[\![d \simeq e]\!]^M_\gamma = 1$ *if* $\gamma(d) = \gamma(e)$ *and* $[\![d \simeq e]\!]^M_\gamma = 0$ *otherwise.*
2. $[\![p \xrightarrow{a} d]\!]^M_\gamma = 1$ *if* $\gamma(p) \xrightarrow{a}_M \gamma(d)$ *and* $[\![p \xrightarrow{a} d]\!]^M_\gamma = 0$ *otherwise.*
3. $[\![t : F]\!]^M_\gamma \stackrel{\text{def}}{=} (\!| F |\!)(\gamma(t))$, *for any process or distribution term* t.

We write $(M, \gamma) \models J$ *if the equality* $[\![J]\!]^M_\gamma = 1$ *holds. We write* $(M, \gamma) \models \Sigma$, *for* $\Sigma = J_1, \ldots, J_m$, *if for all* $i \in \{1, \ldots, m\}$ *it holds that* $(M, \gamma) \models J_i$.

Note that the value $[\![J]\!]^M_\gamma$ of a logical assertion is either 1 or 0, whereas $[\![t : F]\!]^M_\gamma$ lies anywhere in $[0, 1]$, representing, using the informal reading of pLμ discussed in Section 3, the probability of the property expressed by F holding at $\gamma(t)$. In order to extend the semantics from assertions to sequents, we first recall basic notions from *Łukasiewicz* logic [8].

Definition 7 ([8]). *The operations* $\oplus : [0, 1]^2 \to [0, 1]$ *and* $\neg : [0, 1] \to [0, 1]$ *defined as* $x \oplus y = \min\{x + y, 1\}$ *and* $\neg x = 1 - x$, *are known as* Łukasiewicz *disjunction and negation. The induced conjuction* ($\ominus$) *and implication* ($\Rightarrow$) *operations are defined as* $x \ominus y = \neg(\neg x \oplus \neg y)$ *and* $x \Rightarrow y = \neg x \oplus y$. *Note that* $((x \ominus y) \Rightarrow z) = (x \Rightarrow (\neg y \oplus z))$ *and* $(x \Rightarrow y) = 1$ *if and only if* $x \leq y$.

Definition 8 (Semantics of Sequents). *Let* $\Sigma \vdash \Delta$ *be a sequent with* $\Sigma = J_1, \ldots, J_n$ *and* $\Delta = \phi_1, \ldots, \phi_m$. *Given a model* M *and an interpretation* γ *of the variables, we define the semantics* $[\![\Sigma \vdash \Delta]\!]^M_\gamma \in [0, 1]$ *of the sequent as:*

$$[\![\Sigma \vdash \Delta]\!]^M_\gamma \stackrel{\text{def}}{=} ([\![J_1]\!]^M_\gamma \ominus \ldots \ominus [\![J_n]\!]^M_\gamma) \Rightarrow ([\![\phi_1]\!]^M_\gamma \oplus \ldots \oplus [\![\phi_m]\!]^M_\gamma)$$

Note that, since each $[\![J_i]\!]^M_\gamma$ *is in* $\{0, 1\}$, *so is the value of the antecedent. We write* $(M, \gamma) \models \Sigma \vdash \Delta$ *if* $[\![\Sigma \vdash \Delta]\!]^M_\gamma = 1$. *A sequent is* valid, *written* $\models \Sigma \vdash \Delta$, *if* $(M, \gamma) \models \Sigma \vdash \Delta$ *for every pair* (M, γ).

The choice of considering a quantitative semantics of sequents is naturally motivated by the $[0, 1]$-valued semantics of the logic pLμ. Among the many possible choices (several are studied in *fuzzy logic* [8]) for interpreting commas in sequents, Łukasiewicz logic enjoys pleasant properties. Its operators coincide with the ordinary boolean ones when arguments have values in $\{0, 1\}$. Furthermore Ł-negation coincides with pLμ negation (see Proposition 5) and interacts well with Ł-disjunction (as in Definition 7). This allows our one-sided (in the set of logical assertions) formulation of sequents. Lastly, and most importantly, Ł-disjuction validates a sound probabilistic interpretation of some key rules of the proof system (see Proposition 15 below). The validity of a sequent $\Sigma \vdash \Delta$ expresses a form of implication: for every model (M, γ) satisfying all the operational assertions in Σ, the sum of the values of the (interpreted) assertions in Δ is at least 1. Valid sequents express nontrivial relations between the probabilities associated to the logical assertions in Δ. For example, the validity of a sequent of the form $\Sigma \vdash \neg\phi_1, \neg\phi_2, \phi$ can be understood as follows: in every model (M, γ) satisfying the operational assertions in Σ, the value of the assertion ϕ is bounded below by a function ($\ominus$) of the values of ϕ_1 and ϕ_2. We now briefly discuss a few illustrative examples of valid sequents showing how the chosen quantitative interpretation allows the expression of interesting properties.

Example 9. Define $F \stackrel{\text{def}}{=} \mu X.\,[a]\,X$. The following sequents are valid:

$$\mathrm{Seq}_1 \stackrel{\text{def}}{=} x \xrightarrow{a} \alpha \vdash x\!:\!\langle a\rangle\top \qquad \mathrm{Seq}_3 \stackrel{\text{def}}{=} \emptyset \vdash x\!:\!\neg F, y\!:\!\neg F, x|y\!:\!F$$
$$\mathrm{Seq}_2 \stackrel{\text{def}}{=} \emptyset \vdash x|y\!:\!\neg F, x\!:\!F \qquad \mathrm{Seq}_4 \stackrel{\text{def}}{=} \emptyset \vdash !\tfrac{1}{2}(a.0)\!:\!F$$

The first example expresses a trivial property: if a process x can perform an a-labeled transition then it satisfies the pLμ formula $\langle a\rangle\top$ with probability 1. The meaning of the pLμ formula F above can be understood as expressing a termination goal (i.e., the impossibility of producing an infinite sequence of a's) under an adversary environment. Thus the sequent Seq_2 expresses a simple property: the parallel (non-communicating) system $x|y$ with two components has termination probability less than or equal to that of its components. The third sequent Seq_3 expresses a slightly less obvious property, providing a lower bound on the termination probability for $x|y$. Lastly, the fourth sequent Seq_4 expresses the fact that the process $!\frac{1}{2}(a.0)$ terminates with probability 1, i.e., almost surely. All the sequents above can be proven valid by our proof system. In Section 5 we present proofs for Seq_3 and Seq_4.

Before introducing the derivation rules of our system, we introduce an auxiliary kind of judgement useful for expressing entailments between operational assertions. We shall consider *operational judgments* of the form $\Sigma \rhd \{\Sigma_i\}_{0\leq i\leq n}$.

Definition 10. *Given an operational judgment $\Sigma \rhd \{\Sigma_i\}_{0\leq i\leq n}$, we write $(M,\gamma) \models \Sigma \rhd \{\Sigma_i\}_{0\leq i\leq n}$ when the following implication holds: if $(M,\gamma)\models\Sigma$ then there is some $i\in\{0,\ldots,n\}$ such that $(M,\gamma^i)\models\Sigma_i$, for some interpretation γ^i that agrees with γ on all (process and distribution) variables appearing in Σ. We say that $\Sigma \rhd \{\Sigma_i\}_{0\leq i\leq n}$ is* valid, *written $\models \Sigma \rhd \{\Sigma_i\}_{0\leq i\leq n}$, if for every pair (M,γ) it holds that $(M,\gamma)\models \Sigma \rhd \{\Sigma_i\}_{0\leq i\leq n}$. Note that $\models \Sigma \rhd \{\emptyset\}$ holds, $\models \Sigma \vdash \emptyset$ holds iff Σ is not satisfiable and $\emptyset \rhd \{\Sigma\}$ holds iff $(M,\gamma)\models\Sigma$ for all (M,γ). In order to improve readability, we just write $\Sigma \vdash J$ instead of $\Sigma \vdash \{\{J\}\}$.*

Example 11. The following are examples of valid operational judgments:

1. $0 \xrightarrow{a} \alpha \rhd \emptyset$ 2. $\emptyset \rhd \{\{a.x \xrightarrow{a} \alpha, \alpha \simeq \delta(x)\}\}$ 3. $a.x \xrightarrow{b} \alpha \rhd \emptyset$, if $b\neq a$
4. $x|y \xrightarrow{a} \alpha \rhd \{\Sigma_{\mathcal{L}}, \Sigma_{\mathcal{R}}\}$ 5. $!x \xrightarrow{a} \alpha \rhd \big\{\{x \xrightarrow{a} \beta, \alpha \simeq \{\beta \rightsquigarrow y\}\delta(y|!x)\}\big\}$
5. $\emptyset \rhd \alpha \simeq \{\alpha \rightsquigarrow x\}\delta(x)$ 6. $\emptyset \rhd \alpha +_{\frac{1}{4}} \beta \simeq (\alpha +_{\frac{1}{2}} \beta) +_{\frac{1}{2}} \beta$

where $\Sigma_{\mathcal{L}} = x \xrightarrow{a} \beta, \alpha \simeq \{\beta \rightsquigarrow x'\}\delta(x'|y)$ and $\Sigma_{\mathcal{R}} = y \xrightarrow{a} \beta, \alpha \simeq \{\beta \rightsquigarrow y'\}\delta(x|y')$.

The derivation rules for our main proof system for quantitative sequents are presented in Figure 2. The rule Σ-Rule supports reasoning about the operational semantics by means of case analysis, using a side-condition exploiting the semantic validity of operational judgements (Definition 10). In practice, this semantic side-condition can be replaced with a formal proof system for proving validity of operational judgements, which can be constructed following established approaches (see, e.g., "action assertions rules" in [19, p. 18]). For lack of space, we do not go into further details about this, focussing instead on our main proof system for quantitative sequents, whose design is significantly more intricate.

$$\frac{\{\Sigma_i \vdash \Delta\}_{i\in I}}{\Sigma \vdash \Delta}\ \Sigma\text{-Rule}\ \ (\text{proviso: } \Sigma \rhd \{\Sigma_i\}_{i\in I})$$

$$\frac{\Sigma \vdash \Gamma, \psi \qquad \Sigma \vdash \Delta, \neg\psi}{\Sigma \vdash \Gamma, \Delta}\ \text{Cut}$$

$$\frac{\Sigma \vdash \Delta}{\Sigma[q/x] \vdash \Delta[q/x]}\ \text{P-Sub} \qquad \frac{\Sigma \vdash \Delta}{\Sigma[d/\alpha] \vdash \Delta[d/\alpha]}\ \text{D-Sub}$$

$$\frac{\Sigma[e/\alpha] \vdash \Delta[e/\alpha]}{\Sigma[d/\alpha] \vdash \Delta[d/\alpha]}\ \Sigma\text{-Sub}\ \ (\text{proviso: } \Sigma \rhd d \simeq e)$$

$$\frac{\Sigma \vdash p\!:\!F_i, \Delta}{\Sigma \vdash p\!:\!F_1 \vee F_2, \Delta}\ \vee_i\ \ i\in\{1,2\} \qquad \frac{\Sigma \vdash p\!:\!F, \Delta \qquad \Sigma \vdash p\!:\!G, \Delta}{\Sigma \vdash p\!:\!F \wedge G, \Delta}\ \wedge$$

$$\frac{\Sigma \vdash d\!:\!F, \Delta}{\Sigma \vdash p\!:\!\langle a\rangle F, \Delta}\ \langle a\rangle\, (\text{proviso: } \Sigma \rhd p \xrightarrow{a} d) \qquad \frac{\Sigma, p \xrightarrow{a} \alpha \vdash \alpha\!:\!F, \Delta}{\Sigma \vdash p\!:\,[a]\,F, \Delta}\ [a]\ \ \alpha \text{ fresh}$$

$$\frac{\Sigma \vdash p\!:\!F[\mu X.F/X], \Delta}{\Sigma \vdash p\!:\!\mu X.F, \Delta}\ \mu \qquad \frac{\Sigma \vdash p\!:\!F[\nu X.F/X], \Delta}{\Sigma \vdash p\!:\!\nu X.F, \Delta}\ \nu$$

$$\frac{\Sigma \vdash \Delta, p\!:\!F}{\Sigma \vdash \Delta, \delta(p)\!:\!F}\ \delta$$

$$\frac{\Sigma \vdash \Delta, d_1\!:\!F_1, \ldots, d_n\!:\!F_n \qquad \Sigma \vdash \Delta, e_1\!:\!F_1, \ldots, e_n\!:\!F_n}{\Sigma \vdash \Delta, d_1 +_\lambda e_1\!:\!F_1, \ldots, d_n +_\lambda e_n\!:\!F_n}\ +_\lambda\ \ \lambda\in(0,1)$$

$$\frac{\Sigma \vdash \Delta, e_1[y/x_1]\!:\!F_1, \ldots, e_n[y/x_n]\!:\!F_n}{\Sigma \vdash \Delta, \{d \rightsquigarrow x_1\}e_1\!:\!F_1, \ldots, \{d \rightsquigarrow x_n\}e_n\!:\!F_n}\ \{\rightsquigarrow\}\ \ y \text{ fresh.}$$

Fig. 2. Derivation rules

A typical usage of the Σ-Rule is better explained by means of a simple example. Consider the valid sequent $x|y \xrightarrow{a} \alpha \vdash x : \langle a\rangle\top, y : \langle a\rangle\top$ asserting that in every model such that $x|y$ can make an a-transition then either x or y or both can make an a-transition (this qualitative interpretation holds since $(\!|\langle a\rangle\top|\!)(p) \in \{0,1\}$ in every model). The crucial step in proving its validity is:

$$\frac{\Sigma_{\mathcal{L}} \vdash x\!:\!\langle a\rangle\top, y\!:\!\langle a\rangle\top \qquad \Sigma_{\mathcal{R}} \vdash x\!:\!\langle a\rangle\top, y\!:\!\langle a\rangle\top}{x|y \xrightarrow{a} \alpha \vdash x\!:\!\langle a\rangle\top, y\!:\!\langle a\rangle\top}\ \Sigma\text{-Rule: } x|y \xrightarrow{a} \alpha \rhd \{\Sigma_{\mathcal{L}}, \Sigma_{\mathcal{R}}\}$$

where $\Sigma_{\mathcal{L}}$ and $\Sigma_{\mathcal{R}}$ are as in Example 11. This step performs the required case analysis, based on the operational semantics of the (non-communicating) parallel operator, required to distinguish the two relevant cases. Both premises above are easily seen to be valid (see also Seq_1 in Example 9). Note that the only axiom rule (i.e., rule without premises) in the proof system is the instance of the Σ-Rule when the proviso is of the form $\Sigma \rhd \emptyset$, i.e., when Σ is unsatisfiable. We refer to this particular use of this rule as Σ-Axiom. For example, the sequent $\emptyset \vdash 0\!: [a]\,\bot$ can be proved as follows,

$$\frac{\dfrac{}{0 \xrightarrow{a} \alpha \vdash \alpha\!:\!\bot}\ \Sigma\text{-Rule}\ (0 \xrightarrow{a} \alpha \rhd \emptyset)}{\emptyset \vdash 0\!: [a]\,\bot}\ [a]$$

where the axiom is used to reveal the inconsistency in the assumption that the null process 0 could make an a-transition.

Definition 12. *Let $\mathcal{R}$ be a derivation rule. We say that $\mathcal{R}$ is* sound *if, whenever the sequent $\Sigma \vdash \Delta$ is derived using $\mathcal{R}$ from the premises $\{\Sigma_i \vdash \Delta_i\}_{i \in I}$ (for some finite index set I) which are all valid, then also $\Sigma \vdash \Delta$ is valid. We also say that $\mathcal{R}$ is* strongly sound *if, for every (M, γ), the following inequality holds*

$$[\![\Sigma \vdash \Delta]\!]^M_\gamma \geq \min \left\{ [\![\Sigma_i \vdash \Delta_i]\!]^M_{\gamma'} \right\}_{i \in I}$$

for all interpretations γ' that agree with γ on all variables appearing in $\Sigma \vdash \Delta$.

The notion of strong soundness clearly implies the ordinary one. The proposition bellow explains the reason for omitting the contraction rule.

Proposition 13. *The contraction rule* $\dfrac{\Sigma \vdash \Delta, \phi, \phi}{\Sigma \vdash \Delta, \phi}$ *is not sound.*

Proposition 14. *The CUT rule is sound but not strongly sound. All other derivation rules of Figure 2 are strongly sound.*

Proof. Most cases are trivial to verify. The strong soundness of the rules $+_\lambda$ and $\{\leadsto\}$ follows from Proposition 15 below. □

Remark 1. Strong soundness of derivation rules is a technical requirement needed in the proof of our main theorem (see remarks after Theorem 17 below). As stated in Proposition 14, the CUT rule is not strongly sound. This fact requires restrictions to be placed on applications of CUT in proofs (see Definition 16 below). These restrictions are needed for our soundness proof to go through.

The rules {P-Sub, D-Sub, Σ-Sub} are called *substitution rules* and support parametric reasoning [19]. In particular note how using the rule Σ-Sub one can substitute some of the occurrences of a compound distribution term with another equivalent (in all models satisfying Σ) compound distribution term. For example, the term $d +_{\frac{1}{3}} e$ can be rewritten to $(d +_{\frac{2}{3}} e) +_{\frac{1}{2}} e$. Such equational reasoning on distribution terms can be very useful (see, e.g., Remark 2 below). The rules $\{\vee_1, \vee_2, \wedge, \langle a \rangle, [a], \mu, \nu\}$ are called *logical rules.* The rules $\vee_1$, $\vee_2$, $\wedge$, μ, ν are standard and also the rules $\langle a \rangle$, $[a]$ for reasoning about modalities are natural counterparts to the analogous rules adopted in proof systems for modal (fixed point) logics appeared in the literature (see, e.g., [19], [21] and [15]). The rules $\{\delta, +_\lambda, \{\leadsto\}\}$ are called *distribution rules* and constitute a crucial aspect of the system. Together with the rule Σ-Sub, these are the only rules that operate on logical assertions containing distribution terms. All distribution rules can be understood probabilistically as (partially) evaluating the probability distribution terms of the active logical assertions. The simplest rule δ just evaluates a Dirac distribution to the corresponding process. In the rule $+_\lambda$, each active logical assertions (of the form $d_i +_\lambda e_i : F_i$ for a fixed $\lambda \in [0, 1]$) is evaluated to the left (resp. right) sub-term in the left (resp. right) premise of the rule with probability λ (resp. $1 - \lambda$). Note that the evaluation steps of each active probability distribution are not independent of each other. To the contrary, useful dependencies can be established by applications of the rule $+_\lambda$. The rule $\{\leadsto\}$ can be understood, by similar arguments, as a symbolic variant of the rule $+_\lambda$.

Remark 2. Note that the distribution rules $[+_\lambda]$ and $[\{\rightsquigarrow\}]$ may only be applicable once the distribution terms have been rewritten. Consider for example the sequent $\Sigma \vdash (d +_{\frac{1}{3}} e) : F, (d' +_{\frac{1}{2}} e) : G$. The distribution rule $\{+_\lambda\}$ is not directly applicable since the two distribution terms have a different outermost connective. However, by application of the rule Σ-Sub, the distribution term $d +_{\frac{1}{3}} e$ can be rewritten as $(d +_{\frac{2}{3}} e) +_{\frac{1}{2}} e$. This could be used as follows

$$\dfrac{\dfrac{\Sigma \vdash (d +_{\frac{2}{3}} e) : F, d' : G \qquad \Sigma \vdash e : F, e : G}{\Sigma \vdash ((d +_{\frac{2}{3}} e) +_{\frac{1}{2}} e) : F, (d' +_{\frac{1}{2}} e) : G}\, +_{\frac{1}{2}}}{\Sigma \vdash (d +_{\frac{1}{3}} e) : F, (d' +_{\frac{1}{2}} e) : G}\, \Sigma\text{-Sub}$$

to reduce the original problem to the verification of the two new subgoals.

Proposition 15. *Let $\Sigma \vdash \Delta$ be derived by application of the rule $+_\lambda$ from the two premises $\Sigma \vdash \Delta_1$ and $\Sigma \vdash \Delta_2$. Then, for every (M, γ) it holds that*

$$[\![\Sigma \vdash \Delta]\!]^M_\gamma \geq \lambda \cdot [\![\Sigma \vdash \Delta_1]\!]^M_\gamma \; + \; (1-\lambda) \cdot [\![\Sigma \vdash \Delta_2]\!]^M_\gamma .$$

Similarly, let $\Sigma \vdash \Delta$ be derived by application of the rule $\{\rightsquigarrow\}$ and let $\Sigma \vdash \Sigma_1$ be its only premise, as depicted in Figure 2. Then, for every (M, γ), it holds that

$$[\![\Sigma \vdash \Delta]\!]^M_\gamma \geq \sum_{m \in M} \gamma(d)(m) \cdot [\![\Sigma \vdash \Delta_1]\!]^M_{\gamma[m/y]}$$

where $\gamma[m/y]$ updates γ by assigning to the fresh variable y the process $m \in M$.

Proof. Both points follow easily from the following arithmetical inequality (and variants thereof): $\oplus_{i \in I}\{x_i +_\lambda y_i\} \geq (\oplus\{x_i\}_{i\in I}) +_\lambda (\oplus\{y_i\}_{i\in I})$ which is valid for every index set I, reals $x_i, y_i \in [0,1]$, where $x +_\lambda y = \lambda \cdot x + (1-\lambda) \cdot y$. □

4.1 Markov Proofs

As anticipated in the introduction, to enable the proof system to handle the fixed points in the logic pLμ, we allow *cyclic* proof trees (cf. [15,21,3]) in which some leaves of the tree are identified with sequents internal to the tree, with the proof looping back to that point. Technically, it is convenient to view such cyclic trees as the infinite trees they unfold to, and to work with general infinite trees, with the finite cyclic ones corresponding exactly to the *regular* trees (those with only finitely many subtrees). We call a (possibly infinite) tree of rule applications, in which all leaves are instances of the axiom rule Σ-Axiom, a *preproof*. A preproof is *cut-free* if it does not contain occurrences of the CUT rule. Since they may have infinite branches, preproofs are not guaranteed to have valid endsequents even though every rule is (strongly) sound.

In the literature on infinitary proof systems for fixed-point logics (see, e.g., [15,21,3]), valid proofs are defined as those preproofs whose infinite branches all contain at least one legitimate sequence (called a *valid trace*) of fixed-point unfoldings, along which a greatest fixed-point is unfolded infinitely often. This can equivalently be reformulated in terms of a single player game. The aim of the single player, Refuter, is to find an infinite branch along which all traces are invalid. The preproof is then considered a valid proof just in case Refuter cannot win his game.

For the proof system in this paper, we adopt a similar approach, except that we now interpret Refuter's game as a single-player stochastic game $\mathcal{G}(T)$ (i.e., a Markov Decision Process) over the preproof T, and we also need to constrain applications of the CUT rule (see Remark 1 and those following Theorem 17). Once again, in the game $\mathcal{G}(T)$, Refuter is trying to find an infinite branch in T along which all traces are invalid. This time, however, instances of the rule $+_\lambda$ in T are interpreted as probabilistic nodes under the choice of Nature, who extends the branch thus far with the left (resp. right) premise with probability λ (resp. $1-\lambda$). At all other rules, Refuter has the choice of premise. A preproof satisfies the *game condition* just in case Refuter almost surely fails in his goal; that is, no matter what strategy Refuter adopts, the probability of him finding an infinite branch with all traces invalid is 0. We now specify the collection of valid derivations, which we call *Markov proofs*, by the following *inductive* definition.

Definition 16. *A* Markov proof *is a preproof T satisfying the game condition and such that, for that every occurrence of the* CUT *rule in T,*

$$\frac{\begin{array}{c}T_1\\ \Sigma\vdash\Delta,\phi\end{array}\qquad\begin{array}{c}T_2\\ \Sigma\vdash\Gamma,\neg\phi\end{array}}{\Sigma\vdash\Delta,\Gamma}\,CUT$$

either the sub-preproof T_1 or T_2 (or both) is a Markov proof.

Note that a Markov proof can contain infinite branches on which Refuter wins as long as the set of such branches has probability 0 for every Refuter strategy in $\mathcal{G}(T)$. We shall see an example of this kind of Markov proof in Section 5. We remark also that the inductive definition could be replaced with a combinatorial condition. A Markof proof could equivalently be defined as a preproof T satisfying the game condition, for which there exists an assignment of a privileged premise (a 'switching') to every CUT rule such that no infinite path in the proof runs through infinitely many privileged ('switched') CUT premises.

The set of rules of Figure 2 has been kept as small as possible to simplify the proof of Theorem 17 below. Other expected rules are admissible, such as:

$$\frac{}{\Sigma\vdash\Delta,x\!:\!\top}\,\mathrm{Ax}(\top)\qquad\frac{}{\Sigma\vdash\Delta,x\!:\!F,x\!:\!\neg F}\,\mathrm{Ax}(\neg)\qquad\frac{\Sigma\vdash\Delta}{\Sigma,\Sigma'\vdash\Delta,\Gamma}\,\mathrm{Weak}$$

The following result is the main technical contribution of this paper.

Theorem 17 (Soundness). *The endsequent of every Markov proof is valid.*

Proof Sketch. Our proof technique is based on the game semantics of pLμ and, therefore, crucially exploits the equivalence result of [14]. The result is first proved for cut-free Markov proofs and then extended to general Markov proofs. The structure of a Markov proof Π with endsequent $\Sigma\vdash\{\phi_i\}_{i\in I}$ is seen as providing strategies σ_1^i for player Maximizer in the two-player stochastic games associated with the assertions ϕ_i. On the other hand, a Markov play (i.e., a Markov chain) P_Π in $\mathcal{G}(\Pi)$, resolving the choices corresponding to the occurrences of rules $\{\wedge, \Sigma\text{-Rule}\}$ in Π, is seen as providing strategies σ_2^i for Minimizer as well as information about a counter-model M. This allows us to consider P_Π as a *coupling* of Markov chains, i.e., as a non-independent product of the probabilistic pLμ plays $P^i_{\sigma_1^i,\sigma_2^i}$ associated with ϕ_i, whose probabilistic dependencies have been introduced by the rules $[+_\lambda, \{\rightsquigarrow\}]$ in Π. Our proof is by *reductio ad absurdum.* One assumes that M and σ_2^i's constitute a counterexample to the validity of the endsequent of Π, i.e., that the expected probability of victory for Maximizer in the Markov plays $P^i_{\sigma_1^i,\sigma_2^i}$ sum up to a value $\lambda < 1$. We show that this implies that P_Π must assign at least probability $1-\lambda$ (i.e., positive measure) to the set of branches in Π corresponding to plays losing for Maximizer in the pLμ game associated with ϕ_i, for all $i \in I$. These are precisely branches without valid traces. From these assumption it follows that Refuter can win in the game $\mathcal{G}(\Pi)$ with positive probability. Thus Π cannot be a Markov proof, a contradiction. □

The following theorem shows that regular (cyclic) Markov proofs do indeed form an effective proof system. This is essential for the potential applicability of the approach. It is proved along the lines of similar results for non-probabilistic cyclic proofs (see, e.g., [15], [21] and [3]), using decidability results for one-player stochastic parity games established in [5].

Theorem 18. *It is decidable if a regular preproof T is a Markov proof.*

5 Examples of Markov Proofs

In this section we provide Markov proofs of the sequents Seq_3 and Seq_4 discussed in Example 9. Despite the simplicity of the process algebra considered in this paper, these small examples illustrate nontrivial instances of compositional reasoning and verification of infinite state systems. However, due to the space limits, some important features of our proof system, such as the possibility of recombining distribution terms (see Remark 2) and the capability of handling more realistic process algebras (such as those including, e.g, a *communicating* parallel operator), are not illustrated in this paper.

The validity of Seq_3 is proved by the (cut-free) Markov proof Π_3 depicted in Figure 3 (top), where the proviso of Σ-Rule, expressing a case analysis, is as in Example 11. The left sub-Markov proof Π_L, itself containing Π_3 as sub-Markov proof, is depicted as in Figure 3, and Π_R is the similar Markov proof of the sequent $y \xrightarrow{a} \beta, \alpha \simeq \{\beta \rightsquigarrow y'\}\delta(x|y') \vdash \alpha : F, x : \neg F, y : \neg F$. Each infinite play (i.e., branch in Π_3 since there are no probabilistic vertices in $\mathcal{G}(\Pi_3)$, see Section 4), has a valid trace because the greatest fixed point operators are unfolded infinitely many times. Thus Π_3 is a Markov proof as desired.

Compositional reasoning is supported in our system by the CUT rule (cf. [19]). For instance, as we have established the validity of Seq_3, we can reduce the problem of verifying the validity of the sequent $\emptyset \vdash p|q : F$ (i.e., prove that

$$\dfrac{\dfrac{\dfrac{\Pi_L \qquad \Pi_R}{x|y \xrightarrow{a} \alpha \vdash \alpha : F, x : \neg F, y : \neg F}\ \Sigma\text{-Rule: } x|y \xrightarrow{a} \alpha \rhd \{\Sigma_{\mathcal{L}}, \Sigma_{\mathcal{R}}\}}{\emptyset \vdash x|y : [a]\,F, x : \neg F, y : \neg F}\ [a]}{\emptyset \vdash x|y : \mu Z.\,[a]\,Z, x : \nu X.\langle a\rangle X, y : \nu Y.\langle a\rangle Y}\ \mu$$

$$\dfrac{\dfrac{\dfrac{\dfrac{\dfrac{\dfrac{\dfrac{\dfrac{\Pi_3}{\emptyset \vdash x|y : F, x : \neg F, y : \neg F}}{\emptyset \vdash x'|y : F, x' : \neg F, y : \neg F}\ \text{P-SUB: } [x/x']}{\emptyset \vdash \delta(x'|y) : F, \delta(x') : \neg F, y : \neg F}\ \delta,\delta}{\emptyset \vdash \{\beta \rightsquigarrow x'\}\delta(x'|y) : F, \{\beta \rightsquigarrow x'\}\delta(x') : \neg F, y : \neg F}\ \{\rightsquigarrow\}}{\emptyset \vdash \{\beta \rightsquigarrow x'\}\delta(x'|y) : F, \beta : \neg F, y : \neg F}\ \Sigma\text{-Sub}}{x \xrightarrow{a} \beta \vdash \{\beta \rightsquigarrow x'\}\delta(x'|y) : F, x : \langle a\rangle\neg F, y : \neg F}\ \langle a\rangle\text{: } x \xrightarrow{a} \beta \rhd x \xrightarrow{a} \beta}{x \xrightarrow{a} \beta \vdash \{\beta \rightsquigarrow x'\}\delta(x'|y) : F, x : \neg F, y : \neg F}\ \nu}{x \xrightarrow{a} \beta, \alpha \simeq \{\beta \rightsquigarrow x'\}\delta(x'|y) \vdash \alpha : F, x : \neg F, y : \neg F}\ \Sigma\text{-Sub}$$

$$\dfrac{\dfrac{\dfrac{\dfrac{\Pi_3}{\emptyset \vdash x{:}\neg F, y{:}\neg F, x|y{:}F}}{\emptyset \vdash p{:}\neg F, q{:}\neg F, p|q{:}F}\ \text{P-SUB} \qquad \emptyset \vdash p{:}F}{\emptyset \vdash p{:}\neg F, p|q{:}F}\ \text{CUT} \qquad \emptyset \vdash q{:}F}{\emptyset \vdash p|q{:}F}\ \text{CUT}$$

$$\dfrac{\dfrac{\dfrac{\Pi_3}{\emptyset \vdash x'|z : \neg F, x'{:}\neg F, z{:}\neg F}}{\emptyset \vdash (x|y)|z{:}F, x|y{:}\neg F, z{:}\neg F}\ \text{P-SUB } [(x|y)/x'] \qquad \dfrac{\Pi_3}{\emptyset \vdash x|y : F, x : \neg F, y : \neg F}}{\emptyset \vdash (x|y)|z{:}F, x{:}\neg F, y{:}\neg F, z{:}\neg F}\ \text{CUT}$$

$$\dfrac{\dfrac{\dfrac{\dfrac{\dfrac{\dfrac{\Pi_A}{p \xrightarrow{a} \beta \vdash \beta{:}F} \qquad \dfrac{\dfrac{\Pi_4}{\emptyset \vdash !\tfrac{1}{2}p{:}F} \qquad \dfrac{\Pi_B}{p \xrightarrow{a} \beta \vdash \{\beta \rightsquigarrow y\}\delta(y|!\tfrac{1}{2}p){:}F, !\tfrac{1}{2}p{:}\neg F}}{p \xrightarrow{a} \beta \vdash \{\beta \rightsquigarrow y\}\delta(y|\tfrac{1}{2}p\}{:}F}\ \text{CUT}}{p \xrightarrow{a} \beta \vdash \beta +_{\frac{1}{2}} \{\beta \rightsquigarrow y\}\delta(y|!\tfrac{1}{2}p){:}F}\ +_{\frac{1}{2}}}{p \xrightarrow{a} \beta, \alpha \simeq \beta +_{\frac{1}{2}} \{\beta \rightsquigarrow y\}\delta(y|!\tfrac{1}{2}p\} \vdash \alpha{:}F}\ \Sigma\text{-Sub}}{!\tfrac{1}{2}p \xrightarrow{a} \alpha \vdash \alpha{:}F}\ \Sigma\text{-Rule: } P}{\emptyset \vdash !\tfrac{1}{2}p{:}\,[a]\,F}\ [a]}{\emptyset \vdash !\tfrac{1}{2}p{:}\,\mu X.\,[a]\,X}\ \mu$$

Fig. 3. Examples of Markov proofs

the compound system $p|q$ almost surely terminates) to the verification of the two smaller goals $\emptyset \vdash p{:}F$ and $\emptyset \vdash q{:}F$ by means of the Markov proof depicted in Figure 3. Furthermore, other useful results can be proved, without searching for direct proofs, by using already proved lemmas. For example, the validity of the sequent $\emptyset \vdash (x|y)|z{:}F, x{:}\neg F, y{:}\neg F, z{:}\neg F$ can be proved as in Figure 3.

$$\dfrac{\dfrac{\Pi_p}{\emptyset \vdash p:F} \qquad \dfrac{p \xrightarrow{a} \beta \vdash \beta : F, \beta : \neg F}{p \xrightarrow{a} \beta \vdash \beta : F, p : \neg F}\,\nu, \langle a \rangle}{p \xrightarrow{a} \beta \vdash \beta : F}\,\mathrm{CUT}$$

$$\dfrac{\dfrac{\dfrac{\dfrac{\dfrac{\Pi_3}{\vdash y|x:F, x:\neg F, y:\neg F}}{\vdash y|!^{\frac{1}{2}}p:F, !^{\frac{1}{2}}p:\neg F, y:\neg F}\,\text{P-SUB}\,[!^{\frac{1}{2}}p/x]}{\vdash \{\beta \rightsquigarrow y\}\delta(y|!^{\frac{1}{2}}p):F, !^{\frac{1}{2}}p:\neg F, \{\beta \rightsquigarrow y\}\delta(y):\neg F}\,\{\rightsquigarrow\}, \delta}{p \xrightarrow{a} \beta \vdash \{\beta \rightsquigarrow y\}\delta(y|!^{\frac{1}{2}}p):F, !^{\frac{1}{2}}p:\neg F, p:\neg F}\,\nu, \langle a \rangle \qquad \dfrac{\Pi_p}{\emptyset \vdash p:F}}{p \xrightarrow{a} \beta \vdash \{\beta \rightsquigarrow y\}\delta(y|!^{\frac{1}{2}}p):F, !^{\frac{1}{2}}p:\neg F}\,CUT$$

Fig. 4. Sub-Markov proofs Π_A and Π_B of Π_4

Compositional reasoning is the key to the verification of infinite state systems. Consider, for example, a process p that almost surely terminates, i.e., such that the validity of $S_p = \emptyset \vdash p : F$ has been proven by a Markov proof Π_p. It is simple to verify that $!^{\frac{1}{2}}p$ is an infinite state system, even when p is a finite state process such as $a.0$. Nevertheless, it is possible to prove that $!^{\frac{1}{2}}p$ almost surely terminates. This is expressed (for $p{=}a.0$) by the sequent Seq_4 whose validity is witnessed by the regular Markov proof Π_4 depicted in Figure 3 (bottom) where the operational judgment $P \stackrel{\text{def}}{=} !^{\frac{1}{2}}p \xrightarrow{a} \alpha \rhd \{\{p \xrightarrow{a} \beta, \alpha \simeq \beta +_{\frac{1}{2}} \{\beta \rightsquigarrow y\}\delta(y|!^{\frac{1}{2}}p)\}\}$ used in the rule Σ-Rule is defined is valid, and the Markov proofs Π_A and Π_B can be depicted as in Figure 4. Note how the use of the CUT rule in Π_B allows us to make use of the result already proved by the Markov proof Π_3. This time Π_4 contains probabilistic vertices: at occurrences of the rule $+_{\frac{1}{2}}$ the game probabilistically branches. The only infinite branch in Π_4 without valid traces is the one never joining one of the the two sub-Markov proofs Π_A or Π_B. However, the probability that this branch is the outcome of the game $\mathcal{G}(\Pi_4)$ is easily seen to be an event of probability 0. Thus Π_4 satisfies the proof condition and is a Markov proof, as desired.

6 Further Directions

There are numerous directions for improvement to the approach of this paper. One is relax the restrictions on applications of CUT. Is it possible to reconfigure the proof system and soundness proof so that an unrestricted CUT rule is available? Another is to address completeness issues, which we have ignored entirely. Are completeness results available for restricted classes of processes (e.g., finite state)? Yet another is to attempt to extend the proof system to deal with extensions of the probabilistic μ-calculus with other operators, for example those considered in [13], allowing the full expressivity of PCTL to be captured.

It is unclear to us whether or not the approach of this paper is able to scale up to establish useful properties of real-world systems. Nevertheless, we see the main value of our paper as contributing novel techniques towards the challenging problem of compositional verification for concurrent probabilistic systems. In particular, we believe that the use of proofs containing probabilistic branching will generalise to other proof systems for other probabilistic logics.

Acknowledgements. We thank the anonymous referees for helpful suggestions.

References

1. Baier, C., Katoen, J.P.: Principles of Model Checking. The MIT Press (2008)
2. Bartels, F.: GSOS for probabilistic transition systems. Electronic Notes in Theoretical Computer Science 65(1) (2002)
3. Brotherston, J., Simpson, A.: Sequent calculi for induction and infinite descent. Journal of Logic and Computation 21(6), 1177–1216 (2011)
4. Cardelli, L., Larsen, K.G., Mardare, R.: Modular Markovian Logic. In: Aceto, L., Henzinger, M., Sgall, J. (eds.) ICALP 2011, Part II. LNCS, vol. 6756, pp. 380–391. Springer, Heidelberg (2011)
5. Chatterjee, K.: Stochastic ω-Regular Games. PhD thesis, University of California, Berkeley (2007)
6. Delahaye, B., Katoen, J.-P., Larsen, K.G., Legay, A., Pedersen, M.L., Sher, F., Wąsowski, A.: Abstract Probabilistic Automata. In: Jhala, R., Schmidt, D. (eds.) VMCAI 2011. LNCS, vol. 6538, pp. 324–339. Springer, Heidelberg (2011)
7. Forejt, V., Kwiatkowska, M., Norman, G., Parker, D., Qu, H.: Quantitative Multi-objective Verification for Probabilistic Systems. In: Abdulla, P.A., Leino, K.R.M. (eds.) TACAS 2011. LNCS, vol. 6605, pp. 112–127. Springer, Heidelberg (2011)
8. Hájek, P.: Metamathematics of Fuzzy Logic. Trends in Logic. Springer (2001)
9. Henzinger, T.A., Sifakis, J.: The Embedded Systems Design Challenge. In: Misra, J., Nipkow, T., Sekerinski, E. (eds.) FM 2006. LNCS, vol. 4085, pp. 1–15. Springer, Heidelberg (2006)
10. Huth, M., Kwiatkowska, M.: Quantitative analysis and model checking. In: Proc. of 12th LICS (1997)
11. Kwiatkowska, M., Norman, G., Parker, D., Qu, H.: Assume-Guarantee Verification for Probabilistic Systems. In: Esparza, J., Majumdar, R. (eds.) TACAS 2010. LNCS, vol. 6015, pp. 23–37. Springer, Heidelberg (2010)
12. McIver, A., Morgan, C.: Results on the quantitative μ-calculus q$M\mu$. ACM Transactions on Computational Logic 8(1) (2007)
13. Mio, M.: Game Semantics for Probabilistic μ-Calculi. PhD thesis, School of Informatics, University of Edinburgh (2012)
14. Mio, M.: On the equivalence of denotational and game semantics for the probabilistic μ-calculus. Logical Methods in Computer Science 8(2) (2012)
15. Niwinski, D., Walukiewicz, I.: Games for the μ-calculus. Theoretical Computer Science 163, 99–116 (1997)
16. Panangaden, P.: Labelled Markov processes. Imperial College Press (2009)
17. Pnueli, A.: The temporal logic of programs. In: Proc. of 19th FOCS (1977)
18. Segala, R.: Modeling and Verification of Randomized Distributed Real-Time Systems. PhD thesis, Laboratory for Computer Science, M.I.T (1995)
19. Simpson, A.: Sequent calculi for process verification: Hennessy-Milner logic for an arbitrary GSOS. Journal of Logic and Algebraic Programming 60-61, 287 (2004)
20. Stirling, C.: Modal and temporal logics for processes. Springer (2001)
21. Studer, T.: On the proof theory of the modal mu-calculus. Studia Logica 89(3) (2007)

Partiality and Recursion in Higher-Order Logic

Łukasz Czajka

Institute of Informatics, University of Warsaw
Banacha 2, 02-097 Warszawa, Poland
lukaszcz@mimuw.edu.pl

Abstract. We present an illative system $\mathcal{I}_s$ of classical higher-order logic with subtyping and basic inductive types. The system $\mathcal{I}_s$ allows for direct definitions of partial and general recursive functions, and provides means for handling functions whose termination has not been proven. We give examples of how properties of some recursive functions may be established in our system. In a technical appendix to the paper we prove consistency of $\mathcal{I}_s$. The proof is by model construction. We then use this construction to show conservativity of $\mathcal{I}_s$ over classical first-order logic. Conservativity over higher-order logic is conjectured, but not proven.

1 Introduction

We present an illative λ-calculus system $\mathcal{I}_s$ of classical higher-order logic with subtyping and basic inductive types. Being illative means that the system is a combination of higher-order logic with the *untyped* λ-calculus. It therefore allows for unrestricted recursive definitions directly, including definitions of possibly non-terminating partial functions. We believe that this feature of $\mathcal{I}_s$ makes it potentially interesting as a logic for an interactive theorem prover intended to be used for program verification.

In order to ensure consistency, most popular proof assistants allow only total functions, and totality must be ensured by the user, either by very precise specifications of function domains, restricting recursion in a way that guarantees termination, explicit well-foundedness proofs, or other means. There are various indirect ways of dealing with general recursion in popular theorem provers based on total logics. There are also many non-standard logics allowing partial functions directly. We briefly survey some related work in Sect. 5.

In Sect. 2 we introduce the system $\mathcal{I}_s$. Our approach builds on the old tradition of illative combinatory logic [1,2,3]. This tradition dates back to early inconsistent systems of Shönfinkel, Church and Curry proposed in the 1920s and the 1930s [2]. However, after the discovery of paradoxes most logicians abandoned this approach. A notable exception was Haskell Curry and his school, but not much progress was made in establishing consistency of illative systems strong enough to interpret traditional logic. Only in the 1990s some first-order illative system were shown consistent and complete for traditional first-order logic [1,4]. The system $\mathcal{I}_s$, in terms of the features it provides, may be considered an extension of the illative system $\mathcal{I}_\omega$ from [3]. We briefly discuss the relationship between $\mathcal{I}_s$ and $\mathcal{I}_\omega$ in Sect. 5.

F. Pfenning (Ed.): FOSSACS 2013, LNCS 7794, pp. 177–192, 2013.

Because $\mathcal{I}_s$ is based on the untyped λ-calculus, its consistency is obviously open to doubt. In an appendix we give a proof by model construction of consistency of $\mathcal{I}_s$. Unfortunately, the proof is too long to fit within the page limits of a conference paper. In Sect. 3 we give a general overview of the proof. The model construction is similar to the one from [3] for the traditional illative system $\mathcal{I}_\omega$. It is extended and adapted to account for additional features of $\mathcal{I}_s$. To our knowlege $\mathcal{I}_s$ is the first higher-order illative system featuring subtypes and some form of induction, for which there is a consistency proof.

In Sect. 4 we provide examples of proofs in $\mathcal{I}_s$ indicating possible applications of our approach to the problem of dealing with partiality, non-termination and general recursion in higher-order logic. We are mainly interested in partiality arising from non-termination of non-well-founded recursive definitions.

For lack of space we omit proofs of the lemmas and theorems we state. The proofs of non-trivial results are in technical appendices to this paper. The appendices may be found in [5].

2 The Illative System

In this section we present the system $\mathcal{I}_s$ of illative classical higher-order logic with subtyping and derive some of its basic properties.

Definition 1. The system $\mathcal{I}_s$ consists of the following.

- A countably infinite set of variables $V_s = \{x, y, z, \ldots\}$ and a set of constants Σ_s.
- The set of sorts $\mathcal{S} = \{\text{Type}, \text{Prop}\}$.
- The set of *basic inductive types* $\mathcal{T}_I$ is defined inductively by the rule: if $\iota_{1,1}, \ldots, \iota_{1,n_1}, \ldots, \iota_{m,1}, \ldots, \iota_{m,n_m} \in \mathcal{T}_I \cup \{\star\}$ then
$$\mu(\langle \iota_{1,1}, \ldots, \iota_{1,n_1} \rangle, \ldots, \langle \iota_{m,1}, \ldots, \iota_{m,n_m} \rangle) \in \mathcal{T}_I$$
where $m \in \mathbb{N}_+$ and $n_1, \ldots, n_m \in \mathbb{N}$.
- We define the sets of *constructors* $\mathcal{C}$, *destructors* $\mathcal{D}$, and *tests* $\mathcal{O}$ as follows. For each $\iota \in \mathcal{T}_I$ of the form
$$\iota = \mu(\langle \iota_{1,1}, \ldots, \iota_{1,n_1} \rangle, \ldots, \langle \iota_{m,1}, \ldots, \iota_{m,n_m} \rangle) \in \mathcal{T}_I$$
where $\iota_{i,j} \in \mathcal{T}_I \cup \{\star\}$, the set $\mathcal{C}$ contains m distinct constants $c_1^\iota, \ldots, c_m^\iota$. The number n_i is called the *arity* of c_i^ι, and $\langle \iota_{i,1}, \ldots, \iota_{i,n_i} \rangle$ is its *signature*. With each $c_i^\iota \in \mathcal{C}$ of arity n_i we associate n_i distinct destructors $d_{i,1}^\iota, \ldots, d_{i,n_i}^\iota \in \mathcal{D}$ and one test $o_i^\iota \in \mathcal{O}$. When we use the symbols c_i^ι, o_i^ι and $d_{i,j}^\iota$ we implicitly assume that they denote the constructors, tests and destructors associated with ι. When it is clear from the context which type ι is meant, we use the notation $\iota_{i,j}^*$ for $\iota_{i,j}$ if $\iota_{i,j} \neq \star$, or for ι if $\iota_{i,j} = \star$.
- The set of $\mathcal{I}_s$*-terms* $\mathbb{T}$ is defined by the following grammar.
$$\mathbb{T} ::= V_s \mid \Sigma_s \mid \mathcal{S} \mid \mathcal{C} \mid \mathcal{D} \mid \mathcal{O} \mid \mathcal{T}_I \mid \lambda V_s \,.\, \mathbb{T} \mid (\mathbb{T}\mathbb{T}) \mid \text{Is} \mid \text{Subtype} \mid \text{Fun} \mid$$
$$\forall \mid \vee \mid \bot \mid \epsilon \mid \text{Eq} \mid \text{Cond}$$
We assume application associates to the left and omit spurious brackets.

- We identify α-equivalent terms, i.e. terms differing only in the names of bound variables are considered identical. We use the symbol $\equiv$ for identity of terms up to α-equivalence. We also assume that all bound variables in a term are distinct from the free variables, unless indicated otherwise.[1]
- In what follows we use the abbreviations:

$$\begin{aligned}
t_1 : t_2 &\equiv \mathrm{Is}\, t_1\, t_2 \\
\{x : \alpha \mid \varphi\} &\equiv \mathrm{Subtype}\, \alpha\, (\lambda x \,.\, \varphi) \\
\alpha \to \beta &\equiv \mathrm{Fun}\, \alpha\, \beta \\
\forall x : \alpha \,.\, \varphi &\equiv \forall\, \alpha\, (\lambda x \,.\, \varphi) \\
\forall x_1, \ldots, x_n : \alpha \,.\, \varphi &\equiv \forall x_1 : \alpha \,.\, \ldots \forall x_n : \alpha \,.\, \varphi \\
\varphi \supset \psi &\equiv \forall x : \{y : \mathrm{Prop} \mid \varphi\} \,.\, \psi \quad \text{where } x, y \notin FV(\varphi, \psi) \\
\neg\varphi &\equiv \varphi \supset \bot \\
\top &\equiv \bot \supset \bot \\
\varphi \vee \psi &\equiv \vee \varphi \psi \\
\varphi \wedge \psi &\equiv \neg(\neg\varphi \vee \neg\psi) \\
\exists x : \alpha \,.\, \varphi &\equiv \neg\forall x : \alpha \,.\, \neg\varphi
\end{aligned}$$

We assume that $\neg$ has the highest precedence.
- The system $\mathcal{I}_s$ is given by the following rules and axioms, where Γ is a finite set of terms, $t, \varphi, \psi, \alpha, \beta$, etc. are arbitrary terms. The notation Γ, φ is a shorthand for $\Gamma \cup \{\varphi\}$. We use Greek letters φ, ψ, etc. to highlight that a term is to be intuitively interpreted as a proposition, and we use α, β, etc. when it is to be interpreted as a type, but there is no a priori syntactic distinction. All judgements have the form $\Gamma \vdash t$ where Γ is a set of terms and t a term. In particular, $\Gamma \vdash t : \alpha$ is a shorthand for $\Gamma \vdash \mathrm{Is}\, t\, \alpha$.

Axioms

1: $\Gamma, \varphi \vdash \varphi$
2: $\Gamma \vdash \mathrm{Eq}\, t\, t$
3: $\Gamma \vdash \mathrm{Prop} : \mathrm{Type}$
4: $\Gamma \vdash \iota : \mathrm{Type}$ for $\iota \in \mathcal{T}_I$
5: $\Gamma \vdash o_i^\iota(c_i^\iota t_1 \ldots t_{n_i})$ if $c_i^\iota \in \mathcal{C}$ has arity n_i
6: $\Gamma \vdash \neg(o_i^\iota(c_j^\iota t_1 \ldots t_{n_j}))$ if $i \neq j$ and $c_j^\iota \in \mathcal{C}$ has arity n_j
7: $\Gamma \vdash \mathrm{Eq}\, (d_{i,k}^\iota(c_i^\iota t_1 \ldots t_{n_i}))\, t_k$ for $k = 1, \ldots, n_i$, if $c_i^\iota \in \mathcal{C}$ has arity n_i
$\bot_t$: $\Gamma \vdash \bot : \mathrm{Prop}$
c: $\Gamma \vdash \forall p : \mathrm{Prop} \,.\, p \vee \neg p$
β: $\Gamma \vdash \mathrm{Eq}\, ((\lambda x \,.\, t_1) t_2)\, (t_1[x/t_2])$

Rules

$$\forall_i : \frac{\Gamma \vdash \alpha : \mathrm{Type} \qquad \Gamma, x : \alpha \vdash \varphi \qquad x \notin FV(\Gamma, \alpha)}{\Gamma \vdash \forall x : \alpha \,.\, \varphi}$$

[1] So e.g. in the axiom β the free variables of t_2 do not become bound in $t_1[x/t_2]$.

$$\forall_e : \frac{\Gamma \vdash \forall x : \alpha . \varphi \qquad \Gamma \vdash t : \alpha}{\Gamma \vdash \varphi[x/t]}$$

$$\forall_t : \frac{\Gamma \vdash \alpha : \mathrm{Type} \qquad \Gamma, x : \alpha \vdash \varphi : \mathrm{Prop} \qquad x \notin FV(\Gamma, \alpha)}{\Gamma \vdash (\forall x : \alpha . \varphi) : \mathrm{Prop}}$$

$$\exists_i : \frac{\Gamma \vdash \alpha : \mathrm{Type} \qquad \Gamma \vdash t : \alpha \qquad \Gamma \vdash \varphi[x/t]}{\Gamma \vdash \exists x : \alpha . \varphi}$$

$$\exists_e : \frac{\Gamma \vdash \exists x : \alpha . \varphi \qquad \Gamma, x : \alpha, \varphi \vdash \psi \qquad x \notin FV(\Gamma, \psi, \alpha)}{\Gamma \vdash \psi}$$

$$\vee_{i1} : \frac{\Gamma \vdash \varphi}{\Gamma \vdash \varphi \vee \psi} \qquad \vee_{i2} : \frac{\Gamma \vdash \psi}{\Gamma \vdash \varphi \vee \psi}$$

$$\vee_e : \frac{\Gamma \vdash \varphi_1 \vee \varphi_2 \qquad \Gamma, \varphi_1 \vdash \psi \qquad \Gamma, \varphi_2 \vdash \psi}{\Gamma \vdash \psi}$$

$$\vee_t : \frac{\Gamma \vdash \varphi : \mathrm{Prop} \qquad \Gamma \vdash \psi : \mathrm{Prop}}{\Gamma \vdash (\varphi \vee \psi) : \mathrm{Prop}}$$

$$\wedge_{e1} : \frac{\Gamma \vdash \varphi \wedge \psi}{\Gamma \vdash \varphi} \qquad \wedge_{e2} : \frac{\Gamma \vdash \varphi \wedge \psi}{\Gamma \vdash \psi}$$

$$\supset_{t2} : \frac{\Gamma \vdash (\varphi \supset \psi) : \mathrm{Prop}}{\Gamma \vdash \varphi : \mathrm{Prop}} \qquad \bot_e : \frac{\Gamma \vdash \bot}{\Gamma \vdash \varphi}$$

$$\rightarrow_i : \frac{\Gamma \vdash \alpha : \mathrm{Type} \qquad \Gamma, x : \alpha \vdash t : \beta \qquad x \notin FV(\Gamma, \alpha, \beta)}{\Gamma \vdash (\lambda x . t) : \alpha \rightarrow \beta}$$

$$\rightarrow_e : \frac{\Gamma \vdash t_1 : \alpha \rightarrow \beta \qquad \Gamma \vdash t_2 : \alpha}{\Gamma \vdash t_1 t_2 : \beta} \quad \rightarrow_t : \frac{\Gamma \vdash \alpha : \mathrm{Type} \qquad \Gamma \vdash \beta : \mathrm{Type}}{\Gamma \vdash (\alpha \rightarrow \beta) : \mathrm{Type}}$$

$$s_i : \frac{\Gamma \vdash \{x : \alpha \mid \varphi\} : \mathrm{Type} \qquad \Gamma \vdash t : \alpha \qquad \Gamma \vdash (\lambda x . \varphi) t \qquad x \notin FV(\alpha)}{\Gamma \vdash t : \{x : \alpha \mid \varphi\}}$$

$$s_e : \frac{\Gamma \vdash t : \{x : \alpha \mid \varphi\}}{\Gamma \vdash \varphi[x/t]} \qquad s_{et} : \frac{\Gamma \vdash t : \{x : \alpha \mid \varphi\}}{\Gamma \vdash t : \alpha}$$

$$s_t : \frac{\Gamma \vdash \alpha : \mathrm{Type} \qquad \Gamma, x : \alpha \vdash \varphi : \mathrm{Prop} \qquad x \notin FV(\alpha)}{\Gamma \vdash \{x : \alpha \mid \varphi\} : \mathrm{Type}}$$

$$\epsilon_i : \frac{\Gamma \vdash \exists x : \alpha . \top}{\Gamma \vdash (\epsilon \alpha) : \alpha} \qquad p_i : \frac{\Gamma \vdash \varphi}{\Gamma \vdash \varphi : \mathrm{Prop}}$$

$$c_1 : \frac{\Gamma \vdash \varphi}{\Gamma \vdash \mathrm{Eq}\,(\mathrm{Cond}\,\varphi\, t_1\, t_2)\, t_1} \qquad c_2 : \frac{\Gamma \vdash \neg\varphi}{\Gamma \vdash \mathrm{Eq}\,(\mathrm{Cond}\,\varphi\, t_1\, t_2)\, t_2}$$

$$c_3 : \frac{\Gamma, \varphi \vdash \mathrm{Eq}\, t_1\, t_1' \qquad \Gamma \vdash \varphi : \mathrm{Prop}}{\Gamma \vdash \mathrm{Eq}\,(\mathrm{Cond}\,\varphi\, t_1\, t_2)\,(\mathrm{Cond}\,\varphi\, t_1'\, t_2)}$$

$$c_4 : \frac{\Gamma, \neg\varphi \vdash \mathrm{Eq}\, t_2\, t_2' \qquad \Gamma \vdash \varphi : \mathrm{Prop}}{\Gamma \vdash \mathrm{Eq}\,(\mathrm{Cond}\,\varphi\, t_1\, t_2)\,(\mathrm{Cond}\,\varphi\, t_1\, t_2')}$$

$$c_5 : \frac{\Gamma \vdash \varphi : \mathrm{Prop}}{\Gamma \vdash \mathrm{Eq}\,(\mathrm{Cond}\,\varphi\, t\, t)\, t}$$

$$\text{eq}: \frac{\Gamma \vdash \varphi \qquad \Gamma \vdash \mathrm{Eq}\,\varphi\,\varphi'}{\Gamma \vdash \varphi'} \qquad \text{eq-sym}: \frac{\Gamma \vdash \mathrm{Eq}\,t_1\,t_2}{\Gamma \vdash \mathrm{Eq}\,t_2\,t_1}$$

$$\text{eq-trans}: \frac{\Gamma \vdash \mathrm{Eq}\,t_1\,t_2 \qquad \Gamma \vdash \mathrm{Eq}\,t_2\,t_3}{\Gamma \vdash \mathrm{Eq}\,t_1\,t_3}$$

$$\text{eq-cong-app}: \frac{\Gamma \vdash \mathrm{Eq}\,t_1\,t_1' \qquad \Gamma \vdash \mathrm{Eq}\,t_2\,t_2'}{\Gamma \vdash \mathrm{Eq}\,(t_1 t_2)\,(t_1' t_2')}$$

$$\text{eq-}\lambda\text{-}\xi: \frac{\Gamma \vdash \mathrm{Eq}\,t\,t' \qquad x \notin FV(\Gamma)}{\Gamma \vdash \mathrm{Eq}\,(\lambda x\,.\,t)\,(\lambda x\,.\,t')}$$

$$i_i^\iota: \frac{\begin{array}{c}\Gamma, x_1 : \iota_{i,1}^*, \ldots, x_{n_i} : \iota_{i,n_i}^*, t x_{j_{i,1}}, \ldots, t x_{j_{i,k_i}} \vdash t(c_i^\iota x_1 \ldots x_{n_i}) \\ \text{for } i = 1, \ldots, m\end{array}}{\Gamma \vdash \forall x : \iota\,.\,t x}$$

where $x, x_1, \ldots, x_{n_i} \notin FV(\Gamma, t)$, $c_1^\iota, \ldots, c_m^\iota \in \mathcal{C}$ are all constructors associated with $\iota \in \mathcal{T}_I$, and $j_{i,1}, \ldots, j_{i,k_i}$ is an increasing sequence of all indices $1 \le j \le n_i$ such that $\iota_{i,j} = \star$

$$i_t^{\iota,k}: \frac{\Gamma \vdash t_j : \iota_{k,j}^* \text{ for } j = 1, \ldots, n_k}{\Gamma \vdash (c_k^\iota t_1 \ldots t_{n_k}) : \iota}$$

For an arbitrary set of terms Γ, we write $\Gamma \vdash_{\mathcal{I}_s} \varphi$ if there exists a finite subset $\Gamma' \subseteq \Gamma$ such that $\Gamma' \vdash \varphi$ is derivable in the system $\mathcal{I}_s$. We drop the subscript when irrelevant or obvious from the context.

Lemma 1. *If $\Gamma \vdash \varphi$ then $\Gamma, \psi \vdash \varphi$.*

Lemma 2. *If $\Gamma \vdash \varphi$ then $\Gamma[x/t] \vdash \varphi[x/t]$, where $\Gamma[x/t] = \{\psi[x/t] \mid \psi \in \Gamma\}$.*

2.1 Representing Logic

The inference rules of $\mathcal{I}_s$ may be intuitively justified by appealing to an informal many-valued semantics. A term t may be true, false, or something entirely different ("undefined", a program, a natural number, a type, ...). By way of an example, we explain an informal meaning of some terms:

- t : Prop is true iff t is true or false,
- α : Type is true iff α is a type,
- $t : \alpha$ is true iff t has type α, assuming α is a type,
- $\forall x : \alpha.\varphi$ is true iff α is a type and for all t of type α, $\varphi[x/t]$ is true,
- $\forall x : \alpha.\varphi$ is false iff α is a type and there exists t of type α such that $\varphi[x/t]$ is false,
- $t_1 \vee t_2$ is true iff t_1 is true or t_2 is true,
- $t_1 \vee t_2$ is false iff t_1 is false and t_2 is false,

- $t_1 \supset t_2$ is true iff t_1 is false or both t_1 and t_2 are true,
- $t_1 \supset t_2$ is false iff t_1 is true and t_2 is false,
- $\neg t$ is true iff t is false,
- $\neg t$ is false iff t is true.

Obviously, $\Gamma \vdash t$ is then (informally) interpreted as: for all possible substitution instances Γ^*, t^* of Γ, t, [2] if all terms in Γ^* are true, then the term t^* is also true.

Note that the logical connectives are "lazy", e.g. for $t_1 \vee t_2$ to be true it suffices that t_1 is true, but t_2 need not have a truth value at all – it may be something else: a program, a type, "undefined", etc. This laziness allows us to omit many restrictions which would otherwise be needed in inference rules, and would thus make the system less similar to ordinary logic.

The following rules may be derived in $\mathcal{I}_s$.

$$\supset_i: \frac{\Gamma \vdash \varphi : \text{Prop} \qquad \Gamma, \varphi \vdash \psi}{\Gamma \vdash \varphi \supset \psi} \qquad \supset_e: \frac{\Gamma, \varphi \vdash \psi \qquad \Gamma \vdash \varphi}{\Gamma \vdash \psi}$$

$$\supset_t: \frac{\Gamma \vdash \varphi : \text{Prop} \qquad \Gamma, \varphi \vdash \psi : \text{Prop}}{\Gamma \vdash (\varphi \supset \psi) : \text{Prop}} \qquad \wedge_i : \frac{\Gamma \vdash \varphi \qquad \Gamma \vdash \psi}{\Gamma \vdash \varphi \wedge \psi}$$

Note that in general the elimination rules for $\wedge$ and the rules for $\exists$ cannot be derived from the rules for $\vee$ and $\forall$, because we would not be able to prove the premise φ : Prop when trying to apply the rule $\supset_i$. It is instructive to try to derive these rules and see where the proof breaks down.

In $\mathcal{I}_s$ the only non-standard restriction in the usual inference rules for logical connectives is the additional premise $\Gamma \vdash \varphi$: Prop in the rule $\supset_i$. It is certainly unavoidable, as otherwise Curry's paradox may be derived (see e.g. [1,2]). However, we have standard classical higher-order logic if we restrict to terms of type Prop, in the sense that the natural deduction rules then become identical to the rules of ordinary logic. This is made more precise in Sect. 3 where a sound translation from a traditional system of higher-order logic into $\mathcal{I}_s$ is described.

Note that we have the law of excluded middle only in the form $\forall p : \text{Prop} \,.\, p \vee \neg p$. Adding $\Gamma \vdash \varphi \vee \neg \varphi$ as an axiom for an arbitrary term φ gives an inconsistent system.[3]

It is well-known (see e.g. [6, Chapter 11]) that in higher-order logic all logical connectives may be defined from $\forall$ and $\supset$. One may therefore wonder why we take $\vee$ and $\bot$ as primitive. The answer is that if we defined the connectives from $\forall$ and $\supset$, then the inference rules that could be derived for them would need to contain additional restrictions.

2.2 Equality, Recursive Definitions and Extensionality

It is well-known (see e.g. [7, Chapters 2, 6]) that since untyped λ-terms are available together with the axiom β and usual rules for equality, any set of

[2] To be more precise, for every possible substitution of terms for the free variables of Γ, t we perform this substitution on Γ, t, denoting the result by Γ^*, t^*.

[3] By defining (see the next subsection) $\varphi = \neg\varphi$ one could then easily derive $\bot$ using the rule $\vee_e$ applied to $\varphi \vee \neg\varphi$.

equations of the form $\{z_i x_1 \ldots x_m = \Phi_i(z_1, \ldots, z_n, x_1, \ldots, x_m) \mid i = 1, \ldots, n\}$ has a solution for $z_1, \ldots, z_n$, where $\Phi_i(z_1, \ldots, z_n, x_1, \ldots, x_m)$ are arbitrary terms with the free variables listed. In other words, there exist terms $t_1, \ldots, t_n$ such that for any terms $s_1, \ldots, s_m$ we have $\vdash \mathrm{Eq}\ (t_i s_1 \ldots s_m)\ (\Phi_i(t_1, \ldots, t_n, s_1, \ldots, s_m))$ for each $i = 1, \ldots, n$.

We will often define terms by such equations. In what follows we freely use the notation $t_1 = t_2$ for $\vdash \mathrm{Eq}\, t_1\, t_2$, or for $\Gamma \vdash \mathrm{Eq}\, t_1\, t_2$ when it is clear which Γ is meant. We use $t_1 = t_2 = \ldots = t_n$ to indicate that $\mathrm{Eq}\, t_i\, t_{i+1}$ may be derived for $i = 1, \ldots, n-1$. We also write a term of the form $\mathrm{Eq}\, t_1\, t_2$ as $t_1 = t_2$.

In $\mathcal{I}_s$ there is no rule for typing the equality Eq. One consequence is that $\vdash \neg(\mathrm{Eq}\, t_1\, t_2)$ cannot be derived for any terms t_1, t_2.[4] For this reason Eq is more like a meta-level notion of equality.

Definition 2. Leibniz equality Eql is defined as:

$$\mathrm{Eql} \equiv \lambda\alpha\lambda x\lambda y.\forall p : \alpha \to \mathrm{Prop}\,.\, px \supset py$$

As with $=$, we often write $t_1 =_\alpha t_2$ to denote $\vdash \mathrm{Eql}\, \alpha\, t_1\, t_2$ or $\Gamma \vdash \mathrm{Eql}\, \alpha\, t_1\, t_2$, or write $t_1 =_\alpha t_2$ instead of $\mathrm{Eql}\, \alpha\, t_1\, t_2$.

Lemma 3. *If* $\Gamma \vdash \alpha : \mathrm{Type}$ *then*

- $\Gamma \vdash \forall x, y : \alpha\,.\,(x =_\alpha y) : \mathrm{Prop}$,
- $\Gamma \vdash \forall x : \alpha\,.\,(x =_\alpha x)$,
- $\Gamma \vdash \forall x, y : \alpha\,.\,(x =_\alpha y) \supset (y =_\alpha x)$,
- $\Gamma \vdash \forall x, y, z : \alpha\,.\,(x =_\alpha y) \wedge (y =_\alpha z) \supset (x =_\alpha z)$.

The system $\mathcal{I}_s$, as it is stated, is intensional with respect to Leibniz equality. We could add the rules

$$e_f : \frac{\Gamma \vdash \alpha : \mathrm{Type} \qquad \Gamma \vdash \beta : \mathrm{Type}}{\forall f_1, f_2 : \alpha \to \beta\,.\,(\forall x : \alpha\,.\, f_1 x =_\beta f_2 x) \supset (f_1 =_{\alpha\to\beta} f_2)}$$

$$e_b : \frac{\Gamma \vdash \varphi_1 \supset \varphi_2 \qquad \Gamma \vdash \varphi_2 \supset \varphi_1}{\Gamma \vdash \varphi_1 = \varphi_2}$$

to obtain an extensional variant $e\mathcal{I}_s$ of $\mathcal{I}_s$. The system $e\mathcal{I}_s$ is still consistent – the model we construct for $\mathcal{I}_s$ validates the above rules.

Lemma 4. $\vdash_{e\mathcal{I}_s} \forall x, y : \mathrm{Prop}\,.\,(x =_{\mathrm{Prop}} y) \supset (x = y)$

2.3 Induction and Natural Numbers

The system $\mathcal{I}_s$ incorporates basic inductive types. In accordance with the terminology from [8], an inductive type is basic if its constructors have no functional arguments. This class of inductive types includes most simple commonly used inductive types, e.g. natural numbers, lists, finite trees.

---

[4] We mean this in a precise sense. This follows from our model construction.

Lemma 5. *If* $c_i^\iota \in \mathcal{C}$ *of arity* n_i *has signature* $\langle \iota_1, \ldots, \iota_{n_i} \rangle$ *then* $\vdash_{\mathcal{I}_s} c_i^\iota : \iota_1^* \to \ldots \to \iota_{n_i}^* \to \iota$.

Lemma 6. $\vdash_{\mathcal{I}_s} o_i^\iota : \iota \to \text{Prop}$ *and* $\vdash_{\mathcal{I}_s} \forall x : \iota \, . \, o_i^\iota x \supset (d_{i,j}^\iota x : \iota_{i,j}^*)$

Lemma 7. *If* $\iota \in \mathcal{T}_I$ *then* $\vdash_{\mathcal{I}_s} \forall x, y : \iota \, . \, x =_\iota y \supset x = y$.

We may define the type of natural numbers by $\text{Nat} \equiv \mu(\langle\rangle, \langle\star\rangle)$. We use the abbreviations: $0 \equiv c_1^{\text{Nat}}$ (zero), $\mathfrak{o} \equiv o_1^{\text{Nat}}$ (test for zero), $\mathfrak{s} \equiv c_2^{\text{Nat}}$ (successor) and $\mathfrak{p} \equiv \lambda x \, . \, \text{Cond}\,(\mathfrak{o}x)\, 0\, (d_{2,1}^{\text{Nat}} x)$ (predecessor). The rules i_i^{Nat} and $i_t^{\text{Nat},k}$ become:

$$n_i : \frac{\Gamma \vdash t0 \qquad \Gamma, x : \text{Nat}, tx \vdash t(\mathfrak{s}x) \qquad x \notin FV(\Gamma, t)}{\Gamma \vdash \forall x : \text{Nat} \, . \, tx}$$

$$n_t^1 : \frac{}{\Gamma \vdash 0 : \text{Nat}} \qquad\qquad n_t^2 : \frac{\Gamma \vdash t : \text{Nat}}{\Gamma \vdash (\mathfrak{s}t) : \text{Nat}}$$

To simplify the exposition, we discuss some properties of our formulation of inductive types using the example of natural numbers. Much of what we say applies to other basic inductive types, with appropriate modifications.

The rule n_i is an induction principle for natural numbers. An important property of this induction principle is that it places no restrictions on t. This allows us to prove by induction on natural numbers properties of terms about which nothing is known beforehand. In particular, we do not need to know whether t has a β-normal form in order to apply the rule n_i to it. In contrast, an induction principle of the form e.g.

$$n_i' : \; \forall f : \text{Nat} \to \text{Prop} \, . \, ((f0 \wedge (\forall x : \text{Nat} \, . \, fx \supset f(\mathfrak{s}x))) \supset \forall x : \text{Nat} \, . \, fx)$$

would be much less useful, because to apply it to a term t we would have to prove $t : \text{Nat} \to \text{Prop}$ *beforehand.* Examples of the use of the rule n_i for reasoning about possibly nonterminating general recursive programs are given in Sect. 4.

The operations $+$, $-$, $\cdot$, $<$ and $\leq$, usually used in infix notation, may be defined by recursive equations. It is possible to derive all Peano axioms.

Lemma 8. *The following terms are derivable in* $\mathcal{I}_s$*:*

- $\forall x, y : \text{Nat} \, . \, (x + y) : \text{Nat}$, $\forall x, y : \text{Nat} \, . \, (x - y) : \text{Nat}$, $\forall x, y : \text{Nat} \, . \, (x \cdot y) : \text{Nat}$,
- $\forall x, y : \text{Nat} \, . \, (x \leq y) : \text{Prop}$, $\forall x, y : \text{Nat} \, . \, (x < y) : \text{Prop}$.

The next theorem shows that any function for which there exists a measure on its arguments, which may be shown to decrease with every recursive call in each of a finite number of exhaustive cases, is typable in our system.

Theorem 1. *Suppose* $\Gamma \vdash \forall x_1 : \alpha_1 \ldots \forall x_n : \alpha_n \, . \, \varphi_1 \vee \ldots \vee \varphi_m$, $\Gamma \vdash \alpha_j : \text{Type}$ *for* $j = 1, \ldots, n$, *and for* $i = 1, \ldots, m$: $\Gamma \vdash \forall x_1 : \alpha_1 \ldots \forall x_n : \alpha_n \, . \, t_i : \beta \to \ldots \to \beta$ *where* β *occurs* $k_i + 1$ *times,* $\Gamma \vdash \forall x_1 : \alpha_1 \ldots \forall x_n : \alpha_n \, . \, t_{i,j,k} : \alpha_k$ *for* $j = 1, \ldots, k_i$, $k = 1, \ldots, n$, $x_1, \ldots, x_n \notin FV(f, \alpha_1, \ldots, \alpha_n, \beta)$ *and*

$$\Gamma \vdash \forall x_1 : \alpha_1 \ldots \forall x_n : \alpha_n . \varphi_i \supset (f x_1 \ldots x_n = t_i(f t_{i,1,1} \ldots t_{i,1,n}) \ldots (f t_{i,k_i,1} \ldots t_{i,k_i,n})).$$

If there is a term g such that $\Gamma \vdash g : \alpha_1 \to \ldots \to \alpha_n \to \mathrm{Nat}$ and for $i = 1, \ldots, m$

$$\Gamma \vdash \forall x_1 : \alpha_1 \ldots \forall x_n : \alpha_n . \varphi_i \supset (((f x_1 \ldots x_n) : \beta) \vee ((g t_{i,1,1} \ldots t_{i,1,n}) < (g x_1 \ldots x_n) \wedge \ldots \wedge (g t_{i,k_i,1} \ldots t_{i,k_i,n}) < (g x_1 \ldots x_n)))$$

where $x_1, \ldots, x_n \notin FV(g)$, then $\Gamma \vdash f : \alpha_1 \to \ldots \to \alpha_n \to \beta$.

3 Conservativity and Consistency

In this section we show a sound embedding of ordinary classical higher-order logic into $\mathcal{I}_s$, which we also conjecture to be complete. We have a completeness proof only for a restriction of this embedding to first-order logic. We also give a brief overview of the model construction used to establish consistency of $\mathcal{I}_s$.

First, let us define the system CPREDω of classical higher-order logic.

- The *types* of CPREDω are given by $\mathcal{T} ::= o \mid \mathcal{B} \mid \mathcal{T} \to \mathcal{T}$ where $\mathcal{B}$ is a specific finite set of base types. The type o is the type of propositions.
- The set of terms of CPREDω of type τ, denoted T_τ, is defined as follows:
  - $V_\tau, \Sigma_\tau \subseteq T_\tau$,
  - if $t_1 \in T_{\sigma\to\tau}$ and $t_2 \in T_\sigma$ then $t_1 t_2 \in T_\tau$,
  - if $x \in V_{\tau_1}$ and $t \in T_{\tau_2}$ then $\lambda x : \tau_1 . t \in T_{\tau_1 \to \tau_2}$,
  - if $\varphi, \psi \in T_o$ then $\varphi \supset \psi \in T_o$,
  - if $x \in V_\tau$ and $\varphi \in T_o$ then $\forall x : \tau . \varphi \in T_o$,

  where for each $\tau \in \mathcal{T}$ the set V_τ is a set of variables and Σ_τ is a set of constants. We assume that the sets V_τ and Σ_σ are all pairwise disjoint. We write x_τ for a variable $x_\tau \in V_\tau$. Terms of type o are *formulas*.
- The system CPREDω is given by the following rules and axioms, where Δ is a finite set of formulas, φ, ψ are formulas.

  Axioms
  - $\Delta, \varphi \vdash \varphi$
  - $\Delta \vdash \forall p : o . ((p \supset \bot) \supset \bot) \supset p$ where $\bot \equiv \forall p : o . p$

  Rules

$$\supset_i^P : \frac{\Delta, \varphi \vdash \psi}{\Delta \vdash \varphi \supset \psi} \qquad \supset_e^P : \frac{\Delta \vdash \varphi \supset \psi \quad \Delta \vdash \varphi}{\Delta \vdash \psi}$$

$$\forall_i^P : \frac{\Delta \vdash \varphi}{\Delta \vdash \forall x_\tau . \varphi}\ x_\tau \notin FV(\Delta) \qquad \forall_e^P : \frac{\Delta \vdash \forall x_\tau . \varphi}{\Delta \vdash \varphi[x_\tau / t]}\ t \in T_\tau$$

$$\mathrm{conv}^P : \frac{\Delta \vdash \varphi \qquad \varphi =_\beta \psi}{\Delta \vdash \psi}$$

In CPREDω, we define Leibniz equality in type $\tau \in \mathcal{T}$ by

$$t_1 =_\tau t_2 \equiv \forall p : \tau \to o \,.\, pt_1 \supset pt_2.$$

The system CPREDω is intensional. An extensional variant E-CPREDω may be obtained by adding the following axioms for all $\tau, \sigma \in \mathcal{T}$:

$$e_f^P : \forall f_1, f_2 : \tau \to \sigma \,.\, (\forall x : \tau \,.\, f_1 x =_\sigma f_2 x) \supset (f_1 =_{\tau\to\sigma} f_2)$$
$$e_b^P : \forall \varphi_1, \varphi_2 : o \,.\, ((\varphi_1 \supset \varphi_2) \wedge (\varphi_2 \supset \varphi_1)) \supset (\varphi_1 =_o \varphi_2)$$

For an arbitrary set of formulas Δ we write $\Delta \vdash_S \varphi$ if φ is derivable from a subset of Δ in system S.

We now define a mapping $\lceil - \rceil$ from types and terms of CPREDω to terms of $\mathcal{I}_s$, and a mapping $\Gamma(-)$ from sets of terms of CPREDω to sets of terms of $\mathcal{I}_s$. We assume $\mathcal{B} \subseteq \Sigma_s$, $\Sigma_\tau \subseteq \Sigma_s$ and $V_\tau \subseteq V_s$ for $\tau \in \mathcal{T}$.

- $\lceil \tau \rceil = \tau$ for $\tau \in \mathcal{B}$,
- $\lceil o \rceil = \mathrm{Prop}$,
- $\lceil \tau_1 \to \tau_2 \rceil = \lceil \tau_1 \rceil \to \lceil \tau_2 \rceil$ for $\tau_1, \tau_2 \in \mathcal{T}$,
- $\lceil c \rceil = c$ if $c \in \Sigma_\tau$ for some $\tau \in \mathcal{T}$,
- $\lceil x \rceil = x$ if $x \in V_\tau$ for some $\tau \in \mathcal{T}$,
- $\lceil t_1 t_2 \rceil = \lceil t_1 \rceil \lceil t_2 \rceil$,
- $\lceil \lambda x : \tau \,.\, t \rceil = \lambda x \,.\, \lceil t \rceil$,
- $\lceil \varphi \supset \psi \rceil = \lceil \varphi \rceil \supset \lceil \psi \rceil$,
- $\lceil \forall x : \tau \,.\, \varphi \rceil = \forall x : \lceil \tau \rceil \,.\, \lceil \varphi \rceil$.

By $\lceil \Delta \rceil$ we denote the image of $\lceil - \rceil$ on Δ. The set $\Gamma(\Delta)$ is defined to contain:

- $x : \lceil \tau \rceil$ for all $\tau \in \mathcal{T}$ and all $x \in FV(\Delta)$ such that $x \in V_\tau$,
- $c : \lceil \tau \rceil$ for all $\tau \in \mathcal{T}$ and all $c \in \Sigma_\tau$,
- $\tau : \mathrm{Type}$ for all $\tau \in \mathcal{B}$,
- $y : \tau$ for all $\tau \in \mathcal{B}$ and some $y \in V_\tau$ such that $y \notin FV(\Delta)$.

Theorem 2. *If* $\Delta \vdash_{\mathrm{CPRED}\omega} \varphi$ *then* $\Gamma(\Delta, \varphi), \lceil \Delta \rceil \vdash_{\mathcal{I}_s} \lceil \varphi \rceil$. *The same holds if we change CPREDω to E-CPREDω and $\mathcal{I}_s$ to $e\mathcal{I}_s$.*

The above theorem shows that $\mathcal{I}_s$ may be considered an extension of ordinary higher-order logic, obtained by relaxing typing requirements on allowable λ-terms. Type-checking is obviously undecidable in $\mathcal{I}_s$, but the purpose of types in illative systems is not to have a decidable method for syntactic correctness checks, but to provide general means for classifying terms into various categories.

Conjecture 1. If $\Gamma(\Delta, \varphi), \lceil \Delta \rceil \vdash_{\mathcal{I}_s} \lceil \varphi \rceil$ then $\Delta \vdash_{\mathrm{CPRED}\omega} \varphi$. The same holds if we change CPREDω to E-CPREDω and $\mathcal{I}_s$ to $e\mathcal{I}_s$.

We prove this conjecture only for first-order logic. The system of classical first-order logic (FOL) is obtained by restricting CPREDω in obvious ways.

Theorem 3. *If* $\mathcal{I} = \mathcal{I}_s$ *or* $\mathcal{I} = e\mathcal{I}_s$ *then:* $\Delta \vdash_{\mathrm{FOL}} \varphi$ *iff* $\Gamma(\Delta, \varphi), \lceil \Delta \rceil \vdash_{\mathcal{I}} \lceil \varphi \rceil$.

Theorem 4. *The systems $\mathcal{I}_s$ and $e\mathcal{I}_s$ are consistent, i.e.* $\not\vdash_{\mathcal{I}_s} \bot$ *and* $\not\vdash_{e\mathcal{I}_s} \bot$.

We now give an informal overview of the model construction. To simplify the exposition we pretend $\mathcal{I}_s$ allows only function types. Other types add some technicalities, but the general idea of the construction remains the same.

An $\mathcal{I}_s$-model is defined as a λ-model (see e.g. [7, Chapter 5]) with designated elements interpreting the constants of $\mathcal{I}_s$, satisfying certain conditions. By $[\![t]\!]^{\mathcal{M}}$ we denote the interpretation of the $\mathcal{I}_s$-term t in a model $\mathcal{M}$, and $\cdot$ is the application operation in the model. The conditions imposed on an $\mathcal{I}_s$-model express the meaning of each rule of $\mathcal{I}_s$ according to the intuitive semantics. For instance, we have the condition:

($\forall_\top$) for $a \in \mathcal{M}$, if $[\![\mathrm{Is}]\!]^{\mathcal{M}} \cdot a \cdot [\![\mathrm{Type}]\!]^{\mathcal{M}} = [\![\top]\!]^{\mathcal{M}}$ and for all $c \in \mathcal{M}$ such that $[\![\mathrm{Is}]\!]^{\mathcal{M}} \cdot c \cdot a = [\![\top]\!]^{\mathcal{M}}$ we have $b \cdot c = [\![\top]\!]^{\mathcal{M}}$ then $[\![\forall]\!]^{\mathcal{M}} \cdot a \cdot b = [\![\top]\!]^{\mathcal{M}}$.

We show that the semantics based on $\mathcal{I}_s$-models is sound for $\mathcal{I}_s$. Then it suffices to construct a non-trivial $\mathcal{I}_s$-model to establish consistency of $\mathcal{I}_s$. The model will in fact satisfy additional conditions corresponding to the rules e_f and e_b, so we obtain consistency of $e\mathcal{I}_s$ as well.

The model is constructed as the set of equivalence classes of a certain relation $\overset{*}{\Leftrightarrow}$ on the set of so called semantic terms. A semantic term is a well-founded tree whose leaves are labelled with variables or constants, and whose internal nodes are labelled with $\cdot$, λx or $\mathrm{A}\tau$. For semantic terms with the roots labelled with $\cdot$ and λx we use the notation $t_1 t_2$ and $\lambda x.t$. A node labelled with $\mathrm{A}\tau$, where τ is a set of constants, "represents" the statement: for all $c \in \tau$, tc is true. Such a node has one child for each $c \in \tau$. The relation $\overset{*}{\Leftrightarrow}$ is defined as the equivalence relation generated by a certain reduction relation $\Rightarrow$ on semantic terms. The relation $\Rightarrow$ will satisfy[5]: $(\lambda x.t_1)t_2 \Rightarrow t_1[x/t_2]$, $\vee\top t \Rightarrow \top$, $\vee\bot\bot \Rightarrow \bot$, etc. The question is how to define $\Rightarrow$ for $\forall t_1 t_2$ so that the resulting structure satisfies ($\forall_\top$). One could try closing $\Rightarrow$ under the rule: if $\mathrm{Is}\, t_1\, \mathrm{Type} \overset{*}{\Rightarrow} \top$ and for all t such that $\mathrm{Is}\, t\, t_1 \overset{*}{\Rightarrow} \top$ we have $t_2 t \overset{*}{\Rightarrow} \top$, then $\forall t_1 t_2 \Rightarrow \top$. However, there is a negative reference to $\Rightarrow$ here, so the definition would not be monotone, and we would not necessarily reach a fixpoint. This is a major problem. We need to know the range of all quantifiers beforehand. However, the range (i.e. the set of all semantic terms t such that $t_1 t \overset{*}{\Rightarrow} \top$) depends on the definition of $\Rightarrow$, so it is not at all clear how to achieve this.

Fortunately, it is not so difficult to analyze a priori the form of types of $\mathcal{I}_s$. Informally, if $t : \mathrm{Type}$ is true, then t corresponds to a set in $\mathcal{T}$, where $\mathcal{T}$ is defined as follows, ignoring subtypes and inductive types.

- $\mathrm{Bool} \in \mathcal{T}$ where $\mathrm{Bool} = \{\top, \bot\}$.
- If $\tau_1, \tau_2 \in \mathcal{T}$ then $\tau_2^{\tau_1} \in \mathcal{T}$, where $\tau_2^{\tau_1}$ is the set of all set-theoretic functions from τ_1 to τ_2.

We take the elements of $\mathcal{T}$ and $\bigcup \mathcal{T} \setminus \mathrm{Bool}$ as fresh constants, i.e. they may occur as constants in semantic terms. The elements of $\bigcup \mathcal{T}$ are *canonical constants.* If

[5] Substitution is defined for semantic terms in an obvious way, avoiding variable capture.

$c \in \tau_2^{\tau_1}$ and $c_1 \in \tau_1$ then we write $\mathcal{F}(c)(c_1)$ instead of $c(c_1)$ to avoid confusion with the semantic term cc_1. We then define a relation $\succ$ satisfying:

- $c \succ c$ for a canonical constant c,
- if $c \in \tau_2^{\tau_1}$ and for all $c_1 \in \tau_1$ there exists a semantic term t' such that $tc_1 \overset{*}{\Rightarrow} t' \succ \mathcal{F}(c)(c_1)$, then $t \succ c$.

Intuitively, $t \succ c \in \tau$ holds if c "simulates" t in type τ, i.e. t behaves exactly like c in every context where a term of type τ is "expected".

The relation $\Rightarrow$ is then defined by transfinite induction in a monotone way. It will satisfy e.g.:

- if $t \succ c \in \tau \in \mathcal{T}$ then $\mathrm{Is}\, t\, \tau \Rightarrow \top$,
- if $t \succ c_1 \in \tau_1$ and $c \in \tau_2^{\tau_1}$ then $ct \Rightarrow \mathcal{F}(c)(c_1)$,
- $\mathrm{Fun}\, \tau_1\, \tau_2 \Rightarrow \tau_2^{\tau_1}$,
- $\forall \tau t \Rightarrow t'$ where the label at the root of t' is $\mathrm{A}\tau$, and for each $c \in \tau$, t' has a child tc,
- $t \Rightarrow \top$ if the label of the root of t is $\mathrm{A}\tau$, and all children of t are labelled with $\top$,
- if $t_c \overset{*}{\Rightarrow} t'_c$ for all $c \in \tau \in \mathcal{T}$, the label of the root of t is $\mathrm{A}\tau$, and $\{t_c \mid c \in \tau\}$ is the set of children of t, then $t \Rightarrow t'$, where the label of the root of t' is $\mathrm{A}\tau$ and $\{t'_c \mid c \in \tau\}$ is the set of children of t'.

We removed negative references to $\Rightarrow$, but it is not easy to show that the resulting model satisfies the required conditions. Two key properties established in the correctness proof are: 1. $\Rightarrow$ has the Church-Rosser property, and 2. if $t_2 \succ c$ and $t_1 c \overset{*}{\Rightarrow} d \in \{\top, \bot\}$ then $t_1 t_2 \overset{*}{\Rightarrow} d$. The second property shows that quantifying over only canonical constants of type τ is in a sense equivalent to quantifying over all terms of type τ. This is essential for establishing e.g. the condition $(\forall_\top)$.

Both properties have intricate proofs. Essentially, the proofs show certain commutation and postponement properties for $\Rightarrow$, $\succ$ and other auxiliary relations. The proofs proceed by induction on lexicographic products of various ordinals and other parameters associated with the relations and terms involved.

4 Partiality and General Recursion

In this section we give some examples of proofs in $\mathcal{I}_s$ of properties of functions defined by recursion. For lack of space, we give only informal indications of how formal proofs may be obtained, assuming certain basic properties of operations on natural numbers. The transformation of the given informal arguments into formal proofs in $\mathcal{I}_s$ is not difficult. Mostly complete formal proofs may be found in a technical appendix.

Example 1. Consider a term subp satisfying the following recursive equation:

$$\mathrm{subp} = \lambda ij\,.\,\mathrm{Cond}\,(i =_{\mathrm{Nat}} j)\,0\,((\mathrm{subp}\, i\,(j+1)) + 1)\ .$$

If $i \geq j$ then $\mathrm{subp}\, i\, j = i - j$. If $i < j$ then $\mathrm{subp}\, i\, j$ does not terminate. An appropriate specification for subp is $\forall i, j : \mathrm{Nat}\,.\,(i \geq j) \supset (\mathrm{subp}\, x = i - j)$.

Let $\varphi(y) = \forall i : \mathrm{Nat}\,.\,\forall j : \mathrm{Nat}\,.\,(i \geq j \supset y =_{\mathrm{Nat}} i - j \supset \mathrm{subp}\, i\, j = i - j)$. We show by induction on y that $\forall y : \mathrm{Nat}\,.\,\varphi(y)$.

First note that under the assumptions $y : \mathrm{Nat}$, $i : \mathrm{Nat}$, $j : \mathrm{Nat}$ it follows from Lemma 8 that $(i \geq j) : \mathrm{Prop}$ and $(y =_{\mathrm{Nat}} i - j) : \mathrm{Prop}$. Hence, whenever $y : \mathrm{Nat}$, to show $i \geq j \supset y =_{\mathrm{Nat}} i - j \supset \mathrm{subp}\, i\, j = i - j$ it suffices to derive $\mathrm{subp}\, i\, j = i - j$ under the assumptions $i \geq j$ and $y =_{\mathrm{Nat}} i - j$. By Lemma 7 the assumption $y =_{\mathrm{Nat}} i - j$ may be weakened to $y = i - j$.

In the base step it thus suffices to show $\mathrm{subp}\, i\, j = i - j$ under the assumptions $i : \mathrm{Nat}$, $j : \mathrm{Nat}$, $i \geq j$, $i - j = 0$. From $i - j = 0$ we obtain $\mathfrak{o}(i - j)$, so $j \geq i$. From $i \geq j$ and $i \leq j$ we derive $i =_{\mathrm{Nat}} j$. Then $\mathrm{subp}\, i\, j = i - j$ follows by simple computation (i.e. by applying rules for Eq and appropriate rules for the conditional).

In the inductive step we have $\varphi(y)$ for $y : \mathrm{Nat}$ and we need to obtain $\varphi(\mathfrak{s}y)$. It suffices to show $\mathrm{subp}\, i\, j = i - j$ under the assumptions $i : \mathrm{Nat}$, $j : \mathrm{Nat}$ and $\mathfrak{s}y = i - j$. Because $\mathfrak{s}y \neq_{\mathrm{Nat}} 0$ we have $i \neq_{\mathrm{Nat}} j$, hence $\mathrm{subp}\, i\, j = \mathfrak{s}(\mathrm{subp}\, i\, (\mathfrak{s}j))$ follows by computation. Using the inductive hypothesis we now conclude $\mathrm{subp}\, i\, (\mathfrak{s}j) = i - (\mathfrak{s}j)$, and thus $\mathrm{subp}\, i\, (\mathfrak{s}j) =_{\mathrm{Nat}} i - (\mathfrak{s}j)$ by reflexivity of $=_{\mathrm{Nat}}$ on natural numbers. Then it follows by properties of operations on natural numbers that $\mathfrak{s}(\mathrm{subp}\, i\, (\mathfrak{s}j)) =_{\mathrm{Nat}} i - j$. By Lemma 7 we obtain the thesis.

We have thus completed an inductive proof of $\forall y : \mathrm{Nat}\,.\,\varphi(y)$. Now we use this formula to derive $\mathrm{subp}\, i\, j = i - j$ under the assumptions $i : \mathrm{Nat}$, $j : \mathrm{Nat}$, $i \geq j$. Then it remains to apply $\supset_i$ and $\forall_i$ twice.

In the logic of PVS [9] one may define subp by specifying its domain precisely using predicate subtypes and dependent types, somewhat similarly to what is done here. However, an important distinction is that we do not require a domain specification to be a part of the definition. Because of this, we may easily derive $\varphi \equiv \forall i, j : \mathrm{Nat}\,.\,((\mathrm{subp}\, i\, j = i - j) \vee (\mathrm{subp}\, j\, i = j - i))$. This is not possible in PVS because the formula φ translated to PVS generates false proof obligations [9].

Example 2. The next example is a well-known "challenge" posed by McCarthy:

$$f(n) = \mathrm{Cond}\,(n > 100)\,(n - 10)\,(f(f(n + 11)))$$

For $n \leq 101$ we have $f(n) = 91$, which fact may be proven by induction on $101 - n$. This function is interesting because of its use of nested recursion. Termination behavior of a nested recursive function may depend on its functional behavior, which makes reasoning about termination and function value interdependent. Below we give an indication of how a formal proof of $\forall n : \mathrm{Nat}\,.\,n \leq 101 \supset f(n) = 91$ may be derived in $\mathcal{I}_s$. Lemma 8 is used implicitly with implication introduction.

Let $\varphi(y) \equiv \forall n : \mathrm{Nat}\,.\,n \leq 101 \supset 101 - n \leq y \supset f(n) = 91$. We prove $\forall y : \mathrm{Nat}\,.\,\varphi(y)$ by induction on y. In the base step we need to prove $f(n) = 91$ under the assumptions $n : \mathrm{Nat}$, $n \leq 101$ and $101 - n \leq y = 0$. We have $n =_{\mathrm{Nat}} 101$, hence $n = 101$, and the thesis follows by simple computation.

In the inductive step we distinguish three cases: 1. $n+11 > 101$ and $n < 101$, 2. $n + 11 > 101$ and $n \geq 101$, 3. $n + 11 \leq 101$. We need to prove $f(n) = 91$ under the assumptions of the inductive hypothesis $y : \mathrm{Nat}, \forall m : \mathrm{Nat} \,.\, m \leq 101 \supset 101 - m \leq y \supset f(m) = 91$, and of $n : \mathrm{Nat}$, $n \leq 101$ and $101 - n \leq (\mathfrak{s}y)$.

We treat only the third case, other cases being similar. From $101 - n \leq s(y)$ we infer $101 - (n + 11) \leq y$. Since $n + 11 \leq 101$ we conclude by the inductive hypothesis that $f(n + 11) = 91$. Because $n + 11 \leq 101$, so $n \leq 100$, and by definition we infer $f(n) = f(f(n+11)) = f(91)$. Now we simply compute $f(91) = f(f(102)) = f(92) = f(f(103)) = \ldots = f(100) = f(f(111)) = f(101) = 91$ (i.e. we apply rules for Eq and Cond an appropriate number of times).

This concludes the inductive proof of $\forall y : \mathrm{Nat} \,.\, \forall n : \mathrm{Nat} \,.\, n \leq 101 \supset 101 - n \leq y \supset f(n) = 91$. Having this it is easy to show $\forall n : \mathrm{Nat} \,.\, n \leq 101 \supset f(n) = 91$.

Note that the computation of $f(91)$ in the inductive step relies on the fact that in our logic values of functions may always be computed for specific arguments, regardless of what we know about the function.

5 Related Work

In this section we discuss the relationship between $\mathcal{I}_s$ and the traditional illative system $\mathcal{I}_\omega$. We also briefly survey some approaches to dealing with partiality and general recursion in proof assistants. A general overview of the literature relevant to this problem may be found in [10].

5.1 Relationship with Systems of Illative Combinatory Logic

In terms of the features provided, the system $\mathcal{I}_s$ may be considered an extension of $\mathcal{I}_\omega$ from [3], which is a direct extension of $\mathcal{I}\Xi$ from [1] to higher-order logic. The ideas behind $\mathcal{I}_\omega$ date back to [11], or even earlier as far as the general form of inference rules is concerned.

However, there are some technical differences between $\mathcal{I}_s$ and traditional illative systems. For one thing, traditional systems strive to use as few constants and rules as possible. For instance, $\mathcal{I}_\omega$ has only two primitive constants, disregarding constants representing base types. Because of this in $\mathcal{I}_\omega$ e.g. $\mathrm{Is} = \lambda xy \,.\, yx$ and $\mathrm{Prop} = \lambda x \,.\, \mathrm{Type}(\lambda y.x)$, using the notation of the present paper. Moreover, the names of the constants and the notations employed are not in common use today. We will not explain these technicalities in any more detail. The reader may consult [2,1,3] for more information on illative combinatory logic.

Below we briefly describe a system $\mathcal{I}'_\omega$ which is a variant of $\mathcal{I}_\omega$ adapted to our notation. It differs somewhat from $\mathcal{I}_\omega$, mostly by taking more constants as primitive. The terms of $\mathcal{I}'_\omega$ are those of $\mathcal{I}_s$, except that we do not allow subtypes, inductive types, Eq, Cond, $\vee$, $\bot$ and ϵ. There are also additional constants: ω (the type of all terms), ε (the empty type) and $\supset$. The axioms are: $\Gamma, \varphi \vdash \varphi$, $\Gamma \vdash \mathrm{Prop} : \mathrm{Type}$, $\Gamma \vdash \varepsilon : \mathrm{Type}$, $\Gamma \vdash \omega : \mathrm{Type}$, $\Gamma \vdash t : \omega$. The rules are: $\forall_i$, $\forall_e$, $\forall_t$, $\supset_i$, $\supset_e$, $\supset_t$, $\rightarrow_i$, $\rightarrow_e$, $\rightarrow_t$, p_i, and the rules:

$$\text{conv}: \frac{\Gamma \vdash \varphi \qquad \varphi =_\beta \psi}{\Gamma \vdash \psi} \qquad \varepsilon_\bot : \frac{\Gamma \vdash t : \varepsilon}{\Gamma \vdash \bot}$$

$$\rightarrow_p : \frac{\Gamma \vdash \alpha : \text{Type} \qquad \Gamma, x : \alpha \vdash ((tx) : \beta) : \text{Prop} \qquad x \notin FV(\Gamma, t)}{\Gamma \vdash (t : \alpha \to \beta) : \text{Prop}}$$

5.2 Partiality and Recursion in Proof Assistants

Perhaps the most common way of dealing with recursion in interactive theorem provers is to impose certain syntactic restrictions on the form of recursive definitions so as to guarantee well-foundedness. Well-foundedness of definitions is then checked by a built-in automatic syntactic termination checker. Some systems, e.g. ACL2 or PVS, pass the task of proving termination to the user. Such systems require that a well-founded relation or a measure be given with each recursive function definition. Then the system generates so called proof obligations, to be shown by the user, which state that the recursive calls are made on smaller arguments.

The method of restricting possible forms of recursive definitions obviously works only for total functions. If a function does not in fact terminate on some elements of its specified domain, then it cannot be introduced by a well-founded definition. One solution is to use a rich type system, e.g. dependent types combined with predicate subtyping, to precisely specify function domains so as to rule out the arguments on which the function does not terminate. This approach is adopted by PVS [9].

A different approach to dealing with partiality and general recursion is to use a special logic which allows partial functions directly. Systems adopting this approach are often based on variants of the logic of partial terms of Beeson [12,13]. For instance, the IMPS interactive theorem prover [14] uses Farmer's logic PF of partial functions [15], which is essentially a variant of the logic of partial terms adapted to higher-order logic.

The above gives only a very brief overview. There are many approaches to the problem of partiality and general recursion in proof assistants, most of which we didn't mention. We do not attempt here to provide a detailed comparison with a multitude of existing approaches or give in-depth arguments in favor of our system. For such arguments to be entirely convincing, they would need to be backed up by extensive experimentation in proving properties of sizable programs using our logic. No such experimentation has been undertaken. In contrast, our interest is theoretical.

6 Conclusion

We have presented a system $\mathcal{I}_s$ of classical higher-order illative λ-calculus with subtyping and basic inductive types. A distinguishing characteristic of $\mathcal{I}_s$ is that it is based on the untyped λ-calculus. Therefore, it allows recursive definitions of potentially non-terminating functions directly. The inference rules of $\mathcal{I}_s$ are formulated in a way that makes it possible to apply them even when some of the terms used in the premises have not been proven to belong to any type. Additionally, our system may be considered an extension of ordinary higher-order logic, obtained by relaxing the typing restrictions on allowable λ-terms. We believe these facts alone make it relevant to the problem of partiality and recursion in proof assistants, and the system at least deserves some attention.

References

1. Barendregt, H., Bunder, M.W., Dekkers, W.: Systems of illative combinatory logic complete for first-order propositional and predicate calculus. Journal of Symbolic Logic 58(3), 769–788 (1993)
2. Seldin, J.P.: The logic of Church and Curry. In: Gabbay, D.M., Woods, J. (eds.) Logic from Russell to Church. Handbook of the History of Logic, vol. 5, pp. 819–873. North-Holland (2009)
3. Czajka, Ł.: Higher-order illative combinatory logic. Journal of Symbolic Logic (2013), `http://arxiv.org/abs/1202.3672` (accepted)
4. Dekkers, W., Bunder, M.W., Barendregt, H.: Completeness of the propositions-as-types interpretation of intuitionistic logic into illative combinatory logic. Journal of Symbolic Logic 63(3), 869–890 (1998)
5. Czajka, Ł.: Partiality and recursion in higher-order logic. Technical report, University of Warsaw (2013), `http://arxiv.org/abs/1210.2039`
6. Sørensen, M.H., Urzyczyn, P.: Lectures on the Curry-Howard isomorphism. Studies in Logic and the Foundations of Mathematics, vol. 149. Elsevier, Amsterdam (2006)
7. Barendregt, H.P.: The lambda calculus: Its syntax and semantics. Revised edn. North Holland (1984)
8. Blanqui, F., Jouannaud, J., Okada, M.: Inductive-data-type systems. Theoretical Computer Science 272(1), 41–68 (2002)
9. Rushby, J., Owre, S., Shankar, N.: Subtypes for specifications: Predicate subtyping in PVS. IEEE Transactions on Software Engineering 24(9) (1998)
10. Bove, A., Krauss, A., Sozeau, M.: Partiality and recursion in interactive theorem provers: An overview. Mathematical Structures in Computer Science (2012) (to appear)
11. Bunder, M.W.: Predicate calculus of arbitrarily high finite order. Archive for Mathematical Logic 23(1), 1–10 (1983)
12. Feferman, S.: Definedness. Erkenntnis 43, 295–320 (1995)
13. Beeson, M.J.: Proving programs and programming proofs. In: Marcus, R.B., Dorn, G., Weingartner, P. (eds.) Logic, Methodology and Philosophy of Science VII, pp. 51–82. North-Holland (1986)
14. Farmer, W.M., Guttman, J.D., Thayer, F.J.: IMPS: An interactive mathematical proof system. Journal of Automated Reasoning 11(2), 213–248 (1993)
15. Farmer, W.M.: A partial functions version of Church's simple theory of types. Journal of Symbolic Logic 55(3), 1269–1291 (1990)

Some Sahlqvist Completeness Results for Coalgebraic Logics

Fredrik Dahlqvist and Dirk Pattinson

Dept. of Computing, Imperial College London
{f.dahlqvist09,d.pattinson}@imperial.ac.uk

Abstract. This paper presents a first step towards completeness-via-canonicity results for coalgebraic modal logics. Specifically, we consider the relationship between classes of coalgebras for ω-accessible endofunctors and logics defined by Sahlqvist-like frame conditions. Our strategy is based on conjoining two well-known approaches: we represent accessible functors as (equational) quotients of polynomial functors and then use canonicity results for boolean algebras with operators to transport completeness to the coalgebraic setting.

Keywords: Modal logic, coalgebraic modal logic, canonicity, completeness, Sahlqvist formula.

1 Introduction

Coalgebras have gained popularity as an elegant and general framework to study and represent a wide variety of dynamical systems in computer science (see [12]) and even in physics (see [1]). In parallel to this area of research, the field of coalgebraic logic has emerged as a unifying framework for the many types of (modal) logics used to reason about dynamical systems (see [8] for an overview). One of the great insights into the relationship between coalgebras and coalgebraic logics, is that the class of all T-coalgebras for a functor T can always be characterised logically by its one-step behaviour, i.e. axioms and rules with nesting depth of modal operators uniformly equal to 1 (see [13]). However, once the transition type (i.e. the functor T) has been described logically in such a way, one may be interested in subclasses of T-coalgebras which are characterised by more complex axioms (such as transitivity for example) which we will refer to as *frame conditions*. The problem of logically characterising subclasses of the class of all T-coalgebras for an arbitrary functor T is by and large still open ([11] offers a solution for some of the standard frame conditions of classical modal logic).

This paper aims to isolate a large class of frame conditions which can be used to logically characterise proper subclasses of coalgebras, i.e. axioms giving a sound and complete description of certain classes of coalgebras. Our strategy is based on the following observations. Firstly, it is well known that accessible **Set** functors can be represented as (equational) quotients of polynomial functors. Secondly, the coalgebraic logics for polynomial **Set** functors turns out to be very closely related to Boolean Algebras with Operators (BAOs). Thirdly, there is a

F. Pfenning (Ed.): FOSSACS 2013, LNCS 7794, pp. 193–208, 2013.

well developed theory of Sahlqvist formulae for general BAOs (see [6,5,14]). The first step will therefore be to show how Sahlqvist formulae can be imported into the coalgebraic logics of polynomial functors, the second step will be to show how they can then be transported to logics of general functors via the presentation.

The paper in organized as follows: in *Section 2* we will present the basic facts about BAOs and coalgebraic logics that are needed for the rest of the paper. This will be very succinct and the reader is referred to [7,6,5,14,9,8] for further details. The ∇-style of coalgebraic logic requires some notational discipline, and the notation of [9] is presented in detail. The section concludes with our first Sahlqvist-like completeness result for polynomial functors. *Section 3* will first address the idea of presenting a functor T with a polynomial functor S (again we will present the bare minimum and the reader is referred to [2] for all the details), and then explore what this means for coalgebraic logics. In *Section 4* we present the main technical result of the paper, the Translation Theorem, which relates the derivability in the logic associated to a functor T to that in the logic of the functor S presenting it. Finally, in *Section 5* we gather all our results together and present a Sahlqvist completeness theorem for coalgebraic logics.

2 BAOs and Coalgebraic Logics

We start with some notation, basic definitions and facts about BAOs and coalgebraic modal logics. Readers familiar with this material can safely move to Example 1 which should offer a first glimpse at what this paper aims to achieve.

Boolean Algebras with Operators (BAOs). We roughly follow the terminology of [6] which itself is based on the seminal paper [7]. A **Boolean Algebra with Operator (BAO)** is a Boolean Algebra (BA) $\mathfrak{A}$ together with functions $f_\sigma : A^{\mathrm{ar}(\sigma)} \to A$ where A is the set underlying $\mathfrak{A}$ and σ is an element of a signature (Σ, ar) with arity map $\mathrm{ar} : \Sigma \to \mathbb{N}$. The maps f_σ are required to preserve joins in each of their arguments, in which case they are known as **operators**. The BAOs with a given signature Σ, together with the BA-morphism preserving operators in the obvious way, form a category which we will call $\mathbf{BAO}(\Sigma)$. As shown in [7], every BA $\mathfrak{A}$ can be embedded in a unique Complete Atomic Boolean Algebra (CABA) $\mathfrak{A}^\varepsilon$ called its **canonical extension** and which has the property that (1) every atom of $\mathfrak{A}^\varepsilon$ is a meet of elements of $\mathfrak{A}$ and (2) every subset in A (the set underlying $\mathfrak{A}$) whose join in $\mathfrak{A}^\varepsilon$ is $\top$, has a *finite* subset whose join in $\mathfrak{A}$ is also $\top$. This result can be extended to include operators (in fact any monotone map), viz. any BAO $\mathcal{A}$ can be embedded in a BAO $\mathcal{A}^\varepsilon$ - its canonical extension - whose underlying BA is the canonical extension of that of $\mathcal{A}$. This result is of fundamental importance because the category of CABAs is dual to the category **Set** in which models live.

Sahlqvist Formulae in a BAO. Let us fix a BAO $\mathcal{A} = (\mathfrak{A}, \{f_\sigma \mid \sigma \in \Sigma\})$. We define a **$\Sigma$-term** to be an element of the algebra freely generated by the elements of $\mathfrak{A}$ and the operators $f_\sigma, \sigma \in \Sigma$. Following [6] and [5], we define a **Sahlqvist term** to be a Σ-term of the form:

$$u[v_1, \dots, v_n, \neg w_1, \dots, \neg w_m] \tag{1}$$

where (i) u is a *strictly positive* $m+n$-ary term, i.e. contains no negations, (ii) the v_i's are terms of the shape $v_i = \sigma_1^d(\dots(\sigma_k^d(x))\dots)$ where each $\sigma_i^d = \neg\sigma_i\neg$ is the dual of a unary operator $\sigma_i \in \Sigma$, and (iii) the w_k's ($1 \leq k \leq m$) are *positive* terms, i.e. all variables in w_k must occur in the scope of an even number of complementation symbols. A **Sahlqvist equation** is an equation of the type $s = 0$ where s is a Sahlqvist term. A **Sahlqvist inequality** (or Sahlqvist formula in the context of algebras of terms) is an inequality of the type $s \leq t$ where t is positive. As shown in [6], all Sahlqvist identities are canonical, i.e. if $s = 0$ holds in $\mathcal{A}$, then it holds in its canonical extension $\mathcal{A}^\varepsilon$.

Coalgebraic Logics. Coalgebraic logics come in two flavours which we now introduce very succinctly. In both cases V denotes a set of propositional variables.

We start with the **predicate lifting style** of coalgebraic logic. A coalgebraic language L_T has a syntax given by

$$a ::= p \mid \bot \mid \neg a \mid a \wedge b \mid \sigma(a_1, \dots, a_n)$$

where $p \in V$ and $\sigma \in \Sigma$ are modal operators belonging to a signature (Σ, ar). Note that we're using the notational convention of [9] where the lower case Roman letters a, b, c stand for formulae. Such a language is interpreted in terms of coalgebras and predicate liftings. Given a standard **Set**-endofunctor T (we will assume throughout the paper that all functors are standard), a **coalgebra** is a pair (W, γ) where W is a set (of worlds) and $\gamma : W \to TW$ is a transition map (T defines the 'transition type'). Each modal operator σ is interpreted by a **predicate lifting**, i.e. a natural transformation $[\![\sigma]\!] : \mathcal{Q}^n \to \mathcal{Q}T$ where $\mathcal{Q}$ is the contravariant powerset functor. Intuitively, predicate liftings 'lift' n-tuples of predicates (i.e. subsets, hence the powerset functor) to a predicate on transitions (hence $\mathcal{Q}T$). A **coalgebraic model** - or T-model - is a triple $\mathcal{M} = (W, \gamma, \pi)$ where $\pi : W \to \mathcal{P}(V)$ is a valuation. The notion of **truth** of a formula a at a point $w \in W$ is defined inductively in the usual manner for propositional variables and boolean operators, and by

$$\mathcal{M}, w \models \sigma(a_1, \dots, a_n) \text{ iff } \gamma(w) \in [\![\sigma]\!]_W([\![a_1]\!], \dots, [\![a_n]\!])$$

for modal operators, where $[\![a_i]\!]$ is the interpretation of a_i in W. A formula a is **satisfiable** in $\mathcal{M}$ if there exists $w \in W$ such that $\mathcal{M}, w \models a$. A **coalgebraic frame** - or T-frame - is just a T-coalgebra (W, γ) and a formula a is **valid** on the frame if for any valuation π, a is true at every point in the model (W, γ, π).

The **∇-style** of coalgebraic logic (also known as Moss style, or coalgebraic logic for the cover modality) has a very different flavour. We recall the basic definitions and results, and refer to [9] for a very good and very thorough overview of the topic. Since the language involves objects of many types, our notation follows the conventions of [9] very closely to avoid confusion. We start by fixing a weak-pullback preserving functor T and we define $T_\omega = \bigcup\{TY \mid Y \subseteq X \text{ finite}\}$, the finitary version of T. The coalgebraic language $\mathcal{L}_T$ induced by T is given by:

$$a ::= p \mid \neg a \mid \bigwedge \phi \mid \bigvee \phi \mid \nabla \alpha$$

where $p \in V$, $\phi \in \mathcal{P}_\omega \mathcal{L}_T$ and $\alpha \in T_\omega \mathcal{L}_T$. $\bigvee \emptyset$ defines $\bot$. For any $\alpha \in T_\omega \mathcal{L}_T$ we define the **base** of α by $\mathsf{Base}^T(\alpha) = \bigcap \{U \subseteq \mathcal{L}_T \mid \alpha \in TU\}$, i.e. the set of immediate subformulae of $\nabla \alpha$. Given a T-model $\mathcal{M} = (W, \gamma, \pi)$, the truth relation $\models \subseteq W \times \mathcal{L}_T$ is inductively defined for any world $w \in W$ and formula $a \in \mathcal{L}_T$ by the usual clauses for atomic propositions and propositional connectives and

$$\mathcal{M}, w \models \nabla \alpha \text{ iff } \gamma(w) \bar{T}(\models) \alpha$$

where $\bar{T}(\models) \subseteq TW \times T\mathcal{L}_T$ is the *relation lifting* of the truth-relation $\models \subseteq W \times \mathcal{L}_T$ (see [9] for an extensive discussion of relation liftings).

Coalgebraic Logic is weakly complete (see [9]) with respect to the 2-dimensional Hilbert system which we call $\mathsf{KKV}(T)$ and is given by the axioms and rules:

$$\frac{}{a \leq a} \qquad (\text{Cut}) \frac{a \leq c \quad c \leq b}{a \leq b}$$

$$(\bigvee L) \frac{\{a \leq b \mid a \in \phi\}}{\bigvee \phi \leq b} \qquad (\bigvee R) \frac{a \leq b}{a \leq \bigvee \phi}\, b \in \phi$$

$$(\bigwedge L) \frac{a \leq b}{\bigwedge \phi \leq b}\, a \in \phi \qquad (\bigwedge R) \frac{\{a \leq b \mid b \in \phi\}}{a \leq \bigwedge \phi}$$

$$(\neg E) \frac{\bigwedge \{\phi \cup \{\neg a\}\} \leq \bigvee \psi}{\bigwedge \phi \leq \bigvee \{\psi \cup \{a\}\}} \qquad (\neg I) \frac{\bigwedge \{\phi \cup \{a\}\} \leq \bigvee \psi}{\bigwedge \phi \leq \bigvee \{\psi \cup \{\neg a\}\}}$$

$$(\text{Distributivity}) \frac{}{\bigwedge \{\bigvee \phi \mid \phi \in X\} \leq \bigvee \{\bigwedge \mathsf{rng}(\gamma) \mid \gamma \in \mathsf{Choice}(X)\}}$$

$$(\nabla 1) \frac{\{a \leq b \mid (a, b) \in R\}}{\nabla \alpha \leq \nabla \beta}\, (\alpha, \beta) \in \bar{T}R$$

$$(\nabla 2) \frac{\{\nabla (T \bigwedge)(\Phi) \leq b \mid \Phi \in SRD(A)\}}{\bigwedge \{\nabla \alpha \mid \alpha \in A\} \leq b} \qquad (\nabla 3) \frac{\{\nabla \alpha \leq b \mid \alpha \bar{T}{\in} \Phi\}}{\nabla (T \bigvee)(\Phi) \leq b}$$

where $a, b \in \mathcal{L}_T$, $\phi, \psi \in \mathcal{P}_\omega \mathcal{L}_T$, $X \in \mathcal{P}_\omega \mathcal{P}_\omega \mathcal{L}_T$, $\alpha, \beta \in T_\omega \mathcal{L}_T$, $\Phi \in T_\omega \mathcal{P}_\omega \mathcal{L}_T$, $A \in \mathcal{P}_\omega T_\omega \mathcal{L}_T$. The set $\mathsf{Choice}(X)$ is the set of choice functions on X, i.e. the maps $\gamma \colon X \to \mathcal{L}_T$ such that $\gamma(\phi) \in \phi$, and rng denotes the range of the function. $R \subseteq \mathcal{L}_T \times \mathcal{L}_T$ is any relation and $\bar{T}R$ is its lifting. Finally $SRD(A)$ is the set of so-called 'slim redistributions' of A. This last concept is important, and we therefore define it in extenso. A **redistribution** of $A \in \mathcal{P}_\omega T_\omega \mathcal{L}_T$ is an element Φ of $T_\omega \mathcal{P}_\omega \mathcal{L}_T$ which 'contains' all the elements of A as lifted members, i.e. $\alpha \bar{T}{\in} \Phi$ for all $\alpha \in A$. It is called **slim** if it is build from the direct subformulae of the elements of A, i.e. if $\Phi \in T_\omega \mathcal{P}_\omega (\bigcup_{\alpha \in A} \mathsf{Base}(\alpha))$.

To help the reader digest this rather heavy load of definitions, let us look at an example which will cover both BAOs and coalgebraic logics.

Example 1. Given a signature (Σ, ar), we define a functor $S_\Sigma : \mathbf{Set} \to \mathbf{Set}$ by

$$S_\Sigma X = \coprod_{n \in \omega} \Sigma_n \times X^n$$

where Σ_n is the set regrouping all operation symbols $\sigma \in \Sigma$ of arity n. Any functor of this shape is called a **polynomial functor**. We write $U : \mathbf{BA} \to \mathbf{Set}$ for the forgetful functor, and $F : \mathbf{Set} \to \mathbf{BA}$ for its left adjoint (the associated free construction). This allows us to lift set-functors $T\colon \mathbf{Set} \to \mathbf{Set}$ to the category of boolean algebras by putting $\mathsf{T} = FTU : \mathbf{BA} \to \mathbf{BA}$. In particular, every signature Σ induces the functor $\mathsf{S}_\Sigma : \mathbf{BA} \to \mathbf{BA}$ defined by

$$\mathsf{S}_\Sigma \mathfrak{A} = FS_\Sigma U\mathfrak{A} = F\left(\coprod_{\sigma \in \Sigma} A^{\mathrm{ar}(\sigma)}\right)$$

where $A = U\mathfrak{A}$. From now on we will drop the Σ subscript if there is no risk of confusion. It is easy to see by the freeness of the construction of S that the category $\mathbf{Alg}(\mathsf{S})$ of S-algebras in $\mathbf{BA}$ is isomorphic to the category of boolean algebras $\mathfrak{A}$ with maps $f_\sigma : A^n \to A$ where n is the arity of σ. But this is almost the category $\mathbf{BAO}(\Sigma)$ defined above! The only difference is that the maps do not have to be operators, but we will return to this in an instant.

If we now turn our attention to coalgebraic logic, we can use the signature Σ to define a coalgebraic language L_S in the predicate lifting style defined above. It is relatively straightforward to see that the Lindenbaum-Tarski algebra of L_S, which we will denote $\mathcal{A}_S$, is the initial object in $\mathbf{Alg}(\mathsf{S})$, or equivalently, the free BAO of Σ-terms as defined above but with the preservation of joins not being enforced. We would like to interpret L_S in S-models by reading $\mathcal{M}, w \models \sigma(a_1, \ldots, a_n)$ as 'w has a σ-tuple successor and a_i holds at the ith component of this successor'. The predicate liftings are then defined by:

$$[\![\sigma]\!]_W : (\mathcal{P}W)^n \to \mathcal{P}SW, (U_1, \ldots, U_n) \mapsto \{\sigma(x_1, \ldots, x_n) \mid x_i \in U_i, 1 \leq i \leq n\}$$

The obvious question now is: what axioms will give a sound and complete axiomatization of $\mathbf{CoAlg}(S)$, the class of all S-coalgebras? It is not too difficult to find this list of axioms from scratch and then use a canonical model construction to prove completeness, however, the fact that we're using Σ in our syntax and our semantics suggests turning our attention to the ∇-flavour of coalgebraic logic.

There is an obvious one-to-one correspondence between the language L_S defined above and the ∇-style language $\mathcal{L}_S$ induced by S. Since $S = S_\omega$ and since $\sigma(a_1 \ldots, a_n)$ can be seen as an element of $S\mathsf{L}_S$, we can recursively add ∇ in front of every modal operator σ in order to get a ∇-style formula. Conversely, by recursively removing every ∇ from a formula in $\mathcal{L}_S$ we get a formula in L_S (see [10] for a detailed discussion of translations between the two flavours of coalgebraic logic). We now have rules that provide us with a sound and (weakly) complete axiomatization of $\mathbf{CoAlg}(S)$, namely:

$$(\nabla 1)_S \ \frac{\{a_i \leq b_i \mid 1 \leq i \leq n\}}{\nabla\sigma(a_1,\dots,a_n) \leq \nabla\sigma(b_1,\dots,b_n)}$$
$$(\nabla 2)_S \ \bigwedge\{\nabla\sigma(a_1,\dots,a_n) \mid \sigma(a_1,\dots,a_n) \in A\} = \nabla\sigma(\bigwedge \pi_1[A],\dots,\bigwedge \pi_n[A])$$
$$(\nabla 3)_S \ \nabla\sigma(\bigvee \phi_1,\dots,\bigvee \phi_n) = \bigvee\{\nabla\sigma(a_1,\dots,a_n) \mid a_i \in \phi_i\}$$

where $=$ means both $\leq$ and $\geq$, $a_i \in \mathcal{L}_S, A \in \mathcal{P}_\omega S\mathcal{L}_S, \phi_i \in \mathcal{P}_\omega\mathcal{L}_S$. Let us discuss these rules and axioms briefly. The $(\nabla 1)_S$ axiom takes this simplified form because relation lifting by polynomial functors is very simple: if $R \subseteq X \times X$, then $(\alpha, \beta) \in \bar{S}R$ only if α, β lie in the same part of the co-product SX and each component of the tuple α is R-related to the corresponding component of β. For $(\nabla 2)_S$, it is easy to see from the definition of slim redistribution that $SRD(A)$ is empty if A contains elements lying in different parts of the co-products, hence the presence of σ-terms only. Moreover, it is not too hard to check that if $\Phi, \Psi \in SRD(A)$ and $(\Psi, \Phi) \in \bar{S} \subseteq$, then $\nabla S \bigwedge \Phi \leq \nabla S \bigwedge \Psi$, and since $\sigma(\pi_1[A],\dots,\pi_n[A])$ is a lifted subset of all other elements of $SRD(A)$, $\nabla\sigma(\bigwedge \pi_1[A],\dots,\bigwedge \pi_n[A]) \leq b$ implies $\nabla S \bigwedge \Phi \leq b$ for all other $\Phi \in SRD(A)$. Finally, and most importantly for our purpose, $(\nabla 3)_S$ is just another way of saying that ∇ preserves joins in each of its arguments. Of course the number of arguments of ∇ can vary, but by trivially translating into L_S we get that σ - which has a fixed number of arguments - preserves joins in each of its arguments. To see this for the first argument for example, just take $\phi_1 = \{a_1, b_1\}$ and $\phi_i = \{a_i\}$ for $1 < i \leq n$, and note that the premise of the $(\nabla 3)$ rule is a finite set for which we can take the join[1].

Let us denote by K_S the predicate-lifting style logic defined by the trivial translations of the axioms $(\nabla 1)_S - (\nabla 3)_S$ from $\mathcal{L}_S$ to L_S and any axiomatization of propositional logic. It is easy to see that the semantics of $\mathcal{L}_S$ is essentially the same as the semantics we defined for L_S and since we know that $\mathsf{KKV}(S)$ is sound and complete w.r.t. the class of all S-coalgebras and that $\mathsf{KKV}(S)$ and K_S are in bijective correspondence, we can conclude that K_S is also sound and weakly complete w.r.t. the class of all S-coalgebras. The conclusion of this example is therefore that if we look at the *logic* K_S, rather than at the *language* L_S, then the Lindenbaum-Tarski algebra $\mathcal{A}_S$ of K_S is a bona fide BAO since preservation of joins has been enforced by the $(\nabla 3)_S$ axiom.

The previous example suggests our first Sahlqvist completeness theorem. But let us first define a notion of Sahlqvist formula for the coalgebraic logic of a polynomial functor.

Definition 2. Let a be a formula in $\mathsf{KKV}(S)$ for a polynomial functor S, and let a_S be its trivial translation into the predicate-lifting-style logic K_S. Then a (and a_S) will be called a **polynomial Sahlqvist formula** if (the equivalence class of) a_S in the Lindenbaum-Tarski algebra $\mathcal{A}_S$ of K_S is a Sahlqvist formula as defined above (following [6]).

[1] We refer the reader to [9] to see that the converse of $(\nabla 2)$ and $(\nabla 3)$ (i.e. with the inequalities going in the opposite direction) are derivable using $(\nabla 1)$ and the fact that S preserves weak-pullbacks.

Our first Sahlqvist completeness theorem is shown by following the well-trodden path of completeness-via-canonicity proofs (for more details on this technique we refer the reader to Chapter 4 and 5 of the classic [4]). Assume that Σ is a signature defining a polynomial functor S and that C is a set of frame conditions, then we want to endow the set of ultrafilters of the Lindenbaum-Tarski algebra $\mathcal{A}_S(C)$ of $\mathsf{K}_S + C$ (i.e. the BAO of formulae L_S quotiented by equivalence under $\mathsf{K}_S + C$) with the structure of an S-coalgebra. The natural transition map to use is $\gamma_c : \mathsf{Uf}(\mathcal{A}_S(C)) \to S\mathsf{Uf}(\mathcal{A}_S(C))$ defined by

$$\gamma_c(\phi) = \sigma(\psi_1, \ldots, \psi_n) \tag{2}$$

where $a_i \in \psi_i, 1 \leq i \leq n$ if $\sigma(a_1, \ldots, a_n) \in \phi$. However, when Σ is infinite some care must be taken due to the following fact. Consider the set

$$\zeta = \{\neg\sigma(\top, \ldots, \top) \mid \sigma \in \Sigma\}$$

The set ζ is K_S-consistent but γ_c is undefined on any ultrafilter containing it. In particular, any set of frame conditions containing ζ will lead to a situation where γ_c cannot be defined anywhere. The set ζ characterises precisely the set of ultrafilters for which γ_c is well-defined. Note that if Σ is finite, this problem does not arise, moreover ζ cannot be a subset of any finite set of frame conditions. But as we shall see later, infinite signatures and infinite sets of frame conditions will be very useful. This justifies the following definition.

Definition 3. We recursively define the collection $\mathcal{Z}$ of **deadlocking sets of formulae** as follows: $\zeta \in \mathcal{Z}$ and if $\zeta' \in \mathcal{Z}$, then for any operator $\sigma \in \Sigma, \text{ar}(\sigma) = n$ and map $\chi : n \to \{1, 2\}$ s.th. $1 \in \mathsf{rng}(\chi)$

$$\{\sigma(\pi_{\chi(0)}(z, \top), \ldots, \pi_{\chi(n)}(z, \top) \mid z \in \zeta'\} \in \mathcal{Z}$$

where π_1, π_2 are the obvious projections. Intuitively, ζ characterises a deadlock ultrafilter whereas $\mathcal{Z}$ characterises all the ultrafilters from which a deadlock state can be reached in finitely many transitions. We then define an **acceptable** set of frame conditions as a set of L_S-formulae which are K_S-consistent and do not contain any deadlocking set of formulae. This definition is extended via the trivial translation to $\mathsf{KKV}(S)$-frame conditions.

Note that ζ characterizes ultrafilters from which no transition is possible at all, even trivial transitions defined by nullary terms in the signature - which also encodes a notion of 'deadlock' - are forbidden.

Theorem 4 (Sahlqvist Completeness for Polynomial Functors). *Let S be a polynomial functor, $\mathcal{L}_S$ be the ∇-language it defines and let $C \subseteq \mathcal{L}_S$ be an acceptable set of Sahlqvist frame conditions, then $\mathsf{KKV}(S) + C$ is complete w.r.t. the class of S-coalgebras validating C.*

Proof (Sketch). We start with a $\mathsf{KKV}(S) + C$-consistent formula a and we will show that we can find a model in the class of S-coalgebras which validate C, in which a is satisfied. Let Σ be the signature defining S.

Finite Signatures: The map γ_c defined by Eq. (2) is well-defined because we can derive $\top \leq \bigvee_{\sigma \in \Sigma} \sigma(\top, \ldots, \top)$ in K_S, i.e. ultrafilters always have a successor.

Moreover, by the $(\nabla 2)_S$ axioms we cannot have tuples prefixed with different operator symbols in an ultrafilter ϕ since we cannot have $\bot \in \phi$. This guarantees that $\gamma_c(\phi)$ lands in a unique component of the coproduct defining S. It is not too difficult to check that $\gamma_c(\phi)$ is indeed an ultrafilter. The canonical valuation is given as expected by $\pi_c : \mathsf{Uf}(\mathcal{A}_S(C)) \to \mathcal{P}(V), \phi \mapsto V \cap \phi$. Since we have a total function γ_c, rather than a relation like in the traditional Kripke setting, we do not need an Existence Lemma and we can move straight to the Truth Lemma which is easily proven: if we define $\mathcal{M}_c = (\mathsf{Uf}(\mathcal{A}_S(C)), \gamma_c, \pi_c)$, then

$$(\mathcal{M}_c, \phi) \models a \text{ iff } a \in \phi$$

We can now build a model in which a is satisfied: take any ultrafilter ϕ containing a and we have $(\mathcal{M}_c, \phi) \models a$. Now, all we need to do is to show that our canonical model is based on a coalgebra which validates the frame conditions of C. To do this, we need to consider the *complex algebra* associated with $(\mathsf{Uf}(\mathcal{A}(C)), \gamma_c)$. Specifically, we put a Σ-BAO structure on $\mathcal{P}(\mathsf{Uf}(\mathcal{A}_S(C)))$ by defining

$$\sigma^\varepsilon(X_1, \ldots, X_n) = \{\phi \in \mathsf{Uf}(\mathcal{A}_S(C)) \mid \gamma_c(\phi)_i \in X_i, 1 \leq i \leq n\}$$

for all $\sigma \in \Sigma$. The reason for the notation σ^ε, is that the BAO we've just defined is nothing but $\mathcal{A}_S(C)^\varepsilon$, the canonical extension of $\mathcal{A}_S(C)$. Now, since all the formulae in C are Sahlqvist, then they must be canonical (see [6]), i.e. they must all hold in $\mathcal{A}_S(C)^\varepsilon$. It is then easy to check that if a formula of C holds in $\mathcal{A}_S(C)^\varepsilon = \mathcal{P}(\mathsf{Uf}(\mathcal{A}_S(C)))$, it must be valid on the S-frame $(\mathsf{Uf}(\mathcal{A}(C)), \gamma_c)$.

Infinite Signatures: To account for the possibility of deadlock states we start by building a slightly different canonical model. The carrier set is given by

$$W_c = \{\phi \in \mathsf{Uf}(\mathcal{A}(C)) \mid \text{ for all } \zeta' \in \mathcal{Z}, \zeta' \not\subseteq \phi\}$$

The map $\gamma_c : W_c \to SW_c$ is then defined as above and is well-defined by construction and by the comments made in the finite signature case. The Truth lemma holds just as in the finite signature case. So to build a model for a we just need to find an ultrafilter $\phi \in W_c$ containing a. It is quite straightforward to check that this is indeed possible (in fact it is possible for any finite set of formulae). The complex algebra associated with W_c is defined as in the finite signature case and is a subalgebra of the canonical extension $(\mathcal{A}(C))^\varepsilon$ of $\mathcal{A}(C)$. Since C is a set of Sahlqvist formulae, they are canonical and thus $(\mathcal{A}(C))^\varepsilon$ belongs to the variety they define. By Birkhoff's theorem this variety is closed under taking subalgebras and so $\mathcal{P}(W_c)$ is an algebra satisfying the equations of C. The fact that (W_c, γ_c) validates the formulae in C follows.

Remark 5. As was hinted in the proof above, if the polynomial functor S is defined by a signature with only finitely many operation symbols, then the result above can be strengthened to a *strong completeness* result, i.e. any consistent set of formulae is satisfiable. In the case of infinite signatures, only finite sets of consistent formulae are guaranteed to be satisfiable. However, *acceptable* sets of formulae in the sense of Definition (3) are also satisfiable, providing a result which is somewhere between weak and strong completeness.

3 Presentations and Translations

We will make crucial use of the fact that every accessible functor arises as the quotient of a polynomial functor. By a λ-*ary presentation* of a set-endofunctor T we understand a λ-ary signature (Σ, ar) (i.e. arities are bounded by λ) together with an epi natural transformation $q : S_\Sigma \twoheadrightarrow T$. It is well known that every λ-accessible endofunctor has a λ-ary presentation and we refer the reader to [2] for a detailed overview of presentations in the context of coalgebras.

A natural question to ask in this context is: given a natural transformation $q : S \to T$, what can we say about the relationship between the coalgebraic logics associated with S and T? Is there a syntactic relationship? And what happens at the semantic level? These questions seem natural but, as far as we know, have not really been studied systematically in the literature. Let us first look at what happens at the syntactic level.

Definition 6. Let S, T be two weak-pullback preserving standard functors on **Set** and let $q : S \to T$ be a natural transformation. We define the translation map $(\cdot)^q : \mathcal{L}_S \to \mathcal{L}_T$ recursively by

$$(\nabla\alpha)^q = \nabla(q_{\mathcal{L}_T} \circ S(\cdot)^q)(\alpha)$$

We call $(\cdot)^q$ the **translation along** q and will use the following notational conventions for maps associated with $(\cdot)^q : \mathcal{L}_S \to \mathcal{L}_T$

- $\langle . \rangle^q : S\mathcal{L}_S \to T\mathcal{L}_T$ will be shorthand for the map $q_{\mathcal{L}_T} \circ S(\cdot)^q$
- $[\cdot]^q : \mathcal{P}_\omega S\mathcal{L}_S \to \mathcal{P}_\omega T\mathcal{L}_T$ will be shorthand for the map $\mathcal{P}_\omega\langle\cdot\rangle^q$
- $\{\cdot\}^q : S\mathcal{P}_\omega\mathcal{L}_S \to T\mathcal{P}_\omega\mathcal{L}_T$ will be shorthand for the map $q_{\mathcal{P}_\omega\mathcal{L}_T} \circ S\mathcal{P}_\omega(\cdot)^q$

Note that with this notation we have $(\nabla\alpha)^q = \nabla(\langle\alpha\rangle^q)$.

At the level of the semantics, note that a natural transformation $q : S \to T$ induces a functor $Q : \mathbf{CoAlg}(S) \to \mathbf{CoAlg}(T)$ on the corresponding categories of coaglebras, given by $Q(W, \gamma) = (W, q_W \circ \gamma)$. In particular Q turns models for $\mathcal{L}_S$-formulae into models for $\mathcal{L}_T$-formulae and we will now show that the translation along q agrees with the functor Q in the sense that truth is preserved by applying both simultaneously. Formally:

Proposition 7. *Suppose that $q : S \to T$ is a natural transformation and that $a \in \mathcal{L}_S$. Suppose also that we have a model $\mathcal{M} = (W, \gamma, \pi)$ such that*

$$\mathcal{M}, w \models a$$

for some $w \in W$. If we then define $Q(\mathcal{M}) = (W, q_W \circ \gamma, \pi)$ we have

$$Q(\mathcal{M}), w \models (a)^q$$

The following lemma shows how the functors Base^S and Base^T are related by an epi natural transformation $q : S \twoheadrightarrow T$. This lemma will be very useful to relate concepts for S and T which depend on the bases.

Proposition 8. *Let S, T be* **Set** *functors and $q : S \twoheadrightarrow T$ an epi natural transformation between them, let T weakly preserve pullbacks and let $\mathcal{L}_S$ and $\mathcal{L}_T$ be the ∇-languages induced by S and T respectively, then the following diagram commutes:*

$$\begin{array}{ccc} S\mathcal{L}_S & \xrightarrow{\mathsf{Base}^S} & \mathcal{P}\mathcal{L}_S \\ \downarrow{\scriptstyle (\cdot)^q} & & \downarrow{\scriptstyle \mathcal{P}(\cdot)^q} \\ T\mathcal{L}_T & \xrightarrow{\mathsf{Base}^T} & \mathcal{P}\mathcal{L}_T \end{array}$$

We conclude this section with an example.

Example 9. A functor of particular interest for applications is the so-called bag functor which we denote $\mathcal{B}$. Coalgebras for the bag functor are models for **Graded Modal Logic** which is essentially the modal logic version of *cardinality restrictions* in Description Logics. The bag functor can be defined by

$$\mathcal{B}X = \{f : X \to \mathbb{N} \mid supp(f) \text{ is finite }\}$$

Alternatively and equivalently, an element of $\mathcal{B}X$ can be defined as a 'multiset', i.e. a set of pairs (denoted with ':') $\{(x_i : n_i) \mid i \in I\}$ where the elements x_i , $i \in I$ are distinct elements of X and the $n_i, i \in I$ are integers thought of as the multiplicities of the elements x_i. $\mathcal{B}$ has a simple presentation in terms of the list functor $\mathsf{List}X = \coprod_{n\in\omega} X^n$. The presentation is given by:

$$q_X : \mathsf{List}X \to \mathcal{B}X, (a_1, \ldots, a_n) \mapsto \{(a_{p(1)}, \ldots, a_{p(n)}) \mid p \in \mathsf{Perm}(n)\}$$

where $\mathsf{Perm}(n)$ is the group of permutations of n elements. In other words q identifies all permutations of a given tuple, and thus an element of $\mathcal{B}X$ can be represented as a multiset $(a_1 : k_1, \ldots, a_n : k_n)$ where $a_i : k_i$ means a_i appears k_i times in the (equivalence class of) tuple. In this context the translation $(\cdot)^q$ works as follows: a $\mathcal{L}_{\mathsf{List}}$-formula of the form $\nabla n(a_1, \ldots, a_n)$ gets translated into an $\mathcal{L}_{\mathcal{B}}$-formula of the shape $\nabla n(k_1 : (a'_1)^q, \ldots, k_m : (a'_m)^q)$ where the a'_i are the distinct elements of the set $\{a_1, \ldots, a_n\}$ and k_i their multiplicity (and $m \leq n$).

4 The Translation Theorem

We have so far established the following: (1) we have a logic for coalgebras based on polynomial functors which is suitable for defining Sahlqvist formulae and (2) every accessible functor T can be presented from a polynomial functor S and this presentation allows us to move from the language and the coalgebras based on S to those based on T in a sensible way. This section is they key technical contribution of this paper and shows how we can connect the facts that we have established in the two previous sections. The idea will be to show that the translation map $(\cdot)^q$ acts not just on the *language* $\mathcal{L}_S$ but also on the *logic* $\mathsf{KKV}(S)$. Crucially, we will show how derivability in $\mathsf{KKV}(S)$ is related to derivabilty in $\mathsf{KKV}(T)$. This section is rather technical and we must start with a few lemmata. The intended meaning of our first lemma is that $(\cdot)^q$ sends substitution instances of axioms to substitution instances of axioms.

Definition 10. A substitution is a map $\hat{\pi} : \mathcal{L}_T \to \mathcal{L}_T$ defined inductively from a map $\pi : V \to \mathcal{L}_T$ by: $\hat{\pi}(p) = \pi(p)$ for all $p \in V$, $\hat{\pi}(\phi \wedge \psi) = \hat{\pi}(\phi) \wedge \hat{\pi}(\psi)$, $\hat{\pi}(\neg\phi) = \neg\hat{\pi}(\phi)$ and $\hat{\pi}(\nabla\alpha) = \nabla(T\hat{\pi}(\alpha))$.

Lemma 11. *Let S, T be two weak-pullback preserving functors on* **Set**, *let $q : S \twoheadrightarrow T$ be a epi natural transformation and let $\mathcal{L}_S$ and $\mathcal{L}_T$ be the ∇-languages induced by S and T respectively. Let $\pi : V \to \mathcal{L}_S$ define a substitution $\hat{\pi} : \mathcal{L}_S \to \mathcal{L}_S$ and and let ρ be the map $\rho = (\cdot)^q \circ \pi : V \to \mathcal{L}_T$, then for all $a \in \mathcal{L}_S$*

$$(\cdot)^q \circ \hat{\pi}(a) = \hat{\rho} \circ (\cdot)^q(a)$$

There are two important constructions in the KKV axiomatization: the notion of slim redistribution and that of lifted member (used for the (∇2) and (∇3) axiom). The following two lemmata show how these notions interact with the translation map. They are generalisation of Lemmata 5.44 and 5.45 in [10] where a special class of presentations (called 'well-based' presentations) is considered, here we consider arbitrary epi natural transformations between weak-pullback preserving functors.

Lemma 12. *Let S, T be two weak-pullback preserving functors on* **Set**, *let $q : S \twoheadrightarrow T$ be an epi natural transformation, and let $\mathcal{L}_S$ and $\mathcal{L}_T$ be the ∇-languages induced by S and T respectively. For any $A \in \mathcal{P}_\omega S\mathcal{L}_S$ and $\Phi \in T\mathcal{P}_\omega\mathcal{L}_T$, the following two conditions are equivalent:*

(1) $\Phi \in SRD([A]^q)$
(2) there exist $\Phi' \in S\mathcal{P}_\omega\mathcal{L}_S$ such that $\{\Phi'\}^q = \Phi$ and for all $\alpha \in A$ there exist α' such that $\alpha' \bar{S}{\in} \Phi'$ and $\langle\alpha'\rangle^q = \langle\alpha\rangle^q$

Lemma 13. *Let S, T be two weak-pullback preserving functors on* **Set**, *let $q : S \twoheadrightarrow T$ be an epi natural transformation, and let $\mathcal{L}_S$ and $\mathcal{L}_T$ be the ∇-languages induced by S and T respectively. For any $\alpha \in T\mathcal{L}_T$ and $\Phi \in T\mathcal{P}_\omega\mathcal{L}_T$ the following two conditions are equivalent:*

(1) $\alpha \bar{T}{\in} \Phi$ for $\Phi \in T\mathcal{P}_\omega\mathcal{L}_T$
(2) there exist $\Phi' \in T\mathcal{P}_\omega\mathcal{L}_S$ such that $\{\Phi'\}^q = \Phi$ and $\alpha' \in S\mathcal{L}_S$ such that $\langle\alpha'\rangle^q = \alpha$ and $\alpha' \bar{S}{\in} \Phi'$

We are now ready to move to our key technical result. Our main motivation is to get a completeness result for $\mathsf{KKV}(T)+C$ where C is a set of 'Sahlqvist formulae' - we will define what this means precisely in the next section. Following the usual method we'll start with a $\mathsf{KKV}(T)+C$-consistent formula a and try to build a model for it. By Theorem 4, we know how to do this for $\mathsf{KKV}(S)+D$, when $S \twoheadrightarrow T$ is a presentation of T and D is a set of Sahlqvist formulae. So what seems to be required is a result linking $\mathsf{KKV}(T)+C$-consistency to $\mathsf{KKV}(S)+D$-consistency for the right D. More specifically, we want a result relating $\neg a \leq \bot$ not being derivable in $\mathsf{KKV}(T)+C$ to a similar statement in $\mathsf{KKV}(T)+D$ for a certain D. As it turns out, the trick is to look at *all* the pre-images of a and of C, and, using the contrapositive, the result we are looking for is therefore:

Theorem 14 (Translation Theorem). *Let T be a weak-pullback preserving* **Set** *functor, let $q : S \twoheadrightarrow T$ be a presentation of T and let $\mathcal{L}_S$ and $\mathcal{L}_T$ be the ∇-languages defined by S and T. Assume we have a set C of* $\mathsf{KKV}(T)$*-consistent formulae (the frame conditions) and let us define the set $C' \subseteq \mathcal{L}_S$ by:*

$$C' = \{c' \in \mathcal{L}_S \mid (c')^q \in C \text{ and } c' \text{ is } \mathsf{KKV}(S) - \text{consistent}\}$$

Then

$$\mathsf{KKV}(S) + C' \vdash \{a' \leq \bot \mid (a')^q = a\}$$

implies

$$\mathsf{KKV}(T) + C \vdash a \leq \bot$$

Proof (Sketch). We proceed by induction on the depth n of the shortest $\mathsf{KKV}(S)+C'$-proof of $a' \leq \bot$ amongst all a' such that $(a')^q = a$. The base case is if $n = 0$, i.e. if there there exist an inequality $a' \leq \bot$ which is either an axiom of the propositional fragment of the logic or a substitution instance of an axiom in C'. By Lemma 11 and the definition of C', it is clear that if a' is a substitution instance of a formula $c' \in C'$, then its translation $(a')^q = a$ is a substitution instance of a formula in $c \in C$, and we can thus conclude that $\mathsf{KKV}(T) + C \vdash a \leq \bot$. The inductive hypothesis is the following: if we have $\mathsf{KKV}(S) + C'$ proofs that all the pre-images under $(\cdot)^q$ of a formula a are false and if the smallest of these proofs has depth n, then we have a proof that a is false in $\mathsf{KKV}(T) + C$. So let's assume that we have proofs

$$\mathsf{KKV}(S) + C' \vdash \{a' \leq \bot \mid (a')^q = a\}$$

and that the depth of the shortest proof if $n + 1$. We then show that we can always find a set of $\mathsf{KKV}(S) + C'$-proofs of minimal depth n whose conclusion are the pre-images of a premise in $\mathsf{KKV}(T) + C$ whose conclusion is $a \leq \bot$. In other words, we build the last step of a T-proof by using the last steps of existing S-proofs and the inductive hypothesis. This is done by examining, in turn, each of the possible outermost connectives of a, and thus of a', i.e. ∇ or a boolean connective. Each of these possible outermost connectives of a' specifies a small number of rules which could have been the last rule applied to reach $(a' \leq \bot)$. The proof then consists in examining each of these possibilities and show that they all lead to a situation where the induction hypothesis can be applied and lead to a T-rule with conclusion $a \leq \bot$.

5 Sahlqvist Formulae for Coalgebraic Logics

We now have all we need to formulate our Sahlqvist completeness result for coalgebraic modal logic. We start by defining a notion of Sahlqvist formula for a general (i.e. not necessarily polynomial) functor.

Definition 15. Let T be a weak-pullback preserving functor, let $q : S \twoheadrightarrow T$ be a presentation of T and let $\mathcal{L}_T$ and $\mathcal{L}_S$ be the ∇-languages induced by S and T respectively, then $a \in \mathcal{L}_T$ will be called a **coalgebraic Sahlqvist formula** if every pre-image of a under the translation map $(\cdot)^q : \mathcal{L}_S \to \mathcal{L}_T$ is Sahlqvist in the sense of Definition 2. A set $C \subseteq \mathcal{L}_T$ will be called an **acceptable** set of frame conditions if its inverse image under $(\cdot)^q$ is acceptable in the sense of Definition 3.

Theorem 16 (Sahlqvist Completeness Theorem). *Let T be a weak-pullback preserving* **Set** *functor, let $q : S \twoheadrightarrow T$ be a presentation of T and let $\mathcal{L}_T$ and $\mathcal{L}_S$ be the ∇-language induced by S and T respectively. Assume that $C \subseteq \mathcal{L}_T$ is an acceptable set of coalgebraic Sahlqvist formulae, then $\mathsf{KKV}(T) + C$ is complete w.r.t. the class of T-coalgebra validating C.*

Proof. As is customary, we start with a formula $a \in \mathcal{L}_T$ which is $\mathsf{KKV}(T) + C$-consistent, and we will build a model for a in the class of T-coalgebras validating the coalgebraic frame conditions in C. The proof is in four steps.

Firstly, by using the contrapositive of Theorem 14, we know that since a is $\mathsf{KKV}(T) + C$-consistent, then there must exist a pre-image a' of a under $(\cdot)^q$ which is $\mathsf{KKV}(S) + C'$-consistent.

Secondly, since C is a set of coalgebraic Sahlqvist formulae we can apply Theorem 4 and conclude that there exists a model $\mathcal{M}_S$ based on an S-coalgebra (W, γ) which belongs to the class of coalgebraic frames validating the axioms of C' and such that a' is satisfied in $\mathcal{M}_S$, i.e. there exist $w \in W$ such that $\mathcal{M}_S, w \models a'$.

Thirdly, by Proposition 7 if we define $\mathcal{M}_T = Q(\mathcal{M}_S)$ then we have that since $\mathcal{M}_S, w \models a'$ and $(a')^q = a$, $\mathcal{M}_T, w \models a$.

Finally, we need to check that $\mathcal{M}_T$ is a coalgebraic frame validating the formulae in C. Assume that it is not, then there must exist $c \in C$ and $w \in W$ such that $\mathcal{M}_T, w \not\models c$. By the contrapositive of Proposition 7 this means that $\mathcal{M}_S, w \not\models c'$ for all c' in the pre-image of c under $(\cdot)^q$. But by Definition 15 and the second step of the proof we know that we must have $\mathcal{M}_S, w \models c'$ for any such c' and we therefore have a contradiction.

Example 17. We return to the bag functor $\mathcal{B}$ of Example 9 to illustrate what is, as far as we know, the first Sahlqvist completeness result for Graded Modal Logic (GML). An $\mathcal{L}_\mathcal{B}$-formula of the shape $\nabla n(a_1 : k_1, \ldots, a_m : k_m)$ is true at a point w if w has n successors, of which k_i satisfy a_i, for $1 \leq i \leq m$. Note that by construction ($\mathcal{B}$ is presented by List) our $\mathcal{B}$-coalgebras are finitely branching. For this example we will place ourselves in the predicate lifting style logic obtained by the trivial translation removing ∇ operators (see Example 1), and we rewrite $\nabla n(a_1 : k_1, \ldots, a_m : k_m)$ as $\langle n \rangle (a_1 : k_1, \ldots, a_m : k_m)$. For clarity's sake we define the following derived modal operators which are closer to the traditional operators of GML:

$$\Diamond_n p = \bigvee_{i=1}^{n} \langle n \rangle (p : i, \top : (n-i))$$

$$\Diamond_{\leq n} p = \bigvee_{i=1}^{n} \Diamond_n p$$

Thus $\Diamond_n p$ holds at w if w has n successors and p is true at (at least) one of them, whereas $\Diamond_{\leq n} p$ holds at w if w has at most n successors and p is true at (at least) one of them, i.e. p is true at at least one and at most n successors. Using these operators we can define graded versions of the most popular Sahlqvist frame conditions, for example 'transitivity for at most n successors':

$$(4_n): \quad \Diamond_{\leq n} \Diamond_{\leq n} p \rightarrow \Diamond_{\leq n} p$$

To see that (4_n) is Sahlqvist, note first that all the pre-images of $\langle n \rangle (p : i, \top : (n-i))$ under the translation map $(\cdot)^q$ introduced in Example 9 are of the shape (in the predicate-lifting style):

$$n(\pi(\underbrace{p, \ldots, p}_{i \text{ times}}, \underbrace{\top, \ldots, \top}_{(n-i) \text{ times}})) \tag{3}$$

for some $\pi \in \mathsf{Perm}(n)$. i.e. just an operator applied to some variables. So a pre-image of $\Diamond_n p$ is just a join of n formulae of the shape (3) for a choice of n permutations $\pi_i \in \mathsf{Perm}(n), 1 \leq i \leq n$ (or combinations of elements of this shape using meets and joins). In turn, the pre-images of $\Diamond_{\leq n} p$ are joins of n choices of pre-images of $\Diamond_i p, 1 \leq i \leq n$ (or combinations of elements of this shape using meets and joins). Thus the pre-images of the consequent of (4_n) are (strictly) positive. Similarly, the antecedent of (4_n) can be seen to be strictly positive and thus (4_n) is a Sahlqvist formula in the sense of Eq. (1) for any n.

Note that the cardinality restriction leads to a slightly counter-intuitive meaning for the axiom (4_n). Indeed, assume a point w has two successors, that one of these successors has three successors, one of which is the only state to satisfy p, then (4_2) holds, but transitivity doesn't. So (4_n) is transitivity for frames with branching degree at most n. To recover the usual notion of transitivity we need to consider the collection of Sahlqvist formulae $(4) = \{(4_n) \mid n \in \mathbb{N}\}$. It is clear that (4) is acceptable in the sense of Definition 3 and the basic GML + (4) is thus weakly complete w.r.t. finitely branching transitive frames.

Remark 18. We must make two important remarks about the previous example. First, the fact that $\mathcal{L}_{\mathcal{B}}$-formulas count the total number of successors points to an important difference with the traditional language for Graded Modal Logic $\mathcal{L}_{\mathsf{GML}}$ where a formula $\Diamond_k \phi$ is traditionally interpreted as 'ϕ holds at k distinct successors', leaving the total number of successors unspecified. Clearly we cannot express this in a finitary way in $\mathcal{L}_{\mathcal{B}}$, so our Sahlqvist formulae are expressed in a fragment of $\mathcal{L}_{\mathsf{GML}}$. But there is a translation tr from $\mathcal{L}_{\mathcal{B}}$ to $\mathcal{L}_{\mathsf{GML}}$ defined by

$$\mathsf{tr}(\nabla n(a_1 : k_1, \ldots, a_m : k_m)) = \Diamond_n \top \wedge \Box_{n+1} \bot \wedge \bigwedge_{i=1}^{m} \Diamond_{k_i} a_i$$

where $\Diamond_n \top \wedge \Box_{n+1} \bot$ just says that there are exactly n successors. Our second remark is that the $\Diamond_k$ modalities are algebraically ill-behaved as they do not distribute over joins, so there is no way of applying the theory of Sahlqvist formulae in BAOs to GML in the usual setting which may explain why we were unable to find any Sahlqvist completeness result for this logic in the literature.

Example 19. Our next example, is intended to show the relationship between our notion of Sahlqvist formula and the traditional one from relational modal logic. Here we will look at the finite powerset functor $\mathcal{P}_\omega$ which has a very simple presentation $q : \mathsf{List} \twoheadrightarrow \mathcal{P}_\omega$ given by

$$q_X : \mathsf{List}X \twoheadrightarrow \mathcal{P}_\omega X, (a_1, \ldots, a_n) \mapsto \{a_1, \ldots, a_n\}$$

The empty list is sent to the empty set. Thus, the pre-images of a $\mathcal{L}_{\mathcal{P}_\omega}$-formula of the type $\nabla\{a_1, \ldots, a_k\}$ are all the $\mathcal{L}_{\mathsf{List}}$-formula of the type $\nabla n(a'_1, \ldots, a'_n), n \geq k$ where $(a'_1, \ldots, a'_n)$ is any list containing all the elements of $\{a_1, \ldots, a_k\}$. Here we are in a slightly better position than in the graded case as there are semantic-preserving translations of the usual modal language $\mathcal{L}_{ML}$ in terms of $\Diamond$ and $\Box$ into $\mathcal{L}_{\mathcal{P}_\omega}$ and vice-versa (see [9]), in particular $\Diamond p$ is translated by $\nabla\{p, \top\}$ and $\Box p$ by $\nabla\emptyset \vee \nabla\{p\}$. We can check that the traditional Sahlqvist formulae as defined for example in [4] are also Sahlqvist formulae in the sense of this paper. Notice first that the pre-images under $(\cdot)^q$ of $\Diamond p$, or equivalently of $\nabla\{p, \top\}$, are of the shape $\nabla\alpha$ for an $\alpha \in \mathsf{List}\{p, \top\}$, or, in the predicate-lifting style, $\langle n \rangle \alpha$ with $\alpha \in (\{p, \top\})^n$ for some $n \in \mathbb{N}$. It is then quite straightforward to check that positive formulae in $\mathcal{L}_{ML}$ are translated into positive formulae in $\mathcal{L}_{\mathcal{P}_\omega}$ whose inverse images under $(\cdot)^q$ are also positive. This takes care of the consequent of Sahlqvist formulae. Now for the antecedent. As defined in [4], the antecedent must be built from $\bot, \top$, boxed atoms and negative formulae using $\wedge, \vee$ and $\Diamond$. Clearly, $\bot$ and $\top$ pose no problem. Negative $\mathcal{L}_{ML}$-formulae get mapped to negative $\mathcal{L}_{\mathcal{P}_\omega}$-formulae whose inverse image under $(\cdot)^q$ are also negative. The only potentially problematic building block are the boxed atoms. The formula $\Box p$ is translated to $\nabla\emptyset \vee \nabla\{p\}$ whose inverse images under $(\cdot)^q$ are of the shape (in the predicate lifting style) $\langle 0 \rangle \vee \langle n \rangle (p, \ldots, p)$, i.e. *strictly* positive terms. Clearly, the nesting of more boxes doesn't change this and so all inverse images of boxed atoms are strictly positive and we can therefore view them as part of the strictly positive term in Eq. (1) defining Sahlqvist antecedents.

6 Outlook

As illustrated by Example 17, there are instances of logics in the predicate-lifting style which can make statements that cannot be translated in the ∇-style (see [10]). We would like to extend our result to such logics, possibly by enriching the polynomial logics with operators that carry an infinitary meaning but remain algebraically well-behaved. We would also like to extend our result to richer coalgebraic logics such as coalgebraic μ-calculus (see [3] for recent advances in defining Sahlqvist formulae for the μ-calculus) and hybrid coalgebraic modal logic. Finally we would like to find examples and applications of our results to more logics such as probabilistic or coalition logics.

Acknowledgement. We are grateful to Clemens Kupke for his help on some of the finer points of ∇-style logics and to Alexander Kurz for hinting at some parts of [10] which were very useful. We would also like to thank the anonymous referees for their insightful and useful comments.

References

1. Abramsky, S.: Coalgebras, Chu Spaces, and Representations of Physical Systems. CoRR, abs/0910.3959 (2009)
2. Adámek, J., Gumm, H.P., Trnková, V.: Presentation of set functors: A coalgebraic perspective. J. Log. Comput. 20(5), 991–1015 (2010)
3. Bezhanishvili, N., Hodkinson, I.: Sahlqvist theorem for modal fixed point logic. Theoretical Computer Science 424, 1–19 (2012)
4. Blackburn, P., de Rijke, M., Venema, Y.: Modal Logic. Cambridge Tracts in Theoretical Computer Scie., vol. 53. Cambridge University Press (2001)
5. de Rijke, M., Venema, Y.: Sahlqvist's Theorem For Boolean Algebras With Operators With An Application To Cylindric Algebras. Studia Logica (1995)
6. Jónsson, B.: On the canonicity of Sahlqvist identities. Studia Logica 53(4), 473–492 (1994)
7. Jónsson, B., Tarski, A.: Boolean algebras with operators. part 1. Amer. J. Math. 33, 891–937 (1951)
8. Kupke, C., Pattinson, D.: Coalgebraic semantics of modal logics: an overview. Theoretical Computer Science 412(38), 5070–5094 (2011); Special issue CMCS 2010
9. Kupke, C., Kurz, A., Venema, Y.: Completeness for the coalgebraic cover modality. Logical Methods in Computer Science 8(3) (2012)
10. Kurz, A., Leal, R.: Modalities in the Stone age: A comparison of coalgebraic logics. In: MFPS XXV, Oxford (2009)
11. Pattinson, D., Schröder, L.: Beyond Rank 1: Algebraic Semantics and Finite Models for Coalgebraic Logics. In: Amadio, R. (ed.) FOSSACS 2008. LNCS, vol. 4962, pp. 66–80. Springer, Heidelberg (2008)
12. Rutten, J.J.M.M.: Rutten. Universal coalgebra: a theory of systems. Theor. Comput. Sci. 249(1), 3–80 (2000)
13. Schröder, L.: A Finite Model Construction for Coalgebraic Modal Logic. In: Aceto, L., Ingólfsdóttir, A. (eds.) FOSSACS 2006. LNCS, vol. 3921, pp. 157–171. Springer, Heidelberg (2006)
14. Venema, Y.: Algebras and coalgebras. In: van Benthem, J., Blackburn, P., Wolter, F. (eds.) Handbook of Modal Logic. Elsevier (2006)

Cut Elimination in Nested Sequents for Intuitionistic Modal Logics

Lutz Straßburger

INRIA Saclay Île-de-France — Équipe-projet Parsifal
École Polytechnique — LIX — Rue de Saclay — 91128 Palaiseau Cedex — France
http://www.lix.polytechnique.fr/Labo/Lutz.Strassburger/

Abstract. We present cut-free deductive systems without labels for the intuitionistic variants of the modal logics obtained by extending IK with a subset of the axioms d, t, b, 4, and 5. For this, we use the formalism of nested sequents, which allows us to give a uniform cut elimination argument for all 15 logic in the intuitionistic S5 cube.

1 Introduction

Intuitionistic modal logics are intuitionistic propositional logic extended with the modalities $\Box$ and $\Diamond$, obeying some variants of the k-axiom. Unlike for classical modal logic, there is no canonical choice, and many different versions of intuitionistic modal logics have been considered, e.g., [8,23,24,21,25,2,20]. For a survey see [25]. In this paper we consider the variant proposed in [24,21] and studied in detail by Simpson [25], namely, we add the following axioms to intuitionistic propositional logic:

$$
\begin{aligned}
\mathsf{k}_1 &: \Box(A \supset B) \supset (\Box A \supset \Box B) \\
\mathsf{k}_2 &: \Box(A \supset B) \supset (\Diamond A \supset \Diamond B) \\
\mathsf{k}_3 &: \Diamond(A \vee B) \supset (\Diamond A \vee \Diamond B) \\
\mathsf{k}_4 &: (\Diamond A \supset \Box B) \supset \Box(A \supset B) \\
\mathsf{k}_5 &: \neg\Diamond\bot
\end{aligned}
\qquad (1)
$$

In a classical setting the axioms k_2–k_5 would follow from k_1 and the De Morgan laws. Recently, researchers have also studied the variant which allows only k_1 and k_2, and which is sometimes called *constructive modal logic* (e.g., [1,18]). Since this leads to a different proof theory, it will not be discussed here. Independently from the chosen variant for the intuitionistic modal logic K, denoted by IK, one can add an arbitrary subset of the axioms d, t, b, 4, and 5, shown in Figure 1. As in the classical setting, this yields 15 different modal logics. In [25], Simpson presents labeled natural deduction and labeled sequent calculus systems for all of them. In [11], Galmiche and Salhi present label-free natural deduction systems for the ones not using the d-axiom. In this paper we present label-free sequent calculus systems for all 15 logics in the "intuitionistic modal cube" (shown in Figure 2), together with a uniform syntactic cut-elimination proof. For this we use nested sequents [14,3,22] (in a variant already used in [11]).

The motivation for this work is twofold. First, sequent calculus is much better suited for automated proof search than natural deduction, and second, label-free

F. Pfenning (Ed.): FOSSACS 2013, LNCS 7794, pp. 209–224, 2013.

d: $\Box A \supset \Diamond A$	$\forall w.\, \exists v.\, wRv$	(serial)
t: $(A \supset \Diamond A) \wedge (\Box A \supset A)$	$\forall w.\, wRw$	(reflexive)
b: $(A \supset \Box\Diamond A) \wedge (\Diamond\Box A \supset A)$	$\forall w.\, \forall v.\, wRv \supset vRw$	(symmetric)
$\mathsf{4}$: $(\Diamond\Diamond A \supset \Diamond A) \wedge (\Box A \supset \Box\Box A)$	$\forall w.\, \forall v.\, \forall u.\, wRv \wedge vRu \supset wRu$	(transitive)
$\mathsf{5}$: $(\Diamond A \supset \Box\Diamond A) \wedge (\Diamond\Box A \supset \Box A)$	$\forall w.\, \forall v.\, \forall u.\, wRv \wedge wRu \supset vRu$	(euclidean)

Fig. 1. Intuitionistic modal axioms d, t, b, 4, 5, with corresponding frame conditions

systems make it easier to study the theory of proof search and proof normalization. In fact, the sequent systems together with the cut-reduction procedure presented in this paper are the basis for ongoing research on the following two questions: (i) Is it possible to design a focussed system [16,5,17] yielding new normal forms for cut-free proofs and providing proof search mechanisms based on forward-chaining (program-directed search) and backward-chaining (goal-directed search) for intuitionistic modal logics? (ii) Can we give a term calculus (based on the λ-calculus in the style of [19]) for proofs, in order to provide a Curry-Howard-correspondence for intuitionistic modal logics (and not just the constructive modal logics mentioned above)?

There is a close relationship between the labeled and the label-free natural deduction systems of [25] and [11]. In fact, modulo the correspondence between (tree-)labeled systems and nested sequents [10], the basic systems for IK of [25] and [11] are identical. A similar correspondence can be observed between the labeled sequent systems of [25] and our systems, when restricted to the logic IK. However, the rules dealing with the axioms d, t, b, 4, and 5 are very different from [25]. The shape of these rules is crucial for the internal cut-elimination proof.

Furthermore, note that our treatment of the "intuitionistic" in nested sequents is different from the one in [9] (which is two-sided inside each nesting and does not treat modalities), and the one in [13], (which focuses on variants of bi-intuitionistic tense logics, and does not cover all 15 logics in the IS5-cube).

2 Preliminaries

The formulas of intuitionistic modal logic (IML) are generated by:

$$\mathcal{M} ::= \mathcal{A} \mid \bot \mid \mathcal{M} \wedge \mathcal{M} \mid \mathcal{M} \vee \mathcal{M} \mid \mathcal{M} \supset \mathcal{M} \mid \Box\mathcal{M} \mid \Diamond\mathcal{M} \tag{2}$$

where $\mathcal{A} = \{a, b, c, \ldots\}$ is a countable set of *propositional variables* (or *atoms*). We use $A, B, C, \ldots$ to denote formulas. Negation of formulas is defined as $\neg A = A \supset \bot$. The theorems of the intuitionistic modal logic IK are exactly those formulas that are derivable from the axioms of intuitionistic propositional logic and the axioms k_1–k_5 shown in (1) via the rules mp and nec shown below:

$$\mathsf{mp}\,\frac{A \quad A \supset B}{B} \qquad\qquad \mathsf{nec}\,\frac{A}{\Box A} \tag{3}$$

In the following, we recall the *birelational models* [21,7] for IML, which are a combination of the Kripke semantics for propositional intuitionistic logic and the one for classical modal logic. A *frame* $\langle W, \leq, R\rangle$ is a non-empty set W of *worlds* together with two binary relations $\leq, R \subseteq W \times W$, where $\leq$ is a pre-order (i.e., reflexive and transitive), such that the following two conditions hold

(F1) For all worlds w, v, v', if wRv and $v \leq v'$, then there is a w' such that $w \leq w'$ and $w'Rv'$.

(F2) For all worlds w', w, v, if $w \leq w'$ and wRv, then there is a v' such that $w'Rv'$ and $v \leq v'$.

These two conditions can be visualized as follows:

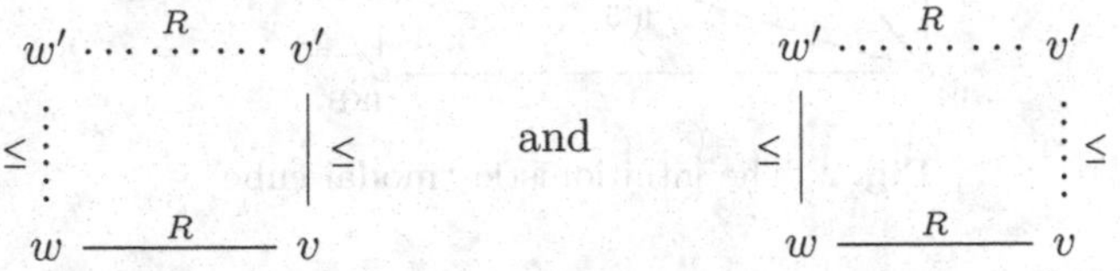

A *model* $\mathfrak{M}$ is a quadruple $\langle W, \leq, R, V\rangle$, where $\langle W, \leq, R\rangle$ is a frame, and V, called the *valuation*, is a monotone function $\langle W, \leq\rangle \to \langle 2^{\mathcal{A}}, \subseteq\rangle$ from the set of worlds to the set of subsets of propositional variables, mapping a world w to the set of propositional variables which are true in w. We write $w \Vdash a$ if $a \in V(w)$. The relation $\Vdash$ is extended to all formulas as follows:

$$
\begin{array}{lll}
w \Vdash A \wedge B & \text{iff} & w \Vdash A \text{ and } w \Vdash B \\
w \Vdash A \vee B & \text{iff} & w \Vdash A \text{ or } w \Vdash B \\
w \Vdash A \supset B & \text{iff} & \text{for all } w' \geq w\colon\ w' \Vdash A \text{ implies } w' \Vdash B \\
w \Vdash \Box A & \text{iff} & \text{for all } w', v' \in W\colon\ \text{if } w' \geq w \text{ and } w'Rv' \text{ then } v' \Vdash A \\
w \Vdash \Diamond A & \text{iff} & \text{there is a } v \in W \text{ such that } wRv \text{ and } v \Vdash A
\end{array}
\qquad (4)
$$

We write $w \nVdash A$ if $w \Vdash A$ does not hold. In particular, note that $w \nVdash \bot$ for all worlds, and that we do *not* have that $w \Vdash \neg A$ iff $w \nVdash A$. However, we get the monotonicity property:

Lemma 2.1 (Monotonicity). *If $w \leq w'$ and $w \Vdash A$ then $w' \Vdash A$.*

Proof. By induction on A, using (4), (F1), and the monotonicity of V. □

We say that a formula A *is valid in a model* $\mathfrak{M} = \langle W, \leq, R, V\rangle$, denoted by $\mathfrak{M} \Vdash A$, if for all $w \in W$ we have $w \Vdash A$. A formula A *is valid in a frame* $\langle W, \leq, R\rangle$, denoted by $\langle W, \leq, R\rangle \Vdash A$, if for all valuations V, we have $\langle W, \leq, R, V\rangle \Vdash A$. Finally, we say a formula is *valid*, if it is valid in all frames. As for classical modal logics, we can consider the axioms $\{\mathsf{d}, \mathsf{t}, \mathsf{b}, \mathsf{4}, \mathsf{5}\}$, whose intuitionistic versions are shown in Figure 1, and that we can add to the logic IK. For $\mathsf{X} \subseteq \{\mathsf{d}, \mathsf{t}, \mathsf{b}, \mathsf{4}, \mathsf{5}\}$ a frame is called an X*-frame* if the relation R obeys the corresponding frame conditions, which are also shown in Figure 1. For example, a $\{\mathsf{b}, \mathsf{4}\}$-frame is one in which R is symmetric and transitive. The following theorem is well-known:

Theorem 2.2. *A formula is derivable from $\mathsf{IK}+\mathsf{X}$ iff it is valid in all X-frames.*

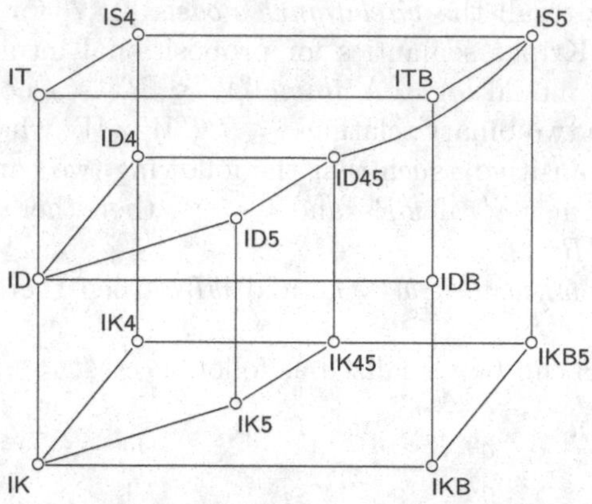

Fig. 2. The intuitionistic "modal cube"

Remark 2.3. Note that we do not have a true correspondence as for classical modal logics. For example, if t is valid in a frame $\langle W, \leq, R\rangle$ then R does not need to be reflexive (see [25,21] for more details).

We will say a formula is X-*valid* iff it is valid in all X-frames. As in classical modal logic, we can, *a priori*, define 32 modal logics with the 5 axioms in Figure 1. But many of them coincide, for example, $\mathsf{IK} + \{\mathsf{t}, \mathsf{b}, \mathsf{4}\}$ and $\mathsf{IK} + \{\mathsf{t}, \mathsf{5}\}$ yield the same logic, called IS5. There are, in fact, 15 different logics, which are shown in Figure 2, the intuitionistic version of the "modal cube" [12].

3 Nested Sequents for Intuitionistic Modal Logics

Let us now turn to nested sequents for IML. The data structure of a nested sequent for intuitionistic modal logics that we employ here has already been used in [11] and is almost the same as for classical modal logics [3,4]: it is a tree whose nodes are multisets of formulas. The only difference is that in the intuitionistic case exactly one formula occurrence in the whole tree is special. We will mark it with a white circle $\circ$, while all other formulas are marked with a black circle $\bullet$. One can see this marking as a polarity assignment: $\bullet$ for *input polarity*, and $\circ$ for *output polarity*.[1] Formally, nested sequents for IML are generated by the grammar (where n and k can both be zero):

$$\Gamma ::= \Lambda, \Pi \qquad \Lambda ::= A_1^\bullet, \ldots, A_n^\bullet, [\Lambda_1], \ldots, [\Lambda_k] \qquad \Pi ::= A^\circ \mid [\Gamma] \tag{5}$$

Thus, a nested sequent consists of two parts: an *LHS-sequent* (denoted by Λ), in which all formulas have input polarity, and an *RHS-sequent* (denoted by Π), which is either a formula with output polarity or a bracketed sequent. A sequent of the shape as Γ in (5) is called a *full sequent*. The letters Δ and Σ

[1] We avoid the use of the "positive/negative" terminology because it is overloaded. For a thorough investigation into polarities as they are used here, see [15].

can stand for full sequents as well as LHS-sequents, depending on the context. Note that any RHS-sequent is also a full sequent, but not the other way around. As usual, we allow sequents to be empty, and we consider sequents to be equal modulo associativity and commutativity of the comma. Sometimes we write $\emptyset$ to denote the empty multiset, allowing us to write $[\emptyset]$, which is a well-formed LHS-sequent. If we forget the polarities, a nested sequent is of the shape $\Gamma = A_1, \ldots, A_k, [\Gamma_1], \ldots, [\Gamma_n]$.

The *corresponding formula* of a nested sequent is defined as follows:

$$\begin{aligned} fm(\Lambda, \Pi) &= fm(\Lambda) \supset fm(\Pi) \\ fm(A_1^\bullet, \ldots, A_n^\bullet, [\Lambda_1], \ldots, [\Lambda_k]) &= A_1 \wedge \cdots \wedge A_n \wedge \Diamond fm(\Lambda_1) \wedge \cdots \wedge \Diamond fm(\Lambda_k) \\ fm(A^\circ) &= A \\ fm([\Gamma]) &= \Box fm(\Gamma) \end{aligned}$$

We say a sequent is X-*valid* if its corresponding formula is.

As in the case of classical modal logics, we need the notion of *context* which is a nested sequent with a hole $\{\ \}$, taking the place of a formula. Since we have two polarities, input and output, there are also two kinds of contexts: *input contexts*, whose holes have to be filled with an input formula for obtaining a full sequent, and *output contexts*, whose holes have to be filled with an output formula for obtaining a full sequent. We also allow the holes in a context to be filled with sequents and not just formulas.

We define the *depth* of a context inductively as follows:

$$\begin{aligned} depth(\{\ \}) &= 0 \\ depth(\Delta, \Gamma\{\ \}) &= depth(\Gamma\{\ \}) \\ depth([\Gamma\{\ \}]) &= 1 + depth(\Gamma\{\ \}) \end{aligned}$$

Example 3.1. *Let $\Gamma_1\{\ \} = C^\bullet, [\{\ \}, [B^\bullet, C^\bullet]]$ and $\Delta_1 = A^\bullet, [B^\circ]$ and $\Gamma_2\{\ \} = C^\bullet, [\{\ \}, [B^\bullet, C^\circ]]$ and $\Delta_2 = A^\bullet, [B^\bullet]$. Then $depth(\Gamma_1\{\ \}) = depth(\Gamma_2\{\ \}) = 1$. Furthermore, $\Gamma_1\{\Delta_2\}$ and $\Gamma_2\{\Delta_1\}$ are not well-formed full sequents, because the former would contain no output formula, and the latter would contain two. However, we can form $\Gamma_1\{\Delta_1\} = C^\bullet, [A^\bullet, [B^\circ], [B^\bullet, C^\bullet]]$ and $\Gamma_2\{\Delta_2\} = C^\bullet, [A^\bullet, [B^\bullet], [B^\bullet, C^\circ]]$. Their corresponding formulas are $fm(\Gamma_1\{\Delta_1\}) = C \supset \Box(A \wedge \Diamond(B \wedge C) \supset \Box B)$ and $fm(\Gamma_2\{\Delta_2\}) = C \supset \Box(A \wedge \Diamond B \supset \Box(B \supset C))$, respectively.*

Observation 3.2. Note that every output context $\Gamma\{\ \}$ is of the shape

$$\Lambda_1, [\Lambda_2, [\ldots, [\Lambda_n, \{\ \}] \ldots]] \tag{6}$$

for some $n \geq 0$, where all Λ_i are LHS-sequents. Filling the hole of an output context with a full sequent yields a full sequent, and filling it with an LHS-sequent yields an LHS-sequent. Every input context $\Gamma\{\ \}$ is of the shape $\Gamma'\{\Lambda\{\ \}, \Pi\}$ where $\Gamma'\{\ \}$ and $\Lambda\{\ \}$ are output contexts (i.e., are of the shape (6) above) and Π is a RHS-sequent. Furthermore, $\Gamma'\{\ \}$ and $\Lambda\{\ \}$ and Π are uniquely defined by the position of the hole $\{\ \}$ in $\Gamma\{\ \}$.

$$\bot^\bullet \frac{}{\Gamma\{\bot^\bullet\}} \qquad \mathsf{id} \frac{}{\Gamma\{a^\bullet, a^\circ\}}$$

$$\wedge^\bullet \frac{\Gamma\{A^\bullet, B^\bullet\}}{\Gamma\{A \wedge B^\bullet\}} \qquad \wedge^\circ \frac{\Gamma\{A^\circ\} \quad \Gamma\{B^\circ\}}{\Gamma\{A \wedge B^\circ\}}$$

$$\vee^\bullet \frac{\Gamma\{A^\bullet\} \quad \Gamma\{B^\bullet\}}{\Gamma\{A \vee B^\bullet\}} \qquad \vee^\circ \frac{\Gamma\{A^\circ\}}{\Gamma\{A \vee B^\circ\}} \qquad \vee^\circ \frac{\Gamma\{B^\circ\}}{\Gamma\{A \vee B^\circ\}}$$

$$\supset^\bullet \frac{\Gamma^\downarrow\{A \supset B^\bullet, A^\circ\} \quad \Gamma\{B^\bullet\}}{\Gamma\{A \supset B^\bullet\}} \qquad \supset^\circ \frac{\Gamma\{A^\bullet, B^\circ\}}{\Gamma\{A \supset B^\circ\}}$$

$$\Box^\bullet \frac{\Gamma\{\Box A^\bullet, [A^\bullet, \Delta]\}}{\Gamma\{\Box A^\bullet, [\Delta]\}} \qquad \Box^\circ \frac{\Gamma\{[A^\circ]\}}{\Gamma\{\Box A^\circ\}}$$

$$\Diamond^\bullet \frac{\Gamma\{[A^\bullet]\}}{\Gamma\{\Diamond A^\bullet\}} \qquad \Diamond^\circ \frac{\Gamma\{[A^\circ, \Delta]\}}{\Gamma\{\Diamond A^\circ, [\Delta]\}}$$

Fig. 3. System NIK

We can chose to fill the hole of a context $\Gamma\{\ \}$ with nothing, which means we simply remove the $\{\ \}$. This is denoted by $\Gamma\{\emptyset\}$. In Example 3.1 above, $\Gamma_1\{\emptyset\} = C^\bullet, [[B^\bullet, C^\bullet]]$ is an LHS-sequent and $\Gamma_2\{\emptyset\} = C^\bullet, [[B^\bullet, C^\circ]]$ is a full sequent. More generally, whenever $\Gamma\{\emptyset\}$ is a full sequent, then $\Gamma\{\ \}$ is an input context. Sometimes we also need a context with many holes, denoted by $\Gamma\{\ \}\cdots\{\ \}$.

Definition 3.3. For every input context $\Gamma\{\ \}$ (resp. full sequent Δ), we define its *output pruning* $\Gamma^\downarrow\{\ \}$ (resp. $\Delta^\downarrow$) to be the same context (resp. sequent) with the unique output formula removed. Thus, $\Gamma^\downarrow\{\ \}$ is an output context (resp. $\Delta^\downarrow$ is an LHS-sequent). If $\Gamma\{\ \}$ is already an output context (resp. if Δ is already an LHS-sequent), then $\Gamma^\downarrow\{\ \} = \Gamma\{\ \}$ (resp. $\Delta^\downarrow = \Delta$).

We are now ready to see the inference rules. Figure 3 shows system NIK, a nested sequent system for intuitionistic modal logic IK. There are more rules than in the classical version [3] because for each connective we need two rules, one for the input polarity, and one for the output polarity. Note how the $\supset^\bullet$-rule makes use of the output pruning. This is necessary because we allow only one output formula in the sequent. Without this restriction, we would collapse into the classical case.

In the course of this paper we will make use of the additional structural rules

$$\mathsf{nec}^{[]} \frac{\Gamma}{[\Gamma]} \quad \mathsf{w} \frac{\Gamma\{\emptyset\}}{\Gamma\{\Lambda\}} \quad \mathsf{c} \frac{\Gamma\{A^\bullet, A^\bullet\}}{\Gamma\{A^\bullet\}} \quad \mathsf{m}^{[]} \frac{\Gamma\{[\Delta_1], [\Delta_2]\}}{\Gamma\{[\Delta_1, \Delta_2]\}} \quad \mathsf{cut} \frac{\Gamma^\downarrow\{A^\circ\} \quad \Gamma\{A^\bullet\}}{\Gamma\{\emptyset\}} \tag{7}$$

called *necessitation*, *weakening*, *contraction*, *box-medial*, and *cut*, respectively. These rules are not part of the system, but we will see later that they are all admissible. Note that in the weakening rule Λ has to be an LHS-sequent, and the contraction rule can only be applied to input formulas. For the $\mathsf{m}^{[]}$-rule it is not relevant where in $\Gamma\{[\Delta_1, \Delta_2]\}$ the output formula is located. The cut-rule makes use of the output pruning, in the same way as the $\supset^\bullet$-rule. Explicit

$$\mathsf{d}^\circ\,\frac{\Gamma\{[A^\circ]\}}{\Gamma\{\Diamond A^\circ\}} \qquad \mathsf{t}^\circ\,\frac{\Gamma\{A^\circ\}}{\Gamma\{\Diamond A^\circ\}} \qquad \mathsf{b}^\circ\,\frac{\Gamma\{[\Delta], A^\circ\}}{\Gamma\{[\Delta, \Diamond A^\circ]\}}$$

$$\mathsf{d}^\bullet\,\frac{\Gamma\{\Box A^\bullet, [A^\bullet]\}}{\Gamma\{\Box A^\bullet\}} \qquad \mathsf{t}^\bullet\,\frac{\Gamma\{\Box A^\bullet, A^\bullet\}}{\Gamma\{\Box A^\bullet\}} \qquad \mathsf{b}^\bullet\,\frac{\Gamma\{[\Delta, \Box A^\bullet], A^\bullet\}}{\Gamma\{[\Delta, \Box A^\bullet]\}}$$

$$\mathsf{4}^\circ\,\frac{\Gamma\{[\Diamond A^\circ, \Delta]\}}{\Gamma\{\Diamond A^\circ, [\Delta]\}} \qquad \mathsf{5}^\circ\,\frac{\Gamma\{\emptyset\}\{\Diamond A^\circ\}}{\Gamma\{\Diamond A^\circ\}\{\emptyset\}}\ \mathrm{depth}(\Gamma\{\ \}\{\emptyset\}) > 0$$

$$\mathsf{4}^\bullet\,\frac{\Gamma\{\Box A^\bullet, [\Box A^\bullet, \Delta]\}}{\Gamma\{\Box A^\bullet, [\Delta]\}} \qquad \mathsf{5}^\bullet\,\frac{\Gamma\{\Box A^\bullet\}\{\Box A^\bullet\}}{\Gamma\{\Box A^\bullet\}\{\emptyset\}}\ \mathrm{depth}(\Gamma\{\ \}\{\emptyset\}) > 0$$

Fig. 4. Intuitionistic $\Diamond^\circ$- and $\Box^\bullet$-rules for the axioms d, t, b, 4, and 5

contraction is not needed in NIK because contraction is implicitly present in the $\supset^\bullet$- and $\Box^\bullet$-rules [6]. Note that the id-rule applies only to atomic formulas. But as usual with sequent style system, the general form is derivable:

Proposition 3.4. *The rule* $\mathsf{id}\,\dfrac{}{\Gamma\{A^\bullet, A^\circ\}}$ *is derivable in* NIK.

Figure 4 shows the intuitionistic versions for the rules for the axioms d, t, b, 4, and 5. They are almost the same as the corresponding rules in the classical case [3]. The only difference is that here we need two rules for each axiom: a $\Diamond^\circ$-rule and a $\Box^\bullet$-rule. Note that contraction is implicitly present in the $\Box^\bullet$-rules but not in the $\Diamond^\circ$-rules. For a subset $\mathsf{X} \subseteq \{\mathsf{d}, \mathsf{t}, \mathsf{b}, \mathsf{4}, \mathsf{5}\}$, we denote by $\mathsf{X}^\bullet$ and X° the corresponding sets of $\Box^\bullet$-rules and $\Diamond^\circ$-rules, respectively.

4 Soundness

In this section we will show that all rules presented in Figures 3 and 4 are indeed sound. More precisely, we prove the following theorem:

Theorem 4.1. *Let* $\mathsf{X} \subseteq \{\mathsf{d}, \mathsf{t}, \mathsf{b}, \mathsf{4}, \mathsf{5}\}$, *and let* $\mathsf{r}\,\dfrac{\Gamma_1 \ \ \dots \ \ \Gamma_n}{\Gamma}$ *(for* $n \in \{0, 1, 2\}$*) be an instance of a rule in* $\mathsf{NIK} + \mathsf{X}^\bullet + \mathsf{X}^\circ$. *Then:*

(i) the formula $\mathit{fm}(\Gamma_1) \wedge \cdots \wedge \mathit{fm}(\Gamma_n) \supset \mathit{fm}(\Gamma)$ *is* X*-valid, and*
(ii) whenever a sequent Γ *is provable in* $\mathsf{NIK} + \mathsf{X}^\bullet + \mathsf{X}^\circ$, *then* Γ *is* X*-valid.*

Clearly, (ii) follows almost immediately from (i). But for proving (i), we need a series of lemmas. We begin by showing that the deep inference principle used in all rules is sound.

Lemma 4.2. *Let* $\mathsf{X} \subseteq \{\mathsf{d}, \mathsf{t}, \mathsf{b}, \mathsf{4}, \mathsf{5}\}$, *and let A, B, and C be formulas.*

(i) If $A \supset B$ *is* X*-valid, then so is* $(C \supset A) \supset (C \supset B)$.
(ii) If $A \supset B$ *is* X*-valid, then so is* $\Box A \supset \Box B$.
(iii) If $A \supset B$ *is* X*-valid, then so is* $(C \wedge A) \supset (C \wedge B)$.
(iv) If $A \supset B$ *is* X*-valid, then so is* $\Diamond A \supset \Diamond B$.
(v) If $A \supset B$ *is* X*-valid, then so is* $(B \supset C) \supset (A \supset C)$.

Proof. This follows immediately from (4) and Lemma 2.1. □

Lemma 4.3. *Let* $\mathsf{X} \subseteq \{\mathsf{d}, \mathsf{t}, \mathsf{b}, 4, 5\}$, *let* Δ *and* Σ *be full sequents, and let* $\Gamma\{\ \}$ *be an output context. If* $\mathit{fm}(\Delta) \supset \mathit{fm}(\Sigma)$ *is* X*-valid, then so is* $\mathit{fm}(\Gamma\{\Delta\}) \supset \mathit{fm}(\Gamma\{\Sigma\})$.

Proof. Induction on $\Gamma\{\ \}$ (see Obs. 3.2), using Lemma 4.2.(i) and (ii). □

Lemma 4.4. *Let* $\mathsf{X} \subseteq \{\mathsf{d}, \mathsf{t}, \mathsf{b}, 4, 5\}$, *let* Δ *and* Σ *be LHS-sequents, and* $\Gamma\{\ \}$ *an input context. If* $\mathit{fm}(\Sigma) \supset \mathit{fm}(\Delta)$ *is* X*-valid, then so is* $\mathit{fm}(\Gamma\{\Delta\}) \supset \mathit{fm}(\Gamma\{\Sigma\})$.

Proof. By Observation 3.2, we have that $\Gamma\{\ \} = \Gamma'\{\Lambda\{\ \}, \Pi\}$ for some $\Gamma'\{\ \}$ and $\Lambda\{\ \}$ and Π. By induction on $\Lambda\{\ \}$, using Lemma 4.2.(iii) and (iv), we get that $\mathit{fm}(\Lambda\{\Sigma\}) \supset \mathit{fm}(\Lambda\{\Delta\})$ is X-valid. From Lemma 4.2.(v) it then follows that $(\mathit{fm}(\Lambda\{\Delta\}) \supset \mathit{fm}(\Pi)) \supset (\mathit{fm}(\Lambda\{\Sigma\}) \supset \mathit{fm}(\Pi))$, i.e., $\mathit{fm}(\Lambda\{\Delta\}, \Pi) \supset \mathit{fm}(\Lambda\{\Sigma\}, \Pi)$ is X-valid. Now the statement follows from Lemma 4.3. □

Lemma 4.5. *Let* $\mathsf{X} \subseteq \{\mathsf{d}, \mathsf{t}, \mathsf{b}, 4, 5\}$. *Then any full sequent of the shape* $\Gamma\{a^\bullet, a^\circ\}$ *or* $\Gamma\{\bot^\bullet\}$ *is* X*-valid.*

Proof. If a formula A is X-valid, then so are $\Box A$ and $C \supset A$ for an arbitrary formula C. Since $a \supset a$ is trivially X-valid, the validity of $\Gamma\{a^\bullet, a^\circ\}$ follows by induction on $\Gamma\{\ \}$ (which is of shape (6)). For $\Gamma\{\bot^\bullet\}$, note that this sequent is of shape $\Gamma'\{\Lambda\{\bot^\bullet\}, \Pi\}$ (by Observation 3.2). By an easy induction on $\Lambda\{\ \}$, we can can show that $\mathit{fm}(\Lambda\{\bot^\bullet\}) \supset \bot$ is X-valid. Since $\bot \supset A$ is X-valid for any formula A, we can conclude that $\mathit{fm}(\Lambda\{\bot^\bullet\}) \supset \mathit{fm}(\Pi)$ is X-valid, and therefore $\mathit{fm}(\Lambda\{\bot^\bullet\}, \Pi)$. Now, X-validity of $\Gamma\{\bot^\bullet\}$ follows by induction on $\Gamma'\{\ \}$. □

Lemma 4.6. *Let* $\mathsf{X} \subseteq \{\mathsf{d}, \mathsf{t}, \mathsf{b}, 4, 5\}$, *and let* $\mathsf{r}\,\dfrac{\Gamma_1}{\Gamma_2}$ *be an instance of* w, c, $\mathsf{m}^{[]}$, $\vee^\circ$, $\Box^\circ$, $\Diamond^\circ$, $\supset^\circ$, $\wedge^\bullet$, $\Diamond^\bullet$, *or* $\Box^\bullet$. *Then* $\mathit{fm}(\Gamma_1) \supset \mathit{fm}(\Gamma_2)$ *is* X*-valid.*

Proof. For the rules $\vee^\circ$, $\Box^\circ$, $\Diamond^\circ$, $\supset^\circ$ this follows immediately from Lemma 4.3, where for $\Diamond^\circ$ we need the k_2-axiom. For the rules $\wedge^\bullet$, $\Diamond^\bullet$, w, and c, the lemma follows immediately from Lemma 4.4. The $\Box^\bullet$-rule can be decomposed into c and the rule $\tilde{\Box}^\bullet\,\dfrac{\Gamma\{[A^\bullet, \Delta]\}}{\Gamma\{\Box A^\bullet, [\Delta]\}}$, for which we need a case distinction: If the output formula occurs inside Δ, then we use the validity of axiom k_1 and Lemma 4.3. If the output formula occurs inside $\Gamma\{\ \}$, then we need the validity of the formula $(\Box A \wedge \Diamond B) \supset \Diamond(A \wedge B)$ for all A and B. This can easily be shown using the definition of $\Vdash$. Then the lemma follows from Lemma 4.4. Finally, for the $\mathsf{m}^{[]}$-rule we also make a case distinction: If the output formula is inside $\Gamma\{\ \}$, we need the validity of the formula $\Diamond(A \wedge B) \supset \Diamond A \wedge \Diamond B$ for all A and B, which can easily be shown using the definition of $\Vdash$. Then the the statement of the lemma follows from Lemma 4.4. If the output formula occurs inside Δ_1 or Δ_2, then we use the validity of axiom k_4 and Lemma 4.3. □

Consider now the rules in Fig. 5, which are special cases of the rules 5° and $5^\bullet$.

Proposition 4.7. *The rule* 5° *is derivable in* $\{5_1^\circ, 5_2^\circ, 5_3^\circ\}$, *and the rule* $5^\bullet$ *is derivable in* $\{5_1^\bullet, 5_2^\bullet, 5_3^\bullet, \mathsf{c}\}$.

$$5_1^\circ\ \frac{\Gamma\{[\Delta],\Diamond A^\circ\}}{\Gamma\{[\Delta,\Diamond A^\circ]\}} \qquad 5_2^\circ\ \frac{\Gamma\{[\Delta],[\Diamond A^\circ,\Sigma]\}}{\Gamma\{[\Delta,\Diamond A^\circ],[\Sigma]\}} \qquad 5_3^\circ\ \frac{\Gamma\{[\Delta,[\Diamond A^\circ,\Sigma]]\}}{\Gamma\{[\Delta,\Diamond A^\circ,[\Sigma]]\}}$$

$$5_1^\bullet\ \frac{\Gamma\{[\Delta],\Box A^\bullet\}}{\Gamma\{[\Delta,\Box A^\bullet]\}} \qquad 5_2^\bullet\ \frac{\Gamma\{[\Delta],[\Box A^\bullet,\Sigma]\}}{\Gamma\{[\Delta,\Box A^\bullet],[\Sigma]\}} \qquad 5_3^\bullet\ \frac{\Gamma\{[\Delta,[\Box A^\bullet,\Sigma]]\}}{\Gamma\{[\Delta,\Box A^\bullet,[\Sigma]]\}}$$

Fig. 5. Variants of the rules for the 5-axiom

Proof. The rule 5° allows to move an output $\Diamond^\circ$-formula from anywhere in the sequent tree, except the root, to any other place in the sequent tree. The same can be achieved with the rules $5_1^\circ, 5_2^\circ, 5_3^\circ$, and similarly for $5^\bullet$. □

Lemma 4.8. *Let* $\mathsf{X} \subseteq \{\mathsf{d},\mathsf{t},\mathsf{b},\mathsf{4},\mathsf{5}\}$, *let* $\mathsf{x} \in \mathsf{X}$, *and let* $\mathsf{r}\,\frac{\Gamma_1}{\Gamma_2}$ *be an instance of* x° *or* $\mathsf{x}^\bullet$. *Then* $fm(\Gamma_1) \supset fm(\Gamma_2)$ *is* X*-valid.*

Proof. For the rules d°, t°, b°, and $\mathsf{4}^\circ$ this follows immediately from Lemma 4.3 and the validity of the corresponding axioms, shown in Fig. 1 (note that b° can be decomposed into $\mathsf{m}^{[]}$ and $\tilde{\mathsf{b}}^\circ\ \frac{\Gamma\{A^\circ\}}{\Gamma\{[\Diamond A^\circ]\}}$, and $\mathsf{4}^\circ$ into $\Diamond^\circ$ and $\tilde{\mathsf{4}}^\circ\ \frac{\Gamma\{\Diamond\Diamond A^\circ\}}{\Gamma\{\Diamond A^\circ\}}$). For 5° we use Proposition 4.7, where soundness of 5_1°, 5_2°, and 5_3° is shown as for b° and $\mathsf{4}^\circ$ (using that axiom 5 implies $\Diamond\cdots\Diamond A \supset \Box\Diamond A$). For the rules $\mathsf{d}^\bullet$, $\mathsf{t}^\bullet$, $\mathsf{b}^\bullet$, $\mathsf{4}^\bullet$, and $5^\bullet$ we proceed similarly, using soundness of the c-rule and Lemma 4.4 instead of Lemma 4.3. □

Let us now turn to showing the soundness of the branching rules $\wedge^\circ$, $\vee^\bullet$, $\supset^\bullet$, and cut. For this, we start with the binary versions of Lemmas 4.2, 4.3, and 4.4.

Lemma 4.9. *Let* $\mathsf{X} \subseteq \{\mathsf{d},\mathsf{t},\mathsf{b},\mathsf{4},\mathsf{5}\}$, *and let* A, B, C, *and* D *be formulas.*
(i) If $A \wedge B \supset C$ *is* X*-valid, then so is* $(D \supset A) \wedge (D \supset B) \supset (D \supset C)$.
(ii) If $A \wedge B \supset C$ *is* X*-valid, then so is* $\Box A \wedge \Box B \supset \Box C$.
(iii) If $C \supset A \vee B$ *is* X*-valid, then so is* $(D \wedge C) \supset (D \wedge A) \vee (D \wedge B)$.
(iv) If $C \supset A \vee B$ *is* X*-valid, then so is* $\Diamond C \supset \Diamond A \vee \Diamond B$.
(v) If $C \supset A \vee B$ *is* X*-valid, then so is* $(A \supset D) \wedge (B \supset D) \supset (C \supset D)$.

Proof. As Lemma 4.2, this follows immediately from (4) and Lemma 2.1. □

Lemma 4.10. *Let* $\mathsf{X} \subseteq \{\mathsf{d},\mathsf{t},\mathsf{b},\mathsf{4},\mathsf{5}\}$, *let* Δ_1, Δ_2, *and* Σ *be full sequents, and let* $\Gamma\{\ \}$ *be an output context. If* $fm(\Delta_1) \wedge fm(\Delta_2) \supset fm(\Sigma)$ *is* X*-valid, then so is* $fm(\Gamma\{\Delta_1\}) \wedge fm(\Gamma\{\Delta_2\}) \supset fm(\Gamma\{\Sigma\})$.

Proof. Induction on $\Gamma\{\ \}$, using Lemma 4.9.(i) and (ii). □

Lemma 4.11. *Let* $\mathsf{X} \subseteq \{\mathsf{d},\mathsf{t},\mathsf{b},\mathsf{4},\mathsf{5}\}$, *let* Δ_1, Δ_2, *and* Σ *be LHS-sequents, and let* $\Gamma\{\ \}$ *be an input context. If* $fm(\Sigma) \supset fm(\Delta_1) \vee fm(\Delta_2)$ *is* X*-valid, then so is* $fm(\Gamma\{\Delta_1\}) \wedge fm(\Gamma\{\Delta_2\}) \supset fm(\Gamma\{\Sigma\})$.

Proof. By Observation 3.2, we have that $\Gamma\{\ \} = \Gamma'\{\Lambda\{\ \},\Pi\}$ for some $\Gamma'\{\ \}$ and $\Lambda\{\ \}$ and Π. By induction on $\Lambda\{\ \}$, using Lemma 4.9.(iii) and (iv), we get that $fm(\Lambda\{\Sigma\}) \supset fm(\Lambda\{\Delta_1\}) \vee fm(\Lambda\{\Delta_2\})$ is X-valid. From Lemma 4.9.(v) it then follows that $fm(\Lambda\{\Delta_1\},\Pi) \wedge fm(\Lambda\{\Delta_2,\Pi\}) \supset fm(\Lambda\{\Sigma\},\Pi)$ is X-valid. Now the statement follows from Lemma 4.10. □

Lemma 4.12. *Let* $\mathsf{X} \subseteq \{\mathsf{d}, \mathsf{t}, \mathsf{b}, 4, 5\}$, *and let* $\mathsf{r}\,\dfrac{\Gamma_1 \quad \Gamma_2}{\Gamma_3}$ *be an instance of* $\wedge^\circ$, $\vee^\bullet$, $\supset^\bullet$, *or* cut. *Then* $fm(\Gamma_1) \wedge fm(\Gamma_2) \supset fm(\Gamma_3)$ *is* X*-valid.*

Proof. For the $\wedge^\circ$- and $\vee^\bullet$-rules, this follows immediately from Lemmas 4.10 and 4.11. For $\supset^\bullet$ and cut, it suffices to show the statement for the rule

$$\tilde{\supset}^\bullet\,\frac{\Gamma^\downarrow\{A^\circ\} \qquad \Gamma\{B^\bullet\}}{\Gamma\{A \supset B^\bullet\}} \tag{8}$$

By Observation 3.2 and Definition 3.3, this rule is of shape

$$\tilde{\supset}^\bullet\,\frac{\Gamma'\{\Lambda\{A^\circ\}, [\Pi\{\emptyset\}]\} \qquad \Gamma'\{\Lambda\{B^\bullet\}, [\Pi\{C^\circ\}]\}}{\Gamma'\{\Lambda\{A \supset B^\bullet\}, [\Pi\{C^\circ\}]\}}$$

where $\Gamma'\{\ \}$, $\Lambda\{\ \}$, and $\Pi\{\ \}$ are output contexts. In particular, let $\Lambda\{\ \} = \Lambda_0, [\Lambda_1, [\ldots, [\Lambda_n, \{\ \}] \ldots]]$ and $\Pi\{\ \} = \Pi_1, [\Pi_2, [\ldots, [\Pi_m, \{\ \}] \ldots]]$. Now let $L_i = fm(\Lambda_i)$ for $i = 0 \ldots n$ and $P_j = fm(\Pi_j)$ for $j = 1 \ldots m$, and let

$$\begin{aligned}
L_X &= fm(\Lambda\{A^\circ\}) = L_0 \supset \Box(L_1 \supset \Box(L_2 \supset \Box(\cdots \supset \Box(L_n \supset A) \cdots))) \\
L_Y &= fm(\Lambda\{B^\bullet\}) = L_0 \wedge \Diamond(L_1 \wedge \Diamond(L_2 \wedge \Diamond(\cdots \wedge \Diamond(L_n \wedge B) \cdots))) \\
L_Z &= fm(\Lambda\{A \supset B^\bullet\}) = L_0 \wedge \Diamond(L_1 \wedge \Diamond(L_2 \wedge \Diamond(\cdots \wedge \Diamond(L_n \wedge (A \supset B)) \cdots))) \\
P_\emptyset &= fm([\Pi\{\emptyset\}]) = \Diamond(P_1 \wedge \Diamond(P_2 \wedge \Diamond(\cdots \wedge \Diamond(P_{m-1} \wedge \Diamond P_m) \cdots))) \\
P_C &= fm([\Pi\{C^\circ\}]) = \Box(P_1 \supset \Box(P_2 \supset \Box(\cdots \supset \Box(P_{m-1} \supset \Box(P_m \supset C)) \cdots)))
\end{aligned}$$

We are first going to show that $(L_X \wedge (L_Y \supset P_C)) \supset (L_Z \supset P_C)$ is X-valid. For this, it suffices to show that for every world w_0 of an arbitrary X-frame, if $w_0 \Vdash L_X$ and $w_0 \Vdash L_Y \supset P_C$ then $w_0 \Vdash L_Z \supset P_C$. So, assume that $w_0 \Vdash L_X$ and $w_0 \Vdash L_Y \supset P_C$. By definition, $w_0 \Vdash L_X$ means that

for all worlds $w'_0, w''_0, w_1, w'_1, w''_1, \ldots, w_n, w'_n$, if $w''_j R w_{j+1}$ and $w_i \leq w'_i \leq w''_i$ and $w'_i \Vdash L_i$ then $w'_n \Vdash A$, (9)

and $w_0 \Vdash L_Y \supset P_C$ means that

for all worlds $\hat{w}_0$ with $w_0 \leq \hat{w}_0$, if there are worlds $\hat{w}_1, \ldots, \hat{w}_n$ with $\hat{w}_i R \hat{w}_{i+1}$ and $\hat{w}_i \Vdash L_i$ and $\hat{w}_n \Vdash B$ then $\hat{w}_0 \Vdash P_C$. (10)

We want to show $w_0 \Vdash L_Z \supset P_C$, which means that

for all worlds $\tilde{w}_0$ with $w_0 \leq \tilde{w}_0$, if there are worlds $\tilde{w}_1, \ldots, \tilde{w}_n$ with $\tilde{w}_i R \tilde{w}_{i+1}$ and $\tilde{w}_i \Vdash L_i$ and $\tilde{w}_n \Vdash A \supset B$ then $\tilde{w}_0 \Vdash P_C$. (11)

So, let us assume we have a chain $\tilde{w}_0 R \tilde{w}_1 R \ldots R \tilde{w}_n$ with $\tilde{w}_i \Vdash L_i$ and $\tilde{w}_n \Vdash A \supset B$. By (9), (F1), and monotonicity (Lemma 2.1), we can conclude that $\tilde{w}_n \Vdash A$. Therefore, we also get $\tilde{w}_n \Vdash B$. Thus, by (10), we get $\tilde{w}_0 \Vdash P_C$, as desired. In a similar way, one can show that $(P_\emptyset \supset P_C) \supset P_C$ is X-valid. Now note that

$$((P_\emptyset \supset P_C) \supset P_C) \wedge (L_X \wedge (L_Y \supset P_C) \supset (L_Z \supset P_C)) \supset \\ ((P_\emptyset \supset L_X) \wedge (L_Y \supset P_C) \supset (L_Z \supset P_C))$$

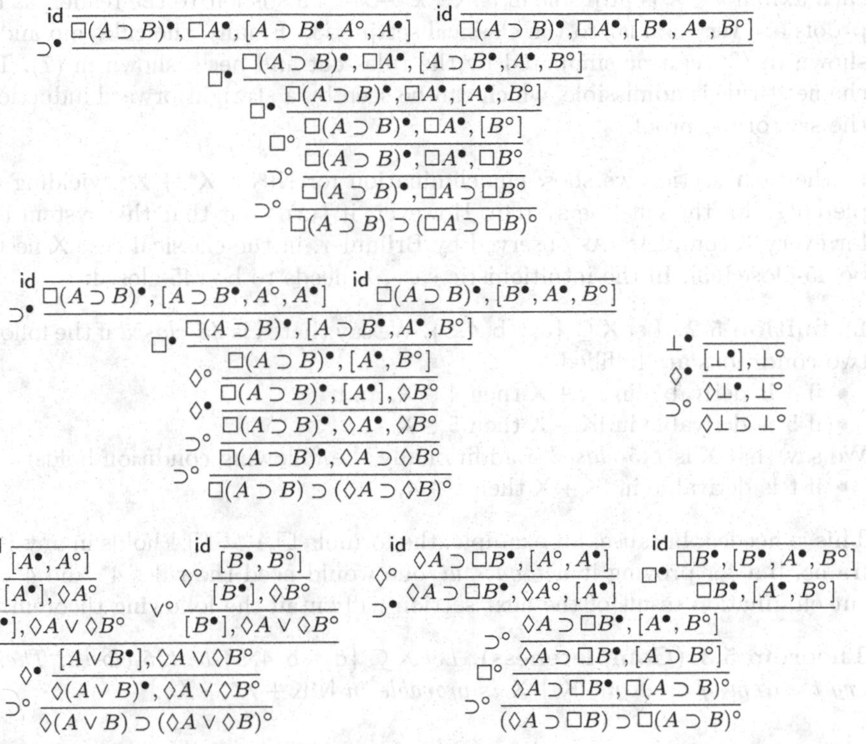

Fig. 6. Proofs of $\mathsf{k}_1, \ldots, \mathsf{k}_5$ in NIK

is a valid intuitionistic formula (for arbitrary $P_\emptyset, P_C, L_X, L_Y, L_Z$). Thus, we can conclude that $(P_\emptyset \supset L_X) \wedge (L_Y \supset P_C) \supset (L_Z \supset P_C)$ is X-valid, and we can apply Lemma 4.10. □

Now we can put everything together to prove Theorem 4.1.

Proof (of Theorem 4.1). Point (i) is just Lemmas 4.5, 4.6, 4.12, and 4.8. Point (ii) follows immediately from (i) using induction on the size of the derivation. □

5 Completeness

For simplifying the presentation, we show completeness with respect to the Hilbert system.

Theorem 5.1. *Let* $\mathsf{X} \subseteq \{\mathsf{d}, \mathsf{t}, \mathsf{b}, 4, 5\}$. *Then every theorem of the logic* $\mathsf{IK} + \mathsf{X}$ *is provable in* $\mathsf{NIK} + \mathsf{X}^\bullet + \mathsf{X}^\circ + \mathsf{cut}$.

Proof. Clearly, all axioms of propositional intuitionistic logic are provable in NIK. The axioms $\mathsf{k}_1, \ldots, \mathsf{k}_5$ are provable in NIK, as shown in Figure 6. Furthermore, each axiom $\mathsf{x} \in \mathsf{X}$ is provable in $\mathsf{NIK} + \mathsf{x}^\bullet + \mathsf{x}^\circ$. This is left to the reader, as these proofs are very similar to the classical setting [3]. Finally, the rules mp and nec, shown in (3), can be simulated by the rules cut and $\mathsf{nec}^{[]}$, shown in (7). Then, the $\mathsf{nec}^{[]}$-rule is admissible, which can be seen by a straightforward induction on the size of the proof. □

In the next section we show cut elimination for $\mathsf{NIK} + \mathsf{X}^\bullet + \mathsf{X}^\circ$, yielding completeness for the cut-free system. However, it turns out that this system is not for every X complete. As observed by Brünnler, in the classical case X needs to be 45-closed [3]. In the intuitionistic case, X needs to be t45-closed:

Definition 5.2. Let $\mathsf{X} \subseteq \{\mathsf{d}, \mathsf{t}, \mathsf{b}, \mathsf{4}, \mathsf{5}\}$. We say that X is *45-closed* if the following two conditions are fulfilled:

- if $\mathsf{4}$ is derivable in $\mathsf{IK} + \mathsf{X}$ then $\mathsf{4} \in \mathsf{X}$, and
- if $\mathsf{5}$ is derivable in $\mathsf{IK} + \mathsf{X}$ then $\mathsf{5} \in \mathsf{X}$.

We say that X is *t45-closed* if additionally the following condition holds:

- if t is derivable in $\mathsf{IK} + \mathsf{X}$ then $\mathsf{t} \in \mathsf{X}$.

This is needed, because, for example, the formula $\Box A \supset \Box\Box A$ holds in any $\{\mathsf{t}, \mathsf{5}\}$-frame, but for proving it *without cut*, one would need the rules $\mathsf{4}^\bullet$ and $\mathsf{4}^\circ$. The cut elimination result of the next section will entail the following theorem:

Theorem 5.3 (Completeness). *Let* $\mathsf{X} \subseteq \{\mathsf{d}, \mathsf{t}, \mathsf{b}, \mathsf{4}, \mathsf{5}\}$ *be t45-closed. Then every theorem of the logic* $\mathsf{IK} + \mathsf{X}$ *is provable in* $\mathsf{NIK} + \mathsf{X}^\bullet + \mathsf{X}^\circ$.

6 Cut Elimination

We define the *depth* of a formula A, denoted by $\mathit{depth}(A)$, inductively as follows:

$$\mathit{depth}(a) = \mathit{depth}(\bot) = 1$$
$$\mathit{depth}(\Box A) = \mathit{depth}(\Diamond A) = \mathit{depth}(A) + 1$$
$$\mathit{depth}(A \wedge B) = \mathit{depth}(A \vee B) = \mathit{depth}(A \supset B) = \max(\mathit{depth}(A), \mathit{depth}(B)) + 1$$

Definition 6.1. Given an instance of cut (as shown in (7)), its *cut formula* is A, and its *cut rank* is $\mathit{depth}(A)$. The *cut rank* of a derivation $\mathcal{D}$, denoted by $\mathit{rank}(\mathcal{D})$, is the maximum of the cut ranks of the cut instances of $\mathcal{D}$. Thus, a derivation with cut rank 0 is cut-free. For $r > 0$, we define the rule cut_r as cut whose cut rank is $\leq r$. As usual, the *height* of a derivation $\mathcal{D}$, denoted by $|\mathcal{D}|$, is defined to be the length of the maximal branch in the derivation tree.

Definition 6.2. We say that a rule r with one premise is *height* (respectively *cut rank*) *preserving admissible* in a system S, if for each derivation $\mathcal{D}$ in S of r's premise there is a derivation $\mathcal{D}'$ of r's conclusion in S, such that $|\mathcal{D}'| \leq |\mathcal{D}|$ (respectively $\mathit{rank}(\mathcal{D}') \leq \mathit{rank}(\mathcal{D})$). Similarly, a rule r is *height* (respectively *cut rank*) *preserving invertible* in a system S, if for every derivation of the conclusion of r there are derivations for each of r's premises with at most the same height (respectively at most the same rank).

$$\mathsf{d}^{[]}\ \frac{\Gamma\{[\emptyset]\}}{\Gamma\{\emptyset\}} \quad \mathsf{t}^{[]}\ \frac{\Gamma\{[\Delta]\}}{\Gamma\{\Delta\}} \quad \mathsf{b}^{[]}\ \frac{\Gamma\{[\Sigma,[\Delta]]\}}{\Gamma\{[\Sigma],\Delta\}} \quad \mathsf{4}^{[]}\ \frac{\Gamma\{[\Delta],[\Sigma]\}}{\Gamma\{[[\Delta],\Sigma]\}} \quad \mathsf{5}^{[]}\ \frac{\Gamma\{[\Delta]\}\{\emptyset\}}{\Gamma\{\emptyset\}\{[\Delta]\}}$$

(where $depth(\Gamma\{\ \}\{\emptyset\}) > 0$)

Fig. 7. Structural rules for the axioms d, t, b, 4, and 5

Figure 7 shows for each axiom in $\{\mathsf{d},\mathsf{t},\mathsf{b},4,5\}$ a corresponding structural rule. They will occur during the cut elimination process. Note that these rules are exactly the same as in the classical case [4]. These rules are admissible for the corresponding system, provided it is t45-closed. This lemma is the only place in the cut elimination proof, where this property is needed. As in the classical case [3], the $\mathsf{d}^{[]}$-rule needs special treatment.

Lemma 6.3. *(i) Let* $\mathsf{X} \subseteq \{\mathsf{t},\mathsf{b},4,5\}$ *be 45-closed, and let* $\mathsf{r} \in \mathsf{X}^{[]}$. *Then the rule* r *is cut-rank preserving admissible for* $\mathsf{NIK} \cup \mathsf{X}^\bullet \cup \mathsf{X}^\circ \cup \{\mathsf{cut}\}$ *as well as for* $\mathsf{NIK} \cup \mathsf{X}^\bullet \cup \mathsf{X}^\circ \cup \{\mathsf{cut}, \mathsf{d}^{[]}\}$.
(ii) Let $\mathsf{X} \subseteq \{\mathsf{d},\mathsf{t},\mathsf{b},4,5\}$ *be t45-closed with* $\mathsf{d} \in \mathsf{X}$. *Then the rule* $\mathsf{d}^{[]}$ *is admissible for* $\mathsf{NIK} \cup \mathsf{X}^\bullet \cup \mathsf{X}^\circ$.

Proof. The proof for (i) is similar to the one in [3]. But in the case analysis every case appears twice, once for the $\mathsf{x}^\bullet$ and once for the x° rule. For (ii), the proof is also almost the same as in [3], except that the rule t° can be introduced when $\{\mathsf{d},\mathsf{b},4\} \subseteq \mathsf{X}$, because there is no contraction available for output formulas. □

Lemma 6.4. *Let* $\mathsf{X} \subseteq \{\mathsf{d},\mathsf{t},\mathsf{b},4,5\}$ *and either* $\mathsf{Z} = \mathsf{NIK} + \mathsf{X}^\bullet + \mathsf{X}^\circ + \mathsf{cut}$ *or* $\mathsf{Z} = \mathsf{NIK} + \mathsf{X}^\bullet + \mathsf{X}^\circ + \mathsf{d}^{[]} + \mathsf{cut}$.

(i) The rules $\mathsf{nec}^{[]}$, w, c, $\mathsf{m}^{[]}$ *are height and cut rank preserving admissible for* Z.
(ii) All rules $\mathsf{r}^\bullet$ *(except* $\bot^\bullet$ *and* $\supset^\bullet$*) in* Z *are height and cut rank preserving invertible.*

Proof. For m, we can proceed by a straightforward induction on the height of the derivation. For all other rules, this proof is exactly the same as in [3]. □

When we eliminate the cut rule from a proof, we will at some point rely on local transformations that reduce the cut rank. However when the cut meets the rules $4^\bullet, 4^\circ$ or $5^\bullet, 5^\circ$ while moving upwards, its rank does not decrease. For this reason, we use the Y-cut-rules [3], defined below for $\mathsf{Y} \subseteq \{4,5\}$:

$$\Diamond\mathsf{Y\text{-}cut}\ \frac{\Gamma^{\downarrow}\{\emptyset\}\{\Diamond A^\circ\} \qquad \Gamma\{\Diamond A^\bullet\}\{\emptyset\}}{\Gamma\{\emptyset\}\{\emptyset\}} \qquad \Box\mathsf{Y\text{-}cut}\ \frac{\Gamma^{\downarrow}\{\Box A^\circ\}\{\emptyset\}^n \qquad \Gamma\{\Box A^\bullet\}\{\Box A^\bullet\}^n}{\Gamma\{\emptyset\}\{\emptyset\}^n}$$

where for $\Diamond$Y-cut there must be a derivation from $\Gamma^{\downarrow}\{\emptyset\}\{\Diamond A^\circ\}$ to $\Gamma^{\downarrow}\{\Diamond A^\circ\}\{\emptyset\}$ in Y°, and for $\Box$Y-cut there must be a derivation from $\Gamma\{\Box A^\bullet\}\{\Box A^\bullet\}^n$ to $\Gamma\{\Box A^\bullet\}\{\emptyset\}^n$ in $\mathsf{Y}^\bullet$. Here, we use the notation $\{\Delta\}^n$ as abbreviation for n holes that are all filled with the same Δ. For $r \geq 0$, the rules $\Diamond\mathsf{Y\text{-}cut}_r$ and $\Box\mathsf{Y\text{-}cut}_r$ are defined analogous to cut_r.

Observation 6.5. If $\mathsf{Y} = \emptyset$ then $\Gamma\{\ \}\{\ \} = \Gamma'\{\{\ \},\{\ \}\}$, for some input context $\Gamma'\{\ \}$, and both $\Diamond\mathsf{Y}$-cut and $\Box\mathsf{Y}$-cut are just ordinary cuts. If $\mathsf{Y} = \{4\}$ then in $\Diamond\mathsf{Y}$-cut we have $\Gamma\{\ \}\{\ \} = \Gamma'\{\{\ \},\Gamma''\{\ \}\}$ for some input contexts $\Gamma'\{\ \}$ and $\Gamma''\{\ \}$, and in $\Box\mathsf{Y}$-cut we have $\Gamma\{\ \}\{\ \}^n = \Gamma'\{\{\ \},\Gamma''\{\ \}^n\}$. If $\mathsf{Y} = \{5\}$ then the first hole must be "inside a box", i.e., in $\Diamond\mathsf{Y}$-cut we have $depth(\Gamma\{\ \}\{\emptyset\}) > 0$ and in $\Box\mathsf{Y}$-cut we have $depth(\Gamma\{\ \}\{\emptyset\}^n) > 0$. If $\mathsf{Y} = \{4,5\}$ then there is no restriction on the context.

Lemma 6.6. *Let* $\mathsf{X} \subseteq \{\mathsf{t},\mathsf{b},4,5\}$ *be 45-closed, let* $\mathsf{Y} \subseteq \{4,5\} \cap \mathsf{X}$, *let either* $\mathsf{Z} = \mathsf{NIK} + \mathsf{X}^\bullet + \mathsf{X}^\circ$ *or* $\mathsf{Z} = \mathsf{NIK} + \mathsf{X}^\bullet + \mathsf{X}^\circ + \mathsf{d}^{[]}$, *and let* $r, n \geq 0$.

(i) If there is a proof of shape

$$\mathsf{cut}_{r+1}\,\frac{\overset{\mathcal{D}_1}{\Gamma^{\downarrow}\{A^\circ\}} \qquad \overset{\mathcal{D}_2}{\Gamma\{A^\bullet\}}}{\Gamma\{\emptyset\}}$$

with $\mathcal{D}_1$ *and* $\mathcal{D}_2$ *in* $\mathsf{Z} + \mathsf{cut}_r$, *then there is a proof of* $\Gamma\{\emptyset\}$ *in* $\mathsf{Z} + \mathsf{cut}_r$.

(ii) If there is a proof of shape

$$\Diamond\mathsf{Y}\text{-}\mathsf{cut}_{r+1}\,\frac{\overset{\mathcal{D}_1}{\Gamma^{\downarrow}\{\emptyset\}\{\Diamond A^\circ\}} \qquad \overset{\mathcal{D}_2}{\Gamma\{\Diamond A^\bullet\}\{\emptyset\}}}{\Gamma\{\emptyset\}\{\emptyset\}}$$

with $\mathcal{D}_1$ *and* $\mathcal{D}_2$ *in* $\mathsf{Z} + \mathsf{cut}_r$, *then there is a proof of* $\Gamma\{\emptyset\}\{\emptyset\}$ *in* $\mathsf{Z} + \mathsf{cut}_r$.

(iii) If there is a proof of shape

$$\Box\mathsf{Y}\text{-}\mathsf{cut}_{r+1}\,\frac{\overset{\mathcal{D}_1}{\Gamma^{\downarrow}\{\Box A^\circ\}\{\emptyset\}^n} \qquad \overset{\mathcal{D}_2}{\Gamma\{\Box A^\bullet\}\{\Box A^\bullet\}^n}}{\Gamma\{\emptyset\}\{\emptyset\}^n}$$

with $\mathcal{D}_1$ *and* $\mathcal{D}_2$ *in* $\mathsf{Z} + \mathsf{cut}_r$, *then there is a proof of* $\Gamma\{\emptyset\}\{\emptyset\}^n$ *in* $\mathsf{Z} + \mathsf{cut}_r$.

Proof (Sketch). This is proved for all three points simultaneously by induction on $|\mathcal{D}_1| + |\mathcal{D}_2|$. If one of $\mathcal{D}_1$ or $\mathcal{D}_2$ is an axiom, the cut disappears. One case is shown below

$$\mathsf{cut}_1\,\frac{\overset{\mathcal{D}_1}{\Gamma^{\downarrow}\{\bot^\circ\}} \qquad \bot^\bullet\,\overline{\Gamma\{\bot^\bullet\}}}{\Gamma\{\emptyset\}} \quad \rightsquigarrow \quad \overset{\mathcal{D}_1'}{\Gamma\{\emptyset\}}$$

where $\mathcal{D}_1'$ is obtained from $\mathcal{D}_1$ by removing the $\bot^\circ$ in every line and keeping the output formula of $\Gamma\{\emptyset\}$ instead. This is possible because there is no rule for $\bot^\circ$. The other axiomatic cases are more standard. If in one of $\mathcal{D}_1$ or $\mathcal{D}_2$ the bottommost rule does not work on the cut formula, we have one of the commutative cases, which are very similar to the standard sequent calculus and make crucial use of the invertability of the $\mathsf{r}^\bullet$-rules. Finally, we have the so called key cases. We show the case involving $\Box\mathsf{Y}$-cut and $\mathsf{b}^\bullet$, in which the derivation

$$
\square\mathsf{Y}\text{-}\mathsf{cut}_{r+1}\;\dfrac{\square^{\circ}\,\dfrac{\overset{\mathcal{D}_1}{\Gamma^{\downarrow}\{[A^{\circ}]\}\{\emptyset\}^{n-1}\{[\Delta^{\downarrow}]\}}}{\Gamma^{\downarrow}\{\square A^{\circ}\}\{\emptyset\}^{n-1}\{[\Delta^{\downarrow}]\}}\qquad \mathsf{b}^{\bullet}\,\dfrac{\overset{\mathcal{D}_2}{\Gamma\{\square A^{\bullet}\}^{n}\{A^{\bullet},[\square A^{\bullet},\Delta]\}}}{\Gamma\{\square A^{\bullet}\}^{n}\{[\square A^{\bullet},\Delta]\}}}{\Gamma\{\emptyset\}^{n}\{[\Delta]\}}
$$

is replaced by

$$
\mathsf{cut}_r\;\dfrac{\mathsf{b}^{[]}\,\dfrac{\mathsf{Y}^{[]}\,\dfrac{\overset{\mathcal{D}_1}{\Gamma^{\downarrow}\{[A^{\circ}]\}\{\emptyset\}^{n-1}\{[\Delta^{\downarrow}]\}}}{\Gamma^{\downarrow}\{\emptyset\}^{n}\{[[A^{\circ}],\Delta^{\downarrow}]\}}}{\Gamma^{\downarrow}\{\emptyset\}^{n}\{A^{\circ},[\Delta^{\downarrow}]\}}\qquad \square\mathsf{Y}\text{-}\mathsf{cut}_{r+1}\,\dfrac{\square^{\circ}\,\dfrac{\mathsf{w}\,\dfrac{\overset{\mathcal{D}_1}{\Gamma^{\downarrow}\{[A^{\circ}]\}\{\emptyset\}^{n-1}\{[\Delta^{\downarrow}]\}}}{\Gamma^{\downarrow}\{[A^{\circ}]\}\{\emptyset\}^{n-1}\{A^{\bullet},[\Delta^{\downarrow}]\}}}{\Gamma^{\downarrow}\{\square A^{\circ}\}\{\emptyset\}^{n-1}\{A^{\bullet},[\Delta^{\downarrow}]\}}\quad \overset{\mathcal{D}_2}{\Gamma\{\square A^{\bullet}\}^{n}\{A^{\bullet},[\square A^{\bullet},\Delta]\}}}{\Gamma\{\emptyset\}^{n}\{A^{\bullet},[\Delta]\}}}{\Gamma\{\emptyset\}^{n}\{[\Delta]\}}
$$

where $\mathsf{Y}^{[]}$ stands for a derivation consisting of $4^{[]}$ and $5^{[]}$, depending on the chosen Y. Then, on the left branch, we use cut rank preserving admissibility of the $\mathsf{b}^{[]}$-, $4^{[]}$-, and $5^{[]}$-rules. On the right branch, we use cut rank and height preserving admissibility of weakening together with the induction hypothesis. The other cases are similar. □

Theorem 6.7. *Let* $\mathsf{X} \subseteq \{\mathsf{d},\mathsf{t},\mathsf{b},4,5\}$ *be t45-closed. If a sequent* Γ *is provable in* $\mathsf{NIK} + \mathsf{X}^{\bullet} + \mathsf{X}^{\circ} + \mathsf{cut}$ *then it is also provable in* $\mathsf{NIK} + \mathsf{X}^{\bullet} + \mathsf{X}^{\circ}$.

Proof. If $\mathsf{d} \notin \mathsf{X}$ the result follows from Lemma 6.6 by a straightforward induction on the cut rank of the derivation. If $\mathsf{d} \in \mathsf{X}$, we first replace all instances of $\mathsf{d}^{\bullet}$ by $\square^{\bullet}$ and $\mathsf{d}^{[]}$, and all instances of d° by $\lozenge^{\circ}$ and $\mathsf{d}^{[]}$. Then we proceed as before, and finally we apply Lemma 6.3.(ii) to remove $\mathsf{d}^{[]}$. □

Finally, we can drop the t45-closed condition and obtain full modularity by also allowing the structural rules of Figure 7 in the system:

Theorem 6.8. *Let* $\mathsf{X} \subseteq \{\mathsf{d},\mathsf{t},\mathsf{b},4,5\}$. *If a sequent* Γ *is provable in* $\mathsf{NIK} + \mathsf{X}^{\bullet} + \mathsf{X}^{\circ} + \mathsf{cut}$ *then it is also provable in* $\mathsf{NIK} + \mathsf{X}^{\bullet} + \mathsf{X}^{\circ} + \mathsf{X}^{[]}$.

Proof (Sketch). We first transform a proof in $\mathsf{NIK} + \mathsf{X}^{\bullet} + \mathsf{X}^{\circ} + \mathsf{cut}$ into one in $\mathsf{NIK} + \mathsf{Y}^{\bullet} + \mathsf{Y}^{\circ}$ by Theorem 6.7, where Y is the t45-closure of X. Trivially, this is also a proof in $\mathsf{NIK} + \mathsf{Y}^{\bullet} + \mathsf{Y}^{\circ} + \mathsf{X}^{[]}$. This is then transformed into a proof in $\mathsf{NIK} + \mathsf{X}^{\bullet} + \mathsf{X}^{\circ} + \mathsf{X}^{[]}$ by showing admissibility of the superfluous rules. □

References

1. Alechina, N., Mendler, M., de Paiva, V., Ritter, E.: Categorical and Kripke Semantics for Constructive S4 Modal Logic. In: Fribourg, L. (ed.) CSL 2001 and EACSL 2001. LNCS, vol. 2142, pp. 292–307. Springer, Heidelberg (2001)
2. Bierman, G.M., de Paiva, V.: On an intuitionistic modal logic. Studia Logica 65(3), 383–416 (2000)

3. Brünnler, K.: Deep sequent systems for modal logic. Archive for Mathematical Logic 48(6), 551–577 (2009)
4. Brünnler, K., Straßburger, L.: Modular Sequent Systems for Modal Logic. In: Giese, M., Waaler, A. (eds.) TABLEAUX 2009. LNCS, vol. 5607, pp. 152–166. Springer, Heidelberg (2009)
5. Chaudhuri, K., Guenot, N., Straßburger, L.: The focused calculus of structures. In: Bezem, M. (ed.) CSL 2011. LIPIcs, vol. 12, pp. 159–173. Schloss Dagstuhl – Leibniz-Zentrum fuer Informatik (2011)
6. Dyckhoff, R.: Contraction-free sequent calculi for intuitionistic logic. J. Symb. Log. 57(3), 795–807 (1992)
7. Ewald, W.B.: Intuitionistic tense and modal logic. J. Symb. Log. 51 (1986)
8. Fitch, F.B.: Intuitionistic modal logic with quantifiers. Portugaliae Mathematica 7(2), 113–118 (1948)
9. Fitting, M.: Nested sequents for intuitionistic logic (2011) (preprint)
10. Fitting, M.: Prefixed tableaus and nested sequents. Annals of Pure and Applied Logic 163, 291–313 (2012)
11. Galmiche, D., Salhi, Y.: Label-free natural deduction systems for intuitionistic and classical modal logics. Journal of Applied Non-Classical Logics 20(4), 373–421 (2010)
12. Garson, J.: Modal logic. In: Zalta, E.N. (ed.) The Stanford Encyclopedia of Philosophy. Stanford University (2008)
13. Goré, R., Postniece, L., Tiu, A.: Cut-elimination and proof search for bi-intuitionistic tense logic. In: Shehtman, V., Beklemishev, L., Goranko, V. (eds.) Advances in Modal Logic, pp. 156–177. College Publications (2010)
14. Kashima, R.: Cut-free sequent calculi for some tense logics. Studia Logica 53(1), 119–136 (1994)
15. Lamarche, F.: On the algebra of structural contexts. Accepted at Mathematical Structures in Computer Science (2001)
16. Liang, C., Miller, D.: Focusing and Polarization in Intuitionistic Logic. In: Duparc, J., Henzinger, T.A. (eds.) CSL 2007. LNCS, vol. 4646, pp. 451–465. Springer, Heidelberg (2007)
17. McLaughlin, S., Pfenning, F.: The focused constraint inverse method for intuitionistic modal logics. Draft manuscript (2010)
18. Mendler, M., Scheele, S.: Cut-free gentzen calculus for multimodal ck. Inf. Comput. 209(12), 1465–1490 (2011)
19. Nanevski, A., Pfenning, F., Pientka, B.: Contextual modal type theory. ACM Transactions on Computational Logic 9(3) (2008)
20. Pfenning, F., Davies, R.: A judgmental reconstruction of modal logic. Mathematical Structures in Computer Science 11(4), 511–540 (2001)
21. Plotkin, G.D., Stirling, C.P.: A framework for intuitionistic modal logic. In: Halpern, J.Y. (ed.) Theoretical Aspects of Reasoning About Knowledge (1986)
22. Poggiolesi, F.: The method of tree-hypersequents for modal propositional logic. In: Makinson, D., Malinowski, J., Wansing, H. (eds.) Towards Mathematical Philosophy. Trends in Logic, vol. 28, pp. 31–51. Springer (2009)
23. Prawitz, D.: Natural Deduction, A Proof-Theoretical. Almquist and Wiksell (1965)
24. Fischer Servi, G.: Axiomatizations for some intuitionistic modal logics. Rend. Sem. Mat. Univers. Politecn. Torino 42(3), 179–194 (1984)
25. Simpson, A.: The Proof Theory and Semantics of Intuitionistic Modal Logic. PhD thesis, University of Edinburgh (1994)

On Monadic Parametricity of Second-Order Functionals

Andrej Bauer[1], Martin Hofmann[2], and Aleksandr Karbyshev[3]

[1] University of Ljubljana
andrej.bauer@andrej.com
[2] Universität München
hofmann@ifi.lmu.de
[3] Technische Universität München
aleksandr.karbyshev@in.tum.de

Abstract. How can one rigorously specify that a given ML functional $f : (\texttt{int} \to \texttt{int}) \to \texttt{int}$ is *pure*, i.e., f produces no computational effects except those produced by evaluation of its functional argument? In this paper, we introduce a semantic notion of *monadic parametricity* for second-order functionals which is a form of purity. We show that every monadically parametric f admits a question-answer strategy tree representation. We discuss possible applications of this notion, e.g., to the verification of generic fixpoint algorithms. The results are presented in two settings: a total set-theoretic setting and a partial domain-theoretic one. All proofs are formalized by means of the proof assistant Coq.

1 Introduction

The problem under consideration is: how do we rigorously specify that a given ML functional $f : (\texttt{int} \to \texttt{int}) \to \texttt{int}$ is pure, i.e., f only produces computational effects (changes store, raises an exceptions, produces output, consumes input, etc.) through calls of its functional argument? Second-order functionals of this type may appear as inputs in various third-order algorithms, such as generic fixpoint solvers [4, 5] and algorithms for exact integration [15, 20]. The algorithms often apply a presumably pure input f to an effectful argument in order to observe the intentional behaviour of f, and to control the computation process.

In a previous paper [8] we addressed the question with regard to functionals that were polymorphic in the state monad and had the type $\forall S.(A \to State_S B) \to State_S C$. The motivation there was rigorous verification of a generic fixpoint algorithm RLD [7] that used state. As it turns out [8], we could not use the standard notion of relational parametricity [17,18] because it is too weak to exclude the snapback functional $f_{\text{snap}} : \forall S.(A \to State_S B) \to State_S B$, defined by

$$f_{\text{snap}}\, S\, k\, s = \textbf{let}\ (b, \_) = k\, a_0\, s\ \textbf{in}\ (b, s) \ .$$

The functional invokes its argument k to compute a result b but then discards the new state and restores the initial one instead. We can show that every functional

F. Pfenning (Ed.): FOSSACS 2013, LNCS 7794, pp. 225–240, 2013.

which is pure in the sense of [8] is represented by a question-answer strategy tree that computes the result by calling its function argument and letting through any effects generated by the calls. The functional f_{snap} is not pure in that sense.

The strategy tree reflects only the "skeleton" of a computation and is not specific to the kind of effects that may be raised. Obviously, we should look for a more general representation theorem that applies to other kinds of effectful computations. Indeed, in this paper we remove the limitation of [8] to state and prove a corresponding theorem for the class of second-order functionals polymorphic with respect to monads from an arbitrary fixed collection *Monad*, i.e., those of type

$$\text{Func} = \prod_{T \in Monad} (A \to TB) \to TC \ ,$$

so long as continuation monads $Cont_R$ are included, for all R. We may think of *Monad* as the class of monads present in a programming language. Every monad can be expressed in terms of continuation and state [6], but we do *not* use this fact and do not require $State \in Monad$. Thus our representation result and that of Filinski [6] are different and do not imply each other directly. An interesting corollary is that a functional F which is pure for the state monads in the sense of [8] has an equivalent implementation which makes no use of state, and is moreover polymorphic in all monads from the given collection *Monad*. Such an implementation is defined by a strategy tree for F.

One possible application of the representation result is formal verification of the above mentioned algorithms. For example, when trying to prove correctness of the local fixpoint solver RLD we assumed without loss of generality that the input constraint system is given in the form of strategy trees. That allowed us to formulate sufficient pre- and post-conditions for the algorithm and complete the proof by induction. The fundamental lemma then allows us to argue that the functional input is indeed pure if it can be defined in some restricted programming language (with recursion) which is often the case in real-life program analysis.

The outline of the paper is as follows. After a preliminary Section 2, we define purity in Section 3 as a semantical notion of monadic parametricity. We also formulate a fundamental lemma for the call-by-value lambda calculus with monadic semantics. In section 4, we define a notion of a strategy tree and show they represent pure functionals of type Func in the total setting. Section 5 provides a similar result in the partial setting. In section 6, we discuss generalizations of purity to other types. In section 7, we discuss some application of purity.

All the proofs have been formalized by means of Coq theorem prover [21] and are available for download at [11]. We used the development of constructive ω-cpos and inverse-limit construction for solution of recursive domain equations by Benton et al. [3]. Our contribution takes around 1500 lines of Coq code.

2 Preliminaries

We study purity in both the total and the partial setting. For the former we interprets types as sets and the latter as continuous posets (cpos), and thus use the notations $a : X$ and $a \in X$ interchangeably. We write $X \times Y$ and $X \to Y$ for Cartesian product and exponential, respectively. We denote pairs by (x, y), and projections by *fst* and *snd*. We use λ, $\circ$ and juxtaposition for function abstraction, composition and applications, correspondingly. For a family of sets or cpos $(X_i)_{i \in I}$ we write $\prod_{i \in I} X_i$ for its Cartesian product.

Definition 1. A *monad* is a triple $(T, \mathsf{val}_T, \mathsf{bind}_T)$ where T is the *monad constructor* assigning to a type X the type TX of computations over X, and

$$\begin{aligned} \mathsf{val}_T^A &: A \to TA \\ \mathsf{bind}_T^{A,B} &: TA \to (A \to TB) \to TB \end{aligned}$$

are the *monadic operators*, satisfying for all a, f, g and t of suitable types

$$\begin{aligned} \mathsf{bind}_T^{A,B}(\mathsf{val}_T^A\, a)(f) &= f\, a \\ \mathsf{bind}_T^{A,A}(t)(\mathsf{val}_T^A) &= t \\ \mathsf{bind}_T^{B,C}(\mathsf{bind}_T^{A,B}(t)(f))(g) &= \mathsf{bind}_T^{A,C}(t)(\lambda x.\, \mathsf{bind}_T^{B,C}(f\, x)(g)) \ . \end{aligned}$$

For the partial case, we require that TX is a pointed cpo (cppo) and that T is *strict*,

$$\mathsf{bind}_T^{A,B} \perp_{TA} f = \perp_{TB} \ .$$

We omit the indices A, B, C when they can be deduced from the context.

The *continuation monad* $Cont_R$ with result type R is defined by $Cont_R X = (X \to R) \to R$ and

$$\begin{aligned} \mathsf{val}_{Cont_R}\, x &= \lambda c.c\, x \\ \mathsf{bind}_{Cont_R}\, t\, f &= \lambda c.t(\lambda x.f\, x\, c) \ . \end{aligned}$$

The *state monad* $State_S$ with the type of states S is defined by $State_S X = S \to X \times S$ and

$$\begin{aligned} \mathsf{val}_{State_S}\, x &= \lambda s.(x, s) \\ \mathsf{bind}_{State_S}\, t\, f &= \lambda s.\mathbf{let}\ (x_1, s_1) \leftarrow t\, s\ \mathbf{in}\ f x_1 s_1 \ . \end{aligned}$$

In the following, we assume that A, B, C, A_i, B_i are sets or cpos, as appropriate. Let *Monad* be a fixed collection of monads such that $Cont_R \in Monad$, for all R, and denote

$$\mathrm{Func} = \prod_{T \subset Monad} (A \to TB) \to TC \ .$$

3 Purity

To define purity in our sense we first introduce several notions and notations. We then provide a relational interpretation of types and terms of call-by-value λ-calculus with monadic semantics, and establish a fundamental lemma of logical relations stating that every well-typed program respects any monadic relation, similar to [8].

Definition 2. If X, X' are types then $\mathrm{Rel}(X, X')$ denotes the type of binary relations between X and X'. Furthermore:

- if X is a type then $\Delta_X \in \mathrm{Rel}(X, X)$ denotes the equality on X;
- if $R \in \mathrm{Rel}(X, X')$ and $S \in \mathrm{Rel}(Y, Y')$ then $R \to S \in \mathrm{Rel}(X \to Y, X' \to Y')$ is given by

$$f\,(R \to S)\,f' \quad \text{iff} \quad \forall x\,x'.\,xRx' \implies (f\,x)\,S\,(f'\,x') \ ;$$

- if $R \in \mathrm{Rel}(X, X')$ and $S \in \mathrm{Rel}(Y, Y')$ then $R \times S \in \mathrm{Rel}(X \times Y, X' \times Y')$ is given by

$$p\,(R \times S)\,p' \quad \text{iff} \quad \mathit{fst}(p)\,R\,\mathit{fst}(p') \wedge \mathit{snd}(p)\,S\,\mathit{snd}(p') \ .$$

Definition 3. For cpos X, X' and $R \in \mathrm{Rel}(X, X')$, R is *admissible* if for any chains $\{c_i\}_{i \in \mathbb{N}}$, $\{c'_i\}_{i \in \mathbb{N}}$ such that $c_i\,R\,c'_i$, for all i, $(\bigsqcup c_i)\,R\,(\bigsqcup c'_i)$ holds.

Definition 4. Fix $T, T' \in \mathit{Monad}$. For every X, X' and $Q \in \mathrm{Rel}(X, X')$ fix a relation $T^{\mathsf{rel}}(Q) \in \mathrm{Rel}(TX, T'X')$. We say that the mapping $(X, X', Q) \mapsto T^{\mathsf{rel}}(Q)$ is an *acceptable monadic relation* if

- for all $X, X', Q \in \mathrm{Rel}(X, X')$, $x \in X$, $x' \in X'$,

$$xQx' \implies (\mathsf{val}_T\ x)\,T^{\mathsf{rel}}(Q)\,(\mathsf{val}_{T'}\ x') \ ;$$

- for all $X, X', Q \in \mathrm{Rel}(X, X')$, $Y, Y', R \in \mathrm{Rel}(Y, Y')$, $t \in TX$, $t' \in T'X'$, $f : X \to TY$, $f' : X' \to T'Y'$,

$$t\,T^{\mathsf{rel}}(Q)\,t' \wedge f(Q \to T^{\mathsf{rel}}(R))f' \implies (\mathsf{bind}_T\ t\,f)\,T^{\mathsf{rel}}(R)\,(\mathsf{bind}_{T'}\ t'\,f') \ .$$

In the domain-theoretic setting, we additionally assume that the monadic relation T^{rel} is

- *admissible*, i.e., $T^{\mathsf{rel}}(Q)$ is admissible for every admissible $Q \in \mathrm{Rel}(X, X')$,
- *strict*, i.e., $(\bot_{TX}, \bot_{T'X'}) \in T^{\mathsf{rel}}(Q)$.

Definition 5. A functional $F \in \mathrm{Func}$ is *pure (monadically parametric)* for the collection *Monad* of monads iff

$$(F_T, F_{T'}) \in (\Delta_A \to T^{\mathsf{rel}}(\Delta_B)) \to T^{\mathsf{rel}}(\Delta_C)$$

holds for all $T, T' \in \mathit{Monad}$ and acceptable monadic relations T^{rel} for T, T'.

Define simple types over a set of base types, ranged over by o, by the grammar

$$\tau ::= o \mid \tau_1 \times \tau_2 \mid \tau_1 \to \tau_2 \,.$$

Fix an assignment of a set or a cpo, as the case may be, $[\![o]\!]_T$ for each base type o and monad $T \in \mathit{Monad}$. We extend $[\![-]\!]_T$ to all types by putting

$$[\![\tau_1 \times \tau_2]\!]_T = [\![\tau_1]\!]_T \times [\![\tau_2]\!]_T \,, \quad [\![\tau_1 \to \tau_2]\!]_T = [\![\tau_1]\!]_T \to T[\![\tau_2]\!]_T \,.$$

Given a set of constants ranged over by c, with corresponding types τ^c, and variables ranged over by x, we define the λ-terms by

$$e ::= x \mid c \mid \lambda x.e \mid e_1\, e_2 \mid e.1 \mid e.2 \mid \langle e_1, e_2\rangle \mid \\ \text{let } x \leftarrow e_1 \text{ in } e_2 \mid \text{let rec } f(x) = e$$

with the last rule for recursive definitions in the partial case only. A typing context Γ is a finite map from variables to types. The typing judgement $\Gamma \vdash e : \tau$ is defined by the usual rules:

$$\frac{x \in \mathrm{dom}(\Gamma)}{\Gamma \vdash x : \Gamma(x)} \qquad \frac{}{\Gamma \vdash c : \tau^c}$$

$$\frac{\Gamma, x : \tau_1 \vdash e : \tau_2}{\Gamma \vdash \lambda x.e : \tau_1 \to \tau_2} \qquad \frac{\Gamma \vdash e_1 : \tau_1 \to \tau_2 \qquad \Gamma \vdash e_2 : \tau_1}{\Gamma \vdash e_1\, e_2 : \tau_2}$$

$$\frac{\Gamma \vdash e_1 : \tau_1 \qquad \Gamma \vdash e_2 : \tau_2}{\Gamma \vdash \langle e_1, e_2\rangle : \tau_1 \times \tau_2} \qquad \frac{\Gamma \vdash e : \tau_1 \times \tau_2}{\Gamma \vdash e.1 : \tau_1} \qquad \frac{\Gamma \vdash e : \tau_1 \times \tau_2}{\Gamma \vdash e.2 : \tau_2}$$

$$\frac{\Gamma \vdash e_1 : \tau_1 \qquad \Gamma, x : \tau_1 \vdash e_2 : \tau_2}{\Gamma \vdash \text{let } x \leftarrow e_1 \text{ in } e_2 : \tau_2} \qquad \frac{\Gamma, f : \tau_1 \to \tau_2, x : \tau_1 \vdash e : \tau_2}{\Gamma \vdash \text{let rec } f(x) = e : \tau_1 \to \tau_2}$$

The term $e : \tau$ is *closed* if $\emptyset \vdash e : \tau$.

For each $T \in \mathit{Monad}$ and constant c fix an interpretation $[\![c]\!]_T \in [\![\tau^c]\!]_T$. An *environment* for a context Γ and $T \in \mathit{Monad}$ is a mapping η such that $x \in \mathrm{dom}(\Gamma)$ implies $\eta(x) \in [\![\Gamma(x)]\!]_T$. If $\Gamma \vdash e : \tau$ and η is such an environment then we define $[\![e]\!]_T(\eta) \in T[\![\tau]\!]_T$ by the following clauses:

$$\begin{aligned}
[\![x]\!]_T(\eta) &= \mathsf{val}_T(\eta(x))\\
[\![c]\!]_T(\eta) &= \mathsf{val}_T([\![c]\!]_T)\\
[\![\lambda x.e]\!]_T(\eta) &= \mathsf{val}_T(\lambda v.[\![e]\!]_T(\eta[x \mapsto v]))\\
[\![e_1\, e_2]\!]_T(\eta) &= \mathsf{bind}_T([\![e_1]\!]_T(\eta))\,(\mathsf{bind}_T([\![e_2]\!]_T(\eta)))\\
[\![e.i]\!]_T(\eta) &= \mathsf{bind}_T([\![e]\!]_T(\eta))\,(\mathsf{val}_T \circ \pi_i),\ i = 1, 2\\
[\![\langle e_1, e_2\rangle]\!]_T(\eta) &= \mathsf{bind}_T([\![e_1]\!]_T(\eta))(\mathsf{bind}_T([\![e_2]\!]_T(\eta)) \circ \mathit{curry}(\mathsf{val}_T))\\
[\![\text{let } x \leftarrow e_1 \text{ in } e_2]\!]_T(\eta) &= \mathsf{bind}_T([\![e_1]\!]_T(\eta))(\lambda v.[\![e_2]\!]_T(\eta[x \mapsto v])))\\
[\![\text{let rec } f(x) = e]\!]_T(\eta) &= \mathsf{val}_T(\mathit{fixp}(\lambda h.\lambda v.[\![e]\!]_T(\eta[f \mapsto h][x \mapsto v])))
\end{aligned}$$

where $\mathit{fixp} : \forall D.(D \to D) \to D$ is the least fixpoint operator for cppos, and *curry* is the currying function.

Definition 6. Fix monads $T, T' \in \mathit{Monad}$ and an acceptable monadic relation T^{rel} for T, T'. Given a binary relation $[\![o]\!]^{\mathsf{rel}} \in \mathrm{Rel}([\![o]\!]_T, [\![o]\!]_{T'})$ for each base type o, we can associate a relation $[\![\tau]\!]^{\mathsf{rel}}_{T^{\mathsf{rel}}} \in \mathrm{Rel}([\![\tau]\!]_T, [\![\tau]\!]_{T'})$ with each type τ by the following clauses:

$$[\![o]\!]^{\mathsf{rel}}_{T^{\mathsf{rel}}} = [\![o]\!]^{\mathsf{rel}}\,, \quad [\![\tau_1 \times \tau_2]\!]^{\mathsf{rel}}_{T^{\mathsf{rel}}} = [\![\tau_1]\!]^{\mathsf{rel}}_{T^{\mathsf{rel}}} \times [\![\tau_2]\!]^{\mathsf{rel}}_{T^{\mathsf{rel}}}\,,$$
$$[\![\tau_1 \to \tau_2]\!]^{\mathsf{rel}}_{T^{\mathsf{rel}}} = [\![\tau_1]\!]^{\mathsf{rel}}_{T^{\mathsf{rel}}} \to T^{\mathsf{rel}}([\![\tau_2]\!]^{\mathsf{rel}}_{T^{\mathsf{rel}}})\,.$$

The following *parametricity theorem* is immediate from the definition of acceptable monadic relation and the previous one.

Theorem 7. *Fix $T, T' \in Monad$, and an acceptable monadic relation T^{rel} for T, T'. Suppose that $[\![c]\!]_T \, [\![\tau^c]\!]^{\mathsf{rel}}_{T^{\mathsf{rel}}} \, [\![c]\!]_{T'}$ holds for all constants c. If $\emptyset \vdash e : \tau$ then*

$$[\![e]\!]_T \; T^{\mathsf{rel}}([\![\tau]\!]^{\mathsf{rel}}_{T^{\mathsf{rel}}}) \; [\![e]\!]_{T'} \; .$$

Proof. One proves the following stronger statement by induction on typing derivations. Given $\Gamma \vdash e : \tau$ and environments η for Γ and T and η' for Γ and T' then

$$\forall x. \, \eta(x) \, [\![\Gamma(x)]\!]^{\mathsf{rel}}_{T^{\mathsf{rel}}} \, \eta'(x) \quad \text{implies} \quad [\![e]\!]_T(\eta) \, T^{\mathsf{rel}}([\![\tau]\!]^{\mathsf{rel}}_{T^{\mathsf{rel}}}) \, [\![e]\!]_{T'}(\eta') \; .$$

The assertion of the theorem follows. □

Every well-typed program $\emptyset \vdash e : \tau$ defines a truly polymorphic function of type $\forall T. [\![\tau]\!]_T$ by taking a product over *Monad*. From theorem 7, we obtain

Corollary 8. *Every truly polymorphic $F \in$ Func implemented in the calculus is monadically parametric.* □

We remark that we could incorporate Theorem 7 into the definition of Func after the definition 1, which would provide "higher-kinded type polymorphism" at that level. The theorem would then turn into a well-definedness assertions to go with the interpretation of type formers. We find the chosen presentation more convenient because it allows for a priori impure functionals whose purity can then be established a posteriori.

4 The Total Case

We first consider the set-theoretic semantics in which all functions are total and there is no general recursion, but we can use structural recursion on inductively defined sets.

Let the set of *strategy trees Tree* be inductively generated by the constructors $\mathsf{Ans} : C \to \mathit{Tree}$ and $\mathsf{Que} : A \to (B \to \mathit{Tree}) \to \mathit{Tree}$. Thus, a strategy tree is either an *answer* leaf $\mathsf{Ans}\, c$ with an answer value $c : C$, or a *question* node $\mathsf{Que}\; a\; f$ with a query $a : A$ and a branching (continuation) function $f : B \to \mathit{Tree}$ that returns a tree for every possible answer of type B.

For a given monad $T \in Monad$, every strategy tree defines a functional. The conversion from trees to functionals is performed by the function $\mathit{tree2fun}_T : \mathit{Tree} \to (A \to TB) \to TC$ defined by structural recursion as

$$\begin{aligned} \mathit{tree2fun}_T(\mathsf{Ans}\, c) &= \lambda k. \, \mathsf{val}_T \; c \\ \mathit{tree2fun}_T(\mathsf{Que}\, a\, f) &= \lambda k. \, \mathsf{bind}_T(k\, a)(\lambda b. \, \mathit{tree2fun}_T\, (f\, b)\, k). \end{aligned}$$

The functional queries and answers its argument k according to the strategy tree, and passes through any effects produces by k. The definition is parametric in the monad T, so we can define the polymorphic version $\mathit{tree2fun}\, t = \Lambda T : Monad. \, \mathit{tree2fun}_T\, t$ whose type is $\mathit{Tree} \to$ Func.

Example 9. For $A = B = C = \mathbb{N}$ and the tree $t = \mathsf{Que}\,0\,(\lambda x.\mathsf{Ans}\,42)$ we have

$$\begin{aligned} \mathit{tree2fun}_T\; t = \mathit{tree2fun}_T\,(\mathsf{Que}\,0\,(\lambda x.\mathsf{Ans}\,42)) = \\ = \lambda k.\,\mathsf{bind}_T\,(k\,0)\,(\lambda b.\,\mathit{tree2fun}_T\,(\mathsf{Ans}\,42)\,k) \\ = \lambda k.\,\mathsf{bind}_T\,(k\,0)\,(\lambda b.\,\mathsf{val}_T\;42). \end{aligned}$$

Thus, the tree $t = \mathsf{Que}\,0\,(\lambda x.\mathsf{Ans}\,42)$ corresponds to a second-order function that queries its argument k at 0 and returns 42. Any effect produced by k is propagated, and no other effects are produced.

The following lemma states that every $t \in \mathit{Tree}$ defines a monadically parametric computation.

Lemma 10. *For any $t \in \mathit{Tree}$, $\mathit{tree2fun}\,t$ is pure.*

Proof. By induction on t, see Appendix. □

It may be a bit surprising that *tree2fun* has an inverse *fun2tree* which is defined with the help of the continuation monad simply as

$$\mathit{fun2tree}\;F = F_{\mathit{Cont}_{\mathit{Tree}}}\;\mathsf{Que}\;\mathsf{Ans}.$$

Let us show that *fun2tree* and *tree2fun* are inverses of each other. As is to be expected, one direction is easier than the other, so we first dispose of the easy one:

Lemma 11. *For any $t \in \mathit{Tree}$, $\mathit{fun2tree}(\mathit{tree2fun}\,t) = t$.*

Proof. We proceed by structural induction on t. The case $t = \mathsf{Ans}\,c$ is easy:

$$\begin{aligned} &\mathit{fun2tree}(\mathit{tree2fun}(\mathsf{Ans}\,c)) = \\ &\qquad \mathit{fun2tree}(\Lambda T.\lambda k.\,\mathsf{val}_T\;c) = (\lambda k.\,\mathsf{val}_{\mathit{Cont}_{\mathit{Tree}}}\;c)\,\mathsf{Que}\,\mathsf{Ans} = \mathsf{Ans}\,c. \end{aligned}$$

To check the case $t = \mathsf{Que}\,a\,f$, assume the induction hypothesis, for all $b \in B$ $\mathit{fun2tree}(\mathit{tree2fun}(f\,b)) = f\,b$, and compute:

$$\begin{aligned} &\mathit{fun2tree}(\mathit{tree2fun}(\mathsf{Que}\,a\,f)) = \\ &\quad = \mathit{fun2tree}(\Lambda T.\lambda k.\,\mathsf{bind}_T(k\,a)(\lambda b.\,\mathit{tree2fun}_T(f\,b)\,k)) \\ &\quad = (\lambda k.\,\mathsf{bind}_{\mathit{Cont}_{\mathit{Tree}}}(k\,a)(\lambda b.\,\mathit{tree2fun}_{\mathit{Cont}_{\mathit{Tree}}}(f\,b)\,k))\,\mathsf{Que}\,\mathsf{Ans} \\ &\quad = (\mathsf{bind}_{\mathit{Cont}_{\mathit{Tree}}}(\mathsf{Que}\,a)(\lambda b.\,\mathit{tree2fun}_{\mathit{Cont}_{\mathit{Tree}}}(f\,b)\,\mathsf{Que}))\,\mathsf{Ans} \\ &\quad = (\mathsf{Que}\,a)(\lambda b.\,\mathit{tree2fun}_{\mathit{Cont}_{\mathit{Tree}}}(f\,b)\,\mathsf{Que}\,\mathsf{Ans}) \\ &\quad = (\mathsf{Que}\,a)(\lambda b.\,\mathit{fun2tree}(\mathit{tree2fun}(f\,b))) \\ &\quad = \mathsf{Que}\,a\,f. \end{aligned}$$

We used the induction hypothesis in the last step. □

Of course, for the other inverse we have to use purity of functionals:

Theorem 12. *For a* pure $F \in \mathrm{Func}$ *and* $T \in$ *Monad,*

$$\mathit{tree2fun}_T(\mathit{fun2tree}\, F) = F_T.$$

We first verify the theorem for the continuation monad.

Lemma 13. *Given a pure* $F \in \mathrm{Func}$, $\mathit{tree2fun}_{\mathit{Cont}_S}(\mathit{fun2tree}\, F) = F_{\mathit{Cont}_S}$ *holds for all* S.

Proof. Given S and functions $q : A \to (B \to S) \to S$ and $a : C \to S$, we define the *conversion* function $\mathit{conv}_{q,a} : \mathit{Tree} \to S$ by $\mathit{conv}_{q,a} = \lambda t.\, \mathit{tree2fun}_{\mathit{Cont}_S}\, t\, q\, a$. We have:

$$\begin{aligned}&\mathit{tree2fun}_{\mathit{Cont}_S}(\mathit{fun2tree}\, F) = F_{\mathit{Cont}_S}\\ &\qquad\iff \forall q, a.(F_{\mathit{Cont}_{\mathit{Tree}}}(\mathsf{Que})(\mathsf{Ans}), F_{\mathit{Cont}_S}\, q\, a) \in \mathcal{G}_{\mathit{conv}_{q,a}}\end{aligned}$$

where $\mathcal{G}_f$ is a *graph* of f, i.e., $(x, y) \in \mathcal{G}_f$ iff $y = f\, x$. We prove the last proposition by constructing an appropriate monadic relation for $\mathit{Cont}_{\mathit{Tree}}$ and Cont_S and utilizing purity of F. Fix some q and a. For X, X' and $R \in \mathrm{Rel}(X, X')$, we define $T_1^{\mathsf{rel}}(R) \in \mathrm{Rel}(\mathit{Cont}_{\mathit{Tree}} X, \mathit{Cont}_S X')$ by

$$(H, H') \in T_1^{\mathsf{rel}}(R) \quad \text{iff} \quad \forall h, h'.(h, h') \in R \to \mathcal{G}_{\mathit{conv}_{q,a}} \implies (Hh, H'h') \in \mathcal{G}_{\mathit{conv}_{q,a}}$$

It is straightforward to show T_1^{rel} is an acceptable monadic relation. Since F is pure, $(F_{\mathit{Cont}_{\mathit{Tree}}}, F_{\mathit{Cont}_S}) \in (\Delta_A \to T_1^{\mathsf{rel}}(\Delta_B)) \to T_1^{\mathsf{rel}}(\Delta_C)$. Thus, it suffices to check that $(\mathsf{Que}, q) \in \Delta_A \to T_1^{\mathsf{rel}}(\Delta_B)$ and $(\mathsf{Ans}, a) \in \Delta_C \to \mathcal{G}_{\mathit{conv}_{q,a}}$. The latter is obvious. For the former, take $a_1 \in A$ and $f : B \to \mathit{Tree}$, $f' : B \to S$ such that $(f, f') \in \Delta_B \to \mathcal{G}_{\mathit{conv}_{q,a}}$. Then

$$\begin{aligned}\mathit{conv}_{q,a}(\mathsf{Que}\, a_1\, f) &= \mathit{tree2fun}_{\mathit{Cont}_S}(\mathsf{Que}\, a_1\, f)\, q\, a\\ &= \mathsf{bind}_{\mathit{Cont}_S}(q\, a_1)(\lambda b.\, \mathit{tree2fun}_{\mathit{Cont}_S}(f\, b)\, q)\, a\\ &= (q\, a_1)(\lambda b.\, \mathit{tree2fun}_{\mathit{Cont}_S}(f\, b)\, q\, a)\\ &= (q\, a_1)(\lambda b. \mathit{conv}_{q,a}(f\, b))\\ &= q\, a_1\, f'\end{aligned}$$

and the former holds. □

Now by the lemma $\mathit{tree2fun}_{\mathit{Cont}_{TC}}(F_{\mathit{Cont}_{\mathit{Tree}}}\, \mathsf{Que}\, \mathsf{Ans}) = F_{\mathit{Cont}_{TC}}$. Let

$$\begin{aligned}\varphi_1 &= \mathsf{bind}_T^{B,C} : TB \to \mathit{Cont}_{TC} B,\\ \varphi_2 &= \lambda g.g\, (\mathsf{val}_T^C) \colon \mathit{Cont}_{TC} C \to TC\end{aligned}$$

and define $\Phi_T : ((A \to \mathit{Cont}_{TC} B) \to \mathit{Cont}_{TC} C) \to (A \to TB) \to TC$ as

$$\Phi_T F = \lambda h.\varphi_2(F(\varphi_1 \circ h)) = \lambda h.F(\mathsf{bind}_T^{B,C} \circ h)(\mathsf{val}_T^C)\ .$$

Lemma 14. *For any pure* $F \in \mathrm{Func}$ *and with* Φ_T *as above,* $\Phi_T(F_{\mathit{Cont}_{TC}}) = F_T$.

Proof. The idea is to construct a suitable acceptable monadic relation and exploit the purity of F. For X, X', $R \in \mathrm{Rel}(X, X')$, we define $T_2^{\mathsf{rel}}(R)$ as an element of $\mathrm{Rel}(\mathit{Cont}_{TC}X, TX')$ by letting $(H, H') \in T_2^{\mathsf{rel}}(R)$ iff

$$\forall h, h'.(h, h') \in R \to \Delta_{TC} \implies (Hh)\,\Delta_{TC}\,(\mathsf{bind}_T\, H'\, h') .$$

It is straightforward to show that T_2^{rel} is an acceptable monadic relation, so we omit the proof. Since F is pure, we have $(F_{\mathit{Cont}_{TC}}, F_T) \in (\Delta_A \to T_2^{\mathsf{rel}}(\Delta_B)) \to T_2^{\mathsf{rel}}(\Delta_C)$. Note that for any $g\colon A \to TB$,

$$\Phi_T(F_{\mathit{Cont}_{TC}})\, g = F_{\mathit{Cont}_{TC}}(\mathsf{bind}_T^{B,C} \circ g)(\mathsf{val}_T^C) \qquad \text{and}$$
$$F_T\, g = \mathsf{bind}_T^{C,C}(F_T\, g)(\mathsf{val}_T^C) .$$

First, we show that $(\mathsf{bind}_T^{B,C} \circ g, g) \in \Delta_A \to T_2^{\mathsf{rel}}(\Delta_B)$. Indeed, for any $a \in A$ and h, h' such that $(h, h') \in \Delta_B \to \Delta_{TC}$ (and thus, $h = h'$) we have $(\mathsf{bind}_T^{B,C} \circ g)\, a\, h = \mathsf{bind}_T^{B,C}(g\, a)\, h'$. Therefore, we conclude

$$(F_{\mathit{Cont}_{TC}}(\mathsf{bind}_T^{B,C} \circ g), \mathsf{bind}_T^{C,C}(F_T\, g)) \in T_2^{\mathsf{rel}}(\Delta_C) .$$

Since $(\mathsf{val}_T^C, \mathsf{val}_T^C) \in \Delta_C \to \Delta_{TC}$, the lemma is proved. □

Proof (of Theorem 12). With the help of lemmas we see that

$$\begin{aligned} F_T &= \Phi_T(F_{\mathit{Cont}_{TC}}) && \text{(by lemma 14)}\\ &= \Phi_T(\mathit{tree2fun}_{\mathit{Cont}_{TC}}(\mathit{fun2tree}\, F)) && \text{(by lemma 13)}\\ &= \mathit{tree2fun}_T(\mathit{fun2tree}\, F) && \text{(by lemmas 10, 14)} \end{aligned}$$

and the other inverse is established. □

We link the present result with that of [8]:

Corollary 15. *Any functional*

$$F : \forall S.(A \to \mathit{State}_S B) \to \mathit{State}_S C$$

which is pure in the sense of [8] may be implemented generically without using state, i.e., there exists a monadically parametric functional $G \in \mathrm{Func}$ *such that* $F_S = G_{\mathit{State}_S}$ *for all* S.

Proof. Take $G = \mathit{tree2fun}\, t_F$ where t_F is the tree representation of F.

5 The Partial Case

In this section, we generalize the characterisation of monadically parametric second-order functionals for the partial case in the domain-theoretic setting. In what follows, we will use the term *acceptable monadic relation* to refer to acceptable monadic relations which are strict and admissible as formulated in Definition 4.

5.1 Domain of Strategy Trees

We construct a cppo of "strategy trees" as a solution of a recursive domain equation $X \simeq \mathcal{F}(X)$ with a locally continuous functor $\mathcal{F} : \mathcal{C} \to \mathcal{C}$ for a suitable category $\mathcal{C}$ of domains.

Let $\eta_X : X \to X_\perp$ and $\mathit{kleisli}_X : (X \to X_\perp) \to (X_\perp \to X_\perp)$ be defined by

$$\eta_X\, x = x \qquad \mathit{kleisli}_X\, f\, x = \begin{cases} \perp & \text{if } x = \perp \\ f\, x & \text{otherwise} \end{cases} .$$

Define the lift monad $T_\perp$ over **Cpo** (category of cpos with continuous functions) by

$$T_\perp X = X_\perp, \quad \mathsf{val}_{T_\perp}^X = \eta_X, \quad \mathsf{bind}_{T_\perp}^{X,Y}\, t\, f = \mathit{kleisli}_X\, f\, t \ .$$

Let $\mathcal{F}(X) = C + B \times (A \to X_\perp)$ be such a functor for the Kleisli category for $T_\perp$ over **Cpo**. Let *Tree* be a cpo such that $\mathit{Tree} \simeq \mathcal{F}(\mathit{Tree})$, together with two (continuous) isomorphism functions

$$\begin{aligned} \mathsf{fold} &: C + B \times (A \to \mathit{Tree}_\perp) \to \mathit{Tree}_\perp \quad \text{and} \\ \mathsf{unfold} &: \mathit{Tree} \to (C + B \times (A \to \mathit{Tree}_\perp))_\perp \ , \end{aligned}$$

i.e., $\mathit{kleisli}(\mathsf{fold}) \circ \mathsf{unfold} = \eta_{\mathit{Tree}}$ and $\mathit{kleisli}(\mathsf{unfold}) \circ \mathsf{fold} = \eta_{\mathcal{F}(\mathit{Tree})}$ hold. For all isomorphisms in the Kleisli category for $T_\perp$, say, $f : X \to Y_\perp$ and $g : Y \to X_\perp$ that $\mathit{kleisli}(f) \circ g = \eta$ and $\mathit{kleisli}(g) \circ f = \eta$, f and g are total functions. Therefore, we can define total

$$\begin{aligned} \mathsf{roll} &: C + B \times (A \to \mathit{Tree}_\perp) \to \mathit{Tree} \quad \text{and} \\ \mathsf{unroll} &: \mathit{Tree} \to C + B \times (A \to \mathit{Tree}_\perp) \end{aligned}$$

using their "partial" counterparts fold and unfold. Moreover, the *minimal invariance* property takes place

$$\mathit{fixp}\, \delta = \eta$$

for $\delta : (\mathit{Tree} \to \mathit{Tree}_\perp) \to (\mathit{Tree} \to \mathit{Tree}_\perp)$ defined by $\delta\, e = \mathsf{fold} \circ F(e) \circ \mathsf{unfold}$. For details on a Coq development of the reverse-limit construction and a formal proof of the minimal invariance, refer to [3].

It is well known that the morphism fold forms an initial F-algebra in the Kleisli category, i.e., for any other F-algebra $\varphi : F(D) \to D$ there exists the *unique* homomorphism $h : \mathit{Tree} \to D_\perp$ such that the $\varphi \circ F(h) = h \circ \mathsf{fold}$.

Definition 16. We call elements of $\mathit{Tree}_\perp$ *strategy trees*. Define continuous "constructor" functions $\mathsf{Ans} : C \to \mathit{Tree}_\perp$ and $\mathsf{Que} : A \to (B \to \mathit{Tree}_\perp) \to \mathit{Tree}_\perp$ by

$$\mathsf{Ans} = \mathsf{fold} \circ \mathit{inl} \quad \text{and} \quad \mathsf{Que} = \mathsf{fold} \circ \mathit{inr} \ .$$

As in the total case, a strategy tree can be extracted by means of the continuation monad $\mathit{Cont}_{\mathit{Tree}_\perp}$. We define the extracting function $\mathit{fun2tree} : \mathrm{Func} \to \mathit{Tree}_\perp$ by

$$\mathit{fun2tree}\, F = F_{\mathit{Cont}_{\mathit{Tree}_\perp}}\, \mathsf{Que}\, \mathsf{Ans} \ .$$

The definition is correct since $Cont_{Tree_\bot}$ is a strict monad. The function *fun2tree* is strict and continuous.

The reverse translation mapping $Tree_\bot$ into Func_T is defined by means of the fixpoint operator $\mathit{fixp} : \forall D.(D \to D) \to D$ for cppos as follows. Given $T \in Monad$, we construct

$$\mathit{tree2fun}_T : Tree_\bot \to \mathrm{Func}_T = \mathit{fixp}\, G_T$$

where

$$G_T : (Tree_\bot \to \mathrm{Func}_T) \to Tree_\bot \to \mathrm{Func}_T = \lambda f.\, \mathit{kleisli}([\phi_T, \psi_T^f] \circ \mathsf{unroll})$$
$$\phi_T : C \to \mathrm{Func}_T = \lambda c.\lambda h.\, \mathsf{val}_T\, c$$
$$\psi_T^f : A \times (B \to Tree_\bot) \to \mathrm{Func}_T = \lambda p.\lambda h.\, \mathsf{bind}_T(h(\pi_1\, p))(\lambda b.(f \circ \pi_2\, p)\, b\, h)\ .$$

For every pointed $T \in Monad$, $\mathit{tree2fun}_T$ is correctly defined (since Func_T is pointed) and is continuous and strict. The parametric version is defined by $\mathit{tree2fun}\, t = \Lambda T.\, \mathit{tree2fun}_T\, t$. The following result is proved in Appendix.

Lemma 17. *For any $t \in Tree_\bot$, tree2fun t is pure.* □

5.2 Representation Theorem

Lemma 18. *For any $t \in Tree_\bot$, $\mathit{fun2tree}(\mathit{tree2fun}\, t) = t$.*

Proof. We note that $\mathit{fun2tree} \circ \mathit{tree2fun}$ is a homomorphism for *Tree*. Thus, the statement follows from initiality of fold. We give a direct formal proof using the minimal invariance property. □

Proofs of the following results are similar to the proofs in the total case.

Theorem 19. *For a pure $F \in \mathrm{Func}$,*

$$\mathit{tree2fun}_T(\mathit{fun2tree}\, F) = F_T$$

holds (extensionally) for any $T \in Monad$.

We first prove that the statement holds for an arbitrary continuation monad with a pointed result domain. See Appendix for the proof.

Lemma 20. *Given pure F, $\mathit{tree2fun}_{Cont_S}(\mathit{fun2tree}\, F) = F_{Cont_S}$ holds for any cppo S.* □

As in the total case, for $T \in Monad$ we define

$$\Phi_T : ((A \to Cont_{TC} B) \to Cont_{TC} C) \to (A \to TB) \to TC$$

and prove

Lemma 21. *For a pure $F \in \mathrm{Func}$ and $T \in Monad$, $\Phi_T(F_{Cont_{TC}}) = F_T$.*

Proof. The proof repeats the one of lemma 14. We only have to check that T_2^{rel} defined as in lemma 14 is a strict, admissible and acceptable monadic relation, which does hold. □

6 Generalizations

In this section, we argue that it is possible to extend the notion of purity to an arbitrary second-order type. Consider a general type n-Func of second-order functionals with n functional arguments

$$n\text{-Func} = \forall T.(A_1 \to TB_1) \to \cdots \to (A_n \to TB_n) \to TC \ .$$

Definition 22. A functional $F \in n$-Func is *pure (monadically parametric)* iff

$$(F_T, F_{T'}) \in (\Delta_{A_1} \to T^{\mathsf{rel}}(\Delta_{B_1})) \to \cdots \to (\Delta_{A_n} \to T^{\mathsf{rel}}(\Delta_{B_n})) \to T^{\mathsf{rel}}(\Delta_C)$$

holds for all $T, T' \in \mathit{Monad}$ and acceptable monadic relations T^{rel} for T, T'.

By theorem 7, any well-typed program of type n-Func is pure in this sense.

Definition 23. The set of *strategy trees n-Tree* is a minimal set generated by constructors

- $\mathsf{Ans} : C \to n\text{-}\mathit{Tree}$
- $\mathsf{Que}_i : A_i \to (B_i \to n\text{-}\mathit{Tree}) \to n\text{-}\mathit{Tree}$, $i = 1, \ldots, n$

Similar to the case of one functional argument, one defines functions

$$\mathit{tree2fun} : n\text{-}\mathit{Tree} \to n\text{-Func} \quad \text{and} \quad \mathit{fun2tree} : n\text{-Func} \to n\text{-}\mathit{Tree} \ .$$

Now, the result of Theorem 12 can be generalized for n-Func.

Theorem 24. *Given a pure $F \in n$-Func, $\mathit{tree2fun}_T(\mathit{fun2tree}\, F) = F_T$ holds (extensionally) for any $T \in \mathit{Monad}$.* □

The formal Coq proof of the theorem is provided in the total setting and uses dependent types.

Characterization for the type n-Func with a parameter type D (equivalently, with finitely many parameter types $D_1, \ldots, D_k$)

$$n\text{-Func}_D = \forall T.D \to (A_1 \to TB_1) \to \cdots \to (A_n \to TB_n) \to TC$$

is similar, with parameterized strategies of type

$$n\text{-}\mathit{Tree}_D = D \to n\text{-}\mathit{Tree} \ .$$

For types of order higher than two it is not that clear yet what corresponding strategies should be let alone how one could characterise their existence by parametricity. It could be, however, that strategies in the sense of game semantics, like in [1, 2, 9], are the right generalization. The another possible approach is in using of Kripke relations of varying arity as in [10]. This might be an interesting question for further investigation.

```
let rec solve (n:int) x s : Maybe S =
  match n with
  | 0 → None
  | _ →
   if is_stable x s then
     Some s
   else
     let s0 = add_stable x s in
     do p ← F (eval (n−1) x) s0;
     let (d, s1) = p in
     let cur = getval s1 x in
     if d ⊑ cur then
       Some s1
     else
       let s2 = setval x (cur ⊔ d) s1 in
       let (w, s3) = extract_work s2 in
         solve_all (n−1) w s3

and solve_all (n:int) w s : Maybe S =
  match n with
  | 0 → None
  | _ →
  match w with
  | [] → Some s
  | x :: xs →
    (solve (n−1) x s) ≫=
      solve_all (n−1) xs

and eval n x y : StateT_S Maybe D =
  match n with
  | 0 → fun s → None
  | _ → fun s →
    let s0 = add_infl y x s in
    do s1 ← solve (n−1) y s0;
    Some (getval s1 y, s1)
```

Fig. 1. The pure functional implementation of totalized RLD

7 Applications

Modulus of Continuity. Recall that a functional $F : \mathbb{B} \to \mathbb{N}$ defined on the Baire space $\mathbb{B} = \mathbb{N} \to \mathbb{N}$ is *continuous at* $f \in \mathbb{B}$ *if* $F\ f$ depends only on finitely many elements of f. A *modulus* for F at f is a number n such that $F\ f$ depends only on the first n terms of f. Suppose F is *pure* functional, i.e., it is given by means of a monadically parametric function $\overline{F} : \prod_T .(\mathbb{N} \to T\mathbb{N}) \to T\mathbb{N}$ such that $F = \overline{F}_{Id}$, where *Id* is the identity monad. Then we can effectively extract a modulus for F at f by means of the functional

$$\mathtt{Mod}\ F\ f = \max(snd\ (\overline{F}_{State_{list\,\mathbb{N}}}(\mathtt{instr}\ f)\ []))$$

where $\mathtt{instr}\ f : \mathbb{N} \to State_{list\,\mathbb{N}} = \lambda a.\lambda l.(f\ a, l{+\!\!+}[a])$ instruments f by means of recording of a list of visited indices. That `Mod` computes what it is supposed to is shown by the following proposition.

Proposition 25. *Let* $F : \mathbb{B} \to \mathbb{N}$ *be pure,* $f : \mathbb{B}$ *and* $m = \mathtt{Mod} F\, f$. *Then for every* $g : \mathbb{B}$ *if* $f\, i = g\, i$ *holds for all* $i \leq m$*, then* $F\, f = F\, g$. □

Proof (sketch). Given $l_f = snd\ (\overline{F}_{State_{list\,\mathbb{N}}}(\mathtt{instr}\ f)\ [])$, one can use l_f to traverse the strategy tree for $\overline{F}$ using values from l_f as corresponding answers at Que-nodes. We show that by so using l_f one reaches a leaf Ans n with $n = F\, f$. By assumption, for all the questions queried when traversing with l_f, f and g must deliver identical answers. We conclude, $l_f = l_g$, and hence $F\, f = F\, g$. □

Verified Fixpoint Algorithms. The provided characterization of pure functionals of type Func can be used for verification of generic off-the-shelf fixpoint algorithms which are used to compute a (local) solution of a constraint system $\mathbf{x} \sqsupseteq F_{\mathbf{x}}$, $\mathbf{x} \in V$, defined over a bounded join-semilattice $\mathbb{D}$ of abstract values and a set of variables V. The local solver RLD, which relies on *self-observation*, applies F to a special stateful function to discover variable dependencies and perform demand-driven evaluations [7]. In order to reason about the algorithm formally, we implement RLD in purely functional manner and model side-effects by means of the state monad. Thus, the pure right-hand side F is assumed to be of type

$$F : \forall S.V \to (V \to \mathit{State}_S\mathbb{D}) \to \mathit{State}_S\mathbb{D}\ .$$

Assuming that all right-hand sides F are pure and hence representable by strategy trees, one can formulate sufficient pre- and post-conditions to verify partial correctness of the algorithm.

Notice that RLD may diverge since we pose no extra restrictions on $\mathbb{D}$ (e.g., ascending chain condition) in general. However, we can define a totalized version of RLD by passing an extra natural parameter to every main function of the algorithm which limits a maximum depth of recursion. Once the limit is reached, the solver terminates with None. Figure 1 gives a pure functional implementation of the totalized version of RLD. Since F is pure, by Corollary 15, a corresponding strategy tree provides a monadically parametric implementation, which can be used as

$$F : V \to (V \to \mathit{StateT}_S\,\mathit{Maybe}\,\mathbb{D}) \to \mathit{StateT}_S\,\mathit{Maybe}\,\mathbb{D}$$

where *Maybe* is an option monad, *StateT* is a state monad transformer, and S is a state structure managed by the solver. The total version can be implemented and proven correct in Coq with the certified code extracted in ML.

The characterization of 2-Func can be applied to verification of local fixpoint algorithms for *side-effecting* constraint systems [19] used for interprocedural analysis and analysis of multithreaded code. The main idea here is that in each constraint $x \sqsupseteq F_{\mathbf{x}}$ the right-hand side $F_{\mathbf{x}}$ is a pure function representable by a strategy tree with two kind of question nodes: QueR for which values of variables are queried using a stateful function `get` and QueW which, when accessed, update current values of some variables by means of a stateful function `set`. Thus, the strategy tree specifies a sequence of *reading* and *writing* accesses to some constraint variables. One version of such a solver although not verified presently is implemented in the program analyzer Goblint [23].

8 Conclusion

We have provided two equivalent characterisations of pure second-order functionals in the presence of nontermination; an extensional one based on preservation of relations and an intensional one based on strategy trees. All verifications have been formalized in Coq.

Our results can be applied to the verification of algorithms that take pure second-order functionals as input. Among these are generic fixpoint algorithms and algorithms for exact real arithmetic. It is generally easier to verify the correctness of such an algorithm assuming the intensional characterisation of purity for its input. On the other hand, for a concretely given input, e.g. in the form of a program in some restricted language it will be easier to establish the extensional characterisation. The techniques developed in this paper were extended to impure higher-order functions enabling modular reasoning about monadic mixin components [13].

We note that a closely related characterisation albeit in a rather different guise has already been given in O'Hearn and Reynolds landmark paper [16]. Our strategy trees appear there as an intensional characterisation of first-order Algol procedures which due to the call-by-name policy are in fact second-order functionals. New aspects of the present work are in particular the monadic formulation, the generalisation of the extensional characterisation to monads other than the state monad, and the complete formalisation in Coq.

Interestingly, our acceptable monadic relations in the total case (Definition 4), also appear in [22] where they are used to derive free theorems in the sense of Wadler [24] for Haskell programs in monadic style. However, the application to the characterisation of pure second-order functionals and the subsequent characterisation with strategy trees do not appear in loc.cit. It is, however, fair to say that the method of [22], being essentially the same as ours, could be used to derive our main result (Representation Theorems), assuming that one adapts it to the partial case which was left open in loc.cit.

As pointed out by an anonymous reviewer, a proof of our results can be given using Katsumata's $\top\top$-lifting construction [12] if one considers stategy trees as free monads. This approach would require $\mathit{Tree} \in \mathit{Monad}$ for all possible result sets C. However, from the practical point of view, we would prefer that F is defined for continuation monads rather than for syntactical monads *Tree*.

A natural question, albeit of mostly academic interest, is the extension of this work to higher than second order. Given that the strategy trees resemble winning strategies in game semantics it would seem natural to attempt to find extensional characterisations of the existence of a winning strategy. Care would have to be taken so as to sidestep the undecidability of lambda definability [14], thus the extensional property would have to be undecidable even if basic types receive a finite interpretation.

Acknowledgements. We thank Alex Simpson, University of Edinburgh, for raising an interesting question and fruitful discussions on the topic. We thank Helmut Seidl and anonymous reviewers for valuable comments on the paper. The third author was supported by GRK 1480.

References

1. Abramsky, S., Malacaria, P., Jagadeesan, R.: Full Abstraction for PCF. In: Hagiya, M., Mitchell, J.C. (eds.) TACS 1994. LNCS, vol. 789, pp. 1–15. Springer, Heidelberg (1994)
2. Abramsky, S., McCusker, G.: Linearity, sharing and state: a fully abstract game semantics for idealized algol with active expressions. Electr. Notes Theor. Comput. Sci. 3 (1996)

3. Benton, N., Kennedy, A., Varming, C.: Formalizing domains, ultrametric spaces and semantics of programming languages. Submitted to Math. Struct. in Comp. Science (2010)
4. Charlier, B.L., Hentenryck, P.V.: A universal top-down fixpoint algorithm. Technical Report CS-92-25, Brown University, Providence, RI 02912 (1992)
5. Fecht, C., Seidl, H.: A faster solver for general systems of equations. Sci. Comput. Program. 35(2), 137–161 (1999)
6. Filinski, A.: Representing monads. In: Boehm, H.-J., Lang, B., Yellin, D.M. (eds.) POPL, pp. 446–457. ACM Press (1994)
7. Hofmann, M., Karbyshev, A., Seidl, H.: Verifying a Local Generic Solver in Coq. In: Cousot, R., Martel, M. (eds.) SAS 2010. LNCS, vol. 6337, pp. 340–355. Springer, Heidelberg (2010)
8. Hofmann, M., Karbyshev, A., Seidl, H.: What Is a Pure Functional? In: Abramsky, S., Gavoille, C., Kirchner, C., Meyer auf der Heide, F., Spirakis, P.G. (eds.) ICALP 2010. LNCS, vol. 6199, pp. 199–210. Springer, Heidelberg (2010)
9. Hyland, J.M.E., Ong, C.-H.L.: On full abstraction for PCF: I, ii, and iii. Inf. Comput. 163(2), 285–408 (2000)
10. Jung, A., Tiuryn, J.: A New Characterization of Lambda Definability. In: Bezem, M., Groote, J.F. (eds.) TLCA 1993. LNCS, vol. 664, pp. 245–257. Springer, Heidelberg (1993)
11. Karbyshev, A.: The accompanying Coq implementation (2013), `https://github.com/karbyshev/purity/`
12. Katsumata, S.-Y.: A Semantic Formulation of ⊤⊤-Lifting and Logical Predicates for Computational Metalanguage. In: Ong, L. (ed.) CSL 2005. LNCS, vol. 3634, pp. 87–102. Springer, Heidelberg (2005)
13. Keuchel, S., Schrijvers, T.: Modular monadic reasoning, a (co-)routine. IFL 2012, pre-proceedings, RR-12-06 (August 2012)
14. Loader, R.: The undecidability of lambda-definability
15. Longley, J.: When is a functional program not a functional program? In: ICFP, pp. 1–7 (1999)
16. O'Hearn, P.W., Reynolds, J.C.: From algol to polymorphic linear lambda-calculus. J. ACM 47(1), 167–223 (2000)
17. Reynolds, J.C.: Types, abstraction and parametric polymorphism. In: IFIP Congress, pp. 513–523 (1983)
18. Reynolds, J.C., Plotkin, G.D.: On functors expressible in the polymorphic typed lambda calculus. Inf. Comput. 105(1), 1–29 (1993)
19. Seidl, H., Vene, V., Müller-Olm, M.: Global invariants for analyzing multithreaded applications. Proc. of the Estonian Academy of Sciences: Phys. Math. 52(4), 413–436 (2003)
20. Simpson, A.K.: Lazy Functional Algorithms for Exact Real Functionals. In: Brim, L., Gruska, J., Zlatuška, J. (eds.) MFCS 1998. LNCS, vol. 1450, pp. 456–464. Springer, Heidelberg (1998)
21. The Coq Development Team. The Coq proof assistant reference manual. TypiCal Project (formerly LogiCal), Version 8.4 (2012)
22. Voigtländer, J.: Free theorems involving type constructor classes: functional pearl. In: Hutton, G., Tolmach, A.P. (eds.) ICFP, pp. 173–184. ACM (2009)
23. Vojdani, V., Vene, V.: Goblint: Path-sensitive data race analysis. In: Annales Univ. Sci. Budapest. Sect. Comp., vol. 30, pp. 141–155 (2009)
24. Wadler, P.: Theorems for free! In: FPCA, pp. 347–359 (1989)

Deconstructing General References via Game Semantics

Andrzej S. Murawski[1] and Nikos Tzevelekos[2,*]

[1] University of Warwick
[2] Queen Mary, University of London

Abstract. We investigate the game semantics of general references through the fully abstract game model of Abramsky, Honda and McCusker (AHM), which demonstrated that the visibility condition in games corresponds to the extra expressivity afforded by higher-order references with respect to integer references.

First, we prove a stronger version of the visible factorisation result from AHM, by decomposing any strategy into a visible one and a single strategy corresponding to a reference cell of type unit $\rightarrow$ unit (AHM accounted only for finite strategies and its result involved unboundedly many cells).

We show that the strengthened version of the theorem implies universality of the model and, consequently, we can rely upon it to provide semantic proofs of program transformation results. In particular, one can prove that any program with general references is equivalent to a purely functional program augmented with a single unit $\rightarrow$ unit reference cell and a single integer cell. We also propose a syntactic method of achieving such a transformation.

Finally, we provide a type-theoretic characterisation of terms in which the use of general references can be simulated with an integer reference cell or through purely functional computation, without any changes to the underlying types.

1 Introduction

In computer science, references are a programming idiom that allows the programmer to manipulate objects in computer memory. The referenced content can be accessed (dereferenced) or overwritten (updated). The most common sort of reference is that to a ground-type value, such as an integer. However, most modern programming languages allow more complicated values to be referenced. For example, the language ML features general references, where memory locations can contain values of any type, in particular of function type. Higher-order references are a very expressive construct. Among others, they can be used to simulate recursion, objects [5] and aspects [10]. In this paper we investigate higher-order references taking inspiration from their game semantic model [1]. In particular, we shall provide both semantic and syntactic accounts of how they can be decomposed, which highlight the fact that their full expressive power is already present in their simplest instance, references of type ref(unit $\rightarrow$ unit). Although in general the inclusion of higher-order references strictly increases expressivity, we also consider the question whether there are circumstances, delineated by types, where there is no such increase. Finally, we shall also answer the question when references

* Supported by a Royal Academy of Engineering research fellowship.

F. Pfenning (Ed.): FOSSACS 2013, LNCS 7794, pp. 241–256, 2013.

in general, integer-valued and higher-order ones, can be replaced altogether by purely functional computation over the same types.

Game semantics [2,7] is a semantic theory that interprets computation as an exchange of moves between two players: O (environment) and P (program). In the Hyland-Ong style of playing [7], moves are equipped with pointers to moves made earlier in the game, giving rise to plays like the one shown below.

$$o_1\ p_1\ o_2\ p_2\ o_3\ p_4$$

Existing literature on modelling integer [3] and higher-order references [1] showed that the expressive gap between the two paradigms can be captured by a property called *visibility*, which restricts the range of targets (earlier moves) for pointers: moves by P can only point at moves from a restricted fragment of the history, called the *view*. For example, the last move in the play above violates visibility, because before it is played, the view is $o_1\ p_1\ o_3$. Intuitively, the visibility constraint captures the intuition that, without higher-order references, the set of values available to a program is limited to the lexical environment (as captured by the notion of view). In particular, function values that are available at one point, cannot be taken for granted later during the course of computation. In contrast, in presence of higher-order references, such values can be recorded and reused at will. Hence, the visibility condition needs to be relaxed for modelling higher-order references.

A fully abstract game model of an ML-like language with references was presented in [1] and founded on plays that need not obey the visibility condition. As part of the full abstraction argument, the authors showed how to decompose every *finite* strategy into a finite strategy satisfying visibility and *several* strategies corresponding to reference cells of type $\mathsf{unit} \to \mathsf{unit}$ (Proposition 5 in [1]).

As our first contribution, we sharpen the result to *arbitrary* strategies as well as showing that *one* strategy corresponding to a $(\mathsf{unit} \to \mathsf{unit})$-valued memory cell is sufficient (Theorem 13). This brings two benefits. On the theoretical side, one can show universality: any recursively presentable strategy corresponds to a program with higher-order references. On a more practical note, the refined factorisation result can be applied to denotations of arbitrary programs to yield a powerful expressivity result: any program with higher-order references is equivalent to one of the shape $\mathsf{let}\ u = \mathsf{ref}(\lambda x^{\mathsf{unit}}.\Omega_{\mathsf{unit}})\ \mathsf{in}\ M$, where the ref-constructor in M is restricted to integers (Theorem 17).

Our first proof of the result is purely semantic and relies on recursion theory. Consequently, it does not offer much insight into how to transform the use of higher-order references into uses of a $(\mathsf{unit} \to \mathsf{unit})$-reference cell. Motivated by this, we try to identify semantics-preserving program transformations that will allow us to reprove the same result through syntactic means. The key element of our approach is the bad-variable constructor mkvar, which enables one to create terms of reference types with non-standard behaviour. Although the translation introduces extra occurrences of mkvar, we show how to eliminate them under certain conditions, namely, when the types associated with its free variables and the type of the term do not contain reference types. Note that this allows for arbitrarily complex private uses of general references inside terms as long as the references are not communicated through the program's type interface. From this perspective, mkvar emerges as a useful intermediate construct for

$$\frac{}{\Gamma \vdash () : \mathsf{unit}} \qquad \frac{i \in \mathbb{Z}}{\Gamma \vdash i : \mathsf{int}} \qquad \frac{(x : \theta) \in \Gamma}{\Gamma \vdash x : \theta} \qquad \frac{\Gamma \vdash M_1 : \mathsf{int} \quad \Gamma \vdash M_2 : \mathsf{int}}{\Gamma \vdash M_1 \oplus M_2 : \mathsf{int}}$$

$$\frac{\Gamma \vdash M : \mathsf{int} \quad \Gamma \vdash N_0 : \theta \quad \Gamma \vdash N_1 : \theta}{\Gamma \vdash \mathsf{if}\ M\ \mathsf{then}\ N_1\ \mathsf{else}\ N_0 : \theta} \qquad \frac{\Gamma \vdash M : \theta}{\Gamma \vdash \mathsf{ref}(M) : \mathsf{ref}(\theta)} \qquad \frac{\Gamma \vdash M : \mathsf{ref}(\theta)}{\Gamma \vdash !M : \theta}$$

$$\frac{\Gamma \vdash M : \mathsf{ref}(\theta) \quad \Gamma \vdash N : \theta}{\Gamma \vdash M := N : \mathsf{unit}} \qquad \frac{\Gamma \vdash M : \mathsf{unit} \to \theta \quad \Gamma \vdash N : \theta \to \mathsf{unit}}{\Gamma \vdash \mathsf{mkvar}(M, N) : \mathsf{ref}(\theta)}$$

$$\frac{\Gamma, x : \theta \vdash M : \theta'}{\Gamma \vdash \lambda x^{\theta}.M : \theta \to \theta'} \qquad \frac{\Gamma \vdash M : \theta \to \theta' \quad \Gamma \vdash N : \theta}{\Gamma \vdash MN : \theta'} \qquad \frac{\Gamma \vdash M : (\theta \to \theta') \to (\theta \to \theta')}{\Gamma \vdash \mathsf{Y}(M) : \theta \to \theta'}$$

Fig. 1. Typing judgments of $\mathcal{L}$

program transformation. Altogether our transformations yield an alternative syntactic proof of Theorem 17.

In the remainder of the paper, we give a type-theoretic characterization of terms in which the use of arbitrary references can be faithfully simulated using integer storage alone. Here is a representative selection of types in typing judgments that turn out to guarantee this property.

$$\cdots, \mathsf{int} \to \cdots \to \mathsf{int}, \cdots \vdash M : \mathsf{int},$$
$$\cdots, (\mathsf{int} \to \cdots \to \mathsf{int}) \to \mathsf{int}, \cdots \vdash M : \mathsf{int} \to \cdots \to \mathsf{int}$$

By highlighting the shape of types in the context, we mean to say that all free identifiers should have types of that form or simpler ones. These are the typing judgments over which there is no distinction in expressive power between integer and higher-order references.

Finally, we show that, as long as terms of the form

$$\cdots, x : \Theta_1, \cdots \vdash M : \beta$$

are considered, where $\beta ::= \mathsf{unit} \mid \mathsf{int}$ and $\Theta_1 ::= \beta \mid \mathsf{ref}(\beta) \mid \beta \to \Theta_1$, the use of higher-order references can be replaced with purely functional computation. That is to say, references do not contribute any expressive power. The last two results are obtained in a semantic way, by referring to game models and associated compositionality and universality results.

2 Syntax of the Language

We shall rely on the programming language $\mathcal{L}$ with general references introduced in [1]. Its types θ are generated from unit and int using the $\to$ and ref type constructors, as shown below.

$$\theta ::= \mathsf{unit} \mid \mathsf{int} \mid \mathsf{ref}(\theta) \mid \theta \to \theta$$

The typing rules are reproduced in Figure 1, where $\oplus$ is meant to cover standard arithmetic operations. The operational semantics of the language relies on a countable set L of typed locations. The values of the language are then the locations themselves, $()$, i, λ-abstractions and $\mathsf{mkvar}(V_1, V_2)$, where V_1, V_2 must be values. The big-step reduction judgments have the form $s, M \Downarrow s', V$, where s, s' are stores (partial functions from L to the set of values) and V is a value. Most reduction rules take the form

$$\frac{M_1 \Downarrow V_1 \quad M_2 \Downarrow V_2 \quad \cdots \quad M_n \Downarrow V_n}{M \Downarrow V}$$

which is meant to abbreviate

$$\frac{s_1, M_1 \Downarrow s_2, V_1 \quad s_2, M_2 \Downarrow s_3, V_2 \quad \cdots \quad s_n, M_n \Downarrow s_{n+1}, V_n}{s_1, M_1 \Downarrow s_{n+1}, V}.$$

In particular, this means that the ordering of the hypotheses is significant.

$$\frac{V \text{ is a value}}{s, V \Downarrow s, V} \qquad \frac{M \Downarrow 0 \quad N_0 \Downarrow V}{\mathsf{if}\ M\ \mathsf{then}\ N_1\ \mathsf{else}\ N_0 \Downarrow V} \qquad \frac{i \neq 0 \quad M \Downarrow i \quad N_1 \Downarrow V}{\mathsf{if}\ M\ \mathsf{then}\ N_1\ \mathsf{else}\ N_0 \Downarrow V}$$

$$\frac{M_1 \Downarrow i_1 \quad M_2 \Downarrow i_2}{M_1 \oplus M_2 \Downarrow i_1 \oplus i_2} \qquad \frac{M \Downarrow \lambda x.M' \quad N \Downarrow V' \quad M'[V'/x] \Downarrow V}{MN \Downarrow V}$$

$$\frac{s, M \Downarrow s', \ell \quad s'(\ell) = V}{s, !M \Downarrow s', V} \qquad \frac{s, M \Downarrow s', \ell \quad s', N \Downarrow s'', V}{s, M := N \Downarrow s''(\ell \mapsto V), ()}$$

$$\frac{M \Downarrow \mathsf{mkvar}(V_1, V_2) \quad V_1() \Downarrow V}{!M \Downarrow V} \qquad \frac{M \Downarrow \mathsf{mkvar}(V_1, V_2) \quad N \Downarrow V \quad V_2 V \Downarrow ()}{M := N \Downarrow ()}$$

$$\frac{M \Downarrow V_1 \quad N \Downarrow V_2}{\mathsf{mkvar}(M, N) \Downarrow \mathsf{mkvar}(V_1, V_2)} \qquad \frac{s, M \Downarrow s', V \quad \ell \notin \mathsf{dom}(s')}{s, \mathsf{ref}(M) \Downarrow s' \cup (\ell \mapsto V), \ell}$$

$$\frac{M \Downarrow \lambda x.M' \quad N \Downarrow V' \quad M'[V'/x] \Downarrow V}{MN \Downarrow V} \qquad \frac{M \Downarrow V}{\mathsf{fix}(M) \Downarrow \lambda x^{\theta}.(V(\mathsf{fix}(V)))x}$$

Given a closed term $\vdash M : \theta$ we write $M \Downarrow$ if there exist s', V such that $\emptyset, M \Downarrow s', V$.

Definition 1. *We shall say that two terms $\Gamma \vdash M_1 : \theta$ and $\Gamma \vdash M_2 : \theta$ are contextually equivalent (written $\Gamma \vdash M_1 \cong M_2$) if, for any context $C[-]$ such that $C[M_1], C[M_2]$ are closed, we have $C[M_1] \Downarrow$ if and only if $C[M_2] \Downarrow$.*

Remark 2. $\mathcal{L}$ features the "bad-reference" constructor mkvar in the style of Reynolds [9]. This makes it possible to construct objects of reference types from arbitrary read and write methods. In general this strengthens the discriminating power of contexts, as terms of ref-type can exhibit non-standard behaviour. However, it can be shown that when there are no ref-types in Γ or θ, this extension is inconsequential. At the technical level, this is due to the fact that the corresponding definability argument [1] need not rely on ref then.

Remark 3. $\mathcal{L}$ does not feature reference-equality testing as a primitive, as in general it would not make sense in a setting with bad references. Still, it is possible to construct a term that can tell two different locations apart by writing different values to them and testing their content. This is of course conditional on the existence of such values and our ability to distinguish them. In our setting, this method will be applicable to all types $\mathsf{ref}(\theta)$ except when $\theta \equiv \mathsf{unit}$.

Remark 4. In earlier work, we considered a language called RefML [8] with general references and equality testing for locations, in which bad references could not be created. The above comments imply that the respective notions of contextual equivalence induced by $\mathcal{L}$ and RefML coincide on mkvar-free $\mathcal{L}$-terms $\Gamma \vdash M : \theta$ such that there are no ref-types in Γ or θ. Similarly, one can also say that they converge for RefML-terms $\Gamma \vdash M : \theta$ such that Γ, θ do not contain ref-types and M does not use equality testing for references of type $\mathsf{ref}(\mathsf{unit})$.

We will now define a number of auxiliary terms that will turn out useful in subsequent arguments. As usual, $\mathsf{let}\, x = M \,\mathsf{in}\, N$ stands for $(\lambda x.N)M$. If x does not occur in N, we may also write $M; N$. We also rely on abbreviated notation for nested let's, e.g. $\mathsf{let}\, x, y = M_x, M_y \,\mathsf{in}\, N$ stands for $\mathsf{let}\, x = M_x \,\mathsf{in}\, \mathsf{let}\, y = M_y \,\mathsf{in}\, N$. We shall write Ω_θ for the divergent term $\mathsf{Y}(\lambda f^{\mathsf{unit}\to\theta}.f)\,()$. Also, for any type θ, we define a term $\vdash \mathsf{new}_\theta : \mathsf{ref}(\theta)$, which creates a suitably initialised reference cell.

$$\mathsf{new}_{\mathsf{unit}} \equiv \mathsf{ref}(()) \qquad \mathsf{new}_{\mathsf{ref}(\theta)} \equiv \mathsf{ref}(\mathsf{mkvar}(\lambda x^{\mathsf{unit}}.\, \Omega_\theta, \lambda x^\theta.\, \Omega_{\mathsf{unit}}))$$
$$\mathsf{new}_{\mathsf{int}} \equiv \mathsf{ref}(0) \qquad \mathsf{new}_{\theta\to\theta'} \equiv \mathsf{ref}(\lambda x^\theta.\Omega_{\theta'})$$

3 Game Model

The following arguments are couched in the game model of general references due to Abramsky, Honda and McCusker [1]. We use a more direct, yet equivalent, presentation due to Honda and Yoshida [6].

Definition 5. *An* **arena** $A = (M_A, I_A, \vdash_A, \lambda_A)$ *is given by*

- *a set M_A of moves, and a subset $I_A \subseteq M_A$ of* initial *moves,*
- *a justification relation $\vdash_A \subseteq M_A \times (M_A \setminus I_A)$, and*
- *a labelling function $\lambda_A : M_A \to \{O, P\} \times \{Q, A\}$*

such that $\lambda_A(I_A) = \{PA\}$. Additionally, whenever $m' \vdash_A m$, we have $(\pi_1\lambda_A)(m) \neq (\pi_1\lambda_A)(m')$, and $(\pi_2\lambda_A)(m') = A$ implies $(\pi_2\lambda_A)(m) = Q$.

The role of λ_A is to label moves as *Opponent* or *Proponent* moves and as *Questions* or *Answers*. We typically write them as $m, n, \dots$, or $o, p, q, a, q_P, q_O, \dots$ when we want to be specific about their kind. The simplest arena is $0 = (\emptyset, \emptyset, \emptyset, \emptyset)$. Other "flat" arenas are 1 and $\mathbb{Z}$, defined by $M_1 = I_1 = \{\star\}$, $M_{\mathbb{Z}} = I_{\mathbb{Z}} = \mathbb{Z}$. The two standard constructions on arenas are presented below, where $\overline{I}_A$ stands for $M_A \setminus I_A$, the domain restriction of a function is denoted by $\restriction$, the OP-complement of λ_A is written as $\overline{\lambda}_A$, and i_A, i_B range over initial moves in the respective arenas.

- $M_{A\Rightarrow B} = \{\star\} \uplus I_A \uplus \overline{I}_A \uplus M_B$, $I_{A\Rightarrow B} = \{\star\}$, $\lambda_{A\Rightarrow B} = [(\star, PA), (i_A, OQ), \overline{\lambda}_A \restriction \overline{I}_A, \lambda_B]$, $\vdash_{A\Rightarrow B} = \{(\star, i_A), (i_A, i_B)\} \cup \vdash_A \cup \vdash_B$.

– $M_{A\otimes B} = (I_A \times I_B) \uplus \overline{I}_A \uplus \overline{I}_B$, $I_{A\otimes B} = I_A \times I_B$, $\lambda_{A\otimes B} = [((i_A, i_B), PA), \lambda_A \upharpoonright \overline{I}_A, \lambda_B \upharpoonright \overline{I}_B]$, $\vdash_{A\otimes B} = \{((i_A, i_B), m) \mid i_A \vdash_A m \vee i_B \vdash_B m\} \cup (\vdash_A \upharpoonright \overline{I}_A{}^2) \cup (\vdash_B \upharpoonright \overline{I}_B{}^2)$.

Types of $\mathcal{L}$ can now be interpreted with arenas in the following way.

$$[\![\mathsf{unit}]\!] = 1 \qquad [\![\mathsf{ref}(\theta)]\!] = (1 \Rightarrow [\![\theta]\!]) \otimes ([\![\theta]\!] \Rightarrow 1)$$
$$[\![\mathsf{int}]\!] = \mathbb{Z} \qquad [\![\theta_1 \to \theta_2]\!] = [\![\theta_1]\!] \Rightarrow [\![\theta_2]\!]$$

Example 6. $[\![\mathsf{ref}(\mathsf{int})]\!]$ and $[\![\mathsf{ref}(\mathsf{unit} \to \mathsf{unit})]\!]$ have the following respective shapes.

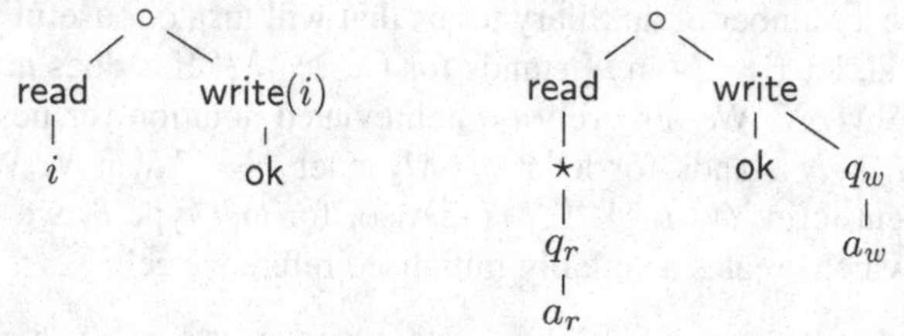

Although arenas model types, the actual games will be played in ***prearenas***, which are defined in the same way as arenas with the exception that initial moves must be O-questions. Given arenas A and B, we can construct the prearena $A \to B$ by setting: $M_{A\to B} = M_A \uplus M_B$, $I_{A\to B} = I_A$, $\lambda_{A\to B} = [(i_A, OQ) \cup (\overline{\lambda}_A \upharpoonright \overline{I}_A), \lambda_B]$ and $\vdash_{A\to B} = \{(i_A, i_B)\} \cup \vdash_A \cup \vdash_B$. A ***justified sequence*** in a prearena A is a finite sequence s of moves of A satisfying the following condition: the first move must be initial, but all other moves m must be equipped with a pointer to an earlier occurrence of a move m' such that $m' \vdash_A m$. We then say that m' *justifies* m. If m is an answer, we also say that m *answers* m'. Given a justified sequence, the last unanswered question will be called *pending*.

Definition 7. *A* **play** *in A is a justified sequence satisfying alternation (players take turns) and well-bracketing (whenever a player plays an answer, it must answer the current pending question). A* **strategy** *in a prearena A is a subset σ of even-length plays in A that is closed under the operation of taking even-length prefixes and satisfies determinacy: if $sp_1, sp_2 \in \sigma$ then $sp_1 = sp_2$.*

Example 8. $\mathsf{cell}_{\mathsf{int}} : 1 \to [\![\mathsf{ref}(\mathsf{int})]\!]$ answers the initial question with $\circ$. Whenever O plays $\mathsf{write}(i)$, it responds with ok. After O plays read, it responds with an integer value present in the latest $\mathsf{write}(i)$ move by O or, if none has been played, with 0. This strategy will model $\vdash \mathsf{ref}(0) : \mathsf{ref}(\mathsf{int})$.

$\mathsf{cell}_{\mathsf{unit}\to\mathsf{unit}} : 1 \to [\![\mathsf{ref}(\mathsf{unit} \to \mathsf{unit})]\!]$ answers the initial question with $\circ$, responds to write and read with ok and $*$ respectively. If O plays q_r justified by an occurrence of $\star$, P plays q_w justified by the last occurrence of ok that precedes the relevant occurrence of $\star$. If none such exists, P has no response. Similarly, if O plays a_w, P will respond with a_r. This strategy will interpret $\vdash \mathsf{ref}(\lambda x^{\mathsf{unit}}.\Omega_{\mathsf{unit}}) : \mathsf{ref}(\mathsf{unit} \to \mathsf{unit})$.

Strategies compose [6], yielding a category of games where objects are arenas and morphisms between objects A and B are strategies in $A \to B$. Let $\Gamma = \{x_1 : \theta_1, \cdots, x_n : \theta_n\}$. We shall write $[\![\Gamma \vdash \theta]\!]$ for the prearena $[\![\theta_1]\!] \otimes \cdots \otimes [\![\theta_n]\!] \to [\![\theta]\!]$ (if $n = 0$ we take the left-hand side to be 1). The game model proposed in [1] interprets a term $\Gamma \vdash M : \theta$ by a strategy in $[\![\Gamma \vdash \theta]\!]$.

We now introduce another condition on plays, known to characterize denotations of terms with ground-type storage only.

Definition 9 (Visibility). *The* view *of a play is inductively defined by:*

$$view(\epsilon) = \epsilon \qquad view(m) = m \qquad view(s_1 \overset{\frown}{m\, s_2\, n}) = view(s_1)\, \overset{\frown}{m\, n}\,.$$

A play s satisfies the **visibility** *condition if, for all even-length prefixes $s'm$ of s, the justifier of m occurs in $view(s')$. A strategy is called* **visible** *if it contains only visible plays.*

It can be shown that in plays the above condition is never violated by answers, because the pending question is always present in the view.

Proposition 10 ([3]). *Let $\Gamma \vdash M : \theta$ be a term in which applications of the* ref(−)*-constructor are restricted to terms of type* unit *and* int*. Then $[\![\Gamma \vdash M : \theta]\!]$ satisfies the visibility condition.*

4 Factorisation

We shall next write $!_A$ for the strategy in $A \to 1$ that responds to the initial move on the left with the unique move on the right. Given strategies $\sigma_i : 1 \to A_i$ that all respond to the initial question, we write $\langle\, \sigma_1, \cdots, \sigma_n \,\rangle$ for the strategy in $1 \to \bigotimes_{i=1}^{n} A_i$ that responds to the initial move with the tuple containing the individual responses of the n strategies and otherwise behaves like σ_i, depending on the component A_i in which O chooses to play.

Let us recall the factorisation result from [1].

Theorem 11 ([1]). *Let $\sigma : A_1 \to A_2$ be a finite strategy and $A = [\![\mathsf{ref}(\mathsf{unit} \to \mathsf{unit})]\!]$. There exists a natural number n and a visible strategy $\overline{\sigma} : (\bigotimes_{i=1}^{n} A) \otimes A_1 \to A_2$ such that $\langle\, \tau, \cdots, \tau, \mathsf{id}_{A_1} \,\rangle; \overline{\sigma} = \sigma$, where $\tau = !_{A_1}; \mathsf{cell}_{\mathsf{unit} \to \mathsf{unit}}$.*

Note that in the result above n may depend on σ. In fact, the proof shows that n can be taken to be (roughly) the length of the longest play in σ.

Remark 12. Violations of visibility describe computational scenarios in which a program attempts to refer to a value that was previously encountered during computation, yet which is not in current scope. The argument from [1] proposes to repair such violations by using free (higher-order) reference variables. Intuitively, they provide an opportunity to record the values currently available to the program. A later attempt to access the reference makes it possible to use the required value. In contrast, our argument will take advantage of a single reference cell. We shall also record the scope at each step, but before doing so we will embed the previous value into the current scope, thus allowing backtracking. In this way, the sought value can be found by backtracking to the desired computational step.

Theorem 13 (Visible Factorisation). *Let $\sigma : A_1 \to A_2$ be a strategy and $A = [\![\mathsf{ref}(\mathsf{unit} \to \mathsf{unit})]\!]$. There exists a visible strategy $\overline{\sigma} : A \otimes A_1 \to A_2$ such that $\langle \tau, \mathsf{id}_{A_1} \rangle ; \overline{\sigma} = \sigma$, where $\tau = !_{A_1}; \mathsf{cell}_{\mathsf{unit}\to\mathsf{unit}}$. If σ is recursively presentable, so is $\overline{\sigma}$.*

Proof. We shall define $\overline{\sigma}$ to be the least strategy containing the plays from $\{\overline{s} \mid s \in \sigma\}$, where $\overline{s}$ will be defined below by induction on the length of a play. Roughly, $\overline{s}$ will consist of s augmented with moves from A.

– In particular, immediately after each O-move of s we shall insert the sequence read $\star$ write ok and, if the move is an answer, it will be followed by a sequence consisting of answers a_w, a_r. Intuitively, each sequence read $\star$ write ok corresponds to reading the current value of the reference (the one modelled in A) and updating it with a new value.

– For P-questions, we shall insert read $\star$ followed by a sequence consisting of questions q_r, q_w *in front of* the P-question. The last q_w will point at the value stored in the reference immediately after the justifier of q was played. P-answers will simply be copied without any extra moves.

We give a precise definition below. The targets of pointers from read, $\star$, write, ok are obvious so, when discussing pointers, we shall focus on those from q_r, q_w, a_r, a_w.

- $\overline{s\, q_O} = \overline{s}\, q_O\, \mathsf{read} \star \mathsf{write}\,\mathsf{ok}$ (if $|s| > 0$); and $\overline{q_O} = q_O\, \mathsf{write}\,\mathsf{ok}$
- $\overline{s\, q_P} = \overline{s}\, \mathsf{read} \star q_r\, (q_w\, q_r)^k\, q_w\, q_P$

  We take k to be the number of O-moves occurring after the justifier o of q_P in s. Let us list them (in order of occurrence) as $o_k, \cdots, o_1$. Then the ith q_w and q_r in $(q_w\, q_r)^k$ are meant to be justified by respectively write and $\star$ from the read $\star$ write ok segment introduced immediately after o_i. The last q_w is justified by write from the read $\star$ write ok segment added after o.
  Note that the resultant sequence will satisfy P-visibility, even if q_P may not have. Additionally, the extra O-moves $\star$, ok, q_w are consistent with the behaviour of the $\mathsf{cell}_{\mathsf{unit}\to\mathsf{unit}}$ strategy.

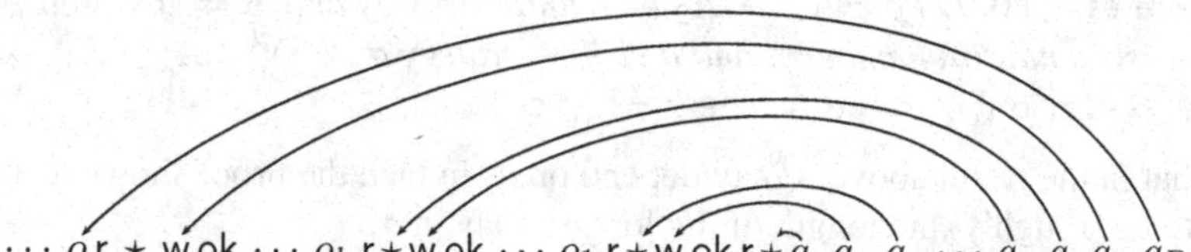

$$\cdots o\, \mathsf{r} \star \mathsf{w}\,\mathsf{ok} \cdots o_k\, \mathsf{r}\star\mathsf{w}\,\mathsf{ok} \cdots o_1\, \mathsf{r}\star\mathsf{w}\,\mathsf{ok}\,\mathsf{r}\star q_r\, q_w\, q_r \cdots q_w\, q_r\, q_w\, q_P$$

- $\overline{s\, a_O} = \overline{s}\, a_O\, \mathsf{read} \star \mathsf{write}\,\mathsf{ok}\, (a_w\, a_r)^{k+1}$
  Suppose a_O answers q_P in s. Then we take k to be the same as in the clause for q_P, i.e. k is the number of O-moves separating q_P's justifier and q_P. The sequence $(a_w\, a_r)^{k+1}$ simply answers all the questions q_w, q_r that were introduced for q_P. Because the pending question of $\overline{s}$ stays the same as that in s, this will yield a valid play. Note also that the O-moves a_r (in response to a_w) are consistent with the $\mathsf{cell}_{\mathsf{unit}\to\mathsf{unit}}$ strategy.

$$\cdots \mathsf{r} \star q_r\, (q_w\, q_r)^k\, q_w\, q_P \cdots a_O\, (a_w\, a_r)^{k+1}$$

– $\overline{s\,a\,p} = \overline{s}\,a\,p$
As we have already mentioned, the construction of $\overline{s}$ from s preserves the pending question. Hence, the above clause leads to a play.

Consequently, $\overline{\sigma}$ is visible and, because the inserted moves are consistent with $\mathsf{cell}_{\mathsf{unit}\to\mathsf{unit}}$, we have $\langle\, !_{A_1}; \mathsf{cell}_{\mathsf{unit}\to\mathsf{unit}}, \mathsf{id}_{A_1}\rangle; \overline{\sigma} = \sigma$. That $\overline{\sigma}$ is recursively presentable follows from our description above. □

In order to apply the Theorem we need two more results. The first of them is classic and concerns decomposing visible strategies into innocent ones. Innocence [7] is a condition even stricter than visibility: responses of innocent strategies are uniquely determined by views.

Theorem 14 (Innocent Factorisation [4]). *Let $\sigma : A_1 \to A_2$ be a visible strategy and $A = [\![\mathsf{ref}(\mathsf{int})]\!]$. There exists an innocent strategy $\hat{\sigma} : A \otimes A_1 \to A_2$ such that $\langle\, !_{A_1}; \mathsf{cell}_{\mathsf{int}}, \mathsf{id}_{A_1}\rangle; \hat{\sigma} = \sigma$.*

Note that this result already applies to arbitrary strategies rather than just finite ones. Also, the construction of $\hat{\sigma}$ is effective and shows that $\hat{\sigma}$ is recursively presentable if σ is.

Finally, we prove a universality result for recursively presentable innocent strategies. Universality results were not necessary in research on full abstraction, because their weaker variants phrased for finite (or finitely generated) strategies sufficed to capture possible separating contexts. Hence, after the initial ones for PCF [2,7], they all but disappeared from subsequent papers. For program transformations, though, we need to be able to express arbitrary recursive strategies, hence the need for universality. Note that there is a huge difference between finite and recursive strategies. For instance, the strategy corresponding to $\lambda x^{\mathsf{int}}.x$ is not finite.

Theorem 15 (Innocent Universality). *Let $\sigma : [\![\Gamma \vdash \theta]\!]$ be a recursively presentable innocent strategy. There exists a* ref*-free term $\Gamma \vdash M : \theta$ such that $[\![\Gamma \vdash M : \theta]\!] = \sigma$. If Γ and θ do not contain occurrences of* ref*-types, then M can be taken to be* mkvar*-free.*

By appealing to Theorems 13, 14 and 15 one can deduce Universality.

Theorem 16 (Universality). *Let $\sigma : [\![\Gamma \vdash \theta]\!]$ be a recursively presentable strategy. Then there exists $\Gamma \vdash M : \theta$ such that $[\![\Gamma \vdash M : \theta]\!] = \sigma$.*

In fact, in the above statement M can be taken to be of the form

$$\mathsf{let}\, f, x = \mathsf{new}_{\mathsf{unit}\to\mathsf{unit}}, \mathsf{new}_{\mathsf{int}}\ \mathsf{in}\ M',$$

where M' is ref-free. Because the game semantics of a term is recursively presentable, we can conclude the following result.

Theorem 17 (Transformation). *Let $\Gamma \vdash M : \theta$. There exists a term $\Gamma, f : \mathsf{ref}(\mathsf{unit} \to \mathsf{unit}), x : \mathsf{ref}(\mathsf{int}) \vdash M' : \theta$ satisfying the following conditions.*

- *$\Gamma \vdash M \cong \mathsf{let}\, f, x = \mathsf{new}_{\mathsf{unit}\to\mathsf{unit}}, \mathsf{new}_{\mathsf{int}}\ \mathsf{in}\ M'$.*
- *M' is* ref*-free.*
- *If there are no occurrences of* ref *in Γ, θ, then M' is* mkvar*-free.*

Thus, general references in $\mathcal{L}$ can be simulated by two memory cells that store values of type $\mathsf{unit} \to \mathsf{unit}$ and int respectively. Our proof was semantic, but the passage from M to M' can be made effective. However, due to reliance on the universality result, we would need to pass through enumerations of partial recursive functions. This is hardly a reasonable way of transforming programs! Next we shall identify several syntactic decomposition principles for general references, which will yield an alternative proof of the Theorem.

5 Syntactic Transformation

Note that $\mathsf{ref}(M)$ is equivalent to $\mathsf{let}\, x = \mathsf{new}_\theta\, \mathsf{in}\, (x := M; x)$ for a suitable θ. Consequently, w.l.o.g. we can assume that the only occurrences of $\mathsf{ref}(\cdots)$ inside terms are those associated with new_θ. Similarly, we assume that terms do not contain fixed-point subterms, as these can be simulated using higher-order reference cells [1].

Next we show new_θ can be decomposed using instances of new at simpler types. Ultimately, this will allow us to replace any occurrences of $\mathsf{ref}(M)$ with $\mathsf{new}_{\mathsf{unit}\to\mathsf{unit}}$ and $\mathsf{new}_{\mathsf{int}}$. The mkvar constructor is central to the transformations.

Lemma 18 (Decomposition of $\mathsf{ref}(\theta_1 \to \theta_2)$). *For all θ_1, θ_2, $\vdash \mathsf{new}_{\theta_1\to\theta_2} \cong \mathsf{let}\, f, x_1, x_2 = \mathsf{new}_{\mathsf{unit}\to\mathsf{unit}}, \mathsf{new}_{\theta_1}, \mathsf{new}_{\theta_2}\, \mathsf{in}\, \mathsf{mkvar}(M_r, M_w)$, where*

$$M_r \equiv \lambda y^{\mathsf{unit}}.\mathsf{let}\, h = !f\, \mathsf{in}\, \lambda z^{\theta_1}.\, (x_1 := z;\, h();\, !x_2),$$
$$M_w \equiv \lambda g^{\theta_1\to\theta_2}.\, f := (\lambda z^{\mathsf{unit}}.\, x_2 := g(!x_1)).$$

We can show the equivalence formally by comparing strategies corresponding to each term. Intuitively, the equivalence is valid because on assignment M_w indirectly records the assigned value g in f. On dereferencing, M_r ensures that the latest value of f is accessed and the corresponding value g applied to the right argument through the internal references x_1 and x_2.

Lemma 19 (Decomposition of $\mathsf{ref}(\mathsf{ref}(\theta))$). *For any θ, $\vdash \mathsf{new}_{\mathsf{ref}(\theta)} \cong \mathsf{let}\, r, w = \mathsf{new}_{\mathsf{unit}\to\theta}, \mathsf{new}_{\theta\to\mathsf{unit}}\, \mathsf{in}\, \mathsf{mkvar}(M_r, M_w)$ for all θ, where*

$$M_r \equiv \lambda z^{\mathsf{unit}}.\, \mathsf{mkvar}(!r, !w),$$
$$M_w \equiv \lambda g^{\mathsf{ref}(\theta)}.\, (r := (\lambda z^{\mathsf{unit}}.\, !g);\;\; w := (\lambda z^{\theta}.\, g := z)).$$

Here, instead of storing a reference of type θ, we store the associated read and write methods, of types $\mathsf{unit} \to \theta$ and $\theta \to \mathsf{unit}$ respectively, which is what references r and w are used for.

Lemma 20. $\vdash \mathsf{new}_{\mathsf{unit}} \cong \mathsf{mkvar}(\lambda x^{\mathsf{unit}}.\, (), \lambda x^{\mathsf{unit}}.\, ())$.

The Lemma is easy to verify by reference to the game model. It illustrates the rather strange status of type $\mathsf{ref}(\mathsf{unit})$ in $\mathcal{L}$, in particular the fact that it is not possible to compare reference names (of type $\mathsf{ref}(\mathsf{unit})$) in the language.

The last three Lemmas imply the following corollary.

Corollary 21. *For any $\Gamma \vdash M : \theta$ there exists $\Gamma \vdash M'' : \theta$ such that $\Gamma \vdash M \cong M'' : \theta$ and occurrences of the* ref *constructor in M'' are restricted to terms of the form* $\mathsf{new}_{\mathsf{unit}\to\mathsf{unit}}$ *or* $\mathsf{new}_{\mathsf{int}}$.

In the result above, $\mathsf{new}_{\mathsf{unit}\to\mathsf{unit}}$ and $\mathsf{new}_{\mathsf{int}}$ are allowed to occur multiple times. In what follows we shall show that one occurrence of each suffices.

Lemma 22. *There exist* ref*-free terms M, N such that*

$$\vdash \lambda x^{\mathsf{unit}}.\,\mathsf{new}_{\mathsf{unit}\to\mathsf{unit}} \cong \mathsf{let}\, f, x = \mathsf{new}_{\mathsf{int}\to\mathsf{unit}}, \mathsf{new}_{\mathsf{int}}\,\mathsf{in}\, M : \mathsf{unit} \to \mathsf{ref}(\mathsf{unit} \to \mathsf{unit}),$$
$$\vdash \lambda x^{\mathsf{unit}}.\,\mathsf{new}_{\mathsf{int}} \cong \mathsf{let}\, x = \mathsf{new}_{\mathsf{int}}\,\mathsf{in}\, N : \mathsf{unit} \to \mathsf{ref}(\mathsf{int}).$$

Proof. We can encode an unbounded number of references of type $\mathsf{ref}(\mathsf{unit} \to \mathsf{unit})$ with a reference f of type $\mathsf{ref}(\mathsf{int} \to \mathsf{unit})$ by giving to each $(\mathsf{unit} \to \mathsf{unit})$-valued reference a unique integer identifier i, and encoding the value of the ith such reference as $\lambda v^{\mathsf{unit}}.\,(!f)i$. We use the internal variable x to count the number of generated references, so as to assign them unique identifiers. Thus, M can be taken to be $\lambda z^{\mathsf{unit}}.\mathsf{let}\, i = \,!x\,\mathsf{in}\,(x := !x + 1); \mathsf{mkvar}(M_r, M_w)$, where $M_r \equiv \lambda u^{\mathsf{unit}}.\mathsf{let}\, h = !f\,\mathsf{in}\,\lambda v^{\mathsf{unit}}.\,h\,i$ and $M_w \equiv \lambda g^{\mathsf{unit}\to\mathsf{unit}}.\,\mathsf{let}\, g' = !f\,\mathsf{in}\, f := (\lambda y^{\mathsf{int}}.\,\mathsf{if}\, y = i\,\mathsf{then}\, g()\,\mathsf{else}\, g'y)$.

For the second part, assume a standard encoding $\mathsf{G}(-) : \mathbb{Z}^* \to \mathbb{Z}$ of lists of integers into integers such that $\mathsf{G}(\epsilon) = 0$. Clearly, one can construct closed PCF terms $\mathsf{len} : \mathsf{int} \to \mathsf{int}$, $\mathsf{add} : \mathsf{int} \to \mathsf{int} \to \mathsf{int}$, $\mathsf{proj} : \mathsf{int} \to \mathsf{int} \to \mathsf{int}$ and $\mathsf{upd} : \mathsf{int} \to \mathsf{int} \to \mathsf{int} \to \mathsf{int}$ such that, for all $s \in \mathbb{Z}^*$ and $i, j \in \mathbb{Z}$:

$$\mathsf{len}\,\mathsf{G}(s) \Downarrow |s|\,, \quad \mathsf{add}\,\mathsf{G}(s)\,i \Downarrow \mathsf{G}(si)\,, \quad \mathsf{proj}\,\mathsf{G}(s)\,j \Downarrow s_j\,, \quad \mathsf{upd}\,\mathsf{G}(s)\,j\,i \Downarrow \mathsf{G}(s[j \mapsto i])\,,$$

where $|s|$ is the length of s, s_j is the jth element of s, and $s[j \mapsto i]$ is the list s with its jth element changed to i. We can then keep track of an unbounded number of integer-valued references by taking N to be

$$\lambda z^{\mathsf{unit}}.\, x := \mathsf{add}\,(!x)\,0; \mathsf{let}\, j{=}\mathsf{len}(!x)\,\mathsf{in}\,\mathsf{mkvar}(\lambda z^{\mathsf{unit}}.\,\mathsf{proj}\,(!x)\,j, \lambda i^{\mathsf{int}}.\, x := \mathsf{upd}\,(!x)\,j\,i).$$

□

Now we are ready to give a new proof of Theorem 17. For a start, we tackle the first two claims therein.

Proof. Given $\Gamma \vdash M : \theta$, from Corollary 21 we can obtain an equivalent term M'', in which occurrences of ref are restricted to $\mathsf{new}_{\mathsf{unit}\to\mathsf{unit}}$ and $\mathsf{new}_{\mathsf{int}}$. Observe that M'' is thus equivalent to $\mathsf{let}\, h = \lambda x^{\mathsf{unit}}.\mathsf{new}_{\mathsf{unit}\to\mathsf{unit}}\,\mathsf{in}\, M_1$, where $M_1 \equiv M''[h()/\mathsf{new}_{\mathsf{unit}\to\mathsf{unit}}]$ and the only occurrences of ref in M_1 are those of $\mathsf{new}_{\mathsf{int}}$. Applying Lemmata 22, 20 and 18, M'' is further equivalent to $\mathsf{let}\, f = \mathsf{new}_{\mathsf{unit}\to\mathsf{unit}}\,\mathsf{in}\, M_2$, where the only occurrences of ref in M_2 are those of $\mathsf{new}_{\mathsf{int}}$. Finally, noting that M_2 is equivalent to $\mathsf{let}\, h' = \lambda x^{\mathsf{unit}}.\mathsf{new}_{\mathsf{int}}\,\mathsf{in}\, M_3$, where M_3 is ref-free, and invoking Lemma 18 we can conclude that M_2 is equivalent to $\mathsf{let}\, x = \mathsf{new}_{\mathsf{int}}\,\mathsf{in}\, M_4$, where M_4 is ref-free. Now we can take M' (from the statement of the Theorem) to be $\mathsf{let}\, f, x = \mathsf{new}_{\mathsf{unit}\to\mathsf{unit}}, \mathsf{new}_{\mathsf{int}}\,\mathsf{in}\, M_4$. □

Note that the decompositions presented in this Section relied on the availability of mkvar and the term M' from the above proof will in general contain many occurrences of mkvar. We devote the remainder of this Section to showing that when Γ and θ are ref-free, all the occurrences of mkvar can actually be eliminated. To that end, we shall rely on a notion of canonical form, defined below.

$$\begin{array}{rl}\mathbb{C} ::= & () \mid x^{\mathsf{int}} \mid \mathsf{mkvar}(\lambda u^{\mathsf{unit}}.\mathbb{C}, \lambda v^{\theta}.\mathbb{C}) \mid \lambda x^{\theta}.\mathbb{C} \mid \mathsf{if}\,\mathbb{C}\,\mathsf{then}\,\mathbb{C}\,\mathsf{else}\,\mathbb{C} \mid \\ & \mathsf{let}\,y = i\,\mathsf{in}\,\mathbb{C} \mid \mathsf{let}\,y = \mathbb{C} \oplus \mathbb{C}\,\mathsf{in}\,\mathbb{C} \mid \mathsf{let}\,y = !x\,\mathsf{in}\,\mathbb{C} \mid \mathsf{let}\,y = (x := \mathbb{C})\,\mathsf{in}\,\mathbb{C} \mid \\ & \mathsf{let}\,y = x\,\mathbb{C}\,\mathsf{in}\,\mathbb{C} \mid \mathsf{let}\,y = \mathsf{ref}(\mathbb{C})\,\mathsf{in}\,\mathbb{C}\end{array}$$

The canonical forms enjoy the following property.

Lemma 23. *For any $\Gamma \vdash M : \theta$ without fixed points, there exists a term $\mathbb{C}_M$ in canonical form such that $\Gamma \vdash M \cong \mathbb{C}_M : \theta$. Moreover, $\mathbb{C}_M$ can be effectively found and the conversion does not add any occurrences of* ref.

It turns out that canonical subterms of canonical terms have types drawn from a rather restricted set. We make this statement precise below.

Definition 24. *Given a type θ, the sets $\mathsf{PST}(\theta)$ (of positive subtypes of θ) and $\mathsf{NST}(\theta)$ (of negative subtypes of θ) are defined respectively as follows. Let us write $\mathsf{ST}(\theta)$ for $\mathsf{PST}(\theta) \cup \mathsf{NST}(\theta)$.*

$$\begin{array}{rlrl}\mathsf{PST}(\mathsf{unit}) = \{\mathsf{unit}\} & & \mathsf{PST}(\mathsf{ref}(\theta)) = \mathsf{ST}(\theta) \cup \{\mathsf{ref}(\theta)\} \\ \mathsf{PST}(\mathsf{int}) = \{\mathsf{int}\} & & \mathsf{PST}(\theta_1 \to \theta_2) = \mathsf{NST}(\theta_1) \cup \mathsf{PST}(\theta_2) \cup \{\theta_1 \to \theta_2\} \\ \mathsf{NST}(\mathsf{unit}) = \emptyset & & \mathsf{NST}(\mathsf{ref}(\theta)) = \mathsf{ST}(\theta) \\ \mathsf{NST}(\mathsf{int}) = \emptyset & & \mathsf{NST}(\theta_1 \to \theta_2) = \mathsf{PST}(\theta_1) \cup \mathsf{NST}(\theta_2)\end{array}$$

Given a canonical form $\mathbb{C}$ such that $\Gamma \vdash \mathbb{C} : \theta$, let $\mathsf{RT}(\mathbb{C})$ stand for the set of types θ' such that $\mathbb{C}$ contains an occurrence of $\mathsf{ref}(\mathbb{C}')$, where $\mathbb{C}'$ of type θ'. It turns out that the types in $\mathsf{RT}(\mathbb{C})$ together with types present in the original typing judgment determine types of canonical subterms, as made precise below.

Lemma 25. *Suppose $\Gamma \vdash \mathbb{C} : \theta$. Let*

$$\begin{array}{l}\mathbf{L} = (\bigcup_{(x:\theta_x)\in\Gamma} \mathsf{PST}(\theta_x)) \cup \mathsf{NST}(\theta) \cup (\bigcup_{\theta_r\in\mathsf{RT}(\mathbb{C})} \mathsf{ST}(\mathsf{ref}(\theta_r))) \cup \{\mathsf{unit}, \mathsf{int}\}, \\ \mathbf{R} = (\bigcup_{(x:\theta_x)\in\Gamma} \mathsf{NST}(\theta_x)) \cup \mathsf{PST}(\theta) \cup (\bigcup_{\theta_r\in\mathsf{RT}(\mathbb{C})} \mathsf{ST}(\theta_r)) \cup \{\mathsf{unit}, \mathsf{int}\}.\end{array}$$

Then, for any subterm $\mathbb{C}'$ of $\mathbb{C}$ which is also in canonical form, we have $\Gamma' \vdash \mathbb{C}' : \theta'$, where $\mathsf{cod}(\Gamma') \subseteq \mathbf{L}$ and $\theta' \in \mathbf{R}$.

Corollary 26. *Suppose $\Gamma \vdash \mathbb{C} : \theta$, $\mathsf{RT}(\mathbb{C}) = \{\mathsf{int}, \mathsf{unit} \to \mathsf{unit}\}$ and Γ, θ are* ref*-free. Then $\mathbb{C}$ does not contain any occurrences of* mkvar.

Proof. Because $\mathsf{RT}(\mathbb{C}) = \{\mathsf{int}, \mathsf{unit} \to \mathsf{unit}\}$, by Lemma 25, $\mathbb{C}$ can only contain mkvar if $(\bigcup_{(x:\theta_x)\in\Gamma} \mathsf{NST}(\theta_x)) \cup \mathsf{PST}(\theta)$ contains a ref-type. Since Γ and θ are ref-free this cannot be the case. □

This completes a syntactic proof of Theorem 17.

Remark 27. Note that Lemma 22 may reintroduce fixed points into the language, because it relies on numerical operations defined in PCF. We can still reduce terms containing such definitions to canonical form by assuming that the required operations are primitive (represented by $\oplus$). If this is not desirable then, after the elimination of mkvar under the above assumption, we can put back the PCF definitions without jeopardizing the result (mkvar is not available in PCF).

6 When Integer References Suffice

Next we shall examine the conditions under which references of type unit $\to$ unit can also be eliminated, i.e. all uses of general references can be replaced with a single integer-valued memory cell. In technical terms, this requires us to characterize the arenas where plays are guaranteed to satisfy visibility.

Definition 28. *Let A be an arena and $m_1, m_2 \in M_A$. We shall say that m_1 and m_2 are* equireachable *if there are paths ms_1m_1 and ms_2m_2 in the graph $(M_A, \vdash_A)$ such that m is initial and, if s_1 and s_2 both start with an answer, say a_1 and a_2 respectively, then $a_1 = a_2$.*

Remark 29. For arenas which are denotations of types, as is the case in Lemma 32, the notion of equireachability trivialises somewhat. In particular, any non-initial O-moves m_1 and m_2 are equireachable. We introduced a more general definition above so as to be able to state Lemma 31.

Definition 30. *An arena A is called* **visible** *if there are no equireachable non-initial moves $m, m' \in M_A$ such that m is an O-question and m' enables a P-question.*

Lemma 31. *Let A be an arena such that each question enables an answer. All plays of A satisfy the visibility condition if and only if A is visible.*

Proof. Let s be a play of A that violates the visibility condition. Suppose further that s ends in the P-move p_2, which breaks visibility for the first time and let o_1 be its justifier. Then, since s breaks visibility at p_2, it must look like:

$$m \cdots p_1 \cdots o_1 \cdots o_2 \cdots p_2$$

for some initial move m, where o_2 appears in the view right before p_2 and where p_2 is a question. Observe also that, since p_2 violates visibility, its justifier o_1 cannot be initial. If o_2 is a question we are done: A is not visible because of $(m, m') = (o_2, o_1)$. So, suppose that o_2 is an answer. Then, p_1 is a question and the move o_2' immediately following it in s is also a question (otherwise it would answer p_1). Moreover o_2' is not initial. Consequently, A is not visible due to $(m, m') = (o_2', o_1)$.

Conversely, suppose that A is not visible and let the latter be witnessed by paths $ms_1p_1o_2$ and $ms_2o_1p_2$ in $(M_A, \vdash_A)$. We form a play s as follows.

- If s_2 does not start with an answer, we set

$$s = m\, s_1\, p_1\, o_2\, s_2\, o_1\, p_2\, o_2\, p_2\,.$$

- If s_1p_1, s_2 both start with an answer, say $s_1p_1 = as_1'$ and $s_2 = as_2'$, we set

$$s = m\, a\, s_1'\, s_2'\, o_1\, p_2\, o_2\, p_2$$

where the leftmost pointer points to the last move of s_1'.

– If s_2 starts with an answer but s_1 does not, we set

$$s = m\, s_1\, p_1\, s_1'\, s_2\, o_1\, p_2\, o_2\, p_2$$

where s_1' is a sequence of moves answering all open questions of $s_1 p_1$.

Now observe that, in each case, the play s breaks visibility at move p_2. □

As a next step we would like to understand what typing judgments give rise to visible arenas. Our answer will be phrased in terms of syntactic shape. For simplicity, we shall now restrict our discussion to types generated from unit (Remark 33 explores the consequences of the results for the full type system). The following two lemmas capture scenarios relevant to verifying visibility for arenas. We write Θ_1 for the collection of first-order types, generated by the grammar $\Theta_1 ::= \mathsf{unit} \mid \mathsf{unit} \to \Theta_1$. Similarly, $\Theta_1 \to \mathsf{unit}$ stands for $\{\theta_1 \to \mathsf{unit} \mid \theta_1 \in \Theta_1\}$.

Lemma 32. *Let $A = [\![\theta_1, \cdots, \theta_k \vdash \theta]\!]$, where $\theta_1, \cdots, \theta_k, \theta$ are generated from* unit.

- *All O-questions in A are initial iff $\theta_i \in \Theta_1$ for all $1 \le i \le k$ and $\theta = \mathsf{unit}$.*
- *A does not contain a P-question enabled by a non-initial O-move iff $\theta_i \in \{\mathsf{unit}\} \cup (\Theta_1 \to \mathsf{unit})$ for $1 \le i \le k$ and $\theta \in \Theta_1$.*

Consequently, A is visible if and only if one of the conditions above is satisfied.

Remark 33. To see whether any occurrences of ref-types generate visible arenas, recall that $[\![\mathsf{ref}(\theta)]\!] = [\![\theta \to \mathsf{unit}]\!] \times [\![\mathsf{unit} \to \theta]\!]$. Consequently, for the purpose of determining visiblity $\mathsf{ref}(\mathsf{unit})$ can be viewed as $\mathsf{unit} \to \mathsf{unit}$. Thus, $\mathsf{ref}(\mathsf{unit})$ can be used whenever $\mathsf{unit} \to \mathsf{unit}$ is allowed. Note also that it is immaterial whether we consider unit or int. The observations yield the following typing constraints for visible arenas: ($\theta_i ::= \beta \mid \mathsf{ref}(\beta) \mid \Theta_1 \to \beta$ and $\theta ::= \Theta_1$) or ($\theta_i ::= \Theta_1$ and $\theta ::= \beta$), where $\beta ::= \mathsf{unit} \mid \mathsf{int}$ and $\Theta_1 ::= \beta \mid \mathsf{ref}(\beta) \mid \beta \to \Theta_1$. Analogously, $\mathsf{ref}(\beta \to \beta)$ should be viewed as a combination of $(\beta \to \beta) \to \beta$ and $\beta \to \beta \to \beta$. The results above do not give us much room for using this type: it cannot occur on the right but, if $\theta \equiv \beta$ we can have $\theta_i \equiv \mathsf{ref}(\beta \to \beta)$.

Thanks to Theorems 14 and 15 we can derive:

Theorem 34. *Let $\Gamma \vdash \theta$ be such that $[\![\Gamma \vdash \theta]\!]$ is visible. For any $\Gamma \vdash M : \theta$, there exists $\Gamma, y : \mathsf{ref}(\mathsf{int}) \vdash M' : \theta$ such that the following conditions are satisfied.*

- $\Gamma \vdash M \cong \mathsf{let}\, y = \mathsf{ref}(0)\, \mathsf{in}\, M'$.
- *M' is* ref*-free.*
- *If $\Gamma \vdash \theta$ does not contain occurrences of* ref*, then M' is* mkvar*-free.*

Next we give several examples of terms in which uses of $\mathsf{ref}(\mathsf{unit} \to \mathsf{unit})$ are definitely not eliminable. This is because the terms generate plays that violate the visibility condition, to be contrasted with Proposition 10.

Example 35. The first example is simply $\vdash \mathsf{new}_{\mathsf{unit}\to\mathsf{unit}} : \mathsf{ref}(\mathsf{unit} \to \mathsf{unit})$. Its semantics contains the play

$$\star\ \circ\ \mathsf{write}\ \mathsf{ok}\ \mathsf{read}\ \star\ q_r\ q_w\,.$$

Other examples are obtained by extending the shape of types from Lemma 32 in various ways.

$$\vdash \mathsf{let}\ x, y = \mathsf{new}_{\mathsf{unit}\to\mathsf{unit}}, \mathsf{new}_{\mathsf{int}}\ \mathsf{in}$$
$$\lambda f^{\mathsf{unit}\to\mathsf{unit}}.\ (\mathsf{if}\ (!y = 0)\ \mathsf{then}\ (y := 1; x := f)\ \mathsf{else}\ ()); (!x)() : (\mathsf{unit} \to \mathsf{unit}) \to \mathsf{unit}$$

$$g : ((\mathsf{unit} \to \mathsf{unit}) \to \mathsf{unit}) \to \mathsf{unit} \vdash \mathsf{let}\ x, y = \mathsf{new}_{\mathsf{unit}\to\mathsf{unit}}, \mathsf{new}_{\mathsf{int}}\ \mathsf{in}$$
$$g(\lambda f^{\mathsf{unit}\to\mathsf{unit}}.\ (\mathsf{if}\ (!y = 0)\ \mathsf{then}\ (y := 1; x := f)\ \mathsf{else}\ ()); (!x)()) : \mathsf{unit}$$

$$g : \mathsf{unit} \to \mathsf{unit} \to \mathsf{unit} \vdash \mathsf{let}\ x, y = \mathsf{new}_{\mathsf{unit}\to\mathsf{unit}}, \mathsf{new}_{\mathsf{int}}\ \mathsf{in}$$
$$\lambda u^{\mathsf{unit}}.(\mathsf{if}\ (!y = 0)\ \mathsf{then}\ (y := 1; x := g())\ \mathsf{else}\ ()); (!x)() : \mathsf{unit} \to \mathsf{unit}$$

$$g : (\mathsf{unit} \to \mathsf{unit}) \to \mathsf{unit} \to \mathsf{unit} \vdash$$
$$\mathsf{let}\ x = \mathsf{new}_{\mathsf{unit}\to\mathsf{unit}}\ \mathsf{in}\ (\ x := g(\lambda z^{\mathsf{unit}}.(!x)())\ );\ (!x)() : \mathsf{unit}$$

7 When All References Are Dispensible

Finally, at some types memory allocation turns out dispensible, i.e. there exist purely functional terms with equivalent observable behaviour. In game-semantic terms, these are types where all strategies are necessarily innocent [7].

Definition 36. *Let A be an arena such that any question enables an answer*[1]. *A is called* **innocent** *if all O-questions are initial.*

Remark 37. Let us observe that $[\![\theta_1, \cdots, \theta_k \vdash \theta]\!]$ is an innocent arena if and only if $\theta_i ::= \Theta_1$ and $\theta ::= \beta$.

Lemma 38. *Let A be an arena such that any question enables an answer. Every strategy $\sigma : A$ is innocent if and only if A is innocent.*

Proof. Suppose A is not innocent, i.e. there exists a non-initial O-question q_O. Let s be the chain of enablers leading from some initial move to q_O and let a_P be an answer to q_O. Then the strategy on A consisting of prefixes of sa_P is not innocent, because it will not contain $sa_P q_O a_P$. Thus, not all strategies in A are innocent.

Now assume that A is innocent. Consequently, all non-initial O-moves are answers. Thus, each odd-length play s in A must have the shape $q(qa)^*$. Consequently, $\mathit{view}(s) = s$ and each strategy on A is thus innocent. □

The following result then follows from Theorem 15.

Theorem 39. *Suppose $\Gamma \vdash M : \theta$ is such that $[\![\Gamma \vdash M : \theta]\!]$ is innocent. Then there exists $\Gamma \vdash M' : \theta$ satisfying all the conditions below.*

- *$\Gamma \vdash M \cong M'$.*
- *M' is* ref*-free.*
- *If there are no occurrences of* ref*-types in $\Gamma \vdash \theta$, then M' is* mkvar*-free.*

[1] All arenas corresponding to types are of this kind.

Example 40. Here are two examples of terms not covered by Theorem 39, i.e. terms that do not have purely functional counterparts, because the corresponding strategies are not innocent.

$$\vdash \mathsf{let}\, y = \mathsf{new}_{\mathsf{int}}\, \mathsf{in}\, \lambda z^{\mathsf{unit}}.\mathsf{if}\ (!y = 0)\ \mathsf{then}\, y := 1\ \mathsf{else}\ \Omega : \mathsf{unit} \to \mathsf{unit}$$

$$g : (\mathsf{unit} \to \mathsf{unit}) \to \mathsf{unit} \vdash \mathsf{let}\, y = \mathsf{new}_{\mathsf{int}}\, \mathsf{in}$$
$$g(\lambda z^{\mathsf{unit}}.\mathsf{if}\ (!y = 0)\ \mathsf{then}\, y := 1\ \mathsf{else}\ \Omega) : \mathsf{unit}$$

8 Conclusion

We showed that general references in $\mathcal{L}$ can be simulated with two reference cells, of types ref(unit $\to$ unit) and ref(int) respectively. This was first demonstrated through a game-semantic argument and subsequently complemented by a syntactic recipe for program transformation. The latter was facilitated by the presence of the mkvar constructor. However, the results apply equally well to the mkvar-free framework, provided no reference types occur in the type of the term or those of its free identifiers (arbitrary internal uses are still allowed). Then the auxiliary occurrences of mkvar can actually be eliminated, so, in this context, mkvar can be viewed as a useful temporary addition to the language.

In the future, we would like to conduct a similar study using the nominal game model of [8]. In the nominal setting, decomposition results such as Lemmata 18 and 19 cannot be expected to hold. Another surprising challenge is that the obvious adaptation of the visibility condition fails to be preserved by composition.

References

1. Abramsky, S., Honda, K., McCusker, G.: Fully abstract game semantics for general references. In: Proceedings of LICS, pp. 334–344. Computer Society Press (1998)
2. Abramsky, S., Jagadeesan, R., Malacaria, P.: Full abstraction for PCF. Information and Computation 163, 409–470 (2000)
3. Abramsky, S., McCusker, G.: Call-by-Value Games. In: Nielsen, M. (ed.) CSL 1997. LNCS, vol. 1414, pp. 1–17. Springer, Heidelberg (1998)
4. Abramsky, S., McCusker, G.: Linearity, sharing and state: a fully abstract game semantics for Idealized Algol with active expressions. In: O'Hearn, P.W., Tennent, R.D. (eds.) Algol-Like Languages, pp. 297–329. Birkhäuser (1997)
5. Bruce, K.B., Cardelli, L., Pierce, B.C.: Comparing object encodings. Inf. Comput. 155(1-2), 108–133 (1999)
6. Honda, K., Yoshida, N.: Game-theoretic analysis of call-by-value computation. Theoretical Computer Science 221(1–2), 393–456 (1999)
7. Hyland, J.M.E., Ong, C.-H.L.: On Full Abstraction for PCF: I. Models, observables and the full abstraction problem, II. Dialogue games and innocent strategies, III. A fully abstract and universal game model. Information and Computation 163(2), 285–408 (2000)
8. Murawski, A.S., Tzevelekos, N.: Game semantics for good general references. In: Proceedings of LICS, pp. 75–84. IEEE Computer Society Press (2011)
9. Reynolds, J.C.: The essence of Algol. In: de Bakker, J.W., van Vliet, J.C. (eds.) Algorithmic Languages, pp. 345–372. North Holland (1981)
10. Sanjabi, S.B., Ong, C.-H.L.: Fully abstract semantics of additive aspects by translation. In: Proceedings of AOSD, pp. 135–148. ACM (2007)

Separation Logic for Non-local Control Flow and Block Scope Variables

Robbert Krebbers and Freek Wiedijk

ICIS, Radboud University Nijmegen, The Netherlands

Abstract. We present an approach for handling non-local control flow (goto and return statements) in the presence of allocation and deallocation of block scope variables in imperative programming languages.

We define a small step operational semantics and an axiomatic semantics (in the form of a separation logic) for a small C-like language that combines these two features, and which also supports pointers to block scope variables. Our operational semantics represents the program state through a generalization of Huet's zipper data structure.

We prove soundness of our axiomatic semantics with respect to our operational semantics. This proof has been fully formalized in Coq.

1 Introduction

There is a gap between programming language features that can be described well by a formal semantics, and those that are available in widely used programming languages. This gap needs to be bridged in order for formal methods to become mainstream. However, interaction between the more 'dirty' features of widely used programming languages tends to make an accurate formal semantics even more difficult. An example of such interaction is the goto statement in the presence of block scope variables in the C programming language.

C allows unrestricted gotos which (unlike break and continue) may not only jump out of blocks, but can also jump into blocks. Orthogonal to this, blocks may contain local variables, which can be "taken out of their block" by keeping a pointer to them. This makes non-local control in C (including break and continue) even more unrestricted, as leaving a block results in the memory of these variables being freed, and thus making pointers to them invalid. Consider:

```
int *p = NULL;
l: if (p) { return (*p); }
   else { int j = 10; p = &j; goto l; }
```

Here, when the label `l` is passed for the first time, the variable `p` is `NULL`. Hence, execution continues in the block containing `j` where `p` is assigned a pointer to `j`. However, after the `goto l` statement, the block containing `j` is left, and the memory of `j` is freed. After this, dereferencing `p` is no longer legal, making the program exhibit *undefined behavior*.

If a program exhibits undefined behavior, the ISO C standard [8] allows it to do literally anything. This is to avoid compilers having to insert (possibly

F. Pfenning (Ed.): FOSSACS 2013, LNCS 7794, pp. 257–272, 2013.

expensive) dynamic checks to handle corner cases. In particular, in the case of non-local control flow this means that an implementation can ignore allocation issues when jumping, but a semantics cannot. Not describing certain undefined behaviors would therefore mean that some programs can be proven correct with respect to the formal semantic whereas they may crash or behave unexpectedly when compiled with an actual C compiler.

It is well known that a small step semantics is more flexible than a big step semantics for modeling more intricate programming language features. In a small step semantics, it is nonetheless intuitive to treat uses of goto as big steps, as executing them makes the program jump to a totally different place in one step. For functional languages, there has been a lot of research on modeling control (`call/cc` and variants thereof) in a purely small step manner (see [6] for example). This indicates that the intuition that uses of non-local control should be treated as big steps is not correct.

We show that a purely small step semantics is also better suited to handle the interaction between gotos and block scope variables in imperative programming languages. Our semantics lets the goto statement traverse in small steps through the program to search for its corresponding label. The required allocations and deallocations are calculated incrementally during this traversal.

Our choice of considering goto at all may seem surprising. Since the revolution of structured programming in the seventies, many people have considered goto as bad programming practice [4]. However, some have disagreed [9], and gotos are still widely used in practice. For example, the current Linux kernel contains about a hundred thousand uses of goto. Goto statements are particularly useful for breaking from multiple nested loops, and for systematically cleaning up resources after an error occurred. Also, gotos can be used to increase performance.

Approach. We define a small step operational semantics for a small C-like language that supports both non-local control flow and block scope variables. To obtain more confidence in this semantics, and to support reasoning about programs in this language, we define an axiomatic semantics for the same fragment, and prove its soundness with respect to the operational semantics.

Our operational semantics uses a *zipper*-like data structure [7] to store both the location of the substatement that is being executed and the program stack. Because we allow pointers to local variables, the stack contains references to the value of each variable instead of the value itself. Execution of the program occurs by traversal through the zipper in one of the following directions: *down* $\searrow$, *up* $\nearrow$, *jump* $\curvearrowright$, or *top* $\Uparrow$. When a `goto` l statement is executed, the direction is changed to $\curvearrowright l$, and the semantics performs a small step traversal through the zipper until the label l has been reached.

Related Work. Goto statements (and other forms of non-local control) are often modeled using continuations. Appel and Blazy provide a small step continuation semantics for Cminor [2] that supports return statements. CompCert extends their approach to support goto statements in Cmedium [12]. Ellison and Rosu [5] also use continuations to model gotos in their C11 semantics, but whereas the

CompCert semantics does not support block scope variables, they do. We further discuss the differences between continuations and our approach in Section 3.

Tews [18] defines a denotational semantics for a C-like language that supports goto and unstructured switch statements. His state includes variants for non-local control corresponding to our directions $\curvearrowright$ and $\upuparrows$.

The most closely related work to our axiomatic semantics is Appel and Blazy's separation logic for Cminor in Coq [2]. Their separation logic supports return statements, but does not support gotos, nor block scope variables. Von Oheimb [15] defines an operational and axiomatic semantics for a Java-like language in Isabelle. His language supports both local variables and mutually recursive function calls. Although his work is fairly different from ours, our approach to mutual recursion is heavily inspired by his. Furthermore, Chlipala [3] gives a separation logic for a low-level language in Coq that supports gotos. His approach to automation is impressive, but he does not give an explicit operational semantics and does not consider block scope variables.

Contribution. Our contribution is threefold:

- We define a small step operational semantics using a novel zipper based data structure to handle the interaction between gotos and block scope variables in a correct way (Section 2 and 3).
- We give an axiomatic semantics that allows reasoning about programs with gotos, pointers to local variables, and mutually recursive function calls. We demonstrate it by verifying Euclid's algorithm (Section 4).
- We prove the soundness of our axiomatic semantics (Section 5). This proof has been fully formalized in the Coq proof assistant (Section 6).

2 The Language

Our memory is a finite partial function from natural numbers to values, where a value is either an unbounded integer, a pointer represented by a natural number corresponding to the index of a memory cell, or the `NULL`-pointer.

Definition 2.1. *A* partial function *from A to B is a (total) function from A to B^{opt}, where A^{opt} is the* option type, *defined as containing either $\bot$ or x for some $x \in A$. A partial function is called* finite *if its domain is finite. The operation $f[x := y]$ stores the value y at index x, and $f[x := \bot]$ deletes the value at index x.* Disjointness, *notation $f_1 \perp f_2$, is defined as $\forall x \,.\, f_1\, x = \bot \vee f_2\, x = \bot$. Given f_1 and f_2 with $f_1 \perp f_2$, the operation $f_1 \cup f_2$ yields their* union. *Moreover, the* inclusion *$f_1 \subseteq f_2$ is defined as $\forall x\, y \,.\, f_1\, x = y \to f_2\, x = y$.*

Definition 2.2. Values *are defined as:*

$$v ::= \mathtt{int}\; n \mid \mathtt{ptr}\; b \mid \mathtt{NULL}$$

Memories *(typically named m) are finite partial functions from natural numbers to values. A value v is* true, *notation* $\mathsf{istrue}\; v$, *if it is of the shape $\mathtt{int}\; n$ with $n \neq 0$, or $\mathtt{ptr}\; b$. It is* false, *notation* $\mathsf{isfalse}\; v$, *otherwise.*

Expressions are side-effect free and will be given a deterministic semantics by an evaluation function. The variables used in expressions are De Bruijn indexes, *i.e.* the variable x_i refers to the ith value on the stack. De Bruijn indexes avoid us from having to deal with shadowing due to block scope variables. Especially in the axiomatic semantics this is useful, as we do not want to lose information by a local variable shadowing an already existing one.

Definition 2.3. Expressions *are defined as:*

$$\circledcirc ::= \; == \mid \leq \mid + \mid * \mid / \mid \%$$
$$e ::= x_i \mid v \mid \texttt{load}\, e \mid e_1 \circledcirc e_2$$

Stacks (typically named ρ) are lists of memory indexes rather than lists of values. This allows us to treat pointers to both local and allocated storage in a uniform way. Evaluation of a variable thus consists of looking up its address in the stack, and returning a pointer to that address.

Definition 2.4. Evaluation $[\![e]\!]_{\rho,m}$ of an expression e *in a stack* ρ *and memory* m *is defined by the following partial function:*

$$\begin{aligned} [\![x_i]\!]_{\rho,m} &:= \texttt{ptr}\, a && \textit{if } \rho\, i = a \\ [\![v]\!]_{\rho,m} &:= v \\ [\![\texttt{load}\, e]\!]_{\rho,m} &:= m\, a && \textit{if } [\![e]\!]_{\rho,m} = \texttt{ptr}\, a \\ [\![e_1 \circledcirc e_2]\!]_{\rho,m} &:= [\![e_1]\!]_{\rho,m} \circledcirc [\![e_2]\!]_{\rho,m} \end{aligned}$$

Lemma 2.5. *If* $m_1 \subseteq m_2$ *and* $[\![e]\!]_{\rho,m_1} = v$*, then* $[\![e]\!]_{\rho,m_2} = v$.

Definition 2.6. Statements *are defined as:*

$$\begin{aligned} s ::= \; &\texttt{block}\, s \mid e_l := e_r \mid f(\vec{e}) \mid \texttt{skip} \mid \texttt{goto}\, l \\ &\mid l : s \mid s_1 \,;\, s_2 \mid \texttt{if}\ (e)\ s_1\ \texttt{else}\ s_2 \mid \texttt{return} \end{aligned}$$

The construct `block` s opens a new scope with one variable. Since we use De Bruijn indexes for variables, it does not contain the name of the variable. For presentation's sake, we have omitted functions that return values (these are however included in our Coq formalization). In the semantics presented here, an additional function parameter with a pointer for the return value can be used instead. Given a statement s, the function labels s collects the labels of labeled statements in s, and the function gotos s collects the labels of gotos in s.

3 Operational Semantics

We define the semantics of statements by a small step operational semantics. That means, computation is defined by the reflexive transitive closure of a reduction relation $\rightarrow$ on *program states*. This reduction relation traverses through the program in small steps by moving the focus on the substatement that is being executed. Uses of non-local control (goto and return) are performed in small steps rather than in big steps as well.

In order to model the concept of focusing on the substatement that is being executed, we need a data structure to capture the location in the program. For this we define *program contexts* as an extension of Huet's zipper data structure [7]. Program contexts extend the zipper data structure by annotating each block scope variable with its associated memory index, and furthermore contain the full call stack of the program. Program contexts can also be seen as a generalization of continuations (as for example being used in CompCert [2,11,12]). However, there are some notable differences.

- Program contexts implicitly contain the stack, whereas a continuation semantics typically stores the stack separately.
- Program contexts also contain the part of the program that has been executed, whereas continuations only contain the part that remains to be done.
- Since the complete program is preserved, looping constructs like the while statement do not have to duplicate code (see the Coq formalization).

The fact that program contexts do not throw away the parts of the statement that have been executed is essential for our treatment of goto. Upon an invocation of a goto, the semantics traverses through the program context until the corresponding label has been found. During this traversal it passes all block scope variables that went out of scope, allowing it to perform required allocations and deallocations in a natural way. Hence, the point of this traversal is not so much to *search* for the label, but much more to incrementally *calculate* the required allocations and deallocations.

In a continuation semantics, upon the use of a goto, one typically computes, or looks up, the statement and continuation corresponding to the target label. However, it is not very natural to reconstruct the required allocations and deallocations from the current and target continuations.

Definition 3.1. Singular statement contexts *are defined as:*

$$E_{\mathcal{S}} ::= \square\,;s_2 \mid s_1\,;\square \mid \texttt{if}\ (e)\ \square\ \texttt{else}\ s_2 \mid \texttt{if}\ (e)\ s_1\ \texttt{else}\ \square \mid l\texttt{:}\square$$

Given a singular statement context $E_{\mathcal{S}}$ and a statement s, substitution of s for the hole in $E_{\mathcal{S}}$, *notation $E_{\mathcal{S}}[\,s\,]$, is defined in the ordinary way.*

A pair $(\vec{E_{\mathcal{S}}}, s)$ consisting of a list of singular statement contexts $\vec{E_{\mathcal{S}}}$ and a statement s forms a zipper for statements without block scope variables. That means, $\vec{E_{\mathcal{S}}}$ is a statement turned inside-out that represents a path from the focused substatement s to the top of the whole statement.

Definition 3.2. Singular program contexts *are defined as:*

$$E ::= E_{\mathcal{S}} \mid \texttt{block}_b\ \square \mid \mathsf{call}\ f\ \vec{e} \mid \mathsf{params}\ \vec{b}$$

Program contexts *(typically named k) are lists of singular program contexts.*

The previously introduced contexts will be used as follows.

- When entering a block, $\texttt{block}\ s$, the context $\texttt{block}_b\ \Box$ is appended to the head of the program context. It associates the block scope variable with its corresponding memory index b.
- Upon a function call, $f(\vec{e})$, the context $\mathsf{call}\ f\ \vec{e}$ is appended to the head of the program context. It contains the location of the caller so that it can be restored when the called function f returns.
- When a function body is entered, the context $\mathsf{params}\ \vec{b}$ is appended to the head of the program context. It contains a list of memory indexes of the function parameters.

As program contexts implicitly contain the stack, we define a function to extract it from them.

Definition 3.3. *The* corresponding stack $\mathsf{getstack}\ k$ *of* k *is defined as:*

$$\begin{aligned}\mathsf{getstack}\ (E_S :: k) &:= \mathsf{getstack}\ k\\ \mathsf{getstack}\ (\texttt{block}_b\ \Box :: k) &:= b :: \mathsf{getstack}\ k\\ \mathsf{getstack}\ (\mathsf{call}\ f\ \vec{e} :: k) &:= []\\ \mathsf{getstack}\ (\mathsf{params}\ \vec{b} :: k) &:= \vec{b} +\!\!+\ \mathsf{getstack}\ k\end{aligned}$$

We will treat getstack *as an implicit coercion and will omit it everywhere.*

We define $\mathsf{getstack}\ (\mathsf{call}\ f\ \vec{e} :: k)$ as $[]$ instead of $\mathsf{getstack}\ k$, as otherwise it would be possible to refer to the local variables of the calling function.

Definition 3.4. Directions, focuses *and* program states *are defined as:*

$$\begin{aligned}d &::= \searrow \mid \nearrow \mid \curvearrowright l \mid \Uparrow\\ \phi &::= (d,\ s) \mid \overline{\mathsf{call}\ f\ \vec{v}} \mid \overline{\mathsf{return}}\\ S &::= \mathbf{S}(k,\ \phi,\ m)\end{aligned}$$

A program state $\mathbf{S}(k, \phi, m)$ consists of a program context k, the part of the program that is focused ϕ, and the memory m. Like Leroy's semantics for Cminor [11], we consider three kinds of states: (a) execution of statements (b) calling a function (c) returning from a function. The CompCert Cmedium semantics [12] also includes a state for execution of expressions and a *stuck* state for undefined behavior. Since our expressions are side-effect free, we do not need an additional expression state. Furthermore, since expressions are deterministic, we can easily capture undefined behavior by letting the reduction get stuck.

Definition 3.5. *The relation* $\texttt{allocparams}\ m_1\ \vec{b}\ \vec{v}\ m_2$ *(non-deterministically)* allocates fresh blocks $\vec{b}$ for function parameters $\vec{v}$. *It is inductively defined as:*

$$\frac{}{\texttt{allocparams}\ m\ []\ []\ m} \qquad \frac{\texttt{allocparams}\ m_1\ \vec{b}\ \vec{v}\ m_2 \qquad m_2\, b = \bot}{\texttt{allocparams}\ m_1\ (b :: \vec{b})\ (v :: \vec{v})\ m_2[b := v]}$$

Definition 3.6. *Given a function δ assigning statements to function names, the* small step reduction relation $S_1 \twoheadrightarrow S_2$ *is defined as:*

1. For simple statements:
   (a) $\mathbf{S}(k, (\searrow, e_1 := e_2), m) \twoheadrightarrow \mathbf{S}(k, (\nearrow, e_1 := e_2), m[a := v])$ *for any a and v such that* $[\![e_1]\!]_{k,m} = \mathbf{ptr}\ a$, $[\![e_2]\!]_{k,m} = v$ *and* $m\,a \neq \bot$.
   (b) $\mathbf{S}(k, (\searrow, f(\vec{e})), m) \twoheadrightarrow \mathbf{S}(\mathsf{call}\ f\ \vec{e} :: k, \overline{\mathsf{call}\ f\ \vec{v}}, m)$ *for any $\vec{v}$ such that* $[\![e_i]\!]_{k,m} = v_i$ *for each i.*
   (c) $\mathbf{S}(k, (\searrow, \mathtt{skip}), m) \twoheadrightarrow \mathbf{S}(k, (\nearrow, \mathtt{skip}), m)$
   (d) $\mathbf{S}(k, (\searrow, \mathtt{goto}\ l), m) \twoheadrightarrow \mathbf{S}(k, (\curvearrowright l, \mathtt{goto}\ l), m)$
   (e) $\mathbf{S}(k, (\searrow, \mathtt{return}), m) \twoheadrightarrow \mathbf{S}(k, (\Uparrow, \mathtt{return}), m)$
2. For compound statements:
   (a) $\mathbf{S}(k, (\searrow, \mathtt{block}\ s), m) \twoheadrightarrow \mathbf{S}((\mathtt{block}_b\ \Box) :: k, (\searrow, s), m[b := v])$ *for any b and v such that* $m\,b = \bot$.
   (b) $\mathbf{S}(k, (\searrow, s_1\,;s_2), m) \twoheadrightarrow \mathbf{S}((\Box\,;s_2) :: k, (\searrow, s_1), m)$
   (c) $\mathbf{S}(k, (\searrow, \mathtt{if}\ (e)\ s_1\ \mathtt{else}\ s_2), m) \twoheadrightarrow \mathbf{S}((\mathtt{if}\ (e)\ \Box\ \mathtt{else}\ s_2) :: k, (\searrow, s_1), m)$ *for any v such that* $[\![e]\!]_{k,m} = v$ *and* $\mathsf{istrue}\ v$.
   (d) $\mathbf{S}(k, (\searrow, \mathtt{if}\ (e)\ s_1\ \mathtt{else}\ s_2), m) \twoheadrightarrow \mathbf{S}((\mathtt{if}\ (e)\ s_1\ \mathtt{else}\ \Box) :: k, (\searrow, s_2), m)$ *for any v such that* $[\![e]\!]_{k,m} = v$ *and* $\mathsf{isfalse}\ v$.
   (e) $\mathbf{S}(k, (\searrow, l : s), m) \twoheadrightarrow \mathbf{S}((l : \Box) :: k, (\searrow, s), m)$
   (f) $\mathbf{S}((\mathtt{block}_b\ \Box) :: k, (\nearrow, s), m) \twoheadrightarrow \mathbf{S}(k, (\nearrow, \mathtt{block}\ s), m[b := \bot])$
   (g) $\mathbf{S}((\Box\,;s_2) :: k, (\nearrow, s_1), m) \twoheadrightarrow \mathbf{S}((s_1\,;\Box) :: k, (\searrow, s_2), m)$
   (h) $\mathbf{S}((s_1\,;\Box) :: k, (\nearrow, s_2), m) \twoheadrightarrow \mathbf{S}(k, (\nearrow, s_1\,;s_2), m)$
   (i) $\mathbf{S}((\mathtt{if}\ (e)\ \Box\ \mathtt{else}\ s_2) :: k, (\nearrow, s_1), m) \twoheadrightarrow \mathbf{S}(k, (\nearrow, \mathtt{if}\ (e)\ s_1\ \mathtt{else}\ s_2), m)$
   (j) $\mathbf{S}((\mathtt{if}\ (e)\ s_1\ \mathtt{else}\ \Box) :: k, (\nearrow, s_2), m) \twoheadrightarrow \mathbf{S}(k, (\nearrow, \mathtt{if}\ (e)\ s_1\ \mathtt{else}\ s_2), m)$
   (k) $\mathbf{S}((l : \Box) :: k, (\nearrow, s), m) \twoheadrightarrow \mathbf{S}(k, (\nearrow, l : s), m)$
3. For function calls:
   (a) $\mathbf{S}(k, \overline{\mathsf{call}\ f\ \vec{v}}, m_1) \twoheadrightarrow \mathbf{S}(\mathsf{params}\ \vec{b} :: k, (\searrow, s), m_2)$ *for any s, $\vec{b}$ and m_2 such that* $\delta\,f = s$ *and* $\mathtt{allocparams}\ m_1\ \vec{b}\ \vec{v}\ m_2$.
   (b) $\mathbf{S}(\mathsf{params}\ \vec{b} :: k, (\nearrow, s), m) \twoheadrightarrow \mathbf{S}(k, \overline{\mathsf{return}}, m[\vec{b} := \vec{\bot}])$
   (c) $\mathbf{S}(\mathsf{params}\ \vec{b} :: k, (\Uparrow, s), m) \twoheadrightarrow \mathbf{S}(k, \overline{\mathsf{return}}, m[\vec{b} := \vec{\bot}])$
   (d) $\mathbf{S}(\mathsf{call}\ f\ \vec{e} :: k, \overline{\mathsf{return}}, m) \twoheadrightarrow \mathbf{S}(k, (\nearrow, f(\vec{e})), m)$
4. For non-local control flow:
   (a) $\mathbf{S}((\mathtt{block}_b\ \Box) :: k, (\Uparrow, s), m) \twoheadrightarrow \mathbf{S}(k, (\Uparrow, \mathtt{block}\ s), m[b := \bot])$
   (b) $\mathbf{S}(E_S :: k, (\Uparrow, s), m) \twoheadrightarrow \mathbf{S}(k, (\Uparrow, E_S[\,s\,]), m)$
   (c) $\mathbf{S}(k, (\curvearrowright l, \mathtt{block}\ s), m) \twoheadrightarrow \mathbf{S}((\mathtt{block}_b\ \Box) :: k, (\curvearrowright l, s), m[b := v])$ *for any b and v such that* $m\,b = \bot$, *and provided that* $l \in \mathsf{labels}\ s$.
   (d) $\mathbf{S}(k, (\curvearrowright l, l : s), m) \twoheadrightarrow \mathbf{S}((l : \Box) :: k, (\searrow, s), m)$
   (e) $\mathbf{S}(k, (\curvearrowright l, E_S[\,s\,]), m) \twoheadrightarrow \mathbf{S}(E_S :: k, (\curvearrowright l, s), m)$ *provided that* $l \in \mathsf{labels}\ s$.
   (f) $\mathbf{S}(\mathtt{block}_b\ \Box :: k, (\curvearrowright l, s), m) \twoheadrightarrow \mathbf{S}(k, (\curvearrowright l, \mathtt{block}\ s), m[b := \bot])$ *provided that* $l \notin \mathsf{labels}\ s$.
   (g) $\mathbf{S}(E_S :: k, (\curvearrowright l, s), m) \twoheadrightarrow \mathbf{S}(k, (\curvearrowright l, E_S[\,s\,]), m)$ *provided that* $l \notin \mathsf{labels}\ s$.

Note that the rules 4d and 4e overlap, and that the splitting into E_S and s in rule 4e is non-deterministic. We let $\twoheadrightarrow^*$ denote the reflexive-transitive closure, and $\twoheadrightarrow^n$ paths of $\leq n$ steps.

Execution of a statement $\mathbf{S}(k, (d, s), m)$ is performed by traversing through the program context k and statement s in direction d. The direction *down* $\searrow$ (respectively *up* $\nearrow$) is used to traverse downwards (respectively upwards) to the next substatement to be executed. Consider the example from the introduction (with the return expression omitted).

```
int *p = NULL;
l: if (p) { return; }
   else { int j = 10; p = &j; goto l; }
```

Figure 1 below displays some states corresponding to execution of this program starting at `p = &j` in downwards direction.

Execution of a function call $\mathbf{S}(k, (\searrow, f(\vec{e})), m)$ consists of two reductions. The reduction to $\mathbf{S}(\mathsf{call}\ f\ \vec{e} :: k, \overline{\mathsf{call}\, f\, \vec{v}}, m)$ evaluates the function parameters $\vec{e}$ to values $\vec{v}$, and stores the location of the calling function on the program context. The subsequent reduction to $\mathbf{S}(\mathsf{params}\ \vec{b} :: \mathsf{call}\ f\ \vec{e} :: k, (\searrow, s), m')$ looks up the called function's body s, allocates storage for the parameters $\vec{v}$, and then performs a transition to execute the called function's body.

We consider two directions for non-local control flow: *jump* $\curvearrowright l$ and *top* $\Uparrow$. After a `goto` l the direction $\curvearrowright l$ is used to traverse to the substatement labeled l. Although this search is non-deterministic, there are some side conditions on it so as to ensure it not going back and forth between the same locations. This is required as it otherwise may impose non-terminating behavior on terminating programs. The non-determinism could be removed entirely by adding additional side conditions. However we omitted doing so in order to ease formalization.

The direction $\Uparrow$ is used to traverse to the *top* of the statement after a `return`. When it reaches the top, there will be two reductions to leave the called function. The first reduction, from $\mathbf{S}(\mathsf{params}\ \vec{b} :: \mathsf{call}\ f\ \vec{e} :: k, (\Uparrow, s), m)$ to $\mathbf{S}(\mathsf{call}\ f\ \vec{e} :: k, \overline{\mathsf{return}}, m[\vec{b} := \bot])$, deallocates the function parameters, and the second, to $\mathbf{S}(k, (\nearrow, f(\vec{e})), m)$, reinstates the calling function.

$k_1 = \square\,;\texttt{goto}\ l$	$k_2 = \square\,;\texttt{goto}\ l$	$k_3 = x_1 := x_0\,;\square$
$::\ x_0 := \texttt{int}\ 10\,;\square$	$::\ x_0 := \texttt{int}\ 10\,;\square$	$::\ x_0 := \texttt{int}\ 10\,;\square$
$::\ \texttt{block}_{b_j}\ \square$	$::\ \texttt{block}_{b_j}\ \square$	$::\ \texttt{block}_{b_j}\ \square$
$::\ \texttt{if (load}\ x_0\texttt{) return else}\ \square$	$::\ \texttt{if (load}\ x_0\texttt{) return else}\ \square$	$::\ \texttt{if (load}\ x_0\texttt{) return else}\ \square$
$::\ l : \square$	$::\ l : \square$	$::\ l : \square$
$::\ x_0 := \texttt{NULL}\,;\square$	$::\ x_0 := \texttt{NULL}\,;\square$	$::\ x_0 := \texttt{NULL}\,;\square$
$::\ \texttt{block}_{b_p}\ \square$	$::\ \texttt{block}_{b_p}\ \square$	$::\ \texttt{block}_{b_p}\ \square$
$\phi_1 = (\searrow, x_1 := x_0)$	$\phi_2 = (\nearrow, x_1 := x_0)$	$\phi_3 = (\searrow, \texttt{goto}\ l)$
$m_1 = \{b_p \mapsto \texttt{NULL}, b_j \mapsto 10\}$	$m_2 = \{b_p \mapsto \texttt{ptr}\ b_j, b_j \mapsto 10\}$	$m_3 = \{b_p \mapsto \texttt{ptr}\ b_j, b_j \mapsto 10\}$
$S_1 = \mathbf{S}(k_1, \phi_1, m_1)$	$S_2 = \mathbf{S}(k_2, \phi_2, m_2)$	$S_3 = \mathbf{S}(k_3, \phi_3, m_3)$

Fig. 1. An example reduction path $S_1 \twoheadrightarrow S_2 \twoheadrightarrow S_3$

When we relate our operational and axiomatic semantics in Section 5, we will have to restrict the traversal through the program to remain below a certain context.

Definition 3.7. *The* k-restricted reduction $S_1 \twoheadrightarrow_k S_2$ *is defined as* $S_1 \twoheadrightarrow S_2$ *provided that* k *is a suffix of the program context of* S_2. *We let* $\twoheadrightarrow_k^*$ *denote the reflexive-transitive closure, and* $\twoheadrightarrow_k^n$ *paths of* $\leq n$ *steps.*

Lemma 3.8. *If* $\mathbf{S}(k, (d, s), m) \twoheadrightarrow_k^* \mathbf{S}(k, (d', s'), m')$, *then* $s = s'$.

The previous lemma shows that the small step semantics indeed behaves as traversing through a zipper. Its proof is not entirely trivial due to the presence of function calls, as these add the statement of the called function to the state.

4 Axiomatic Semantics

Judgments of Hoare logic are triples $\{P\}\, s\, \{Q\}$, where s is a statement, and P and Q are *assertions* called the pre- and postcondition. The intuitive reading of such a triple is: if P holds for the state before executing s, and execution of s terminates, then Q holds afterwards. To deal with non-local control flow and function calls, our judgments become sextuples $\Delta;\ J;\ R \vdash \{P\}\, s\, \{Q\}$, where:

- Δ is a finite function from function names to their pre- and postconditions. This environment is used to cope with (mutually) recursive functions.
- J is a function that gives the jumping condition for each `goto`. When executing a `goto` l, the assertion $J\, l$ has to hold.
- R is the assertion that has to hold when executing a `return`.

The assertions P, Q, J and R correspond to the four directions $\searrow$, $\nearrow$, $\curvearrowright$ and $\Uparrow$ in which traversal through a statement can be performed. We therefore often treat the sextuple as a triple $\Delta;\ \bar{P} \vdash s$, where $\bar{P}$ is a function from directions to assertions such that $\bar{P} \searrow = P$, $\bar{P} \nearrow = Q$, $\bar{P}\,(\curvearrowright l) = J\, l$ and $\bar{P} \Uparrow = R$.

We use a shallow embedding for the representation of assertions. This treatment is similar to that of Appel and Blazy [2] and Von Oheimb [15].

Definition 4.1. Assertions *are predicates over the the stack and the memory. We define the following connectives on assertions.*

$$\begin{aligned}
P \to Q &:= \lambda \rho\, m\,.\, P\, \rho\, m \to Q\, \rho\, m & \ulcorner P \urcorner &:= \lambda \rho\, m\,.\, P \\
P \wedge Q &:= \lambda \rho\, m\,.\, P\, \rho\, m \wedge Q\, \rho\, m & e \Downarrow v &:= \lambda \rho\, m\,.\, [\![e]\!]_{\rho,m} = v \\
P \vee Q &:= \lambda \rho\, m\,.\, P\, \rho\, m \vee Q\, \rho\, m & e \Downarrow - &:= \exists v\,.\, e \Downarrow v \\
\neg P &:= \lambda \rho\, m\,.\, \neg P\, \rho\, m & e \Downarrow \top &:= \exists v\,.\, \ulcorner \mathsf{istrue}\, v \urcorner \wedge e \Downarrow v \\
\forall x\,.\, P\, x &:= \lambda \rho\, m\,.\, \forall x\,.\, P\, x\, \rho\, m & e \Downarrow \bot &:= \exists v\,.\, \ulcorner \mathsf{isfalse}\, v \urcorner \wedge e \Downarrow v \\
\exists x\,.\, P\, x &:= \lambda \rho\, m\,.\, \exists x\,.\, P\, x\, \rho\, m & P[a := v] &:= \lambda \rho\, m\,.\, P\, \rho\, m[a := v]
\end{aligned}$$

We treat $\ulcorner \_ \urcorner$ *as an implicit coercion, for example, we write* True *instead of* $\ulcorner$True$\urcorner$. *Also, we often lift the above connectives to functions to assertions, for example, we write* $P \wedge Q$ *instead of* $\lambda v\,.\, P\, v \wedge Q\, v$.

Definition 4.2. *An assertion P is* stack independent *if $P\,\rho_1\,m \to P\,\rho_2\,m$ for each memory m and stacks ρ_1 and ρ_2, and similarly is* memory independent *if $P\,\rho\,m_1 \to P\,\rho\,m_2$ for each stack ρ and memories m_1 and m_2.*

Next, we define the assertions of separation logic [14]. The assertion emp asserts that the memory is empty. The *separating conjunction* $P * Q$ asserts that the memory can be split into two disjoint parts such that P holds in the one part, and Q in the other. Finally, $e_1 \mapsto e_2$ asserts that the memory consists of exactly one cell at address e_1 with contents e_2, and $e_1 \hookrightarrow e_2$ asserts that the memory contains at least a cell at address e_1 with contents e_2.

Definition 4.3. *The* connectives of separation logic *are defined as follows.*

$$
\begin{aligned}
\mathsf{emp} &:= \lambda \rho\, m \,.\, m = \emptyset \\
P * Q &:= \lambda \rho\, m \,.\, \exists m_1\, m_2 \,.\, m = m_1 \cup m_2 \wedge m_1 \perp m_2 \wedge P\,\rho\, m_1 \wedge Q\,\rho\, m_2 \\
e_1 \mapsto e_2 &:= \lambda \rho\, m \,.\, \exists b\, v \,.\, [\![\, e_1 \,]\!]_{\rho,m} = \mathtt{ptr}\; b \wedge [\![\, e_2 \,]\!]_{\rho,m} = v \wedge m = \{(b,\, v)\} \\
e_1 \mapsto - &:= \exists e_2 \,.\, e_1 \mapsto e_2 \\
e_1 \hookrightarrow e_2 &:= \lambda \rho\, m \,.\, \exists b\, v \,.\, [\![\, e_1 \,]\!]_{\rho,m} = \mathtt{ptr}\; b \wedge [\![\, e_2 \,]\!]_{\rho,m} = v \wedge m\, b = v \\
e_1 \hookrightarrow - &:= \exists e_2 \,.\, e_1 \hookrightarrow e_2
\end{aligned}
$$

To deal with block scope variables we need to lift an assertion such that the De Bruijn indexes of its variables are increased. We define the lifting $P\!\uparrow$ of an assertion P semantically, and prove that it indeed behaves as expected.

Definition 4.4. *The assertion $P\!\uparrow$ is defined as $\lambda \rho\, m \,.\, P\,(\mathsf{tail}\,\rho)\, m$.*

Lemma 4.5. *We have $(e \Downarrow v)\!\uparrow\, = (e\!\uparrow) \Downarrow v$ and $(e_1 \mapsto e_2)\!\uparrow\, = (e_1\!\uparrow) \mapsto (e_2\!\uparrow)$, where the operation $e\!\uparrow$ replaces each variable x_i in e by x_{i+1}. Furthermore, $(\_)\!\uparrow$ distributes over the connectives $\to$, $\wedge$, $\vee$, $\neg$, $\forall$, $\exists$, and $*$.*

In order to relate the pre- and postcondition of a function, we allow universal quantification over arbitrary logical variables $\vec{y}$. The specification of a function with parameters $\vec{v}$ consists therefore of a precondition $P\,\vec{y}\,\vec{v}$ and postcondition $Q\,\vec{y}\,\vec{v}$. These should be stack independent because local variables will have a different meaning at the calling function than in the called function's body. We will write such a specification as $\forall \vec{y} \forall \vec{v} \,.\, \{P\,\vec{y}\,\vec{v}\}\,\{Q\,\vec{y}\,\vec{v}\}$.

Definition 4.6. *Given a function δ assigning statements to function names, the rules of the axiomatic semantics are defined as:*

$$
\frac{\Delta;\, J;\, R \vdash \{P\}\, s\, \{Q\}}{\Delta;\, A * J;\, A * R \vdash \{A * P\}\, s\, \{A * Q\}}
\qquad
\frac{\forall x. (\Delta;\, J;\, R \vdash \{P\,x\}\, s\, \{Q\})}{\Delta;\, J;\, R \vdash \{\exists x \,.\, P\,x\}\, s\, \{Q\}}
\quad \text{(frame \& exists)}
$$

$$
\frac{(\forall l \in \mathsf{labels}\; s \,.\, J'\,l \to J\,l) \quad (\forall l \notin \mathsf{labels}\; s \,.\, J\,l \to J'\,l) \quad R \to R' \qquad P' \to P \quad \Delta;\, J;\, R \vdash \{P\}\, s\, \{Q\} \quad Q \to Q'}{\Delta;\, J';\, R' \vdash \{P'\}\, s\, \{Q'\}}
\quad \text{(weaken)}
$$

$$
\frac{}{\Delta;\, J;\, R \vdash \{P\}\, \mathtt{skip}\, \{P\}}
\qquad
\frac{}{\Delta;\, J;\, R \vdash \{R\}\, \mathtt{return}\, \{Q\}}
\quad \text{(skip \& return)}
$$

$$\frac{}{\Delta;\, J;\, R \vdash \{\exists a\, v\,.\, e_1 \Downarrow a \wedge e_2 \Downarrow v \wedge \mathtt{ptr}\; a \hookrightarrow - \wedge P[a := v]\}\, e_1 := e_2\, \{P\}} \quad \text{(assign)}$$

$$\frac{\Delta;\, J;\, R \vdash \{J\, l\}\, s\, \{Q\}}{\Delta;\, J;\, R \vdash \{J\, l\}\, l : s\, \{Q\}} \qquad \frac{}{\Delta;\, J;\, R \vdash \{J\, l\}\, \mathtt{goto}\; l\, \{Q\}} \quad \text{(label \& goto)}$$

$$\frac{\Delta;\, x_0 \mapsto - * J \uparrow;\, x_0 \mapsto - * R \uparrow \vdash \{x_0 \mapsto - * P \uparrow\}\, s\, \{x_0 \mapsto - * Q \uparrow\}}{\Delta;\, J;\, R \vdash \{P\}\, \mathtt{block}\; s\, \{Q\}} \quad \text{(block)}$$

$$\frac{\Delta;\, J;\, R \vdash \{P\}\, s_1\, \{P'\} \qquad \Delta;\, J;\, R \vdash \{P'\}\, s_2\, \{Q\}}{\Delta;\, J;\, R \vdash \{P\}\, s_1\, ;\, s_2\, \{Q\}} \quad \text{(comp)}$$

$$\frac{\Delta;\, J;\, R \vdash \{e \Downarrow \top \wedge P\}\, s_1\, \{Q\} \qquad \Delta;\, J;\, R \vdash \{e \Downarrow \bot \wedge P\}\, s_2\, \{Q\}}{\Delta;\, J;\, R \vdash \{e \Downarrow - \wedge P\}\, \mathtt{if}\; (e)\; s_1\; \mathtt{else}\; s_2\, \{Q\}} \quad \text{(cond)}$$

$$\frac{\Delta\, f = \{P\}\, \{Q\} \qquad \vec{e} \Downarrow \vec{v} \wedge P\, \vec{y}\, \vec{v} \to A \qquad A \textit{ memory independent}}{\Delta;\, J;\, R \vdash \{\vec{e} \Downarrow \vec{v} \wedge P\, \vec{y}\, \vec{v}\}\, f(\vec{e})\, \{A \wedge Q\, \vec{y}\, \vec{v}\}} \quad \text{(call)}$$

$$\frac{\begin{array}{l}\forall f\, P'\, Q'\,.\, \Delta' f = (\forall \vec{z} \forall \vec{w}\,.\, \{P\, \vec{z}\, \vec{w}\}\, \{Q\, \vec{z}\, \vec{w}\}) \to \forall \vec{y}\, \vec{v}\,. \\ (\Delta' \cup \Delta;\, \lambda l.\mathsf{False};\, \Pi_*[x_i \mapsto -] * Q' \vec{y}\, \vec{v} \vdash \{\Pi_*[x_i \mapsto v_i] * P' \vec{y}\, \vec{v}\}\, \delta\, f\, \{\Pi_*[x_i \mapsto -] * Q' \vec{y}\, \vec{v}\}) \\ \Delta' \cup \Delta;\, J;\, R \vdash \{P\}\, s\, \{Q\} \qquad\qquad \mathsf{dom}\; \Delta' \subseteq \mathsf{dom}\; \delta \end{array}}{\Delta;\, J;\, R \vdash \{P\}\, s\, \{Q\}} \quad \text{(add funs)}$$

The traditional *frame rule* of separation logic [14] includes the side-condition $\mathsf{vars}\, s \cap \mathsf{vars}\, A = \emptyset$ on the free variables in the statement s and assertion A. However, as our local variables are just (immutable) references into the memory, we do not need this side-condition. Also, the (frame) and (block) rule are uniform in all assertions, allowing us to write:

$$\frac{\Delta;\, \bar{P} \vdash s}{\Delta;\, A * \bar{P} \vdash s} \qquad \frac{\Delta;\, \bar{P} \vdash \mathtt{block}\; s}{\Delta;\, x_0 \mapsto - * \bar{P} \uparrow \vdash s}$$

Since the return and goto statements leave the normal control flow, the postconditions of the (goto) and (return) rules are arbitrary.

Our rules for function calls are similar to those by Von Oheimb [15]. The (call) rule is to call a function that is already in Δ. It is important to notice that its postcondition is not $\vec{e} \Downarrow \vec{v} \wedge Q\, \vec{y}\, \vec{v}$, as after calling f evaluation of $\vec{e}$ may be different after all. However, in case $\vec{e}$ contains no load expressions, we have that $\vec{e} \Downarrow \vec{v}$ is memory independent, and we can simply take A to be $\vec{e} \Downarrow \vec{v}$.

The (add funs) rule can be used to add an arbitrary family Δ' of specifications of (possibly mutually recursive) functions to Δ. For each function f in Δ' with precondition P' and postcondition Q', it has to be verified that the function body $\delta\, f$ is correct for all instantiations of the logical variables $\vec{y}$ and input values $\vec{v}$. The precondition $\Pi_*[x_i \mapsto v_i] * P' \vec{y}\, \vec{v}$, where $\Pi_*[x_i \mapsto v_i]$ denotes the assertion $x_i \mapsto v_i * \ldots x_n \mapsto v_n$, states that the function parameters $\vec{x}$ are allocated with values $\vec{v}$ for which the precondition P' of the function holds. The post- and returning condition $\Pi_*[x_i \mapsto -] * Q' \vec{y}\, \vec{v}$ ensure that the parameters have not been deallocated while executing the function body and that the postcondition P' of the function holds on a return. The jumping condition $\lambda l.\mathsf{False}$ ensures that all gotos jump to a label that occurs in the function body.

Euclid's algorithm in C:

```
void swap(int *p, int *q) {
  int z = *p; *p = *q; *q = z;
}

int gcd(int y, int z) {
  l: if (z) {
    y = y % z; swap(&y, &z); goto l;
  }
  return y;
}
```

Verification of the body of swap:

$\{x_0 \mapsto p * x_1 \mapsto q * p \mapsto y * q \mapsto z\}$
  block (
$\{x_0 \mapsto - * x_1 \mapsto p * x_2 \mapsto q * p \mapsto y * q \mapsto z\}$
    $x_0 := \texttt{load}\ (\texttt{load}\ x_1)$;
$\{x_0 \mapsto y * x_1 \mapsto p * x_2 \mapsto q * p \mapsto y * q \mapsto z\}$
    $\texttt{load}\ x_1 := \texttt{load}\ (\texttt{load}\ x_2)$;
$\{x_0 \mapsto y * x_1 \mapsto p * x_2 \mapsto q * p \mapsto z * q \mapsto z\}$
    $\texttt{load}\ x_2 := \texttt{load}\ x_0$
$\{x_0 \mapsto y * x_1 \mapsto p * x_2 \mapsto q * p \mapsto z * q \mapsto y\}$
)
$\{x_0 \mapsto p * x_1 \mapsto q * p \mapsto z * q \mapsto y\}$

Verification of the body of gcd:

$\{x_0 \mapsto \texttt{int}\ y * x_1 \mapsto \texttt{int}\ z\}$
  l :
$\{J\,l\}$
  if (load x_1) (
$\{\texttt{load}\ x_1 \Downarrow \top \wedge J\,l\}$
$\{x_0 \mapsto \texttt{int}\ y' * x_1 \mapsto \texttt{int}\ z' \wedge$
$z' \neq 0 \wedge \mathsf{gcd}\ y\ z = \mathsf{gcd}\ y'\ z'\}$
    $x_0 := \texttt{load}\ x_0\ \%\ \texttt{load}\ x_1$;
$\{x_0 \mapsto \texttt{int}\ (y'\ \%\ z') * x_1 \mapsto \texttt{int}\ z' *$
$(z' \neq 0 \wedge \mathsf{gcd}\ y\ z = \mathsf{gcd}\ y'\ z' \wedge \mathsf{emp})\}$
    swap(x_0, x_1) ;
$\{x_0 \mapsto \texttt{int}\ z' * x_1 \mapsto \texttt{int}\ (y'\ \%\ z') *$
$(z' \neq 0 \wedge \mathsf{gcd}\ y\ z = \mathsf{gcd}\ y'\ z' \wedge \mathsf{emp})\}$
$\{J\,l\}$
    goto l
$\{x_0 \mapsto \texttt{int}\ (\mathsf{gcd}\ y\ z) * x_1 \mapsto \texttt{int}\ 0\}$
) else
$\{\texttt{load}\ x_1 \Downarrow \bot \wedge J\,l\}$
$\{x_0 \mapsto \texttt{int}\ y' * x_1 \mapsto \texttt{int}\ 0 \wedge \mathsf{gcd}\ y\ z = \mathsf{gcd}\ y'\ 0\}$
    skip
$\{x_0 \mapsto \texttt{int}\ (\mathsf{gcd}\ y\ z) * x_1 \mapsto \texttt{int}\ 0\}$

Fig. 2. Verification of Euclid's algorithm

As an example we verify Euclid's algorithm for computing the greatest common divisor. We first verify the swap function, which takes two pointers p and q and swaps their contents. Its specification is as follows:

$$\forall y\, z\, \forall p\, q\,.\, \{p \mapsto y * q \mapsto z\}\, \{p \mapsto z * q \mapsto y\}$$

Here, universal quantification over the logical variables y and z is used to relate the contents of the pointers p and q in the pre- and postcondition. In order to add this function to the context Δ of verified functions using the (add funs) rule, we have to prove that the body satisfies the above specification. An outline of this proof (with implicit uses of weakening) is displayed in Figure 2.

To verify the body of the gcd function, we use a jumping environment J that assigns $\exists y' z'\,.\, x_0 \mapsto \texttt{int}\ y' * x_1 \mapsto \texttt{int}\ z' \wedge \mathsf{gcd}\ y\ z = \mathsf{gcd}\ y'\ z'$ to the label l. For the function call to swap, we use the (frame) rule with the framing condition $z' \neq 0 \wedge \mathsf{gcd}\ y\ z = \mathsf{gcd}\ y'\ z' \wedge \mathsf{emp}$ as displayed in Figure 2. We refer to the Coq formalization for the full details of these proofs.

5 Soundness of the Axiomatic Semantics

We define $\Delta;\, J;\, R \vDash \{P\}\, s\, \{Q\}$ for Hoare sextuples in terms of our operational semantics. Proving *soundness* of the axiomatic semantics then consists of showing that $\Delta;\, J;\, R \vdash \{P\}\, s\, \{Q\}$ implies that $\Delta;\, J;\, R \vDash \{P\}\, s\, \{Q\}$.

We want $\Delta;\, J;\, R \models \{P\}\, s\, \{Q\}$ to ensure partial program correctness. Intuitively, that means: if $P\,k\,m$ and $\mathbf{S}(k,\,(s,\searrow),\,m) \twoheadrightarrow^* \mathbf{S}(k,\,(s,\nearrow),\,m')$, then $Q\,k\,m'$. However, due to the additional features, this is too simple.

1. We also have to account for reductions starting in $\Uparrow$ or $\curvearrowright$ direction. Hence, we take the four assertions J, R, P and Q together as one function $\bar{P}$ and write $\Delta;\, J;\, R \models \{P\}\, s\, \{Q\}$ as $\Delta;\, \bar{P} \models s$. The intuitive meaning of $\Delta;\, \bar{P} \models s$ is: if $\bar{P}\,d\,k\,m$ and $\mathbf{S}(k,\,(s,\,d),\,m) \twoheadrightarrow^* \mathbf{S}(k,\,(s,\,d'),\,m')$, then $\bar{P}\,d'\,k\,m'$.
2. We have to enforce the reduction $\mathbf{S}(k,\,(s,\,d),\,m) \twoheadrightarrow^* \mathbf{S}(k,\,(s,\,d'),\,m')$ to remain below k as it could otherwise take too many items off the context.
3. Our language has various kinds of undefined behavior (*e.g.* invalid pointer dereferences). We therefore also want $\Delta;\, \bar{P} \models s$ to guarantee that s does not exhibit such undefined behaviors. Hence, $\Delta;\, \bar{P} \models s$ should at least guarantee that if $\bar{P}\,d\,k\,m$ and $\mathbf{S}(k,\,(s,\,d),\,m) \twoheadrightarrow_k^* S$, then S is either:
   – reducible (no undefined behavior has occurred); or:
   – of the shape $\mathbf{S}(k,\,(s,\,d'),\,m')$ with $\bar{P}\,d'\,k\,m'$ (execution is finished).
4. The program should satisfy a form of memory safety so as the make the frame rule derivable. Hence, if before execution the memory can be extended with a disjoint part, that part should not be modified during the execution.
5. We take a step indexed approach in order to relate the assertions of functions in Δ to the statement s.

Together this leads to the following definitions:

Definition 5.1. *Given a predicate $\bar{P}$ over stacks, focuses and memories, specifying valid ending states, the relation $\bar{P} \models_n \hat{\mathbf{S}}(k,\,\phi,\,m \cup \Box)$ is defined as: for each reduction $\mathbf{S}(k,\,\phi,\,m \cup m_f) \twoheadrightarrow_k^n \mathbf{S}(k',\,\phi',\,m')$, we have that m' is of the shape $m' = m'' \cup m_f$ for some memory m'', and either:*
1. *there is a state S such that $\mathbf{S}(k',\,\phi',\,m') \twoheadrightarrow_k S$; or:*
2. *$k' = k$ and $\bar{P}\,k'\,\phi'\,m''$.*

Definition 5.2. Validity of the environment Δ, *notation* $\models_n \Delta$ *is defined as: if* $\Delta\, f = (\forall \vec{y}\,.\, \forall \vec{v}\,.\, \{P\,\vec{y}\,\vec{v}\}\, \{Q\,\vec{y}\,\vec{v}\})$ *and* $P\,\vec{y}\,\vec{v}\,k\,m$, *then* $Q' \models_n \hat{\mathbf{S}}(k,\,\overline{\mathsf{call}\, f\,\vec{v}},\,m \cup \Box)$, *where* $Q' := \lambda \rho\, \phi\, m'\,.\,(\phi = \overline{\mathsf{return}}) \wedge Q\,\vec{y}\,\vec{v}\,\rho\, m'$.

Definition 5.3. Validity of a statement s, *notation* $\Delta;\, \bar{P} \models s$ *is defined as: if* $\models_n \Delta$, $\mathsf{down}\; d\; s$, *and* $\bar{P}\,d\,k\,m$, *then* $Q' \models_n \hat{\mathbf{S}}(k,\,(d,\,s),\,m \cup \Box)$, *where*

$$Q' := \lambda \rho\, \phi\, m'\,.\, \exists d'\, s'\,.\, \phi = (d',\,s') \wedge \neg \mathsf{down}\; d'\; s' \wedge \bar{P}\,d'\,\rho\, m'$$

The predicate down *holds if* $\mathsf{down} \searrow s'$ *or* $\mathsf{down}\; (\curvearrowright l)\; s'$ *with* $l \in \mathsf{labels}\; s'$.

Proposition 5.4 (Soundness). $\Delta;\, \bar{P} \vdash s$ *implies* $\Delta;\, \bar{P} \models s$.

This proposition is proven by induction on the derivation of $\Delta;\, \bar{P} \vdash s$. Thus, for each rule of the axiomatic semantics, we have to show that it holds in the model. The rules for the skip, return, assignment, goto and function calls are proven by chasing all possible reduction paths. In the case of the assignment statement, we need weakening of expression evaluation (Lemma 2.5).

All structural rules are proven by induction on the length of the reduction. These proofs involve chasing all possible reduction paths. We refer to the Coq formalization for the proofs of these rules.

6 Formalization in Coq

All proofs in this paper have been fully formalized using the Coq proof assistant. Formalization has been of great help in order to develop and debug the semantics. We used Coq's notation mechanism combined with unicode symbols and type classes for overloading to let the Coq development correspond as well as possible to the definitions in this paper.

There are some small differences between the Coq development and this paper. Firstly, we omitted while statements and functions with return values here, whereas they are included in the Coq development. Secondly, in this paper, we presented the axiomatic semantics as an inference system, and then showed that it has a model. Since we did not consider completeness, in Coq we directly proved all rules to be derivable with respect to the model.

We used type classes to provide abstract interfaces for commonly used structures like finite sets and finite functions, so we were able to prove theory and implement automation in an abstract way. Our approach is greatly inspired by the *unbundled* approach of Spitters and van der Weegen [17]. However, whereas their work heavily relies on *setoids* (types equipped with an equivalence relation), we tried to avoid that, and used Leibniz equality wherever possible. In particular, our interface for finite functions requires *extensionality* with respect to Leibniz equality. That means $m_1 = m_2 \leftrightarrow \forall x \,.\, m_1\, x = m_2\, x$.

Intensional type theories like Coq do not satisfy extensionality. However, finite functions indexed by a countable type can still be implemented in a way that extensionality holds. For the memory we used finite functions indexed by binary natural numbers implemented as radix-2 search trees. This implementation is based on the implementation in CompCert [12]. But whereas CompCert's implementation does not satisfy canonicity, and thus allows different trees for the same finite function, we have equipped our trees with a proof of canonicity. This way, equality on these finite functions as trees becomes extensional.

Extensional equality on finite functions is particularly useful for dealing with assertions, which are defined as predicates on the stack and memory (Definition 4.1). Due to extensionality, we did not have to equip assertions with a proof of well-definedness with respect to extensional equality on memories.

Although the semantics described in this paper is not extremely big, it is still quite cumbersome to be treated without automation in a proof assistant. In particular, the operational semantics is defined as an inductive type consisting of 32 constructors. To this end, we have automated many steps of the proofs. For example, we implemented the tactic `do_cstep` to automatically perform reduction steps and to solve the required side-conditions, and the tactic `inv_cstep` to perform case analyzes on reductions and to automatically discharge impossible cases. Ongoing experiments show that this approach is successful, as the semantics can be extended easily without having to redo many proofs.

Our Coq code, available at `http://robbertkrebbers.nl/research/ch2o/`, is about 3500 lines of code including comments and white space. Apart from that, the library on general purpose theory (finite sets, finite functions, lists, *etc.*) is about 7000 lines, and the `gcd` example is about 250 lines.

7 Conclusions and Further Research

The further reaching goal of this work is to develop an operational semantics for a large part of the C11 programming language [8] as part of the Formalin project [10]. In order to get there, support for non-local control flow is a necessary step. The operational semantics in this paper extends easily to most other forms of non-local control in C: the break and continue statement, and non-structured switch statements (*e.g.* Duff's device). To support these, we just have to add an additional *direction* and its corresponding reduction rules.

In this paper we have also defined an axiomatic semantics. The purpose of this axiomatic semantics is twofold. Firstly, it gives us more confidence in the correctness and usability of our operational semantics. Secondly, in order to reason about actual C programs, non-local control flow and pointers to block scope variables have to be supported by the axiomatic semantics.

Unfortunately, the current version of our axiomatic semantics is a bit cumbersome to be used for actual program verification. The foremost reason is that our way of handling local variables introduces some overhead. In traditional separation logic [14], there is a strict separation between local variables and allocated storage: the values of local variables are stored directly on the stack, whereas the memory is only used for allocated storage. To that end, the separating conjunction does not deal with local variables, and many assertions can be written down in a shorter way. For example, even though we do not use pointers to the local variables of the `swap` function (Figure 2), we still have to deal with two levels of indirection.

It seems not too hard to make allocation of local variables in the memory optional, so that it can be used only for variables that actually have pointers to them. Ordinary variables then correspond nicely to those with the `register` keyword in C. Alternatively, the work of Parkinson *et al.* [16] on variables as resources may be useful.

Another requirement to conveniently use an axiomatic semantics for program verification is strong automation. Specific to the Coq proof assistant there has been work on this by for example Appel [1] and Chlipala [3]. As our main purpose is to develop an operational semantics for a large part of C11, we consider automation a problem for future work.

In order to get closer to a semantics for C11 we are currently investigating the following additional features of the C11 standard.

- Expressions with side effects and *sequence points*.
- The C type system including structs, unions, arrays and integer types.
- The non-aliasing restrictions (*effective types* in particular).

We intend to support these features in both our operational and axiomatic semantics. Ongoing work shows that our current operational semantics can easily be extended with non-deterministic expressions using a similar approach as Norrish [13] and Leroy [12]. As non-determinism in expressions is closely related to concurrency, we use separation logic for a Hoare logic for expressions.

In this paper we have not considered completeness of the axiomatic semantics as it is not essential for program verification. Also, our future extension for non-deterministic expressions with side-effects will likely be incomplete.

Another direction for future research is to relate our semantics to the CompCert semantics [12] (by eliminating block scope variables). That way we can link it to actual non-local jumps in assembly.

Acknowledgments. We thank Erik Poll for bringing up the idea of an axiomatic semantics for gotos, and thank Herman Geuvers and the anonymous referees for providing several helpful suggestions. This work is financed by the Netherlands Organisation for Scientific Research (NWO).

References

1. Appel, A.W.: Tactics for Separation Logic (2006), `http://www.cs.princeton.edu/~appel/papers/septacs.pdf`
2. Appel, A.W., Blazy, S.: Separation Logic for Small-Step Cminor. In: Schneider, K., Brandt, J. (eds.) TPHOLs 2007. LNCS, vol. 4732, pp. 5–21. Springer, Heidelberg (2007)
3. Chlipala, A.: Mostly-automated verification of low-level programs in computational separation logic. In: PLDI, pp. 234–245. ACM (2011)
4. Dijkstra, E.W.: Go To statement considered harmful. Communications of the ACM 11(3), 147–148 (1968); Letter to the Editor
5. Ellison, C., Rosu, G.: An executable formal semantics of C with applications. In: POPL, pp. 533–544 (2012)
6. Felleisen, M., Hieb, R.: The Revised Report on the Syntactic Theories of Sequential Control and State. Theoretical Computer Science 103(2), 235–271 (1992)
7. Huet, G.P.: The Zipper. Journal of Functional Programming 7(5), 549–554 (1997)
8. International Organization for Standardization. ISO/IEC 9899-2011: Programming languages – C. ISO Working Group 14 (2012)
9. Knuth, D.: Structured programming with go to statements. In: Classics in software engineering, pp. 257–321. Yourdon Press (1979)
10. Krebbers, R., Wiedijk, F.: A Formalization of the C99 Standard in HOL, Isabelle and Coq. In: Davenport, J.H., Farmer, W.M., Urban, J., Rabe, F. (eds.) MKM 2011 and Calculemus 2011. LNCS (LNAI), vol. 6824, pp. 301–303. Springer, Heidelberg (2011)
11. Leroy, X.: A formally verified compiler back-end. Journal of Automated Reasoning 43(4), 363–446 (2009)
12. Leroy, X.: The CompCert verified compiler, software and commented proof (2012), `http://compcert.inria.fr/`
13. Norrish, M.: C formalised in HOL. PhD thesis, University of Cambridge (1998)
14. O'Hearn, P.W., Reynolds, J.C., Yang, H.: Local Reasoning about Programs that Alter Data Structures. In: Fribourg, L. (ed.) CSL 2001 and EACSL 2001. LNCS, vol. 2142, pp. 1–19. Springer, Heidelberg (2001)
15. von Oheimb, D.: Hoare Logic for Mutual Recursion and Local Variables. In: Pandu Rangan, C., Raman, V., Sarukkai, S. (eds.) FSTTCS 1999. LNCS, vol. 1738, pp. 168–180. Springer, Heidelberg (1999)
16. Parkinson, M.J., Bornat, R., Calcagno, C.: Variables as Resource in Hoare Logics. In: LICS, pp. 137–146 (2006)
17. Spitters, B., van der Weegen, E.: Type classes for mathematics in type theory. Mathematical Structures in Computer Science 21(4), 795–825 (2011)
18. Tews, H.: Verifying Duff's device: A simple compositional denotational semantics for Goto and computed jumps (2004)

The Parametric Ordinal-Recursive Complexity of Post Embedding Problems*

Prateek Karandikar[1,2] and Sylvain Schmitz[2]

[1] CMI, Chennai, India
[2] LSV, ENS Cachan & CNRS, Cachan, France

Abstract. *Post Embedding Problems* are a family of decision problems based on the interaction of a rational relation with the subword embedding ordering, and are used in the literature to prove non multiply-recursive complexity lower bounds. We refine the construction of Chambart and Schnoebelen (LICS 2008) and prove parametric lower bounds depending on the size of the alphabet.

1 Introduction

Ordinal Recursive functions and subrecursive hierarchies [24, 12] are employed in computability theory, proof theory, Ramsey theory, rewriting theory, etc. as tools for bounding derivation sizes and other objects of very high combinatory complexity. A standard example is the ordinal-indexed *extended Grzegorczyk hierarchy* $\mathscr{F}_\alpha$ [21], which characterizes classical classes of functions: for instance, $\mathscr{F}_2$ is the class of elementary functions, $\bigcup_{k<\omega} \mathscr{F}_k$ of primitive-recursive ones, and $\bigcup_{k<\omega} \mathscr{F}_{\omega^k}$ of multiply-recursive ones. Similar tools are required for the classification of decision problems arising with verification algorithms and logics, prompting the investigation of "natural" decision problems complete for *fast-growing complexity* classes $\mathbf{F}_\alpha$ [14, 27].

Post Embedding Problems. (PEPs) have been introduced by Chambart and Schnoebelen [7] as a tool to prove the decidability of safety and termination problems in unreliable channel systems. The most classical instance of a PEP is called "regular" by Chambart and Schnoebelen [7], but we will follow Barceló et al. [4] and rather call it *rational* in this paper:

Rational Embedding Problem (EP[Rat])

Input. A rational relation R in $\Sigma^* \times \Sigma^*$.

Question. Is the relation $R \cap \sqsubseteq$ empty?

Here, the $\sqsubseteq$ relation denotes the *subword embedding* ordering, which relates two words w and w' if $w = c_1 \cdots c_n$ and $w' = w_0 c_1 w_1 \cdots w_n c_n w_{n+1}$ for some symbols

* Research partially funded by the ANR ReacHard project (ANR 11 BS02 001 01). The first author is partially funded by Tata Consultancy Services. Part of this research was conducted while the second author was visiting the Department of Computer Science at Oxford University thanks to a grant from the ESF *Games for Design and Verification* activity.

F. Pfenning (Ed.): FOSSACS 2013, LNCS 7794, pp. 273–288, 2013.

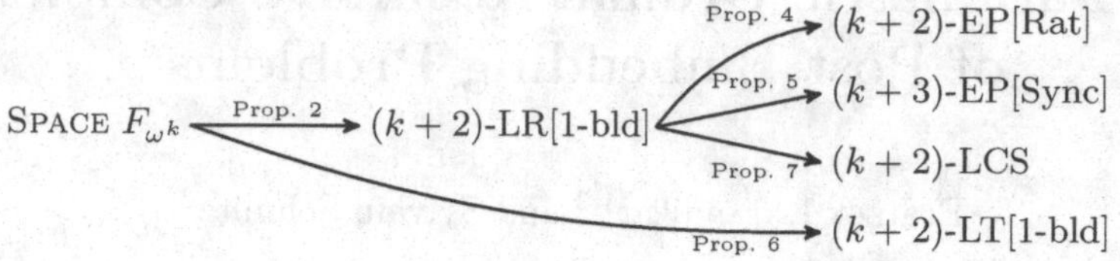

Fig. 1. Relationships between PEPs and similar decision problems

c_i in Σ and words w_i in Σ^*; in other words, w can be obtained from w' by "losing" some symbol occurrences (maybe none).

Although PEPs appear naturally in relation with channel systems [7, 8, 16] and queries on graph databases [4], their main interest lies in their use in lower bound proofs for other, sometimes seemingly distantly related problems [23, 19, 3]: in spite of their simple formulation, they are known to be of non multiply-recursive complexity in general. In fact, this motivation has been present from their inception in [7]: find a "master" decision problem complete for $\mathbf{F}_{\omega^\omega}$, the class of *hyper-Ackermannian* problems, solvable with non multiply-recursive complexity, but no less—much like SAT is often taken as the canonical NPTIME-complete problem, or the Post Correspondence Problem for Σ_1^0. This has also prompted a wealth of research into variants and related questions [10, 4, 18].

In this paper, we revisit and simplify the original proof of Chambart and Schnoebelen [9] that established the hardness of PEPs, and prove tight *parameterized* lower bounds when the size of the alphabet Σ is fixed. More precisely, we show that the $(k+2)$-rational embedding problem, i.e. the restriction of EP[Rat] to alphabets Σ of size at most $k+2$, is hard for $\mathbf{F}_{\omega^k}$ the class of *k-Ackermannian problems* if $k \geq 2$. As the problem can be shown to be in $\mathbf{F}_{\omega^{k+1}+1}$ [26, 18], we argue this to be a rather tight bound. The hyper-Ackermannian lower bound of $\mathbf{F}_{\omega^\omega}$ first proven by Chambart and Schnoebelen then arises when $|\Sigma|$ is not fixed but depends on the instance.

Our main tool to this end is another problem that involves a rational relation together with the subword embedding:

Lossy Rewriting (LR[Rat])

Input. A rational relation R in $\Sigma^* \times \Sigma^*$ and two words w and w' in Σ^*.
Question. Does (w, w') belong to the reflexive transitive closure $R_{\sqsupseteq}^{\circledast}$?

Here $R_{\sqsupseteq}$ denotes the "lossy version" of the relation R, defined formally as the composition $\sqsupseteq ; R ; \sqsupseteq$. We prove our lower bounds on this variant of EP[Rat] and then use them to prove lower bounds for EP[Rat] and other embedding problems; Fig. 1 summarizes the lower bounds presented in this paper. In a sense, LR is our own champion for the title of "master" problem for $\mathbf{F}_{\omega^\omega}$. Besides its rather simple statement, note that the related question of whether (w, w') belongs to $R^{\circledast}$ is undecidable by an easy reduction from the acceptance problem for Turing machines.

Overview. Technically, our results rely on an implementation of the computations for the *Hardy functions* $H^{\omega^{\omega^k}}$ and their inverses by successive applications of a relation with a fixed *bounded length discrepancy.* The main difficulty here is that this implementation should be *robust* for the symbol losses associated with the embedding relation. It requires in particular a robust encoding of ordinals below ω^{ω^k} as sequences over an alphabet of $k+2$ symbols, for which we adapt the constructions of [9, 15]; see Sec. 3. Compared with previous work, we make the most of the rational relations framework, leading to simpler and more detailed proofs of robustness.

This allows us to show in Sec. 4 that for $k \geq 2$, $(k+2)$-LR[1-bld], i.e. a version of LR[Rat] over an alphabet of size $|\Sigma| = k+2$ and with a relation R with bounded length discrepancy of 1, is $\mathbf{F}_{\omega^k}$-hard. We also show that this lower bound is quite tight, as $(k+2)$-LR[Rat] is in $\mathbf{F}_{\omega^{k+1}}$.

We then show in Sec. 5 that LR[1-bld] can easily be reduced to EP[Rat] and other (parameterized) embedding problems—including EP[Sync], a restriction of EP[Rat] introduced by Barceló et al. [4] where the relation R is *synchronous* (aka *regular*), and which required a complex lower bound proof.

Let us now turn to the necessary formal background on PEPs in Sec. 2. Due to space constraints, some proof details will be found in the full version of this paper, available as arXiv:1211.5259 [cs.LO].

2 Post Embedding Problems

Rational Relations [11] play an important role in the following, as they provide a notion of finitely presentable relations over strings more powerful than string rewrite systems, and come with a large body of theory and results [see e.g. 25, Chap. IV]. Let us quickly skim over the notations and definitions that will be needed in this paper.

We assume the reader to be familiar with the basic characterizations of *rational relations* R between two finite alphabets Σ and Δ by

closure of the finite relations in $\Sigma^* \times \Delta^*$ under union, concatenation, and Kleene star,[1]

finite transductions defined by normalized transducers $\mathcal{T} = \langle Q, \Sigma, \Delta, \delta, I, F\rangle$ where Q is a finite set of states, $\delta \subseteq Q \times ((\Sigma \times \{\varepsilon\}) \cup (\{\varepsilon\} \times \Delta)) \times Q$ is a transition relation—where ε denotes the empty word, of length $|\varepsilon| = 0$—, initial set of states $I \subseteq Q$, and final set of states $F \subseteq Q$,

decomposition into a regular language L over some finite alphabet Γ and two morphisms $u\colon \Gamma^* \to \Sigma^*$ and $v\colon \Gamma^* \to \Delta^*$ s.t. $R = u^{-1} \mathbin{;} \mathrm{Id}_L \mathbin{;} v$, where Id_L is the identity function over the restricted domain L.

[1] We use different symbols "$*$" and "$+$" for Kleene star and Kleene plus, i.e. iteration of concatenation "$\cdot$" on the one hand, and "$\circledast$" and "$\oplus$" for reflexive transitive closure and transitive closure, i.e. iteration of composition "$;$" on the other hand. Rational relations and length-preserving relations are closed under Kleene star, but none of the classes of relations we consider is closed under reflexive transitive closure.

This last characterization is known as Nivat's Theorem, and shows that EP[Rat] can be stated alternatively as taking as input a regular language L in Γ^* and two morphisms u and v from Γ^* to Σ^* and asking whether there exists some word x in L s.t. $u(x) \sqsubseteq v(x)$ [7]. This justifies the name of "Post Embedding Problem", as the well-known, undecidable *Post Correspondence Problem* asks instead given u and v whether there exists x in Γ^+ s.t. $u(x) = v(x)$.

Synchronous Relations are a restricted class of rational relations, and are closed under intersection and complement, in addition to e.g. the closure under composition and inverse that all rational relations enjoy. A rational relation has *b-bounded length discrepancy* if the absolute value of $|u| - |v|$ is at most b for all (u, v) in R, and has *bounded length discrepancy* (bld) if there exists such a finite b. In particular, it is *length-preserving* if $|u| = |v|$, i.e. if it has bld 0. A *synchronous relation* is a finite union of relations of form $\{(u, vw) \mid (u, v) \in R \wedge w \in L\}$ and $\{(uw, v) \mid (u, v) \in R \wedge w \in L\}$ where R ranges over length-preserving rational relations and L over regular languages. In terms of classes of relations in $\Sigma^* \times \Delta^*$, we have the strict inclusions [25]:

$$\text{lp} = 0\text{-bld} \subsetneq \cdots \subsetneq b\text{-bld} \subsetneq (b+1)\text{-bld} \subsetneq \cdots \subsetneq \text{bld} \subsetneq \text{Sync} \subsetneq \text{Rat}\,. \qquad (1)$$

Post Embedding Problems, as we have seen in the introduction, are concerned with the interplay of a rational relation R in $\Sigma^* \times \Sigma^*$ with the subword embedding ordering $\sqsubseteq$. The latter is a particular case of a (deterministic) rational relation that is not synchronous. Both EP[Rat] and LR[Rat] are particular instances of more general, undecidable problems: the emptiness of intersection of two rational relations for EP[Rat], and the word problem in the reflexive transitive closure of a rational relation for LR[Rat]. We can add another natural problem to the set of PEPs:

Lossy Termination (LT[Rat])

Input. A rational relation R over Σ and a word w in Σ^*.

Question. Does $R_{\sqsupseteq}^{\circledast}$ terminate from w, i.e. is every sequence $w = w_0 \, R_{\sqsupseteq} \, w_1 \, R_{\sqsupseteq} \cdots R_{\sqsupseteq} \, w_i \, R_{\sqsupseteq} \cdots$ with $w_0, w_1, \ldots, w_i, \ldots$ in Σ^* finite?

Again, this is a variant of the termination problem, which is in general undecidable when the relation is not lossy.

Restrictions. We parameterize PEPs with the subclass of rational relations under consideration for R and the cardinal of the alphabet Σ; for instance, $(k+2)$-EP[Sync] is the variant of EP[Rat] where the relation is synchronous and $|\Sigma| = k+2$. We are interested in this paper in providing $\mathbf{F}_{\omega^k}$ lower bounds with the smallest possible class of relations and smallest possible alphabet size, but we should also mention that some (rather strong) restrictions become tractable:

- Barceló et al. [4] show that EP[Rec]—where a *recognizable relation* is a finite union of products $L \times L'$ where L and L' range over regular languages—is in NLogSpace, because the intersection $R \cap \sqsubseteq$ is rational, and can effectively be constructed and tested for emptiness on the fly,

- Chambart and Schnoebelen [7] show that EP[2Morph]—where a *2-morphic relation* [20] is the composition $R = (u^{-1} \mathbin{;} v) \setminus \{(\varepsilon, \varepsilon)\}$ of two morphisms u and v from Γ^* to Σ^*—is in LogSpace, because it reduces to checking whether there exists a in Γ s.t. $u(a) \sqsubseteq v(a)$,
- the case EP[Rewr] of *rewrite relations* is similarly in LogSpace: a rewrite relation R is defined from a *semi-Thue system*, i.e. a finite set Υ of rules (u, v) in $\Sigma^* \times \Sigma^*$, as $\to_\Upsilon = \{(wuw', wvw') \mid w, w' \in \Sigma^* \wedge (u, v) \in \Upsilon\}$, and EP[Rewr] reduces to checking whether $u \sqsubseteq v$ for some rule (u, v) of Υ,
- the unary alphabet case of 1-EP[Rat] is in NLogSpace: this can be seen using Parikh images and Presburger arithmetic:

Proposition 1. *The problem 1-EP[Rat] is in* NLogSpace.

3 Hardy Computations

We use the *Hardy hierarchy* as our main subrecursive hierarchy [21, 24, 12]. Although we will only use the lower levels of this hierarchy, its general definition is worth knowing, as it is archetypal of ordinal-indexed *subrecursive hierarchies*; see [27] for a self-contained presentation.

3.1 The Hardy Hierarchy

Ordinal Terms. Let ε_0 be the smallest solution of the equation $\omega^x = x$. It is well-known that any ordinal $\alpha < \varepsilon_0$ can be written uniquely in Cantor Normal Form (CNF) as a sum

$$\alpha = \omega^{\beta_1} \dotplus \cdots \dotplus \omega^{\beta_n} \tag{2}$$

where $\beta_n \leq \cdots \leq \beta_1 < \alpha$ and each β_i is itself in CNF. This ordinal α is 0 if $n = 0$ in (2), a *successor ordinal* if β_n is 0, and a *limit ordinal* otherwise. In the following, we write $\alpha \dotplus \beta$ to denote a direct sum $\alpha + \beta$ where $\alpha > \beta$ or $\alpha = 0$.

Subrecursive hierarchies are defined through assignments of *fundamental sequences* $(\lambda_n)_{n<\omega}$ for limit ordinals $\lambda < \varepsilon_0$, satisfying $\lambda_n < \lambda$ for all n and $\lambda = \sup_n \lambda_n$. A standard assignment on terms in CNF is defined by:

$$\left(\gamma \dotplus \omega^{\alpha+1}\right)_n \stackrel{\text{def}}{=} \gamma \dotplus \omega^\alpha \cdot n, \qquad \left(\gamma \dotplus \omega^\lambda\right)_n \stackrel{\text{def}}{=} \gamma \dotplus \omega^{\lambda_n}, \tag{3}$$

thus verifying $\omega_n = n$. Let $\Omega \stackrel{\text{def}}{=} \omega^{\omega^\omega}$; this yields for instance $\Omega_k = \omega^{\omega^k}$ and, if $k > 0$, $(\Omega_k)_n = \omega^{\omega^{k-1} \cdot n}$.

Hardy Hierarchy. The *Hardy hierarchy* $(H^\alpha)_{\alpha<\varepsilon_0}$ is an ordinal-indexed hierarchy of functions $H^\alpha \colon \mathbb{N} \to \mathbb{N}$ defined by

$$H^0(n) \stackrel{\text{def}}{=} n \qquad H^{\alpha+1}(n) \stackrel{\text{def}}{=} H^\alpha(n+1) \qquad H^\lambda(n) \stackrel{\text{def}}{=} H^{\lambda_n}(n)\,. \tag{4}$$

Observe that H^1 is simply the successor function, and more generally H^α is the αth iterate of the successor function, using diagonalisation to treat limit ordinals. A related hierarchy is the *fast growing hierarchy* $(F_\alpha)_{\alpha<\varepsilon_0}$, which can be defined by $F_\alpha \stackrel{\text{def}}{=} H^{\omega^\alpha}$, resulting in $F_0(n) = H^1(n) = n+1$, $F_1(n) = H^\omega(n) = H^n(n) = 2n$, $F_2(n) = H^{\omega^2}(n) = 2^n n$ being exponential, $F_3 = H^{\omega^3}$ being non-elementary, $F_\omega = H^{\omega^\omega} = H^{\Omega_1}$ being an Ackermannian function, $F_{\omega^k} = H^{\Omega_k}$ a k-Ackermannian function, and $F_{\omega^\omega} = H^\Omega$ an hyper-Ackermannian function.

Fast-Growing Complexity Classes. Our intention is to establish the "F_{ω^k} hardness" of Post embedding problems. In order to make this statement more precise, we define the class $\mathbf{F}_{\omega^k}$ of *k-Ackermannian problems* as a specific instance of the *fast-growing complexity classes* defined for $\alpha \geq 3$ by

$$\mathbf{F}_\alpha \stackrel{\text{def}}{=} \bigcup_{p\in\bigcup_{\beta<\alpha}\mathscr{F}_\beta} \mathrm{DTIME}(F_\alpha(p(n)))\;, \qquad \mathscr{F}_\alpha = \bigcup_{c<\omega} \mathrm{FDTIME}(F_\alpha^c(n))\;, \tag{5}$$

where $\mathscr{F}_\alpha$ defined above is the αth level of the *extended Grzegorczyk hierarchy* [21] when $\alpha \geq 2$. The classes $\mathbf{F}_\alpha$ are naturally equipped with $\bigcup_{\beta<\alpha}\mathscr{F}_\beta$ as class of reductions. For instance, because $\bigcup_{k<\omega}\mathscr{F}_{\omega^k}$ is exactly the set of multiply-recursive functions, $\mathbf{F}_{\omega^\omega}$ captures the intuitive notion of hyper-Ackermannian problems closed under multiply-recursive reductions.[2]

Hardy Computations. The fast-growing and Hardy hierarchies have been used in several publications to establish Ackermannian and higher complexity bounds [9, 26, 15, 27]. The principle in their use for lower bounds is to view (4), read left-to-right, as a rewrite system over $\varepsilon_0 \times \mathbb{N}$, and later implement it in the targeted formalism. Formally, a (forward) *Hardy computation* is a sequence

$$\alpha_0, n_0 \to \alpha_1, n_1 \to \alpha_2, n_2 \to \cdots \to \alpha_\ell, n_\ell \tag{6}$$

of evaluation steps implementing the equations in (4) seen as left-to-right rewrite rules over *Hardy configurations* α, n. It guarantees $\alpha_0 > \alpha_1 > \alpha_2 > \cdots$ and keeps $H^{\alpha_i}(n_i)$ invariant. We say it is *complete* when $\alpha_\ell = 0$ and then $n_\ell = H^{\alpha_0}(n_0)$ (we also consider incomplete computations). A *backward Hardy computation* is obtained by using (4) as right-to-left rules. For instance,

$$\omega^{\omega^k}, n \to \omega^{\omega^{k-1}\cdot n}, n \to \omega^{\omega^{k-1}\cdot(n-1)+\omega^{k-2}\cdot n}, n \tag{7}$$

constitute the first three steps of the forward Hardy computation starting from Ω_k, n if $k > 1$ and $n > 0$.

---

[2] Note that, at such high complexities, the usual distinctions between deterministic vs. nondeterministic, or time-bounded vs. space-bounded computations become irrelevant. In particular, $\mathscr{F}_2$ is the set of elementary functions, and $\mathbf{F}_3$ the class of problems with a tower of exponentials of height bounded by some elementary function of the input as an upper bound.

Termination of Hardy Computations. Because $\alpha_0 > \alpha_1 > \cdots > \alpha_\ell$ in a forward Hardy computation like (6), it necessarily terminates. For inverse computations, this is less immediate, and we introduce for this a *norm* $\|\alpha\|$ of an ordinal α in ε_0 as its count of "ω" symbols when written as an ordinal term: formally, $\|.\|\colon \varepsilon_0 \to \mathbb{N}$ is defined by

$$\|0\| \stackrel{\text{def}}{=} 0 \qquad \|\omega^\alpha\| \stackrel{\text{def}}{=} 1 + \|\alpha\| \qquad \|\alpha \dotplus \alpha'\| \stackrel{\text{def}}{=} \|\alpha\| + \|\alpha'\| \,. \tag{8}$$

We can check that, for any limit ordinal λ, $\|\lambda_n\| > \|\lambda\|$ whenever $n > 1$. Therefore, in a backward Hardy computation, the pair $(n, \|\alpha\|)$ decreases for the lexicographic ordering over $\mathbb{N}^2$. As this is a well-founded ordering, we see that backward computations terminate if n remains larger than 1—which is a reasonable hypothesis for the following.

3.2 Encoding Hardy Configurations

Our purpose is now to encode Hardy computations as relations over Σ^*. This entails in particular (1) encoding configurations α, n in $\Omega_k \times \mathbb{N}$ of a Hardy computation as finite sequences using *cumulative ordinal descriptions* or "*codes*", which we do in this subsection, and (2) later in Sec. 3.3 designing a 1-bld relation that implements Hardy computation steps over codes. A constraint on codes is that they should be *robust* against losses, i.e. if $\pi(x)$ and $\pi(x')$ are the ordinals associated to the codes x and x' and $\pi(x) \sqsubseteq \pi(x')$, then $H^{\pi(x)}(n) \le H^{\pi(x')}(n)$—pending some hygienic conditions on x and x', see Lem. 2.

Finite Ordinals below k can be represented as single symbols $a_0, \ldots, a_{k-1}$ of an alphabet Σ_k along with a bijection

$$\varphi(a_i) \stackrel{\text{def}}{=} i \,. \tag{9}$$

Small Ordinals below ω^k are then easily encoded as finite words over Σ_k: given a word $w = b_1 \cdots b_n$ over Σ_k, we define its associated ordinal in ω^k as

$$\beta(w) \stackrel{\text{def}}{=} \omega^{\varphi(b_1)} + \cdots + \omega^{\varphi(b_n)} \,. \tag{10}$$

Note that β is surjective but not injective: for instance, $\beta(a_0 a_1) = \beta(a_1) = \omega$. By restricting ourselves to *pure* words over Σ_k, i.e. words satisfying $\varphi(b_j) \ge \varphi(b_{j+1})$ for all $1 \le j < n$, we obtain a bijection between ω^k and $\mathsf{p}(\Sigma_k^*)$ the set of pure finite words in Σ_k^*, because then (10) is the CNF of $\beta(w)$.

Large Ordinals below Ω_k are denoted by *codes* [9, 15], which are #-separated words over the extended alphabet $\Sigma_{k\#} \stackrel{\text{def}}{=} \Sigma_k \uplus \{\#\}$. A code x can be seen as a concatenation $w_1 \# w_2 \# \cdots \# w_p \# w_{p+1}$ where each w_i is a word over Σ_k. Its associated ordinal $\pi(x)$ in Ω_k is then defined as

$$\pi(x) \stackrel{\text{def}}{=} \omega^{\beta(w_1 w_2 \cdots w_p)} \dotplus \cdots \dotplus \omega^{\beta(w_1 w_2)} \dotplus \omega^{\beta(w_1)} \,, \tag{11}$$

or inductively by

$$\pi(w) \stackrel{\text{def}}{=} 0, \qquad \pi(w\#x) \stackrel{\text{def}}{=} \omega^{\beta(w)} \cdot \pi(x) \dotplus \omega^{\beta(w)} \tag{12}$$

for w a word in Σ_k^* and x a code. For instance, $\pi(a_1a_0\#) = \omega^{\omega+1} = \pi(a_0a_1a_0\#a_3)$, or, closer to our concerns, the initial ordinal in our computations is $\pi(a_{k-1}^n\#) = (\Omega_k)_n$ when $k > 0$.

Observe that π is surjective, but not injective. We can mend this by defining a *pure* code $x = w_1\# \cdots \#w_p\#w_{p+1}$ as one where $w_{p+1} = \varepsilon$ and every word w_i for $1 \leq i \leq p$ is pure—note that it does not force the concatenation of two successive words w_iw_{i+1} of x to be pure. This is intended, as this is the very mechanism that allows π to be a bijection between Ω_k and $\mathsf{p}(\Sigma_{k\#}^*)$

Lemma 1. *The function π is a bijection from $\mathsf{p}(\Sigma_{k\#}^*)$ to Ω_k.*

We also define $\mathsf{p}(x)$ to be the unique pure code x' verifying $\pi(x) = \pi(x')$; then $\mathsf{p}(x) \sqsubseteq x$, and $x \sqsubseteq x'$ implies $\mathsf{p}(x) \sqsubseteq \mathsf{p}(x')$.

Hardy Configurations α, n are finally encoded as sequences $c = \pi^{-1}(\alpha) \mid \#^n$ using a separator "$\mid$", i.e. as sequences in the language Confs $\stackrel{\text{def}}{=} \mathsf{p}(\Sigma_{k\#}^*)\cdot\{\mid\}\cdot\{\#\}^*$. This is a regular language over $\Sigma_{k\#} \uplus \{\mid\}$, but the most important fact about this encoding is that it is *robust* against symbol losses as far as the corresponding computed Hardy values are concerned. Robustness is a critical part of hardness proofs based on Hardy functions. The main difficulty rises from the fact that the Hardy functions are not monotone in their ordinal parameter: for instance, $H^\omega(n) = H^n(n) = 2n$ is less than $H^{n+1}(n) = 2n + 1$. Code robustness is addressed in [9, Prop. 4.3]. Robustness is the limiting factor that prevents us from reducing languages in $\mathbf{F}_\alpha$ for $\alpha > \Omega$ into PEPs.

Lemma 2 (Robustness). *Let $c = x \mid \#^n$ and $c' = x' \mid \#^{n'}$ be two sequences in* Confs. *If $c \sqsubseteq c'$, then $H^{\pi(x)}(n) \leq H^{\pi(x')}(n')$.*

3.3 Encoding Hardy Computations

It remains to present a 1-bld relation that implements Hardy computations over Hardy configurations encoded as sequences in Confs. We translate the equations from (4) into a relation $R_H = (R_0 \cup R_1 \cup R_2) \cap (\text{Confs} \times \text{Confs})$, which can be reversed for backward computations:

$$R_0 \stackrel{\text{def}}{=} \{(\#x \mid \#^n, x \mid \#^{n+1}) \mid n \geq 0, x \in \Sigma_{k\#}^*\} \tag{13}$$

$$R_1 \stackrel{\text{def}}{=} \{(wa_0\#x \mid \#^n, w\#^n\mathsf{p}(a_0x) \mid \#^n) \mid n > 1, w \in \Sigma_k^*, x \in \Sigma_{k\#}^*\} \tag{14}$$

$$R_2 \stackrel{\text{def}}{=} \{(wa_i\#x \mid \#^n, wa_{i-1}^n\#\mathsf{p}(a_ix) \mid \#^n) \mid n > 1, i > 0, w \in \Sigma_k^*, x \in \Sigma_{k\#}^*\} \tag{15}$$

The relation R_0 implements the successor case, while R_1 and R_2 implement the limit case of (3) for ordinals of form $\gamma \dotplus \omega^{\alpha+1}$ and $\gamma \dotplus \omega^\lambda$ respectively. The restriction to $n > 1$ in R_1 and R_2 enforces termination for backward computations; it is not required for correctness. Because R_H is a direct translation of (4) over Confs:

Lemma 3 (Correctness). *For all* α, α' *in* Ω_k *and* $n, n' > 1$, $(\pi^{-1}(\alpha) \shortmid \#^n)$ $(R_H \cup R_H^{-1})^{\circledast}(\pi^{-1}(\alpha') \shortmid \#^{n'})$ *iff* $H^\alpha(n) = H^{\alpha'}(n')$.

Unfortunately, although R_0 is a length-preserving rational relation, R_1 and R_2 are not 1-bld, nor even rational. However, they can easily be broken into smaller steps, which are rational—as we are applying a reflexive transitive closure, this is at no expense in generality. This requires more complex encodings of Hardy configurations, with some "finite state control" and a working space in order to keep track of where we are in our small steps. Because we do not want to spend new symbols in this encoding, given some finite set Q of states, we work on sequences in

$$\mathrm{Seqs} \stackrel{\text{def}}{=} \{a_0, a_1\}^{\lceil \log |Q| \rceil} \cdot \{\shortmid\} \cdot \mathsf{p}(\Sigma_k^*) \cdot \{\#\}^* \cdot \{\shortmid\} \cdot \mathsf{p}(\Sigma_{k\#}^*) \cdot \{\shortmid\} \cdot \{\#, a_0, a_1\}^* \; . \quad (16)$$

with four segments separated by "$\shortmid$": a state, a working segment, an ordinal encoding, and a counter. Given a state q in Q, we use implicitly its binary encoding as a sequence of fixed length over $\{a_0, a_1\}$.

We define two relations Fw and Bw with domain and range Seqs that implement forward and backward computations with R_H. A typical case is that of computations with R_1, which can be implemented as the closure of the union:

$$q_{\mathsf{Fw}} \shortmid\shortmid w a_0 \# x \shortmid \#^{n+2} \;\mathsf{Fw}_1\; q_{\mathsf{Fw}_1} \shortmid w\# \shortmid \mathsf{p}(a_0 x) \shortmid \#^{n+1} a_0 \quad (17)$$

$$q_{\mathsf{Fw}_1} \shortmid w\#^m \shortmid x \shortmid \#^{n+1} a_0^{p+1} \;\mathsf{Fw}_1\; q_{\mathsf{Fw}_1} \shortmid w\#^{m+1} \shortmid x \shortmid \#^n a_0^{p+2} \quad (18)$$

$$q_{\mathsf{Fw}_1} \shortmid w\#^{m+1} \shortmid x \shortmid a_0^{n+2} \;\mathsf{Fw}_1\; q_{\mathsf{Fw}_1} \shortmid\shortmid w\#^{m+1} x \shortmid \#^{n+2} \quad (19)$$

for m, n, p in $\mathbb{N}$, w in $\mathsf{p}(\Sigma_k^*)$, and x in $\mathsf{p}(\Sigma_{k\#}^*)$. Note that $\mathsf{p}(a_0 x)$ returns $a_0 x$ if x begins with $\#$ or a_0, and x otherwise. The corresponding backward computation for R_1 inverses the relations in (17–19) and substitutes q_{Bw} and q_{Bw_1} for q_{Fw} and q_{Fw_1}. The reader should be able to convince herself that this is indeed feasible in a rational 1-bld fashion; for instance, (18) can be written as a rational expression

$$\begin{bmatrix} q_{\mathsf{Fw}_1} \shortmid \\ q_{\mathsf{Fw}_1} \shortmid \end{bmatrix} \cdot \mathrm{Id}_{\Sigma_k^*} \cdot \begin{bmatrix} \# \\ \# \end{bmatrix}^* \cdot \begin{bmatrix} \varepsilon \\ \# \end{bmatrix} \cdot \begin{bmatrix} \shortmid \\ \shortmid \end{bmatrix} \cdot \mathrm{Id}_{\Sigma_{k\#}^*} \cdot \begin{bmatrix} \shortmid \\ \shortmid \end{bmatrix} \cdot \begin{bmatrix} \# \\ \# \end{bmatrix}^* \cdot \begin{bmatrix} \# \\ \varepsilon \end{bmatrix} \cdot \begin{bmatrix} a_0 \\ a_0 \end{bmatrix}^+ \cdot \begin{bmatrix} \varepsilon \\ a_0 \end{bmatrix} \cdot \quad (20)$$

Observe that separators "$\shortmid$" are reliable, and that losses cannot pass unnoticed in the constant-sized state segment of a sequence in Seqs; thus we can use lemmas 2 and 3 to prove that $\mathsf{Fw}_{\sqsupseteq}^{\circledast}$ and $\mathsf{Bw}_{\sqsupseteq}^{\circledast}$ are "weak" implementations of H^α and its inverse when α is in Ω_k. Not any reformulation of R_H as the closure of a rational relation would work here: our relation also needs to be robust to losses; see the full paper for details.

Lemma 4 (Weak Implementation). *The relations* Fw *and* Bw *are 1-bld and terminating. Furthermore, if* $k \geq 1$, $m, n > 1$ *and* $\alpha \in \Omega_k$,

$$(q_{\mathsf{Fw}} \shortmid\shortmid \pi^{-1}(\alpha) \shortmid \#^n) \;\mathsf{Fw}_{\sqsupseteq}^{\circledast}\; (q_{\mathsf{Fw}} \shortmid\shortmid\shortmid \#^m) \qquad \textit{implies } m \leq H^\alpha(n)$$

$$(q_{\mathsf{Bw}} \shortmid\shortmid\shortmid \#^m) \;\mathsf{Bw}_{\sqsupseteq}^{\circledast}\; (q_{\mathsf{Bw}} \shortmid\shortmid \pi^{-1}(\alpha) \shortmid \#^n) \qquad \textit{implies } m \geq H^\alpha(n)$$

and there exists rewrites verifying $m = H^\alpha(n)$ *in both of the above cases.*

4 The Parametric Complexity of LR[1-bld]

Now equipped with suitable encodings for Hardy computations, we can turn to the main result of the paper: Prop. 2 below shows the $\mathbf{F}_{\omega^k}$-hardness of $(k+2)$-LR[1-bld]. As we obtain almost matching upper bounds in Sec. 4.2, we deem this to be rather tight.

4.1 Lower Bound

Thanks to the relations over $\Sigma_{k\#} \uplus \{\shortmid\}$ defined in Sec. 3, we know that we can weakly compute with Fw a "budget space" as a unary counter of size $F_{\omega^k}(n)$, and later check that this budget has been maintained by running through Bw. We are going to insert the simulation of an $\mathbf{F}_{\omega^k}$-hard problem between these two phases of budget construction and budget verification, thereby constructing $\mathbf{F}_{\omega^k}$-hard instances of $(k+2)$-LR[1-bld].

Proposition 2. *Let $k \geq 2$. Then $(k+2)$-LR[1-bld] is $\mathbf{F}_{\omega^k}$-hard.*

Bounded Semi-Thue Reachability. The problem we reduce from is a space-bounded variant of the *semi-Thue reachability problem* (aka *semi-Thue word problem*): as already mentioned in Sec. 2, a *semi-Thue system* Υ over an alphabet is a finite set of rules (u, v) in $\Sigma^* \times \Sigma^*$, defining a *rewrite relation* $\to_\Upsilon$. The semi-Thue reachability problem, or R[Rewr], is a reliable version of the lossy reachability problem. This problem is in general undecidable, as one can express the "next configuration" relation of a Turing machine as a semi-Thue system. Its *F_{ω^k}-bounded* version for some $k \geq 1$ takes as input an instance $\langle \Upsilon, y, y' \rangle$ of size n where, if $y \to_\Upsilon^{\circledast} x$, then $|x| \leq F_{\omega^k}(n)$. This is easily seen to be hard for $\mathbf{F}_{\omega^k}$, even for a binary alphabet Σ.

Reduction. Let $\langle \Upsilon, y, y' \rangle$ be an instance of size $n > 1$ of F_{ω^k}-bounded R[Rewr] over the two-letters alphabet $\{a_0, a_1\}$. We build a $(k+2)$-LR[1-bld] instance in which the rewrite relation R performs the following sequence:

1. Weakly compute a budget of size $F_{\omega^k}(n)$, using Fw described in Sec. 3.
2. In this allocated space, simulate the rewrite steps of Υ starting from y.
3. Upon reaching y', perform a reverse Hardy computation using Bw and check that we obtain back the initial Hardy configuration. This check ensures that the lossy rewrites were in fact reliable (i.e., no symbols were lost).

For Phase 2, we define a #-padded version Sim of $\to_\Upsilon$ that works over Seqs:

$$\mathsf{Sim} \stackrel{\text{def}}{=} \{(q_{\mathsf{Sim}} \shortmid\shortmid\shortmid u\#^p, q_{\mathsf{Sim}} \shortmid\shortmid\shortmid v\#^q) \mid u \to_\Upsilon v, |u| + p = |v| + q\}\,. \tag{21}$$

This is a length-preserving rational relation. We define two more length-preserving rational relations Init and Fin that initialize the simulation with y on the budget space, and launch the verification phase if y' appears there, allowing to move from Phase 1 to Phase 2 and from Phase 2 to Phase 3, respectively:

$$\mathsf{Init} \stackrel{\text{def}}{=} \{(q_{\mathsf{Fw}} \shortmid\shortmid\shortmid \#^{\ell+|y|}, q_{\mathsf{Sim}} \shortmid\shortmid\shortmid y\#^{\ell}) \mid \ell \geq 0\}\,, \tag{22}$$

$$\mathsf{Fin} \stackrel{\text{def}}{=} \{(q_{\mathsf{Sim}} \shortmid\shortmid\shortmid y'\#^{\ell}, q_{\mathsf{Bw}} \shortmid\shortmid\shortmid \#^{\ell+|y'|}) \mid \ell \geq 0\}\,. \tag{23}$$

Finally, because $F_{\omega^k}(n) = H^{(\Omega_k)_n}(n)$, we define our source and target by

$$w \stackrel{\text{def}}{=} q_{\mathsf{Fw}} \shortparallel a_{k-1}^n \# \mid \#^n \;, \qquad w' \stackrel{\text{def}}{=} q_{\mathsf{Bw}} \shortparallel a_{k-1}^n \# \mid \#^n \;, \tag{24}$$

and we let R be the 1-bld rational relation $\mathsf{Fw} \cup \mathsf{Init} \cup \mathsf{Sim} \cup \mathsf{Fin} \cup \mathsf{Bw}$.

Claim. The given R[Rewr] instance is positive if and only if $\langle R, w, w' \rangle$ is a positive instance of the $(k+2)$-LR[1-bld] problem.

Proof. Suppose $w \; R_{\sqsupseteq}^{\circledast} \; w'$. It is easy to see that the separator symbol "$\mid$" and the encodings of states from Q are reliable. Because of the way the relations treat the states, we in fact get

$$w \; \mathsf{Fw}_{\sqsupseteq}^{\circledast} \; (q_{\mathsf{Fw}} \mid\mid\mid \#^{\ell_1}) \; \mathsf{Init}_{\sqsupseteq} \; (q_{\mathsf{Sim}} \mid\mid\mid z_1) \; \mathsf{Sim}_{\sqsupseteq}^{\circledast} \; (q_{\mathsf{Sim}} \mid\mid\mid z_2) \; \mathsf{Fin}_{\sqsupseteq} \; (q_{\mathsf{Sim}} \mid\mid\mid \#^{\ell_2}) \; \mathsf{Bw}_{\sqsupseteq}^{\circledast} \; w'$$

for some strings z_1, z_2 and naturals $\ell_1, \ell_2 \in \mathbb{N}$. By Lem. 4, we have $\ell_1 \leq F_{\omega^k}(n)$ and $\ell_2 \geq F_{\omega^k}(n)$. Since Init, Sim, and Fin are length-preserving, we get

$$F_{\omega^k(n)} \geq \ell_1 \geq |z_1| \geq |z_2| \geq \ell_2 \geq F_{\omega^k(n)} \tag{25}$$

Thus equality holds throughout, and therefore the lossy steps of $\mathsf{Sim}_{\sqsupseteq}$ in Phase 2 were actually reliable, i.e. were steps of Sim. This allows us to conclude that the original R[Rewr] instance was positive.

Suppose conversely that the R[Rewr] instance is positive. We can translate this into a witnessing run for $w \; R_{\sqsupseteq}^{\circledast} \; w'$, in particular, for $w \; \mathsf{Fw}^{\circledast} \,\fatsemi\, \mathsf{Init} \,\fatsemi\, \mathsf{Sim}^{\circledast} \,\fatsemi\, \mathsf{Fin} \,\fatsemi\, \mathsf{Bw}^{\circledast} \; w'$, because any successful run from the R[Rewr] instance can be plugged into the $\mathsf{Sim}^{\circledast}$ phase; Lem. 4 and the fact that the configurations of Υ are bounded by $F_{\omega^k}(n)$ together ensure that this can be done. □

4.2 Upper Bound

Well-Structured Transition Systems. As a preliminary, let us show that the lossy rewriting problem is decidable. Indeed, the relation $R_{\sqsupseteq}$ can be viewed as the transition relation of an infinite transition system over the state space Σ^*. Furthermore, by Higman's Lemma, the subword embedding ordering $\sqsubseteq$ is a *well quasi ordering* (wqo) over Σ^*, and the relation $R_{\sqsupseteq}$ is *compatible* with it: if $u \, R_{\sqsupseteq} \, v$ and $u \sqsubseteq u'$ for some u, v, u' in Σ^*, then there exists v' in Σ^* s.t. $u' \, R_{\sqsupseteq} \, v'$: here it suffices to use $v' = v$ by transitivity of $\sqsupseteq$.

A transition system $\mathcal{S} = \langle S, \rightarrow, \leq \rangle$ with a wqo $(S, \leq)$ as state space and a compatible transition relation $\rightarrow$ is called a *well-structured transition system* (WSTS), and several problems are decidable on such systems under very light effectiveness assumptions [1, 13], among which the *coverability problem*, which asks given a WSTS $\mathcal{S}$ and two states s and s' in S whether there exists $s'' \geq s'$ s.t. $s \rightarrow^{\circledast} s''$. The lossy rewrite problem when $w \not\sqsupseteq w'$ can be restated as a coverability problem for the WSTS $\langle \Sigma^*, R_{\sqsupseteq}, \sqsubseteq \rangle$ and w and w', since if there exists $w'' \sqsupseteq w'$ with $w \; R_{\sqsupseteq}^{\circledast} \; w''$, then $w \; R_{\sqsupseteq}^{\circledast} \; w'$ also holds by transitivity of $\sqsupseteq$.

Parameterized Upper Bound. In many cases, a *combinatory algorithm* can be employed instead of the classical backward coverability algorithm for WSTS: we can find a particular coverability witness $w' = w_0 \sqsubseteq ; R^{-1} w_1 \cdots w_{\ell-1} \sqsubseteq ; R^{-1} w_\ell \sqsubseteq w$ of length ℓ *bounded* by a function akin to $F_{\omega^{k-1}}$ using the Length Function Theorem of [26]. This is a generic technique for coverability explained in [27], and the reader will find it instantiated for $(k+2)$-LR[Rat] in the long version of this paper:

Proposition 3 (Upper Bound). *The problem $(k+2)$-LR[Rat] is in $\mathbf{F}_{\omega^{k+1}}$.*

The small gap of complexity we witness here with Prop. 2 stems from the encoding apparatus, which charges us with one extra symbol. We have not been able to close this gap; for instance, the encoding breaks if we try to work without our separator symbol "|".

5 Applications

We apply in this section the proof of Prop. 2 to prove parametric complexity lower bounds for several problems. In three cases (propositions 4, 5, and 7 below), we proceed by a reduction from the LR problem, but take advantage of the specifics of the instances constructed in the proof Prop. 2 to obtain tighter parameterized bounds. The hardness proof for the LT problem in Prop. 6 requires more machinery, which needs to be incorporated to the construction of Sec. 4.1 in order to obtain a reduction.

Rational Embedding. We first deal with the classical embedding problem: We reduce from a $(k+2)$-LR[Rat] instance and use Prop. 2. The issue is to somehow convert an iterated composition into an iterated concatenation—the idea is similar to the one typically used for proving the undecidability of PCP.

Proposition 4. *Let $k \geq 2$. Then $(k+2)$-EP[Rat] is $\mathbf{F}_{\omega^k}$-hard.*

Proof. Assume without loss of generality that $w \neq w'$ in a $(k+2)$-LR[Rat] instance $\langle R, w, w' \rangle$. We consider sequences of consecutive configurations of $\sqsupseteq ; (R ; \sqsupseteq)^{\oplus}$ of form

$$w = v_0 \sqsupseteq u_0 \, R \, v_1 \sqsupseteq u_1 \, R \, v_2 \sqsupseteq \cdots R \, v_n \sqsupseteq u_n = w' \tag{26}$$

that prove the LR instance to be positive. Let \$ be a fresh symbol; we construct a new relation R' that attempts to read the u_i's on its first component and the v_i's on the second, using the \$'s for synchronization:

$$R' \stackrel{\text{def}}{=} \begin{bmatrix} \$w'\$ \\ \$ \end{bmatrix} \cdot \left(R \cdot \begin{bmatrix} \$ \\ \$ \end{bmatrix} \right)^+ \cdot \begin{bmatrix} \varepsilon \\ w\$ \end{bmatrix} \tag{27}$$

Observe that in any pair of words (u, v) of R', one finds the same number of occurrences of the separator \$ in u and v, i.e. we can write $u = \$u_n\$ \cdots \$u_0\$$ and

$v = \$v_n\$\cdots\$v_0\$$ with $n > 0$, verifying $v_0 = w$, $u_n = w'$, and $u_i \mathrel{R} v_{i+1}$ for all i. Assume $u \sqsubseteq v$: the embedding ordering is restricted by the \$ symbols to the factors $u_i \sqsubseteq v_i$. We can therefore exhibit a sequence of form (26). Conversely, given a sequence of form (26), the corresponding pair (u, v) belongs to $R' \cap \sqsubseteq$.

In order to conclude, observe that we can set $\$ \stackrel{\text{def}}{=} \shortmid$ in the proof of Prop. 2 and adapt the previous arguments accordingly, since "$\shortmid$" is preserved by R and appears in both w and w' in the particular instances we build. □

Synchronous Embedding. Turning now to the case of synchronous relations, we proceed as in the previous proof, but employ an extra padding symbol $\perp$ to construct a length-preserving version of the relation R in an instance of $(k+2)$-LR[Sync], allowing us to apply the Kleene star operator while remaining regular.

Proposition 5. *Let $k \geq 2$. Then $(k+3)$-EP[Sync] is $\mathbf{F}_{\omega^k}$-hard.*

Proof. Let $\langle R, w, w' \rangle$ be an instance of $(k+2)$-LR[Sync] with $w \neq w'$ and let \$ and $\perp$ be two fresh symbols. Because $R \cdot \{(\$, \$)\}$ is a synchronous relation, we can construct a padded length-preserving relation

$$R_\perp \stackrel{\text{def}}{=} \{(u\$\perp^m, v\$\perp^p) \mid m, p \geq 0 \wedge (u, v) \in R \wedge |u\$\perp^m| = |v\$\perp^p|\} \tag{28}$$

and define a relation similar to (27):

$$R'_\perp \stackrel{\text{def}}{=} \begin{bmatrix} \$w'\$ \\ \$ \end{bmatrix} \cdot R_\perp^+ \cdot \begin{bmatrix} \varepsilon \\ w\$ \end{bmatrix} \cdot \begin{bmatrix} \varepsilon \\ \perp \end{bmatrix}^* . \tag{29}$$

Let us show that $R'_\perp$ is regular: $\{(\$w'\$, \$)\}$ and $\{(\varepsilon, w\$)\}$ are relations with bounded length discrepancy and $R_\perp^*$ is length preserving, thus their concatenation has bounded length discrepancy, and can be effectively computed by *resynchronization* [25]. Suffixing $\{(\varepsilon, \perp)\}^*$ thus yields a synchronous relation.

As in the proof of Prop. 4, $R'_\perp$ preserves the \$ separators, thus if (u, v) belongs to $R'_\perp$, then we can write

$$\begin{aligned} u &= \$\, u_n \,\$\, \perp^{m_n}\, u_{n-1} \,\$\, \perp^{m_{n-1}} \cdots \$\, \perp^{m_1}\, u_0 \,\$\, \perp^{m_0}, \\ v &= \$\, v_n \,\$\, \perp^{p_n}\, v_{n-1} \,\$\, \perp^{p_{n-1}} \cdots \$\, \perp^{p_1}\, v_0 \,\$\, \perp^{p_0}. \end{aligned} \tag{30}$$

with $n > 0$ and $m_n = 0$. Furthermore, $v_0 = w$, $u_n = w'$, and $(u_i\$\perp^{m_i}, v_{i+1}\$\perp^{p_{i+1}})$ belongs to $R_\perp$, thus $u_i \mathrel{R} v_{i+1}$ for all i. If the EP instance is positive, i.e. if $u \sqsubseteq v$, then $u_i \sqsubseteq v_i$ and $m_i < p_i$ for all i, and we can build a sequence of form (26) proving the LR instance to be positive. Conversely, if the LR instance is positive, there exists a sequence of form (26), and we can construct a pair (u, v) as in (30) above by guessing a sufficient padding amount p_0 that will allow to carry the entire rewriting. Finally, as in the proof of Prop. 4, we can set $\$ \stackrel{\text{def}}{=} \shortmid$. □

Lossy Termination. In contrast with the previous cases, our hardness proof for the LT problem does not reduce from LR but directly from a semi-Thue word problem, by adapting the proof of Prop. 2 in such a way that $R^{\circledast}_{\sqsupseteq}$ is *guaranteed* to terminate. The main difference is that we reduce from a semi-Thue system where the length of *derivations* is bounded, rather than the length of configurations—this is still $\mathbf{F}_{\omega^k}$-hard since the distinction between time and space complexities is insignificant at such high complexities. The simulation of such a system then builds two copies of the initial budget in Phase 1: a *space* budget, where the derivation simulation takes place, and a *time* budget, which gets decremented with each new rewrite of Phase 2, and enforces its termination even in case of losses. See the full paper for details.

Proposition 6. *Let $k \geq 2$. Then $(k+2)$-LT[1-bld] is $\mathbf{F}_{\omega^k}$-hard.*

Lossy Channel Systems. By over-approximating the behaviours of a channel system by allowing uncontrolled, arbitrary message losses, Abdulla, Cécé, et al. [6, 2] obtain decidability results on an otherwise Turing-complete model. Many variants of this model have been studied in the literature [7, 8, 16], but our interest here is that LCSs were originally used as the formal model for the $\mathbf{F}_{\omega^\omega}$ lower bound proof of Chambart and Schnoebelen [9], rather than a PEP.

Formally, a *lossy channel system* (LCS) is a finite labeled transition system $\langle Q, \Sigma, \delta \rangle$ where transitions in $\delta \subseteq Q \times \{?, !\} \times \Sigma \times Q$ read and write on an unbounded channel. An channel system defines an infinite transition system over its set of configurations $Q \times \Sigma^*$—holding the current state and channel content—, with transition relation $q, x \to q', x'$ if either δ holds a read $(q, ?m, q')$ and $x = mx'$, or if it holds a write $(q, !m, q')$ and $xm = x'$. The operational semantics of an LCS then use the lossy version $\to_{\sqsupseteq}$ of this transition relation. In the following, we consider a slightly extended model, where transitions carry sequences of instructions instead, i.e. δ is a finite set included in $Q \times (\{?, !\} \times \Sigma)^* \times Q$. The natural decision problem associated with a LCS is its reachability problem:

Lossy Channel System Reachability (LCS)
Input. A LCS $\mathcal{C}$ and two configurations (q, x) and (q', x') of $\mathcal{C}$.
Question. Is (q', x') reachable from (q, x) in $\mathcal{C}$, i.e. does $q, x \to_{\sqsupseteq}^{\circledast} q', x'$?

The lossy rewriting problem easily reduces to a reachability problem in a LCS: the LCS *cycles* through the channel contents thanks to a distinguished symbol, and applies the rational relation at each cycle; see the full version for details.

Proposition 7. *Let $k \geq 2$. Then $(k+2)$-LCS is $\mathbf{F}_{\omega^k}$-hard.*

6 Concluding Remarks

Post embedding problems provide a high-level packaging of hyper-Ackermannian decision problems—and thanks to our parametric bounds, for k-Ackermannian problems—, compared to e.g. reachability in lossy channel systems (as used in [9]). The lossy rewriting problem is a prominent example: because it is stated in terms of a rational relation instead of a machine definition, it benefits automatically from the theoretical toolkit and multiple characterizations associated with rational relations. For a simple example, the *increasing* rewriting problem, which employs $R_{\sqsubseteq} \stackrel{\text{def}}{=} \sqsubseteq ; R ; \sqsubseteq$ instead of $R_{\sqsupseteq}$, is immediately seen to be equivalent to LR, by substituting R^{-1} for R and exchanging w and w'.

Interestingly, this inversion trick allows to show the equivalence of the lossy and increasing variants of all our problems, except for lossy termination:

Increasing Termination (IT[Rat])
Input. A rational relation R over Σ and a word w in Σ^*.
Question. Does $R_{\sqsubseteq}^{\circledast}$ terminate from w?

A related problem, termination of increasing channel systems with emptiness tests, is known to be in $\mathbf{F}_3$ [5] instead of $\mathbf{F}_{\omega^\omega}$ for LCS termination, but IT[Rat] is more involved. Like LR[Rat] or EP[Rat], it provides a high-level description, this time of *fair termination* problems in increasing channel systems, which are known to be equivalent to satisfiability of *safety metric temporal logic* [23, 22, 17]. The exact complexity of IT[Rat] is open, with a gigantic gap between the $\mathbf{F}_{\omega^\omega}$ upper bound provided by WSTS theory, and an $\mathbf{F}_4$ lower bound by Jenkins [17].

Acknowledgements. The authors thank Philippe Schnoebelen and the anonymous reviewers for their insightful comments.

References

1. Abdulla, P.A., Čerāns, K., Jonsson, B., Tsay, Y.K.: Algorithmic analysis of programs with well quasi-ordered domains. Inform. and Comput. 160, 109–127 (2000)
2. Abdulla, P.A., Jonsson, B.: Verifying programs with unreliable channels. Inform. and Comput. 127(2), 91–101 (1996)
3. Atig, M.F., Bouajjani, A., Burckhardt, S., Musuvathi, M.: On the verification problem for weak memory models. In: POPL 2010, pp. 7–18. ACM (2010)
4. Barceló, P., Figueira, D., Libkin, L.: Graph logics with rational relations and the generalized intersection problem. In: LICS 2012, pp. 115–124. IEEE Press (2012)
5. Bouyer, P., Markey, N., Ouaknine, J., Schnoebelen, P., Worrell, J.: On termination and invariance for faulty channel machines. Form. Asp. Comput. 24(4), 595–607 (2012)
6. Cécé, G., Finkel, A., Purushothaman Iyer, S.: Unreliable channels are easier to verify than perfect channels. Inform. and Comput. 124(1), 20–31 (1996)
7. Chambart, P., Schnoebelen, P.: Post Embedding Problem Is Not Primitive Recursive, with Applications to Channel Systems. In: Arvind, V., Prasad, S. (eds.) FSTTCS 2007. LNCS, vol. 4855, pp. 265–276. Springer, Heidelberg (2007)
8. Chambart, P., Schnoebelen, P.: Mixing Lossy and Perfect Fifo Channels. In: van Breugel, F., Chechik, M. (eds.) CONCUR 2008. LNCS, vol. 5201, pp. 340–355. Springer, Heidelberg (2008)
9. Chambart, P., Schnoebelen, P.: The ordinal recursive complexity of lossy channel systems. In: LICS 2008, pp. 205–216. IEEE Press (2008)
10. Chambart, P., Schnoebelen, P.: Pumping and Counting on the Regular Post Embedding Problem. In: Abramsky, S., Gavoille, C., Kirchner, C., Meyer auf der Heide, F., Spirakis, P.G. (eds.) ICALP 2010. LNCS, vol. 6199, pp. 64–75. Springer, Heidelberg (2010)

11. Elgot, C.C., Mezei, J.E.: On relations defined by generalized finite automata. IBM Journal of Research and Development 9(1), 47–68 (1965)
12. Fairtlough, M., Wainer, S.S.: Hierarchies of provably recursive functions. In: Handbook of Proof Theory, ch. III, pp. 149–207. Elsevier (1998)
13. Finkel, A., Schnoebelen, P.: Well-structured transition systems everywhere! Theor. Comput. Sci. 256(1-2), 63–92 (2001)
14. Friedman, H.M.: Some decision problems of enormous complexity. In: LICS 1999, pp. 2–13. IEEE Press (1999)
15. Haddad, S., Schmitz, S., Schnoebelen, P.: The ordinal-recursive complexity of timed-arc Petri nets, data nets, and other enriched nets. In: LICS 2012, pp. 355–364. IEEE Press (2012)
16. Jančar, P., Karandikar, P., Schnoebelen, P.: Unidirectional Channel Systems Can Be Tested. In: Baeten, J.C.M., Ball, T., de Boer, F.S. (eds.) TCS 2012. LNCS, vol. 7604, pp. 149–163. Springer, Heidelberg (2012)
17. Jenkins, M.: Synthesis and Alternating Automata over Real Time. Ph.D. thesis, Oxford University (2012)
18. Karandikar, P., Schnoebelen, P.: Cutting through Regular Post Embedding Problems. In: Hirsch, E.A., Karhumäki, J., Lepistö, A., Prilutskii, M. (eds.) CSR 2012. LNCS, vol. 7353, pp. 229–240. Springer, Heidelberg (2012)
19. Lasota, S., Walukiewicz, I.: Alternating timed automata. ACM Trans. Comput. Logic 9(2), 10 (2008)
20. Latteux, M., Leguy, J.: On the Composition of Morphisms and Inverse Morphisms. In: Díaz, J. (ed.) ICALP 1983. LNCS, vol. 154, pp. 420–432. Springer, Heidelberg (1983)
21. Löb, M., Wainer, S.: Hierarchies of number theoretic functions, I. Arch. Math. Logic 13, 39–51 (1970)
22. Ouaknine, J., Worrell, J.B.: Safety Metric Temporal Logic Is Fully Decidable. In: Hermanns, H., Palsberg, J. (eds.) TACAS 2006. LNCS, vol. 3920, pp. 411–425. Springer, Heidelberg (2006)
23. Ouaknine, J.O., Worrell, J.B.: On the decidability and complexity of Metric Temporal Logic over finite words. Logic. Meth. in Comput. Sci. 3(1), 8 (2007)
24. Rose, H.E.: Subrecursion: Functions and Hierarchies, Oxford Logic Guides, vol. 9. Clarendon Press (1984)
25. Sakarovitch, J.: Elements of Automata Theory. Cambridge University Press (2009)
26. Schmitz, S., Schnoebelen, P.: Multiply-Recursive Upper Bounds with Higman's Lemma. In: Aceto, L., Henzinger, M., Sgall, J. (eds.) ICALP 2011, Part II. LNCS, vol. 6756, pp. 441–452. Springer, Heidelberg (2011)
27. Schmitz, S., Schnoebelen, P.: Algorithmic Aspects of WQO Theory. Lecture Notes (2012), `http://cel.archives-ouvertes.fr/cel-00727025`

Deciding Definability by Deterministic Regular Expressions*

Wojciech Czerwiński[1], Claire David[2], Katja Losemann[1], and Wim Martens[1]

[1] Universität Bayreuth
[2] Université Paris-Est Marne-la-Vallée

Abstract. We investigate the complexity of deciding whether a given regular language can be defined with a deterministic regular expression. Our main technical result shows that the problem is PSPACE-complete if the input language is represented as a regular expression or nondeterministic finite automaton. The problem becomes EXPSPACE-complete if the language is represented as a regular expression with counters.

1 Introduction

Schema information is highly advantageous when managing and exchanging XML data. Primarily, schema information is crucial for automatic error detection in the data itself (which is called validation, see, e.g., [5,26,2,20]) or in the procedures that transform the data [24,23,22]. Furthermore, schemas provide information for optimization of XML querying and processing [25,28], they are inevitable when integrating data through schema matching [1], and they provide users with a high-level overview of the structure of the data. From a software development point of view, schemas are very useful to precisely specify pre- and post-conditions of software routines that process XML data.

In their core, XML schemas specify the structure of well-formed XML documents through a set of constraints which are very similar to extended context-free grammar productions. Such schema are usually abstracted as a set of rules of the form

$$\textit{Type} \rightarrow \textit{Content},$$

where *Content* is a regular expression that defines the allowed content inside the element type specified in the left-hand side. As such, regular expressions are a central component of schema languages for XML.

The two most prevalent schema languages for XML data, *Document Type Definition (DTD)* [5] and *XML Schema Definition (XSD)* [10], both developed by the World Wide Web Consortium, do not allow arbitrary regular expressions to define *Content*. Instead, they require these expressions to be *deterministic*. We refer to such deterministic regular expressions as *DREs*. In order to get a good understanding of schema languages for XML, it is thus important to develop a

* This work was supported by grant number MA 4938/2-1 of the Deutsche Forschungsgemeinschaft (Emmy Noether Nachwuchsgruppe).

F. Pfenning (Ed.): FOSSACS 2013, LNCS 7794, pp. 289–304, 2013.

good understanding on DREs. Furthermore, since the concept of determinism in regular expressions is a rather foundational, we believe our results to be relevant in a larger scope as well.

Intuitively, a regular expression is deterministic if, when reading a word from left to right without looking ahead, it is always clear where in the expression the next symbol can be matched. For example, the expression $(a+b)^*b(a+b)$ is not deterministic, because if we read a word that starts with b, it is not clear whether this b should be matched in the expression if we do not know what the remainder of the word will be. As such, determinism in regular expressions is very similar to determinism in finite automata: When we consider each alphabet symbol in an expression as a state and consider transitions between positions in the expression that can be matched by successive symbols, then the expression is deterministic if and only if the thus obtained automaton (which is known as the Glushkov automaton of the expression) is deterministic.

Deterministic regular expressions or DREs have therefore been a subject of research since their foundations were laid in a seminal paper by Brüggemann-Klein and Wood [6,7]. The most important contribution of this paper is a characterization of languages definable by DREs in terms of structural properties on the minimal DFA. In particular, this characterization showed that some regular languages cannot be defined with a DRE. One such language is defined by the expression $(a+b)^*b(a+b)$. Furthermore, Brüggemann-Klein and Wood showed that it is decidable whether a given regular language is definable by a DRE. Since then, DREs have been studied in the context of language approximations [3], learning [4], descriptional complexity [13,21] and static analysis [8,9]. Recently, it was shown that testing if a regular expression is deterministic can be done in linear time [14].

Determinism has also been studied for a more general class of regular expressions which allows a *counting operator* [19,11,16]. This operator allows to write the expression $a^{10,100}$ defining the language that contains strings of length 10 to 100 and labeled with only a's. The motivation for the counting operator again comes from schema languages, because the operator can be used to define expressions in XML Schema. Determinism for expressions with counters seems to pose more challenges than without the counting operator. For example, already testing whether an expression with counters is deterministic is non-trivial [18].

In this paper we study the following problem:

Given a regular expression, can it be determinized?

This problem seems to be very foundational and has first been studied around 20 years ago [6] but the precise complexity was still open, despite the rich body of research discussed above. The best known upper bound is from Brüggeman-Klein and Wood, who showed that the problem is in EXPTIME (by exhibiting an algorithm that works in polynomial time on the minimal DFA [7]) and the best known lower bound is PSPACE-hardness [3]. The main result of this paper settles this question and proves that this problem is PSPACE-complete. Our proof is rather technical and provides deeper insights in the decision algorithm of Brüggemann-Klein and Wood. A central insight, which is a cornerstone of our proof, is that the recursion depth of the algorithm is only polynomial in the size of the smallest NFA for the given regular language.

Since regular expressions with counters are important in the context of W3C XML Schema, we also study the complexity of deciding if a given expressions with counters can be written as a DRE. This problem turns out to be EXPSPACE-complete. We complement these completeness results by proving that it is NLOGSPACE-hard to decide if a given DFA can be written as a DRE. At the moment, it is not clear to us whether this lower bound can be improved. The problem is known to be in polynomial time by [7].

Organisation: We give the basic definitions in Section 2. In Section 3 we present the algorithm of Brüggeman-Klein and Wood and prove preliminary results. Complexity results are presented in Section 4. Due to space limits, some proofs are not presented or only sketched.

2 Definitions

For a finite set S, we denote its cardinality by $|S|$. By Σ we always denote an alphabet, i.e., a finite set of symbols. A (Σ-)*word* w over alphabet Σ is a finite sequence of symbols $a_1 \cdots a_n$, where $a_i \in \Sigma$ for each $i = 1, \ldots, n$. The set of all Σ-words is denoted by Σ^*. The *length* of a word $w = a_1 \cdots a_n$ is n and is denoted by $|w|$. The empty word is denoted by ε. A *language* is a set of words.

A *(nondeterministic) finite automaton* (or *NFA*) N is a tuple $(Q, \Sigma, \delta, q_0, F)$, where Q is a finite set of states, $\delta : Q \times \Sigma \to 2^Q$ is the transition function, q_0 is the initial state, and $F \subseteq Q$ is the set of accepting states. We sometimes denote that $q_2 \in \delta(q_1, a)$ as $q_1 \xrightarrow{a} q_2 \in \delta$ to emphasize that, when N is in state q_1, it can go to state q_2 when reading an a. A *run of N on word* $w = a_1 \cdots a_n$ is a sequence $q_0 \cdots q_n$ where, for each $i = 1, \ldots, n$, we have $q_{i-1} \xrightarrow{a_i} q_i \in \delta$. Word w is *accepted* by N if there is such a run which is *accepting*, i.e., if $q_n \in F$. The *language of* N, also denoted $L(N)$, is the set of words accepted by N. By δ^* we denote the extension of δ to words, i.e., $\delta^*(q, w)$ is the set of states which can be reached from q by reading w. The *size* $|N|$ of an NFA is the total number of transitions, i.e., $\sum_{q,a} |\delta(q, a)|$. An NFA is *deterministic*, or a *DFA*, when every $\delta(q, a)$ has at most one element. Throughout the paper, we will use the notation $\mathcal{P}_N$ for the power set automaton of N and $[N]$ for the minimal DFA for $L(N)$. It is well-known that $[N]$ is unique for N and that it can be obtained by merging states of $\mathcal{P}_N$ [15]. In this paper we assume that all states of automata are *useful* unless mentioned otherwise, that is, every state can appear in some accepting run. This implies that every state can be reached from the initial state and that, from each state in an automaton, an accepting state can be reached. This also implies that we use minimal DFAs without sink state and that $\mathcal{P}_N$ by default only contains the useful subsets of states of N. We sometimes abuse notation and also denote by $\emptyset$ the minimal DFA with no states.

Furthermore, we will often see an NFA as a *graph*, which is obtained by considering its states as nodes and its transitions as (labeled) directed edges. Then, we also refer to a connected sequence of transitions in N as a *path*.

The *regular expressions (RE)* over Σ are defined as follows: ε and every Σ-symbol is a regular expression; and whenever r and s are regular expressions then so are $(r \cdot s)$, $(r + s)$, and $(s)^*$. In addition, we allow $\emptyset$ as a regular expression, but we do not allow $\emptyset$ to occur in any other regular expression. For readability, we usually omit concatenation operators and parentheses in examples. The *language* defined by an RE r, denoted by $L(r)$, is defined as usual. Whenever we say that expressions or automata are *equivalent*, we mean that they define the same language. The *size* $|r|$ of r is defined to be the total number of occurrences of alphabet symbols, epsilons, and operators, i.e., the number of nodes in its parse tree.

2.1 Variations of Regular Expressions

The *regular expressions with counters (RE(#))* extend the REs with a *counting operator*. That is, each RE-expression is an RE(#)-expression. Furthermore, when r and s are RE(#)-expressions then so are $(r \cdot s)$, $(r + s)$, and $r^{k,\ell}$ for $k \in \mathbb{N}$ and $\ell \in \mathbb{N}^+ \cup \{\infty\}$ with $k \leq \ell$. Here, $\mathbb{N}^+$ denotes $\mathbb{N} \setminus \{0\}$. For a language L, define $L^{k,\ell}$ as $\bigcup_{i=k}^{\ell} L^i$. Then, $L(r^{k,\ell}) = \bigcup_{i=k}^{\ell} L(r)^i$. Notice that r^* is equivalent to $r^{0,\infty}$. The size of an expression in RE(#) is the number of nodes in its parse tree, plus the sizes of the counters, where a counter $k \in \mathbb{N}$ has size $\log k$.

Deterministic regular expressions (DREs) put a restriction on the class of REs. Let $\bar{r}$ stand for the RE obtained from r by replacing, for every i and a, the i-th occurrence of alphabet symbol a in r (counting from left to right) by a_i. For example, for $r = b^*a(b^*a)^*$ we have $\bar{r} = b_1^*a_1(b_2^*a_2)^*$. A regular expression r is *deterministic* (or *one-unambiguous* [7] or a *DRE*) if there are no words wa_iv and wa_jv' in $L(\bar{r})$ such that $i \neq j$. The expression $(a+b)^*a$ is not deterministic since both strings a_2 and a_1a_2 are in $L((a_1+b_1)^*a_2)$. The equivalent expression $b^*a(b^*a)^*$ is deterministic. Brüggemann-Klein and Wood showed that not every regular expression is equivalent to a deterministic one [7]. We call a regular language *DRE-definable* if there exists a DRE that defines it. The canonical example for a language that is not DRE-definable is $(a+b)^*b(a+b)$ [7]. We therefore have that the set of DREs forms a strict subset of the REs, which in turn are a strict subset of the RE(#)s.

2.2 Problems of Interest

In this paper, we investigate variants of the following problem.

DRE-DEFINABILITY: Given a regular language L, is L DRE-definable?

We consider this problem for various representations of regular languages: regular expressions, regular expressions with counters, NFAs, and DFAs. Whenever we consider such a variation, we put the respective representation between braces. For example, DRE-DEFINABILITY(RE) is the problem: Given a regular expression r, is $L(r)$ DRE-definable?

3 The BKW Algorithm

DRE-DEFINABILITY was first studied by Brüggemann-Klein and Wood who showed that the problem can be solved in polynomial time in the size of the minimal DFA of a language [7]. Their algorithm (henceforth referred to as the BKW-Algorithm) is not at all trivial and gives good insight in DRE-definable regular languages. We recall the BKW-Algorithm together with some definitions and known results and then we prove deeper properties of the BKW-Algorithm which will be the basis of further results in the paper.

Orbits and Gates: For a state q in an NFA N, the *orbit of* q, denoted $\mathcal{O}(q)$, is the maximal strongly connected component of N that contains q. We call q a *gate* of $\mathcal{O}(q)$ if q is accepting or q has an outgoing transition that leaves $\mathcal{O}(q)$. If an orbit consists only of one state q and q has no self-loops, we say that it is a *trivial* orbit. We say that a transition $q_1 \xrightarrow{a} q_2$ is an *inter-orbit* transition if q_1 and q_2 belong to different orbits. The *orbit automaton of state* q is the sub-automaton of N consisting of $\mathcal{O}(q)$ in which the initial state is q and the accepting states are the gates of $\mathcal{O}(q)$. We denote the orbit automaton of q by N_q. The *orbit language of* q is $L(N_q)$. The *orbit languages of* N are the orbit languages of states of N.

Orbit Property: An NFA N has the *orbit property* if, for every pair of gates q_1, q_2 in the same orbit in N, the following properties hold:

1. q_1 is accepting if and only if q_2 is accepting; and,
2. for all states q outside the orbit of q_1 and q_2, there is a transition $q_1 \xrightarrow{a} q$ if and only if there is a transition $q_2 \xrightarrow{a} q$.

Consistent Symbols: A symbol $a \in \Sigma$ is *N-consistent* if there is a state $f(a)$, such that every accepting state q of N has a transition $q \xrightarrow{a} f(a)$. We refer to the corresponding transitions as *consistent transitions* of N. A set $S \subseteq \Sigma$ is *N-consistent* if every symbol in S is N-consistent. Whenever we consider N-consistent sets S in the remainder of the paper we assume that they are maximal, i.e., there does not exist an $a \in \Sigma$ that is not in S and is N-consistent. Henceforth, we will therefore refer to *the* N-consistent set. For the set S of N-consistent symbols, the *S-cut* of N, denoted N_S, is obtained by removing all consistent transitions from N. Using these notions, Brüggemann-Klein and Wood give the following characterization of the class of DRE-definable languages.

Theorem 1 (Brüggemann-Klein and Wood [7]). *Let D be a minimal DFA and S be the set of D-consistent symbols. Then the following are equivalent:*

1. *$L(D)$ is DRE-definable;*
2. *D has the orbit property and all orbit languages of D are DRE-definable;*
3. *D_S has the orbit property and all orbit languages of D_S are DRE-definable.*

Furthermore, if D consists of a single, nontrivial orbit and $L(D)$ is DRE-definable, then there is at least one D-consistent symbol.

Algorithm 1. The BKW-Algorithm [7].

Algorithm BKW
2: **Input:** Minimal DFA $D = (Q, \Sigma, \delta, q_0, F)$
Output: *true* if $L(D)$ is DRE-definable, else *false*
4: $S \leftarrow$ the maximal set of D-consistent symbols
if D has only one trivial orbit **then return** *true*
6: **if** D has precisely one orbit and $S = \emptyset$ **then return** *false*
compute the orbits of D_S
8: **if** D_S does not have the orbit property **then return** *false*
for each orbit $\mathcal{O}$ in D_S **do**
10: choose a state q in $\mathcal{O}$
if not BKW($(D_S)_q$) **then return** *false*
12: **return** *true*

They also show this result about the orbit property and the orbit languages.

Lemma 2 (Brüggemann-Klein and Wood [7]). *Let D be a minimal DFA and S be the set of D-consistent symbols.*

1. *If D_S has the orbit property, then $(D_S)_q$ is minimal for each state q in D.*
2. *If p and q are states in the same orbit of D_S, then $L((D_S)_p)$ is DRE-definable if and only if $L((D_S)_q)$ is DRE-definable.*

Point 1 of the above lemma is immediate from combining Lemmas 5.9 and 5.10 from [7]. Point 2 is immediate from the fact that DRE-definable regular languages are closed under derivatives [7]. Notice that, in general, D_S does not have to be a minimal DFA. In particular, it can have states that are not reachable from the initial state. These results lead to a recursive test that decides whether the language of a minimal DFA is DRE-definable. We present this test in Algorithm 1. Notice that Lemma 2 ensures that we never have to minimize the DFA that we give to the recursive call in line 11 of the algorithm.

In the remainder of this article, D always denotes a minimal DFA. In the following we investigate the recursion depth of Algorithm 1. Therefore, we examine how, for a state q of D, the orbit of q evolves during the recursion. In one iteration of Algorithm 1 we always delete two kinds of transitions, if they are present: the consistent transitions (which we delete to obtain D_S from D) and the inter-orbit transitions in D_S (which we delete to obtain $(D_S)_q$).

Level Automata: For a state q of a minimal DFA D and $k \in \mathbb{N}$ we inductively define the *level k automaton of D for the state q*, denoted $\mathrm{lev}_k(D, q)$, as follows:

- $\mathrm{lev}_0(D, q) = D$.
- Let S be the maximal set of consistent symbols in D. Then

$$\mathrm{lev}_1(D, q) = \begin{cases} (D_S)_q & \text{if } D \text{ has more than one orbit and } D_S \text{ has the orbit property;} \\ (D_S)_q & \text{if } S \neq \emptyset \text{ and } D_S \text{ has the orbit property;} \\ \emptyset & \text{otherwise.} \end{cases}$$

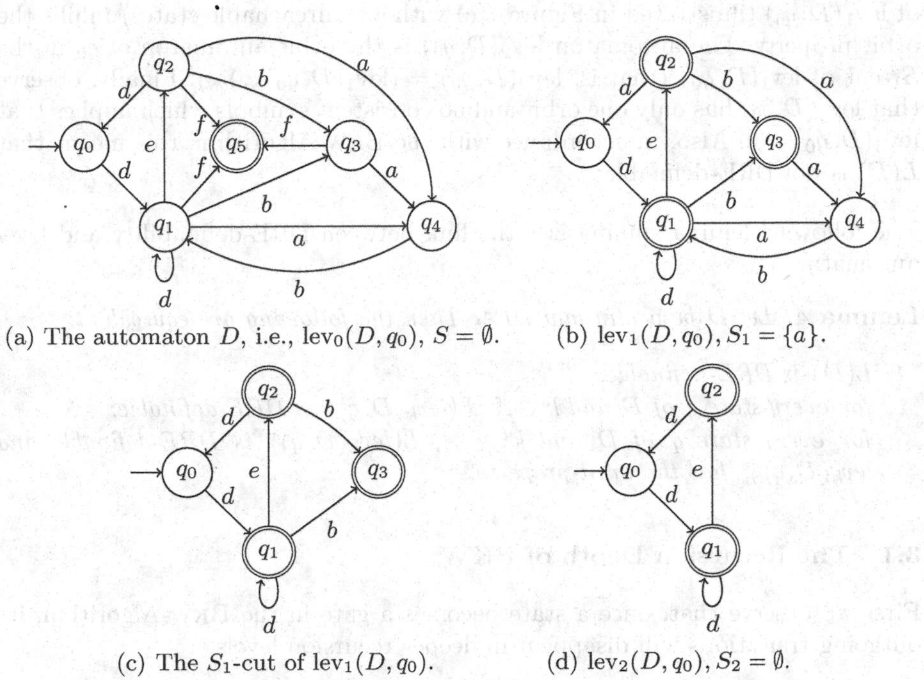

(a) The automaton D, i.e., $\mathrm{lev}_0(D, q_0)$, $S = \emptyset$. (b) $\mathrm{lev}_1(D, q_0), S_1 = \{a\}$.

(c) The S_1-cut of $\mathrm{lev}_1(D, q_0)$. (d) $\mathrm{lev}_2(D, q_0), S_2 = \emptyset$.

Fig. 1. An example of level automata for a minimal DFA D

– Let $k > 1$ and S_{k-1} be the maximal set of consistent symbols in $B := \mathrm{lev}_{k-1}(D, q)$. Then

$$\mathrm{lev}_k(D, q) = \begin{cases} (B_{S_{k-1}})_q & \text{if } S_{k-1} \neq \emptyset \text{ and } B_{S_{k-1}} \text{ fulfills the orbit property} \\ \emptyset & \text{otherwise.} \end{cases}$$

The above definition actually precisely follows the construction in Algorithm 1 if state q is chosen every time in line 10. The definition makes clear that the top level recursion of the BKW-Algorithm (in which we construct $\mathrm{lev}_1(D, q)$) is slightly different from the others: the input DFA D of the top level can have multiple orbits, whereas this is not the case for deeper recursive levels. According to Lemma 2, $\mathrm{lev}_k(D, q)$ is always minimal.

Example 3. Figure 1 provides an example to illustrate the notion of level automata. Consider the minimal DFA D from Figure 1(a) and its state q_0. By definition, $\mathrm{lev}_0(D, q_0)$ is the automaton D itself. In order to build $\mathrm{lev}_1(D, q_0)$ (see Figure 1(b)), observe that D has two orbits and its set of consistent symbols S is empty since no transitions leave state q_5. Furthermore, D_S, which equals D, fulfills the orbit property since all transitions that leave $\mathcal{O}(q_0)$ go to state q_5. As such, $\mathrm{lev}_1(D, q_0)$ equals $(D_\emptyset)_{q_0}$, the orbit automaton of q_0 in D. We now explain how to obtain $\mathrm{lev}_2(D, q_0)$ (see Figure 1(d)). First notice that $S_1 = \{a\}$ is the maximal set of consistent symbols in $\mathrm{lev}_1(D, q_0)$. Furthermore, the S_1-cut of $\text{lev}_1(D, q_0)$ (illustrated in Figure 1(c) without unreachable states) fulfills the orbit property. The automaton $\text{lev}_2(D, q_0)$ is the orbit automaton of q_0 in the S_1-cut of $\text{lev}_1(D, q_0)$ (that is, $\text{lev}_2(D, q_0) = (\text{lev}_1(D, q_0)_{S_1})_{q_0}$). Finally, observe that $\text{lev}_2(D, q_0)$ has only one orbit and no consistent symbols which implies that $\text{lev}_3(D, q_0) = \emptyset$. Also, in accordance with the BKW-Algorithm, this means that $L(D)$ is not DRE-definable.

The following lemma summarizes the link between DRE-definability and level automata.

Lemma 4. *Let D be a minimal DFA. Then the following are equivalent:*

1. *$L(D)$ is DRE-definable;*
2. *for every state q of D and $k \in \mathbb{N}$, $L(\text{lev}_k(D, q))$ is DRE-definable;*
3. *for every state q of D and $k \in \mathbb{N}$, $L(\text{lev}_k(D, q))$ is DRE-definable and $\text{lev}_k(D, q)_{S_k}$ has the orbit-property.*

3.1 The Recursion Depth of BKW

First we observe that, once a state becomes a gate in the BKW-Algorithm, its outgoing transitions will disappear in deeper recursion levels.

Lemma 5. *Let D be a minimal DFA and q be a gate in $\text{lev}_k(D, q)$ for some $k > 0$. Then either $\text{lev}_{k+1}(D, q) = \emptyset$ or q has strictly less outgoing transitions in $\text{lev}_{k+1}(D, q)$. In the latter case, q is also a gate in $\text{lev}_{k+1}(D, q)$.*

From Lemma 5, we can infer how long it takes for a state p to become a gate.

Lemma 6. *Let D be a minimal DFA and p be a state of $\text{lev}_k(D, p)$ for some $k \in \mathbb{N}$. Let ℓ be the length of the shortest path from p to a gate in $\text{lev}_k(D, p)$. Then either $\text{lev}_{k+|\Sigma|\cdot\ell+1}(D, p) = \emptyset$ or p is a gate in $\text{lev}_{k+|\Sigma|\cdot\ell+1}(D, p)$.*

Next, we want to combine Lemma 6 with an observation about NFAs versus their minimal DFAs, namely that paths to accepting states are short.

Lemma 7. *Let N be an NFA with size n. Then for every state of $[N]$ there is a path leading to some accepting state of length at most $n - 1$.*

Combining Lemma 6 and 7 tells us how long states can be present in the recursion of the BKW-Algorithm, compared to the size of an NFA for the language.

Lemma 8. *Let N be an NFA with size n. Then it holds that $\text{lev}_{n\cdot|\Sigma|+2}([N], p) = \emptyset$ for every state p of $[N]$.*

Summarized, we know that the recursion depth of the BKW-Algorithm is polynomial in the size of the minimal NFA for a language.

Theorem 9. *Let N be an NFA with size n. The recursion depth of Algorithm 1 on $[N]$ is at most $n \cdot |\Sigma| + 2$.*

3.2 Consistency Violations

In the following, we analyze the possible causes of failure for the BKW-Algorithm. We identify three properties such that the BKW-Algorithm fails if and only if one of them holds for some orbit automaton at some level k. This will be a tool for our PSPACE algorithm that will search for one of these violations.

From the BKW-Algorithm, we can immediately see that there are two situations in which it can reject: (in line 6) at some point in the recursion, the automaton consists only of one orbit which has no consistent symbols or (in line 8) at some point in the recursion, the S-cut of the automaton does not have the orbit property. The latter means that there exist two gates of the same orbit in the S-cut, such that either they do not have the same transitions to the outside or one of them is accepting while the other one is not. We now formalize these different types of violations and we then prove that the BKW-Algorithm fails if and only if one of these violations is found at some point in the recursion. Let D be a minimal DFA and S be its set of consistent symbols. Then D has

- an OUT-CONSISTENCY VIOLATION, if there exist gates q_1 and q_2 in the same orbit $\mathcal{O}$ of D_S and there exists a state q outside $\mathcal{O}$ such that there is a transition $q_1 \xrightarrow{a} q$ and no transition $q_2 \xrightarrow{a} q$;
- an ACCEPTANCE CONSISTENCY VIOLATION, if there exist gates q_1 and q_2 in the same orbit of D_S such that q_1 is accepting and q_2 is not; and
- an ORBIT CONSISTENCY VIOLATION, if there exists an accepting state q_1 such that, for every symbol a, there exists another accepting state q_2 in $\mathcal{O}(q_1)$ in D, such that for every state q, at most one of the transitions $q_1 \xrightarrow{a} q$ and $q_2 \xrightarrow{a} q$ exists.[1]

Notice that the first two violations focus on D_S and the last one on D, as in the BKW-Algorithm. In summary, we will also say that a DFA D *has a violation* if and only if it has at least one of the above violations. We show that these violations are a valid characterization of DRE-definable languages.

Theorem 10. *Let D be a minimal DFA. Then it holds that $L(D)$ is not DRE-definable if and only if there exist a state q of D and $k \in \mathbb{N}$ such that $lev_k(D, q)$ has a violation.*

4 The Definability Problem

We are now ready to prove results about the complexity of DRE-DEFINABILITY for different formalisms. We first show that the problem is PSPACE-complete for NFAs and REs. Then we look at RE(#)s which are natural extensions in the context of W3C XML Schema. In that setting, the problem becomes EXPSPACE-complete. Finally, we look at the class of DFAs.

[1] Notice that $\delta(q_1, a)$ may be empty if D only has useful states.

4.1 Definability for REs and NFAs

Our PSPACE algorithm for DRE-definability for REs and NFAs exploits Theorem 10 in the following way. Given an NFA N, we search for a level k and a state p of $[N]$ such that $\text{lev}_k([N], p)$ has a violation. As PSPACE is closed under complement, the result follows.

Notice that, in general $[N]$ can be exponentially larger than N and therefore we cannot simply compute $[N]$ in space polynomial in $|N|$. To overcome this difficulty, we use the following two ideas:

1. Use the fact that the maximal recursion depth of Algorithm 1 on $[N]$ is polynomial in the size of the NFA N (Theorem 9);
2. Adapt Algorithm 1 using Theorem 10 and apply it on the minimal DFA by only partially constructing it on-the-fly from the NFA.

In the following we explain how we can detect if there occurs a violation in the minimal DFA for some NFA on the fly, i.e., without constructing the DFA explicitly. To this end, we fix the following notations for the remainder of the section. By $N = (Q_N, \Sigma, \delta_N, q_N^0, F_N)$ we always denote an NFA. So, in particular, we always denote by Q_N the state set of N. For a set of states $q \subseteq Q_N$, we denote by $[q]$ the corresponding state in the minimal DFA $[N]$. More formally, $[q]$ is the set of words $\{w \mid \exists t \in q \text{ s.t. } \delta_N^*(t, w) \cap F_N \neq \emptyset\}$, i.e., the Myhill-Nerode class of q. Also, whenever we talk about $\text{lev}_k([N], [q])_{S_k}$, the set S_k is the set of consistent symbols in $\text{lev}_k([N], [q])$. The key result (Lemma 17) is to show that we can detect if a violation occurs in a level k for $[N]$ in space polynomial in k and $|N|$. Here are the precise problems we consider. For each of them the input is an NFA N and $k \in \mathbb{N}$:

OUT-CONS-VIOLATION:	Given N and k, is there a $q \subseteq Q_N$ such that $\text{lev}_k([N], [q])_{S_k}$ has an out-consistency violation?
ACC-CONS-VIOLATION:	Given N and k, is there a $q \subseteq Q_N$ such that $\text{lev}_k([N], [q])_{S_k}$ has an acceptance consistency violation?
ORBIT-CONS-VIOLATION:	Given N and k, is there a $q \subseteq Q_N$ such that $\text{lev}_k([N], [q])$ has an orbit consistency violation?

We first study the complexity of the following subproblems which will be used in the proof of Lemma 17. The input is always a subset of an NFA N, sets $p, q \subseteq Q_N$, $a \in \Sigma$, $k \in \mathbb{N}$ that is relevant to the problem.

EDGE:	Given (N, p, q, a, k), is $[p] \xrightarrow{a} [q]$ a transition in $\text{lev}_k([N], [p])$?
REACHABILITY:	Given (N, p, q, k), is $[q]$ reachable from $[p]$ in $\text{lev}_k([N], [p])$?
SAMEORBIT:	Given (N, p, q, k), are $[p]$ and $[q]$ in the same orbit of $\text{lev}_k([N], [p])$?
INTERORBIT:	Given (N, p, k), is there an inter-orbit transition $[p] \xrightarrow{a} [q]$ for some label a and q in $\text{lev}_k([N], [p])$?
ACCEPTANCE:	Given (N, p, k), is $[p]$ accepting in $\text{lev}_k([N], [p])$?
ISGATE:	Given (N, p, k), is $[p]$ a gate in $\text{lev}_k([N], [p])$?

Notice that SAMEORBIT and INTERORBIT are only non-trivial if $k = 0$. Furthermore, for some of the above problems X we consider a variation called X-CUT in which, with the same input, we want to decide if the problem X is true for automaton $\text{lev}_k([N],[p])_{S_k}$ (instead of $\text{lev}_k([N],[p])$).
We will heavily use the following result:

Theorem 11 (Corollary to Savitch's Theorem). *Let $f(n) \geq \log n$ be a non-decreasing polynomial function. Then NSPACE($f(n)$) $\subseteq$ SPACE($f^2(n)$).*

Our proof is a careful mutual induction on the above defined problems. First we show that EDGE, EDGE-CUT, and ACCEPTANCE can be computed in polynomial space on level 0 and then we prove a set of implications of the sort *if we can solve X on level k, then we can solve Y on level k (or level $k+1$).* All the lemmas have to be carefully put together in the right order.

Lemma 12. *Given N and $p, q \subseteq Q_N$, we can test whether $[p] = [q]$ in space $O(|N|^2)$.*

Proof (Sketch). A non-deterministic Turing Machine can test in non-deterministic space $O(|N|)$ whether $[p] \neq [q]$. The lemma then follows from Theorem 11 and the fact that deterministic complexity classes are closed under complement. □

In the following, whenever we say that we solve a problem *for N at level k*, we mean that we solve it for the NFA N and arbitrary sets of states $p, q \subseteq Q_N$, $a \in \Sigma$, and k. The basis of our entire mutual induction is being able to test if a certain transition is present in the minimal DFA equivalent to N, if it is present in its S-cut, or if some state is accepting.

Lemma 13. EDGE, ACCEPTANCE *and* EDGE-CUT *for N at level* 0 *can be solved in space $O(|N|^2)$.*

Proof. We first show how to check, for a given set $p \subseteq Q_N$, whether $[p]$ is a state in $[N]$, i.e., whether it is useful. As N only has useful states, there exists a path from $[p]$ to some accepting state in $[N]$. Thus it is enough to check whether there is a path from $[\{q_N^0\}]$ to $[p]$. This clearly can be done by a nondeterministic algorithm working in space $O(|N|)$. This algorithm would guess a word symbol by symbol, simulate $\mathcal{P}_N$ on the fly, starting from $\{q_N^0\}$ and test at each step whether the reached state q is equivalent to p, i.e., if $[p] = [q]$ (proof of Lemma 12). Thus, by Theorem 11, it can also be done by a deterministic algorithm working in space $O(|N|^2)$.

We now turn our attention to EDGE. Let $(N, p, q, a, 0)$ be the input for EDGE. We must decide if $[p] \xrightarrow{a} [q]$ is a transition in $[N]$. Since $[N]$ does not have useless states, this means that we should test two things:

- both $[p]$ and $[q]$ are states in $[N]$;
- $[\delta_{\mathcal{P}_N}(p, a)] = [q]$, where $\mathcal{P}_N$ is the power set automaton of N.

The former can be solved in $O(|N|^2)$, as we mentioned before. It remains to prove the latter. Given p and a, we can easily compute $\delta_{\mathcal{P}_N}(p, a)$ in space $O(|N|)$. According to Lemma 12, we can then decide if $[\delta_{\mathcal{P}_N}(p, a)] = [q]$ in space $O(|N|^2)$.

We omit the proof how to solve ACCEPTANCE and EDGE-CUT. □

When we can solve EDGE or EDGE-CUT at a certain level, we can use it to make more complex tests.

Lemma 14. *Assume that we can solve* EDGE *for* N *at level* k *in space* $f(k,|N|)$. *Then we can solve*

- REACHABILITY *for* N *at level* k *in space* $f(k,|N|)+O(|N|^2)$.
- SAMEORBIT *for* N *at level* k *in space* $f(k,|N|)+O(|N|^2)$.
- INTERORBIT *for* N *at level* k *in space* $f(k,|N|)+O(|N|^2)$.

Analogously, if we can solve EDGE-CUT *for* N *at level* k *in space* $f(k,|N|)$, *then we can solve* REACHABILITY-CUT, SAMEORBIT-CUT, *and* INTERORBIT-CUT *for* N *at level* k *in space* $f(k,|N|)+O(|N|^2)$.

The next lemma allows us to do a single induction step that allows us to compute the structure of an automaton at level $k+1$ if we can compute the automaton at level k.

Lemma 15. *Assume that we can solve* EDGE *and* ACCEPTANCE *for* N *at level* k *in space* $f(k,|N|)$. *Furthermore, assume that* $\mathit{lev}_k([N],[p])$ *has no violation. Then we can solve* EDGE *and* ACCEPTANCE *for* N *at level* $k+1$ *in space* $f(k,|N|)+O(|N|^2)$.

The following lemma shows that we can compute properties of the level k automaton if we can decide some properties for all smaller levels. For technical reasons, we need the assumption that all lower level automata $\mathrm{lev}_i([N],[p])$ have no violation. This is because, otherwise, the level k automaton would be empty. The proof is basically a careful induction that puts together the previous lemmas.

Lemma 16. *Assume that for* $0 \le i \le k-1$ *all automata* $\mathit{lev}_i([N],[p])$ *have no violation. Then all problems* EDGE, REACHABILITY, SAMEORBIT, INTERORBIT, ACCEPTANCE *and* EDGE-CUT, REACHABILITY-CUT, SAMEORBIT-CUT, INTERORBIT-CUT, *and* ISGATE-CUT *for* N *at level* k *are in space* $O((k+1)|N|^2)$.

Lemma 16 states that we can decide on-the-fly which transitions are present and which states are accepting in level k automata (if no violations occur in more shallow levels). Since these properties give the entire structure of the level k automata, we can now also test for violations on level k.

Lemma 17. *Assume that all automata* $\mathit{lev}_i([N],[p])$ *for* $0 \le i \le k-1$ *have no violation. We can solve* OUT-CONS-VIOLATION, ACC-CONS-VIOLATION *and* ORBIT-CONS-VIOLATION *for* N *at level* k *in space* $O((k+1)|N|^2)$.

Proof. Let the input for OUT-CONS-VIOLATION be N and k. Then, an out-consistency violation occurs at level k of N if and only if there exist $p,q \subseteq Q_N$ such that all of the following hold:

- all automata $\mathrm{lev}_i([N],[p])$ for $0 \le i \le k-1$ have no violation;
- both $[p]$ and $[q]$ are gates in $\mathrm{lev}_k([N],[p])_{S_k}$;
- both $[p]$ and $[q]$ are in the same orbit of $\mathrm{lev}_k([N],[p])_{S_k}$;

– there exist a symbol $a \in \Sigma$ and $[q']$ outside the orbit of $[p]$ in $\text{lev}_k([N],[p])_{S_k}$ such that the transition $[p] \xrightarrow{a} [q']$ exists in $\text{lev}_k([N],[p])_{S_k}$, but $[q] \stackrel{a}{\nrightarrow} [q']$ does not.

By the lemma statement we know that all automata $\text{lev}_i([N],[p])$ for $0 \le i \le k-1$ have no violation. According to Lemma 16 we can solve IsGate-Cut and SameOrbit-Cut for N at level k in space $O((k+1)|N|^2)$. We solve the last point by enumerating all $a \in \Sigma$ and states $q' \subseteq Q_N$ and testing whether

– a transition $[p] \xrightarrow{a} [q']$ exists and $[q] \xrightarrow{a} [q']$ does not exist;
– $[p]$ and $[q']$ are in different orbits.

Again by Lemma 16 Edge-Cut and SameOrbit-Cut for N at level k may be done in space $O((k+1)|N|^2)$. This concludes the proof for Out-Cons-Violation.

The proofs for Acc-Cons-Violation and Orbit-Cons-Violation are similar. □

We now have all the elements to prove our main result.

Theorem 18. DRE-Definability(*NFA*) *and* DRE-Definability(*RE*) *are PSPACE-complete.*

Proof. DRE-Definability(RE) is known to be PSPACE-hard [3]. Since an RE can be translated in polynomial time into an equivalent NFA, the lower bound also holds for NFAs.

Furthermore, the upper bound for REs follows from the upper bound for NFAs. We therefore show that DRE-Definability for an NFA N is in space $O(|N|^4)$, which proves the theorem. We assume w.l.o.g. that $|\Sigma| \le |N|$. According to Theorem 10, a language $L(N)$ is not DRE-definable if and only if one of Out-Cons-Violation, Acc-Cons-Violation and Orbit-Cons-Violation occurs at some level k for N. Due to Theorem 9 we need to check this only for levels up to $|N|^2 + 2$, since the recursion depth of Algorithm 1 is never bigger than this.

We test whether there exists a violation at some level k for N, starting from level 0 and moving into higher and higher levels, up to level $|N|^2+2$. For every single level k the test can be done in space $O((k+1)|N|^2)$, according to Lemma 17. Note that during the application of the above lemma we know that the smaller levels do not contain any violation, since it was checked before, thus the assumptions in the lemma statements are fulfilled. Therefore the space needed for solving DRE-Definability(NFA) for N is bounded by $O((k+1)|N|^2)$ for $k = |N|^2 + 2$, i.e., bounded by $O(|N|^4)$, which finalises the proof for NFAs. □

4.2 Definability for RE(#)s

Next we show that testing DRE-Definability is EXPSPACE-complete when the input is given as a regular expression with counters. The upper bound is immediate from Theorem 18 and the fact that we can translate an RE(#) into an RE of exponential size by unfolding the counters.

For the lower bound, we reduce from the exponential corridor tiling problem, which is defined as follows. An *exponential tiling instance* is a tuple $\mathcal{I} = (T, H, V, x_\perp, x_\top, n)$ where T is a finite set of tiles, $H, V \subseteq T \times T$ are the horizontal and vertical constraints, $x_\perp, x_\top \in T$, and n is a natural number in unary notation. A *correct exponential corridor tiling for* $\mathcal{I}$ is a mapping $\lambda : \{1, \dots, m\} \times \{1, \dots, 2^n\} \to T$ for some $m \in \mathbb{N}$, such that every of the following constraints is satisfied:

- the first tile of the first row is $x_\perp$: $\lambda(1,1) = x_\perp$;
- the first tile of the m-th row is $x_\top$: $\lambda(m,1) = x_\top$;
- all vertical constraints are satisfied: $\forall i < m, \forall j \leq 2^n, (\lambda(i,j), \lambda(i+1,j)) \in V$; and,
- all horizontal constraints are satisfied: $\forall i \leq m, \forall j < 2^n, (\lambda(i,j), \lambda(i,j+1)) \in H$.

Then, the EXP-TILING problem asks, given an exponential tiling instance, whether there exists a correct exponential corridor tiling. The latter problem is known to be EXPSPACE-complete [27].

The full proof of the lower bound combines elements from the proof that DRE-definability for REs is PSPACE-complete [3] and that language universality for RE(#)s is EXPSPACE-complete [12].

Theorem 19. *Given a regular expression with counters r, the problem of deciding whether L(r) is DRE-definable is EXPSPACE-complete.*

4.3 Definability for DFAs

As explained before, the DRE-definability problem is in polynomial time if the input is a minimal DFA [7]. As a final, minor result, we give an NLOGSPACE lower bound in this case.

Theorem 20. DRE-DEFINABILITY(*minDFA*) *is NLOGSPACE-hard.*

Proof (sketch). The reduction for the lower bound is from the reachability problem in directed acyclic graphs. This problem asks, given a DAG $G = (V, E)$, a source node s, and a target node t, whether t is reachable from s by a directed path. The DAG reachability problem is well-known to be NLOGSPACE-complete [17].

In this proof, we will use the fact that finite languages are always DRE-definable. (This can be checked through the BKW-Algorithm which discovers immediately that all orbits are trivial.) For the reduction, let $G = (V, E)$, and nodes $s, t \in V$ be an instance of DAG reachability. We construct a minimal DFA D such that $L(D)$ is DRE-definable if and only if vertex t is reachable from vertex s in graph G.

We build $D = (Q, \Sigma, \delta, q_0, \{q_f\})$ from G as follows. The set Q of states is the disjoint union of the vertices V of G, plus two distinguished states q_0 (which is D's initial state) and q_f (which is D's only accepting state). The alphabet

Σ is defined as $(V \uplus \{q_0, q_f\})^2$. The transitions of D are defined as follows. Let $V = \{v_1, \ldots, v_n\}$ be the vertices of G.

- For each directed edge $(v_i, v_j) \in E$, the automaton D has the transition $\delta(v_i, (v_i, v_j)) = v_j$;
- for each vertex v_i, the automaton D has the transitions $\delta(q_0, (q_0, v_i)) = v_i$ and $\delta(v_i, (v_i, q_f)) = q_f$; and
- $\delta(t, (t, s)) = s$ is a transition of the automaton D.

As such, every transition has its unique label. This concludes the reduction. The reduction can be conducted in logarithmic space. □

5 Conclusions

We have pinned down the exact complexity of testing whether a regular expression can be determinized and considered additional variants of this problem. Our proof provides additional insights on such DRE-definable languages and on the decision algorithm of Brüggemann-Klein and Wood. An important open question is about the possible blow-up in the determinization process: Given a regular expression, what is the worst-case (unavoidable) blow-up when converting it to an equivalent deterministic one? At the moment, we know that an exponential blow-up cannot be avoided [21] but the best known upper bound is double exponential. While our proofs seem to give some insight in how to improve this upper bound, testing if a language is DRE-definable and actually constructing a minimal equivalent DRE are quite different matters. It is not yet clear to us how our techniques can be leveraged to obtain better upper bounds for this question.

Acknowledgement. We thank Wouter Gelade for bringing this problem to our attention.

References

1. Arenas, M., Barceló, P., Libkin, L., Murlak, F.: Foundations of Data Exchange. Book (to appear, 2013)
2. Balmin, A., Papakonstantinou, Y., Vianu, V.: Incremental validation of XML documents. ACM Trans. on Datab. Syst. 29(4), 710–751 (2004)
3. Bex, G.J., Gelade, W., Martens, W., Neven, F.: Simplifying XML Schema: effortless handling of nondeterministic regular expressions. In: ACM SIGMOD International Conference on Management of Data (SIGMOD), pp. 731–744 (2009)
4. Bex, G.J., Gelade, W., Neven, F., Vansummeren, S.: Learning deterministic regular expressions for the inference of schemas from XML data. In: World Wide Web Conference (WWW), pp. 825–834 (2008)
5. Bray, T., Paoli, J., Sperberg-McQueen, C.M., Maler, E., Yergeau, F.: Extensible Markup Language XML 1.0, 5th edn. W3C Recommendation (November 2008)
6. Brüggemann-Klein, A., Wood, D.: Deterministic Regular Languages. In: Finkel, A., Jantzen, M. (eds.) STACS 1992. LNCS, vol. 577, pp. 173–184. Springer, Heidelberg (1992)

7. Brüggemann-Klein, A., Wood, D.: One-unambiguous regular languages. Inf. and Comput. 142(2), 182–206 (1998)
8. Chen, H., Chen, L.: Inclusion Test Algorithms for One-Unambiguous Regular Expressions. In: Fitzgerald, J.S., Haxthausen, A.E., Yenigun, H. (eds.) ICTAC 2008. LNCS, vol. 5160, pp. 96–110. Springer, Heidelberg (2008)
9. Colazzo, D., Ghelli, G., Sartiani, C.: Efficient inclusion for a class of XML types with interleaving and counting. Inform. Syst. 34(7), 643–656 (2009)
10. Fallside, D., Walmsley, P.: XML Schema Part 0: Primer, 2nd edn. World Wide Web Consortium (October 2004)
11. Gelade, W., Gyssens, M., Martens, W.: Regular expressions with counting: Weak versus strong determinism. SIAM J. Comput. 41(1), 160–190 (2012)
12. Gelade, W., Martens, W., Neven, F.: Optimizing schema languages for XML: Numerical constraints and interleaving. SIAM J. Comput. 38(5), 2021–2043 (2009)
13. Gelade, W., Neven, F.: Succinctness of the complement and intersection of regular expressions. ACM Trans. on Comput. Logic 4, 1–19 (2012)
14. Groz, B., Maneth, S., Staworko, S.: Deterministic regular expressions in linear time. In: ACM Symposium on Principles of Database Systems (PODS), pp. 49–60 (2012)
15. Hopcroft, J.E., Motwani, R., Ullman, J.D.: Introduction to Automata Theory, Languages, and Computation. Addison Wesley (2007)
16. Hovland, D.: Regular Expressions with Numerical Constraints and Automata with Counters. In: Leucker, M., Morgan, C. (eds.) ICTAC 2009. LNCS, vol. 5684, pp. 231–245. Springer, Heidelberg (2009)
17. Jones, N.D.: Space-bounded reducibility among combinatorial problems. J. Comput. Syst. Sci. 11(1), 68–85 (1975)
18. Kilpeläinen, P.: Checking determinism of XML Schema content models in optimal time. Inform. Syst. 36(3), 596–617 (2011)
19. Kilpeläinen, P., Tuhkanen, R.: One-unambiguity of regular expressions with numeric occurrence indicators. Inform. and Comput. 205(6), 890–916 (2007)
20. Konrad, C., Magniez, F.: Validating XML documents in the streaming model with external memory. In: International Conference on Database Theory (ICDT), pp. 34–45 (2012)
21. Losemann, K., Martens, W., Niewerth, M.: Descriptional Complexity of Deterministic Regular Expressions. In: Rovan, B., Sassone, V., Widmayer, P. (eds.) MFCS 2012. LNCS, vol. 7464, pp. 643–654. Springer, Heidelberg (2012)
22. Maneth, S., Berlea, A., Perst, T., Seidl, H.: XML type checking with macro tree transducers. In: ACM Symposium on Principles of Database Systems (PODS), pp. 283–294 (2005)
23. Martens, W., Neven, F.: On the complexity of typechecking top-down XML transformations. Theor. Comp. Sc. 336(1), 153–180 (2005)
24. Milo, T., Suciu, D., Vianu, V.: Typechecking for XML transformers. J. Comput. Syst. Sci. 66(1), 66–97 (2003)
25. Neven, F., Schwentick, T.: On the complexity of XPath containment in the presence of disjunction, DTDs, and variables. Log. Meth. in Comp. Sc. 2(3) (2006)
26. Segoufin, L., Vianu, V.: Validating streaming XML documents. In: ACM Symposium on Principles of Database Systems (PODS), pp. 53–64 (2002)
27. van Emde Boas, P.: The convenience of tilings. In: Complexity, Logic and Recursion Theory, pp. 331–363. Marcel Dekker Inc. (1997)
28. Wood, P.T.: Containment for XPath Fragments under DTD Constraints. In: Calvanese, D., Lenzerini, M., Motwani, R. (eds.) ICDT 2003. LNCS, vol. 2572, pp. 297–311. Springer, Heidelberg (2002)

Type-Based Complexity Analysis for Fork Processes

Emmanuel Hainry, Jean-Yves Marion, and Romain Péchoux

Université de Lorraine and LORIA
{emmanuel.hainry,jean-yves.marion,romain.pechoux}@loria.fr

Abstract. We introduce a type system for concurrent programs described as a parallel imperative language using while-loops and fork/wait instructions, in which processes do not share a global memory, in order to analyze computational complexity. The type system provides an analysis of the data-flow based both on a data ramification principle related to tiering discipline and on secure typed languages. The main result states that well-typed processes characterize exactly the set of functions computable in polynomial space under termination, confluence and lock-freedom assumptions. More precisely, each process computes in polynomial time so that the evaluation of a process may be performed in polynomial time on a parallel model of computation. Type inference of the presented analysis is decidable in linear time provided that basic operator semantics is known.

Keywords: Implicit Computational Complexity, Tiering, Secure Information Flow, Concurrent Programming, PSpace.

1 Introduction

We propose a type system for an imperative language with while loops and forks calls, which provides an upper bound on the complexity of a process by controlling its data flow. Threads are created dynamically and they do not share a global memory. Each `fork` call creates a new child process with a distinct identifier (id) and duplicates the execution context including the program counter. The parent process keeps the ids of its children (but not the converse). Communications between children and their parent are performed through the use of a `wait` instruction that allows a returning child to pass a value to its parent process. The domain of computation is, in essence, the set of strings, on which various admissible operators can be defined. Thus, we are able to use arrays of strings with respect to the typing limitation as in Example 1.

We demonstrate that each subprocess generated by a well-typed process (under some restrictions) runs in polynomial time (Proposition 2) and that the number of process offsprings is bounded by a polynomial (Proposition 3). As a result, the amount of interactions between two processes is also bounded by a polynomial. Thanks to Savitch's Theorem [21], we show that a program runs in polynomial space on a sequential machine (Theorem 2), and conversely

F. Pfenning (Ed.): FOSSACS 2013, LNCS 7794, pp. 305–320, 2013.

(Theorem 3). This result is expressed in the central Theorem 1. As far as we know, it is the first characterization of FPSpace based on a typed system for an imperative language. The type inference procedure is computable in linear time in the program size (Proposition 1). As a result, applications about the complexity measure of process calculi may be considered. We refer to the conclusion for a discussion about practical issues.

That said, one of our main motivation is to understand the relationship between data flow control and computational complexity. In [17], a type system was introduced for a sequential imperative programming language, characterizing the set of polynomial time computable functions. The idea behind such a typing system is to control the information flow in an execution by enforcing a data tiering discipline. The idea is that data are ramified into tiers in such a way that data at a given tier may only produce information at the same tier or at a lower tier. Thus, the type system prevents the alteration of upper tier data by lower tier data, following the principle of an integrity policy [3]. From this observation, there is only a small step to do in order to devise a type system based on works on type-based information flow analysis. We refer to the survey [20] for further explanations. The type system assigns tiers (which are similar to security levels) to variables. As in [23,22], the type system prevents information to flow from lower tier variables to higher tier variables by direct or indirect assignments. From a complexity point of view, the main novelty here is that we have to control the information that flows between processes so that the termination of a father process does not depend on the return values of its children. For this, we suggest a three-tier lattice, $\{\mathbf{-1}, \mathbf{0}, \mathbf{1}\}$, where tier $\mathbf{1}$ variables control the while loops, tier $\mathbf{0}$ variables are working data, and tier $\mathbf{-1}$ variables are output values. Consequently, we prevent an unsafe declassification due to the termination of a process through the wait/return mechanism. Lastly, it is worth noticing that we establish a non-interference property (Proposition 2), which expresses the relationship between information flow and complexity.

There are several works, which are directly related to our results. On function algebra, a characterization of PSPACE has been provided by ramified recursion with parameter substitutions [14] and by ramified recurrence over functions [15]. On functional languages, several characterizations of PSPACE have been provided in [8] consisting in restrictions on data manipulation and on recursion schemes. Moreover, a typed lambda calculus based on Soft Linear Logic [11], which identifies exactly PSPACE, is introduced in [7]. The completeness proofs of the aforementioned works simulate alternating polynomial time Turing machines [5]. Another approach is the one of Pola [6] which relates complexity and category theory. On imperative languages, an approach consists in analyzing the growth of the data flow by means of a matrix calculus propagating constraints between variables [19,9,18]. Thus, Niggl and Wunderlich gave a characterization of PSPACE [19]. On the complexity of message passing languages based on implicit computational complexity, Amadio and Dabrowski show how to obtain an upper bound on instants for synchronous languages by using quasi-interpretation based tools [1,2]. More recently, Dal Lago *et al* proposed to polynomially bound process interactions by type systems based on Light Linear Logic [13,12]. Notice that the fragment of the process calculi is too weak to establish a completeness result. Lastly, the recent work of Madet [16] proposes a multithreaded program with side effects also based on light linear logic which are polynomial.

2 Imperative Language with Forks

2.1 Syntax of Processes

Expressions, instructions, commands and processes are defined by the following grammar, where $\mathcal{V}$ is the set of variables and $\mathbb{O}$ is the set of operators. The size of an expression is $|X| = 1$ and $|op(E_1, \ldots, E_n)| = \sum_{i=1}^{n} |E_i| + 1$. We note $\mathcal{V}(K)$, $K \in \text{Exp} \cup \text{Proc}$ the set of variables occurring in K.

$$
\begin{array}{lll}
E, E_1, \ldots, E_n \in \text{Exp} & ::= X \mid op(E_1, \ldots, E_n) & X \in \mathcal{V}, op \in \mathbb{O} \\
I \in \text{Inst} & ::= \texttt{fork}() \mid \texttt{wait}(E) & \\
C, C' \in \text{Cmd} & ::= X{:=}E \mid C;\ C' \mid \texttt{while}\ E\ \texttt{do}\ C \mid & \\
 & \quad \texttt{skip} \mid X{:=}I \mid \texttt{if}\ E\ \texttt{then}\ C\ \texttt{else}\ C' & \\
\text{P} \in \text{Proc} & ::= \texttt{return}\ X \mid C;\ \text{P} &
\end{array}
$$

2.2 Informal Semantics

The semantics is similar to the one of `C` programs and `Unix` processes. Each process has an id (a pid). When a fork command is executed, a new child process is created, that will run concurrently with its parent process with its own duplicated memory. The parent process knows the child id, whereas a child has no knowledge of its parent process id. A process P is evaluated inside a *configuration*, a triple $(\text{P}, \mu)_\rho$ consisting of a process, a *store* μ mapping each variable to a value of the computational domain and a set of ids ρ. In a configuration, the process can be viewed as the code that remains to be executed, and ρ stores the children process id's. At the creation of a child process, the store μ and the program counter are duplicated, and ρ is $\emptyset$. The main process id is always 1. When a new child is created by a `fork` its id is automatically set to a new integer and it is recorded in the set ρ of the father's configuration. All the configurations corresponding to the main process and its subprocesses are stored inside an *environment* $\mathscr{E}$, a partial function mapping the process id to a configuration. Processes do not share a global memory. Consequently, the only way for a process to communicate with its children is through the use of a `wait`(E) instruction, which provides a one-way communication. The following program illustrates how to define distinct code that will be executed by the child but not the parent process and conversely. For the parent, X contains the pid of its child, for the child, X contains 0.

```
P:   X := fork();           // X contains the child pid
     if X > 0 then {        // father's code (X>0)
          Y := wait(X);
          Y := "father"
```

```
} else {                    // child's code (X=0)
    Y := "child"
}
return Y   // The father returns "father" and the child returns "child"
```

2.3 Semantics of Expressions, Configurations and Environments

Domain. Let $\mathbb{W}$ be the set of words over a finite alphabet Σ including two symbols `tt` and `ff` that denote truth values true and false. Let ε be the empty word. The length of a word d is denoted $|d|$. As usual, we set $|\varepsilon| = 0$. Define $\trianglelefteq$ as the sub-word relation over $\mathbb{W}$, by $v \trianglelefteq w$, iff there are u and u' of $\mathbb{W}$ s.t. $w = u.v.u'$, where '.' is the concatenation. We write $\underline{n}$ to mean the binary word encoding the natural number n.

Store. A *store* μ is a total function from process variables in $\mathcal{V}$ to words in $\mathbb{W}$. Let $\mu\{X_1 \leftarrow d_1, \ldots, X_n \leftarrow d_n\}$, with X_i pairwise distinct, denote the store μ where the value stored by X_i is updated to d_i, for each $i \in \{1, \ldots, n\}$. The size of a store μ, denoted $|\mu|$ is defined by $|\mu| = \sum_{X \in \mathcal{V}} |\mu(X)|$.

Configuration. Given a store μ and a process P, the triplet $c = (\mathrm{P}, \mu)_\rho$, where ρ is an element of $\mathcal{P}(\mathbb{N})$, is called a *configuration*. Let $\bot$ be a special erased configuration that will be used to replace the content of a configuration once it has terminated: it no longer uses space. The size of a configuration is defined by $|\bot| = 0$ and $|(\mathrm{P}, \mu)_\rho| = |\mu| + \sharp\rho$, where $\sharp\rho$ is the cardinality of ρ.

Expressions and Configurations. Each operator of arity n is interpreted by a total function $[\![op]\!] : \mathbb{W}^n \mapsto \mathbb{W}$. The expression evaluation relation $\overset{\mathsf{e}}{\to}$ and the sequential command evaluation relation $\overset{\mathsf{c}}{\to}$ are described in Figure 1.

$(X, \mu) \overset{\mathsf{e}}{\to} \mu(X)$		(Variable)
$(op(E_1, \ldots, E_n), \mu) \overset{\mathsf{e}}{\to} [\![op]\!](d_1, \ldots, d_n),$	if $\forall i, (E_i, \mu) \overset{\mathsf{e}}{\to} d_i$	(Operator)
$(\mathtt{skip};\ \mathrm{P}, \mu)_\rho \overset{\mathsf{c}}{\to} (\mathrm{P}, \mu)_\rho$		(Skip)
$(X{:=}E;\ \mathrm{P}, \mu)_\rho \overset{\mathsf{c}}{\to} (\mathrm{P}, \mu\{X \leftarrow d\})_\rho$	if $(E, \mu) \overset{\mathsf{e}}{\to} d$	(Assign)
$(\mathtt{if}\ E\ \mathtt{then}\ C_{\mathtt{tt}}\ \mathtt{else}\ C_{\mathtt{ff}};\ \mathrm{P}, \mu)_\rho \overset{\mathsf{c}}{\to} (C_{\mathtt{tt}};\ \mathrm{P}, \mu)_\rho$	if $(E, \mu) \overset{\mathsf{e}}{\to} \mathtt{tt}$	($\mathrm{If}_{\mathtt{tt}}$)
$(\mathtt{if}\ E\ \mathtt{then}\ C_{\mathtt{tt}}\ \mathtt{else}\ C_{\mathtt{ff}};\ \mathrm{P}, \mu)_\rho \overset{\mathsf{c}}{\to} (C_{\mathtt{ff}};\ \mathrm{P}, \mu)_\rho$	if $(E, \mu) \overset{\mathsf{e}}{\to} \mathtt{ff}$	($\mathrm{If}_{\mathtt{ff}}$)
$(\mathtt{while}\ E\ \mathtt{do}\ C\ ;\ \mathrm{P}, \mu)_\rho \overset{\mathsf{c}}{\to} (\mathrm{P}, \mu)_\rho$	if $(E, \mu) \overset{\mathsf{e}}{\to} \mathtt{ff}$	($\mathrm{While}_{\mathtt{ff}}$)
$(\mathtt{while}\ E\ \mathtt{do}\ C\ ;\ \mathrm{P}, \mu)_\rho \overset{\mathsf{c}}{\to} (C;\ \mathtt{while}\ E\ \mathtt{do}\ C\ ;\ \mathrm{P}, \mu)_\rho$	if $(E, \mu) \overset{\mathsf{e}}{\to} \mathtt{tt}$	($\mathrm{While}_{\mathtt{tt}}$)

Fig. 1. Small step semantics of expressions and configurations

Environments. An *environment* $\mathscr{E}$ is a partial function from $\mathbb{N}$ to configurations. The domain of $\mathscr{E}$ is denoted $dom(\mathscr{E})$ and we denote $\sharp\mathscr{E}$ its cardinal when it is finite. We abbreviate $\mathscr{E}(n)$ by $\mathscr{E}_n$. The size of a finite environment $\|\mathscr{E}\|$ is defined by $\|\mathscr{E}\| = \sum_{i \in dom(\mathscr{E})} |\mathscr{E}_i|$. The notation $\mathscr{E}[i := c]$ is the environment $\mathscr{E}'$ defined by $\mathscr{E}'(j) = \mathscr{E}(j)$ for all $j \neq i \in dom(\mathscr{E})$ and $\mathscr{E}'(i) = c$. As usual $\mathscr{E}[i_1 := c_1, \ldots, i_k := c_k]$ is a shortcut for $\mathscr{E}[i_1 := c_1] \ldots [i_k := c_k]$. The *initial* environment is noted $\mathscr{E}_{init}[\mathrm{P}, \mu]$ and consists in the main process with no child. That is $\mathscr{E}_{init}[\mathrm{P}, \mu](1) = (\mathrm{P}, \mu)_\emptyset$ and $dom(\mathscr{E}_{init}[\mathrm{P}, \mu]) = \{1\}$. An environment $\mathscr{E}$ is *terminal* if the root process satisfies $\mathscr{E}_1 = (\mathtt{return}\ X, \mu)_\rho$.

Semantics. The transition $\rightarrow$ for process evaluation is provided in Figure 2. The (Fork) rule creates a new configuration, a new process, say of id n, with a new store, and adds it to the environment. The parent process records its child id $\underline{n}$ into the variable X on which the fork instruction has been called and the child id set of the parent is updated to $\rho \cup \{n\}$. Note also that X is assigned to 0 in the child configuration. The (Wait) commands Z:=$\mathtt{wait}(E)$ evaluates the expression E to some binary numeral $\underline{n}$. If the process $\underline{n}$ is a terminating configuration $\mathscr{E}_n$ with $n \in \rho$, then the output value $\mu'(Y)$ is transmitted and stored in the variable Z. Finally, the returning process n is killed by erasing it through the following operation $\mathscr{E}[n := \perp]$. Note that the side condition $n \in \rho$ prevents locks.

$$\mathscr{E}[i := c] \rightarrow \mathscr{E}[i := c'] \quad \text{if } c \xrightarrow{c} c' \qquad \text{(Conf)}$$

$$\mathscr{E}[i := (X\text{:=}\mathtt{fork}();\ \mathrm{P}, \mu)_\rho] \rightarrow \mathscr{E}[i := (\mathrm{P}, \mu\{X \leftarrow \underline{n}\})_{\rho \cup \{n\}}, n := (\mathrm{P}, \mu\{X \leftarrow \underline{0}\})_\emptyset] \qquad \text{(Fork)}$$
with $n = \sharp\mathscr{E} + 1$

$$\mathscr{E}[i := (X\text{:=}\mathtt{wait}(E);\ \mathrm{P}, \mu)_\rho] \rightarrow \mathscr{E}[i := (\mathrm{P}, \mu\{X \leftarrow \mu'(Y)\})_\rho, n := \perp] \qquad \text{(Wait)}$$
if $(E, \mu) \xrightarrow{e} \underline{n}$, $n \in \rho$ and $\mathscr{E}_n = (\mathtt{return}\ Y, \mu')$

Fig. 2. Semantics of Environments

2.4 Strong Normalization, Lock-Freedom and Confluence

Throughout the paper, given a relation $\mapsto$, let $\mapsto^*$ be the reflexive and transitive closure of $\mapsto$ and let $\mapsto^k$ denote the k-fold self composition of $\mapsto$. A process P is *strongly normalizing* if there is no infinite reduction starting from the initial environment $\mathscr{E}_{init}[\mathrm{P}, \mu]$ through the relation $\rightarrow$, for any store μ. Given an initial environment $\mathscr{E}_{init}[\mathrm{P}, \mu]$, for some strongly normalizing process P and some store μ, if $\mathscr{E}_{init}[\mathrm{P}, \mu] \rightarrow^* \mathscr{E}'$, for some environment $\mathscr{E}'$ such that there is no environment $\mathscr{E}''$, $\mathscr{E}' \rightarrow \mathscr{E}''$, then either $\mathscr{E}'$ is terminal, i.e. $\mathscr{E}'_1 = (\mathtt{return}\ X, \mu')_\rho$ (the *main process is returning*) or $\mathscr{E}'_1 = (X\text{:=}\mathtt{wait}(E);\ C', \mu')_\rho$(we say that the environment $\mathscr{E}'$ is locked). A process $\mathrm{P} = C;\ \mathtt{return}\ X$ is *lock-free* if for any initial environment $\mathscr{E}_{init}[\mathrm{P}, \mu]$, there is no locked environment $\mathscr{E}'$ such that $\mathscr{E}_{init}[\mathrm{P}, \mu] \xrightarrow{*} \mathscr{E}'$.

A process P is *confluent* if for each initial environment $\mathscr{E}_{init}[\mathrm{P}, \mu]$ and any reductions $\mathscr{E}_{init}[\mathrm{P}, \mu] \rightarrow^* \mathscr{E}'$ and $\mathscr{E}_{init}[\mathrm{P}, \mu] \rightarrow^* \mathscr{E}''$ there exists an environment $\mathscr{E}^3$ such that $\mathscr{E}' \rightarrow^* \mathscr{E}^3$ and $\mathscr{E}'' \rightarrow^* \mathscr{E}^3$. A strongly normalizing, lock free and confluent process P computes a total function $f : \mathbb{W}^n \rightarrow \mathbb{W}$ defined:

$$\forall d_1, \ldots, d_n \in \mathbb{W},\ f(d_1, \ldots, d_n) = w$$

if $\mathscr{E}_{init}[\mathrm{P}, \mu[X_i \leftarrow d_i]] \rightarrow^* \mathscr{E}$, for some terminal environment $\mathscr{E}$ with $\mathscr{E}_1 = (\mathtt{return}\ X, \mu')_\rho$ and $\mu'(X) = w$.

3 Type System

3.1 Tiers and Typing Environments

Tiers are elements of the lattice $(\{\mathbf{-1}, \mathbf{0}, \mathbf{1}\}, \vee, \wedge)$ where $\vee$ and $\wedge$ are the least upper bound operator and the greatest lower bound operator, respectively. The induced order, denoted $\preceq$, is such that $\mathbf{-1} \preceq \mathbf{0} \preceq \mathbf{1}$. In what follows, let $\alpha, \beta, \ldots$ denote tiers in $\{\mathbf{-1}, \mathbf{0}, \mathbf{1}\}$. Tiers will be used to type both expressions and commands. Operator types τ are defined by $\tau ::= \alpha \mid \alpha \longrightarrow \tau$. As usual, we will use right associativity for $\longrightarrow$. A *variable typing environment* Γ maps each variable in $\mathcal{V}$ to a tier. An *operator typing environment* Δ maps each operator *op* to a set of operator types $\Delta(op)$, where the operator types corresponding to an operator of arity n are of the shape $\tau = \alpha_1 \longrightarrow \ldots \longrightarrow \alpha_n \longrightarrow \alpha$.

Intuitively, each tier variable has a specific role to play:

- tier $\mathbf{1}$ variables will be used as guards of while loops. They should not be allowed to take more than a polynomial number of distinct values;
- tier $\mathbf{0}$ variables may increase and cannot be used as while loop guards;
- tier $\mathbf{-1}$ variables will store values returned by child processes and cannot increase. Intuitively, they play the role of an output tape.

3.2 Well-Typed Processes

Typing Rules. Figure 3 provides the typing rules for expressions, commands and processes. Typing rules consist in judgments of the shape $\Gamma, \Delta \vdash K : \alpha$, $K \in \mathrm{Exp} \cup \mathrm{Cmd}$, meaning that K has type α under variable and operator typing environments Γ and Δ, respectively.

There are some important points to explain in this type system. First, the typing discipline precludes values from flowing from tier α to tier β, whenever $\alpha \preceq \beta$ and $\alpha \neq \beta$. Consequently, the guards of while loops are enforced to be of tier $\mathbf{1}$ in rule (CW). Moreover, in a (CB) rule, we enforce the tier of the guard to be equal to the tier of both branches. Also note that the subtyping rule (CSub) is restricted to commands in order not to break this preclusion. On the opposite, information may flow from tier $\mathbf{1}$ to tier $\mathbf{0}$ then to tier $\mathbf{-1}$. This point is underlined by the side condition of the (CA) rule. The (F) rule enforces the tier of the variable storing the process id to be of tier $\mathbf{0}$ since the value stored

$$\frac{\Gamma(X)=\alpha}{\Gamma,\Delta\vdash X:\alpha}\ (EV)\qquad \frac{\Gamma,\Delta\vdash E_i:\alpha_i \quad \alpha_1\longrightarrow\ldots\longrightarrow\alpha_n\longrightarrow\alpha\in\Delta(op)}{\Gamma,\Delta\vdash op(E_1,\ldots,E_n):\alpha}\ (EO)$$

$$\frac{\Gamma,\Delta\vdash X:\mathbf{0}}{\Gamma,\Delta\vdash X{:=}\texttt{fork()}:\mathbf{0}}\ (F)\qquad \frac{\Gamma,\Delta\vdash E:\mathbf{0}\quad \Gamma,\Delta\vdash X:\mathbf{-1}}{\Gamma,\Delta\vdash X{:=}\texttt{wait}(E):\mathbf{-1}}\ (W)$$

$$\frac{\Gamma,\Delta\vdash X:\alpha \qquad \Gamma,\Delta\vdash E:\alpha' \qquad E\in\text{Exp}}{\Gamma,\Delta\vdash X{:=}E:\alpha}\ \alpha\preceq\alpha'\ \ (CA)$$

$$\frac{\Gamma,\Delta\vdash C:\alpha \qquad \Gamma,\Delta\vdash C':\alpha'}{\Gamma,\Delta\vdash C;\ C':\alpha\vee\alpha'}\ (CC)\qquad \frac{\Gamma,\Delta\vdash E:\mathbf{1} \qquad \Gamma,\Delta\vdash C:\alpha}{\Gamma,\Delta\vdash \texttt{while}(E)\texttt{do}\{C\}:\mathbf{1}}\ (CW)$$

$$\frac{\Gamma,\Delta\vdash E:\alpha \qquad \Gamma,\Delta\vdash C:\alpha \qquad \Gamma,\Delta\vdash C':\alpha}{\Gamma,\Delta\vdash \texttt{if}\ E\ \texttt{then}\ C\ \texttt{else}\ C':\alpha}\ (CB)\qquad \frac{}{\Gamma,\Delta\vdash \texttt{skip}:\alpha}\ (CS)$$

$$\frac{\Gamma,\Delta\vdash C:\alpha}{\Gamma,\Delta\vdash C:\beta}\ \alpha\preceq\beta\quad (CSub)\qquad \frac{\Gamma,\Delta\vdash X:\beta}{\Gamma,\Delta\vdash \texttt{return}\ X:\beta}\ (R)$$

Fig. 3. Type system for expressions, instructions, commands and processes

will increase dynamically during the process execution. Finally, the tier of the variable storing the result returned by a child process (rule (W)) has to be of tier $\mathbf{-1}$, which means that no information may flow from a variable of a child process to tier $\mathbf{0}$ and tier $\mathbf{1}$ variables of its parent process.

Notations. In practice, we write $C:\alpha$ to say that C is of type α, and E^{α} to says that E is of type α under the considered environments.

Example 1. The following example illustrates how we can program a reduce operation used in parallel prefix sum [10]. The problem consists in computing the greatest element among n integers in an array-like structure $(A[0],\ldots,A[n-1])$.

```
max_reduce(n^1, A^0) ::=
 r^0 := 0: 0;
 f^-1 := A[r]^0: -1;
 flag^0 := tt: 0;
 while (n^1 ≠ 1)^1 do {
     if flag^0 then {                              // not finished
         pidl^0 := fork(): 0
         if (pidl>0)^0 then {                      // father process
             r^0 := 2*r+2: 0;
             pidr^0 := fork(): 0}
             else { r^0 := 2*r+1: 0 }              // left son
         if (pidr==0)^0 or (pidl==0)^0 then { f^-1 := A[r]^0: 0; }
             else {
                flag^0 := ff: 0;                   // father
                xl^-1 := wait(pidl): 0;
                xr^-1 := wait(pidr): 0;
```

```
            f⁻¹ := max(f⁻¹, max(xl,xr)): 0; } }
   n¹ := half(n)¹: 1 }               // end of while
return f:-1
```

The notation $A[r]$ simply denotes the access to cell r of the array A.

The program queries the first element of the array $A[0]$ then will spawn two children subprocesses, each one exploring one half of the array (the left son will compute the max of $\{A[i]; 2^k - 1 \leq i \leq 2^k + 2^{k-1} - 2, k \geq 1\} = \{A[1], A[3], A[4], A[7], \ldots\}$, the right son computing the max of $\{A[i]; 2^k + 2^{k-1} - 1 \leq i \leq 2^{k+1} - 2, k \geq 1\} = \{A[2], A[5], A[6], A[11], \ldots\}$). When labelling the nodes of the subprocesses tree with the character each process checks by itself, it is such that the array is a breadth first search of it.

Notice that the typing of commands uses the subtyping rule *CSub*.

We find convenient to describe now the nature of operators used, that will be presented in the next section (Definitions 1 and 2). The get operation on arrays, expressed above by the assignment $f := A[r]$, is a neutral operation of type $\mathbf{0} \longrightarrow \mathbf{0} \longrightarrow \mathbf{0}$ and it can be in *Ntr* because of its type. The length n and the index r are identified by their binary notations. So the operators $2 * r + 2$ and $2*r+1$ increase the length by 1 and are positive operators of type $\mathbf{0}, \mathbf{0} \longrightarrow \mathbf{0}$, and so are in *Pos*. Similarly, the operator `half`(n) divides the length n by two, that is it deletes a letter. So, it is a neutral operator of type $\mathbf{1} \longrightarrow \mathbf{1}$ and is in *Ntr*. The predicates `or`, $\neq$ are typical neutral operators of types resp. $\mathbf{0} \longrightarrow \mathbf{0} \longrightarrow \mathbf{0}$ and $\mathbf{1} \longrightarrow \mathbf{1} \longrightarrow \mathbf{1}$. However the typing implies that `or` $\in Ntr \cup Pos$, and $\neq \in Ntr$. Lastly, `max` is a max operator of type $-\mathbf{1} \longrightarrow -\mathbf{1} \longrightarrow -\mathbf{1}$ and is in *Max*.

4 Safe Processes, Type Inference and Complexity

4.1 Neutral, Max, and Positive Operators

The type system guarantees that information flow goes from tier $\mathbf{1}$ to tier $-\mathbf{1}$, and prevents any flow in the other way from a lower tier to higher tier. But this is not sufficient to bound process resources. We need to fix the class of operator interpretations based on their typing. For that we define neutral operators (that make the variables decrease and preserve tier), max operators (that do not make variables grow and preserve lower tiers but not tier $\mathbf{1}$), and positive operators (that can only produce a result of tier $\mathbf{0}$ as they can create something bigger than their arguments).

Definition 1.

1. *An operator op is* neutral *if:*
   (a) *either op computes a binary predicate (i.e. the codomain of op is* $\{\mathtt{tt}, \mathtt{ff}\}$*).*
   (b) *or* $\forall d_1, \ldots, d_n,\ \exists i \in \{1, \ldots, n\},\ [\![op]\!](d_1, \ldots, d_n) \trianglelefteq d_i$.
2. *An operator op is* max *if* $\forall (d_i)_{1,n}\ |[\![op]\!](d_1, \ldots, d_n)| \leq \max_{i \in [1,n]} |d_i|$.
3. *An operator op is* positive *if* $\forall (d_i)_{1,n},\ |[\![op]\!](d_1, \ldots, d_n)| \leq \max_{i \in [1,n]} |d_i| + c$, *for a constant* $c \geq 0$.

Say that a type $\alpha_1 \longrightarrow \ldots \longrightarrow \alpha_n \longrightarrow \alpha$ is decreasing if $\alpha \preceq \wedge_{i=1,n}\alpha_i$. We now give a partition of operators into three classes which depend both on their types and on their growth rates.

Definition 2 (Safe operator typing environment). *An operator typing environment Δ is* safe *if each type given by Δ is decreasing and there exist three disjoint classes of operators Ntr, Max and Pos such that for any operator op and $\forall \alpha_1 \longrightarrow \ldots \longrightarrow \alpha_n \longrightarrow \alpha \in \Delta(op)$, the following conditions hold:*

- *If op $\in$ Ntr then op is a neutral operator.*
- *If op $\in$ Max then op is a max operator and $\alpha \neq \mathbf{1}$.*
- *If op $\in$ Pos then op is a positive operator and $\alpha = \mathbf{0}$.*

Intuitively, expressions in while guards are of tier $\mathbf{1}$ and so the iteration length just depends on the number of possible tier $\mathbf{1}$ configurations. Inside a while loop, we can perform operations on variables of other tiers. The tier $-\mathbf{1}$ values are return values and processes are confined in the sense that the information flow of a process does not depend on a return value.

4.2 Main Result

Proposition 1 (Type inference). *Given a safe operator typing environment Δ, deciding if there exists a variable typing environment Γ such that the typing rules are satisfied can be done in time linear in the size of the program.*

Proof. We encode the tier of each variable X by 3 boolean variables $x_{\mathbf{1}}$, $x_{\mathbf{0}}$ and $x_{-\mathbf{1}}$. We enforce that variable X is of exactly one tier will be encoded by $(\neg x_{\mathbf{0}} \vee \neg x_{\mathbf{1}}) \wedge (\neg x_{\mathbf{0}} \vee \neg x_{-\mathbf{1}}) \wedge (\neg x_{\mathbf{1}} \vee \neg x_{-\mathbf{1}})$. Each command gives some constraints. Thus, in the case of an assignment $X := op(Y)$, we need to encode $\Gamma(Y) \geq \Gamma(X)$, which can be represented as $(\neg y_{-\mathbf{1}} \vee x_{-\mathbf{1}}) \wedge (\neg y_{\mathbf{0}} \vee \neg x_{\mathbf{1}}) \wedge (\neg x_{\mathbf{1}} \vee y_{\mathbf{1}})$. As a result, the type inference problem is reduced to 2-SAT.

Definition 3 (Safe process). *Given Γ a variable typing environment and Δ an operator typing environment, we say that a process* P *is a* safe process *if:*

- P *is well-typed with respect to Γ and Δ, i.e. $\Gamma, \Delta \vdash \mathrm{P} : \beta$;*
- *Δ is safe.*

The main result below is a consequence of the soundness Theorem 2 and the completeness Theorem 3

Theorem 1. *The set of polynomial space computable functions is exactly the set of functions computed by strongly normalizing, lock-free, confluent and safe processes, where a unit-cost is taken as the cost of an operator computation.*

Therefore, the process in example 1 can run within a polynomial space. It is worth noticing that the demonstration below says more about process interaction and runtime, which are, in fact, polynomially bounded as we shall see.

5 Complexity Soundness

5.1 Process Runtime

We establish that each process runs in polynomial time if the time measure is the number of reductions. We follow the line of the soundness proof of [23] to demonstrate a non-interference property. We first prove two lemmas 1 and 2 in order to express that an expression of tier **1** just depends on tier **1** variables, and not on lower tier variables. As a result, there is only a polynomial number of tier **1** configurations, because of neutral operator growth rate. From the security point of view, this means that a variable of tier (level) **1** can be updated by information of the same tier of integrity. Proposition 2 corresponds to the non-interference property. Intuitively, it says that the complexity of a process depends only on tier **1** variables, and so it is not modified if values of lower tiers are updated and so the process runtime does not interfere with lower tier values.

Lemma 1 (Simple security). *Given a safe process* P *wrt typing environments* Γ *and* Δ, *if* $\Gamma, \Delta \vdash E : \mathbf{1}$, *for an expression* E *in* P, *then for each* $X \in \mathcal{V}(E)$, $\Gamma(X) = \mathbf{1}$ *and all operators in* E *are neutral.*

Proof. By induction on E.

Given a typing environment Γ say that $|\mu|_{\mathrm{i}} = \sum_{\Gamma(X)=\mathrm{i}} |\mu(X)|$. We will first prove that that the values taken by each tier **1** expression are at most in polynomial number.

Lemma 2. *Given a safe process* P *wrt typing environments* Γ *and* Δ, *for each expression* E *in* P *such that* $\Gamma, \Delta \vdash E : \mathbf{1}$, *the number of distinct values taken by* E *during the evaluation of the initial environment* $\mathscr{E}_{init}[\mathrm{P}, \mu]$ *is bounded polynomially in* $|\mu|_{\mathbf{1}}$.

Proof. A store can be updated either (i) by using a `fork` instruction, (ii) by using a `wait` instruction or (iii) by assigning the result of an expression. The typing discipline prevents variables assigned to in (i) and (ii) to be of tier **1**. For (iii), we have $E = op(E_1, \ldots, E_n)$, for some neutral operator op, by Lemma 1. Consequently, the result of the computation is either `tt` or `ff` or a subword of the initial values. The number of sub-words being quadratic in the size, the number of distinct values of tier **1** variables is at most quadratic in $|\mu|_{\mathbf{1}}$.

The relations $\Rightarrow_i$ are defined to express the computation of the process of id i.

Definition 4 ($\Rightarrow_i$). *Given two environments* $\mathscr{E}$ *and* $\mathscr{E}'$ *such that* $\mathscr{E} \to \mathscr{E}'$, *we write* $\mathscr{E} \to_i \mathscr{E}'$ *if the transition* $\to$ *corresponds to the evaluation of the process in the configuration* $\mathscr{E}_i$ *of* $\mathscr{E}$. *In the same way define the transition* $\mathscr{E} \to_{\neq i} \mathscr{E}'$ *if there exists* $k \neq i$ *such that* $\mathscr{E} \to_k \mathscr{E}'$. *Now define* $\Rightarrow_i$ *by* $\mathscr{E} \Rightarrow_i \mathscr{E}'$ *if there are environments* $\mathscr{E}^1, \mathscr{E}^2$ *such that* $\mathscr{E} \to^*_{\neq i} \mathscr{E}^1 \to_i \mathscr{E}^2 \to^*_{\neq i} \mathscr{E}'$.

The proposition below expresses that the number of instructions performed by a single process is bounded by a polynomial in the tier **1** initial values. Intuitively, this means that each process runs in polynomial time, in the tier **1** initial values, if we do not count the waiting time due to forks.

Proposition 2 (Non-interference wrt time). *Given a strongly normalizing and safe process* P*, there is a polynomial* Q *such that, for each initial environment* $\mathscr{E}_{init}[\mathrm{P}, \mu]$*,* $\forall i \in \mathbb{N}$*, if* $\mathscr{E}_{init}[\mathrm{P}, \mu] \Rightarrow_i^k \mathscr{E}$ *then* $k \leq Q(|\mu|_{\mathbf{1}})$*.*

Proof. The typing discipline enforces all the expressions in while-loop guards of a safe process to be of tier **1**. The evaluation of tier **1** values does not depend on lower tier values. By Lemma 2, the number of values taken by tier **1** variables during the evaluation is bounded by $Q(|\mu|_{\mathbf{1}})$ for some polynomial Q. If a process enters twice in the same configuration with the same tier **1** values, then the computation loops forever. Consequently, if P is strongly normalizing then all the while loops are executed at most $Q(|\mu|_{\mathbf{1}})$ times.

5.2 Process Spawning

Now, we demonstrate that given a strongly normalizing and safe process the number of children of a process (Lemma 3), the number of process generations (Lemma 4) and the size of a configuration (Lemma 5) are polynomially bounded.

Definition 5. *Given a finite environment* $\mathscr{E}$*, the* process tree $T(\mathscr{E})$ *is defined:*

- *the nodes are the configurations* $\{\mathscr{E}_1, \ldots, \mathscr{E}_{\sharp\mathscr{E}}\}$ *and the root is* $\mathscr{E}_1$*;*
- *for each* $l \in [1, \sharp\mathscr{E}]$*, there is an edge from* $\mathscr{E}_l = (\mathrm{P}, \mu)_\rho$ *to* $\mathscr{E}_k$*, if* $k \in \rho$*.*

The degree $d(T)$ corresponds to the number of children generated by fork instructions of a given process. The height $h(T)$ is the number of nested processes.

Lemma 3. *Given a strongly normalizing and safe process* P*, there exists a polynomial* Q *such that, for each initial environment* $\mathscr{E}_{init}[\mathrm{P}, \mu]$*, if* $\mathscr{E}_{init}[\mathrm{P}, \mu] \to^* \mathscr{E}$ *then* $d(T(\mathscr{E})) \leq Q(|\mu|_{\mathbf{1}})$*. In other words, for each* $\mathscr{E}_i = (\mathrm{P}_i, \mu_i)_{\rho_i}$*,* $i \leq \sharp\mathscr{E}$*, the number of subprocesses is bounded by* $Q(|\mu|_{\mathbf{1}})$*, i.e.* $\sharp\rho_i \leq Q(|\mu|_{\mathbf{1}})$*.*

Proof. By Proposition 2, there is a polynomial Q such that the transition $\to_i$ is taken at most $Q(|\mu|_{\mathbf{1}})$ times, which bounds the number of executed `fork` instructions.

Lemma 4. *Given a strongly normalizing and safe process* P*, there exists a polynomial* Q *such that, for each initial environment* $\mathscr{E}_{init}[\mathrm{P}, \mu]$*, if* $\mathscr{E}_{init}[\mathrm{P}, \mu] \to^* \mathscr{E}$ *then* $h(T(\mathscr{E})) \leq Q(|\mu|_{\mathbf{1}})$*.*

Proof. By Proposition 2, there is a polynomial Q which bounds the number of executed instructions and which just depends on the size of tier **1** initial values. Now, when $X :=$ `fork`() is performed, the parent process store is duplicated in the child process store, except for the value of X. But X is of tier 0 and so has no impact on the computational time of both processes. Next, a fork command has been executed in the parent process. So, the runtime of the child generated is strictly less than the runtime of its father. Therefore, we deduce that Q bounds the height of the process tree.

We end by showing that subprocess stores are polynomially bounded in the size of the initial store, which is a consequence of the non-interference property, as stated in Proposition 2.

Lemma 5. *Given a strongly normalizing and safe process* P*, there exists a polynomial* S *such that, for each initial environment* $\mathscr{E}_{init}[\mathrm{P},\mu]$*, if* $\mathscr{E}_{init}[\mathrm{P},\mu] \to^* \mathscr{E}$ *then* $\forall i \leq \sharp\mathscr{E}$*, if* $\mathscr{E}_i = (\mathrm{P}_i, \mu_i)_{\rho_i}$ *then* $|\mu_i| \leq S(|\mu|_{\mathbf{1}}) + |\mu|_{\mathbf{0}} + |\mu|_{-\mathbf{1}}$.

Proof. There are three cases to consider. First, a variable of tier **1** is updated by an expression of tier **1**. By Lemma 1, a tier **1** expression just consists in neutral operators and tier **1** variables. So, tier **1** variables are always sub-words of tier **1** initial values. Thus, the size of a variable of tier **1** is always bounded by $|\mu|_{\mathbf{1}}$. Second, take a tier **0** variable X, which is either updated inside a process by a composition of operators or updated by a pid return of a fork. By combining Proposition 2 and lemmas 3 and 4, we obtain a polynomial R which depends on operators and on the program such that the size of X is bounded by $R(Q(|\mu|_{\mathbf{1}}))+|\mu|_{\mathbf{0}}$ in each process computation. Third, take a tier $-\mathbf{1}$ variable Y. Suppose that the variable Y is assigned to max operators and variables of any tier. Observe that the case when $Y := \mathtt{wait}(E')$ is a particular case of an assignment if we see globally all processes together. Hence Y is bounded by $R(Q(|\mu|_{\mathbf{1}}) + |\mu|_{\mathbf{0}} + |\mu|_{-\mathbf{1}}$.

5.3 PSpace Abiding Evaluation Strategy

We define a deterministic evaluation strategy $\twoheadrightarrow$ starting from the initial environment and evaluating it until it reaches a wait instruction for a process n. Then, the strategy runs the process n until it returns a value. Formally, define a state $s = (\mathscr{E}, l)$ to be a pair of an environment $\mathscr{E}$ and a stack l of ids representing the queued processes. The initial state is $(\mathscr{E}_{init}[\mathrm{P},\mu], [1])$. We denote :: the stack constructor.

1. If $\mathscr{E} \to_h \mathscr{E}'$ for some rule (R) of Figure 2, $R \neq Wait$, then $(\mathscr{E}, h :: q) \twoheadrightarrow (\mathscr{E}', h :: q)$
2. If $\mathscr{E}_h = (X := \mathtt{wait}(E); \mathrm{P}, \mu)_\rho$ and $(E,\mu) \overset{\mathtt{e}}{\to} \underline{n}$ then $(\mathscr{E}, h :: q) \twoheadrightarrow (\mathscr{E}, n :: h :: q)$
3. If $\mathscr{E}_n = (\mathtt{return}\ X, \mu)_\rho$ and $\mathscr{E}_h = (Y{:=}\mathtt{wait}(E); \mathrm{P}, \mu')_{\rho'}$ then $(\mathscr{E}, n :: h :: q) \twoheadrightarrow (\mathscr{E}', h :: q)$ where $\mathscr{E} \to_h \mathscr{E}'$ for the (Wait) rule of Figure 2.

Notice that the rule (2) implies that $(E,\mu') \overset{\mathtt{e}}{\to} \underline{n}$ in rule (3).

Lemma 6 (Correction Lemma). *Given a strongly normalizing and confluent process* P *and an initial environment* $\mathscr{E}_{init}[\mathrm{P},\mu]$*, if there exists* $\mathscr{E}$ *such that* $\mathscr{E}_{init}[\mathrm{P},\mu] \to^* \mathscr{E}[1 := (\mathtt{return}\ X, \mu')_\rho]$ *then there exists* $\mathscr{E}'$ *s.t.* $(\mathscr{E}_{init}[\mathrm{P},\mu], [1]) \twoheadrightarrow^* (\mathscr{E}'[1 := (\mathtt{return}\ X, \mu'')_{\rho'}], [1])$ *and* $\mu''(X) = \mu'(X)$*, where* $[1]$ *is the stack containing the single process* 1.

The size of a stack $|l|$ is the number of elements in it (e.g. $|[1]| = 1$).

Lemma 7. *Given a strongly normalizing and safe process* P, *there is a polynomial* Q *such that for each initial environment* $\mathscr{E}_{init}[\mathrm{P},\mu]$ *and each stack* l, *if* $(\mathscr{E}_{init}[\mathrm{P},\mu],[1]) \twoheadrightarrow^* (\mathscr{E},l)$ *then (i)* $|l| \leq Q(|\mu|)$ *and (ii)* $\|\mathscr{E}\| \leq Q(|\mu|)$.

Proof. The $\twoheadrightarrow$ strategy explores the process tree using a stack. The height of the stack corresponds to the number of nested processes. The first inequality follows Lemma 4. Next, by combining Lemmata 3 and 5, we obtain the second inequality.

Theorem 2 (Soundness). *A strongly normalizing, confluent and safe process* P *can be evaluated in polynomially bounded space using strategy* $\twoheadrightarrow$, *where a unit-cost is taken as the cost of an operator computation.*

6 Completeness

Let us show that each polynomial space computable function can be computed by a strongly normalizing and safe process.

Theorem 3. *Every polynomial space function is computable by a strongly normalizing, lock-free and safe process* P.

Proof. We present how to compute the value of a quantified boolean formula (QBF). The program below may be seen as a skeleton from which we may simulate, for example, an alternating Turing machine running in polynomial time. Consequently, the computation of a polynomial space function necessitates to compute its output bit by bit, which may be uniformly performed by generating each address and querying the output bit. The program below generates 2^n forks where n is the number of variables. We suppose that the concrete syntax of a QBF is implemented as a string `phi` of tier **1**. Two neutral operators `kind` and `variable` of type $\mathbf{1} \longrightarrow \mathbf{1}$ return the quantifier kind and the variable bound at the root of `phi`. The neutral operator `next` of type $\mathbf{1} \longrightarrow \mathbf{1}$ erases the root quantifier and moves to the next one. The positive operator `evaluate` of type $\mathbf{1}, \mathbf{0} \longrightarrow \mathbf{0}$ computes a boolean formula with respect to an evaluation encoded as an array of booleans, which is an NC^1 complete problem [4]. As in example 1, arrays are words of type **0**, whose length is bounded by `phi`. The set array operator, that we conveniently note `vartab[x] := tt`, is a positive operator of type $\mathbf{0}, \mathbf{0} \longrightarrow \mathbf{0}$. We consider the array to be pre-allocated using a `calloc` operator that can be implemented using a positive operator in a loop.

This program has two big while loops: the first one generates 2^n subprocesses, one for each possible variable configuration, for each variable, the father will assign `tt` and the son will assign `ff`; the second one gathers the results of the children, making a disjonction with the parent if they correspond to a $\exists$, a conjonction for a $\forall$.

qbf(phi) ::=

```
    psi^1 := phi^1 : 1;
    q^1 := kind(phi^1) : 1;
```

```
  x¹ := variable(phi¹) : 1;
  i⁰ := 0⁰ : 0; // son number of a process
  pidtab⁰ := calloc(phi¹, phi¹) : 0;
  vartab⁰ := calloc(phi¹, tt) : 0;
  while (q¹ == '∃') or (q¹ == '∀') do {
      phi¹ := next(phi¹) : 1;
      pid⁰ := fork() : 0;
      if pid⁰>0 then {                          // father process
          pidtab[i] := (pid,q) : 0;
          vartab[x] := tt : 0;
          i⁰ := i + 1⁰ :0
      } else {                                  // son process
          vartab[x] := ff : 0;
          i⁰ := 0⁰ : 0} // end of else
      q¹ := kind(phi¹) : 1;
      x¹ := variable(phi¹) : 1;} // end of while
  res⁻¹ := evaluate(phi¹, vartab)⁰ : 0;
  while(state(psi¹) == '∃') or (state(psi¹) == '∀') do {
      psi¹ := next(psi¹) : 1;
      if (i>0) then {// for each son
          i⁰ := i − 1⁰ : 0;
          (pid⁰,op⁰) := pidtab[i]: 0;
          res_son⁻¹ := wait(pid⁰)⁻¹ : −1
          if op⁰ == '∃' then
              res⁻¹ := or(res, res_son)⁻¹ :0;
          else
              res⁻¹ := and(res, res_son)⁻¹ :0; } }
  return res⁻¹
```

7 Conclusion

We established that a typed process will generate only a polynomial number of offspring processes and that each process runs within polynomial time bounds. This work may have practical applications to determine whether or not a process runs within certain limited computing resources. Indeed, it may be a tool to control the spawning mechanism, and so to prevent, for example, denial of service attacks.

Let us come back to Example 1 describing a max-reduce algorithm. The runtime is $O(log^2(n))$. The reason of this apparent shortcoming is that we can not allow a break instruction inside a while-loop. In this case, a return instruction would "disrupt" the information flow and would interfere with tier **1** values. As a result, the runtime would not be anymore bounded. However, if the goal is to analyze program complexity, then we can devise a program transformation Θ, which, approximately, moves break instructions outside while loops, whenever it is possible. Such program transformation may be efficiently performed by a tree transducer. Moreover, there is no reason to preserve the program semantics because we are just interested in resource bounds. So, we may omit implementation details, which may increase drastically the expressivity of the considered programs. Therefore, we may think of this overall scenario: A system receives a process P to run. First, it computes an abstraction of P using a program transformation Θ. Second, it checks that the abstraction is well-typed. As a result, and even if the system does not know precisely an upper-bound on the complexity, it gets efficiently some confidence on the resource usage of P.

The program solving QBF gives another limitation because we are not allowed to make a loop bounded by the number of subprocesses, which is a tier **0** value. However, we know that this number is bounded by a tier **1** value. To solve this, Hofmann suggested to manage the garbage collector thanks to a type system. We may think to allow iterations on controlled unsafe values by a kind of declassification rule.

The above questions suggest the need to delve deeper into the question of determining the amount of information, which can be declassified while guarantying complexity. In regards to future research, the relationship between information, information flow control and complexity should be better understood.

References

1. Amadio, R.M., Dabrowski, F.: Feasible reactivity for synchronous cooperative threads. Electron. Notes Theor. Comput. Sci. 154(3), 33–43 (2006)
2. Amadio, R.M., Dabrowski, F.: Feasible reactivity in a synchronous pi-calculus. In: Proceedings of the 9th ACM SIGPLAN International Conference on Principles and Practice of Declarative Programming, PPDP 2007, pp. 221–230. ACM, New York (2007)
3. Biba, K.: Integrity considerations for secure computer systems. Technical report, Mitre corp Rep. (1977)
4. Buss, S.R.: The boolean formula value problem is in ALOGTIME. In: Proceedings of the Nineteenth Annual ACM Symposium on Theory of Computing, STOC 1987, pp. 123–131. ACM, New York (1987)
5. Chandra, A., Kozen, D., Stockmeyer, L.: Alternation. J. ACM 28(1), 114–133 (1981)
6. Cockett, R., Redmond, B.: A categorical setting for lower complexity. Electron. Notes Theor. Comput. Sci. 265, 277–300 (2010)
7. Gaboardi, M., Marion, J.-Y., Ronchi Della Rocca, S.: A logical account of PSPACE. In: POPL 2008, pp. 121–131. ACM (2008)
8. Jones, N.D.: The expressive power of higher-order types or, life without cons. J. Funct. Program. 11(1), 5–94 (2001)
9. Jones, N.D., Kristiansen, L.: A flow calculus of *wp*-bounds for complexity analysis. ACM Trans. Comput. Log. 10(4) (2009)
10. Ladner, R.E., Fischer, M.J.: Parallel prefix computation. J. ACM 27(4), 831–838 (1980)
11. Lafont, Y.: Soft linear logic and polynomial time. Theor. Comput. Sci. 318(1-2), 163–180 (2004)
12. Dal Lago, U., Di Giamberardino, P.: Soft session types. In: EXPRESS 2011. EPTCS, vol. 64, pp. 59–73 (2011)
13. Dal Lago, U., Martini, S., Sangiorgi, D.: Light logics and higher-order processes. In: EXPRESS 2010. EPTCS, vol. 41, pp. 46–60 (2010)

14. Leivant, D., Marion, J.-Y.: Ramified Recurrence and Computational Complexity II: Substitution and Poly-Space. In: Pacholski, L., Tiuryn, J. (eds.) CSL 1994. LNCS, vol. 933, pp. 486–500. Springer, Heidelberg (1995)
15. Leivant, D., Marion, J.-Y.: Predicative Functional Recurrence and Poly-Space. In: Bidoit, M., Dauchet, M. (eds.) TAPSOFT 1997. LNCS, vol. 1214, pp. 369–380. Springer, Heidelberg (1997)
16. Madet, A.: A polynomial time lambda-calculus with multithreading and side effects. In: PPDP (2012)
17. Marion, J.-Y.: A type system for complexity flow analysis. In: LICS, pp. 123–132 (2011)
18. Moyen, J.-Y.: Resource control graphs. ACM Trans. Comput. Logic 10(4), 29:1–29:44 (2009)
19. Niggl, K.-H., Wunderlich, H.: Certifying polynomial time and linear/polynomial space for imperative programs. SIAM J. Comput. 35(5), 1122–1147 (2006)
20. Sabelfeld, A., Myers, A.C.: Language-based information-flow security. IEEE J. Selected Areas in Communications 21(1), 5–19 (2003)
21. Savitch, W.J.: Relationship between nondeterministic and deterministic tape classes. JCSS 4, 177–192 (1970)
22. Smith, G., Volpano, D.: Secure information flow in a multi-threaded imperative language. In: POPL, pp. 355–364. ACM (1998)
23. Volpano, D., Irvine, C., Smith, G.: A sound type system for secure flow analysis. Journal of Computer Security 4(2/3), 167–188 (1996)

Pure Pointer Programs and Tree Isomorphism*

Martin Hofmann, Ramyaa Ramyaa, and Ulrich Schöpp

Ludwig-Maximilians Universität München
Oettingenstraße 67, 80538 Munich, Germany
{mhofmann,ramyaa,schoepp}@tcs.ifi.lmu.de

Abstract. In a previous work, Hofmann and Schöpp have introduced the programming language PURPLE to formalise the common intuition of LOGSPACE-algorithms as pure pointer programs that take as input some structured data (e.g. a graph) and store in memory only a constant number of pointers to the input (e.g. to the graph nodes). It was shown that PURPLE is strictly contained in LOGSPACE, being unable to decide *st*-connectivity in undirected graphs.

In this paper we study the options of strengthening PURPLE as a manageable idealisation of computation with logarithmic space that may be used to give some evidence that PTIME-problems such as Horn satisfiability cannot be solved in logarithmic space.

We show that with counting, PURPLE captures all of LOGSPACE on locally ordered graphs. Our main result is that without a local ordering, even with counting and nondeterminism, PURPLE cannot solve tree isomorphism. This generalises the same result for Transitive Closure Logic with counting, to a formalism that can iterate over the input structure, furnishing a new proof as a by-product.

1 Introduction

In a previous work we have introduced the Pure Pointer Language PURPLE [9], which captures the intuitive idea of accessing read only graph-like input data via pointers as an abstract datatype. In addition to the usual control structures PURPLE provides a way of iterating over all nodes of the input graph in an unspecified order. PURPLE programs can be evaluated in LOGSPACE, but are strictly weaker, for example because counting cannot be defined [9]. One can thus hope to prove strict separations between PURPLE and PTIME, which could then be seen as additional evidence for the assumed inequality LOGSPACE $\neq$ PTIME. We have already shown that PURPLE strictly subsumes deterministic transitive closure (DTC) logic with local ordering and that it cannot in general decide connectivity of undirected (locally ordered) graphs [8]. As a consequence, the same follows for DTC with local ordering, which was open until then. This trivially shows that PURPLE programs cannot solve PTIME-complete problems, such as Horn satisfiability, because reachability is an instance thereof.

Unfortunately, such a result cannot be seen as evidence for LOGSPACE $\neq$ PTIME because reachability in undirected graphs is in LOGSPACE by Reingold's theorem. Thus, the way forward would consist of strengthening PURPLE so as to still remain within LOGSPACE or at least POLY-LOGSPACE, yet in such a way that reachability does become

* This work was supported by Deutsche Forschungsgemeinschaft (DFG) under grant PURPLE.

F. Pfenning (Ed.): FOSSACS 2013, LNCS 7794, pp. 321–336, 2013.

definable, whereas Horn satisfiability demonstrably is not. While we are not yet in a position to offer such an extension, we report in this paper considerable progress with pinning down its possible form. A reasonable attempt borrowed from finite model theory consists of adding arithmetic variables ranging over the size of the input ("counting"). We show, however, that with a local-ordering (edges of input graph navigable bidirectionally) such an extension captures all of LOGSPACE. If the edges of the input graph can be followed only in one direction (an analogue of "1LO" in finite model theory) then counting, even in the presence of nondeterminism, does not capture LOGSPACE, because the tree isomorphism problem, which is in LOGSPACE, cannot be solved. This constitutes the main technical result of this paper. It implies a similar result for TC logic, which was known, but proved in a completely different fashion. Our method is based on bisimulation, while the proof for TC logic uses Ehrenfeucht-Fraïssé (EF) games [5].

The iteration construct of PURPLE subsumes first-order quantification but seems to be stronger. Despite some effort on our side it was not possible to extend the EF-method to PURPLE. The method of bisimulation, on the other hand, did do the trick.

While we believe that PURPLE with nondeterminism and counting is strictly stronger than TC logic with counting, it is difficult to come up with an explicit example, for this would have to be between LOGSPACE and TC logic with counting, but tree isomorphism is the only known such example.

We can, however, notice that PURPLE is able to traverse the nodes of a tree in some (non reproducible) breadth-first ordering whereas TC apparently can not. Whether this results in a clean separation we do not presently know, but we may note that, if we replace breadth-first with depth-first then, being able to traverse *is* strictly stronger than being able to quantify which means that PURPLE-style traversal is not a mere reformulation of quantification: If we assume that we can traverse the nodes of a tree in depth first order for some (not necessarily reproducible order between siblings) then we can implement Lindell's algorithm and thus solve tree isomorphism, and this is in fact what happens in [7]. On the other hand, TC logic is able to, for example, count the number of nodes that lie between two given nodes in *any* depth-first ordering.

Related Work. The first attempt at a "relativised separation" by permitting access to input data only via an abstract interface are Cook & Rackoff's results on jumping automata on graphs (JAGs) [3], which show that no finite automaton being able to place pebbles on a graph and moving these along edges can solve undirected st-connectivity. At that time this problem was probably believed to be not in LOGSPACE unless NLOGSPACE $=$ LOGSPACE, so that such result might have been seen as evidence for NLOGSPACE $\neq$ LOGSPACE. Reingold's LOGSPACE algorithm [15] for undirected st-connectivity has changed this picture and consequently it has been shown that it suffices to extend JAGs with counting in order to encode this algorithm [14]. In our own work on PURPLE we extended Cook & Rackoff's result by allowing such JAGs to visit every node of the input albeit in an unspecified order [9]. We [8] were able to show that st-connectivity remains unsolvable in such a framework thus solving an open question about the strength of transitive closure logic on locally ordered structures [6]. The latter is an alternative to the automata-theoretic approach to restricting the access to the input. There, and more generally in finite model theory [4,10], "programs" are formulas of some logic to be evaluated over the input represented as a logical structure.

When it comes to relativisation many people first think of oracles; however just as in the case of P vs. NP also the oracle-relativised versions of LOGSPACE vs. PTIME can go either way depending on the choice of oracle [12], which, as usual, shows that techniques such as diagonalization that work in the presence of oracles cannot possibly lead to a full separation of LOGSPACE and PTIME. Incidentally, we may remark here that the proof techniques we use for PURPLE, namely Cook and Rackoff's [3] radius bounds that we generalized in [8] and bisimulation as in this paper do *not* "relativize" in the classical sense. But, of course, we nevertheless do not claim that we achieve a full separation with them! Another approach from general complexity theory that should be mentioned is [2], which shows that the existence of P-complete sparse sets implies LOGSPACE = PTIME.

Finite model theory has a related but slightly different goal to ours. One also seeks to access the input, e.g. a graph, in an abstract fashion; yet the ultimate goal is to capture complexity classes, i.e. to define a logic in which despite the limited access all and only the properties decidable in a given complexity class are definable. In general, this is possible if a total ordering on the elements (nodes) of the input is available or definable.

Once a complexity class is "captured" by a formalism it becomes next to useless for relativised separation, for then the relativised separation is as hard as full unrelativised separation. This also applies to several alternative characterisations of LOGSPACE by programming language methods, such as [1,11]. As far as we know PURPLE is the first programming formalism for a proper, yet nontrivial subset of LOGSPACE which can therefore be used as a vehicle for relativised separation.

More specific to the technical contribution of this paper as opposed to PURPLE as a whole, we mention that the addition of arithmetical variables ranging over the size of the input structure called "counting" is a popular device for strengthening such logics and it came as a surprise that despite the obvious power to simulate log-sized work-tapes, counting still does not render (deterministic) transitive closure logic equivalent to (N)LOGSPACE. This is because despite Lindell's intriguing LOGSPACE algorithm [13], isomorphism of unordered trees has been shown not to be definable in transitive closure logic with counting [5]. Recently, this has led to the introduction of a new recursion principle allowing the implementation of Lindell's algorithm in logic [7].

2 PURPLE and Extensions

PURPLE programs are parameterised by a finite set L of labels and a finite set S of predicate symbols. Each predicate symbol p is assumed to have a finite arity $ar(p) \in \mathbb{N}$.

The input of a program is a pointer structure, which interprets the labels and predicates: A *pointer structure* on L and S ((L, S)-model, for short) specifies a finite set U as a universe, a function $[\![l]\!]\colon U \to U$ for each label $l \in L$ and a set $[\![p]\!] \subseteq U^{ar(p)}$ for each predicate symbol p.

Relational structures, such as graphs, are special cases of such pointer structures. For graphs one would use an empty set of labels and a single binary predicate $edge(x, y)$. In some cases, pointer structures are arguably more natural than relational structures, however. For example bounded degree graphs with a one-way local ordering (1LO), in which the edges emanating from each node are ordered, are represented naturally using one sort V and labels $\mathtt{succ}_i$, where i ranges from 1 to the degree of the graph. For a two-way local ordering (2LO), which also orders the incoming edges at each node, one may in addition use labels of the form $\mathtt{pred}_i$.

A program with labels L and predicate symbols S can access its input structure through the following terms for pointers to elements of the universe and for booleans.

$$t^U ::= x^U \mid t^U.l \text{ for any label } l \in L$$
$$t^{\mathtt{bool}} ::= x^{\mathtt{bool}} \mid \neg t^{\mathtt{bool}} \mid t_1^{\mathtt{bool}} \wedge t_2^{\mathtt{bool}} \mid t_1^U = t_2^U \mid p(x_1^U, \dots, x_{ar(p)}^U) \text{ for any } p \in S$$

We call x^U pointer variables. The intention is that $t^U.l$ is interpreted by $[\![l]\!](t^U)$.

The programs themselves are given by the grammar.

$$\begin{aligned} \mathit{Prg} ::= {} & \mathtt{skip} \mid \mathit{Prg}_1; \mathit{Prg}_2 \mid x^U := t^U \mid x^{\mathtt{bool}} := t^{\mathtt{bool}} \\ & \mid \mathtt{if}\ t^{\mathtt{bool}}\ \mathtt{then}\ \mathit{Prg}_1\ \mathtt{else}\ \mathit{Prg}_2 \mid \mathtt{forall}\ x^U\ \mathtt{do}\ \mathit{Prg} \end{aligned}$$

We do not include a `while`-loop, as it can be defined from the `forall`-loop, see [9]. We write $\mathtt{if}\ t^{\mathtt{bool}}\ \mathtt{then}\ \mathit{Prg}$ for $\mathtt{if}\ t^{\mathtt{bool}}\ \mathtt{then}\ \mathit{Prg}\ \mathtt{else}\ \mathtt{skip}$.

A *configuration* $\langle \rho, q \rangle$ consists of a *pebbling* ρ and a *state* q. The pebbling ρ maps pointer variables (which we also call pebbles) to elements of the universe U. The state q is a function from boolean variables to booleans. Given a configuration I, we define an interpretation of the terms $[\![t^{\mathtt{bool}}]\!]_I \in \{\mathtt{true}, \mathtt{false}\}$ and $[\![t^U]\!]_I \in U$ in the usual way.

A big-step reduction relation $\mathit{Prg} \vdash_M I \longrightarrow O$ between configurations I and O on some (L, S)-model M and a program Prg is defined inductively by the following clauses:

- $\mathtt{skip} \vdash_M I \longrightarrow I$.
- $\mathit{Prg}_1; \mathit{Prg}_2 \vdash_M I \longrightarrow O$ if $\mathit{Prg}_1 \vdash_M I \longrightarrow R$ and $\mathit{Prg}_2 \vdash_M R \longrightarrow O$ for some R.
- $x^U := t^U \vdash_M \langle \rho, q \rangle \longrightarrow \langle \rho[x \mapsto [\![t]\!]_{\langle \rho, q \rangle}], q \rangle$.
- $x^{\mathtt{bool}} := t^{\mathtt{bool}} \vdash_M \langle \rho, q \rangle \longrightarrow \langle \rho, q[x \mapsto [\![t]\!]_{\langle \rho, q \rangle}] \rangle$.
- $\mathtt{if}\ t\ \mathtt{then}\ \mathit{Prg}_1\ \mathtt{else}\ \mathit{Prg}_2 \vdash_M I \longrightarrow O$ if $[\![t]\!]_I = \mathtt{true}$ and $\mathit{Prg}_1 \vdash_M I \longrightarrow O$.
- $\mathtt{if}\ t\ \mathtt{then}\ \mathit{Prg}_1\ \mathtt{else}\ \mathit{Prg}_2 \vdash_M I \longrightarrow O$ if $[\![t]\!]_I = \mathtt{false}$ and $\mathit{Prg}_2 \vdash_M I \longrightarrow O$.
- $\mathtt{forall}\ x^U\ \mathtt{do}\ \mathit{Prg}_1 \vdash_M I \longrightarrow O$ if there exists an enumeration $u_1, u_2, \dots, u_n$ of $[\![U]\!]$ and configurations $I = \langle \rho_1, q_1 \rangle, \langle \rho_2, q_2 \rangle, \dots, \langle \rho_{n+1}, q_{n+1} \rangle = O$, such that $\mathit{Prg}_1 \vdash_M \langle \rho_k[x \mapsto u_k], q_k \rangle \longrightarrow \langle \rho_{k+1}, q_{k+1} \rangle$ holds for all $k \in \{1, \dots, n\}$.

When the model M is clear from the context we may omit the subscript.

In order for a PURPLE program to accept (resp. reject) an input, it must do so no matter what enumerations are chosen for its `forall`-loops. This is defined formally in the next definition, in which we use a boolean variable *result* to indicate acceptance.

Definition 1. *A program Prg* accepts *(resp. rejects) an (L, S)-model M if $\mathit{Prg} \vdash_M \langle \rho, q \rangle \longrightarrow \langle \rho', q' \rangle$ implies $q'(\mathit{result}) = \mathtt{true}$ (resp. $q'(\mathit{result}) = \mathtt{false}$) for all ρ, ρ', q and q'.*

A program *Prg recognises* a set X of (L, S)-models if it accepts any model in X and rejects all others. Note that a program may neither accept nor reject its input, namely if for some runs it returns `true` and for others it returns `false`. To put it simply, a PURPLE program for some problem X should give the correct answer, be it `true` or `false` for any given input and independent of the run, i.e. the traversal sequences chosen.

We also note that predicate symbols, which were not part of the original definition of PURPLE [9], are there just for notational convenience and do not add expressive power. Unary predicates can be modelled with an extra pointer that points to designated nodes for "true" and "false". A binary relation can be modelled by introducing an extra node for each pair of related nodes with pointers *fst* and *snd* pointing to the latter two nodes. One uses a unary predicate to differentiate between actual nodes and these helper nodes.

2.1 Counting

PURPLE cannot solve counting problems, as shown in [9]. So, one obvious addition to PURPLE is counting. Here we extend PURPLE by counting variables ("counters"), each of which can hold a number from 0 to the size of the input structure's universe, and we extend the terms with arithmetic operations:

$$t^{\texttt{bool}} ::= \cdots \mid \mathit{iszero}(t^{\texttt{count}}) \qquad t^{\texttt{count}} ::= x^{\texttt{count}} \mid \mathit{max} \mid \mathit{pred}(t^{\texttt{count}})$$

We extend the operational semantics such that the state q now not only maps boolean variables to booleans, but also counting variables to numbers.

PURPLE with counters (PURPLE_c) can do arithmetic: the complement of a counter can be computed using *max* and repeated decrement; the increment can be implemented using double complementation and decrement; The rest of the operations follow by repeated applications of these. PURPLE_c can count the number of tuples of nodes satisfying any PURPLE-definable property, and so can simulate counting quantifiers used in (D)TC logics.

Lemma 1. PURPLE_c *captures* LOGSPACE *on graphs with a two-way local ordering, represented as pointer structures as described above.*

Proof (Outline). PURPLE captures all of LOGSPACE on ordered graphs. So, it suffices to show that a total ordering can be defined on any input graph.

To this end we note that given any graph node n, PURPLE_c can define a total ordering of the weakly connected component containing n. We use Reingold's algorithm for undirected st-connectivity, which checks for connectivity by enumerating all nodes in the weakly connected component of s and checking if t appears therein. This algorithm can be implemented by RAMJAGs [14], and so, by an easy translation, also in PURPLE_c. In this way, PURPLE_c can order the nodes of the weakly connected component according to their order of their first appearance in the enumeration.

In their proof that TC-logic with counting captures NLOGSPACE on graphs with a two-way local ordering [5], Etessami and Immerman have shown how a total ordering can be defined using counting from such orderings of the weakly connected components. This argument can be adapted to PURPLE_c to complete the proof.

2.2 Nondeterminism

Another way to extend the power of PURPLE is to add nondeterminism. This gives PURPLE the power to decide st-connectivity (reachability) even for directed graphs and in the presence of counting also the power of deciding non-reachability by Immerman-Szelepcsenyi. PURPLE-programs with non-determinism can be evaluated in NLOGSPACE but (probably) not LOGSPACE.

Adding nondeterminism is not completely straightforward, as we must separate nondeterministic choices from the choices made in the evaluation of `forall`-loops. We would like to allow programs to make nondeterministic choices, while still maintaining that their acceptance behaviour is independent of the enumerations chosen in the `forall`-loops.

PURPLE with nondeterminism (PURPLE_{nd}) has a command for nondeterministic choice:

$$\mathit{Prg} ::= \ldots \mid \texttt{choose}\ \mathit{Prg}_1\ \texttt{or}\ \mathit{Prg}_2$$

To define the semantics of PURPLE with nondeterminism, we amend the notion of configuration so that it now consists of a triple $\langle \rho, q, \sigma \rangle$, where ρ and q are a pebbling and a state as before and σ is an infinite list enumerations of the universe U. This new component σ specifies the runs of all future `forall` loops. Therefore, in the definition of $(\texttt{forall}\ x^U\ \texttt{do}\ \mathit{Prg}) \vdash \langle \rho, q, \sigma \rangle \longrightarrow \langle \rho', q', \sigma' \rangle$ we do not use an arbitrary enumeration of U, but we take the first enumeration $u_1, \ldots, u_n$ from σ. The subsequent computation uses the list $\mathit{tail}(\sigma)$ of the remaining enumerations from σ. That is, we require there to be configurations $\langle \rho, q, \mathit{tail}(\sigma) \rangle = \langle \rho_1, q_1, \sigma_1 \rangle, \ldots, \langle \rho_{n+1}, q_{n+1}, \sigma_{n+1} \rangle = \langle \rho', q', \sigma' \rangle$ such that $\mathit{Prg} \vdash \langle \rho_k[x \mapsto u_k], q_k, \sigma_k \rangle \longrightarrow \langle \rho_{k+1}, q_{k+1}, \sigma_{k+1} \rangle$ holds for all $k \in \{1, \ldots, n\}$. For the semantics of the new term, we stipulate $\texttt{choose}\ \mathit{Prg}_1\ \texttt{or}\ \mathit{Prg}_2 \vdash I \longrightarrow O$ if $\mathit{Prg}_1 \vdash I \longrightarrow O$ or $\mathit{Prg}_2 \vdash I \longrightarrow O$. In all other cases, σ is merely passed on. E.g. $x := t \vdash \langle \rho, q, \sigma \rangle \longrightarrow \langle \rho[x \mapsto [\![t]\!]_I], q, \sigma \rangle$.

With these provisos, we can make the role of the two kinds of choices precise and define when a nondeterministic program accepts an input:

Definition 2. *A nondeterministic program Prg* accepts *an (L, S)-model M if for all I there exists O with $\mathit{Prg} \vdash_M I \longrightarrow O$ and $O(\mathit{result}) = \texttt{true}$. It rejects M if for all I and for all O with $\mathit{Prg} \vdash_M I \longrightarrow O$ one has $O(\mathit{result}) = \texttt{false}$.*

Thus, in the positive case, M must find, for all traversals of the `forall`-loops, appropriate nondeterministic choices leading to result `true`. In the negative case, however, the program must yield result `false` no matter how the nondeterministic choices are made and how the `forall`-loops are being traversed.

Note that for programs without `choose`, this definition agrees with the one for PURPLE above. For programs without `forall`-loop it agrees with the standard definition of nondeterminism.

In [9] we have shown that PURPLE can evaluate the formulae of DTC-logic. One may expect that with nondeterminism, this result generalises to TC-logic. Indeed, the TC-operator itself can be evaluated by nondeterministic PURPLE much like the DTC-operator by deterministic PURPLE. However, with nondeterminism it becomes harder to evaluate negations of TC-formulae. To be able to do so, we add counting as well, so that we can implement Immerman-Szelepcsenyi's algorithm for complementation in NL. We refer to PURPLE with counting and nondeterminism by $\text{PURPLE}_{c,nd}$.

We obtain the following proposition. Recall that any relational structure M can be understood as a pointer structure.

Lemma 2. *For each closed TC-formula φ on a relational signature Σ, there exists a program P_φ in PURPLE$_{c,nd}$ such that, for any Σ-structure of Σ, $M \models \varphi$ holds if and only if P_φ recognises M.*

3 Preliminaries

We first define AB-trees, a family of leaf-coloured trees. (These are modified versions of the trees presented in [5]). AB-trees are defined without any ordering, in particular there is no ordering on the children of any node. For each k, there are exactly two AB-trees A_{2k} and B_{2k} of height $2k$, and exactly three AB-trees C_{2k+1}, D_{2k+1} and E_{2k+1} of height $2k+1$. We define these trees by induction on k.

- The tree A_0 is a leaf with colour A and B_0 is a leaf with colour B.
- For any k, the tree C_{2k+1} (resp. E_{2k+1}) has six immediate subtrees: four of A_{2k} (resp. B_{2k}) and two of B_{2k} (resp. A_{2k}).
- For any k, the tree D_{2k+1} has six immediate subtrees: three each of A_{2k} and B_{2k}.
- For $k > 0$, the tree A_{2k} has eight immediate subtrees: three each of E_{2k-1} and C_{2k-1}, and two of D_{2k-1}.
- For $k > 0$, the tree B_{2k} has eight immediate subtrees: two each of E_{2k-1} and C_{2k-1}, and four of D_{2k-1}.

While an AB-tree only comes with a colouring of its leaves, we can consider it as a coloured graph in which each node is assigned a colour from the set $\{A,B,C,D,E\}$. Write $ht(n)$ for the height of a node n, measured from leaf nodes, which have height 0. A node n has colour A, B, C, D or E respectively if the subtree rooted at this node is $A_{ht(n)}$, $B_{ht(n)}$, $C_{ht(n)}$, $D_{ht(n)}$ or $E_{ht(n)}$ respectively. With this definition, each node n in an AB-tree has a uniquely defined colour; we denote it by $colour(n)$.

A *structural isomorphism* between two AB-trees is a bijection between the nodes of the two trees that preserves the tree structure, but that need not preserve colours. In contrast, a *colour isomorphism* must preserve both structure and colours. For any $X,Y \in \{A,B,C,D,E\}$ and any k the trees X_k and Y_k are structurally isomorphic. They are (colour) isomorphic iff $X = Y$. AB-trees have many isomorphic subtrees, and in particular, have the following properties.

Lemma 3. *For all nodes n, m with children $n_1,\dots,n_t$ and $m_1,\dots,m_t$:*
1. For any $i,j,k : \{1,\dots,t\}$ with $i \neq j$ there is a bijection $f : \{1,\dots,t\} \to \{1,\dots,t\}$ with $f(i) = k$ and $colour(n_j) = colour(m_{f(j)})$.
2. For any $i,j : \{1,\dots,t\}$, there is a bijection $f : \{1,\dots,t\} \to \{1,\dots,t\}$ with $colour(n_i) = colour(m_{f(i)})$ and $colour(n_j) = colour(m_{f(j)})$.
3. If n, m are of even height (i.e., of colour A or B), there is a colour preserving bijection from the grandchildren of n to the grandchildren of m.

An AB-tree T is presented to a PURPLE programs as a pointer structure over the set of nodes of T with the label *parent* (which maps $root(T)$ to itself and other nodes to their parents), and the predicate A_0 (true only for leaf nodes with colour A). The colours of internal nodes are not available to PURPLE programs. PURPLE can solve isomorphism of AB-trees if and only if there is a program that can determine, for inputs of any height, the colour of the root. The following example shows that PURPLE can determine the colour of internal nodes up to fixed height. W.l.o.g., the initial configuration of any program places its pebbles on the root of its input, and initializes all boolean variables to `false`.

Example: The following PURPLE program accepts tree C_1 but rejects D_1 and E_1. It uses boolean variables *seen1*, *seen2*, *seen3* and *result* and pointers *rt* and *chld*.

```
forall chld do
    if (chld.parent = rt) ∧ A_0(chld) then
        if seen3 then result := true;
        if seen2 then seen3 := true;
        if ¬seen1 then seen1 := true else seen2 := true
```

Using this repeatedly, for any height h, there is a PURPLE program Prg_h that determines the colour of nodes of height up to h. However, the number of variables in Prg_h increases with h. Our main result asserts that no program can do this for all heights.

Definition 3. *A PURPLE (PURPLE$_{c,nd}$) program Prg is* simple *if it does not have any Boolean variables and only contains compositions of the form $Prg_1; Prg_2$ where Prg_1 does not contain loops, and in every* `forall` p `do` *Prg_1, Prg_1 does not modify the variable p. Output is represented by the equality of dedicated pointers* res_1 *and* res_2.

Lemma 4. *For every PURPLE (PURPLE$_{c,nd}$) program Prg, there is a simple PURPLE (PURPLE$_{c,nd}$) program Prg' such that an input with more than two nodes is accepted (resp. rejected) by Prg' iff it is accepted (resp. rejected) by Prg.*

A pebbling I *fixes* a node n, if I places a pebble p on a node of the subtree rooted at n. Fixed nodes must be treated like pebbled ones, as PURPLE programs can navigate the path from pebbled nodes to the root.

Pebblings I, J with pebbles P *distance-match* ($I \sim J$) if the relative placement of pebbles is the same, i.e. $\forall p, q \in P.\ ht(I(p)) = ht(J(p))$ and $dist(I(p), I(q)) = dist(J(p), J(q))$, where, $dist(n, m) = \langle h - ht(n), h - ht(m) \rangle$ where h is the height of the closest common ancestor of n and m. If $I \sim J$ we define $match_{(I,J)}$ as a function from nodes fixed by I to nodes fixed by J such that $\forall p \in P.\ match_{(I,J)}(I(p)) = J(p)$ and $\forall n.\ (match_{(I,J)}(n) = m) \rightarrow (match_{(I,J)}(parent(n)) = parent(m))$. The pebblings I, J *colour-match till* h, written $I \sim_h J$, if $I \sim J$ and $colour(n) = colour(match_{(I,J)}(n))$ for any node n with $ht(n) \leq h$. Given pebblings I_1 and I_2 with pebbles P, we write $I_1 I_2$ for their disjoint combination, defined to be the pebbling with pebbles $\{1, 2\} \times P$, such that $I_1 I_2(\langle i, p \rangle) = I_i(p)$ for any $\langle i, p \rangle \in \{1, 2\} \times P$.

Lemma 5. *Given pebblings I_1, I_2, J_1, and J_2 with pebbles P, $I_1 I_2 \sim_h J_1 J_2$ iff $I_1 \sim_h J_1$ and $I_2 \sim_h J_2$ and $\forall p, q \in P.\ dist(I_1(p), I_2(q)) = dist(J_1(p), J_2(q))$.*

A structural isomorphism F *witnesses* $I \sim J$ if $J(r) = F(I(r))$ for every pebble r. We say that J is induced by F and I. Note that for any node n fixed by I, $match_{(I,J)}(n) = F(n)$. Given a node n of a tree T, pebblings I, J and a structural isomorphism F, we denote by $T|_n$ the subtree rooted at the node n; we denote by $I|_n$ and $F|_n$ the restrictions of I and F to $T|_n$ respectively. Likewise, given a height h, we write $I|^h$ and $F|^h$ for the restrictions of I and F to nodes above h respectively. Finally, given a height h, we write $F\|_h$ to mean that for any node n with height $\leq h$ and $colour(n) = colour(F(n))$, $F|_n$ is a colour isomorphism; analogously, we write $I\|_h J$ if for any nodes n, m with height $\leq h$ and $match_{(I,J)}(n) = m$, and $colour(m) = colour(n)$, we have $I|_n \sim_h J|_m$.

Lemma 6. *For all pebblings I, J with $I \sim_h J$, there is a structural isomorphism F with $F\|_h$ that witnesses it.*

4 Main Result

In this section we state and prove our main result – the impossibility of distinguishing between different tree colours at all heights. For simplicity and space reasons, we only give the proof for plain PURPLE as defined in Sec. 2; the extension with nondeterminism and counting is not given for space reasons; its structure and main invariants are identical to the proof presented in the main part.

Configurations of simple PURPLE program are just pebblings. We show that every simple PURPLE program that halts on input A_H in some pebbling O_A, will halt on input B_H (for some traversal) in some pebbling O_B such that $O_A(res_1) = O_A(res_2)$ iff $O_B(res_1) = O_B(res_2)$.

Theorem 1. *For any simple* PURPLE *program Prg there exists a height h with the following property: Whenever $h' \geq h$ and $H \geq h'$ (both h' and H even), then for any pebbling I_1, there exists a pebbling O_1 with $Prg \vdash_{A_H} I_1 \longrightarrow O_1$ such that whenever $I_1 \sim_{h'} I_2$ for some I_2, then we have*

- *$INV1(h)$: there is a pebbling O_2 with $I_1O_1 \sim I_2O_2$ and $Prg \vdash_{B_H} I_2 \longrightarrow O_2$;*
- *$INV2(h)$: for any pebbling O_2 with $I_1O_1 \sim_{h'} I_2O_2$ we have $Prg \vdash_{B_H} I_2 \longrightarrow O_2$.*

We take the properties $INV1(h)$ and $INV2(h)$ as defined by the statement of the theorem and denote their conjunction by $INV(h)$. The proof of the theorem is by induction on *Prg* and broken down into lemmas, one for each constructor, the most interesting being of course the forall loop. Before doing so, we note that this (together with Lemma 4) implies that PURPLE cannot solve tree isomorphism.

Corollary 1. *No* PURPLE *program can recognize the set $\{A_{2n} \mid n \in \mathbb{N}\}$.*

Proof. Assume, for a contradiction, that *Prg* recognizes $\{A_{2n} \mid n \in \mathbb{N}\}$. By Theorem 1 there exists an h such that *Prg* satisfies $INV(h)$. Choose even $h' \geq h$ and $H > h'$ and choose I_1 and I_2 to be the configurations on A_H and B_H respectively that place all pebbles of *Prg* on the roots. Clearly, we have $I_1 \sim_{h'} I_2$. Theorem 1 gives us O_1 with $Prg \vdash_{A_H} I_1 \longrightarrow O_1$. Since *Prg* accepts A_H, configuration O_1 must be accepting. Property $INV1(h)$ then furnishes O_2 with $I_1O_1 \sim I_2O_2$ and $Prg \vdash_{B_H} I_2 \longrightarrow O_2$. By $I_1O_1 \sim I_2O_2$, configuration O_2 must also be accepting. Hence, *Prg* does not reject B_H in the sense of Def. 1. Therefore it cannot recognize $\{A_{2n} \mid n \in \mathbb{N}\}$.

We now come to the inductive cases.

Lemma 7. *Let Prg be* `skip` *or an assignment. Then Prg satisfies INV(0).*

Lemma 8. *Let Prg be* `if` *c* `then` *Prg_1* `else` *Prg_2 where Prg_i satisfies $INV(h_i)$ for $i = 1, 2$. Then Prg satisfies $INV(\max(h_1, h_2))$.*

Proof. On pebblings I and J with $I \sim_h J$ $(h \geq 0)$, c evaluates identically since atomic conditions do. As this involves no pebble movements, the result follows.

Lemma 9. *Let Prg be $Prg_1; Prg_2$ where Prg_1 contains no loops and Prg_2 satisfies $INV(h)$. Then Prg satisfies $INV(h)$.*

Proof. On pebblings I_1, I_2 with $I_1 \sim_{h'} I_2$ $(h' \geq h)$, conditions evaluate identically (as in Lemma 8). Since Prg_1 can only move pebbles to already pebbled nodes, it halts with final pebblings J_1 and J_2 with $J_1 \sim_{h'} J_2$. The result follows since Prg_2 satisfies $INV(h)$.

Let *Prg* be `forall` r `do` Prg_1. Let $X = X_1, \ldots, X_{|A_H|}$ be the list of pebblings generated at the beginning of each iteration of *Prg*. Thus, $X_i(r) = E[i]$ for some enumeration E of A_H. Then, $\forall i < |A_H|.\ Prg_1 \vdash X_i \longrightarrow X_{i+1}[r \mapsto E[i]]$ and $Prg \vdash X_1 \longrightarrow X_{|A_H|+1}$ where $Prg_1 \vdash X_{|A_H|} \longrightarrow X_{|A_H|+1}$. Consider a list $Y = Y_1, \ldots, Y_{|B_H|+1}$ of pebblings of B_H such that $\forall i < |A_H|, X_i X_{i+1} \sim_h Y_i Y_{i+1}$. Then, assuming $INV2$ for Prg_1, $\forall i < |A_H| + 1.\ Prg_1 \vdash Y_i \longrightarrow Y_{i+1}$. Further, if $Y_i(r)$ gives an enumeration of B_H, then $Prg \vdash Y_1 \longrightarrow Y_{|B_H|+1}$. So, to prove the invariant for *Prg*, we need to construct such a list Y for some list X as above.

We first construct this matching list for a list X in which $X[i](r)$ is not necessarily an enumeration of the tree. The construction appears in Lemma 11 below. Its proof (not given for space reasons) relies on the following lemma, which shows that to colour match up to some height, it suffices to match what we call single children.

Definition 4. *A (non-root) node n is a* single child *at pebbling I, written in short as $singlechild(n, I)$, if I fixes n, but does not fix any of the siblings of n.*

Lemma 10. *For all pebblings I and J of trees of height H with pebbles P, if $(I \|_{H-|P|} J)$ and $\forall n.singlechild(n, I) \rightarrow colour(n) = colour(match_{(I,J)}(n))$, then $(I \sim_{H-|P|} J)$.*

Proof. Towards a contradiction, assume I fixes a node n of height $h \leq H - |P|$ and $colour(n) \neq colour(match_{I,J}(n))$. Since $(I \|_{H-|P|} J)$, for every ancestor m of n, $colour(m) \neq colour(match_{I,J}(m))$. So, I fixes some sibling of every ancestor of n (including n) i.e., I pebbles more than $|P|$ pairwise disjoint trees. □

The following lemma states that given a list of pebblings of A_H, it is possible to construct a list of pebblings over B_H such that successive pebbling pairs will colour match till desired height. The required list of pebblings is constructed recursively to colour match single children.

Lemma 11. *Let $X = X_0, \ldots, X_k$ be a list of pebblings of A_H with pebbles P, and Y_0 and Y_k be pebblings of B_H with $X_0 X_k \sim_h Y_0 Y_k$ witnessed by F. There exists a list of pebblings $Y_1, \ldots, Y_{k-1}$ such that for $0 \leq i < k$, $X_i X_{i+1} \sim_{h-|P|} Y_i Y_{i+1}$, and $Y_i|^h = F \circ X_i|^h$ and $Y_i|_n = F \circ X_i|_n$ for every node n at height h with $colour(n) = colour(F(n))$.*

To use these results for a program *Prg* defined as `forall` r `do` Prg_1, we need to extend the proof for a list of pebblings X where the list of nodes $X[i](r)$ is some enumeration of the A_H and construct a list Y where the list of nodes $Y[i](r)$ is some enumeration of the B_H. Since PURPLE programs behave the same for any enumeration, we choose for $X[i](r)$ an enumeration of A_H called "special run" (SR) such that there is an enumeration "match run" (MR) of B_H with $ht(SR[i]) = ht(MR[i])$ and $dist(SR[i], SR[i+1]) = dist(MR[i], MR[i+1])$ and $colour(SR[i]) = colour(MR[i])$ if $ht(SR[i]) \leq h$ (determined by Prg_1).

We define the enumeration of the tree A_H called the *special run* $SR_{H,h}$ parameterized by an even height h. We fix a function gc that maps the nodes of A_H to an enumeration of their grandchildren. Int is a list of sets of nodes of A_H such that

$$Int[0] = \{m \mid ht(m) \geq h-1\},$$
$$Int[k] = \{m \mid m \in A_H|_{gc(n)[k]} \text{ for some } n \text{ with } ht(n) = h\} \text{ for } 1 \leq k \leq 48.$$

Define *interval boundary* $IB_{H,h}(j) = \sum_{i=0}^{j-1} |Int[i]|$ for $0 \leq j \leq 49$. Then, for $0 \leq i \leq 48$ the sublist of $SR_{H,h}$ from $IB_{H,h}(i)+1$ till $IB_{H,h}(i+1)$ is called the ith interval of $SR_{H,h}$ and is defined to be an arbitrary enumeration of $Int[i]$.

Lemma 12. *For all even heights H and h, the jth interval of $SR_{H,h}$ ($1 \leq j < 49$) contains, for every node n with $ht(n) = h$, the subtree rooted at exactly one grandchild of n; the 0th interval contains nodes with height $\geq h-1$.*

Given a structural isomorphism $F : A_H \to B_H$, we define the enumeration *match run* $MR_{H,h,F}$ of B_H. For each n of A_H of even height, we fix some colour preserving bijection f_n from the grandchildren of n to the grandchildren of $F(n)$ and let GC_n be the union of some colour isomorphisms from $A_H|_{n'}$ to $B_H|_{f_n(n')}$ for each grandchild n' of n. (GC_n may not preserve distances.) With these we define the enumeration *match run* $MR_{H,h,F}$ of B_H by: if $ht(SR_{H,h}[i]) \geq h-1$, then $MR_{H,h,F}[i] = F(SR_{H,h}[i])$; otherwise let n be the ancestor of $SR_{H,h}[i]$ at height h. If $colour(n) = colour(F(n))$, then, $MR_{H,h,F}[i] = F(SR_{H,h}[i])$; else $MR_{H,h,F}[i] = GC_n(SR_{H,h}[i])$.

Lemma 13. *Let H and h be even heights, and $F : A_H \to B_H$ be a structural isomorphism with $F \| h$. Let IB, SR and MR be $IB_{H,h}$, $SR_{H,h}$ and $MR_{H,h,F}$ respectively.*

1. *for all $i < |A_H|$, we have $ht(SR[i]) = ht(MR[i])$ and $ht(SR[i]) \leq h-2 \to colour(SR[i]) = colour(MR[i])$;*
2. *if $0 \leq j \leq 49$ and $IB[j] \leq i < IB[j+1]$, then $dist(SR[i],\ SR[i+1]) = dist(MR[i], MR[i+1])$;*
3. *if n is the ancestor of $SR[IB[j]]$ at height h and $colour(n) = colour(F(n))$, then we have $dist(SR[IB[j]-1],\ SR[IB[j]]) = dist(MR[IB[j]-1], MR[IB[j]])$ for $0 < j \leq 49$.*

Since GC_n may not preserve distances, we needed the extra conditions at the interval boundaries to ensure MR is defined using F. Lemma 11 shows that given a list X of pebblings and an initial structural isomorphism F, we can construct a list Y whose corresponding elements colour-match with X, such that under some conditions, $Y[i](r) = F \circ X[i](r)$. When we add the constraint that $X[i] = SR(i)$, if we have $F(SR_{H,h}(i)) = MR_{H,h,G}(i)$ for some G, and the conditions required to ensure $Y[i](r) = F \circ X[i](r)$, the result follows. However, no structural (let alone colour) isomorphism satisfies $Y[i](r) = F \circ X[i](r)$. So, we divide X into 49 sublists, one for each interval of $SR_{H,h}$, such that there are 49 structural isomorphisms with the property that successive ones agree on the nodes fixed at the boundaries of the intervals, and $MR_{H,h,F_0}[i] = F_j(SR[i])$ for if i is the in j^{th} interval of SR.

Lemma 14. *Let H, h be even heights, P be a set of pebbles, $h' = h - 50|P|$, $IB = IB_{H,h'}$, $SR = SR_{H,h'}$, $|A_H| = N$. Let $X = X_1, \ldots, X_N$ be a list of pebblings with $X_i(r) = SR[i]$. Let Y_1, Y_N be pebblings of B_H with $X_1 X_{A_H} \sim_h Y_1 Y_N$. Then, there are structural isomorphisms F_j $(0 \leq j \leq 48)$ with*

1. *$MR_{H,h',F_0}[i] = F_j(SR[i])$ for $IB[j] \leq i \leq IB[j+1]$;*
2. *$F_0(n) = F_j(n)$ on all nodes n fixed by $X_{IB[j]}$ and $X_{IB[j+1]}$;*
3. *for all i, either $ht(SR[i]) > h' - 2$ or the ancestor n of $SR[i]$ at height $h' - 2$ satisfies $colour(n) = colour(F_j(n))$ where $IB[j] \leq i \leq IB[j+1]$.*

Proof. We first want to define an F_0 such that $F_0||_{h'}$ and $colour(n) = colour(F_0(n))$ for any node n at height h' fixed by $X_{IB[j]}$ for some j $(0 \leq j \leq 48)$. Define F_0 as follows: Let $S = \{IB[i] \mid 0 \leq i \leq 49\}$, and R_1, R', R_N be pebbling with pebbles $S \times P$, defined as $R_1(\langle i, p\rangle) = X_1(p)$, $R_N(\langle i, p\rangle) = X_N(p)$and $R'(\langle i, p\rangle) = X_i(p)$ for each $i \in S, p \in P$. Using Lemma 11 applied to the list $R_1 : R' : R_2$, Y_1 and Y_N with any structural isomorphism witnessing $X_1 X_{A_H} \sim_h Y_1 Y_{B_H}$ we can construct a Y' with $R_0 R' \sim_{h'} Y_0 Y'$. Then, there is a structural isomorphism F_0 witnessing this with the required properties.

Define F_j as follows: Take $F_j|^{h'} = F_0|^{h'}$, and $F_j|_n = F_0|_n$ for any node n of height h', with $colour(n) = colour(F_0(n))$ (this satisfies (1) for nodes below such n; Since F_0 is such that all nodes fixed by $X_{IB[j]}$ for $0 \leq j \leq 48$, (2) is satisfied). To satisfy (1) below any node n with height h' and $colour(n) \neq colour(F_0(n))$, define $F_j|_n$ as follows: By Lemma 12, there is at most one grandchild of n in j^{th} interval of $IB[j], \ldots, IB[j+1]$ of SR - say n'. So the only nodes that (1) enforces are in $A_H|_{n'}$. Set $F_j(n'') = MR[l'']$ for each $n'' \in A_H|_{n'}$ (including n')) where $SR[l''] = n''$. To preserve distances, set $F_j(parent(n')) = parent(F_j(n'))$. Elsewhere in $A_H|_n$, F_j is any structural isomorphism. Thus, $MR[i] = F_j(X_i(r))$ for $IB[j] < i \leq IB[j+1]$. (3) follows from the construction. □

The following lemma shows that given a list of pebblings of A_H, with pebble i enumerating it, these structural isomorphisms can be used to construct a list of pebblings B_H such that the successive pairs of pebblings colour match to the desired heights, and pebble i enumerates B_H.

Lemma 15. *Let H, h be even heights, P be a set of pebbles, $h' = h - 51|P|$, $IB = IB_{H,h'}$, $SR = SR_{H,h'}$, $|A_H| = N$. Let $X = X_1, \ldots, X_N$ be a list of pebblings with $X_i(r) = SR[i]$. Let Y_1, Y_N be pebblings of B_H such that $X_1 X_N \sim_h Y_1 Y_N$. Then there is a list of pebblings $Y = Y_1, \ldots, Y_N$ such that $Y_i \sim_{h'-|P|-2} X_i$ and a structural isomorphism F_0 such that $Y_i(r) = MR_{H,h',F_0}[i]$.*

Proof. By Lemma 14, for all $0 \leq j \leq 48$ there exists F_j with properties 1, 2 and 3 as in Lemma 14. Define $Y_{IB[j]} = F_0 \circ X_{IB[j]}$ for $0 \leq j \leq 49$. By Lemma 14(2), F_j agrees with F_0 on nodes fixed by pebblings $X_{IB[j]}$ and $X_{IB[j+1]}$, i.e., F_j witnesses $X_{IB[j]}X_{IB[j+1]} \sim_{h'} Y_{IB[j]}Y_{IB[j+1]}$. In order to make use of Lemma 14(3), we weaken this to $X_{IB[j]}X_{IB[j+1]} \sim_{h'-2} Y_{IB[j]}Y_{IB[j+1]}$. By Lemma 11, applied to $X_{IB[j]}, \ldots, X_{IB[j+1]}, Y_{IB[j]}$ and $Y_{IB[j+1]}$ with F_j witnessing $X_{IB[j]}X_{IB[j+1]} \sim_{h'-2} Y_{IB[j]}Y_{IB[j+1]}$, construct $Y_{IB[j]+1}, \ldots, Y_{IB[j+1]-1}$ such that for (a) $IB[j] \leq i \leq IB[j+1]$, $X_iX_{i+1} \sim_{h'-2-|P|} Y_iY_{i+1}$ and (b) $Y_i|^{h'} = F_j \circ X_i|^{h'}$ and (c) $Y_i|_n = F_j \circ X_i|_n$ for every node n at height h with $colour(n) = colour(F_j(n))$. By (b), (c) and 14 (3), $Y_i(r) = F_j \circ X_i(r)$ since $X_i(r) = SR[i]$. By Lemma 14 (1) we have $Y_i(r) = MR_{H,h',F_0}[i]$. □

Using this, the INV for the forall loop can be proved directly.

Lemma 16. *If* $Prg = \texttt{forall}\ r\ \texttt{do}\ Prg_1$ *and* Prg_1 *satisfies* $INV(h_1)$ *then* Prg *satisfies* $INV(h)$ *where* $h = h_1 + 52|P| + 2$ *and* P *is set of pointers in* Prg.

Proof. Consider an even height $h' > h$. Let $gph = h' - 51|P|$, $gch = gph - 2$, $h_1' = gch - |P|$, and $H \geq h'$ be even, $IB = IB_{H,gph}$, $SR = SR_{H,gph}$, and $|A_H| = N$.

INV1: Let pebblings I_1 of A_H and I_2 of B_H satisfy $I_1 \sim_{h'} I_2$. Let $X = X_1, \ldots, X_N$ be a list of pebblings with $X_i(r) = SR[i]$, $X_1 = I_1[r \mapsto SR[1]]$, and for all $1 \leq i < N$, $Prg_1 \vdash X_i \longrightarrow X_{i+1}[r \mapsto SR[i]]$ such that for any pebblings J_1 and J_2 with $X_iX_{i+1}[r \mapsto SR[i]] \sim_{h_1'} J_1J_2$ we have $Prg_1 \vdash J_1 \longrightarrow J_2$. (Since Prg_1 does not modify r, by IH ($INV2$), such an X exists). By Lemma 15 with $Y_1 = I_2$ and $Y_N = F \circ (X_N)$ for some F that witnesses $I_1 \sim_{h'} I_2$, we have a list Y with $Prg_1 \vdash Y_i \longrightarrow Y_{i+1}[r \mapsto MR[i]]$ for all $i \leq N$. Applying IH again to get $Prg_1 \vdash X_N \longrightarrow O_1$ and $Prg_1 \vdash Y_N \longrightarrow O_2$, the result follows.

INV2: Let pebblings I_1, O_1 of A_H and I_2, O_2 of B_H satisfy $I_1O_1 \sim_{h'} I_2O_2$. Let X be as above. By Lemma 11 applied to $I_1 : X_N : O_1$ and I_2 and O_2, construct Y_N.

By Lemma 15 construct list Y with $Prg_1 \vdash Y_i \longrightarrow Y_{i+1}[r \mapsto MR[i]]$ for all $i \leq N$. So the result follows. □

The proof of Theorem 1 is now a straightforward induction on program structure.

The key to generalizing our main result for nondeterminism and counting lies in the correct formulation of the invariants, in particular the quantification for runs and nondeterministic choices:

Theorem 2. *For any simple* PURPLE$_{c,nd}$ *program* Prg *there exists a height* h *with the following property: Whenever* $h' \geq h$ *and* $H > h'$ *(both* h' *and* H *even), then for any pebblings* $\rho_{i1} \sim_{h'} \rho_{i2}$ *and any state* q*, there exists* σ_1 *(pre-selected traversal sequences) such that* $Prg \vdash_{A_H} \langle \rho_{i1}, q, \sigma_1 \rangle \longrightarrow \langle \rho_{o1}, q', \sigma_1' \rangle$ *implies*

- $INV1(h)$ *: there is a pebbling* ρ_{o2} *with* $\rho_{i1}\rho_{o1} \sim \rho_{i2}\rho_{o2}$ *and* σ_2 *and* σ_2' *such that* $Prg \vdash_{B_H} \langle \rho_{i2}, q, \sigma_2 \rangle \longrightarrow \langle \rho_{o2}, q', \sigma_2' \rangle$*;*
- $INV2(h)$ *: for any pebbling* ρ_{o2} *with* $\rho_{i1}\rho_{o1} \sim_{h'} \rho_{i2}\rho_{o2}$ *there exist* σ_2 *and* σ_2' *with* $Prg \vdash_{B_H} \langle \rho_{i2}, q, \sigma_2 \rangle \longrightarrow \langle \rho_{o2}, q', \sigma_2' \rangle$*.*

The proof follows the proof for PURPLE almost verbatim.

Corollary 2. *No* PURPLE$_{c,nd}$ *program can recognize the set* $\{A_{2n} \mid n \in \mathbb{N}\}$.

Proof. Assume, for a contradiction, that *Prg* recognizes $\{A_{2n} \mid n \in \mathbb{N}\}$. Choose h such that *Prg* satisfies *INV*(h). Let $H > h$ be even and choose ρ_{i1} and ρ_{i2} to be the pebblings on A_H and B_H respectively that place all pebbles of *Prg* on the roots. Let q be the state that maps all counters to 0. Clearly, we have $\rho_{i1} \sim_{h'} \rho_{i2}$. Let σ_1 be the pre-selected traversal sequences, as provided by Theorem 2. Since *Prg* accepts A_H, we must have $Prg \vdash_{A_H} \langle \rho_{i1}, q, \sigma_1 \rangle \longrightarrow \langle \rho_{o1}, q', \sigma_1' \rangle$ for some accepting configuration $\langle \rho_{o1}, q', \sigma_1' \rangle$. *INV1*$(h)$: then furnishes ρ_{o2} and σ_2 and σ_2' with $\rho_{i1}\rho_{o1} \sim \rho_{i2}\rho_{o2}$ and $Prg \vdash_{B_H} \langle \rho_{i2}, q, \sigma_2 \rangle \longrightarrow \langle \rho_{o2}, q', \sigma_2' \rangle$. As $\langle \rho_{o2}, q', \sigma_2' \rangle$ must be accepting by $\rho_{i1}\rho_{o1} \sim \rho_{i2}\rho_{o2}$, program *Prg* does not reject B_H. Therefore *Prg* cannot recognize $\{A_{2n} \mid n \in \mathbb{N}\}$.

5 Horn Satisfiability

For any AB-tree T, PURPLE can construct a Horn satisfiability problem S_T that is unsatisfiable iff T is A_H or C_H: The problem S_T has one variable X_n for each node n in T. The goal clause is $\neg X_{root(T)}$. For each node n with $A_0(n)$ there is a clause X_n making X_n a fact. Finally, for each node n with 8 (resp. 6) children and any choice of $t = 3$ (resp. $t = 4$) pairwise distinct children $n_1, \ldots, n_t$ of n, there is a clause $X_n, \neg X_{n_1}, \ldots, \neg X_{n_t}$. As a result, S_T is satisfiable iff T is an B, D or an E tree. So, PURPLE cannot solve Horn satisfiability problems presented as sets of clauses.

6 Conclusion and Future Work

We have introduced an extension of PURPLE (imperative language with statically allocated pointers and iteration) with non-determinism and counting. Our main result shows that even in this extension isomorphism of unordered trees cannot be decided even though this problem is in LOGSPACE. This generalises an analogous result for transitive closure logic with counting [5]. A completely new proof method based on bisimulation was necessary to prove this result. This then furnishes a new proof of the result in [5], but more importantly sheds new light on the strength and weakness of extensions to PURPLE.

This whole line of work is motivated by the quest for a meaningful rigorization of the intuition that a constant number of pointers cannot be sufficient to check satisfiability of a set of Horn clauses. It seems so obvious that a non-constant number of "already proven" facts must be kept in memory no matter which algorithm is used. On the other hand, since Horn satisfiability is complete for PTIME and only a "constant number of pointers" can be done in LOGSPACE a comprehensive proof of such a statement would separate LOGSPACE from PTIME. Thus, until the techniques become strong enough to prove such a result we seek meaningful relativisations of such a separation. To this end we introduced the programming language PURPLE that provides the ability to iterate over all input elements in an unspecified order. We now seek extensions of this formalism that while remaining within LOGSPACE enhance the power so as to get more expressive relativised separations.

The work reported here narrows down the design space for such an extension considerably. We have proved that PURPLE extended with nondeterminism and counting cannot decide isomorphism of trees where the children of a node are unordered.

For the above research programme this has the important implication that an encoding of Horn satisfiability where clauses and variables are unordered, i.e., we merely have a relation that tells whether a variable appears as a premise or conclusion of a Horn clause, is not appropriate for a meaningful relativised separation because tree isomorphism can be—within PURPLE—reduced to this version of Horn satisfiability. So, Horn satisfiability cannot be decided in PURPLE augmented with nondeterminism and counting either, but for the very reason that a particular LOGSPACE problem, namely tree isomorphism cannot.

This narrowing of the design space leaves the following two options to be investigated: Further extend PURPLE, for instance with the LREC-recursor by Grohe et al. [7] and try to prove that unordered Horn satisfiability cannot be decided in that system. While this would be a very interesting result, we feel that this path does not really get to the essence of the issue because it is too dependent on the precise presentation of the input structure. Our preferred solution is to remove counting, but merely provide nondeterminism and possible restricted patterns of recursion. In view of Prop. 1 this would then allow us to present instances of Horn satisfiability with as much local ordering as desired. E.g. by ordering the premises of any clause and giving for each variable a list of clauses which conclude with that variable. Unordered tree isomorphism can then no longer be encoded as Horn satisfiability.

Thus, in summary, the results of this paper give us robust evidence that in order to stay clear of full LOGSPACE while at the same time not being trivially weak one should stick to a constant number of variables be they pointer or boolean variables.

References

1. Bonfante, G.: Some Programming Languages for LOGSPACE and PTIME. In: Johnson, M., Vene, V. (eds.) AMAST 2006. LNCS, vol. 4019, pp. 66–80. Springer, Heidelberg (2006)
2. Cai, J.-Y., Sivakumar, D.: Sparse hard sets for P: Resolution of a conjecture of Hartmanis. J. Comput. Syst. Sci. 58(2), 280–296 (1999)
3. Cook, S.A., Rackoff, C.: Space lower bounds for maze threadability on restricted machines. SIAM J. Comput. 9(3), 636–652 (1980)
4. Ebbinghaus, H.-D., Flum, J.: Finite model theory. Springer (1995)
5. Etessami, K., Immerman, N.: Tree canonization and transitive closure. In: IEEE Symp. Logic in Comput. Sci., pp. 331–341 (1995)
6. Grädel, E., McColm, G.L.: On the power of deterministic transitive closures. Inf. Comput. 119(1), 129–135 (1995)
7. Grohe, M., Grußien, B., Hernich, A., Laubner, B.: L-recursion and a new logic for logarithmic space. In: CSL, pp. 277–291 (2011)
8. Hofmann, M., Schöpp, U.: Pointer programs and undirected reachability. In: LICS, pp. 133–142 (2009)
9. Hofmann, M., Schöpp, U.: Pure pointer programs with iteration. ACM Trans. Comput. Log. 11(4) (2010)
10. Immerman, N.: Progress in descriptive complexity. In: Curr. Trends in Th. Comp. Sci., pp. 71–82 (2001)

11. Jones, N.D.: LOGSPACE and PTIME characterized by programming languages. Theor. Comput. Sci. 228(1-2), 151–174 (1999)
12. Richard, E.: Ladner and Nancy A. Lynch. Relativization of questions about log space computability. Mathematical Systems Theory 10, 19–32 (1976)
13. Lindell, S.: A logspace algorithm for tree canonization (extended abstract). In: STOC 1992, pp. 400–404. ACM, New York (1992)
14. Lu, P., Zhang, J., Poon, C.K., Cai, J.-Y.: Simulating Undirected *st*-Connectivity Algorithms on Uniform JAGs and NNJAGs. In: Deng, X., Du, D.-Z. (eds.) ISAAC 2005. LNCS, vol. 3827, pp. 767–776. Springer, Heidelberg (2005)
15. Reingold, O.: Undirected connectivity in log-space. J. ACM 55(4) (2008)

A Language for Differentiable Functions

Pietro Di Gianantonio[1] and Abbas Edalat[2]

[1] Dip. di Matematica e Informatica
Università di Udine, 33100 Udine, Italy
pietro.digianatonio@uniud.it
[2] Department of Computing, Imperial College London
ae@ic.ac.uk

Abstract. We introduce a typed lambda calculus in which real numbers, real functions, and in particular continuously differentiable and more generally Lipschitz functions can be defined. Given an expression representing a real-valued function of a real variable in this calculus, we are able to evaluate the expression on an argument but also evaluate the L-derivative of the expression on an argument. The language is an extension of PCF with a real number data-type but is equipped with primitives for min and weighted average to capture computable continuously differentiable or Lipschitz functions on real numbers. We present an operational semantics and a denotational semantics based on continuous Scott domains and several logical relations on these domains. We then prove an adequacy result for the two semantics. The denotational semantics also provides denotational semantics for Automatic Differentiation. We derive a definability result showing that for any computable Lipschitz function there is a closed term in the language whose evaluation on any real number coincides with the value of the function and whose derivative expression also evaluates on the argument to the value of the L-derivative of the function.

1 Introduction

Real-valued locally Lipschitz maps on finite dimensional Euclidean spaces enjoy a number of fundamental properties which make them the appropriate choice of functions in many different areas of applied mathematics and computation. They contain the class of continuously differentiable functions, are closed under composition and the absolute value, min and max operations, and contain the important class of piecewise polynomial functions, which are widely used in geometric modelling, approximation and interpolation and are supported in MatLab [4]. Lipschitz maps with uniformly bounded Lipschitz constants are closed under convergence with respect to the sup norm. Another fundamental property of these maps is that a Lipschitz vector field in $\mathbb{R}^n$ has a unique solution in the initial value problem [3].

In the past thirty years, motivated by applications in control theory and optimisation and using an infinitary double limit superior operation, the notion

F. Pfenning (Ed.): FOSSACS 2013, LNCS 7794, pp. 337–352, 2013.

of Clarke gradient has been developed as a convex and compact set-valued generalised derivative for real-valued locally Lipschitz maps [2]. For example, the absolute value function, which is not classically differentiable at zero, is a Lipschitz map which has Clarke gradient $[-1,1]$ at zero. The Clarke gradient extends the classical derivative for continuously differentiable functions.

Independently, a domain-theoretic Scott continuous Lipschitz derivative, later called the L-derivative, was introduced in [8] for interval-valued functions of an interval variable and was used to construct a domain for locally Lipschitz maps; these results were then extended to higher dimensions [9]. It was later shown that on finite dimensional Euclidean spaces the L-derivative actually coincides with the Clarke gradient [6]. In finite dimensions, therefore, the L-derivative provides a simple and finitary representation for the Clarke gradient.

Since the mid 1990's, a number of typed lambda calculi, namely extensions of PCF with a real number data type, have been constructed, including Real PCF, RL and LPR [11,5,19], which are essentially equivalent and in which computable continuous functions can be defined. Moreover, IC-Reals, a variant of LPR with seven digits, has been implemented with reasonable efficiency in *C* and Haskell [15].

It was relatively straightforward in [7] to equip Real PCF with the integral operator, which is in fact a continuous functional. However, adding a derivative operator to the language has proved to be non-trivial since classic differentiation may not be defined on continuous functions and even when defined it may not result in a continuous function. The development of the Scott continuous L-derivative, defined in a finitary manner, has therefore been essential for construction of a language with a derivative operator.

The aim of this work is to take the current extensions of PCF with a real number data type into a new category and define a typed lambda calculus, in which real numbers, real functions and in particular continuously differentiable and Lipschitz functions are definable objects. Given an expression e representing a function from real numbers to real numbers in this language, we would be able to evaluate both e and its L-derivative on an argument. In this paper we will only be concerned with the theoretical feasibility of such a language and not with questions of efficiency.

To develop such a language, we need to find a suitable replacement for the test for positiveness $(\mathsf{(0<)})$, which is used in the current extensions of PCF with real numbers to define functions by cases. In fact, a function defined using the conditional with this constructor will not be differentiable at zero even if the two outputs of the conditional are both differentiable: Suppose we have two real computable functions f and g whose derivatives $D\,f$ and $D\,g$ are also computable, and consider $l = \lambda x.\ \mathsf{if}\ (\mathsf{0}<)\ x\ \mathsf{then}\ f\,x\ \mathsf{else}\ g\,x$. The function l is computable and there is an effective way to obtain approximations of the value of $l(x)$ including at 0. However, there is no effective way to generate any approximation for the derivative of l, i.e., $D\,l$, at the point 0. In fact, it is correct to generate an approximation of Dl on 0 only if $f(0) = g(0)$, but this equality is undecidable, i.e., it cannot be established by observing the computation of f and g at 0 for any finite time.

In this paper, instead of the test $(0<)$, we will use the functions minimum, negation and weighted average when defining continuously differentiable or Lipschitz maps. These primitives are of course definable in Real PCF, RL and LPR, but the definitions are based on the test $(0<)$, which means that the information about the derivative is lost.

By a simple transfer of the origin and a rescaling of coordinates we can take the interval $[-1,1]$ as the domain of definition of Lipschitz maps. Furthermore, by a rescaling of the values of Lipschitz maps (i.e., multiplying them with the reciprocal of their Lipschitz constant) we can convert them to non-expansive maps, i.e., we can take their Lipschitz constant to be one. Concretely, we take digits similar to those in Real PCF as constructors and develop an operational semantics and a denotational semantics based on three logical relations, and prove an adequacy result. The denotational semantics for first order types is closely related but different from the domain constructed in [8] in that we capture approximations to the function part and to the derivative part regarded as a sublinear map on the tangent space. Finally, we prove a definability result and show that every computable Lipschitz map is definable in the language as the limit of a sequence of piecewise linear maps with the convergence of their L-derivatives.

We note that all our proofs can be found in the extended version of the present paper [13].

1.1 Related Work

Given a program to evaluate the values of a function defined in terms of a number of basic primitives, Automatic Differentiation (also called Algorithmic Differentiation) seeks to use the chain rule to compute the derivative of the function [14]. AD is distinct from symbolic differentiation and from numerical differentiation. Our work can be regarded as providing denotational semantics for *forward* Automatic Differentiation and can be used to extend AD to computation of the generalised derivative of Lipschitz functions.

In [10], the *differential* λ-*calculus*, and in [17], the perturbative λ-calculus that integrates the latter with AD, have been introduced which syntactically model the derivative operation on power series in a typed λ-calculus or a full linear logic. Although apparently similar, our calculus and these two λ-calculi differ in almost every aspect: motivation, syntax, semantics, and the class of definable real functions. (i) These λ-calculi have been presented to analyse linear substitution and formal differentiation, (ii) the syntax is quite structured and contains constructors that have no correspondence in our setting, (iii) the semantics is based on differential categories and not on domain theory, and (iv) the definable real functions are limited to analytical maps which have power series expansion.

On the other hand, Computable Analysis [20,21] and Constructive Analysis [1] are not directly concerned with computation of the derivative and both only deal with continuously differentiable functions. In fact, a computable real-valued function with a continuous derivative has a computable derivative if and only if the derivative has a recursive modulus of uniform continuity [16, p. 191], [20, p. 53], which is precisely the definition of a differentiable function in constructive mathematics [1, p. 44].

2 Syntax

We denote the new language with PCDF (Programming language for Computable and Differentiable Functions).

The types of PCDF are the types of a slightly modified version of PCF where natural numbers are replaced by integers, together with a new type ι, an expression e of type ι denotes a real number in the interval $[-1,1]$ or a partial approximation of a real number, represented by a closed intervals contained in $[-1,1]$. The set T of type expressions is defined by the grammar:

$$\sigma ::= o \mid \nu \mid \iota \mid \sigma \to \sigma$$

where o is the type of booleans and ν is the type of integer numbers.

The expressions of PCDF are the expressions of PCF together with a new set of constants for dealing with real numbers. This set of constants is composed by the following elements:

(i) A constructor for real numbers given by: $\mathsf{dig} : \nu \to \nu \to \nu \to \iota \to \iota$. It is used to build affine transformations, and real numbers are obtained by a limiting process. The expression $\mathsf{dig}\, l\, m\, n$ represents the affine transformation $\lambda x.(l+m\cdot x)/n$, if $0 \le m < n$ and $|l| \le n-m$, or the constant function $\lambda x.0$ otherwise. The above condition on l, m, n implies that $\mathsf{dig}\, l\, m\, n$ represents a rational affine transformation mapping the interval $[-1,1]$ strictly into itself with a non-negative slope or derivative $0 \le m/n < 1$.
In this way we use three integers to encode a rational affine transformation; of course it is possible to devise other encodings where just natural numbers or a single natural number is used, however these alternative encodings will be more complex.
The affine transformations definable by dig are also called generalised digits. Since there is no constant having type ι, an expression e having type ι can never normalise and its evaluation proceeds by producing expressions in the form $\mathsf{dig}\, l\, m\, n\, e'$. These expressions give partial information about the value represented by e, namely they state that e represents a real number contained in the interval $[(l+m)/n,\ (l-m)/n]$, which is the range of the function $\lambda x.(l+m\cdot x)/n$. During the reduction process, this interval is repeatedly refined and the exact result, a completely defined real number, can be obtained as the limit of this sequence.

(ii) The opposite sign function (negation) $\mathsf{opp} : \iota \to \iota$.

(iii) $\mathsf{add} : \nu \to \nu \to \iota \to \iota$, representing the function $\lambda p\, q\, x\,.\ \min((p/q + x), 1)$, if $0 < p < 2\cdot q$, and the constant function $\lambda x.\, 0$ otherwise.

We define $\mathsf{sub}\, p\, q\, x$ as syntactic sugar for the expression $\mathsf{opp}\,(\mathsf{add}\, p\, q\,(\mathsf{opp}\, x))$, which returns the value $\max((x - (p/q), -1))$.

(iv) A weighted average function $\mathsf{av} : \nu \to \nu \to \iota \to \iota \to \iota$. The expression $\mathsf{av}\, p\, q$ represents the function $\lambda x\, y\,.\,(p/q) \cdot x + (1 - p/q) \cdot y$, if $0 < p < q$, and the constant function $\lambda x\, y\,.\,0$ otherwise.

(v) The minimum function

$$\mathsf{min} : \iota \to \iota \to \iota$$

with the obvious action on pairs of real numbers. We define $\mathsf{max}\; x\, y$ as syntactic sugar for the expression $\mathsf{opp}\,(\mathsf{min}\,(\mathsf{opp}\, x)(\mathsf{opp}\, y))$.

(vi) A test function $(\mathsf{0}\!<) : \iota \to o$, which returns true if the argument is strictly greater than zero, and false if the argument is strictly smaller that zero. The test function can be used for constructing functions that are not differentiable, an example being the function $\lambda x.\ \mathsf{if}\ (\mathsf{0}\!<)\,(x)\ \mathsf{then}\ 1\ \mathsf{else}\ 0$; as a consequence we impose some restriction in its use.

(vii) The if-then-else constructor on reals, $\mathsf{if}_\iota : o \to \iota \to \iota \to \iota$, and the parallel if-then-else constructor $\mathsf{pif}_\iota : o \to \iota \to \iota \to \iota$.

The main use for the parallel if operator is to evaluate, without loss of information, derivative of expressions containing the min operator. However, the parallel if operator can be completely avoided in defining non-expansive functions on real numbers. In fact in the constructive proof of our definability result, the parallel if operator is never used.

(viii) A new binding operator D. The operator D can bind only variables of type ι and can be applied only to expressions of type ι. In our language, $\mathsf{D}x.e$ represents the derivative of the real function $\lambda x.e$.

The differential operator D can be applied only to expressions that contain neither the constant $(\mathsf{0}\!<)$ nor the differential operator D itself.

We note that, with the exception of the test functions $(\mathsf{0}\!<)$, all the new constants represent functions on reals that are non-expansive; the if-then-else constructors are also non-expansive if the distance between true (tt) and false (ff) is defined to be equal to two, while the test function $(\mathsf{0}\!<)$ cannot be non-expansive, whatever metric is defined on the Boolean values. The expressions containing neither the constant $(\mathsf{0}\!<)$ nor the differential operator D are called *non-expansive* since they denote functions on real numbers that are non-expansive. This fact, intuitively true, is formally proved by Proposition 2. The possibility to syntactically characterise a sufficiently rich set of expressions representing non-expansive functions is a key ingredient in our approach that allows us to obtain information about the derivative of a function expression without completely evaluating it. For example, from the fact that $e : \iota$ is a non-expansive expression, one can establish that the derivative of $\lambda x.e$, at any point, is contained in the interval $[-1, 1]$ and that the derivative of $\lambda x.\mathsf{dig}\, l\, m\, n\, e$ is contained in the smaller interval $[-m/n, m/n]$.

3 Operational Semantics

The operational semantics is given by a *small step reduction relation*, $\rightarrow$, which is obtained by adding to the PCF reduction rules the following set of extra rules for the new constants.

The operational semantics of add and min operators uses an extra constant $\mathsf{aff} : \nu \rightarrow \nu \rightarrow \nu \rightarrow \iota \rightarrow \iota$. The expression $\mathsf{aff}\, l\, m\, n$ is intended to represent general affine transformations (including expansive ones) with a non-negative derivative, i.e., the affine transformation $\lambda x.(l + mx)/n$ with $m \geq 0, n > 0$. A property preserved (i.e., invariant) by the reduction rules is that the constant aff appears only as the head of one of the arguments of min or as the head of the fourth argument of aff. It follows that in any expression e' in the reduction chain of a standard expression e (without the extra constants aff), the constant aff can appear only in the above positions.

The generalised digit $\mathsf{dig}\, l\, m\, n$ is a special case of an affine transformation. Therefore, in applying the reduction rules, we use the convention that any reduction rule containing, on the left hand side, a general affine transformation aff can be applied also to terms where the affine transformation aff is substituted by the constructor dig.

On affine transformations we will use the following notations:

- $(\mathsf{aff}\, l_1\, m_1\, n_1) \circ (\mathsf{aff}\, l_2\, m_2\, n_2)$ stands for $\mathsf{aff}\, (l_1 \cdot n_2 + m_1 \cdot l_2)\, (m_1 \cdot m_2)\, (n_1 \cdot n_2)$, i.e., the composition of affine transformations.
- If $m \neq 0$, $(\mathsf{aff}\, l\, m\, n)^{-1}$ stands for $(\mathsf{aff}\, (-l)\, n\, m)$, i.e., the inverse affine transformation; if $m = 0$, the expression $(\mathsf{aff}\, l\, m\, n)^{-1}$ is undefined.
- The symbols l, m, n, p, q stand for values of integer type.

The reduction rules arè the PCF reduction rules together the following set of extra rules. First we have three simple reductions:

- $\mathsf{dig}\, l\, m\, n\, e \rightarrow \mathsf{dig}\, 0\, 0\, 1\, e$ if $m < 0$ or $n \leq 0$ or $|l| > n - m$.
- $\mathsf{add}\, p\, q\, e \rightarrow \mathsf{dig}\, 0\, 0\, 1\, e$ if $p \leq 0$ or $q \leq p$.
- $\mathsf{av}\, p\, q\, e_1\, e_2 \rightarrow \mathsf{dig}\, 0\, 0\, 1\, e_1$ if $p \leq 0$ or $q \leq p$.

The above rules deal with those instances of dig, add, av with integer arguments that reduce to the constant zero digit. An implicit condition on the following set of rules is that they apply only if none of the above three rules can be applied.

1. $\mathsf{dig}\, l_1\, m_1\, n_1 (\mathsf{dig}\, l_2\, m_2\, n_2\, e) \rightarrow ((\mathsf{dig}\, l_1\, m_1\, n_1) \circ (\mathsf{dig}\, l_2\, m_2\, n_2))\, e$
2. $\mathsf{opp}\, (\mathsf{dig}\, l\, m\, n\, e) \rightarrow \mathsf{dig}\, (-l)\, m\, n\, (\mathsf{opp}\, e)$
3. $\mathsf{add}\, p\, q\, e \rightarrow \mathsf{min}\, (\mathsf{aff}\, p\, q\, q\, e) (\mathsf{dig}\, 1\, 0\, 1\, e)$
   note that $(\mathsf{aff}\, p\, q\, q)$ and $(\mathsf{dig}\, 1\, 0\, 1)$ represent the functions $\lambda x.p/q + x$ and $\lambda x.1$ respectively.
4. $\mathsf{av}\, p\, q\, (\mathsf{dig}\, l\, m\, n\, e_1)\, e_2 \rightarrow \mathsf{dig}\, l'\, m'\, n' (\mathsf{av}\, p'\, q'\, e_1\, e_2)$
   where $l' = l \cdot p$, $m' = q' = m \cdot p + n \cdot q - n \cdot p$, $n' = n \cdot q$ and $p' = m \cdot p$.
   By a straightforward calculation, one can check that the left and the right parts of the reduction rules represent the same affine transformation on the arguments e_1, e_2.

5. $\mathsf{av}\,p\,q\,e_1\,(\mathsf{dig}\,l\,m\,n\,e_2) \;\rightarrow\; \mathsf{dig}\,l'\,m'\,n'(\mathsf{av}\,p'\,q'\,e_1\,e_2)$
where $l' = l(q-p)$, $m' = q' = np + mq - mp$, $n' = nq$ and $p' = np$.
6. $\mathsf{min}\,(\mathsf{dig}\,l_1\,m_1\,n_1\,e_1)(\mathsf{aff}\,l_2\,m_2\,n_2\,e_2) \rightarrow \mathsf{dig}\,l_1\,m_1\,n_1\,e_1$
if $(l_1 + m_1)/n_1 \leq (l_2 - m_2)/n_2$.
The above condition states that every point in the image of $(\mathsf{dig}\,l_1\,m_1\,n_1)$ is smaller, in the usual Euclidean order, than every point in the image of $(\mathsf{aff}\,l_2\,m_2\,n_2)$, i.e., the first argument of min is certainly smaller that the second.
7. $\mathsf{min}\,(\mathsf{aff}\,l_1\,m_1\,n_1\,e_1)(\mathsf{dig}\,l_2\,m_2\,n_2\,e_2) \rightarrow \mathsf{dig}\,l_2\,m_2\,n_2\,e_2$
if $(l_2 + m_2)/n_2 \leq (l_1 - m_1)/n_1$
The symmetric version of the previous rule.
8. $\mathsf{min}\,(\mathsf{dig}\,l\,m\,n\,e_1)\,e_2 \rightarrow \mathsf{dig}\,l'\,m'\,n'$
$(\mathsf{min}\,((\mathsf{dig}\,l'\,m'\,n')^{-1} \circ (\mathsf{dig}\,l\,m\,n)\,e_1)\,((\mathsf{dig}\,l'\,m'\,n')^{-1}\,e_2))$
if $l + m < n$ and $l' = l + m - n$, $m' = l + m + n \neq 0$, $n' = 2 \cdot n$.
The above equations imply that if $(\mathsf{dig}\,l\,m\,n)$ has image $[a, b]$ then $(\mathsf{dig}\,l'\,m'\,n')$ has image $[-1, b]$. The rule is justified by the fact if the first argument of min are smaller than b then the value of min is also smaller than b.
9. $\mathsf{min}\,e_1\,(\mathsf{dig}\,l\,m\,n\,e_2) \rightarrow \mathsf{dig}\,l'\,m'\,n'$
$(\mathsf{min}\,((\mathsf{dig}\,l'\,m'\,n')^{-1}\,e_1)\,((\mathsf{dig}\,l'\,m'\,n')^{-1} \circ (\mathsf{dig}\,l\,m\,n)\,e_2))$
if $l + m < n$ and $l' = l + m - n$, $m' = l + m + n \neq 0$, $n' = 2 \circ n$.
The symmetric version of the previous rule.
10. $\mathsf{min}\,(\mathsf{aff}\,l_1\,m_1\,n_1\,e_1)(\mathsf{aff}\,l_2\,m_2\,n_2\,e_2) \rightarrow \mathsf{dig}\,l'\,m'\,n'$
$(\mathsf{min}\,((\mathsf{dig}\,l'\,m'\,n')^{-1} \circ (\mathsf{aff}\,l_1\,m_1\,n_1)\,e_1)((\mathsf{dig}\,l'\,m'\,n')^{-1} \circ (\mathsf{aff}\,l_2\,m_2\,n_2)\,e_2))$
if $-1 < (l_1 + m_1)/n_1 \leq (l_2 - m_2)/n_2$ and $l' = l_1 - m_1 + n_1$, $m' = m_1 - l_1 + m_1$, $n' = 2 \cdot n_1$.
The above equation implies that if $(\mathsf{dig}\,l_1\,m_1\,n_1)$ has image $[a, b]$ then $(\mathsf{dig}\,l'\,m'\,n')$ has image $[a, 1]$. The rule is justified by the fact if both arguments of min are greater that a then the value of min is also greater than a.
11. $\mathsf{min}\,(\mathsf{aff}\,l_1\,m_1\,n_1\,e_1)(\mathsf{aff}\,l_2\,m_2\,n_2\,e_2) \rightarrow \mathsf{dig}\,l'\,m'\,n'$
$(\mathsf{min}\,((\mathsf{dig}\,l'\,m'\,n')^{-1} \circ (\mathsf{aff}\,l_1\,m_1\,n_1)\,e_1)((\mathsf{dig}\,l'\,m'\,n')^{-1} \circ (\mathsf{aff}\,l_2\,m_2\,n_2)\,e_2))$
if $-1 < (l_2 - m_2)/n_2 < (l_1 + m_1)/n_1$, and $l' = l_2 - m_2 + n_2$, $m' = m_2 - l_2 + m_2$, $n' = 2 \cdot n_2$.
The symmetric version of the previous rule.
12. $\mathsf{aff}\,l_1\,m_1\,n_1(\mathsf{aff}\,l_2\,m_2\,n_2\,e) \rightarrow ((\mathsf{aff}\,l_1\,m_1\,n_1) \circ (\mathsf{aff}\,l_2\,m_2\,n_2))\,e$
13. $\mathsf{aff}\,l\,m\,n\,e \rightarrow \mathsf{dig}\,l\,m\,n\,e$
if $-1 \leq (l - m)/n$, $(l + m)/n \leq 1$ and e is not in the form $\mathsf{aff}\,a'b'e$.
14. $(0\mathsf{<})\,(\mathsf{dig}\,l\,m\,n\,e) \rightarrow \mathsf{tt}$ if $(l - m)/n > 0$
15. $(0\mathsf{<})\,(\mathsf{dig}\,l\,m\,n\,e) \rightarrow \mathsf{ff}$ if $(l - m)/n < 0$
16. $\mathsf{pif}\;e\;\mathsf{then}\;\mathsf{dig}\,l_1\,m_1\,n_1\,e_1\;\mathsf{else}\;\mathsf{dig}\,l_2\,m_2\,n_2\,e_2 \rightarrow$
$\mathsf{dig}\,l'\,m'\,n'(\mathsf{pif}\;e\;\mathsf{then}\;(\mathsf{dig}\,l'\,m'\,n')^{-1} \circ (\mathsf{dig}\,l_1\,m_1\,n_1)\,e_1$
$\mathsf{else}\;(\mathsf{dig}\,l'\,m'\,n')^{-1} \circ (\mathsf{dig}\,l_2\,m_2\,n_2)\,e_2)$
where $n' = 2 \cdot n_1 \cdot n_2$,
$m' = \max((l_1 + m_1) \cdot n_2, (l_2 + m_2) \cdot n_1) - \min((l_1 - m_1) \cdot n_2, (l_2 - m_2) \cdot n_1)$,
$l' = 2 \cdot \min((l_1 - m_1) \cdot n_2, (l_2 - m_2) + m'$.
Here the values l', m', n' are defined in such a way that if $(\mathsf{dig}\,l_1\,m_1\,n_1)$ has image $[a_1, b_1]$ and $(\mathsf{dig}\,l_2\,m_2\,n_2)$ has image $[a_2, b_2]$, then $(\mathsf{dig}\,l'\,m'\,n')$ has image the convex closure of the set $[a_1, b_1] \cup [a_2, b_2]$.

The remaining rules for pif, if are included in the reduction rules for PCF and therefore are omitted from the present list.

17. $\dfrac{N \to N'}{MN \to MN'}$ if M is a constant different from the combinator Y or is an expression in the form $\mathsf{min}\, n_1$, $\mathsf{min}\, n_1\, n_2$, $\mathsf{min}\, n_1\, n_2\, n_3$, $\mathsf{min}\, n_1\, n_2\, n_3\, M'$, $\mathsf{av}\, n_1$, $\mathsf{av}\, n_1\, n_2$, $\mathsf{av}\, n_1\, n_2\, M'$, $\mathsf{add}\, n_1$, $\mathsf{add}\, n_1\, n_2$, $\mathsf{pif}_\iota\ M'$ then, $\mathsf{pif}_\iota\ M'$ then M'' else, where n_1, n_2, n_3 are values.

The reduction rules for the derivative operator are:

1. $\mathsf{D}x.\, x \to \lambda y.\, \mathsf{dig}\, 0\, 0\, 1\, y$
2. $\mathsf{D}x.\, \mathsf{dig}\, l\, m\, n\, e \to \lambda y.\, \mathsf{dig}\, 0\, m\, n\, (\mathsf{D}x.\, e)y$
3. $\mathsf{D}x.\, \mathsf{opp}\, e \to \lambda y.\, \mathsf{opp}\, (\mathsf{D}x.\, e)y$
4. $\mathsf{D}x.\, \mathsf{add}\, p\, e\, q \to \lambda y.\, \mathsf{pif}_\iota\ (0{<})\, (\mathsf{add}\, (q{-}p)\, q\, (\mathsf{opp}\, e))\ \mathsf{then}\ (\mathsf{D}x.\, e)y\ \mathsf{else}\ \mathsf{dig}\, 0\, 0\, 1\, y$
5. $\mathsf{D}x.\, \mathsf{av}\, p\, q\, e_1\, e_2 \to \lambda y.\, \mathsf{av}\, p\, q\, ((\mathsf{D}x.\, e_1)y)\, ((\mathsf{D}x.\, e_2)y)$
6. $\mathsf{D}x.\, \mathsf{min}\ e_1 e_2 \to$
   $\lambda y.\, \mathsf{pif}\ (\lambda x.\, (0{<})\, (\mathsf{av}\, 1\, 2\, (\mathsf{opp}\, e_1) e_2))y\ \mathsf{then}\ (\mathsf{D}x.\, e_1)y\ \mathsf{else}\ (\mathsf{D}x.\, e_2)y$
7. $\mathsf{D}x.\, \mathsf{pif}_\iota\ e_1\ \mathsf{then}\ e_2\ \mathsf{else}\ e_3 \to\ \lambda y.\, \mathsf{pif}_\iota\ (\lambda x.\, e_1)y\ \mathsf{then}\ (\mathsf{D}x.\, e_1)y\ \mathsf{else}\ (\mathsf{D}x.\, e_2)y$
8. $\mathsf{D}x.\, \mathsf{if}\ e_1\ \mathsf{then}\ e_2\ \mathsf{else}\ e_3 \to\ \lambda y.\, \mathsf{if}\ (\lambda x.\, e_1)y\ \mathsf{then}\ (\mathsf{D}x.\, e_1)y\ \mathsf{else}\ (\mathsf{D}x.\, e_2)y$
9. $\mathsf{D}x.\, Y e \to \mathsf{D}x.\, e(Ye)$
10. $\mathsf{D}x.\, (\lambda y.\, e)e_1 \ldots e_n \to \mathsf{D}x.\, e[e_1/y]e_2 \ldots e_n$

Note that the rules for the derivative operator are a direct derivation of the usual rules for the symbolic computation of the derivative of a function.

3.1 Examples

We will give some examples for defining non-analytic functions in later sections; in particular we will show in the proof of definability how easily piecewise linear maps with rational coefficients are defined in the language. A useful technique to define analytic functions and real constants is to consider their Taylor series expansions and reduce the Taylor series to a sequence of applications of affine transformations. For example the value $e - 2$, where e is the Euler constant, is given by the Taylor series $1/2!+1/3!+1/4! \ldots$. Denoting the affine transformation $\lambda x.(1+x)/n$ as f, the above series can be expressed as $f(2)(f(3)(f(4)(\ldots)\ldots)$. It follows that in PCDF $e - 2$ can be expressed as

$$(Y\, \lambda f : \nu \to \iota.\, \lambda n : \nu.\, f\, n.\, \mathsf{dig}\, 1\, 1\, n\, (f(n+1)))\, 2.$$

Given an expression to represent product in PCDF, it is possible to use the above technique to express analytic functions. For example, suppose hp defines the half-product function $\lambda xy.\, x{\cdot}y/2$. Then, one can express the function $\lambda x.\, e^{x/2} - 1 - x/2$ by the PCDF.

$$\lambda x : \iota.\, \mathsf{hp}\, x\, ((Y \lambda f : \nu \to \iota \to \iota.\, \lambda n : \nu.\, \lambda x : \iota.\, \mathsf{hp}\, x\, (\mathsf{dig}\, 1\, 1\, n\, (f(n+1))))\, 2\, x)$$

The half product hp is definable in PCDF by reducing product to a series of applications of the average and minimum function. The actual definition of the function hp is lengthy and we will not present it here. As a simpler example of the technique involved, we present the definition of the function $\lambda x.x^2/2$.

Consider the following mutual recursive definition of the terms $g, h : \nu \to \iota \to \iota$

$g\,0\,x = \mathsf{max}\,x\,(\mathsf{opp}\,x)$

$h\,n\,x = \mathsf{add}\,1\,(2^{n+1})\,(g\,n\,(\mathsf{add}\,1\,(2^{n+1})\,x))$

$g\,(n+1)\,x = \mathsf{min}\,(h\,n\,x)(h\,n\,(\mathsf{opp}\,x))$

By standard techniques, one can derive a PCDF expression g satisfying the above recursive definition. The careful reader can check that the term:

$\lambda x.\,(\mathsf{sub}\,1\,2\,(\mathsf{av}\,1\,2\,(g\,0\,x)(\mathsf{av}\,1\,2\,(g\,1\,x)(\mathsf{av}\,1\,2\,(g\,2\,x)\ldots(\mathsf{av}\;1\,2\,(g\,n\,x)\,(\mathsf{dig}\,101x)\ldots)$

represents the step-wise linear interpolation of the function $\lambda x.x^2/2$ on the points of the set $\{i/2^n | i \in Z, -2^n \leq i \leq 2^n\}$. It follows that the function $\lambda x.x^2/2$ is defined by the term:

$\lambda x.\,(\mathsf{sub}\,1\,2((Y\lambda f : \nu \to \iota \to \iota.\,\lambda n : \nu.\,\lambda x : \iota.\,(\mathsf{av}\,1\,2\,(g\,n\,x)\,(f\,(n+1)\,x)))\,0\,x)$.

4 Denotational Semantics

The denotational semantics for PCDF is given in the standard way as a family of continuous Scott domains, $UD := \{\mathcal{D}_\sigma \mid \sigma \in T\}$. The basic types are interpreted using the standard flat domains of integers and booleans. The domain associated to real numbers is the product domain $\mathcal{D}_\iota = \mathcal{I} \times \mathcal{I}$, where $\mathcal{I}$ is the continuous Scott domain consisting of the non-empty compact subintervals of the interval $I = [-1, 1]$ partially ordered with reverse inclusion. Elements of $\mathcal{I}$ can represent either a real number x, i.e., the degenerated interval $[x, x]$, or some partial information about a real number x, i.e., an interval $[a, b]$, with $x \in [a, b]$. On the elements of $\mathcal{I}$, we consider both the set-theoretic operation of intersection ($\cap$), the pointwise extensions of the arithmetic operations, and the lattice operations on the domain information order ($\sqcap, \sqcup$), [11]. Function types have the usual interpretation of call-by-name programming languages: $\mathcal{D}_{\sigma\to\tau} = \mathcal{D}_\sigma \to \mathcal{D}_\tau$.

A hand waiving explanation for the definition of the domain $\mathcal{D}_\iota = \mathcal{I} \times \mathcal{I}$, is that the first component is used to define the value part of the function while the second component is used to define the derivative part. More precisely, a (non-expansive) function f from I to I, is described, in the domain, by the product of two functions $\langle f_1, f_2 \rangle : (\mathcal{I} \times \mathcal{I}) \to (\mathcal{I} \times \mathcal{I})$: the function $f_1 : (\mathcal{I} \times \mathcal{I}) \to \mathcal{I}$ represents the value part of f, in particular $f_1(i, j)$ is the image of the interval i under f for all intervals j, i.e., f_1 depends only on the first argument. The second function $f_2 : (\mathcal{I} \times \mathcal{I}) \to \mathcal{I}$ represents the derivative part. If $D\,f$ denotes the derivative of f, then $f_2(i, j)$ is the image of the intervals i and j under the function $\lambda x, y.D\,f\,(x) \cdot y$. Thus, f_2 is linear in its second component and $f_2(\{x\}, \{1\})$ is the derivative of f at the point x.

Note that with respect to the above interpretation, composition behaves correctly, that is if the pair $\langle f_1, f_2 \rangle : (\mathcal{I} \times \mathcal{I}) \to (\mathcal{I} \times \mathcal{I})$ describes the value part and the derivative part of a function $f : I \to I$ and $\langle g_1, g_2 \rangle : (\mathcal{I} \times \mathcal{I}) \to (\mathcal{I} \times \mathcal{I})$ describes a function $g : I \to I$ then $\langle h_1, h_2 \rangle$ describes, by the chain rule, the function $f \circ g$ with $h_1(i, j) = f_1(g_1(i, j), g_2(i, j))$ and $h_2(i, j) = f_2(g_1(i, j), g_2(i, j))$.

The L-derivative of the non-expansive map $f : I \to I$ is the Scott continuous function $\mathcal{L}(f) : I \to \mathcal{I}$ defined by [6]:

$$\mathcal{L}(f)(x) = \bigcap\{b \in \mathcal{I} : \exists \text{ open interval } O \subset I,\, x \in O \text{ with } \frac{f(u)-f(v)}{u-v} \in b \text{ for all } u, v \in O, u \neq v\}.$$

Consider now the case of functions in two arguments. Given a function $g : I \to I \to I$, its domain description will be an element in $(\mathcal{I}\times\mathcal{I}) \to (\mathcal{I}\times\mathcal{I}) \to (\mathcal{I}\times\mathcal{I})$, which is isomorphic to $((\mathcal{I}\times\mathcal{I})\times(\mathcal{I}\times\mathcal{I})) \to (\mathcal{I}\times\mathcal{I})$. Thus again, the domain description of g consists of a pair of functions $\langle g_1, g_2\rangle$, with g_1 describing the value part. If $D\,g\,(x_1, x_2)$ is the linear transformation representing the derivative of g at (x_1, x_2), then the function g_2 is a domain extension of the real function $\lambda x_1, y_1, x_2, y_2. D\,g\,(x_1, x_2)\cdot(y_1, y_2)$.

This approach for describing functions on reals is also used in (forward mode) Automatic Differentiation [14]. While Automatic Differentiation is different from our method in that it does not consider the domain of real numbers and the notion of partial reals, it is similar to our approach in that it uses two real numbers as input and a pair of functions to describe the derivative of functions on reals. Automatic differentiation is also used in [17], while the idea of using two separated components to describe the value part and the derivative part in the domain-theoretic setting can be found also in [8].

The semantic interpretation function $\mathcal{E}$ is defined, by structural induction, in the standard way:

$$\begin{aligned}
\mathcal{E}[\![c]\!]_\rho &= \mathcal{B}[\![c]\!]\\
\mathcal{E}[\![x]\!]_\rho &= \rho(x)\\
\mathcal{E}[\![e_1 e_2]\!]_\rho &= \mathcal{E}[\![e_1]\!]_\rho(\mathcal{E}[\![e_2]\!]_\rho)\\
\mathcal{E}[\![\lambda x^\sigma.e]\!]_\rho &= \lambda d \in \mathcal{D}_\sigma.\mathcal{E}[\![e]\!]_{(\rho[d/x])}
\end{aligned}$$

The semantic interpretation of any PCF constant is the usual one, while the semantic interpretation of the new constants on reals is given by:

$$\mathcal{B}[\![\mathsf{dig}]\!](l, m, n, \langle i, j\rangle) = \begin{cases}\bot & \text{if } l = \bot \vee m = \bot \vee n = \bot\\ \langle[0,0],[0,0]\rangle & \text{if } \neg(0 \leq m < n \wedge |l| \leq n - m)\\ \langle l/n + m/n\cdot i,\; m/n\cdot j\rangle & \text{otherwise}\end{cases}$$

$$\mathcal{B}[\![\mathsf{opp}]\!](\langle i, j\rangle) = \langle -i, -j\rangle$$

$$\mathcal{B}[\![\mathsf{add}]\!](p, q, \langle i, j\rangle) = \begin{cases}\bot & \text{if } p = \bot \vee q = \bot\\ \langle[0,0],[0,0]\rangle & \text{if } \neg(0 < 2\cdot p < q)\\ \langle i + p/q, j\rangle & \text{if } i + p/q < 1\\ \langle[1,1],[0,0]\rangle & \text{if } i + p/q > 1\\ \langle i + p/q \sqcap [-1,1],\; j \sqcap [0,0]\rangle & \text{otherwise}\end{cases}$$

$$\mathcal{B}[\![\mathsf{av}]\!](p, q, \langle i_1, j_1\rangle, \langle i_2, j_2\rangle)$$
$$= \begin{cases}\bot & \text{if } p = \bot \vee q = \bot\\ \langle[0,0],[0,0]\rangle & \text{if } \neg(0 < p < q)\\ \langle p/q\cdot i_1 + (1-p/q)\cdot i_2,\; p/q\cdot j_1 + (1-p/q)\cdot j_2\rangle & \text{otherwise}\end{cases}$$

$$\mathcal{B}[\![\mathsf{min}]\!](\langle i_1, j_1\rangle, \langle i_2, j_2\rangle) = \begin{cases} \langle i_1, j_1\rangle & \text{if } i_1 < i_2 \\ \langle i_2, j_2\rangle & \text{if } i_1 > i_2 \\ \langle i_1 \mathsf{min}\, i_2, j_1 \sqcap j_2\rangle & \text{otherwise} \end{cases}$$

$$\mathcal{B}[\![(0<)]\!](\langle i, j\rangle) = \begin{cases} tt & \text{if } i > 0 \\ ff & \text{if } i < 0 \\ \bot & \text{otherwise} \end{cases}$$

The interpretation of the derivative operator is given by:

$$\mathcal{E}[\![\mathsf{D}x.e]\!]_\rho = \lambda d \in \mathcal{I} \times \mathcal{I} \,.\, \langle \pi_2(\mathcal{E}[\![e]\!]_\rho[\langle \pi_1 d, 1\rangle / x]), \bot\rangle$$

Note that the function $\mathcal{B}[\![(0<)]\!]$ loses the information given by the derivative part, while the function $\mathcal{E}[\![\mathsf{D}x.e]\!]_\rho$, is a sort of translation of the function $\mathcal{E}[\![\lambda x.e]\!]_\rho$: The value of $\mathcal{E}[\![\mathsf{D}x.e]\!]_\rho$ is obtained from the derivative part of $\mathcal{E}[\![\lambda x.e]\!]_\rho$, while the derivative part of $\mathcal{E}[\![\mathsf{D}x.e]\!]_\rho$ is set to $\bot$.

Consider some examples. The absolute value function can be implemented through the term $\mathrm{Ab} = \lambda x.\mathsf{max}\,(\mathsf{opp}\, x) x$ with the following semantic interpretation:

$$\mathcal{E}[\![\mathrm{Ab}]\!]_\rho(\langle i, j\rangle) = \begin{cases} \langle i, j\rangle & \text{if } i > 0 \\ \langle -i, -j\rangle & \text{if } i < 0 \\ \langle [k^-, k^+], [-1, 1] j\rangle & \text{otherwise,} \end{cases}$$

where $k^- = \max(i^-, -i^+)$, $k^+ = \max(i^+, -i^-)$ with $i = [i^-, i^+]$.

When the absolute value function is evaluated at 0, where it is not differentiable, the derivative part of the semantic interpretation returns a partial value: $\pi_2(\mathcal{E}[\![\mathrm{Ab}]\!]_\rho(\{0\}, \{1\}) = [-1, 1]$. This partial value coincides with the Clarke gradient, equivalently the L-derivative, of the absolute value function.

The function $\frac{|x-y|}{2}$, is represented by the expression

$$\text{Ab-dif} = \lambda x.y.\mathsf{max}\,(x \mathsf{av}\, 1/2 (\mathsf{opp}\, y))((\mathsf{opp}\, x) \mathsf{av}\, 1/2 y)$$

whose semantics is the function:

$$\mathcal{E}[\![\text{Ab-dif}]\!]_\rho(\langle i_1, j_1\rangle, \langle i_2, j_2\rangle) =$$

$$\begin{cases} \langle \frac{i_1 - i_2}{2}, \frac{j_1 - j_2}{2}\rangle & \text{if } i_1 > i_2 \\ \langle \frac{i_2 - i_1}{2}, \frac{j_2 - j_1}{2}\rangle & \text{if } i_1 < j_1 \\ \langle [k^-, k^+], [-1/2, 1/2](j_1 - j_2)\rangle & \text{otherwise,} \end{cases}$$

where $k^- = \max(i_1^- - i_2^+, i_2^- - i_1^+)$ and $k^+ = \max(i_1^+ - i_2^-, i_2^+ - i_1^-)$.

From $[\![\text{Ab-dif}]\!]$ it is possible to evaluate the partial derivative of the function $\frac{|x-y|}{2}$, not only along the axes x and y, but along any direction. Considering the Euclidean distance, the derivative of the function at $(0, 0)$ in the direction of the unit vector $(u/\sqrt{u^2+v^2}, v/\sqrt{u^2+v^2})$ is given by
$\mathcal{E}[\![\text{Ab-dif}]\!]_\rho(\langle \{0\}, \{u/\sqrt{u^2+v^2}\}\rangle, \langle \{0\}, \{v/\sqrt{u^2+v^2}\}\rangle)$, that is the the interval $[-1/2, 1/2]\frac{u-v}{\sqrt{u^2+v^2}}$. Again this value coincides with the value of the Clarke gradient of the function $\frac{|x-y|}{2}$ at $(0, 0)$ in the direction $(u/\sqrt{u^2+v^2}, v/\sqrt{u^2+v^2})$.

4.1 Logical Relations Characterisation

In the present approach we choose to define the semantic domains in the simplest possible way. As a consequence, our domains contain also points that are not consistent with the intended meaning, for example, the domain $\mathcal{D}_{\iota\to\iota} = (\mathcal{I}\times\mathcal{I}) \to (\mathcal{I}\times\mathcal{I})$ contains also the product of two functions $\langle f_1, f_2\rangle$ where the derivative part f_2 is not necessarily linear in its second argument and is not necessarily consistent with the value part, i.e., the function f_1; moreover the value part f_1 can be a function depending also on its second argument.

However the semantic interpretation of (non-expansive) PCDF expressions will not have this pathological behaviour. A proof of this fact and a more precise characterisation of the semantic interpretation of expressions can be obtained through the technique of logical relations [18]. In particular we define a set of logical relations on the semantic domains and prove that, for any non-expansive PCDF expression e, the semantic interpretation of e satisfies these relations. Using this method, we can establish a list of properties for the semantic interpretation of PCDF expressions.

Definition 1. *The following list of relations are defined on the domain $\mathcal{D}_\iota$.*

- **Independence**: *A binary relation R^i_ι consisting of the pairs of the form $(\langle i, j_1\rangle, \langle i, j_2\rangle)$. The relation R^i_ι is used to establish that, for a given function, the value part of the result is independent from the derivative part of the argument: $f_1(i, j_1) = f_1(i, j_2)$.*
- **Sub-linearity**: *A family of relations $R^{l,r}_\iota$ indexed by a rational number $r \in [-1, 1]$. The family $R^{l,r}_\iota$ consists of pairs of the form $(\langle i, j_1\rangle, \langle i, j_2\rangle)$ where $j_1 \sqsubseteq r \cdot j_2$. These relations are used to establish the sublinearity of the derivative part: $f_2(i, r\cdot j) \sqsubseteq r \cdot f_2(i, j)$.*
- **Consistency**: *A family of ternary relation $R^{d,r}_\iota$ indexed by a rational number $r \in (0, 2]$, consisting of triples of the form $(\langle i_1, j_1\rangle, \langle i_2, j_2\rangle, \langle i_3, j_3\rangle)$ with $i_3 \sqsubseteq i_1 \sqcap i_2$ and $(r \cdot j_3)$ consistent with $(i_1 - i_2)$, that is the intervals $(r\cdot j_3)$ and $(i_1 - i_2)$ have a non-empty intersection. This relation is used to establish the consistency of the derivative part of a function with respect to the value part.*

The above relations are defined on the other ground domains $\mathcal{D}_o$ and $\mathcal{D}_\nu$ as the diagonal relations in two or three arguments, e.g., $R^{d,r}_\nu(l, m, n)$ iff $l = m = n$. The relations are extended inductively to higher order domains by the usual definition on logical relations: $R^i_{\sigma\to\tau}(f, g)$ iff for every $d_1, d_2 \in \mathcal{D}_\sigma$, $R^i_\sigma(d_1, d_2)$ implies $R^i_\tau(f(d_1), g(d_2))$, and similarly for the other relations.

Proposition 1. *For any closed expression $e : \sigma$, for any rational number $r \in [-1, 1]$, the semantic interpretation $\mathcal{E}[\![e]\!]_\rho$ of e, is self-related by R^i_σ, $R^{l,r}_\sigma$, i.e. $R^i_\sigma(\mathcal{E}[\![e]\!]_\rho, \mathcal{E}[\![e]\!]_\rho)$, and similarly for $R^{l,r}_\sigma$. Moreover, if the expression $e : \sigma$ is non-expansive, the semantic interpretation $\mathcal{E}[\![e]\!]_\rho$, is self-related by $R^{d,r}_\sigma$.*

We now show how the three relations ensure the three properties of independence, sublinearity and consistency. To any element $f = \langle f_1, f_2\rangle$ in the domain $\mathcal{D}_{\iota\to\iota} = (\mathcal{I}\times\mathcal{I}) \to (\mathcal{I}\times\mathcal{I})$ we associate a partial function $f_v : I \to I$ with

$$f_v(x) = \begin{cases} y & \text{if } f_1(\langle\{x\},\bot\rangle) = \{y\} \\ \text{undefined} & \text{if } f_1(\langle\{x\},\bot\rangle) \text{is a proper interval} \end{cases}$$

and a total function

$$f_d : I \to \mathcal{I} = \lambda x. f_2(\langle\{x\},\{1\}\rangle)$$

The preservation of the relations R^i_ι, $R^{l,r}_\iota$ has the following straightforward consequences:

Proposition 2. *(i) For any function $f = \langle f_1, f_2\rangle$ in $\mathcal{D}_{\iota\to\iota}$ self-related by $R^i_{\iota\to\iota}$, for every i, j_1, j_2, $f_1(\langle i, j_1\rangle) = f_1(\langle i, j_2\rangle)$, the return value part is independent from the derivative argument.*

(ii) For any function $f = \langle f_1, f_2\rangle$ in $\mathcal{D}_{\iota\to\iota}$ self-related by $R^{l,r}_{\iota\to\iota}$ for every i, j, and for every rational $r \in [-1,1]$, $f_2(\langle i, r\cdot j\rangle) \sqsubseteq r \cdot f_2(\langle i, j\rangle)$. It follows that:

- *$(f_2(\langle i, \{r\}\rangle))/r \sqsubseteq f_2(\langle i, \{1\}\rangle)$, i.e., the most precise approximation of the L-derivative is obtained by evaluating the function with* 1 *as its second argument,*
- *for every i, j, $f_2(\langle i, -j\rangle) = -f_2(\langle i, j\rangle)$, i.e., the derivative part is an odd function.*

The preservation of the relation $R^{d,r}_\iota$ induces the following properties (see [13] for the proof):

Proposition 3. *For any function f in $\mathcal{D}_{\iota\to\iota}$ self-related by $R^{d,r}_{\iota\to\iota}$:*

(i) the function f_v is non-expansive;

(ii) on the open sets where the functions f_v is defined, the function f_d is an approximation to the L-derivative of the function f_v;

(iii) if f is a maximal element of $\mathcal{D}_{\iota\to\iota}$ then f_v is a total function and f_d is the associated L-derivative.

4.2 Subdomains

By definition, the logical relations are closed under directed lubs, and as a consequence the sets of elements self-related by them are also closed under directed lubs.

For any ground type σ the relations R^i_σ, $R^{l,r}_\sigma$, $R^{d,r}_\sigma$ are closed under arbitrary meets, meaning that if $\forall j \in J \,.\, R^i_\sigma(d_j, e_j)$ then $R^i_\sigma(\bigsqcap_{j\in J} d_j, \bigsqcap_{j\in J} e_j)$ and similarly for the other relations $R^{l,r}_\sigma$, $R^{d,r}_\sigma$. The proof is immediate for $\sigma = o, \nu$, and is a simple check for $\sigma = \iota$. The following result shows that this closure property holds also for $\sigma = \iota \to \iota$.

Proposition 4. *The set of elements in $\mathcal{D}_{\iota\to\iota}$ self-related by any of the three relations $R^i_{\iota\to\iota}$, $R^{l,r}_{\iota\to\iota}$, and $R^{d,r}_{\iota\to\iota}$ is closed under arbitrary meets.*

Proof. For the independence relation $R^i_{\iota\to\iota}$, the closure property is trivial to check. For the consistency relation $R^{d,r}_{\iota\to\iota}$, the closure under non-empty meets follows immediately from the fact that this relation is downward closed. The closure property for the sublinearity relation $R^{l,r}_{\iota\to\iota}$ is given in [13].

We now employ the following result whose proof can be found in [13].

Proposition 5. *In a continuous Scott domain, a non-empty subset closed under lubs of directed subsets and closed under non-empty meets is a continuous Scott subdomain.*

Corollary 1. *If σ is a ground type or first order type, then the set of elements in $\mathcal{D}_\sigma$ self-related by the three logical relations is a continuous Scott subdomain of $\mathcal{D}_\sigma$.*

As we do not deal with second or higher order real types in this extended abstract, we will not discuss the corresponding subdomains here.

4.3 Adequacy

As usual once an operational and denotational semantics are defined, it is necessary to present an adequacy theorem stating that the two semantics agree.

Let us denote by $[a,b] \ll \mathit{Eval}(e)$ the fact that there exits three integers l, m, n such that $e \to^* \mathsf{dig}\, l\, m\, n\, e'$ and $[(l-m)/n, (l+m)/n] \subset (a,b)$. The proof of the following theorem is presented in [13].

Theorem 1 (Adequacy). *For every closed term e with type ι, interval $[a,b]$ and environment ρ, we have:*

$$[a,b] \ll \mathit{Eval}(e) \textit{ iff } [a,b] \ll \pi_1(\mathcal{E}[\![e]\!]_\rho)$$

In the operational semantics that we have proposed, the calculus of the derivative is performed through a sort of symbolic computation: the rewriting rules specify how to evaluate the derivative of the primitive functions and the application of the derivative rules essentially transforms a function expression into the function expression representing the derivative. The denotational semantics provides an alternative approach to the computation of the derivative, which almost exactly coincides with the computation performed by Automatic Differentiation. We can interpret our adequacy result as a proof that symbolic computation of the derivative and the computation of the derivative through Automatic Differentiation coincide. We remark in passing that, inspired by the denotational semantics, it is possible to define an alternative operational semantics that will perform the computation of the derivative in the same way that is performed by Automatic Differentiation.

4.4 Function Definability

We will show in the following theorem that any computable Lipschitz function can be obtained in our framework as the limit, in the sup norm, of a sequence of piecewise linear maps definable in PCDF such that every piecewise linear map in the sequence gives lower and upper bounds for the function and the L-derivative of the function is contained in the L-derivatives of the piecewise linear maps, which converge to the classical derivative of the function wherever it is continuously differentiable.

Theorem 2. *For any maximal computable function f in $\mathcal{D}_{\iota \to \iota}$ preserving the logical relations $R^{i}_{\iota \to \iota}$, $R^{l,r}_{\iota \to \iota}$, $R^{d,r}_{\iota \to \iota}$, there exists a closed PCDF expression* f *such that:*

$$\forall x \in I.\ f_v(x) = (\mathcal{E}[\![\mathsf{f}]\!]_\rho)_v(x) \ \wedge\ f_d(x) = (\mathcal{E}[\![\mathsf{f}]\!]_\rho)_d(x)$$

The above definability result states that if we consider only the behaviour of the domain functions on the total elements of $\mathcal{D}_\iota$ (i.e. the elements representing completely defined real numbers) then PCDF is sufficiently rich to represent the computable elements of $\mathcal{D}_{\iota \to \iota}$.

We do not consider the problem of defining PCDF expressions whose semantics coincides with domain functions also on partial elements. The reason for this choice is that this later problem is technically more difficult and less interesting from a practical point of view.

The proof of the above result is quite lengthy: we define a general methodology to transform the information that can be extracted from a domain function into a PCDF expression. The extended version of the present paper, [13], contains a description of the construction.

5 Conclusion

We have integrated, in a single language, exact real number computation with the evaluation of the derivatives of function expressions.

The language has been designed using a minimal set of primitives sufficient to define any computable (and differentiable) function. It can be seen as a theoretical basis for the implementation of exact real number computation in a programming language. In a practical implementation, however, one needs both to extend the set of primitive functions and to carefully redesign the reduction strategy to increase both the usability of the language and the efficiency of the computation. In this respect, the approach used in [12] to define analytic functions can provide useful ideas.

The main result presented here is an adequate denotational semantics for differentiable functions, which has required original ideas in developing the semantics domains, and a definability result showing the expressivity of the language.

The present research can be extended in several directions. Some possible future works are the following.

- An obvious problem to consider is whether the definability result presented in the paper can be extended to a larger class of function domains. We claim that the techniques presented here can be easily adapted to functions with

several arguments. This is not however the case when considering higher order functions, whose definability is an open problem.
- A second direction for possible further research is the treatment of the second derivative and more generally derivative of arbitrary order.

References

1. Bishop, E., Bridges, D.: Constructive Analysis. Springer (1985)
2. Clarke, F.H.: Optimization and Nonsmooth Analysis. Wiley (1983)
3. Coddington, E.A., Levinson, N.: Theory of Ordinary Differential Equations. McGraw Hill (1955)
4. Davis, T.A., Sigmon, K.: MATLAB Primer, 7th edn. CRC Press (2005)
5. Di Gianantonio, P.: An abstract data type for real numbers. Theoretical Computer Science 221, 295–326 (1999)
6. Edalat, A.: A continuous derivative for real-valued functions. In: New Computational Paradigms, Changing Conceptions of What is Computable, pp. 493–519. Springer (2008)
7. Edalat, A., Escardó, M.: Integration in real PCF. Information and Computation 160, 128–166 (2000)
8. Edalat, A., Lieutier, A.: Domain theory and differential calculus (functions of one variable). Mathematical Structures in Computer Science 14(6), 771–802 (2004)
9. Edalat, A., Lieutier, A., Pattinson, D.: A computational model for multi-variable differential calculus. Information and Computation 224, 22–45 (2013)
10. Ehrhard, T., Regnier, L.: The differential lambda-calculus. Theoretical Computer Science 309 (2003)
11. Escardó, M.: PCF extended with real numbers. Theoretical Computer Science 162(1), 79–115 (1996)
12. Escardó, M., Simpson, A.: A universal characterization of the closed Euclidean interval. In: LICS, pp. 115–125. IEEE Computer Society (2001)
13. Di Gianantonio, P., Edalat, A.: A language for differentiable functions (extended version), `http://www.dimi.uniud.it/pietro/papers/pcdf.pdf`
14. Griewank, A., Walther, A.: Evaluating Derivatives: Principles and Techniques of Algorithmic Differentiation, 2nd edn. SIAM (2008)
15. `http://www.doc.ic.ac.uk/exact-computation/`
16. Ko, K.: Complexity Theory of Real Numbers. Birkhäuser (1991)
17. Manzyuk, O.: A simply typed -calculus of forward automatic differentiation. Electr. Notes Theor. Comput. Sci. 286, 257–272 (2012)
18. Mitchell, J.C.: Foundations of Programming Languages. MIT Press (1996)
19. Potts, P., Edalat, A., Escardó, M.: Semantics of exact real arithmetic. In: LICS, pp. 248–257 (1997)
20. Pour-El, M.B., Richards, J.I.: Computability in Analysis and Physics. Springer (1988)
21. Weihrauch, K.: Computable Analysis (An Introduction). Springer (2000)

Computing Quantiles in Markov Reward Models*

Michael Ummels[1] and Christel Baier[2]

[1] Institute of Transportation Systems, German Aerospace Center
michael.ummels@dlr.de
[2] Technische Universität Dresden
baier@tcs.inf.tu-dresden.de

Abstract. Probabilistic model checking mainly concentrates on techniques for reasoning about the probabilities of certain path properties or expected values of certain random variables. For the quantitative system analysis, however, there is also another type of interesting performance measure, namely *quantiles*. A typical quantile query takes as input a lower probability bound $p \in]0,1]$ and a reachability property. The task is then to compute the minimal reward bound r such that with probability at least p the target set will be reached before the accumulated reward exceeds r. Quantiles are well-known from mathematical statistics, but to the best of our knowledge they have not been addressed by the model checking community so far.

In this paper, we study the complexity of quantile queries for until properties in discrete-time finite-state Markov decision processes with nonnegative rewards on states. We show that qualitative quantile queries can be evaluated in polynomial time and present an exponential algorithm for the evaluation of quantitative quantile queries. For the special case of Markov chains, we show that quantitative quantile queries can be evaluated in pseudo-polynomial time.

1 Introduction

Markov models with reward (or cost) functions are widely used for the quantitative system analysis. We focus here on the discrete-time or time-abstract case. Discrete-time Markov decision processes, MDPs for short, can be used, for instance, as an operational model for randomised distributed algorithms and rewards might serve to reason, e.g., about the size of the buffer of a communication channel or about the number of rounds that a leader election protocol might take until a leader has been elected.

Several authors considered variants of probabilistic computation tree logic (PCTL) [12,4] for specifying quantitative constraints on the behaviour of Markov

* This work was supported by the DFG project QuaOS and the collaborative research centre HAEC (SFB 912) funded by the DFG. This work was partly supported by the European Union Seventh Framework Programme under grant agreement no. 295261 (MEALS), the DFG/NWO project ROCKS and the cluster of excellence cfAED.

F. Pfenning (Ed.): FOSSACS 2013, LNCS 7794, pp. 353–368, 2013.

models with reward functions. Such extensions, briefly called PRCTL here, permit to specify constraints on the probabilities of reward-bounded reachability conditions, on the expected accumulated rewards until a certain set of target states is reached or expected instantaneous rewards after some fixed number of steps [7,6,9,1,15], or on long-run averages [8]. An example for a typical PRCTL formula with PCTL's probability operator and the reward-bounded until operator is the formula $\mathsf{P}_{>p}(a\,\mathsf{U}_{\le r}\,b)$ where p is a lower probability bound in $[0,1[$ and r is an upper bound for the accumulated reward earned by path fragments that lead via states where a holds to a b-state. From a practical point of view, more important than checking whether a given PRCTL formula φ holds for (the initial state of) a Markov model $\mathcal{M}$ are PRCTL *queries* of the form $\mathsf{P}_{=?}\,\psi$ where the task is to calculate the (minimum or maximum) probability for the path formula ψ. Indeed, the standard PRCTL model checking algorithm checks whether a given formula $\mathsf{P}_{\bowtie p}\,\psi$ holds in $\mathcal{M}$ by evaluating the PRCTL query $\mathsf{P}_{=?}\,\psi$ and comparing the computed value q with the given probability bound p according to the comparison predicate $\bowtie$. The standard procedure for dealing with PRCTL formulas that refer to expected (instantaneous or accumulated) rewards relies on an analogous scheme; see e.g. [10]. An exception can be made for qualitative PRCTL properties $\mathsf{P}_{\bowtie p}\,\psi$ where the probability bound p is either 0 or 1, and the path formula ψ is a plain until formula without reward bound (or any ω-regular path property without reward constraints): in this case, a graph analysis suffices to check whether $\mathsf{P}_{\bowtie p}\,\psi$ holds for $\mathcal{M}$ [16,5].

In a common project with the operating system group of our department, we learned that a natural question for the systems community is to swap the given and unknown parameters in PRCTL queries and to ask for the computation of a *quantile* (see [2]). For instance, if $\mathcal{M}$ models a mutual exclusion protocol for competing processes $P_1,\ldots,P_n$ and rewards are used to represent the time spent by process P_i in its waiting location, then the *quantile query* $\mathsf{P}_{>0.9}(wait_i\,\mathsf{U}_{\le ?}\,crit_i)$ asks for the minimal time bound r such that in all scenarios (i.e., under all schedulers) with probability greater than 0.9 process P_i will wait no longer than r time units before entering its critical section. For another example, suppose $\mathcal{M}$ models the management system of a service execution platform. Then the query $\mathsf{P}_{>0.98}(true\,\mathsf{U}_{\le ?}\,tasks\_completed)$ might ask for the minimal initial energy budget r that is required to ensure that even in the worst-case there is more than 98% chance to reach a state where all tasks have been completed successfully.

To the best of our knowledge, quantile queries have not yet been addressed directly in the model checking community. What is known from the literature is that for finite Markov chains with nonnegative rewards the task of checking whether a PRCTL formula $\mathsf{P}_{>p}(a\,\mathsf{U}_{\le r}\,b)$ or $\mathsf{P}_{\ge p}(a\,\mathsf{U}_{\le r}\,b)$ holds for some given state is NP-hard [14] when p and r are represented in binary. Since such a formula holds in state s if and only if the value of the corresponding quantile query at s is $\le r$, this implies that evaluating quantile queries is also NP-hard.

The purpose of this paper is to study quantile queries for Markov decision processes with nonnegative rewards in more details. We consider quantile queries for reward-bounded until formulas in combination with the standard PRCTL quantifier $\mathsf{P}_{\bowtie p}$ (in this paper denoted by $\forall\mathsf{P}_{\bowtie p}$), where universal quantification over all schedulers is inherent in the semantics, and its dual $\exists\mathsf{P}_{\bowtie p}$ that asks for the existence of some scheduler enjoying a certain property. By duality, our results carry over to reward-bounded release properties.

Contributions. First, we address *qualitative* quantile queries, i.e. quantile queries where the probability bound is either 0 or 1, and we show that such queries can be evaluated in strongly polynomial time. Our algorithm is surprisingly simple and does not rely on value iteration or linear programming techniques (as it is e.g. the case for extremal expected reachability times and stochastic shortest-paths problems in MDPs [9]). Instead, our algorithm relies on the greedy method and borrows ideas from Dijkstra's shortest-path algorithm. In particular, our algorithm can be used for checking PRCTL formulas of the form $\forall\mathsf{P}_{\bowtie p}(a\,\mathsf{U}_{\leq r}\,b)$ or $\exists\mathsf{P}_{\bowtie p}(a\,\mathsf{U}_{\leq r}\,b)$ with $p \in \{0,1\}$ in polynomial time. Previously, a polynomial-time algorithm was known only for the special case of MDPs where every loop contains a state with nonzero reward [13].

Second, we consider *quantitative* quantile queries. The standard way to compute the maximal or minimal probabilities for reward-bounded until properties, say $a\,\mathsf{U}_{\leq r}\,b$, relies on the iterative computation of the extremal probabilities $a\,\mathsf{U}_{\leq i}\,b$ for increasing reward bound i. We use here a reformulation of this computation scheme as a linear program whose size is polynomial in the number of states of $\mathcal{M}$ and the given reward bound r. The crux to derive from this linear program an algorithm for the evaluation of quantile queries is to provide a bound for the sought value, which is our second contribution. This bound then permits to perform a sequential search for the quantile, which yields an exponentially time-bounded algorithm for evaluating quantitative quantile queries. Finally, in the special case of Markov chains with integer rewards, we show that this algorithm can be improved to run in time polynomial in the size of the query, the size of the chain, and the largest reward, i.e. in *pseudo-polynomial* time.

Outline. The structure of the paper is as follows. Section 2 summarises the relevant concepts of Markov decision processes and briefly recalls the logic PRCTL. Quantile queries are introduced in Sect. 3. Our polynomial-time algorithms for qualitative quantile queries is presented in Sect. 4, whereas the quantitative case is addressed in Sect. 5. The paper ends with some concluding remarks in Sect. 6.

2 Preliminaries

In the following, we assume a countably infinite set AP of *atomic propositions*. A Markov decision process (MDP) $\mathcal{M} = (S, Act, \gamma, \lambda, rew, \delta)$ with nonnegative rewards consists of a finite set S of states, a finite set Act of actions, a function $\gamma\colon S \to 2^{Act} \setminus \{\emptyset\}$ describing the set of *enabled actions* in each state, a labelling function $\lambda\colon S \to 2^{\mathsf{AP}}$, a reward function $rew\colon S \to \mathbb{R}^{\geq 0}$, and a transition function $\delta\colon S \times Act \times S \to [0,1]$ such that $\sum_{t \in S} \delta(s,\alpha,t) = 1$ for all $s \in S$ and $\alpha \in Act$. If the set Act of actions is just a singleton, we call $\mathcal{M}$ a Markov chain.

Given an MDP $\mathcal{M}$, we say that a state s of $\mathcal{M}$ is *absorbing* if $\delta(s, \alpha, s) = 1$ for all $\alpha \in \gamma(s)$. Moreover, for $a \in \mathsf{AP}$ we denote by $\lambda^{-1}(a)$ the set of states s such that $a \in \lambda(s)$, and for $x = s_0 s_1 \ldots s_k \in S^*$ we denote by $rew(x)$ the accumulated reward after x, i.e. $rew(x) = \sum_{i=0}^{k} rew(s_i)$. Finally, we denote by $|\delta|$ the number of *nontrivial* transitions in $\mathcal{M}$, i.e. $|\delta| = |\{(s, \alpha, t) : \alpha \in \gamma(s) \text{ and } \delta(s, \alpha, t) > 0\}|$.

Schedulers are used to resolve the nondeterminism that arises from the possibility that more than one action might be enabled in a given state. Formally, a scheduler for $\mathcal{M}$ is a mapping $\sigma\colon S^+ \to Act$ such that $\sigma(xs) \in \gamma(s)$ for all $x \in S^*$ and $s \in S$. Such a scheduler σ is *memoryless* if $\sigma(xs) = \sigma(s)$ for all $x \in S^*$ and $s \in S$. Given a scheduler σ and an initial state $s = s_0$, there is a unique probability measure $\Pr_s^\sigma$ on the Borel σ-algebra over S^ω such that $\Pr_s^\sigma(s_0 s_1 \ldots s_k \cdot S^\omega) = \prod_{i=0}^{k-1} \delta(s_i, \sigma(s_0 \ldots s_i), s_{i+1})$; see [3].

Several logics have been introduced in order to reason about the probability measures $\Pr_s^\sigma$. In particular, the logics PCTL and PCTL* replace the path quantifiers of CTL and CTL* by a single probabilistic quantifier $\mathsf{P}_{\bowtie p}$, where $\bowtie \in \{<, \leq, \geq, >\}$ and $p \in [0, 1]$. In these logics, the formula $\varphi = \mathsf{P}_{\bowtie p}\, \psi$ holds in state s (written $s \models \varphi$) if under *all* schedulers σ the probability $\Pr_s^\sigma(\psi)$ of the path property ψ compares positively with p wrt. the comparison operator $\bowtie$, i.e. if $\Pr_s^\sigma(\psi) \bowtie \psi$. A dual existential quantifier $\exists\mathsf{P}_{\bowtie p}$ that asks for the existence of a scheduler can be introduced using the equivalence $\exists\mathsf{P}_{\bowtie p}\, \psi \equiv \neg\mathsf{P}_{\overline{\bowtie} p}\, \psi$, where $\overline{\bowtie}$ denotes the dual inequality. Since many properties of MDPs can be expressed more naturally using the $\exists\mathsf{P}$ quantifier, we consider this quantifier an equal citizen of the logic, and we denote the universal quantifier P by $\forall\mathsf{P}$ in order to stress its universal semantics.

In order to be able to reason about accumulated rewards, we amend the until operator U by a reward constraint of the form $\sim r$, where $\sim$ is a comparison operator and $r \in \mathbb{R} \cup \{\pm\infty\}$. Since we adopt the convention that a reward is earned upon *leaving* a state, a path $\pi = s_0 s_1 \ldots$ fulfils the formula $\psi_1 \mathsf{U}_{\sim r} \psi_2$ if there exists a point $k \in \mathbb{N}$ such that 1. $s_k s_{k+1} \ldots \models \psi_2$, 2. $s_i s_{i+1} \ldots \models \psi_1$ for all $i < k$, and 3. $rew(s_0 \ldots s_{k-1}) \sim r$. Even though our logic is only a subset of the logics PRCTL and PRCTL* defined in [1], we use the same names for the extension of PCTL and PCTL* with the amended until operator. The following proposition states that extremal probabilities for PRCTL* are attainable. This follows, for instance, from the fact that PRCTL* can only describe ω-regular path properties.

Proposition 1. *Let $\mathcal{M}$ be an MDP and ψ a* PRCTL* *path formula. Then there exist schedulers σ^* and τ^* such that $\Pr_s^{\sigma^*}(\psi) = \sup_\sigma \Pr_s^\sigma(\psi)$ and $\Pr_s^{\tau^*}(\psi) = \inf_\tau \Pr_s^\sigma(\psi)$ for all states s of $\mathcal{M}$.*

3 Quantile Queries

A *quantile query* is of the form $\varphi = \forall\mathsf{P}_{\bowtie p}(a\, \mathsf{U}_{\leq ?}\, b)$ or $\varphi = \exists\mathsf{P}_{\bowtie p}(a\, \mathsf{U}_{\leq ?}\, b)$, where $a, b \in \mathsf{AP}$, $p \in [0, 1]$ and $\bowtie \in \{<, \leq, \geq, >\}$. We call queries of the former type *universal* and queries of the latter type *existential*. If $r \in \mathbb{R} \cup \{\pm\infty\}$, we write $\varphi[r]$ for the PRCTL formula that is obtained from φ by replacing ? with r.

Given an MDP $\mathcal{M}$ with rewards, evaluating φ on $\mathcal{M}$ amounts to computing, for each state s of $\mathcal{M}$, the least or the largest $r \in \mathbb{R}$ such that $s \models \varphi[r]$. Formally, if $\varphi = \forall \mathsf{P}_{\bowtie p}(a \,\mathsf{U}_{\leq ?}\, b)$ or $\varphi = \exists \mathsf{P}_{\bowtie p}(a \,\mathsf{U}_{\leq ?}\, b)$ then the *value* of a state s of $\mathcal{M}$ with respect to φ is $\mathrm{val}_\varphi^{\mathcal{M}}(s) := \mathrm{opt}\{r \in \mathbb{R} : s \models \varphi[r]\}$, where opt = inf if $\bowtie \in \{\geq, >\}$ and opt = sup otherwise.[1] Depending on whether $\mathrm{val}_\varphi^{\mathcal{M}}(s)$ is defined as an infimum or a supremum, we call φ a *minimising* or a *maximising* query, respectively. In the following, we will omit the superscript $\mathcal{M}$ when the underlying MDP is clear from the context.

Given a query φ, we define the *dual query* to be the unique quantile query $\overline{\varphi}$ such that $\overline{\varphi}[r] \equiv \neg\varphi[r]$ for all $r \in \mathbb{R} \cup \{\pm\infty\}$. Hence, to form the dual of a query, one only needs to replace the quantifier $\forall \mathsf{P}_{\bowtie p}$ by $\exists \mathsf{P}_{\overline{\bowtie} p}$ and vice versa. For instance, the dual of $\forall \mathsf{P}_{<p}(a \,\mathsf{U}_{\leq ?}\, b)$ is $\exists \mathsf{P}_{\geq p}(a \,\mathsf{U}_{\leq ?}\, b)$. Note that the dual of a universal or minimising query is an existential or maximising query, respectively, and vice versa.

Proposition 2. *Let $\mathcal{M}$ be an MDP and φ a quantile query. Then $\mathrm{val}_\varphi(s) = \mathrm{val}_{\overline{\varphi}}(s)$ for all states s of $\mathcal{M}$.*

Proof. Without loss of generality, assume that φ is a minimising query. Let $s \in S$, $v = \mathrm{val}_\varphi(s)$ and $v' = \mathrm{val}_{\overline{\varphi}}(s)$. On the one hand, for all $r < v$ we have $s \not\models \varphi[r]$, i.e. $s \models \overline{\varphi}[r]$, and therefore $v' \geq v$. On the other hand, since $\varphi[r]$ implies $\varphi[r']$ for $r' \geq r$, for all $r > v$ we have $s \models \varphi[r]$, i.e. $s \not\models \overline{\varphi}[r]$, and therefore also $v' \leq v$. □

Assume that we have computed the value $\mathrm{val}_\varphi(s)$ of a state s with respect to a quantile query φ. Then, for any $r \in \mathbb{R}$, to decide whether $s \models \varphi[r]$, we just need to compare r to $\mathrm{val}_\varphi(s)$.

Proposition 3. *Let $\mathcal{M}$ be an MDP, s a state of $\mathcal{M}$, φ a minimising or maximising quantile query, and $r \in \mathbb{R}$. Then $s \models \varphi[r]$ if and only if $\mathrm{val}_\varphi(s) \leq r$ or $\mathrm{val}_\varphi(s) > r$, respectively.*

Proof. First assume that $\varphi = Q(a \,\mathsf{U}_{\leq ?}\, b)$ is a minimizing query. Clearly, if $s \models \varphi[r]$, then $\mathrm{val}_\varphi(s) \leq r$. On the other hand, assume that $\mathrm{val}_\varphi(s) \leq r$ and denote by R the set of numbers $x \in \mathbb{R}$ of the form $x = \sum_{i=0}^{k} rew(s_i)$ for a finite sequence $s_0 s_1 \ldots s_k$ of states. Since the set $\{x \in R : x \leq n\}$ is finite for all $n \in \mathbb{N}$, we can fix some $\varepsilon > 0$ such that $r + \delta \notin R$ for all $0 < \delta \leq \varepsilon$. Hence, the set of paths that fulfil $a \,\mathsf{U}_{\leq r}\, b$ agrees with the set of paths that fulfil $a \,\mathsf{U}_{\leq r+\varepsilon}\, b$. Since $\mathrm{val}_\varphi(s) < r + \varepsilon$ and φ is a minimising query, we know that $s \models \varphi[r + \varepsilon]$. Since replacing $r+\varepsilon$ by r does not affect the path property, this implies that $s \models \varphi[r]$. Finally, if φ is a maximising query, then $\overline{\varphi}$ is a minimising query, and $s \models \overline{\varphi}[r]$ if and only if $\mathrm{val}_{\overline{\varphi}}(s) = \mathrm{val}_\varphi(s) \leq r$, i.e. $s \models \varphi[r]$ if and only if $\mathrm{val}_\varphi(s) > r$. □

Proposition 3 does not hold when we allow r to take an infinite value. In fact, if φ is a minimizing query and $s \not\models \varphi[\infty]$, then $\mathrm{val}_\varphi(s) = \infty$. Analagously, if φ is a maximising query and $s \not\models \varphi[-\infty]$, then $\mathrm{val}_\varphi(s) = -\infty$.

To conclude this section, let us remark that queries using the reward-bounded *release* operator R can easily be accommodated in our framework. For instance, the query $\forall \mathsf{P}_{\geq p}(a \,\mathsf{R}_{\leq ?}\, b)$ is equivalent to the query $\forall \mathsf{P}_{\leq 1-p}(\neg a \,\mathsf{U}_{\leq ?}\, \neg b)$.

[1] As usual, we assume that $\inf \emptyset = \infty$ and $\sup \emptyset = -\infty$.

Algorithm 1. Solving qualitative queries of the form $Q(a \, \mathsf{U}_{\leq ?} \, b)$

Input: MDP $\mathcal{M} = (S, Act, \gamma, \lambda, rew, \delta)$, $\varphi = Q(a \, \mathsf{U}_{\leq ?} \, b)$
for each $s \in S$ **do**
  if $s \models b$ **then** $v(s) \leftarrow 0$ **else** $v(s) \leftarrow \infty$
$X \leftarrow \{s \in S : v(s) = 0\}$; $R \leftarrow \{0\}$
$Z \leftarrow \{s \in S : s \models a \wedge \neg b$ and $rew(s) = 0\}$
while $R \neq \emptyset$ **do**
  $r \leftarrow \min R$; $Y \leftarrow \{s \in X : v(s) \leq r\} \setminus Z$
  for each $s \in S \setminus X$ with $s \models a \wedge Q \mathsf{X}(Z \, \mathsf{U} \, Y)$ **do**
    $v(s) \leftarrow r + rew(s)$
    $X \leftarrow X \cup \{s\}$; $R \leftarrow R \cup \{v(s)\}$
  $R \leftarrow R \setminus \{r\}$
return v

4 Evaluating Qualitative Queries

In this section, we give a strongly polynomial-time algorithm for evaluating *qualitative queries*, i.e. queries where the probability bound p is either 0 or 1. Throughout this section, let $\mathcal{M} = (S, Act, \gamma, \lambda, rew, \delta)$ be an MDP with non-negative rewards. By Proposition 2, we can restrict to queries using one of the quantifiers $\forall \mathsf{P}_{>0}$, $\exists \mathsf{P}_{>0}$, $\forall \mathsf{P}_{=1}$ and $\exists \mathsf{P}_{=1}$. The following lemma allows to give a unified treatment of all cases. (X denotes the next-step operator).

Lemma 4. *The equivalence* $Q \mathsf{X}(a \, \mathsf{U} \, (\neg a \wedge \psi)) \equiv Q \mathsf{X}(a \, \mathsf{U} \, (\neg a \wedge Q \, \psi))$ *holds in* PRCTL* *for all* $Q \in \{\forall \mathsf{P}_{>0}, \exists \mathsf{P}_{>0}, \forall \mathsf{P}_{=1}, \exists \mathsf{P}_{=1}\}$, $a \in \mathsf{AP}$, *and all path formulas* ψ.

Algorithm 1 is our algorithm for computing the values of a quantile query where we look for an upper bound on the accumulated reward. The algorithm maintains a set X of states, a set R of real numbers, and a table v mapping states to non-negative real numbers or infinity. The algorithm works by discovering states with finite value repeatedly until only the states with infinite value remain. Whenever a new state is discovered, it is put into X and its value is put into R. In the initialisation phase, the algorithm discovers all states labelled with b, which have value 0. In every iteration of the main loop, new states are discovered by picking the least value r that has not been fully processed (i.e. the least element of R) and checking which undiscovered a-labelled states fulfil the PCTL* formula $Q \mathsf{X}(Z \, \mathsf{U} \, Y)$, where Y is the set of already discovered states whose value is at most r and Z is the set of states labelled with a but not with b and having reward 0. Any such newly discovered state s must have value $r + rew(s)$, and r can be deleted from R at the end of the current iteration. The termination of the algorithm follows from the fact that in every iteration of the main loop either the set X increases or it remains constant and one element is removed from R.

Lemma 5. *Let* $\mathcal{M}$ *be an MDP,* $\varphi = Q(a \, \mathsf{U}_{\leq ?} \, b)$ *a qualitative query, and let* v *be the result of Algorithm 1 on* $\mathcal{M}$ *and* φ. *Then* $v(s) = \mathrm{val}_{\varphi}(s)$ *for all states* s.

Proof. We first prove that $s \models \varphi[v(s)]$ for all states s with $v(s) < \infty$. Hence, v is an upper bound on val_φ. We prove this by induction on the number of iterations the while loop has performed before assigning a finite value to $v(s)$. Note that this is the same iteration when s is put into X and that $v(s)$ never changes afterwards. If s is put into X before the first iteration, then $s \models b$ and therefore also $s \models \varphi[0] = \varphi[v(s)]$. Now assume that the while loop has already completed i iterations and is about to add s to X in the current iteration; let X, r and Y be as at the beginning of this iteration (after r and Y have been assigned, but before any new state is added to X). By the induction hypothesis, $t \models \varphi[r]$ for all $t \in Y$. Since s is added to X, we have that $s \models a \wedge Q\mathsf{X}(Z \cup Y)$. Using Lemma 4 and some basic PRCTL* laws, we can conclude that $s \models \varphi[v(s)]$ as follows:

$$\begin{aligned}
& s \models a \wedge Q\mathsf{X}(Z \cup Y) \\
\implies\ & s \models a \wedge Q\mathsf{X}(Z \cup (\neg Z \wedge Q(a \mathsf{U}_{\leq r} b))) \\
\implies\ & s \models a \wedge Q\mathsf{X}(Z \cup (\neg Z \wedge (a \mathsf{U}_{\leq r} b))) \\
\implies\ & s \models a \wedge Q\mathsf{X}(a \mathsf{U}_{\leq r} b) \\
\implies\ & s \models Q(a \mathsf{U}_{\leq r+rew(s)} b) \\
\implies\ & s \models \varphi[v(s)]
\end{aligned}$$

To complete the proof, we need to show that v is also a lower bound on val_φ. We define a strict partial order $\prec$ on states by setting $s \prec t$ if one of the following conditions holds:

1. $s \models b$ and $t \not\models b$,
2. $\mathrm{val}_\varphi(s) < \mathrm{val}_\varphi(t)$, or
3. $\mathrm{val}_\varphi(s) = \mathrm{val}_\varphi(t)$ and $rew(s) > rew(t)$.

Towards a contradiction, assume that the set C of states s with $\mathrm{val}_\varphi(s) < v(s)$ is non-empty, and pick a state $s \in C$ that is minimal with respect to $\prec$ (in particular, $\mathrm{val}_\varphi(s) < \infty$). Since $s \models \varphi[\infty]$ and the algorithm correctly sets $v(s)$ to 0 if $s \models b$, we know that $s \models a \wedge \neg b$ and $\mathrm{val}_\varphi(s) \geq rew(s)$. Moreover, by Proposition 3, $s \models \varphi[\mathrm{val}_\varphi(s)]$. Let T be the set of all states $t \in S \setminus Z$ such that $\mathrm{val}_\varphi(t) + rew(s) \leq \mathrm{val}_\varphi(s)$, i.e. $t \models \varphi[\mathrm{val}_\varphi(s) - rew(s)]$. Note that $T \neq \emptyset$ (because every state labelled with b is in T) and that $t \prec s$ for all $t \in T$. Since s is a minimal counter-example, we know that $v(t) \leq \mathrm{val}_\varphi(t) < \infty$ for all $t \in T$. Consequently, after some number of iterations of the while loop all elements of T have been added to X and the numbers $v(t)$ have been added to R. Since R is empty upon termination, in a following iteration we have that $r = \max\{v(t) : t \in T\}$ and that $T \subseteq Y$. Let $x := \mathrm{val}_\varphi(s) - rew(s)$. Using Lemma 4 and some basic PRCTL* laws, we can conclude that $s \models Q\mathsf{X}(Z \cup Y)$ as follows:

$$\begin{aligned}
& s \models \neg b \wedge \varphi[\mathrm{val}_\varphi(s)] \\
\implies\ & s \models Q(\neg b \wedge (a \mathsf{U}_{\leq x+rew(s)} b)) \\
\implies\ & s \models Q\mathsf{X}(a \mathsf{U}_{\leq x} b) \\
\implies\ & s \models Q\mathsf{X}(Z \cup (\neg Z \wedge (a \mathsf{U}_{\leq x} b))) \\
\implies\ & s \models Q\mathsf{X}(Z \cup (\neg Z \wedge Q(a \mathsf{U}_{\leq x} b)))
\end{aligned}$$

$$\implies s \models Q\,\mathsf{X}(Z \,\mathsf{U}\, T)$$
$$\implies s \models Q\,\mathsf{X}(Z \,\mathsf{U}\, Y)$$

Since also $s \models a$, this means that s is added to X no later than in the current iteration. Hence, $v(s) \leq r + rew(s) \leq \mathrm{val}_\varphi(s)$, which contradicts our assumption that $s \in C$. □

Theorem 6. *Qualitative queries of the form $Q(a \,\mathsf{U}_{\leq ?}\, b)$ can be evaluated in strongly polynomial time.*

Proof. By Lemma 5, Algorithm 1 can be used to compute the values of $Q(a \,\mathsf{U}_{\leq ?}\, b)$. During the execution of the algorithm, the running time of one iteration of the while loop is dominated by computing the set of states that fulfil the PCTL* formula $Q\,\mathsf{X}(Z \,\mathsf{U}\, Y)$, which can be done in time $\mathrm{O}(|\delta|)$ for $Q \in \{\forall\mathsf{P}_{>0}, \exists\mathsf{P}_{>0}, \forall\mathsf{P}_{=1}\}$ and in time $\mathrm{O}(|S| \cdot |\delta|)$ for $Q = \exists\mathsf{P}_{=1}$ (see [3, Chapter 10]). In each iteration of the while loop, one element of R is removed, and the number of elements that are put into R in total is bounded by the number of states in the given MDP. Hence, the number of iterations is also bounded by the number of states, and the algorithm runs in time $\mathrm{O}(|S| \cdot |\delta|)$ or $\mathrm{O}(|S|^2 \cdot |\delta|)$, depending on Q. Finally, since the only arithmetic operation used by the algorithm is addition, the algorithm is strongly polynomial. □

Of course, queries of the form $\exists\mathsf{P}_{>0}(a \,\mathsf{U}_{\leq ?}\, b)$ can actually be evaluated in time $\mathrm{O}(|S|^2 + |\delta|)$ using Dijkstra's algorithm since the value of a state with respect to such a query is just the weight of a shortest path from s via a-labeled states to a b-labelled state.

Algorithm 1 also gives us a useful upper bound on the value of a state with respect to a qualitative query.

Proposition 7. *Let $\mathcal{M}$ be an MDP, $\varphi = Q(a \,\mathsf{U}_{\leq ?}\, b)$ a qualitative quantile query, $n = |\lambda^{-1}(a)|$, and $c = \max\{rew(s) : s \in \lambda^{-1}(a)\}$. Then $\mathrm{val}_\varphi(s) \leq nc$ for all states s with $\mathrm{val}_\varphi(s) < \infty$.*

Proof. By induction on the number of iterations Algorithm 1 performs before assigning a finite number to $v(s)$. □

Finally, let us remark that our algorithm can be extended to handle queries of the form $Q(a \,\mathsf{U}_{\geq ?}\, b)$, where a *lower bound* on the accumulated reward is sought. To this end, the initialisation step has to be extended to identify states with value $-\infty$ and the rule for discovering new states has to be modified slightly. We invite the reader to make the necessary modifications and to verify the correctness of the resulting algorithm. This proves that the fragment of PRCTL with probability thresholds 0 and 1 and without reward constraints of the form $= r$ can be model-checked in polynomial time. Previously, a polynomial-time algorithm was only known for the special case where the models are restricted to MDPs in which every loop contains a state with nonzero reward [13].

5 Evaluating Quantitative Queries

In the following, we assume that all state rewards are natural numbers. This does not limit the applicability of our results since any MDP $\mathcal{M}$ with non-negative rational numbers as state rewards can be converted efficiently to an MDP $\mathcal{M}'$ with natural rewards by multiplying all state rewards with the least common multiple K of all denominators occurring in state rewards. It follows that $\mathrm{val}_\varphi^{\mathcal{M}'}(s) = K \cdot \mathrm{val}_\varphi^{\mathcal{M}}(s)$ for any quantile query φ and any state s of $\mathcal{M}$, so in order to evaluate a quantile query on $\mathcal{M}$ we can evaluate it on $\mathcal{M}'$ and divide by K. Throughout this section, we also assume that any transition probability and any probability threshold p occurring in a quantile query is rational. Finally, we define the *size* of an MDP $\mathcal{M} = (S, Act, \gamma, \lambda, rew, \delta)$ to be $|M| := \sum_{s\in S} \|rew(s)\| + \sum_{(s,\alpha,t)\in\delta,\alpha\in\gamma(s)} \|\delta(s,\alpha,t)\|$, where $\|x\|$ denotes the length of the binary representation of x.

5.1 Existential Queries

In order to solve queries of the form $\exists\mathsf{P}_{\geq p}(a \,\mathsf{U}_{\leq ?}\, b)$ or $\exists\mathsf{P}_{>p}(a \,\mathsf{U}_{\leq ?}\, b)$, we first show how to compute the *maximal* probabilities for fulfilling the path formula $a \,\mathsf{U}_{\leq r}\, b$ when we are given the reward bound r. Given an MDP $\mathcal{M}$, $a, b \in \mathsf{AP}$ and $r \in \mathbb{N}$, consider the following linear program over the variables $x_{s,i}$ for $s \in S$ and $i \in \{0, 1, \ldots, r\}$:

Minimise $\sum x_{s,i}$ subject to

$$x_{s,i} \geq 0 \qquad \text{for all } s \in S \text{ and } i \leq r,$$

$$x_{s,i} = 1 \qquad \text{for all } s \in \lambda^{-1}(b) \text{ and } i \leq r,$$

$$x_{s,i} \geq \textstyle\sum_{t\in S} \delta(s,\alpha,t) \cdot x_{t,i-rew(s)}$$

$$\text{for all } s \in \lambda^{-1}(a),\ \alpha \in Act \text{ and } rew(s) \leq i \leq r.$$

This linear program is of size $r \cdot |\mathcal{M}|$, and it can be shown that setting $x_{i,s}$ to $\max_\sigma \Pr_s^\sigma(a \,\mathsf{U}_{\leq i}\, b)$ yields the optimal solution. Hence, we can compute the numbers $\max_\sigma \Pr_s^\sigma(a \,\mathsf{U}_{\leq i}\, b)$ in time $\mathsf{poly}(r \cdot |\mathcal{M}|)$.

Our algorithm for computing the value of a state s wrt. a query of the form $\exists\mathsf{P}_{>p}(a \,\mathsf{U}_{\leq ?}\, b)$ just computes the numbers $\max_\sigma \Pr_s^\sigma(a \,\mathsf{U}_{\leq i}\, b)$ for increasing i and stops as soon as this probability exceeds p. However, in order to make this algorithm work and to show that it does not take too much time, we need a bound on the value of s provided this value is not infinite. Such a bound can be derived from the following lemma, which resembles a result by Hansen et al., who gave a bound on the convergence rate of *value iteration* in *concurrent reachability games* [11]. Our proof is technically more involved though, since we have to deal with paths that from some point onwards do not earn any more rewards.

Lemma 8. *Let $\mathcal{M}$ be an MDP where the denominator of each transition probability is at most m, and let $n = |\lambda^{-1}(a)|$, $c = \max\{rew(s) : s \in \lambda^{-1}(a)\}$ and $r = kncm^{-n}$ for some $k \in \mathbb{N}^+$. Then $\max_\sigma \Pr_s^\sigma(a \,\mathsf{U}\, b) < \max_\sigma \Pr_s^\sigma(a \,\mathsf{U}_{\leq r}\, b) + \mathrm{e}^{-k}$ for all $s \in S$.*

Proof. Without loss of generality, assume that all b-labelled states are absorbing. Let us call a state s of $\mathcal{M}$ *dead* if $s \models \forall \mathsf{P}_{=0}(a \mathsf{U} b)$, and denote by D the set of dead states. Note that $s \in D$ for all states s with $s \models \neg a \wedge \neg b$. Finally, let τ be a memoryless scheduler such that $\Pr_s^\tau(a \mathsf{U} b) = \max_\sigma \Pr_s^\sigma(a \mathsf{U} b)$ for all states s, and denote by Z the set of all states s with $s \models a \wedge \neg b$ and $rew(s) = 0$. By the definition of D and Z, we have that $\Pr_s^\tau(a \mathsf{U}_{\leq r} (D \vee \mathsf{G}Z) \wedge a \mathsf{U} b) = 0$ for all $s \in S$. Moreover, if s is not dead, then there must be a simple path from s to a b-labelled state via a-labelled states in the Markov chain induced by τ. Since any a-labelled state has reward at most c, this implies that $\Pr_s^\tau(a \mathsf{U}_{\leq nc} b) \geq m^{-n}$ for all non-dead states s. Now let ψ be the path formula $b \vee D \vee \mathsf{G}Z$. We claim that $\Pr_s^\tau(\neg(a \mathsf{U}_{\leq r} \psi)) < \mathrm{e}^{-k}$ for all states s. To prove this, let $s \in S$. We first show that $\Pr_s^\tau(a \mathsf{U}_{\leq i+nc} \psi \mid \neg(a \mathsf{U}_{\leq i} \psi)) \geq m^{-n}$ for all $i \in \mathbb{N}$ with $\Pr_s^\tau(a \mathsf{U}_{\leq i} \psi) < 1$. Let X be the set of sequences $xt \in S^* \cdot S$ such that $xt \in \{s \in S \setminus D : s \models a \wedge \neg b\}^*$, $rew(x) \leq i$ and $rew(xt) > i$. It is easy to see that the set $\{xt \cdot S^\omega : xt \in X\}$ is a partition of the set of infinite sequences over S that violate $a \mathsf{U}_{\leq i} \psi$. Using the fact that τ is memoryless, we can conclude that

$$\begin{aligned}
&\Pr_s^\tau(a \mathsf{U}_{\leq i+nc}\, \psi \mid \neg(a \mathsf{U}_{\leq i}\, \psi)) \\
&\geq \Pr_s^\tau(a \mathsf{U}_{\leq i+nc}\, b \mid \neg(a \mathsf{U}_{\leq i}\, \psi)) \\
&= \Pr_s^\tau(a \mathsf{U}_{\leq i+nc}\, b \cap X \cdot S^\omega) / \Pr_s^\tau(X \cdot S^\omega) \\
&= \sum_{xt \in X} \Pr_s^\tau(a \mathsf{U}_{\leq i+nc}\, b \cap xt \cdot S^\omega) / \Pr_s^\tau(X \cdot S^\omega) \\
&= \sum_{xt \in X} \Pr_t^\tau(a \mathsf{U}_{\leq i-rew(x)+nc}\, b) \cdot \Pr_s^\tau(xt \cdot S^\omega) / \Pr_s^\tau(X \cdot S^\omega) \\
&\geq \sum_{xt \in X} \Pr_t^\tau(a \mathsf{U}_{\leq nc}\, b) \cdot \Pr_s^\tau(xt \cdot S^\omega) / \Pr_s^\tau(X \cdot S^\omega) \\
&\geq \sum_{xt \in X} m^{-n} \cdot \Pr_s^\tau(xt \cdot S^\omega) / \Pr_s^\tau(X \cdot S^\omega) \\
&= m^{-n}\,.
\end{aligned}$$

Now, applying this inequality successively, we get that $\Pr_s^\tau(\neg(a \mathsf{U}_{\leq r} \psi)) \leq (1 - m^{-n})^{\frac{r}{nc}} = (1 - m^{-n})^{km^n} < \mathrm{e}^{-k}$. Finally,

$$\begin{aligned}
\Pr_s^\tau(a \mathsf{U} b) &= \Pr_s^\tau(a \mathsf{U} b \wedge \neg(a \mathsf{U}_{\leq r} (D \vee \mathsf{G}Z))) \\
&\leq \Pr_s^\tau(\neg(a \mathsf{U}_{\leq r} (D \vee \mathsf{G}Z))) \\
&\leq \Pr_s^\tau(\neg(a \mathsf{U}_{\leq r} \psi) \vee (a \mathsf{U}_{\leq r} b)) \\
&\leq \Pr_s^\tau(\neg(a \mathsf{U}_{\leq r} \psi)) + \Pr_s^\tau(a \mathsf{U}_{\leq r} b) \\
&< \mathrm{e}^{-k} + \max\nolimits_\sigma \Pr_s^\sigma(a \mathsf{U}_{\leq r} b)
\end{aligned}$$

for all $s \in S$. Since $\Pr_s^\tau(a \mathsf{U} b) = \max_\sigma \Pr_s^\sigma(a \mathsf{U} b)$, this inequality proves the lemma. □

Given an MDP $\mathcal{M}$ and $a, b \in \mathsf{AP}$, we denote by $\tilde{\mathcal{M}}$ the MDP that arises from $\mathcal{M}$ by performing the following transformation:

1. In each state s, remove all actions α with $\sum_{t\in S} \delta(s,\alpha,t) \cdot \max_\sigma \Pr_t^\sigma(a \,\mathsf{U}\, b) < \max_\sigma \Pr_s^\sigma(a \,\mathsf{U}\, b)$ from the set $\gamma(s)$ of enabled actions.
2. Label all states s such that $s \models \mathsf{P}_{=0}(a \,\mathsf{U}\, b)$ with b.

The following lemma, whose proof is rather technical, allows us to reduce the query $\exists\mathsf{P}_{\geq p}(a \,\mathsf{U}_{\leq ?}\, b)$ to the qualitative query $\exists\mathsf{P}_{=1}(a \,\mathsf{U}_{\leq ?}\, b)$ in the special case that p equals the optimal probability of fulfilling $a \,\mathsf{U}\, b$.

Lemma 9. *Let $\mathcal{M}$ be an MDP, $\varphi = \exists\mathsf{P}_{\geq p}(a \,\mathsf{U}_{\leq ?}\, b)$ and $\tilde{\varphi} = \exists\mathsf{P}_{=1}(a \,\mathsf{U}_{\leq ?}\, b)$. Then $\mathrm{val}_\varphi^{\mathcal{M}}(s) = \mathrm{val}_{\tilde{\varphi}}^{\tilde{\mathcal{M}}}(s)$ for all states s of $\mathcal{M}$ with $p = \max_\sigma \Pr_s^\sigma(a \,\mathsf{U}\, b)$.*

With the help of Lemmas 8 and 9, we can devise an upper bound for the value of any query whose value is finite.

Lemma 10. *Let $\mathcal{M}$ be an MDP where the denominator of each transition probability is at most m, $\varphi = \exists\mathsf{P}_{\rhd p}(a \,\mathsf{U}_{\leq ?}\, b)$ for $\rhd \in \{\geq, >\}$, $n = |\lambda^{-1}(a)|$, $c = \max\{rew(s) : s \in \lambda^{-1}(a)\}$, $s \in S$, and $q = \max_\sigma \Pr_s^\sigma(a \,\mathsf{U}\, b)$. Then at least one of the following statements holds:*

1. *$p \geq q$ and $\mathrm{val}_\varphi(s) = \infty$.*
2. *$p = q$, $\rhd = \geq$ and $\mathrm{val}_\varphi(s) \leq nc$.*
3. *$p < q$ and $\mathrm{val}_\varphi(s) \leq kncm^n$, where $k = \max\{-\lfloor \ln(q-p) \rfloor, 1\}$.*

Proof. Clearly, if either $\rhd = >$ and $p \geq q$ or $\rhd = \geq$ and $p > q$, then $\mathrm{val}_\varphi(s) = \infty$, and 1. holds. Now assume that $p = q$ and $\rhd = \geq$. By Lemma 9, we have that $\mathrm{val}_\varphi^{\mathcal{M}}(s) = \mathrm{val}_{\tilde{\varphi}}^{\tilde{\mathcal{M}}}(s)$. Hence, if $\mathrm{val}_{\tilde{\varphi}}^{\tilde{\mathcal{M}}}(s) = \infty$, then 1. holds. On the other hand, if $\mathrm{val}_{\tilde{\varphi}}^{\tilde{\mathcal{M}}}(s) < \infty$, then Proposition 7 gives us that $\mathrm{val}_{\tilde{\varphi}}^{\tilde{\mathcal{M}}}(s) \leq nc$, and 2. holds. Finally, if $p < q$, then let $r := kncm^n$. By Lemma 8, we have that $\max_\sigma \Pr_s(a \,\mathsf{U}_{\leq r}\, b) > q - e^{-k} \geq q - e^{\lfloor \ln(q-p) \rfloor} \geq q - (q-p) = p$, i.e. $s \models \exists\mathsf{P}_{\rhd p}(a \,\mathsf{U}_{\leq r}\, b)$. Hence, $\mathrm{val}_\varphi(s) \leq r$, and 3. holds. □

It follows from Lemma 10 that we can compute the value of a state s wrt. a query φ of the form $\exists\mathsf{P}_{>p}(a \,\mathsf{U}_{\leq ?}\, b)$ as follows: First compute the maximal probability q of fulfilling $a \,\mathsf{U}\, b$ from s, which can be done in polynomial time. If $p \geq q$, we know that the value of s wrt. φ must be infinite. Otherwise, $\mathrm{val}_\varphi(s) \leq r := kncm^n$, where $k = \max\{-\lfloor \ln(q-p) \rfloor, 1\}$, and we can find the least i such that $\max_\sigma \Pr_s^\sigma(a \,\mathsf{U}_{\leq i}\, b) > p$ by computing $\max_\sigma \Pr_s^\sigma(a \,\mathsf{U}_{\leq i}\, b)$ for all $i \in \{0, 1, \dots, r\}$, which can be done in time $\mathsf{poly}(r \cdot |\mathcal{M}|)$. Since r is exponential in the number of states of the given MDP $\mathcal{M}$, the running time of this algorithm is exponential in the size of $\mathcal{M}$. If φ is of the form $\exists\mathsf{P}_{\geq p}(a \,\mathsf{U}_{\leq ?}\, b)$, the algorithm is similar, but in the case that $p = q$, we compute $\max_\sigma \Pr_s^\sigma(a \,\mathsf{U}_{\leq i}\, b)$ for all $i \in \{0, 1, \dots, nc\}$ in order to determine whether the value is infinite or one of these numbers i.

Theorem 11. *Queries of the form $\exists\mathsf{P}_{\geq p}(a \,\mathsf{U}_{\leq ?}\, b)$ or $\exists\mathsf{P}_{>p}(a \,\mathsf{U}_{\leq ?}\, b)$ can be evaluated in exponential time.*

5.2 Universal Queries

In order to solve queries of the form $\forall \mathsf{P}_{>p}(a \,\mathsf{U}_{\leq ?}\, b)$, we first show how to compute the *minimal* probabilities for fulfilling the path formula $a \,\mathsf{U}_{\leq r}\, b$ when we are given the reward bound r. Given an MDP $\mathcal{M}$, $a, b \in \mathsf{AP}$ and $r \in \mathbb{N}$, consider the following linear program over the variables $x_{s,i}$ for $s \in S$ and $i \in \{0, 1, \ldots, r\}$:

$$
\begin{aligned}
&\text{Maximise } \textstyle\sum x_{s,i} \text{ subject to} \\
&x_{s,i} \leq 1 && \text{for all } s \in S \text{ and } i \leq r, \\
&x_{s,i} = 0 && \text{for all } s \in S \text{ with } s \not\models \forall \mathsf{P}_{>0}(a \,\mathsf{U}_{\leq i}\, b) \text{ and } i \leq r, \\
&x_{s,i} \leq \textstyle\sum_{t \in S} \delta(s, \alpha, t) \cdot x_{t, i - rew(s)} \\
& && \text{for all } s \in S \setminus \lambda^{-1}(b),\ \alpha \in Act \text{ and } rew(s) \leq i \leq r.
\end{aligned}
$$

This program is of size $r \cdot |\mathcal{M}|$, and it can be shown that setting $x_{i,s}$ to $\min_\sigma \Pr_s^\sigma(a \,\mathsf{U}_{\leq i}\, b)$ yields the optimal solution. Since the set of states s with $s \models \forall \mathsf{P}_{>0}(a \,\mathsf{U}_{\leq i}\, b)$ can be computed in polynomial time (Theorem 6), this means that we can compute the numbers $\min_\sigma \Pr_s^\sigma(a \,\mathsf{U}_{\leq i}\, b)$ in time $\mathsf{poly}(r \cdot |\mathcal{M}|)$. The following lemma is the analogue of Lemma 8 for minimal probabilities.

Lemma 12. *Let $\mathcal{M}$ be an MDP where the denominator of each transition probability is at most m, and let $n = |\lambda^{-1}(a)|$, $c = \max\{rew(s) : s \in \lambda^{-1}(a)\}$ and $r = kncm^{-n}$ for some $k \in \mathbb{N}^+$. Then* $\min_\sigma \Pr_s^\sigma(a \,\mathsf{U}\, b) < \min_\sigma \Pr_s^\sigma(a \,\mathsf{U}_{\leq r}\, b) + \mathrm{e}^{-k}$ *for all $s \in S$.*

Proof. Without loss of generality, assume that all b-labelled states are absorbing. Let us call a state s of $\mathcal{M}$ *dull* if $s \models \exists \mathsf{P}_{=0}(a \,\mathsf{U}\, b)$, and denote by D the set of dull states. Note that $s \in D$ for all states s with $s \models \neg a \wedge \neg b$. If s is not dull, then it is easy to see that, for any scheduler σ, the probability of reaching a b-labelled state from s in *at most n steps* (while seeing only a-labelled states before reaching a b-labelled state) is at least m^{-n}. Since any a-labelled state has reward at most c, we get that $\Pr_s^\sigma(a \,\mathsf{U}_{\leq nc}\, b) \geq m^{-n}$ for all non-dull states s and all schedulers σ. In the following, denote by Z the set $\{s \in S : s \models a \wedge \neg b \text{ and } rew(s) = 0\}$, and let ψ be the path formula $b \vee D \vee \mathsf{G} Z$. In the same way as in the proof of Lemma 8, we can infer that $\Pr_s^\sigma(\neg(a \,\mathsf{U}_{\leq r}\, \psi)) < \mathrm{e}^{-k}$ for all states s and all schedulers σ. Now fix a scheduler τ that minimises $\Pr_s^\tau(a \,\mathsf{U}_{\leq r}\, b)$ for all $s \in S$ and a scheduler σ such that $\Pr_s^\sigma(a \,\mathsf{U}\, b) = 0$ for all $s \in D$. From τ and σ, we devise another scheduler τ^* by setting

$$
\tau^*(x) = \begin{cases} \tau(x) & \text{if } x \in (S \setminus D)^*, \\ \sigma(x_2) & \text{if } x = x_1 \cdot x_2 \text{ where } x_1 \in (S \setminus D)^* \text{ and } x_2 \in D \cdot S^*. \end{cases}
$$

Note that $\Pr_s^{\tau^*}(a \,\mathsf{U}_{\leq r}\, (D \vee \mathsf{G} Z) \wedge a \,\mathsf{U}\, b) = 0$ and $\Pr_s^{\tau^*}(a \,\mathsf{U}_{\leq r}\, (D \vee \mathsf{G} Z)) = \Pr_s^\tau(a \,\mathsf{U}_{\leq r}\, (D \vee \mathsf{G} Z))$ for all $s \in S$. Hence,

$$\begin{aligned}
\Pr_s^{\tau^*}(a \mathsf{U} b) &= \Pr_s^{\tau^*}(a \mathsf{U} b \wedge \neg(a \mathsf{U}_{\le r} (D \vee \mathsf{G}Z))) \\
&\le \Pr_s^{\tau^*}(\neg(a \mathsf{U}_{\le r} (D \vee \mathsf{G}Z))) \\
&= \Pr_s^{\tau}(\neg(a \mathsf{U}_{\le r} (D \vee \mathsf{G}Z))) \\
&\le \Pr_s^{\tau}(\neg(a \mathsf{U}_{\le r} \psi) \vee (a \mathsf{U}_{\le r} b)) \\
&\le \Pr_s^{\tau}(\neg(a \mathsf{U}_{\le r} \psi)) + \Pr_s^{\tau}(a \mathsf{U}_{\le r} b) \\
&< e^{-k} + \Pr_s^{\tau}(a \mathsf{U}_{\le r} b) \\
&= e^{-k} + \min_\sigma \Pr_s^{\sigma}(a \mathsf{U}_{\le r} b)
\end{aligned}$$

for all $s \in S$. Since $\min_\sigma \Pr_s^\sigma(a \mathsf{U} b) \le \Pr_s^{\tau^*}(a \mathsf{U} b)$, this inequality proves the lemma. □

With the help of Lemma 12, we can devise an upper bound for the value of a query of the form $\forall \mathsf{P}_{>p}(a \mathsf{U}_{\le ?} b)$ in case this value is finite.

Lemma 13. *Let $\mathcal{M}$ be an MDP where the denominator of each transition probability is $\le m$, $\varphi = \forall \mathsf{P}_{>p}(a \mathsf{U}_{\le ?} b)$, $n = |\lambda^{-1}(a)|$, $c = \max\{rew(s) : s \in \lambda^{-1}(a)\}$, $s \in S$, and $q = \min_\sigma \Pr_s^\sigma(a \mathsf{U} b)$. Then one of the following statements holds:*

1. *$p \ge q$ and $\mathrm{val}_\varphi(s) = \infty$.*
2. *$p < q$ and $\mathrm{val}_\varphi(s) \le kncm^n$, where $k = \max\{-\lfloor \ln(q-p) \rfloor, 1\}$.*

Proof. Clearly, if $p \ge q$, then $\mathrm{val}_\varphi(s) = \infty$, and 1. holds. On the other hand, if $p < q$, then let $r := kncm^n$. By Lemma 12, we have that $\min_\sigma \Pr_s(a \mathsf{U}_{\le r} b) > q - e^{-k} \ge q - e^{\lfloor \ln(q-p) \rfloor} \ge q - (q-p) = p$, i.e. $s \models \forall \mathsf{P}_{>p}(a \mathsf{U}_{\le r} b)$. Hence, $\mathrm{val}_\varphi(s) \le r$, and 3. holds. □

As in the last section, Lemma 13 can be used to derive an exponential algorithm for computing the value of a state wrt. a query of the form $\forall \mathsf{P}_{>p}(a \mathsf{U}_{\le ?} b)$.

Theorem 14. *Queries of the form $\forall \mathsf{P}_{>p}(a \mathsf{U}_{\le ?} b)$ can be evaluated in exponential time.*

Regarding queries of the form $\forall \mathsf{P}_{\ge p}(a \mathsf{U}_{\le ?} b)$, we can compute the value of a state s whenever the probability $\min_\sigma \Pr_s^\sigma(a \mathsf{U} b)$ differs from p using the same algorithm. However, in the case that $p = \min_\sigma \Pr_s^\sigma(a \mathsf{U} b)$ it is not clear how to bound the value of s. As the following example shows, the analogous bound of nc for existential queries from Lemma 10 does not apply in this case.

Example 15. Consider the MDP depicted in Fig. 1, where $Act = \{\flat, \natural\}$ and $q \in [0,1[$ is an arbitrary probability. A state's reward is depicted in its bottom half, and a transition from s to t labelled with α, p indicates that $\delta(s, \alpha, t) = p$. Only transitions from non-absorbing states with nonzero probability and corresponding to enabled actions are shown. Assuming that every state is labelled with a but only s_3 and s_5 are labelled with b, it is easy to see that $\min_\sigma \Pr_{s_0}^\sigma(a \mathsf{U} b) = \frac{1}{2}$. Moreover, a quick calculation reveals that the value of state s_0 with respect to the query $\forall \mathsf{P}_{\ge 1/2}(a \mathsf{U}_{\le ?} b)$ equals $-\lfloor 1/\log_2 q \rfloor$. Since q can be chosen arbitrarily close to 1, this value can be made arbitrarily high.

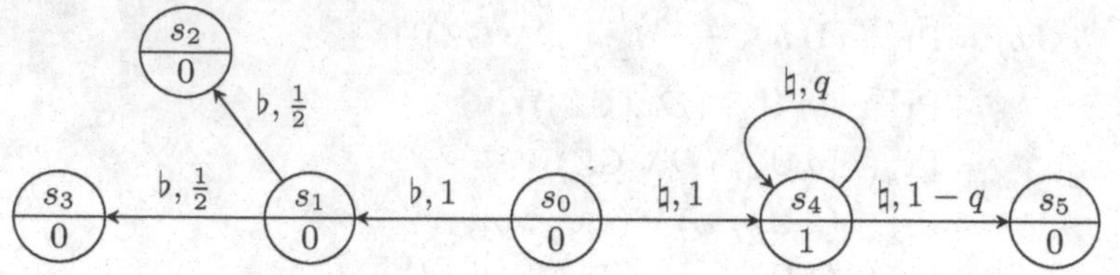

Fig. 1. An MDP with nonnegative rewards

5.3 A Pseudo-polynomial Algorithm for Markov Chains

In this section, we give a *pseudo-polynomial* algorithm for evaluating quantile queries of the form $\mathsf{P}_{\rhd p}(a \,\mathsf{U}_{\leq ?}\, b)$ on Markov chains. (Note that the quantifiers $\exists\mathsf{P}$ and $\forall\mathsf{P}$ coincide for Markov chains.) More precisely, our algorithm runs in time $\mathsf{poly}(c \cdot |\mathcal{M}| \cdot \|p\|)$ if c is the largest reward in $\mathcal{M}$. As an important special case, our algorithm runs in polynomial time on Markov chains where each state has reward 0 or 1.

Our polynomial-time algorithm relies on the following equations for computing the probability of the event $a \,\mathsf{U}_{=i}\, b$ in a Markov chain with rewards 0 and 1. Given such a Markov chain $\mathcal{M}$ and $a \in \mathsf{AP}$, we denote by Z the set of states s such that $\mathit{rew}(s) = 0$ and $s \models a \wedge \neg b$. Then the following equations hold for all $s \in S$, $a, b \in \mathsf{AP}$ and $r \in \mathbb{N}$:

- $\Pr_s(a \,\mathsf{U}_{=0}\, b) = \Pr_s(Z \,\mathsf{U}\, b)$,
- $\Pr_s(a \,\mathsf{U}_{=2r}\, b) = \sum_{t \in S \setminus Z} \Pr_s(a \,\mathsf{U}_{=r}\, \{t\}) \cdot \Pr_t(a \,\mathsf{U}_{=r}\, b)$,
- $\Pr_s(a \,\mathsf{U}_{=2r+1}\, b) = \sum_{t \in \lambda^{-1}(a) \setminus Z} \sum_{u \in S} \Pr_s(a \,\mathsf{U}_{=r}\, \{t\}) \cdot \delta(t,u) \cdot \Pr_u(a \,\mathsf{U}_{=r}\, b)$,

Using these equations, we can compute the numbers $\Pr_s(a \mathsf{U}_{=r} b)$ along the binary representation of r in time $\mathsf{O}(\mathsf{poly}(|\mathcal{M}|) \cdot \log r)$ for Markov chains with rewards 0 and 1 (see also [12]). Since any Markov chain $\mathcal{M}$ with rewards $0, 1, \ldots, c$ can easily be transformed into an equivalent Markov chain of size $c \cdot |\mathcal{M}|$ with rewards 0 and 1, the same numbers can be computed in time $\mathsf{O}(\mathsf{poly}(c \cdot |\mathcal{M}|) \cdot \log r)$ for general Markov chains. Finally, we can compute the numbers $\Pr_s(a \,\mathsf{U}_{\leq r}\, b)$ in the same time by first applying the following operations to each b-labelled state s: Make s absorbing, add a to $\lambda(s)$, and set $\mathit{rew}(s) = 1$; in the resulting Markov chain each state s fulfils $\Pr_s(a \,\mathsf{U}_{\leq r}\, b) = \Pr_s(a \,\mathsf{U}_{=r}\, b)$.

Now let $\varphi = \mathsf{P}_{\rhd p}(a \,\mathsf{U}_{\leq ?}\, b)$. Our algorithm for evaluating φ at state s of a Markov chain $\mathcal{M}$ is essentially the same algorithm as for MDPs. Hence, we first compute the probability $q := \Pr_s(a \,\mathsf{U}\, b)$. If either $p > q$ or $p = q$ and $\rhd = >$, then $\mathrm{val}_\varphi(s) = \infty$, by Lemma 10. If $p < q$, then the same lemma entails that $\mathrm{val}_\varphi(s) \leq r := kncm^n$, where $n = |\lambda^{-1}(a)|$, m is the least denominator of any transition probability, and $k = \max\{-\lfloor \ln(q-p) \rfloor, 1\} \leq \mathsf{poly}(\mathcal{M}) + \|p\|$. Hence, we can determine $\mathrm{val}_\varphi(s)$ using an ordinary binary search in time $\mathsf{O}(\mathsf{poly}(c \cdot |\mathcal{M}|) \cdot \log^2 r) = \mathsf{O}(\mathsf{poly}(c \cdot |\mathcal{M}| \cdot \|p\|))$. Finally, the same method can be applied if $p = q$ and $\rhd = \geq$ since Lemma 10 tells us that $\mathrm{val}_\varphi(s) \leq nc$ in this case.

Theorem 16. *Queries of the form* $\mathsf{P}_{\geq p}(a \mathsf{U}_{\leq ?} b)$ *or* $\mathsf{P}_{>p}(a \mathsf{U}_{\leq ?} b)$ *can be evaluated in pseudo-polynomial time on Markov chains.*

6 Conclusions

Although many researchers presented algorithms and several sophisticated techniques for the PCTL model checking problem and to solve PCTL and PRCTL queries, the class of quantile-based queries has not yet been addressed in the model checking community. In this paper, we presented algorithms for qualitative and quantitative quantile queries of the form $\mathsf{P}_{\bowtie p}(a\, \mathsf{U}_{\leq ?}\, b)$ and their duals $\exists \mathsf{P}_{\bowtie p}(a\, \mathsf{U}_{\leq ?}\, b)$. We established a polynomial algorithms for the qualitative case and exponential algorithms for all but one of the quantitative cases. Although the algorithms for the quantitative cases rely on a simple search algorithm for the quantile, the crucial feature is the bound we presented in Lemmas 8 and 12. These bounds might be interesting also for other purposes. There are several open problems to be studied in future work. First, the precise complexity of quantitative quantile queries is unknown and more efficient algorithms might exist, despite the NP-hardness shown in [14]. Second, we concentrated here on reward-bounded until properties, and by duality our results also apply to reward-bounded release properties. But quantile queries can also be derived from other PCTL-like formulas, such as formulas reasoning about expected rewards, e.g. in combination with step bounds.

Acknowledgments. We would like to thank Manuela Berg, Joachim Klein, Sascha Klüppelholz and Dominik Wojtczak for helpful discussions and the anonymous reviewers for their valuable remarks and suggestions.

References

1. Andova, S., Hermanns, H., Katoen, J.-P.: Discrete-Time Rewards Model-Checked. In: Larsen, K.G., Niebert, P. (eds.) FORMATS 2003. LNCS, vol. 2791, pp. 88–104. Springer, Heidelberg (2004)
2. Baier, C., Daum, M., Engel, B., Härtig, H., Klein, J., Klüppelholz, S., Märcker, S., Tews, H., Völp, M.: Waiting for Locks: How Long Does It Usually Take? In: Stoelinga, M., Pinger, R. (eds.) FMICS 2012. LNCS, vol. 7437, pp. 47–62. Springer, Heidelberg (2012)
3. Baier, C., Katoen, J.-P.: Principles of Model Checking. MIT Press (2008)
4. Bianco, A., de Alfaro, L.: Model Checking of Probabilistic and Nondeterministic Systems. In: Thiagarajan, P.S. (ed.) FSTTCS 1995. LNCS, vol. 1026, pp. 499–513. Springer, Heidelberg (1995)
5. Courcoubetis, C.A., Yannakakis, M.: The complexity of probabilistic verification. Journal of the ACM 42(4), 857–907 (1995)
6. de Alfaro, L.: Formal Verification of Probabilistic Systems. PhD thesis, Stanford University (1997)
7. de Alfaro, L.: Temporal Logics for the Specification of Performance and Reliability. In: Reischuk, R., Morvan, M. (eds.) STACS 1997. LNCS, vol. 1200, pp. 165–176. Springer, Heidelberg (1997)
8. de Alfaro, L.: How to specify and verify the long-run average behavior of probabilistic systems. In: Proceedings of the 13th Annual IEEE Symposium on Logic in Computer Science, LICS, pp. 454–465. IEEE Press (1998)

9. de Alfaro, L.: Computing Minimum and Maximum Reachability Times in Probabilistic Systems. In: Baeten, J.C.M., Mauw, S. (eds.) CONCUR 1999. LNCS, vol. 1664, pp. 66–81. Springer, Heidelberg (1999)
10. Forejt, V., Kwiatkowska, M., Norman, G., Parker, D.: Automated Verification Techniques for Probabilistic Systems. In: Bernardo, M., Issarny, V. (eds.) SFM 2011. LNCS, vol. 6659, pp. 53–113. Springer, Heidelberg (2011)
11. Hansen, K.A., Ibsen-Jensen, R., Miltersen, P.B.: The Complexity of Solving Reachability Games Using Value and Strategy Iteration. In: Kulikov, A., Vereshchagin, N. (eds.) CSR 2011. LNCS, vol. 6651, pp. 77–90. Springer, Heidelberg (2011)
12. Hansson, H., Jonsson, B.: A logic for reasoning about time and reliability. Formal Aspects of Computing 6(5), 512–535 (1994)
13. Jurdziński, M., Sproston, J., Laroussinie, F.: Model checking probabilistic timed automata with one or two clocks. Logical Methods in Computer Science 4(3) (2008)
14. Laroussinie, F., Sproston, J.: Model Checking Durational Probabilistic Systems. In: Sassone, V. (ed.) FOSSACS 2005. LNCS, vol. 3441, pp. 140–154. Springer, Heidelberg (2005)
15. Pekergin, N., Younès, S.: Stochastic Model Checking with Stochastic Comparison. In: Bravetti, M., Kloul, L., Zavattaro, G. (eds.) EPEW/WS-FM 2005. LNCS, vol. 3670, pp. 109–123. Springer, Heidelberg (2005)
16. Vardi, M.: Automatic verification of probabilistic concurrent finite-state programs. In: Proceedings of the 26th IEEE Symposium on Foundations of Computer Science, FOCS, pp. 327–338. IEEE Press (1985)

Parameterized Weighted Containment

Guy Avni and Orna Kupferman

School of Computer Science and Engineering, Hebrew University, Israel

Abstract. Partially-specified systems and specifications are used in formal methods such as stepwise design and query checking. Existing methods consider a setting in which the systems and their correctness are Boolean. In recent years there has been growing interest and need for quantitative formal methods, where systems may be weighted and specifications may be multi valued. Weighted automata, which map input words to a numerical value, play a key role in quantitative reasoning. Technically, every transition in a weighted automaton $\mathcal{A}$ has a cost, and the value $\mathcal{A}$ assigns to a finite word w is the sum of the costs on the transitions participating in the most expensive accepting run of $\mathcal{A}$ on w. We study *parameterized weighted containment*: given three weighted automata $\mathcal{A}$, $\mathcal{B}$, and $\mathcal{C}$, with $\mathcal{B}$ being partial, the goal is to find an assignment to the missing costs in $\mathcal{B}$ so that we end up with $\mathcal{B}'$ for which $\mathcal{A} \leq \mathcal{B}' \leq \mathcal{C}$, where $\leq$ is the weighted counterpart of containment. We also consider a one-sided version of the problem, where only $\mathcal{A}$ or only $\mathcal{C}$ are given in addition to $\mathcal{B}$, and the goal is to find a minimal assignment with which $\mathcal{A} \leq \mathcal{B}'$ or, respectively, a maximal one with which $\mathcal{B}' \leq \mathcal{C}$. We argue that both problems are useful in stepwise design of weighted systems as well as approximated minimization of weighted automata.

We show that when the automata are deterministic, we can solve the problems in polynomial time. Our solution is based on the observation that the set of legal assignments to k missing costs forms a k-dimensional polytope. The technical challenge is to find an assignment in polynomial time even though the polytope is defined by means of exponentially many inequalities. We do so by using a powerful mathematical tool that enables us to develop a divide-and-conquer algorithm based on a separation oracle for polytopes. For nondeterministic automata, the weighted setting is much more complex, and in fact even non-parameterized containment is undecidable. We are still able to study variants of the problems, where containment is replaced by simulation.

1 Introduction

The *automata-theoretic* approach uses the theory of automata as a unifying paradigm for system specification and verification [24,26]. By viewing computations as words (over the alphabet of possible assignments to variables of the system), we can view both the system and its specification as languages. Questions like satisfiability of specifications or their satisfaction can then be reduced to questions about automata and their languages.

The automata-theoretic approach has proven useful also in reasoning about *partially-specified* systems and specifications, where some components are not known or hidden. Partially-specified systems are used mainly in *stepwise design*: One starts with a system with "holes" and iteratively completes them in a way that satisfies some specification

F. Pfenning (Ed.): FOSSACS 2013, LNCS 7794, pp. 369–384, 2013.

[9,10]. Reasoning about partially-specified systems is useful also in automatic partial synthesis [23] and program repair [15]. From the other direction, partially-specified specifications are used for system exploration. In particular, in *query checking* [5], the specification contains variables, and the goal is to find an assignment to the variables with which the explored system satisfies the specification. For example, solutions to the query $\text{ALWAYS}(X_1 \rightarrow \text{EVENTUALLY}\,grant)$ assign to X_1 events that trigger a generation of a grant in the system. Missing information in the system or the specification can be easily encoded in an automaton that models it, and indeed algorithms for the above problems are based on partially specified automata (c.f., [4]).

Traditional automata accept or reject their input, and are therefore Boolean. In recent years, there is growing need and interest in quantitative reasoning. *Weighted finite automaton* (WFA, for short) map words to numerical values. Technically, every transition in a weighted automaton $\mathcal{A}$ has a cost, and the value that $\mathcal{A}$ assigns to a finite word w, denoted $val(\mathcal{A}, w)$, is the sum of the costs of the transitions participating in the most expensive accepting run of $\mathcal{A}$ on w. [1] Applications of weighted automata include formal verification, where they are used for the verification of quantitative properties [6], as well as text, speech, and image processing, where the weights of the automaton are used in order to account for the variability of the data and to rank alternative hypotheses [8,20].

In the Boolean setting, formal verification amounts to checking containment of the language of the system by the language of the specification. This makes the *language-containment* problem of great theoretical and practical interest. In the weighted setting, the analogous problem gets as input two weighted automata $\mathcal{A}$ and $\mathcal{B}$, and decides whether all the words w that are accepted by $\mathcal{A}$ are also accepted by $\mathcal{B}$ and $val(\mathcal{A}, w) \leq val(\mathcal{B}, w)$. We denote this by $\mathcal{A} \subseteq \mathcal{B}$. Weighted automata are much more complicated than Boolean ones. The source of the difficulty is the infinite domain of values that the automata may assign to words. In particular, the problem of weighted containment is in general undecidable [1,18]. Given the importance of the problem, researchers have studied decidable fragments and approximations of weighted containment. We know, for example, that weighted containment is decidable, in fact polynomial, for deterministic WFAs (DWFAs, for short). For general WFA, researchers have suggested a weighted variant of the simulation relation, which approximates weighted containment and is decidable [3,7].

In this paper, we introduce and study *parameterized weighted containment*: given three weighted automata $\mathcal{A}$, $\mathcal{B}$, and $\mathcal{C}$, with $\mathcal{B}$ being partial, the goal is to find an assignment to the missing costs in $\mathcal{B}$ so that we end up with $\mathcal{B}'$ for which $\mathcal{A} \subseteq \mathcal{B}' \subseteq \mathcal{C}$. We also consider a one-bounded version of the problem, where only $\mathcal{A}$ or only $\mathcal{C}$ are given in addition to $\mathcal{B}$, and the goal is to find a minimal assignment with which $\mathcal{A} \subseteq \mathcal{B}'$ or, respectively, a maximal one with which $\mathcal{B}' \subseteq \mathcal{C}$. [2]

[1] In general, weighted automata may be defined with respect to all semirings. For our application here, we consider WFAs over $\mathbb{Q}$, with the sum of the semi-ring being $+$ and its product being max.

[2] An orthogonal research direction is that of *parametric real-time reasoning* [2]. There, the quantitative nature of the automata origins from real-time constraints, the semantics is very different, and the goal is to find restrictions on the behavior of the clocks such that the automata satisfy certain properties.

Before we describe the technical details of the problems and their solutions, let us argue for their usefulness with two applications.

Example 1. Stepwise design Assume we have a weighted specification $\mathcal{C}$. Refining the specification to an implementation involves a refinement of its Boolean behavior, possibly extending its alphabet, and an assignment of values to the refined computations. When the values in $\mathcal{C}$ exhibit upper bounds on costs, we want the implementation $\mathcal{B}$ to satisfy $\mathcal{B} \subseteq \mathcal{C}$. It is relatively easy to refine the Boolean behavior of $\mathcal{C}$ and get an automaton whose language, when restricted to the joint alphabet, is contained in the language of $\mathcal{C}$. It is much harder to design the weighted behavior of $\mathcal{B}$. For this, we apply one-bound parameterized weighted containment: $\mathcal{C}$ is the specification, $\mathcal{B}$ is its Boolean refinement, we label its costs by variables, and we are looking for a maximal assignment for the variables with which $\mathcal{B}$ is contained in $\mathcal{C}$.

For a specific example, consider the problem of ranking contributors to user-generated sites (e.g., Wikipedia). A big challenge for these sites is to develop trust in users. We seek a WFA that distinguishes between good and bad edits. After a user performs an edit on the site, the WFA gives it a score, and decisions on blocking and promotion of users are based on these scores.

We assume that an edit is a sequence of words – these added by the user. We also assume we have a tool, which we refer to as the *mapper*, that, intuitively, performs a pre-processing that abstracts the edit the user performed. More formally, the mapper maps words to some fixed alphabet, which is the alphabet of the WFA. For example, a mapper might map the sentence "The dog bent uver." to the word "$the \cdot noun \cdot verb \cdot misspelledword$."

The WFA combines heuristics, each of which either identifies a positive linguistic feature of a sentence or a negative one. An example of a positive heuristic is: "a sentence in which the multiplicity of the subject matches that of the verb should get a score greater than $1/4$". An example of a negative heuristic is: "a sentence in which *the* appears before a verb should not get a score above $1/2$".

Devising a WFA that takes care of a single heuristic is simple. However, since the automata are weighted, combining them is complicated. Some variants of parameterized weighted containment are useful here: when we want to combine two positive heuristics, modeled by WFAs $\mathcal{A}_1$ and $\mathcal{A}_2$, we seek a minimal WFA $\mathcal{B}$ such that both $\mathcal{A}_1 \subseteq \mathcal{B}$ and $\mathcal{A}_2 \subseteq \mathcal{B}$. This variant of the one-bound problem is useful also when both heuristics are negative. Combining a negative heuristic $\mathcal{A}$ and a positive one $\mathcal{C}$ then corresponds to the problem of finding a WFA $\mathcal{B}$ such that $\mathcal{A} \subseteq \mathcal{B} \subseteq \mathcal{C}$.

Example 2. DWFA approximated minimization Minimization of Boolean deterministic automata is a well-studied problem. For DWFAs, Mohri described a (complicated yet polynomial) minimization algorithm [19]. We argue that two-bound parameterized weighted containment can be used in order to simplify Mohri's algorithm and, which we find more exciting, enables also *approximated minimization*. There, given a DWFA $\mathcal{A}$ and a factor $t \in \mathbb{Q}$, we would like to construct a minimal automaton $\mathcal{B}$ that has the same language as $\mathcal{A}$ and assigns values within a factor of t from $\mathcal{A}$. Given $\mathcal{A}$, we first construct the DWFAs $reduce(\mathcal{A}, t)$ and $increase(\mathcal{A}, t)$, for whatever definitions of reduce and increase we are after; for example, we can take $-t$ and $+t$ as additive factors to the value, or we can take $\frac{1}{t}$ and t as multiplicative ones. We then use parameterized weighted containment in order to find $\mathcal{B}$ such that $reduce(\mathcal{A}, t) \subseteq \mathcal{B} \subseteq increase(\mathcal{A}, t)$.

In both examples above, we left all the components of the generated WFA $\mathcal{B}$ unspecified. When the user has an idea about $\mathcal{B}$'s Boolean behavior, as is typically the case in step-wise refinement, this Boolean behavior is a natural starting point. In Section 3.2, we study the case only $\mathcal{A}$ and $\mathcal{C}$ are given, and we seek a minimal $\mathcal{B}$ such that $\mathcal{A} \subseteq \mathcal{B} \subseteq \mathcal{C}$. We show that the problem is NP-complete for DWFAs, and suggest a heuristic for finding $\mathcal{B}$ that is based on viewing the Boolean product of $\mathcal{A}$ and $\mathcal{C}$ as a partially specified WFA.

Let us now return to parameterized weighted containment where a partial WFA $\mathcal{B}$ is given. Our solution to the problem is based on strong mathematical tools. We explain here briefly the general idea for the two-bound problem for DWFAs. Consider an input $\mathcal{A}$, $\mathcal{B}$, and $\mathcal{C}$ to the problem. Assume that transitions in $\mathcal{B}$ are parameterized by variables from a set $\mathcal{X}$ of size k. Recall that we are looking for a legal assignment $f : \mathcal{X} \to \mathbb{Q}$; that is, one with which $\mathcal{A} \subseteq \mathcal{B}^f \subseteq \mathcal{C}$, where $\mathcal{B}^f$ is the DWFA obtained by replacing each variable $X \in \mathcal{X}$ by $f(X)$. We first show that the products $\mathcal{A} \times \mathcal{B}$ and $\mathcal{B} \times \mathcal{C}$ can be used in order to generate a set of inequalities that the variables have to satisfy. For that, we characterize *critical paths* in the products – it is necessary and sufficient to restrict the assignment of the variables in transitions along these paths in order to guarantee that f is legal. Each critical path induces an inequality and together the inequalities induces a convex polytope $P \subseteq \mathbb{R}^k$ that includes exactly all the legal assignments. Khachiyan's *Ellipsoid's method* [17] then enables us to find a point in this polytope or conclude that no legal assignment exists.

This is, however, not the end of the story. Unfortunately, the number of critical paths we have to consider is exponential, making a naive search for the solution exponential too. Examining Khachiyan's method one can see that it is not necessary to have an implicit list of inequalities that define the polytope P. Indeed, it was shown in [12,16,21] that it is sufficient to have a *separation oracle* for the polytope. That is, instead of a list of inequalities that define P, the input to the problem is an oracle that, given a point $p \in \mathbb{Q}^k$, either says that $p \in P$ or returns a half-space $H \subseteq \mathbb{Q}^k$ such that $p \notin H$ and $P \subseteq H$. We show that we can use the products $\mathcal{A} \times \mathcal{B}$ and $\mathcal{B} \times \mathcal{C}$ in order to define such a separation oracle, leading to polynomial-time a solution to the problem.

For the one-bound variant, we show that the induced polytope is *pointed*, and that the solution we are after is a *vertex* of it, leading to an actually simpler algorithm. For the case the automata are nondeterministic, we argue that the one-bound problem is not interesting, as a minimal/maximal solution need not exist. For the two bound problem, we approximate containment by simulation, and show that the problem is NP-hard. Also, a polynomial algorithm for deciding weighted simulation would imply that it is NP-complete. [3] Given the computational difficulty of handling nondeterministic WFAs in general, we view these results as good news: parameterized language containment can be solved in polynomial time for the deterministic setting, and its approximation by simulation is decidable in the nondeterministic one.

[3] The best algorithm currently known for weighted simulation is in NP $\cap$ co-NP [3].

Due to the lack of space, some of the proofs and examples are omitted in this version and can be found in the full version, in the authors' homepages.

2 Preliminaries

2.1 Weighted Automata

A nondeterministic finite weighted automaton on finite words (WFA, for short) is a tuple $\mathcal{A} = \langle \Sigma, Q, \Delta, Q_0, F, \tau \rangle$, where Σ is an alphabet, Q is a set of states, $\Delta \subseteq Q \times \Sigma \times Q$ is a transition relation, $Q_0 \subseteq Q$ is a set of initial states, $F \subseteq Q$ is a set of accepting states, and $\tau : \Delta \to \mathbb{Q}$ is a function that maps each transition to a rational value, which is the cost of traversing this transition. We assume that there are no redundant states in $\mathcal{A}$. That is, all states are not empty (an accepting state is reachable from them) and accessible (reachable from an initial state).

A run of $\mathcal{A}$ on a word $w = w_1, \ldots, w_n \in \Sigma^*$ is a sequence of states $r = r_0, r_1, \ldots, r_n$ such that $r_0 \in Q_0$ and for every $0 \leq i < n$ we have $\Delta(r_i, w_{i+1}, r_{i+1})$. The run r is accepting iff $r_n \in F$. The value of the run, denoted $val(r, w)$, is the sum of costs of transitions it traverses. That is, $val(r, w) = \sum_{0 \leq i < n} \tau(\langle r_i, w_{i+1}, r_{i+1} \rangle)$. Similarly, for a path π, which is a sequence of transitions, we define $val(\pi) = \sum_{e \in \pi} \tau(e)$. Since $\mathcal{A}$ is non-deterministic, there can be more than one run on a word. We define the value that $\mathcal{A}$ assigns to $w \in \Sigma^*$, denoted $val(\mathcal{A}, w)$, as the value of the maximal-valued accepting run of $\mathcal{A}$ on w. That is, $val(\mathcal{A}, w) = max\{val(r, w) : r \text{ is an accepting run of } \mathcal{A} \text{ on } w\}$. As in NFAs, the language of $\mathcal{A}$, denoted $L(\mathcal{A})$, is the set of words in Σ^* that $\mathcal{A}$ accepts.

We say that $\mathcal{A}$ is *deterministic* if $|Q_0| = 1$ and for every $q \in Q$ and $\sigma \in \Sigma$, there is at most one state $q' \in Q$ such that $\Delta(q, \sigma, q')$. Note that a deterministic WFA (DWFA, for short) has at most one run on every word in Σ^*.

Weighted Containment. For two WFAs $\mathcal{A}$ and $\mathcal{B}$, we say that $\mathcal{A}$ is contained in $\mathcal{B}$, denoted $\mathcal{A} \subseteq \mathcal{B}$, iff $L(\mathcal{A}) \subseteq L(\mathcal{B})$ and for every word $w \in L(\mathcal{A})$ we have that $val(\mathcal{A}, w) \leq val(\mathcal{B}, w)$. It is shown in [1,18] that deciding containment for WFAs is undecidable.

Negativeness. We say that a WFA $\mathcal{A}$ is *negative* if $val(\mathcal{A}, w) \leq 0$ for every word $w \in L(\mathcal{A})$. We say that a path π in $\mathcal{A}$ is a *critical* path iff it is either a simple path from an initial state to an accepting state or a simple cycle. Keeping in mind that all states in $\mathcal{A}$ are not empty and reachable from an initial state, it is not hard to prove the following characterization of negative DWFAs.

Proposition 1. *A DWFA $\mathcal{A}$ is negative iff $val(\pi) \leq 0$ for every critical path π in $\mathcal{A}$.*

Let $\mathcal{A} = \langle \Sigma, Q_\mathcal{A}, \Delta_\mathcal{A}, q_{0_\mathcal{A}}, F_\mathcal{A} \rangle$ and $\mathcal{B} = \langle \Sigma, Q_\mathcal{B}, \Delta_\mathcal{B}, q_{0_\mathcal{B}}, F_\mathcal{B} \rangle$ be two DWFAs. Consider the product $\mathcal{S}_{\mathcal{A},\mathcal{B}} = \langle \Sigma, Q_\mathcal{B} \times Q_\mathcal{A}, \Delta_{\mathcal{A},\mathcal{B}}, \langle q_{0_\mathcal{A}}, q_{0_\mathcal{B}} \rangle, F_\mathcal{A} \times F_\mathcal{B}, \tau_{\mathcal{A},\mathcal{B}} \rangle$, where $\Delta_{\mathcal{A},\mathcal{B}}$ is such that $t = \langle \langle u, v \rangle, \sigma, \langle u', v' \rangle \rangle \in \Delta_{\mathcal{A},\mathcal{B}}$ iff $t_\mathcal{A} = \langle u, \sigma, u' \rangle \in \Delta_\mathcal{A}$ and $t_\mathcal{B} = \langle v, \sigma, v' \rangle \in \Delta_\mathcal{B}$. We refer to the transitions $t_\mathcal{A}$ and $t_\mathcal{B}$ as the transitions that are mapped to t. Then, for every $t \in \Delta_{\mathcal{A},\mathcal{B}}$, we define $\tau_{\mathcal{A},\mathcal{B}}(t) = \tau_\mathcal{B}(t_\mathcal{B}) - \tau_\mathcal{A}(t_\mathcal{A})$, where $t_\mathcal{A}$ and $t_\mathcal{B}$ are mapped to t.

Assume that $L(\mathcal{A}) \subseteq L(\mathcal{B})$ and consider a word $w \in \Sigma^*$. Let $r = \langle r_0^{\mathcal{A}}, r_0^{\mathcal{B}} \rangle, \ldots, \langle r_{|w|}^{\mathcal{A}}, r_{|w|}^{\mathcal{B}} \rangle$ be the run of $\mathcal{S}_{\mathcal{A},\mathcal{B}}$ on w. It is easy to see that $r_0^{\mathcal{A}}, \ldots, r_{|w|}^{\mathcal{A}}$ is the run of $\mathcal{A}$ on w and $r_0^{\mathcal{B}}, \ldots, r_{|w|}^{\mathcal{B}}$ is the run of $\mathcal{B}$ on w. Thus, by the definition of the weight function of $\mathcal{S}_{\mathcal{A},\mathcal{B}}$, it follows that $val(\mathcal{S}_{\mathcal{A},\mathcal{B}}, w) = val(\mathcal{A}, w) - val(\mathcal{B}, w)$. Hence, we have the following proposition:

Proposition 2. *Let $\mathcal{A}$ and $\mathcal{B}$ be two DWFAs such that $L(\mathcal{A}) \subseteq L(\mathcal{B})$. Then, $\mathcal{A} \subseteq \mathcal{B}$ iff $\mathcal{S}_{\mathcal{A},\mathcal{B}}$ is not negative.*

2.2 Parameterized Weighted Containment

Consider a set of variables $\mathcal{X} = \{X_1, \ldots, X_k\}$. An *$\mathcal{X}$-parameterized WFA* is a WFA in which some of the costs are replaced by variables from $\mathcal{X}$. Thus, the weight function is of the form $\tau : \Delta \to \mathbb{Q} \cup \mathcal{X}$. Given an $\mathcal{X}$-parameterized WFA $\mathcal{A}$ and an assignment $f : \mathcal{X} \to \mathbb{Q}$ to the variables in $\mathcal{X}$, we obtain the WFA $\mathcal{A}^f$ by replacing every variable $X \in \mathcal{X}$ with the value $f(X)$. Formally, the components of the WFA $\mathcal{A}^f$ agree with these of $\mathcal{A}$ except for the weight function τ^f, which agrees with τ on all transitions $t \in \Delta$ with $\tau(t) \in \mathbb{Q}$, and is such that $\tau^f(t) = f(X)$ for all $t \in \Delta$ with $\tau(t) = X$, for some $X \in \mathcal{X}$. Note that a variable $X \in \mathcal{X}$ may appear in more than one transition of $\mathcal{A}$.

Definition 1. *We consider the following three variants of parameterized weighted containment (PWC, for short).*

- **Two bound PWC:** *Given WFAs $\mathcal{A}$ and $\mathcal{C}$, and an $\mathcal{X}$-parameterized WFA $\mathcal{B}$, find an assignment $f : \mathcal{X} \to \mathbb{Q}$ such that $\mathcal{A} \subseteq \mathcal{B}^f \subseteq \mathcal{C}$.*
- **Least upper bound PWC:** *Given a WFA $\mathcal{A}$ and an $\mathcal{X}$-parameterized WFA $\mathcal{B}$, find a minimal assignment $f : \mathcal{X} \to \mathbb{Q}$ such that $\mathcal{A} \subseteq \mathcal{B}^f$.*
- **Greatest lower bound PWC:** *Given a WFA $\mathcal{C}$ and an $\mathcal{X}$-parameterized WFA $\mathcal{B}$, find a maximal assignment $f : \mathcal{X} \to \mathbb{Q}$ such that $\mathcal{B}^f \subseteq \mathcal{C}$.*

The least-upper and greatest-lower bound variants are dual and we refer to them as *one-bound* PWC. Their definition uses the terms minimal and maximal, with the expected interpretation: an assignment f is minimal if decreasing the value it assigns to a variable results in a violation of the requirement that $\mathcal{A} \subseteq \mathcal{B}^f$. Formally, f is minimal if for every variable $X \in \mathcal{X}$ and every $\epsilon > 0$, the assignment f' that agrees with f on all variables $X \neq X' \in \mathcal{X}$ and $f'(X) = f(X) - \epsilon$ is such that $\mathcal{A} \not\subseteq \mathcal{B}^{f'}$. Note that without the minimality requirement, the upper-bound variant is trivial: for every variable $X \in \mathcal{X}$ we set $f(X)$ to be a very high value, for example, the maximal cost appearing in $\mathcal{A}$ times the size of $|\mathcal{A}| \cdot |\mathcal{B}|$. The definition of a maximal assignment is dual.

Solving parameterized weighted containment is clearly harder than solving weighted containment, and is therefore undecidable in general. We study two restrictions of the problem. In Section 3, we study the PWC problem where the automata are deterministic. As hinted in Proposition 2, containment is decidable for DWFAs. In Section 4 we study the PWC problem where the automata are nondeterministic, but we replace the containment relation with its approximating relation of *simulation* [3], which is decidable.

2.3 Geometry in $\mathbb{R}^k$

We briefly review some definitions on polytopes. For more details and intuition, see [22].

Polytopes. A *convex polytope* is a set in $\mathbb{R}^k$ that is the intersection of a finite number of *half-spaces*. Thus, it can be defined as the set of points $p \in \mathbb{R}^k$ that are solutions to a system of linear inequalities $Ax \leq b$, where $A \in \mathbb{Q}^{m,k}$ is an $m \times k$ matrix of rationals, $b \in \mathbb{Q}^m$, and $m \in \mathbb{N}$ is the number of inequalities. For example, the system of inequalities $2x_1 + 3x_2 \leq 7$, $5x_1 \leq 3$, and $4x_2 \leq 0$ corresponds to the following representation:

$$\begin{pmatrix} 2 & 3 \\ 5 & 0 \\ 0 & 4 \end{pmatrix} \begin{pmatrix} x_1 \\ x_2 \end{pmatrix} \leq \begin{pmatrix} 7 \\ 3 \\ 0 \end{pmatrix}$$

Dimension. We say that the points $p_1, \ldots, p_m \in \mathbb{R}^n$ are *affinely independent* iff the vectors $p_2 - p_1, p_3 - p_1, \ldots, p_m - p_1$ are independent. The dimension of a convex polytope $P \subseteq \mathbb{R}^k$, denoted $dim(P)$, is defined to be $l \leq k$ iff the maximal number of affinely independent points in P is $l + 1$.

For example, consider the line in Figure 1, which is a convex polytope in $\mathbb{R}^2$. We claim that the dimension of the polytope is 1. Indeed, points a and b in the polytope are affinely independent, as the single vector $b - a$ is linearly independent. On the other hand, points a, b, and c are not affinely independent, as $b - a$ and $c - a$ are linearly dependent.

We say that a polytope $P \subseteq \mathbb{R}^k$ is *full-dimensional* if its dimension is k. When P is not full-dimensional, it is contained in a hyper-plane of dimension less than k.

Vertices. In 2-dimensions, a vertex is the meeting point of two edges. In k-dimensions, a vertex is the meeting point of k *faces*, which are the k-dimensional generalization of edges. We say that a polytope P is *pointed* if it has a vertex. In the full version we define these notions formally.

Geometrical Objects. A k-dimensional *ball* is a generalization of the 2-dimensional circle. For $c \in \mathbb{R}^k$ (the center) and $r \in \mathbb{R}$ (the radius) we define the ball $B(c, r) = \{p \in \mathbb{R}^k : \sum_{1 \leq i \leq k} (p_i - c_i)^2 \leq r^2\}$. Consider a polytope $P = \{x \in \mathbb{R}^k : Ax \leq b\}$. We say that P is *bounded* iff there is a ball with a finite radius that contains it.

Consider an *invertible linear transformation* $L : \mathbb{R}^k \to \mathbb{R}^k$. For example, *rotation* is invertible but the transformation $L(p) = 0$, for every $p \in \mathbb{R}^k$, is not. An *ellipsoid* with center $0 \in \mathbb{R}^k$, is the implication of L on the unit ball $B(0, 1)$. That is, it is the set $L(B(0, 1)) = \{L(p) \in \mathbb{R}^k : p \in B(0, 1)\}$. An ellipsoid centered at $c \in \mathbb{R}^k$, is the translation of the set $L(B(0, 1))$ by c. That is, $Ell(c, L) = c + L(B(0, 1))$[4].

[4] In [22], an ellipsoid is defined as follows: $Ell(z, D) = \{p \in \mathbb{R}^k : (p - z)^T \cdot D^{-}1 \cdot (p - z) \leq 1\}$, where D is a positive definite matrix and $z \in \mathbb{R}^k$. The definition we use is equivalent to this definition.

Volume. Consider a set $S \subseteq \mathbb{R}^k$. We define the volume of S, denoted $vol(S)$, using the *Lebesgue measure*. The volume of a k-dimensional box $B = \{p \in \mathbb{R}^k : a_1 \leq p_1 \leq b_1, \ldots, a_k \leq p_k \leq b_k\}$ is $vol(B) = \prod_{1 \leq i \leq k}(b_i - a_i)$. Consider a collection of countably many boxes C such that $S \subseteq \bigcup_{B \in C} B$. We define $vol(S) = inf_C\{\sum_{B \in C} vol(B)\}$. An important observation is that the volume of a polytope that is not full-dimensional is 0. Generally, there are sets that are not Lebesque measurable. In this work, however, we only use convex sets, which are measurable.

Size of Representation. Consider a number $p/q \in \mathbb{Q}$. The *size* of p/q is, intuitively, the number of bits that are needed to represent it. Thus, we define the size of p/q, denoted $size(p/q)$, to be $1 + \lceil log(|p|+1) \rceil + \lceil log(|q|+1) \rceil$. We define the size of an inequality $\sum_{1 \leq i \leq k} a_i \cdot X_i \leq b$ to be $1 + \sum_{1 \leq i \leq k} size(a_i) + size(b)$.

The Ellipsoid Method. In 1979, Khachiyan [17] introduced the first polynomial time algorithm for feasibility of linear programming. In this problem we get as input a polytope $P = \{x \in \mathbb{R}^k : Ax \leq b\}$, where $A \in \mathbb{Q}^{m \times k}$ and $b \in \mathbb{Q}^m$. Our goal is to find a point $p \in P$ or determine that P is empty. Let φ be the size of the maximal inequality that defines P, and let $\nu = 4k^2\varphi$. Also, let $R = 2^\nu$. We sketch the algorithm referred to as the *ellipsoid method* for bounded full-dimensional polytopes. We find a sequence of ellipsoids $\mathcal{E}^0, \mathcal{E}^1, \ldots, \mathcal{E}^N$ of decreasing volumes, such that for every $1 \leq i \leq N$ the ellipsoid $\mathcal{E}^i$ satisfies $P \subseteq \mathcal{E}^i$.

The initial ellipsoid $\mathcal{E}^0$ is the ball with center $0 \in \mathbb{R}^k$ and radius $R \in \mathbb{N}$. Using the radius R ensures that indeed $P \subseteq \mathcal{E}^0$. Assume that we found the ellipsoid $\mathcal{E}^i = Ell(z^i, L^i)$. We describe the $(i+1)$-th iteration of the algorithm. We test if $z^i \in P$. If it is, we are done. Otherwise, we find an inequality that z^i violates. Let $H \subseteq \mathbb{R}^k$ be the half-space that corresponds to the inequality we find. Next, using the half-space H, and the ellipsoid $\mathcal{E}^i$, we generate the ellipsoid $\mathcal{E}^{i+1}$ with the following properties: $\mathcal{E}^i \cap H \subseteq \mathcal{E}^{i+1}$ and $\mathcal{E}^{i+1}$ has a minimal volume. Moreover, we have that $vol(\mathcal{E}^{i+1})/vol(\mathcal{E}^i) \leq 1/e$. For a 2-dimensional example, consider Figure 2. Finally, it is guaranteed that if all the equations generated by the separation oracle are over $\mathbb{Q}$, so are all the points generated by the algorithm.

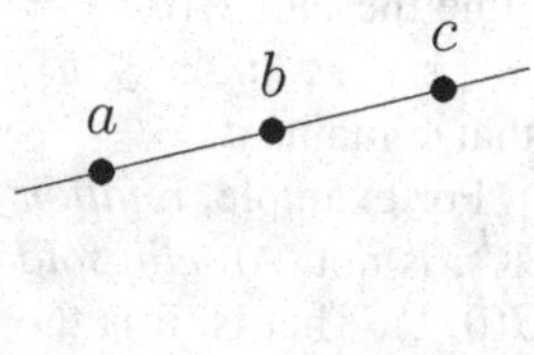

Fig. 1. A 1-dimensional polytope in $\mathbb{R}^2$

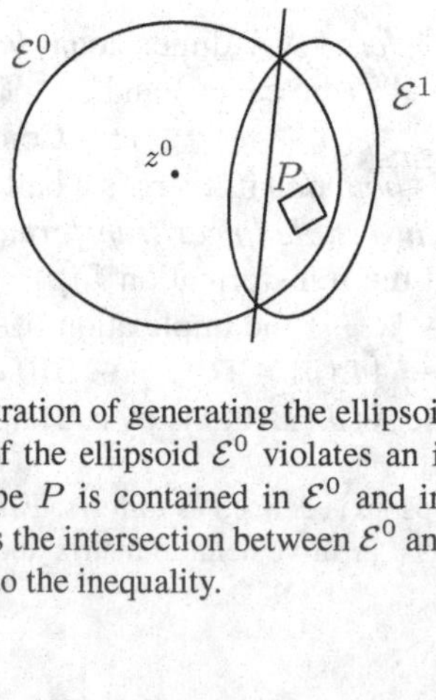

Fig. 2. An illustration of generating the ellipsoid $\mathcal{E}^1$, given that the center z^0 of the ellipsoid $\mathcal{E}^0$ violates an inequality. Note that the polytope P is contained in $\mathcal{E}^0$ and in $\mathcal{E}^1$. Also note that $\mathcal{E}^1$ contains the intersection between $\mathcal{E}^0$ and the half-space corresponding to the inequality.

The termination criterion also depends on φ as above. If P is not empty, then since it is full-dimensional, its volume is at least $2^{-\nu}$, where ν is polynomial in the representation size of φ. Since the volumes of the ellipsoids decrease exponentially, by selecting $N = poly(k, \nu)$, we have that $vol(\mathcal{E}^N) < 2^{-\nu}$. Thus, if $z^N \notin P$, we can conclude that P is empty, and we terminate.

In order to drop the assumption that the polytope is bounded, we use the following property. If the polytope P is not empty, there is a point $p \in P$ with $size(p) \leq 2^{\nu}$. Thus, we can use the polytope that is the intersection of P with the box $\{p \in \mathbb{R}^k : \forall 1 \leq i \leq k, 2^{\nu} \leq p_i \leq 2^{\nu}\}$, which is clearly bounded.

Symbolic Ellipsoid Method. Examining Khachiyan's method one can see that it is not necessary to have the implicit list of inequalities that define the given polytope P. Indeed, it was shown in [12,16,21] that it is sufficient to have a *separation oracle* for P when P is full-dimensional. That is, instead of a full list of inequalities that define P, the input to the problem is an oracle that, given a point $p \in \mathbb{R}^k$, either says that $p \in P$ or returns a half-space $H \subseteq \mathbb{R}^k$ such that $p \notin H$ and $P \subseteq H$. Assuming we found the ellipsoid $\mathcal{E}^i = Ell(z^i, L^i)$, we use the oracle to check if z^i is in P. If it is, we are done, and otherwise, we get a half-space with which we construct the ellipsoid $\mathcal{E}^{i+1}$. Since we construct only polynomially many ellipsoids, we perform only polynomial many calls to the oracle. The runtime is thus polynomial in the runtime of the separation oracle, in k, and in the maximal representation size (aka φ) of the inequalities that define P. In the full version we explain how we can work with a separation oracle even when the polytope is not full-dimensional. To conclude, we have the following.

Theorem 1. *[22] Consider a polytope $P \subseteq \mathbb{R}^k$, defined by linear inequalities over $\mathbb{Q}$ of size at most φ. Given a separation oracle SEP for P, it is possible to find a point in $P \cap \mathbb{Q}^k$ in time that is polynomial in k, φ, and the running time of SEP.*

3 The PWC Problem for Deterministic WFAs

In this section we show that both the two- and one-bound PWC problems can be solved in polynomial time when the input WFAs are deterministic.

3.1 The Two-Bound PWC Problem for Deterministic Automata

Recall that the input to the two-bound PWC problem are DWFAs $\mathcal{A}$ and $\mathcal{C}$ and an $\mathcal{X}$-parametrized DWFA $\mathcal{B}$. Our goal is to find a *legal assignment* for the variables in $\mathcal{X}$. That is, an assignment f such that $\mathcal{A} \subseteq \mathcal{B}^f \subseteq \mathcal{C}$.

From Parameterized Containment to a Convex Polytope. Consider an input $\mathcal{A}$, $\mathcal{B}$, and $\mathcal{C}$ to the two-bound problem. When the automata are deterministic, checking whether $L(\mathcal{A}) \subseteq L(\mathcal{B}) \subseteq L(\mathcal{C})$ can be done in polynomial time. If Boolean containment does not hold, there is clearly no assignment as required. Thus, we assume that $L(\mathcal{A}) \subseteq L(\mathcal{B}) \subseteq L(\mathcal{C})$.

Consider an assignment $f : \mathcal{X} \to \mathbb{Q}$. By Proposition 2, we have that f is legal iff $\mathcal{S}_{\mathcal{A},\mathcal{B}^f}$ and $\mathcal{S}_{\mathcal{B}^f,\mathcal{C}}$ are negative. Moreover, by Proposition 1, the latter holds iff all the critical paths in $\mathcal{S}_{\mathcal{A},\mathcal{B}^f}$ and $\mathcal{S}_{\mathcal{B}^f,\mathcal{C}}$ have a non-positive value. Thus, the set of critical paths in $\mathcal{S}_{\mathcal{A},\mathcal{B}}$ and $\mathcal{S}_{\mathcal{B},\mathcal{C}}$ induce necessary and sufficient restrictions on the possible values the variables can get in a legal assignment. Each critical path induces an inequality over the variables in $\mathcal{X}$, and together all critical paths induce a convex polytope that includes exactly all the legal assignments.

The above observation is the key to our algorithm, and we describe its details below for the product $\mathcal{S}_{\mathcal{A},\mathcal{B}}$. The construction of inequalities induced by $\mathcal{S}_{\mathcal{B},\mathcal{C}}$ is similar. Consider a critical path π in $\mathcal{S}_{\mathcal{A},\mathcal{B}}$. We generate an inequality from π that corresponds to a restriction on legal assignment to the variables. Inequalities for $\mathcal{S}_{\mathcal{B},\mathcal{C}}$ are generated in a similar manner. Recall that the path π is a sequence of transitions in a DWFA that is the product of two DWFAs. For every $e = \langle e_\mathcal{A}, e_\mathcal{B} \rangle \in \pi$, let $c_e = \tau_\mathcal{A}(e_\mathcal{A}) - \tau_\mathcal{B}(e_\mathcal{B})$. Recall that $\tau_\mathcal{B}(e_\mathcal{B})$ can either be a number, in which case c_e is a number too, or a variable $X \in \mathcal{X}$, in which case c_e is of the form $c - X$ with $c = \tau_\mathcal{A}(e_\mathcal{A}) \in \mathbb{Q}$. We define the inequality $(\sum_{e \in \pi} c_e) \leq 0$. Clearly, it is possible to rewrite the inequality as $\sum_{1 \leq i \leq k} -l_i \cdot X_i + c \leq 0$, where $l_i \in \mathbb{N}$ is the number of times that $X_i \in \mathcal{X}$ appears in π and $c \in \mathbb{Q}$.

Remark 1. Let $n = max\{|Q_\mathcal{A}| \cdot |Q_\mathcal{B}|, |Q_\mathcal{B}| \cdot |Q_\mathcal{C}|\}$ and $M = max\{|\tau_\mathcal{A}(e_\mathcal{A}) - \tau_\mathcal{B}(e_\mathcal{B})|, |\tau_\mathcal{B}(e_\mathcal{B})|, |\tau_\mathcal{B}(e_\mathcal{B}) - \tau_\mathcal{C}(e_\mathcal{C})| : e_\mathcal{A} \in \Delta_\mathcal{A}, e_\mathcal{B} \in \Delta_\mathcal{B}, e_\mathcal{C} \in \Delta_\mathcal{C}, \tau(e_\mathcal{B}) \in \mathbb{Q}\}$. Since π is either acyclic or a simple cycle, its length is at most n. Since for every $1 \leq i \leq k$, we have that l_i is the number of times $X_i \in \mathcal{X}$ appears in π, then $0 \leq l_i \leq n$. Clearly, $|c| \leq Mn$. Thus, the size of every inequality we generate is at most $1 + \sum_{1 \leq i \leq k} size(n) + size(Mn) = O(log(nM))$.

Let $|\mathcal{X}| = k$. By the above, we think of an assignment to the variables as a point in $\mathbb{R}^k$, think of the inequalities as half-spaces in $\mathbb{R}^k$, and think of the set of legal assignments as a convex polytope in $\mathbb{R}^k$, namely the intersection of all the half-spaces that are generated from the critical paths in $\mathcal{S}_{\mathcal{A},\mathcal{B}}$ and $\mathcal{S}_{\mathcal{B},\mathcal{C}}$. We denote this polytope by $\mathcal{P} \subseteq \mathbb{R}^k$.

Efficient Reasoning About the Convex Polytope. A naive way to solve the two-bound problem is to generate all the inequalities from $\mathcal{S}_{\mathcal{A},\mathcal{B}}$ and $\mathcal{S}_{\mathcal{B},\mathcal{C}}$ and solve the system of inequalities. Since, however, there can be exponentially many critical paths, the running time of such an algorithm would be at least exponential. In order to overcome this difficulty, we do not construct the induced polytope $\mathcal{P}$ implicitly. Instead, we devise a separation oracle for $\mathcal{P}$. By Theorem 1, this would enable us to find a point in $\mathcal{P} \cap \mathbb{Q}^k$ (or decide that $\mathcal{P}$ is empty) with only polynomial many calls to the oracle.

Recall that a separation oracle for $\mathcal{P}$ is an algorithm that, given a point $p \in \mathbb{R}^k$, either returns that $p \in \mathcal{P}$ or returns a half-space $H \subseteq \mathbb{R}^k$, represented by an inequality, that separates p from $\mathcal{P}$. That is, $\mathcal{P} \subseteq H$ and $p \notin H$.

We describe the separation oracle for $\mathcal{P}$. Given a point $p \in \mathbb{R}^k$, we check if $\mathcal{A} \subseteq \mathcal{B}^p \subseteq \mathcal{C}$. If the latter holds, we conclude that p is a legal assignment, and we are done. Otherwise, there is a word $w \in \Sigma^*$ such that $w \in L(\mathcal{A})$ and $val(\mathcal{A}, w) > val(\mathcal{B}^p, w)$, or $w \in L(\mathcal{B})$ and $val(\mathcal{B}^p, w) > val(\mathcal{C}, w)$. Using w, we find a critical path that p

violates and we return the inequality induced by this path. Note that the run on w may not be a critical path: we know it is a path form an initial state to an accepting state, but this path may not be simple. We describe how to detect a critical path from w. Assume that w is such that $val(\mathcal{A}, w) > val(\mathcal{B}^p, w)$. The other case is similar. Since $\mathcal{A}$ and $\mathcal{B}$ are deterministic there is a single accepting run r of $\mathcal{S}_{\mathcal{A},\mathcal{B}^p}$ on w. If r is acyclic, then it is a critical path. Otherwise, we remove every non-positive cycle from r. Let r' be the obtained path in $\mathcal{S}_{\mathcal{A},\mathcal{B}^p}$. Clearly, $val(r') \geq val(r)$. If r' is acyclic, we found a critical path. Otherwise, since $val(r') \geq val(r) > 0$, there must be a positive valued cycle in r'. This cycle is a critical path, and we are done.

We can now use Theorem 1 and conclude with the following.

Theorem 2. *The two-bound PWC problem for DWFAs can be solved in polynomial time.*

Remark 2 (Speeding up the Separation Oracle). Reasoning about critical paths involves a calculation of distances in the graphs corresponding to the product automata and is done by solving the All-Pairs Shortest Path problem. As we update the ellipsoids, we also update costs in the product automata. There is much research in the field of *dynamic graph algorithms* (specifically, [25] suggests a fully-dynamic data-structure to solve the All-Pairs Shortest Path problem) that we can use here in order to speed up the running time of the separation oracle so that the time required for solving a distance query is proportional to the updates rather than to the automata.

3.2 When $\mathcal{B}$ Is Not Given

An interesting variant of the two-bound PWC problem is one in which we are not given $\mathcal{B}$ and we seek a DWFA of a minimal size such that $\mathcal{A} \subseteq \mathcal{B} \subseteq \mathcal{C}$. One may start the search for $\mathcal{B}$ with a non-weighted version of the problem. That is, seek a minimal DFA $\mathcal{B}$ such that $L(\mathcal{A}) \subseteq L(\mathcal{B}) \subseteq L(\mathcal{C})$. We can then turn $\mathcal{B}$ into a candidate DWFA by labeling all its transitions by variables. The corresponding decision problem, which we refer to as the *Boolean sandwich* problem, gets as input two DFAs $\mathcal{A}$ and $\mathcal{C}$ and index $k \in \mathbb{N}$ and decides whether there is a DFA $\mathcal{B}$ with k states such that $L(\mathcal{A}) \subseteq L(\mathcal{B}) \subseteq L(\mathcal{C})$. It is easy to see that the weighted sandwich problem is at least as hard as the Boolean one. Indeed, by defining all costs to be 0, we get an easy reduction between the two. In order to neutralize the difficulty of the Boolean aspect of the language, we define a pure-weighted sandwich problem, where $\mathcal{A}$ and $\mathcal{C}$ are such that $L(\mathcal{A}) = L(\mathcal{C})$, and we are looking for a minimal $\mathcal{B}$ such that $\mathcal{A} \subseteq \mathcal{B} \subseteq \mathcal{C}$. Note that such a WFA $\mathcal{B}$ exists iff $\mathcal{A} \subseteq \mathcal{C}$. As we show now, all sandwich problems are difficult.

Theorem 3. *The Boolean, weighted, and pure-weighted sandwich problems are NP-complete.*

Proof: Since the automata are deterministic, checking whether a given k-state automaton satisfies the Boolean or weighted sandwich requirements can be done in polynomial time. Thus, membership in NP in easy.

For the lower bound, we start with the Boolean case and show a reduction from the vertex coloring problem (VC, for short). Recall that the input to the VC problem is

$\langle G, k \rangle$, where G is a graph and $k \in \mathbb{N}$ is an index. We say that $\langle G, k \rangle \in$ VC iff there is a coloring of G's vertices in k colors such that two adjacent vertices are not colored in the same color.

Consider an input $\langle G, k \rangle$ to the VC problem. We construct an input $\langle \mathcal{A}, \mathcal{C}, k+2 \rangle$ to the Boolean-sandwich problem. The idea is that $\mathcal{A}$ has a state representing every vertex in G. In order to construct $\mathcal{B}$ one must, intuitively, *merge* different states of $\mathcal{A}$. The automaton $\mathcal{C}$ enforces that merging two states can only be done if the corresponding vertices do not share a common edge, and thus the states of $\mathcal{B}$ correspond to a legal coloring of the vertices of G. For the details of the proof, see the full version.

As noted above, hardness in the Boolean setting implies hardness in the weighted variant. For the pure-weighted variant, we go through the t-approximation problem for DWFAs, defined as follows: given a DWFA $\mathcal{A}$, a factor $t \in \mathbb{Q}$ such that $t > 1$, and an index $k \in \mathbb{N}$, we ask whether there is a DWFA $\mathcal{A}'$ with k states such that $L(\mathcal{A}) = L(\mathcal{A}')$ and for every word $w \in L(\mathcal{A})$, we have $1/t \cdot val(\mathcal{A}, w) \leq val(\mathcal{A}', w) \leq t \cdot val(\mathcal{A}, w)$. We say that $\mathcal{A}'$ t-approximates $\mathcal{A}$. As detailed in Section 1, t-approximation can be easily reduced to the pure-weighted sandwich problem. We prove that t-approximation is NP-hard by a reduction from VC. Given an input $\langle G, k \rangle$ to the VC problem, we construct a DWFA $\mathcal{A}$ with a state corresponding to every vertex in G. Constructing an approximating automaton for $\mathcal{A}$ is done by merging states. We construct $\mathcal{A}$ so that by merging two states whose corresponding vertices share an edge, there is a word in $L(\mathcal{A})$ that violates the t-approximation requirement. Thus, an approximating automaton for $\mathcal{A}$ corresponds to a legal coloring of G. For the details of the proof, see the full version. □

We suggest a heuristic to cope with the complexity of the pure-weighted sandwich problem. Consider DWFAs $\mathcal{A}$ and $\mathcal{C}$ such that $L(\mathcal{A}) = L(\mathcal{C})$ and $\mathcal{A} \subseteq \mathcal{C}$. We start the search for $\mathcal{B}$ with a DFA $\mathcal{D}$ that has the state space $Q_{\mathcal{A}} \times Q_{\mathcal{C}}$. Note that since $L(\mathcal{A}) = L(\mathcal{C})$ and the automata are deterministic, we have that $(F_{\mathcal{A}} \times Q_{\mathcal{C}}) \cup (Q_{\mathcal{A}} \times F_{\mathcal{C}}) = F_{\mathcal{A}} \times F_{\mathcal{C}}$, which we define as the set of $\mathcal{D}$'s accepting states. We try to minimize $\mathcal{D}$ by iteratively searching for states that can be merged. Clearly, we cannot hope to obtain an automaton with fewer states than the one required for $L(\mathcal{B})$, thus candidates for merging are states that are merged in the standard DFA minimization algorithm. Consider two such states $q, q' \in Q_{\mathcal{D}}$. Let $\mathcal{D}'$ be the DFW obtained by merging q and q'. We label each transition of $\mathcal{D}'$ with a different variable and use parametric weighted containment in order to find a DWFA $\mathcal{B}$ on the structure of $\mathcal{D}'$ that satisfies $\mathcal{A} \subseteq \mathcal{B} \subseteq \mathcal{C}$. If we find, we continue with further mergings in $\mathcal{D}'$. Otherwise, we un-merge the states, and look for new candidates.

3.3 The One-Bound PWC Problem for Deterministic Automata

Recall that the input to the one-bound PWC problem is a DWFA $\mathcal{A}$ and an $\mathcal{X}$-parameterized automaton $\mathcal{B}$, which we want to complete to a DWFW that either upper bounds $\mathcal{A}$ in a minimal way or lower bounds $\mathcal{A}$ in a maximal way. We focus here on the case we seek a least upper bound. The second case is similar. As in the two-bound case, we say that an assignment f is legal if it satisfies $\mathcal{A} \subseteq \mathcal{B}^f$.

As detailed below, it is technically simpler to assume that all the states in the DWFAs are accepting. Thus, in this model, the language of a DWFA is Σ^*. We can, however, use weights and encode rejecting states. For example, we can add to the alphabet a letter # that leaves all states to some state with either a bottom value, when we do not want the origin state to be considered accepting, or with value 0 when we want it to be accepting. We then restrict attention to prefixes of words that end after the first #.

As in the two-bound problem, we view assignments as points in $\mathbb{R}^k$ and use inequalities induced by critical paths in order to define a polytope $\mathcal{P} \subseteq \mathbb{R}^k$ of legal assignments. We show that the polytope generated in this case is in full-dimensional. Intuitively, it follows from the fact that increasing a point by ϵ results in a point that is still in $\mathcal{P}$.

Lemma 1. *If $\mathcal{P} \neq \emptyset$, then $\mathcal{P}$ is full-dimensional.*

Unlike the two-bound case, here we are not looking for an arbitrary point in $\mathcal{P}$, but one that is a minimal assignment. We show that a vertex of $\mathcal{P}$ is such an assignment. Intuitively, it follows from the fact that points on a face F of $\mathcal{P}$ are minimal assignments with respect to the variables participating in the inequality corresponding to F. A vertex is the intersection of k faces, and thus, it corresponds to an assignment that is minimal with respect to all faces and hence also with respect to all variables.

Lemma 2. *A vertex in $\mathcal{P}$ is a minimal assignment.*

Recall that some of the inequalities that define $\mathcal{P}$ are induced by critical paths that are simple paths from the initial vertex to an accepting state. Since we assume that all states are accepting, prefixes of such critical paths are also critical. From the geometrical point of view, this implies the following.

Lemma 3. *If $\mathcal{P} \neq \emptyset$, then $\mathcal{P}$ is pointed.*

For full-dimensional pointed polytopes, Schrijver shows a strengthening of Theorem 1 that enables us to find vertices:

Theorem 4. *[22] Consider a full-dimensional pointed polytope $P \subseteq \mathbb{R}^k$, defined by linear inequalities over $\mathbb{Q}$ of size at most φ. Given a separation oracle* SEP *for P, it is possible to find a vertex of P in $\mathbb{Q}^k$, in time that is polynomial in k, φ, and the running time of* SEP.

By the lemmas above, Theorem 4 is applicable in the one-bound PWC problem and we conclude with the following.

Theorem 5. *The one-bounded problem can be solved in polynomial time.*

Remark 3. In [11], the authors define *functional weighted automata*, which are nondeterministic weighted automata in which all the accepting runs on a word have the same value. The authors show that in this model, containment is decidable: given two functional weighted automata $\mathcal{A}$ and $\mathcal{B}$, we check if $L(\mathcal{A}) \subseteq L(\mathcal{B})$, and then we check if for every word $w \in \Sigma^*$, we have $val(\mathcal{A}, w) \leq val(\mathcal{B}, w)$ by reasoning on the automaton $\mathcal{S}_{\mathcal{A},\mathcal{B}}$. It is easy to extend the technique presented in this section to functional automata. Essentially, as in the case of deterministic automata, the construction of the separation oracle can be based on $\mathcal{S}_{\mathcal{A},\mathcal{B}}$. Accordingly, the computational bottleneck is the Boolean $L(\mathcal{A}) \subseteq L(\mathcal{B})$ check, making the PWC problem for functional automata PSPACE-complete.

4 The PWC Problem for WFAs

In this section we study the one- and two-bound problems for WFAs. Recall that containment for WFAs is undecidable, making the decidability of the PWC hopeless. Consequently, we replace the containment order for WFAs by *weighted simulation* [3,7]. Simulation has been extensively used in order to approximate containment in the Boolean setting, and was recently used as a decidable approximation of containment in the weighted setting.

Let us explain the idea behind weighted simulation. Given two WFAs $\mathcal{A}$ and $\mathcal{B}$, deciding whether $\mathcal{A} \subseteq \mathcal{B}$ can be thought of as a two-player game of one round: Player 1, the Player whose goal it is to show that there is no containment, chooses a word w and a run r_1 of $\mathcal{A}$ on w. Player 2 then replies by choosing a run r_2 of $\mathcal{B}$ on w. Player 1 wins if r_1 is accepting and r_2 is not or if $val(r_1, w) > val(r_2, w)$. While this game clearly captures containment, it does not lead to interesting insights or algorithmic ideas about checking containment. A useful way to view simulation is as a "step-wise" version of the above game in which in each round the players proceed according to a single transition of the WFAs. More formally, $\mathcal{B}$ simulates $\mathcal{A}$, denoted $\mathcal{A} \leq \mathcal{B}$, if Player 2 has a strategy that wins against all strategies of Player 1: no matter how Player 1 proceeds in the WFA $\mathcal{A}$, Player 2 can respond in a transition so that whenever the run generated so far in $\mathcal{A}$ by Player 1 is accepting, so is the run generated by Player 2 in $\mathcal{B}$. Moreover, the cost of the run in $\mathcal{A}$ is smaller than the one in $\mathcal{B}$. For full details, see [3].

So, in the nondeterministic setting, we replace containment with simulation and seek, in the two-bound case, a valuation f such that $\mathcal{A} \leq \mathcal{B}^f \leq \mathcal{C}$, and in the one bound cases minimal and maximal assignments so that $\mathcal{A} \leq \mathcal{B}^f$ or $\mathcal{B}^f \leq \mathcal{C}$, respectively.

4.1 The One-Bound PWC Problem for Nondeterministic Automata

We argue that this version of the PWC problem is not very interesting; in fact it is not well defined as is, as there are cases in which we do not have even a minimal (or maximal) assignment.

Consider for example the WFAs in Figure 4.1. The candidates for minimal assignments for the variables in $\mathcal{B}$ are the ones that assign the value 0 to X_1 or X_2. However, an assignment f with $f(X_1) = 0$ can assign to X_2 an arbitrarily low value, and, symmetrically, an assignment with $f(X_2) = 0$ can assign an arbitrarily low value to X_1. Hence, there is no minimal assignment for the variables in $\mathcal{B}$.

Thus, the nondeterministic setting calls for a different definition of the one-bound PWC problem – one that considers alternative sets of variables whose values should be minimized. We do not find such definitions well motivated.[5]

[5] Having said that, the setting does suggest some very interesting theoretical problems, like deciding when a solution exists, and the relation between the observation above and the fact WFAs cannot always be determined.

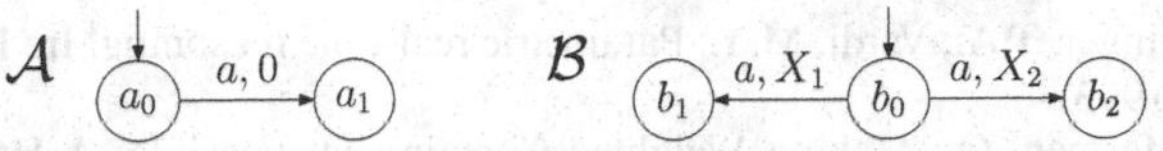

Fig. 3. An example in which there is no minimal assignment

4.2 The Two-Bound PWC Problem for Nondeterministic Automata

We now turn to study the two-bound problem. As we shall show, here the set of legal assignments corresponds to a vertex polytope, so it is either empty or not and the problem is well defined. On the other hand, the complexity of the problem depends on the complexity of deciding weighted simulation, and is NP-complete even if one finds a polynomial algorithm for deciding weighted simulation (the best known algorithm for weighted simulation positions it in NP $\cap$ co-NP). We first show that the problem is NP-hard.

Theorem 6. *The two-bound PWC problem for WFAs is NP-hard.*

Proof: We prove that finding a satisfying assignment to a 3-SAT formula is easier than solving the two-bound PWC problem for WFAs. Consider an input formula ψ to the 3-SAT problem. The intuition behind the construction is as follows. For every variable in ψ there are two variables in $\mathcal{X}$. We construct $\mathcal{B}$ and $\mathcal{C}$, so that the simulation game corresponding to them guarantees that only one of the two variables in $\mathcal{X}$ that correspond to a variable in ψ can get a value greater than or equal to 1. Thus, an assignment to the variables in $\mathcal{X}$ corresponds to an assignment to the variables in ψ. The idea of the game that corresponds to $\mathcal{A}$ and $\mathcal{B}$ is to force the assignment to variables in ψ to be satisfying. That is, Player 1 challenges Player 2 with a clause. Player 2 in his turn chooses a literal that is satisfied in this clause. For the full proof see the full version. □

For the upper bound, we first need the following result by Schrijver, relating the size of the inequalities that define a polytope P and the size of its vertices.

Theorem 7. *[22] Let $P \subseteq \mathbb{R}^k$ be a polytope that is defined by inequalities of size at most φ. Then, the size of each of its vertices is at most $4k^2\varphi$.*

We now rely on the fact that when Player 2 wins the weighted simulation game, he has a *memoryless winning strategy* [3]. Consequently, we can trim the two arenas and, as in the deterministic case, represent the set of assignments that are legal for these specific strategies by a k-dimensional polytope. Thus, we use Theorem 7 to show that if there is a legal assignment, then there is one of polynomial size. Now, if we assume that deciding weighted simulation can be done in polynomial time, we can conclude with the following. For the full proof see the full version.

Theorem 8. *A polynomial algorithm for solving simulation games implies that solving the two-bound nondeterministic PWC is in NP.*

Acknowledgments. We thank Nati Linial for the helpful discussions and pointers.

References

1. Almagor, S., Boker, U., Kupferman, O.: What's Decidable about Weighted Automata? In: Bultan, T., Hsiung, P.-A. (eds.) ATVA 2011. LNCS, vol. 6996, pp. 482–491. Springer, Heidelberg (2011)

2. Alur, R., Henzinger, T.A., Vardi, M.Y.: Parametric real-time reasoning. In: Proc. 25th STOC, pp. 592–601 (1993)
3. Avni, G., Kupferman, O.: Making Weighted Containment Feasible: A Heuristic Based on Simulation and Abstraction. In: Koutny, M., Ulidowski, I. (eds.) CONCUR 2012. LNCS, vol. 7454, pp. 84–99. Springer, Heidelberg (2012)
4. Bruns, G., Godefroid, P.: Temporal logic query checking. In: Proc. 16th LICS, pp. 409–420 (2001)
5. Chan, W.: Temporal-logic Queries. In: Emerson, E.A., Sistla, A.P. (eds.) CAV 2000. LNCS, vol. 1855, pp. 450–463. Springer, Heidelberg (2000)
6. Chatterjee, K., Doyen, L., Henzinger, T.: Quantitative Languages. In: Kaminski, M., Martini, S. (eds.) CSL 2008. LNCS, vol. 5213, pp. 385–400. Springer, Heidelberg (2008)
7. Chatterjee, K., Doyen, L., Henzinger, T.A.: Expressiveness and closure properties for quantitative languages. LMCS 6(3) (2010)
8. Culik, K., Kari, J.: Digital images and formal languages. In: Handbook of Formal Languages: Beyond words, vol. 3, pp. 599–616 (1997)
9. Dijkstra, E.W.: A Discipline of Programming. Prentice Hall (1976)
10. Fix, L., Francez, N., Grumberg, O.: Program Composition and Modular Verification. In: Leach Albert, J., Monien, B., Rodríguez-Artalejo, M. (eds.) ICALP 1991. LNCS, vol. 510, pp. 93–114. Springer, Heidelberg (1991)
11. Filiot, E., Gentilini, R., Raskin, J.-F.: Quantitative Languages Defined by Functional Automata. In: Koutny, M., Ulidowski, I. (eds.) CONCUR 2012. LNCS, vol. 7454, pp. 132–146. Springer, Heidelberg (2012)
12. Grötschel, M., Lovász, L., Schrijver, A.: The ellipsoid method and its consequences in combinatorial optimization. Combinatorica 1(2), 169–197 (1981)
13. Grötschel, M., Lovász, L., Schrijver, A.: Corrigendum to our paper "the ellipsoid method and its consequences in combinatorial optimization". Combinatorica 4(4) (1984)
14. Grötschel, M., Lovász, L., Schrijver, A.: Geometric Algorithms and Combinatorial Optimization. Springer (1988)
15. Jobstmann, B., Griesmayer, A., Bloem, R.: Program Repair as a Game. In: Etessami, K., Rajamani, S.K. (eds.) CAV 2005. LNCS, vol. 3576, pp. 226–238. Springer, Heidelberg (2005)
16. Karp, R., Papadimitriou, C.: On linear characterizations of combinatorial optimization problems. In: Proc. 21st FOCS, pp. 1–9 (1980)
17. Khachiyan, L.G.: A polynomial algorithm in linear programming. Doklady Akademii Nauk SSSR 244, 1093–1096 (1979)
18. Krob, D.: The equality problem for rational series with multiplicities in the tropical semiring is undecidable. International Journal of Algebra and Computation 4(3), 405–425 (1994)
19. Mohri, M.: Finite-state transducers in language and speech processing. Computational Linguistics 23(2), 269–311 (1997)
20. Mohri, M., Pereira, F.C.N., Riley, M.: Weighted finite-state transducers in speech recognition. Computer Speech and Language 16(1), 69–88 (2002)
21. Padberg, M.W., Rao, M.R.: The Russian Method and Integer Programming. Working paper series. Salomon Brothers Center for the Study of Financial Institutions (1980)
22. Schrijver, A.: Theory of linear and integer programming. Wiley-Interscience series in discrete mathematics and optimization. Wiley (1999)
23. Solar-Lezama, A., Rabbah, R.M., Bodík, R., Ebcioglu, K.: Programming by sketching for bit-streaming programs. In: PLDI, pp. 281–294 (2005)
24. Thomas, W.: Automata on infinite objects. Handbook of Theoretical Computer Science, 133–191 (1990)
25. Thorup, M.: Fully-Dynamic All-Pairs Shortest Paths: Faster and Allowing Negative Cycles. In: Hagerup, T., Katajainen, J. (eds.) SWAT 2004. LNCS, vol. 3111, pp. 384–396. Springer, Heidelberg (2004)
26. Vardi, M.Y., Wolper, P.: Reasoning about infinite computations. I& C 115(1), 1–37 (1994)

Weighted Specifications over Nested Words*

Benedikt Bollig, Paul Gastin, and Benjamin Monmege

LSV, ENS Cachan, CNRS & Inria, France
`firstname.lastname@lsv.ens-cachan.fr`

Abstract. This paper studies several formalisms to specify quantitative properties of finite nested words (or equivalently finite unranked trees). These can be used for XML documents or recursive programs: for instance, counting how often a given entry occurs in an XML document, or computing the memory required for a recursive program execution. Our main interest is to translate these properties, as efficiently as possible, into an automaton, and to use this computational device to decide problems related to the properties (e.g., emptiness, model checking, simulation) or to compute the value of a quantitative specification over a given nested word. The specification formalisms are weighted regular expressions (with forward and backward moves following linear edges or call-return edges), weighted first-order logic, and weighted temporal logics. We introduce weighted automata walking in nested words, possibly dropping/lifting (reusable) pebbles during the traversal. We prove that the evaluation problem for such automata can be done very efficiently if the number of pebble names is small, and we also consider the emptiness problem.

1 Introduction

In this paper, we develop a denotational formalism to express quantitative properties of nested words. Nested words, introduced in [1], are strings equipped with a binary nesting relation. Just like trees, they have been used as a model of XML documents or recursive programs. Though nested words can indeed be encoded in trees (and vice versa), they are often more convenient to work with, e.g., in streaming applications, as they come with a linear order that is naturally given by an XML document [2]. Moreover, nested words better reflect system runs of recursive programs where the nesting relation matches a procedure call with its corresponding return. There is indeed a wide range of works that address logics and automata over nested words to process XML documents or to model recursive programs e.g., [2,3,4].

Most previous approaches to nested words (or unranked trees) consider Boolean properties: logical formulae are evaluated to either true or false, or to a set of word positions if the formula at hand represents a unary query. Now, given an XML document in terms of a nested word, one can imagine a number of *quantitative* properties that one would like to compute: What is the number of books

* This work was partially supported by LIA INFORMEL.

F. Pfenning (Ed.): FOSSACS 2013, LNCS 7794, pp. 385–400, 2013.

of a certain author? Are there more fiction than non-fiction books? What is the total number of entries? So, we would like to have flexible and versatile languages allowing us to evaluate arithmetic expressions, possibly guarded by logical conditions written in a standard language (e.g., first-order logic or XPath). To this aim, we introduce (1) weighted regular expressions, which can indeed be seen as a quantitative extension of XPath, (2) weighted first-order logic as already studied in [5] over words, and (3) weighted nested word temporal logic in the flavor of [3]. Their application is not restricted to XML documents, though. For instance, when nested words model recursive function calls, these specification languages can be used to quantify the call-depth of a given system run, i.e., the maximal number of open calls.

Thus, when one considers expressiveness and algorithmic issues, a natural question arises: Is there a robust automata model able to compile specifications written in these languages? Our answer will be positive: we actually obtain a Kleene-Schützenberger correspondence, stating the equivalence of weighted regular expressions with a model of automata. Not only do we obtain this correspondence, but we also give similar complexity results concerning the size of the automaton derived from a regular expression, and the time and space used to construct it. We also prove that we can evaluate these automata efficiently and that emptiness problem is decidable (in case the underlying weight structure has no zero divisors). Towards a suitable automata device, we consider navigational automata with pebbles, for two reasons. First, weighted automata, the classical quantitative extensions of finite automata [6], are not expressive enough to encode powerful quantitative expressions, neither for words [7] nor for nested words or trees [8,9]. Second, we are looking for a model that conforms with common query languages for nested words or trees, such as XPath or equivalent variants of first-order logic, aiming at a quantitative version of the latter and a suitable algorithmic framework. Indeed, tree-walking automata are an appropriate machine model for compiling XPath queries [10].

Contribution. In Section 3, we introduce *weighted expressions with pebbles* over nested words, mixing navigational constructs and rational arithmetic expressions. As an operational counterpart of weighted expressions, we then introduce *weighted automata with pebbles* in Section 4, which can traverse a nested word along nesting edges and direct successors in both directions, and occasionally place reusable pebbles. In a sense, these are extensions of the *tree-walking automata with invisible pebbles*, introduced in [11], to the weighted setting and to nested words. We extend results over words stated in [12], namely a Kleene-Schützenberger theorem showing correspondence between weighted expressions with pebbles and *layered* weighted automata with pebbles (i.e., those that can only use a bounded number of pebbles). We also show how to efficiently compute the value associated to a given nested word in a weighted automaton with pebbles, and prove decidability (not surprinsingly, with non-elementary complexity) of the emptiness problem in case the underlying weight structure has no zero divisor.

In order to allow more flexibility, we also discuss, in Section 5, more logical quantitative formalisms like first-order logic and temporal logics, and show how to compile them efficiently into weighted automata with pebbles.

For lack of space, proof details are given in the full version [13].

2 Preliminaries

Nested Words. We fix a finite alphabet A. For $n \in \mathbb{N}$, we let $[n] = \{0, 1, \ldots, n-1\}$. A *nested word* over A is a pair $W = (w, \curvearrowright)$ where $w = a_0 \cdots a_{n-1} \in A^+$ is a nonempty string and $\curvearrowright \subseteq ([n] \times [n]) \cap <$ is a *nesting relation*: for all $(i,j), (i',j') \in \curvearrowright$, we have (1) $i = i'$ iff. $j = j'$, and (2) $i < i'$ implies ($j < i'$ or $j > j'$). We will more often denote $(i,j) \in \curvearrowright$ as $i \curvearrowright j$. Moreover, the inverse of relation $\curvearrowright$ will be denoted as $\curvearrowleft$, so that $i \curvearrowright j$ iff. $j \curvearrowleft i$. For uniformity reasons, we denote as $i \rightarrow j$ the fact that j is the successor of i, i.e., $j = i + 1$. In case we want to stress that i is the predecessor of j, we rather denote it $j \leftarrow i$. The *length* n of W is denoted $|W|$, and $\mathrm{pos}(W) = [n]$ is the set of *positions* of W. In order to ease some definitions of the paper, a virtual position n can be added to the positions: we will then denote $\overline{\mathrm{pos}}(W) = [n] \cup \{n\}$ the extended set of positions.

Let $\mathcal{T} = \{\mathsf{first}, \mathsf{last}, \mathsf{call}, \mathsf{ret}, \mathsf{int}\}$. Each position $i \in \overline{\mathrm{pos}}(W)$ in a nested word $W = (w, \curvearrowright)$ has a *type* $\tau(i) \subseteq \mathcal{T}$: $\mathsf{first} \in \tau(i)$ iff. $i = 0$; $\mathsf{last} \in \tau(i)$ iff. $i = |W|$; $\mathsf{call} \in \tau(i)$ iff. there exists j such that $i \curvearrowright j$; dually, $\mathsf{ret} \in \tau(i)$ iff. there exists j such that $j \curvearrowright i$; finally, $\mathsf{int} \in \tau(i)$ iff. $i < |W|$ and $\tau(i) \cap \{\mathsf{call}, \mathsf{ret}\} = \emptyset$.

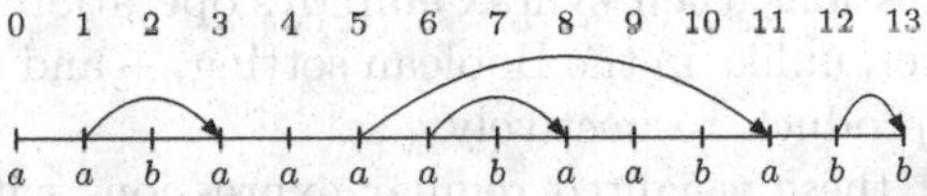

Fig. 1. A nested word

It is convenient to represent the pairs of the relation $\curvearrowright$ pictorially by curved lines which do not cross. Fig. 1 shows a nested word $W = (w, \curvearrowright)$ over $A = \{a, b\}$ of length $|W| = 14$. We have $w = aabaaaabaababb$ and the nesting relation defined by $1 \curvearrowright 3$, $5 \curvearrowright 11$, $6 \curvearrowright 8$ and $12 \curvearrowright 13$. Moreover, $\tau(0) = \{\mathsf{first}, \mathsf{int}\}$, $\tau(13) = \{\mathsf{ret}\}$ and $\tau(14) = \{\mathsf{last}\}$. In the examples of this paper, we will consider the *call-depth* c-d(j) of a position $j \in \overline{\mathrm{pos}}(W)$, i.e., the number of contexts in which position j lies. Formally, c-d$(j) = |\{(i,k) \mid i \curvearrowright k \wedge i < j < k\}|$. For example, position 7 has call-depth 2, whereas position 6 has call-depth 1. The call-depth of a nested word is the maximal call-depth among the positions: the call-depth of W is here 2.

Weights. A semiring is a set $\mathbb{S}$ equipped with two binary internal operations denoted $+$ and $\times$, and two neutral elements 0 and 1 such that $(\mathbb{S}, +, 0)$ is a commutative monoid, $(\mathbb{S}, \times, 1)$ is a monoid, $\times$ distributes over $+$ and $0 \times s = s \times 0 = 0$ for every $s \in \mathbb{S}$. If the monoid $(\mathbb{S}, \times, 1)$ is commutative, the semiring itself is called commutative. In this paper, we only consider continuous semirings: intuitively, these are semirings in which sums of infinite families are always well-defined, and such that these sums can be approximated by finite partial sums. This allows us to define in particular a star operation: for every $s \in \mathbb{S}$, the element $s^* = \sum_{i\in\mathbb{N}} s^i$ exists (where s^i is defined inductively by $s^0 = 1$ and $s^{i+1} = s^i \times s$). See [14] for more discussions about semirings, especially continuous ones. In all the rest, $\mathbb{S}$ will denote a continuous semiring. Here are some examples of continuous semirings.

- The Boolean semiring $(\{0,1\}, \vee, \wedge, 0, 1)$ with $\sum$ defined as an infinite disjunction.
- $(\mathbb{R}_{\geq 0} \cup \{\infty\}, +, \times, 0, 1)$ with $\sum$ defined as usual for positive (not necessarily convergent) series: in particular, $s^* = \infty$ if $s \geq 1$ and $s^* = 1/(1-s)$ if $0 \leq s < 1$.
- $(\mathbb{N} \cup \{\infty\}, +, \times, 0, 1)$ as a continuous subsemiring of the previous one.
- $(\mathbb{R} \cup \{-\infty\}, \min, +, -\infty, 0)$ with $\sum = \inf$ and $(\mathbb{R} \cup \{\infty\}, \max, +, \infty, 0)$ with $\sum = \sup$.
- Complete lattices such as $([0,1], \min, \max, 0, 1)$.
- The semiring of languages over an alphabet A: $(2^{A^*}, \cup, \cdot, \emptyset, \{\varepsilon\})$ with $\sum$ defined as (infinite) union.

3 Weighted Regular Expressions with Pebbles

In this section, we introduce weighted regular expressions with pebbles. Like classical regular expressions, their syntax employs operations $+$ and $\cdot$, as well as a Kleene star. However, unlike in the Boolean setting, $+$ and $\cdot$ will be interpreted as sum and Cauchy product, respectively.

We introduce first these weighted regular expressions with an example. We consider a nested word over the alphabet $\{a, b\}$. The classical regular expression $(a+b)^*b(a+b)^*$ *checks* that the given nested word contains an occurrence of letter b. We will rather use the shortcut $\rightarrow$ to denote the non-guarded move to the right encoded by the choice $(a+b)$, and use a weighted semantics in the semiring $(\mathbb{N} \cup \{\infty\}, +, \times, 0, 1)$: hence, expression $\rightarrow^* b \rightarrow^*$ *counts* the number of occurrences of letter b in the nested word. We now turn to the more complex task of counting the total number of occurrences of the letter b *inside a context with a call position labelled* a: more formally we want to sum over all possible call positions labelled a, the number of occurrences of b that appear strictly in-between this position and the matching return. For the nested word of Fig. 1, we must count 4 (in particular, position 7 must count for both call positions 5 and 6). In our formalism, we will achieve this task using expression:

$$E = \rightarrow^* (a? \wedge \mathsf{call}?)\, x! \big[\rightarrow^* x? \curvearrowright (\neg x? \leftarrow)^+ b? \rightarrow^*\big] \rightarrow^* .$$

First, we search for a call position labelled with a: there we use the test $a?$ to check the label without moving. Then, we mark the call position of the interesting context with a pebble named x: this permits us to compute independently the subexpression between brackets on the nested word, restarting from the first position. The latter subexpression first searches for the pebble, follows the call-return edge and then moves backward inside the context to pick non-deterministically a position carrying letter b.

We turn to the formal syntax of our expressions. We let $\text{Peb} = \{x, y, \ldots\}$ be an infinite set of pebble names. Weighted expressions are built upon simple Boolean tests from a set Test. The syntax of these basic tests is given as follows:

$$\alpha ::= a? \mid \tau? \mid x? \mid \neg\alpha \mid \alpha \wedge \alpha \mid \alpha \vee \alpha$$

where $a \in A$, $\tau \in \mathcal{T}$ and $x \in \text{Peb}$. Thus, a test is a Boolean combination of atomic checks allowing one to verify whether a given position in a nested word has label a; whether it is the first or last position (which is useful since we deal with 2-way expressions); whether it is a call, a return or an internal action; or whether it carries a pebble with name x, respectively. Given a nested word W, a position $i \in \overline{\text{pos}}(W)$ and an assignment of free pebble names given by a partial mapping $\sigma\colon \text{Peb} \to \text{pos}(W)$, we denote $W, i, \sigma \models \alpha$ if test α is verified over the given model: this semantics is defined as expected. Next, we present weighted regular expressions.

Definition 1. *The set* pebWE *of* weighted expressions with pebbles *is given by the following grammar:*

$$E ::= s \mid \alpha \mid \rightarrow \mid \leftarrow \mid \curvearrowright \mid \curvearrowleft \mid x!E \mid E + E \mid E \cdot E \mid E^+$$

where $s \in \mathbb{S}$, $\alpha \in$ Test, *and* $x \in$ Peb.

We get the classical Kleene star as an abbreviation: $E^* \stackrel{\text{def}}{=} 1 + E^+$. It is also convenient to introduce macros for "check-and-move": $a \stackrel{\text{def}}{=} a? \cdot \rightarrow$. This allows us to use common syntax such as $(ab)^+abc$, or to write $\rightarrow^* abba \leftarrow^+ \mathsf{first?} \rightarrow^* baab \rightarrow^*$ to identify words having both $abba$ and $baab$ as subwords.

A pebWE is interpreted over a nested word W with a marked initial position i and a marked final position j (as is the case in rational expressions or path expressions from XPath) and an assignment of free pebbles (as is the case in logics with free variables), given as a partial mapping $\sigma\colon \text{Peb} \to \text{pos}(W)$. The atomic expressions $\rightarrow$, $\leftarrow$, $\curvearrowright$, $\curvearrowleft$ have their natural interpretation as a binary relation R and are evaluated 1 or 0 depending on whether or not $(i, j) \in R$. On the contrary, formulae s, α, $x!E$ are non-progressing and require $i = j$. In particular, $x!E$ evaluates E in W with the current position marked with pebble x. The formal semantics of pebWE is given in Table 1. By default, i and j are the first and the last position of W, i.e., 0 and $|W|$, so that we use $[\![E]\!](W, \sigma)$ as a shortcut for $[\![E]\!](W, \sigma, 0, |W|)$. In the following, we call *pebble-depth* of an expression in pebWE its maximal number of nested $x!E$ operators.

Example 2. Over $(\mathbb{N} \cup \{-\infty\}, \max, +, -\infty, 0)$, consider the pebWE

$$E = \big((1\,\mathsf{call?} \rightarrow \neg\mathsf{ret?}) + (\mathsf{int?} \rightarrow \neg\mathsf{ret?}) + \curvearrowright + (\mathsf{ret?} \rightarrow \neg\mathsf{ret?})\big)^* x? \rightarrow^* .$$

Table 1. Semantics of pebWE

$$[\![s]\!](W,\sigma,i,j) = \begin{cases} s & \text{if } j=i \\ 0 & \text{otherwise} \end{cases} \qquad [\![\alpha]\!](W,\sigma,i,j) = \begin{cases} 1 & \text{if } j=i \wedge W,\sigma,i \models \alpha \\ 0 & \text{otherwise} \end{cases}$$

$$[\![d]\!](W,\sigma,i,j) = \begin{cases} 1 & \text{if } i\,d\,j \\ 0 & \text{otherwise} \end{cases} \qquad (\text{with } d \in \{\leftarrow,\rightarrow,\curvearrowright,\curvearrowleft\})$$

$$[\![x!E]\!](W,\sigma,i,j) = \begin{cases} [\![E]\!](W,\sigma[x\mapsto i],0,|W|) & \text{if } j=i<|W| \\ 0 & \text{otherwise} \end{cases}$$

$$[\![E\cdot F]\!](W,\sigma,i,j) = \sum_{k\in\overline{\mathrm{pos}}(W)} [\![E]\!](W,\sigma,i,k)\times[\![F]\!](W,\sigma,k,j)$$

$$[\![E+F]\!] = [\![E]\!]+[\![F]\!] \qquad [\![E^+]\!] = \sum_{n>0}[\![E^n]\!]$$

Notice the use of $1 \in \mathbb{N}$ which is not the unit of the semiring. Moreover, operations + are resolved by the max operator, whereas concatenation implies the use of addition in $\mathbb{N}\cup\{-\infty\}$. For every nested word W, and every position $i \in \mathrm{pos}(W)$, $[\![E]\!](W,[x\mapsto i])$ computes the call-depth of position i: indeed the first Kleene star is unambiguous, meaning that only one path starting from position 0 will lead to x in this iteration; along this path – the shortest one – we only count the number of times we enter inside the context of a call position. Hence, the call-depth of W can be computed with expression $E' = \rightarrow^*(x!E)\rightarrow^*$.

4 Weighted Automata with Pebbles

We define an automaton that walks in a nested word W. Whether a transition is applicable depends on the current control state and the current position i in W, i.e., its letter and type $\tau(i)$. A transition then either moves to a successor/predecessor position (following the linear order or the nesting relation), or drops/lifts a pebble whose name is taken from Peb. The effect of a transition is described by a move from the set $\mathrm{Move} = \{\rightarrow,\leftarrow,\curvearrowright,\curvearrowleft,\uparrow\}\cup\{\downarrow_x \mid x\in\mathrm{Peb}\}$.

Definition 3. *A* pebble weighted automaton *(or shortly* pebWA*) is a tuple* $\mathcal{A} = (Q,A,I,\delta,T)$ *where* Q *is a finite set of* states, A *is the input alphabet,* $I\in\mathbb{S}^Q$ *is the* vector of initial weights, $T\in\mathbb{S}^Q$ *is the* vector of final weights, *and* $\delta\colon Q\times\mathrm{Test}\times\mathrm{Move}\times Q\to\mathbb{S}$ *is a transition function with finite support*[1].

Informally, I assigns to any state $q\in Q$ the weight I_q of entering a run in q. Similarly, T determines the exit weight T_q at q. Finally, $\delta(p,\alpha,m,q)$, determines the weight of going from state p to state q depending on the move $m\in\mathrm{Move}$ and on the outcome of a test $\alpha\in\mathrm{Test}$. The set of *pebbles names of* $\mathcal{A}$ is defined to be the set of pebble names that appear either in drop transitions $\downarrow_x$ or in tests $x?$ of $\mathcal{A}$.

[1] The support of δ is the set of tuples (q,α,m,q') such that $\delta(q,\alpha,m,q')\neq 0$.

Let us turn to the formal semantics of a pebWA $\mathcal{A} = (Q, A, I, \delta, T)$. A run is described as a sequence of *configurations* of $\mathcal{A}$. A configuration is a tuple (W, σ, q, i, π). Here, W is the nested word at hand, $\sigma\colon \mathrm{Peb} \to \mathrm{pos}(W)$ is a valuation, $q \in Q$ indicates the current state, $i \in \overline{\mathrm{pos}}(W)$ the current position, and $\pi \in (\mathrm{Peb} \times \mathrm{pos}(W))^*$. The valuation σ indicates the position of *free pebbles* x, which may be tested successfully using $x?$ even before being dropped with $\downarrow_x$. It can be omitted when there are no free pebbles. The string π may be interpreted as the contents of a stack (its top being the rightmost symbol of π) that keeps track of the positions where pebbles have been dropped, and in which order. Pebbles are reusable (or invisible as introduced in [11]): this means that we have an unbounded supply of pebbles each marked by a pebble name in Peb. More than one pebble with name x can be placed at the same time, but only the last dropped is *visible* in the configuration. However, when the latter will be lifted, the previous occurrence of pebble x will become visible again. Formally, this means that a pebble name can occur at several places in π, but only its topmost occurrence is visible. Having this in mind, we define, given σ and π, a new valuation $\sigma_\pi\colon \mathrm{Peb} \to \mathrm{pos}(W)$ by $\sigma_\varepsilon = \sigma$ and $\sigma_{\pi(x,i)}(x) = i$, $\sigma_{\pi(x,i)}(y) = \sigma_\pi(y)$ if $y \neq x$.

Any two configurations with fixed W and σ give rise to a *concrete transition* $(W, \sigma, p, i, \pi) \rightsquigarrow (W, \sigma, q, j, \pi')$. Its *weight* is defined by

$$\sum_{\substack{d \in \{\rightarrow, \leftarrow, \curvearrowright, \curvearrowleft\} \mid i\, d\, j \\ \alpha \in \mathrm{Test} \mid W, \sigma_\pi, i \models \alpha}} \delta(p, \alpha, d, q) \quad \text{if } \pi' = \pi$$

$$\sum_{\alpha \in \mathrm{Test} \mid W, \sigma_\pi, i \models \alpha} \delta(p, \alpha, \downarrow_x, q) \quad \text{if } j = 0,\ i < |W| \text{ and } \pi' = \pi(x, i)$$

$$\sum_{\alpha \in \mathrm{Test} \mid W, \sigma_\pi, i \models \alpha} \delta(p, \alpha, \uparrow, q) \quad \text{if } \pi = \pi'(y, j) \text{ for some } y \in \mathrm{Peb}$$

and 0, otherwise. In particular, this implies that a pebble cannot be dropped on position $|W|$ in agreement with the convention adopted for weighted expressions.

A *run* of $\mathcal{A}$ is a sequence of consecutive transitions. Its weight is the product of transition weights, multiplied from left to right. We are interested in runs that start at some position i, in state p, and end in some configuration with position j and state q. So, let $[\![\mathcal{A}_{p,q}]\!](W, \sigma, i, j)$ be defined as the sum of the weights of all runs from $(W, \sigma, p, i, \varepsilon)$ to $(W, \sigma, q, j, \varepsilon)$. Since the semiring is continuous, $[\![\mathcal{A}_{p,q}]\!](W, \sigma, i, j)$ is well defined.

The semantics of $\mathcal{A}$ wrt. the nested word W and the initial assignment σ includes the initial and terminal weights, and we let

$$[\![\mathcal{A}]\!](W, \sigma) = \sum_{p,q \in Q} I_p \times [\![\mathcal{A}_{p,q}]\!](W, \sigma, 0, |W|) \times T_q\,.$$

In order to evaluate automata, or prove some expressiveness results, we consider the natural subclass of pebWA that cannot drop an unbounded number of pebbles. We will hence identify *K-layered* automata, for $K \geq 0$, where a

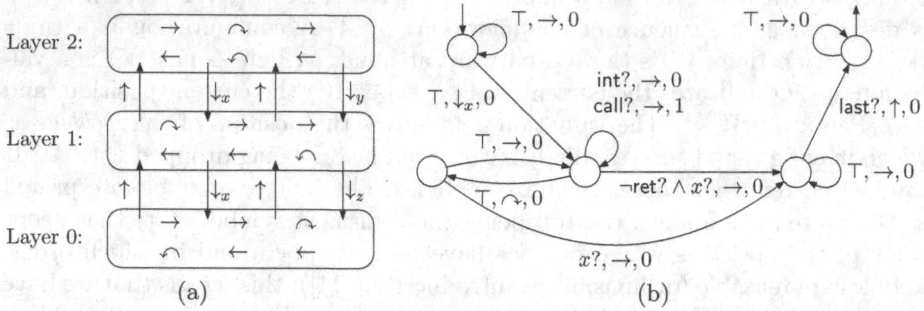

Fig. 2. (a) A 2-layered pebWA, (b) A pebWA computing the call-depth

state contains information about the number $n \in \{0, \dots, K\}$ of currently available pebbles. Formally, a pebWA $\mathcal{A} = (Q, A, I, M, T)$ is K-*layered* if there is a mapping $\ell\colon Q \to \{0, \dots, K\}$ satisfying, for all $p, q \in Q$, (i) if $I_q \neq 0$ or $T_q \neq 0$ then $\ell(q) = K$; (ii) if there is $\alpha \in$ Test and $d \in \{\leftarrow, \rightarrow, \curvearrowright, \curvearrowleft\}$ such that $\delta(p, \alpha, d, q) \neq 0$ then $\ell(q) = \ell(p)$; (iii) if there is $\alpha \in$ Test such that $\delta(p, \alpha, \uparrow, q) \neq 0$ then $\ell(q) = \ell(p) + 1$; and (iv) if there is $\alpha \in$ Test and $x \in$ Peb such that $\delta(p, \alpha, \downarrow_x, q) \neq 0$ then $\ell(q) = \ell(p) - 1$. Fig. 2(a) schematizes a 2-layered pebWA.

Example 4. We depict in Fig. 2(b) a pebWA which computes the call-depth of a nested word: it has the same semantics as expression E' of Example 2. Notice that this automaton is 1-layered.

We now extend the Kleene theorem to our setting. In order to express the complexity of the construction, we define the *literal-length* $\ell\ell(E)$ of an expression as the number of occurrences of moves in $\{\rightarrow, \leftarrow, \curvearrowright, \curvearrowleft\}$ plus twice the number of occurrences of ! (in $x!-$).

Theorem 5. *For each* pebWE E, *we can construct a* layered pebWA $\mathcal{A}(E)$ *equivalent to* E, *i.e., for all nested words* W *and for all assignments* σ *we have:* $[\![\mathcal{A}(E)]\!](W, \sigma) = [\![E]\!](W, \sigma)$. *Moreover, the number of layers in* $\mathcal{A}(E)$ *is the pebble-depth of* E, *and its number of states is* $1 + \ell\ell(E)$. *Conversely, for each* layered pebWA *we can construct an equivalent* pebWE.

Such extensions of Kleene's theorem have been proved for various weighted models. In [15], Sakarovitch gives a survey about different constructions establishing Schützenberger's theorem, namely Kleene's theorem for weighted one-way automata over finite words. An efficient algorithm constructing an automaton from an expression uses standard automata (which has as variants Berry-Sethi algorithm, or Glushkov algorithm). In [12], we extended this algorithm to deal with pebbles and two-way navigation in (non-nested) words. It is not difficult – and not surprising – to see that this extension holds in the context of nested words too. For the converse translation, weighted versions of the state elimination method of Brzozowski-McCluskey, or the procedure of McNaughton-Yamada can easily be applied to our two-way/pebble setting as previously stated in [12] over (non-nested) words.

We now study the evaluation problem of a K-layered pebWA $\mathcal{A}$: given a nested word W over A and a valuation $\sigma\colon \mathrm{Peb} \to \mathrm{pos}(W)$, compute $[\![\mathcal{A}]\!](W, \sigma)$. The problem is non-trivial since, even if the nested word is fixed, the number of accepting runs may be infinite.

Our evaluation algorithm requires the computation of the matrix N^* given a square matrix $N \in \mathbb{S}^{n\times n}$. By definition, N^* is defined as the infinite sum $\sum_{k\geq 0} N^k$, which is well defined since the semiring is continuous, but may seem difficult to compute. However, using Conway's decomposition of the star of a matrix (see [16] for more details), we can compute N^* with $\mathcal{O}(n)$ scalar star operations and $\mathcal{O}(n^3)$ scalar sum and product operations.

Theorem 6. *Given a layered* pebWA *with ρ pebble names and a nested word W, we can compute with $\mathcal{O}((\rho+1)|Q|^3|W|^{\rho+1})$ scalar operations (sum, product, star) the values $[\![\mathcal{A}_{p,q}]\!](W,\sigma)$ for all states $p, q \in Q$ and valuations $\sigma\colon \mathrm{Peb} \to \mathrm{pos}(W)$.*

Proof (Sketch). In the whole proof, we fix a nested word $W = (w, \curvearrowright)$. We follow the same basic idea used to evaluate weighted automata over words, namely computing matrices of weights for partial runs [12]. The 2-way navigation is resolved by computing simultaneously matrices of weights of the back and forth loops, whereas we deal with layers inductively. Finally, we deal with call-return edges by using a hierarchical order based on the call-depth to compute the different matrices. Hence, for every position $i \in \overline{\mathrm{pos}}(W)$ we consider the pair $(\mathrm{start}(i), \mathrm{end}(i))$ of start and end positions as follows:

$$\mathrm{start}(i) = \min\{j \in \mathrm{pos}(W) \mid j \leq i \wedge \forall k \quad j \leq k \leq i \implies \text{c-d}(k) \geq \text{c-d}(i)\}$$
$$\mathrm{end}(i) = \max\{j \in \overline{\mathrm{pos}}(W) \mid i \leq j \wedge \forall k \quad i \leq k \leq j \implies \text{c-d}(k) \geq \text{c-d}(i)\}$$

For positions of call-depth 0, the start position is 0 whereas the end position is $|W|$. For positions of call-depth at least 1 (see Fig. 3), the start position is the linear successor of the closest call in the past such that its matched return is after position i, whereas its end position is the linear predecessor of this return position.

Let $\mathcal{A} = (Q, A, I, M, T)$ be a K-layered pebWA. For $k \leq K$, we let $Q^{(k)} = \ell^{-1}(k)$ be the set of states in layer k. For every layer $k \in \{0, \dots, K\}$ and all states $p, q \in Q^{(k)}$ we denote by $B^{(k)}_{\sigma,p,q}$ the sum of weights of the runs from configuration $(W, \sigma, p, 0, \varepsilon)$ to configuration $(W, \sigma, q, |W|, \varepsilon)$: observe that the stack of pebbles is empty at the beginning of these runs, hence they stay in layers $k, k-1, \dots, 0$. Notice that $B^{(k)}_{\sigma,p,q} = [\![\mathcal{A}_{p,q}]\!](W, \sigma)$. In the following, these coefficients (and others that will be defined later) will be grouped into matrices: for example, we denote by $B^{(k)}_\sigma$ the $(Q^{(k)} \times Q^{(k)})$-matrix containing all coefficients $(B^{(k)}_{\sigma,p,q})_{p,q\in Q^{(k)}}$.

Fix a layer $k \in \{0, \dots, K\}$ of the automaton. Suppose by induction that we have already computed matrices $B^{(k-1)}_\sigma$ for every valuation $\sigma : \mathrm{Peb} \to \mathrm{pos}(W)$. For a valuation $\sigma : \mathrm{Peb} \to \mathrm{pos}(W)$, the matrix $B^{(k)}_\sigma$ will be obtained by the

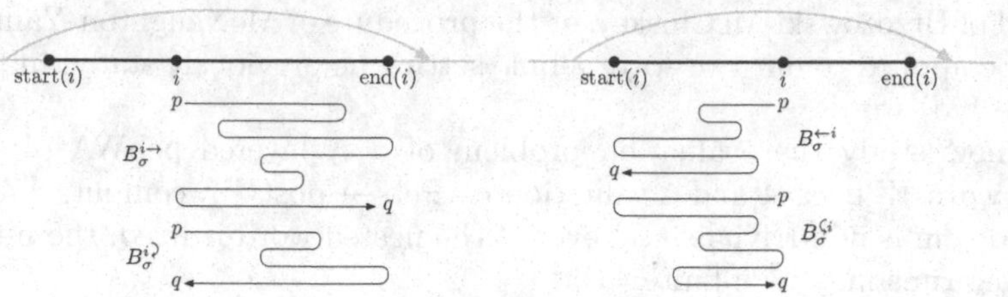

Fig. 3. Representation of the four types of matrices

computation of four types of matrices for every position (see Fig. 3). For example, $B^{i\rightarrow}_{\sigma,p,q}$ (resp. $B^{i\circlearrowright}_{\sigma,p,q}$) is the sum of weights of the runs from configuration $(W,\sigma,p,i,\varepsilon)$ to $(W,\sigma,q,\mathrm{end}(i),\varepsilon)$ (resp. $(W,\sigma,q,i,\varepsilon)$) with intermediary configurations of the form (W,σ,r,j,π) with $\pi \neq \varepsilon$ or $i \leq j \leq \mathrm{end}(i)$. These runs are those which stay between their starting position and the corresponding end position (except when they drop pebbles, allowing the automaton to scan the whole nested word) stopping at the corresponding end position (resp. their starting position). Note that $B^{(k)}_\sigma = B^{0\rightarrow}_\sigma$.

We compute these four types of matrices for every position i by decreasing value of call-depth. Suppose this has been done for every position of call-depth greater than c. We describe how to compute matrices $B^{i\rightarrow}_\sigma$ and $B^{i\circlearrowright}_\sigma$ for every position i of call-depth c, by decreasing values of i. Similarly, matrices $B^{\leftarrow i}_\sigma$ and $B^{\circlearrowleft i}_\sigma$ can be computed by increasing values of positions i having call-depth c.

We let $M^d_{\sigma,i}$ be the matrix with p,q-coefficient $\sum_{\alpha\in\mathrm{Test}\mid W,\sigma,i\models\alpha} \delta(p,\alpha,d,q)$ for $d \in \{\leftarrow,\rightarrow,\curvearrowright,\curvearrowleft\}$: this coefficient denotes the weight of taking a transition with move d from state p to state q on position i with current valuation σ. We similarly define matrices for drop, denoted $M^{\downarrow_x}_{\sigma,i}$, and lift moves: without loss of generality, we assume that lift moves only occur on position $|W|$, so that it may be denoted as $M^{\uparrow}$ since it does not depend on the valuation σ.

We only explain how to compute matrices $B^{i\rightarrow}_\sigma$ and $B^{i\circlearrowright}_\sigma$ when $i < \mathrm{end}(i)$ and i is not a call. Other cases may be found in the long version [13]. A loop on the right of i either starts by dropping a pebble over i, or starts with a right move, followed by a loop on the right of $i+1$ and a left move. All of this may be iterated using a star operation:

$$B^{i\circlearrowright}_\sigma = \Big(\sum_{x\in\mathrm{Peb}} M^{\downarrow_x}_{\sigma,i} \times B^{(k-1)}_{\sigma[x\mapsto i]} \times M^{\uparrow} + M^{\rightarrow}_{\sigma,i} \times B^{i+1\circlearrowright}_\sigma \times M^{\leftarrow}_{\sigma,i+1}\Big)^* .$$

Moving to the right of position i, until $\mathrm{end}(i)$, can be decomposed as a loop on the right of i, followed by a right move (from that point, we will not reach position i anymore) and a run from $i+1$ to $\mathrm{end}(i+1) = \mathrm{end}(i)$:

$$B^{i\rightarrow}_\sigma = B^{i\circlearrowright}_\sigma \times M^{\rightarrow}_{\sigma,i} \times B^{i+1\rightarrow}_\sigma . \qquad \square$$

Notice that if the nested word is in fact a word, our algorithm only needs to compute the two sets of matrices $B^{i\rightarrow}_\sigma$ and $B^{i\circlearrowright}_\sigma$ with a backward visit of the

Table 2. Semantics of wFO

$$[\![s]\!](W,\sigma) = s \qquad [\![\varphi]\!](W,\sigma) = \begin{cases} 1 & \text{if } W,\sigma \models \varphi \\ 0 & \text{otherwise} \end{cases}$$

$$[\![\Phi + \Psi]\!] = [\![\Phi]\!] + [\![\Psi]\!] \qquad \left[\!\left[\textstyle\sum_x \Phi\right]\!\right](W,\sigma) = \sum_{u \in \mathrm{pos}(W)} [\![\Phi]\!](W,\sigma[x \mapsto u])$$

$$[\![\Phi \times \Psi]\!] = [\![\Phi]\!] \times [\![\Psi]\!] \qquad \left[\!\left[\textstyle\prod_x \Phi\right]\!\right](W,\sigma) = \prod_{u \in \mathrm{pos}(W)} [\![\Phi]\!](W,\sigma[x \mapsto u])$$

positions of the word. This is indeed a different algorithm than the one presented in [12] where the positions are visited in a forward manner.

Classical decision problems over finite state automata have natural counterparts in the weighted setting. For example, the emptiness problem takes as input a pebWA $\mathcal{A}$, and asks whether there exists a nested word W such that $[\![\mathcal{A}]\!](W) \neq 0$.

Theorem 7. *The emptiness problem is decidable, with non-elementary complexity, for layered* pebWA *over a continuous semiring* $\mathbb{S}$ *with no zero divisor, i.e., such that* $s \times s' = 0$ *implies* $s = 0$ *or* $s' = 0$.

5 Weighted Logical Specifications over Nested Words

5.1 Weighted First-Order Logic

We fix an infinite supply of first-order variables $\mathcal{V} = \{x, y, \ldots\}$. We suppose known the fragment of (Boolean) first-order formulae, denoted as FO, over nested words, defined by the grammar

$$\varphi ::= \top \mid P_a(x) \mid x \le y \mid x \curvearrowright y \mid \neg\varphi \mid \varphi \vee \varphi \mid \varphi \wedge \varphi \mid \exists x\, \varphi \mid \forall x\, \varphi$$

where $a \in A$ and $x, y \in \mathcal{V}$.

The weighted extension is based on sums and products as for FO with counting (see [17], e.g.). The class of weighted first-order formulae, denoted as wFO, is defined by:

$$\Phi ::= s \mid \varphi \mid \Phi + \Phi \mid \Phi \times \Phi \mid \textstyle\sum_x \Phi \mid \prod_x \Phi$$

where $s \in \mathbb{S}$, $\varphi \in$ FO, and $x \in \mathcal{V}$.

The semantics of a wFO formula is a map from nested words to the semiring. For the inductive definition, we need to consider formulae with free variables. So let $\Phi \in$ wFO. Then, $[\![\Phi]\!]$ maps to a value in $\mathbb{S}$ each pair (W, σ) where W is a nested word and $\sigma : V \to \mathrm{pos}(W)$ is a valuation of a subset $V \subseteq \mathcal{V}$ containing the free variables of Φ. The inductive definition is given in Table 2. It is possible to define a quantitative implication: given $\varphi \in$ FO and $\Phi \in$ wFO, we let $\varphi \stackrel{+}{\Longrightarrow} \Phi$ be a macro for the formula $\neg\varphi + \varphi \times \Phi$. Then, its semantics coincides with the semantics of Φ if φ holds, and its semantics is 1 (the unit of the semiring) if φ does not hold.

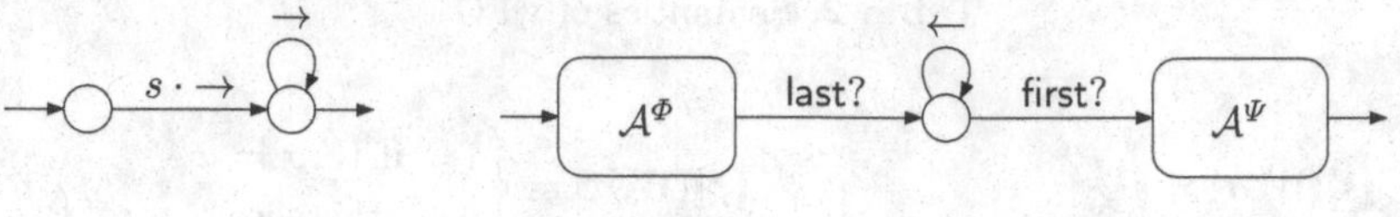

Fig. 4. Automata for $\Xi = s$ and $\Xi = \Phi \times \Psi$

Example 8. In the semiring of natural numbers, the formula $\sum_y (x \leq y \wedge P_a(y))$ computes the number of a's after the position of variable x in the nested word.

In a probabilistic setting, the formula $\Phi(x) = \prod_y (y \leq x \overset{+}{\Longrightarrow} 1/2)$ maps a position i to $(1/2)^{i+1}$, hence defining a geometric probability distribution over the positions of the nested word. We can then compute the expectation of a formula $\Psi(x)$ by the formula $\sum_x \Phi(x) \times \Psi(x)$.

Finally, the formula $\sum_x \prod_y \big[(\exists z\ (y \curvearrowright z \wedge y < x < z)) \overset{+}{\Longrightarrow} 1\big]$ computes the call-depth of a nested word, as it was already presented for expressions and automata in Examples 2 and 4. Notice again that here, sum stands for max, product for +, and the unit of the semiring is 0.

Theorem 9. *Let $\Phi \in$ wFO be a formula and $V \subseteq \mathcal{V}$ be a finite set containing the variables occurring in Φ (free or bound). We can effectively construct an equivalent layered* pebWA *$\mathcal{A}^\Phi$ with $\mathcal{O}(|\Phi|)$ states and set V of pebble names: $[\![\mathcal{A}^\Phi]\!](W,\sigma) = [\![\Phi]\!](W,\sigma)$ for every nested word W and valuation $\sigma \colon V \to \mathrm{pos}(W)$.*

Proof. This is achieved by structural induction on the formula. We deal with Boolean formulae below. For $\Xi = s \in \mathbb{S}$, the automaton is given on the left of Fig. 4. For the sum $\Xi = \Phi + \Psi$, we use a non-deterministic choice as usual. For the product $\Xi = \Phi \times \Psi$, the construction is described on the right of Fig. 4. Constructions for $\Xi = \sum_x \Phi$ and $\Xi = \prod_x \Phi$ are schematized in Fig. 5.

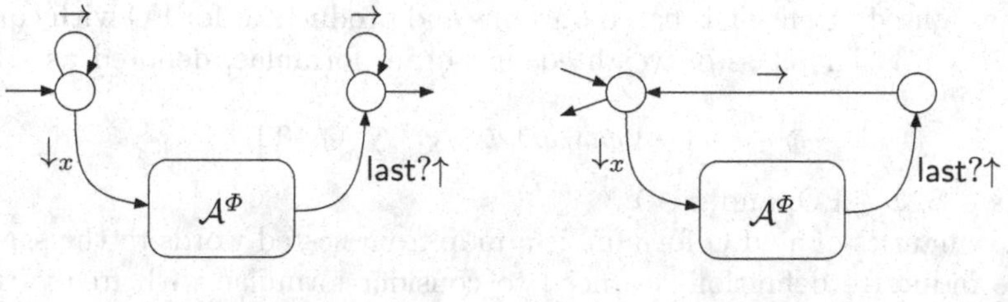

Fig. 5. Automata for $\sum_x \Phi$ and $\prod_x \Phi$

We explain now the construction of pebWA for Boolean formulae. The values computed by such automata should be in $\{0, 1\}$ so we cannot freely use non-determinism. Also, to achieve the complexity stated in the theorem, we cannot use classical constructions yielding deterministic automata. Instead, we build

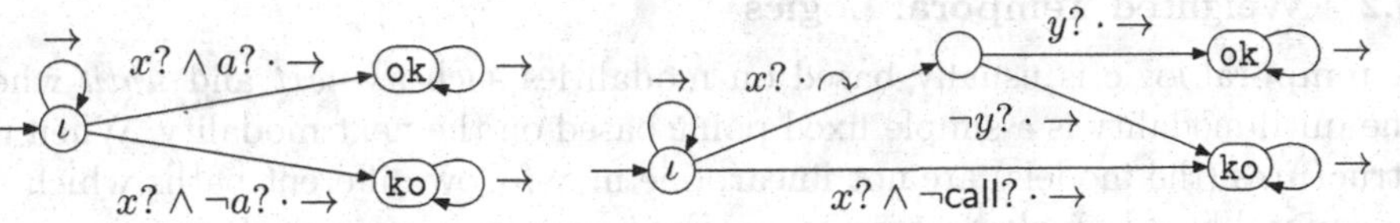

Fig. 6. Automata for $P_a(x)$ and $x \curvearrowright y$

unambiguous automata. Hence, for every $\varphi \in$ FO and $V \subseteq \mathcal{V}$ containing the free variables of φ, we construct a pebWA $\mathcal{B}^\varphi$ having one initial state ι and two (final) states ok and ko such that for all nested words W and valuations $\sigma \colon V \to \mathrm{pos}(W)$ we have:

$$[\![\mathcal{B}^\varphi_{\iota,\mathsf{ok}}]\!](W,\sigma) = \begin{cases} 1 & \text{if } W,\sigma \models \varphi \\ 0 & \text{otherwise,} \end{cases} \qquad [\![\mathcal{B}^\varphi_{\iota,\mathsf{ko}}]\!](W,\sigma) = \begin{cases} 0 & \text{if } W,\sigma \models \varphi \\ 1 & \text{otherwise.} \end{cases}$$

We obtain automaton $\mathcal{A}^\varphi$ by considering $\mathcal{B}^\varphi$ with ι (resp. ok) having initial (resp. final) weight 1. To get an automaton for the negation of a formula, we simply exchange states ok and ko. Automata for atoms $P_a(x)$ and $x \curvearrowright y$ are given in Fig. 6 and the automaton for $x \le y$ is left to the reader.

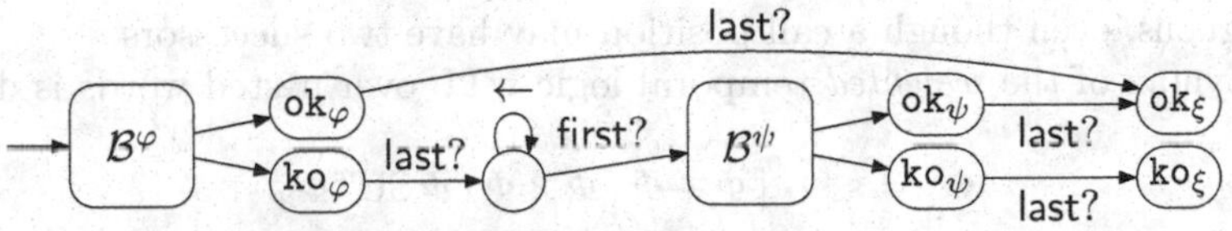

The construction for disjunction $\xi = \varphi \vee \psi$ is described above. It is similar to the one used for the product: we start computing φ and stop if it is verified, otherwise, we reset to the beginning of the nested word and check formula ψ. The construction for conjunction is obtained dually.

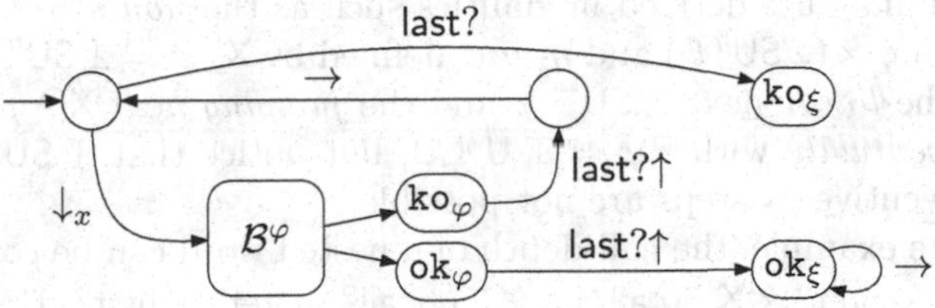

Finally, the construction for existential quantification $\xi = \exists x\, \varphi$ is described above. Again, the construction for universal quantification is dual. ☐

5.2 Weighted Temporal Logics

A temporal logic is usually based on modalities such as *next* and *until* where the until modality is a simple fixed point based on the next modality. When the structures (the models) are not linear, one may follow different paths which are in general based on elementary *steps*. For instance, in unranked trees, one may move vertically down to a child or up to the father, or horizontally to the right or left brother. Similarly, for nested words, several types of paths were introduced in [3] yielding various until modalities, some of them will be discussed below.

Here, we adopt a generic definition of temporal logics where until modalities are based on various elementary steps. Formally, an elementary step η is an *unambiguous* regular expression following the syntax:

$$\eta ::= \alpha \mid \rightarrow \mid \leftarrow \mid \curvearrowright \mid \curvearrowleft \mid \eta + \eta \mid \eta \cdot \eta \mid \eta^+$$
$$\alpha ::= a? \mid \tau? \mid \neg\alpha \mid \alpha \wedge \alpha \mid \alpha \vee \alpha \tag{1}$$

with $a \in A$ and $\tau \in \mathcal{T}$. By *unambiguous* we mean that the quantitative semantics $[\![\eta]\!]$ as defined in Section 3 coincides with the Boolean semantics: $[\![\eta]\!](W, i, j) \in \{0, 1\}$ for all nested words W and positions $i, j \in \overline{\mathrm{pos}}(W)$.

For instance, the (classical) linear until is based on the linear step $\eta = \rightarrow$. The *summary-up* until is based on the summary-up step σ^u defined as $\sigma^u = \curvearrowright + \neg\mathsf{call}? \cdot \rightarrow$ which may move directly from a call to the matching return, or go to the successor, but cannot "enter" a call. The *summary-down* until is based on the summary-down step defined as $\sigma^d = \curvearrowright + \rightarrow \cdot \neg\mathsf{ret}?$. Notice that σ^d is unambiguous, even though a call position may have two successors.

The syntax of the *weighted* temporal logic wTL over nested words is defined by

$$\Phi ::= s \mid \alpha \mid \Phi + \Phi \mid \Phi \times \Phi \mid \Phi \,\mathsf{SU}^\eta\, \Phi$$

with $s \in \mathbb{S}$, α simple tests as defined in (1), and η elementary steps. Since a (weighted) temporal logic formula has an implicit free variable, the quantitative semantics $[\![\Phi]\!](W, i) \in \mathbb{S}$ maps a nested word W and a position $i \in \mathrm{pos}(W)$ to a value in the semiring. It is defined in Table 3. Given a nested word W, we say that two positions $i, j \in \mathrm{pos}(W)$ form an η-step if $[\![\eta]\!](W, i, j) = 1$. Moreover, an η-path is a sequence $i_0, \dots, i_n \in \mathrm{pos}(W)$ such that (i_k, i_{k+1}) is an η-step for all $0 \leq k < n$.

As usual, we may use derived modalities such as the *non strict until* defined by $\varphi \,\mathsf{U}^\eta\, \psi \stackrel{\text{def}}{=} \psi + \varphi \times (\varphi \,\mathsf{SU}^\eta\, \psi)$ and η-*next* defined by $\mathsf{X}^\eta\, \varphi \stackrel{\text{def}}{=} \bot \,\mathsf{SU}^\eta\, \varphi$. As special cases, we get the *linear* next $\bot \,\mathsf{SU}^\rightarrow\, \varphi$ and the *jumping* next $\mathsf{X}^\curvearrowright \varphi = \bot \,\mathsf{SU}^\curvearrowright\, \varphi$. We also get *eventually* with $\mathsf{F}\varphi = \top \,\mathsf{U}^\rightarrow\, \varphi$, but notice that $\top \,\mathsf{SU}^\curvearrowright\, \varphi = \mathsf{X}^\curvearrowright \varphi$ since two consecutive $\curvearrowright$-steps are not possible.

As a concrete example, the call-depth of a nested word can be computed with the formula $(1 \times \neg\mathsf{ret}? \times \mathsf{X}^\leftarrow(\mathsf{call}?) + \mathsf{X}^\leftarrow(\neg\mathsf{call}?) + \mathsf{ret}? + \mathsf{first}?) \,\mathsf{U}^{\sigma^d}\, \top$.

Notice that the sum in (2) may be infinite for some step expressions such as $\eta = \leftarrow + \rightarrow$. On the other hand, if a step expression only moves forward (resp. backward) then it defines a *future* (resp. *past*) modality and the sum in (2) is finite. The following theorem shows that wTL formulae can be translated into equivalent layered pebWA.

Table 3. Semantics of wTL

$$[\![s]\!](W,i) = s \qquad [\![\alpha]\!](W,i) = \begin{cases} 1 & \text{if } W,i \models \alpha \\ 0 & \text{otherwise} \end{cases}$$

$$[\![\Phi + \Psi]\!] = [\![\Phi]\!] + [\![\Psi]\!] \qquad [\![\Phi \times \Psi]\!] = [\![\Phi]\!] \times [\![\Psi]\!]$$

$$[\![\Phi \,\mathsf{SU}^\eta\, \Psi]\!](W,i) = \sum_{i=i_0,i_1,\ldots,i_n\ \eta\text{-path}} \left(\prod_{0<k<n} [\![\Phi]\!](W,i_k) \right) \times [\![\Psi]\!](W,i_n) . \tag{2}$$

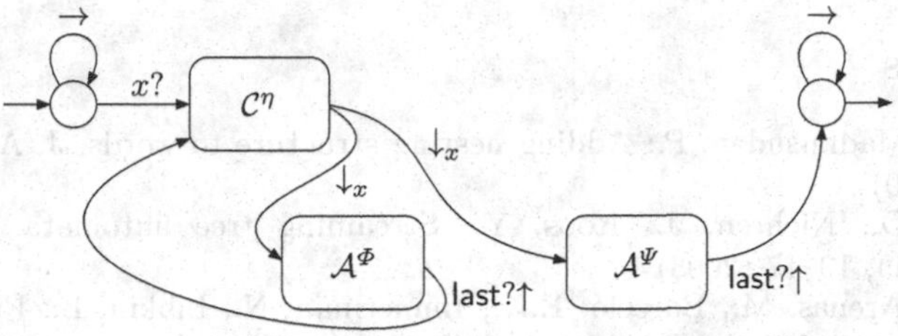

Fig. 7. Automaton for $\Xi = \Phi \,\mathsf{SU}^\eta\, \Psi$

Theorem 10. *For each* wTL *formula* Φ *we can effectively construct an equivalent layered* pebWA $\mathcal{A}^\Phi$ *with a single pebble name and* $\mathcal{O}(|\Phi|)$ *states: for all nested words* W *and positions* $i \in \mathrm{pos}(W)$ *we have* $[\![\mathcal{A}^\Phi]\!](W, x \mapsto i) = [\![\Phi]\!](W,i)$.

Proof (Sketch). We explain the construction for SU^η where η is a step expression. Using Theorem 5, from the pebWE η we obtain an equivalent pebWA $\mathcal{C}^\eta$: $[\![\mathcal{C}^\eta]\!](W,i,j) = [\![\eta]\!](W,i,j)$ for all nested words W and positions $i,j \in \overline{\mathrm{pos}}(W)$.

Consider the wTL formula $\Xi = \Phi \,\mathsf{SU}^\eta\, \Psi$ and assume we have already constructed the pebWA $\mathcal{A}^\Phi$ and $\mathcal{A}^\Psi$. The pebWA $\mathcal{A}^\Xi$ is given in Fig. 7. Observe that we have added one layer and a constant number of states. □

Notice that we can extend the wTL in several ways. First, instead of simple (pebble-free) tests α we may allow arbitrary Boolean formula $\varphi(x) \in \mathrm{FO}$ having a single free variable. Second, we may allow *weighted* regular expressions for steps, in which case we have to include the weights of η-steps in the semantics of the weighted until in (2).

6 Conclusion and Perspectives

We have presented a general framework to specify quantitative properties of nested words, and compile them into automata.

Several improvements can be considered. First, concerning our procedure for evaluation, [12] also presented an improved algorithm in the strongly layered case, namely when at every layer, only one pebble name can be dropped. We did not consider this case in this paper for lack of space, but we believe it can be adapted in the nested word case, and leave it for future work. Second, we would like to extend the decidability of emptiness in the general case, where the semiring may have zero divisors.

Finally, notice that, contrary to [3], wFO is strictly less expressive than our temporal logics. This is due to the power of η steps, which gives to wTL a flavor of weighted transitive closure (see [5]). As other directions of research, we would like to study this transitive closure operator in order to find a logical fragment expressively equivalent to pebWA and pebWE.

We thank the anonymous referees for their valuable comments.

References

1. Alur, R., Madhusudan, P.: Adding nesting structure to words. J. ACM 56, 16:1–16:43 (2009)
2. Gauwin, O., Niehren, J., Roos, Y.: Streaming tree automata. Inf. Process. Lett. 109(1), 13–17 (2008)
3. Alur, R., Arenas, M., Barceló, E.K., Immerman, N., Libkin, L.: First-order and temporal logics for nested words. LMCS 4(4) (2008)
4. Madhusudan, P., Viswanathan, M.: Query Automata for Nested Words. In: Královič, R., Niwiński, D. (eds.) MFCS 2009. LNCS, vol. 5734, pp. 561–573. Springer, Heidelberg (2009)
5. Bollig, B., Gastin, P., Monmege, B., Zeitoun, M.: Pebble Weighted Automata and Transitive Closure Logics. In: Abramsky, S., Gavoille, C., Kirchner, C., Meyer auf der Heide, F., Spirakis, P.G. (eds.) ICALP 2010. LNCS, vol. 6199, pp. 587–598. Springer, Heidelberg (2010)
6. Schützenberger, M.P.: On the definition of a family of automata. Information and Control 4, 245–270 (1961)
7. Droste, M., Gastin, P.: Weighted automata and weighted logics. In: Handbook of Weighted Automata, pp. 175–211. Springer (2009)
8. Mathissen, C.: Weighted logics for nested words and algebraic formal power series. Logical Methods in Computer Science 6(1) (2010)
9. Droste, M., Vogler, H.: Weighted logics for unranked tree automata. Theory Comput. Syst. 48(1), 23–47 (2011)
10. Bojańczyk, M.: Tree-Walking Automata. In: Martín-Vide, C., Otto, F., Fernau, H. (eds.) LATA 2008. LNCS, vol. 5196, pp. 1–2. Springer, Heidelberg (2008)
11. Engelfriet, J., Hoogeboom, H.J., Samwel, B.: XML transformation by tree-walking transducers with invisible pebbles. In: Proceedings of PODS 2007, pp. 63–72 (2007)
12. Gastin, P., Monmege, B.: Adding Pebbles to Weighted Automata. In: Moreira, N., Reis, R. (eds.) CIAA 2012. LNCS, vol. 7381, pp. 28–51. Springer, Heidelberg (2012)
13. Bollig, B., Gastin, P., Monmege, B.: Weighted specifications over nested words. Technical report (2013), `http://www.lsv.ens-cachan.fr/Publis/RAPPORTS_LSV/newrapports`
14. Droste, M., Kuich, W.: Semirings and formal power series. In: Handbook of Weighted Automata, pp. 3–27. Springer (2009)
15. Sakarovitch, J.: Automata and expressions. In: AutoMathA Handbook (2012) (to appear)
16. Conway, J.: Regular Algebra and Finite Machines. Chapman & Hall (1971)
17. Libkin, L.: Elements of Finite Model Theory. In: EATCS. Springer (2004)

An Algebraic Presentation of Predicate Logic
(Extended Abstract)

Sam Staton*

Computer Laboratory, University of Cambridge

Abstract. We present an algebraic theory for a fragment of predicate logic. The fragment has disjunction, existential quantification and equality. It is not an algebraic theory in the classical sense, but rather within a new framework that we call 'parameterized algebraic theories'.

We demonstrate the relevance of this algebraic presentation to computer science by identifying a programming language in which every type carries a model of the algebraic theory. The result is a simple functional logic programming language.

We provide a syntax-free representation theorem which places terms in bijection with sieves, a concept from category theory.

We study presentation-invariance for general parameterized algebraic theories by providing a theory of clones. We show that parameterized algebraic theories characterize a class of enriched monads.

1 Introduction

This paper is about the following fragment of predicate logic:

$$P, Q ::= \bot \mid P \vee Q \mid (t = u) \wedge P \mid \exists a. P(a) \mid x[t_1, \ldots, t_n]$$

where t, u range over the domain of discourse and x is an n-ary predicate symbol. We provide an algebraic presentation of logical equivalence using a new algebraic framework that we call 'parameterized algebraic theories'. This syntactic framework comes with a straightforward deduction system.

Having introduced the new algebraic framework and presented the theory of predicate logic, we make three further contributions.

1. We consider a programming language in which every type is equipped with the structure of a model of the theory. This yields a simple functional logic programming language (in the sense of [7]). In doing this, we add weight to the slogan of Plotkin and Power [35]: 'algebraic theories determine computational effects'. We demonstrate our language by providing a simple implementation[1].

2. We give a representation theorem for terms in our algebraic theory. There is nothing canonical about a presentation of an algebraic theory: which function

* Research partially supported by a grant from the Isaac Newton Trust and ERC Project ECSYM.

[1] The implementation is available at `http://www.cl.cam.ac.uk/users/ss368/flp`.

F. Pfenning (Ed.): FOSSACS 2013, LNCS 7794, pp. 401–417, 2013.

symbols should be used? which assortment of equations? We show that our algebraic presentation of predicate logic is correct by representing terms up-to equivalence as mathematical objects. More precisely, we show that terms up-to equivalence can be understood as *sieves*, a kind of generalized subset.

3. The idea of presentation-invariance of theories is an important one, and so we introduce a general notion of 'clone' for parameterized algebraic theories. Via enriched category theory we obtain a semantic status for the syntactic framework of parameterized algebraic theories. In particular, we show that the parameterized algebraic theories can be understood as a class of enriched monads.

This work thus provides a principled foundation for program equations, with anticipated consequences for program verification and compiler design (see [24]).

In many ways our algebraic presentation of predicate logic is an elaboration of the algebraic theory of semilattices. Recall that the theory of semilattices has a constant $\bot$ and a binary function symbol $\vee$, and the following equations

$$(x \vee y) \vee z \equiv x \vee (y \vee z) \qquad x \vee y \equiv y \vee x \qquad x \vee x \equiv x \qquad \bot \vee x \equiv x. \quad (1)$$

Predicate logic combines these equations with others, such as the equation $x[b] \vee \exists a.\, x[a] \equiv \exists a.\, x[a]$. We introduce the article by considering the three contributions from the simpler perspective of semilattices.

Extending the Theory to a Programming Language (§3). To extend the theory of semilattices to a functional language, we add a constant $\bot$ and a binary operation $\vee$ at each type. The result is a language that is declarative in two ways: first as a functional language, and second in that the semilattice structure provides a kind of non-determinism.

The additional constructs of predicate logic provide further techniques for declarative programming: $(t = u) \wedge P$ can be understood as unification and $\exists a.\, P(a)$ can be understood as introducing a new logic variable. Thus the functional language gains an abstract type `param`, representing the domain of discourse. If the domain of discourse is the Peano natural numbers, we have expressions `z:param` and `s:param→param`. Consider the following recursive program `add` of type `param→param→param→unit`:

$$\mathtt{add}\, a\, b\, c \stackrel{\text{def}}{=} \big((a = \mathtt{z}) \wedge (b = c) \wedge ()\big) \vee \big(\exists a'.\, \exists b'.\, (a = \mathtt{s}(a')) \wedge (c = \mathtt{s}(c')) \wedge \mathtt{add}\, a'\, b\, c'\big)$$

The program returns if $a+b = c$, and fails if not. Thus functions into `unit` are like predicates. To experiment in more depth we provide a simple implementation in Standard ML.

Representation Theorem (§4). Any presentation of an algebraic theory is somewhat arbitrary. We could have presented semilattices (1) using a ternary disjunction, or by replacing associativity with mediality $((v\vee x)\vee(y\vee z) \equiv (v\vee y)\vee(x\vee z))$. What really matters about the theory of semilattices is that to give a term is to give the set of variables that appear in it. For instance, $(x \vee z) \vee v$ and $v \vee ((z \vee x) \vee \bot)$ both use the same variables, and they are provably equal.

When we move to the algebraic theory of disjunctive predicate logic, a representation result is even more desirable, since the axiomatization is more complicated. We no longer have a characterization in terms of sets of variables. Rather, we generalize that analysis by understanding a set of variables as a kind of weakening principle: in any algebraic theory, any term involving those variables can also be considered as a term involving more variables. In the setting of parameterized algebraic theories, the notions of substitution and weakening are more sophisticated. Nonetheless our representation theorem characterizes terms of disjunctive predicate logic as classes of substitutions that satisfy a closure condition. We set this up using category theory, by defining a category whose objects are contexts and whose morphisms are substitutions, so that a class of substitutions is a 'sieve'.

This representation theorem shows that our algebraic theory of disjunctive predicate logic is not just an ad-hoc syntactic gadget: it is mathematically natural. This corroborates the following general hypothesis: algebraic theories for computational effects should have elegant characterizations of terms and free models [35,30].

Clones (§5). In our third contribution we stay with the theme of presentation invariance, but we study it for parameterized algebraic theories in general. In classical universal algebra, presentation-invariance is studied via clones: closed sets of operations. Recall that an abstract clone is given by a set $T(n)$ for each natural number n, a tuple $\eta_n \in T(n)^n$ for each number n, and a family of functions $\{*_{m,n} : Tm \times (Tn)^m \to Tn\}_{m,n}$, all satisfying some conditions (e.g. [14, Ch. III]).

The terms in the theory of semilattices form a clone: $T(n)$ is the set of terms in n variables, η picks out the terms that are merely variables; and $*$ describes simultaneous substitution. Similarly the subsets form a clone: $T(n)$ is the set of subsets of n, η picks out the singleton subsets, and $*$ is a union construction. These two clones are isomorphic.

Abstract clones can be equivalently described as monoids in a suitable monoidal category, and moreover equivalently described as finitary monads on the category of sets. This provides the connection between Moggi's work on computational monads [32] and the assertion of Plotkin and Power [35] that computational effects determine algebraic theories.

In Section 5, we revisit this situation in the context of parameterized algebraic theories. We provide a general notion of enriched clone. Specialized to enrichment in a presheaf category, this provides a presentation-invariant description of parameterized algebraic theories. For general reasons, enriched clones can be equivalently described as sifted-colimit-preserving enriched monads on a presheaf category. Thus our syntactic framework of parameterized algebraic theories is given a canonical semantic status and a connection with Moggi's work [32].

2 Presentations of Parameterized Algebraic Theories

We will present disjunctive predicate logic as an algebraic theory within a new framework of parameterized algebraic theories. It is an algebraic theory that takes parameters from another algebraic theory. When we write

$$((a = a) \wedge x) \equiv x \tag{2}$$

this is a judgement of equality between predicates, and while x is a variable standing for a predicate, a is not: it stands for a term (e.g. a natural number). This phenomenon is common across mathematics. For example, in linear algebra, when we write $a(x + y) \equiv ax + ay$ the variables x and y have a different status to the scalar parameter a.

Parameterized algebraic theories are not merely 2-sorted theories. In equation (2), the variable x stands for a predicate which might itself have parameters. For example, we can describe the substitutive nature of equality like this:

$$(a = b) \wedge x[a] \equiv (a = b) \wedge x[b]. \tag{3}$$

The substitution of predicates for variables is now quite elaborate. One instance of equation (3) is $(a = b) \wedge (a = a) \wedge y \equiv (a = b) \wedge (b = a) \wedge y$, under the assignment $x[c] \mapsto (c = a) \wedge y$.

Quantifiers bind parameters, requiring equations like this:

$$x[b] \vee \exists a.\, x[a] \equiv \exists a.\, x[a] \tag{4}$$

in which the parameter b is free while the parameter a is bound. We work up-to α-equivalence ($\exists a.\, x[a] = \exists b.\, x[b]$) and substitution must avoid variable capture: we must change bound variables before substituting $x[c] \mapsto (c = a) \wedge y$ in (4).

In this section we give a technical account of what constitutes a signature of parameters (§2.1) and an algebraic theory parameterized in that signature (§2.2). In the example of predicate logic, terms of the signature of parameters describe the domain of discourse. Terms of the parameterized theory are predicates over the domain of discourse, or alternatively simple logic programs over the domain of discourse.

The general framework is essentially a single-sorted version of the 'effect theories' developed in joint work with Møgelberg [31], based on proposals by Plotkin and Pretnar [34,37].

2.1 Signatures of Parameters

Recall the notion of signature used in universal algebra. A signature is given by a set of function symbols, f, g, ..., each equipped with an arity, which is a natural number specifying how many arguments it takes. From a signature we can build terms-in-context:

$$\frac{}{\vec{a} \vdash a_i}\,(i \leqslant |\vec{a}|) \qquad\qquad \frac{\vec{a} \vdash t_1 \;\dots\; \vec{a} \vdash t_n}{\vec{a} \vdash \mathsf{f}(t_1, \dots, t_n)}\,(\mathsf{f}\text{ has arity } n)$$

(We write $\vec{a}$ for a list of variables $a_1, a_2, \ldots$ and $|\vec{a}|$ for the length of the list.)

Here are some simple examples of signatures:

- The empty signature has no function symbols.
- The signature of natural numbers has a constant symbol z (i.e. a function symbol with arity 0) and a unary function symbol s.
- The domain of a database can be described by a signature with a constant symbol for each element of the domain.

2.2 Parameterized Algebraic Theories

Let $\mathbb{S}$ be a signature as in Sect. 2.1. A *signature* $\mathbb{T}$ *parameterized by* $\mathbb{S}$ is given by a set of function symbols $\mathrm{F}, \mathrm{G}, \ldots$ each equipped with an arity, written $\mathrm{F} : (p \mid \vec{n})$, where p is a natural number and $\vec{n}$ a list of natural numbers. We distinguish between parameters and arguments. The number p determines how many parameters (from $\mathbb{S}$) the function symbol takes. The length of the list $\vec{n}$ determines how many arguments (from $\mathbb{T}$) the function symbol takes, and each component of the list determines the valence or binding depth of that argument. For instance, the arity of binary join is $\vee : (0 \mid [0, 0])$, since it takes two arguments with no variable binding; the arity of the equality predicate $(a = b) \wedge x$ is $(2 \mid [0])$ as it takes two parameters (a and b) and one argument (x); the arity of the quantifier is $\exists : (0 \mid [1])$ since it takes one argument with a bound variable.

From a parameterized signature we can build parameterized terms in context. A context $\Gamma \mid \Delta$ for a parameterized term has two parts, comprising two kinds of variable. The first part Γ is a finite set of variables ranging over parameters. The second part Δ is a finite set of variables ranging over terms, each equipped with a natural number, which specifies how many parameters the variable takes. As usual, we write a set of variables as a list with the convention that all variables are different. The terms-in-context are built from variables, function symbols and terms from the signature of parameters, using the following two rules.

$$\frac{\Gamma \vdash t_1 \ \ldots \ \Gamma \vdash t_{n_i}}{\Gamma \mid x_1 : n_1 \ldots x_k : n_k \vdash x_i[t_1 \ldots t_{n_i}]}$$

$$\frac{\Gamma \vdash t_1 \ \ldots \ \Gamma \vdash t_p \quad \Gamma, a_{1,1} \ldots a_{1,n_1} \mid \Delta \vdash u_1 \ \ldots \ \Gamma, a_{k,1} \ldots a_{k,n_k} \mid \Delta \vdash u_k}{\Gamma \mid \Delta \vdash \mathrm{F}(t_1, \ldots, t_p, \vec{a}_1.\, u_1, \ldots, \vec{a}_k.\, u_k)} \tag{5}$$

where $\mathrm{F} : (p \mid [n_1 \ldots n_k])$. We work up-to α-renaming the binders $\vec{a}$. Notice our distinction between judgements of parameters $(\Gamma \vdash t)$ and of terms $(\Gamma \mid \Delta \vdash t)$.

Definition 1. *A presentation of a parameterized algebraic theory is a parameterized signature together with a collection of equations, where an equation is a pair of two terms in the same context.*

We define substitution for the two kinds of variable in a standard way, so as to give the following derived rules.

$$\frac{\Gamma \vdash t \quad \Gamma, a \mid \Delta \vdash u}{\Gamma \mid \Delta \vdash u[{}^{t}\!/\!{}_{a}]} \qquad \frac{\Gamma, a_1 \ldots a_n \mid \Delta \vdash t \quad \Gamma \mid \Delta, x : n \vdash u}{\Gamma \mid \Delta \vdash u[{}^{a_1 \ldots a_n . t}\!/\!{}_{x}]} \tag{6}$$

Recall that all variables in a context are distinct. Thus substitution is capture-avoiding unless the capture is explicit.

We form a deductive system from a presentation by combining all substitution-instances of the equations with the usual laws of reflexivity, symmetry, transitivity, and the following congruence rule:

$$\frac{\Gamma \vdash t_1 \ \dots \ \Gamma \vdash t_p \qquad \Gamma, \vec{a}_1 \mid \Delta \vdash u_1 \equiv u_1' \ \dots \ \Gamma, \vec{a}_k \mid \Delta \vdash u_k \equiv u_k'}{\Gamma \mid \Delta \vdash \mathrm{F}(t_1, \dots, t_p, \vec{a}_1.\, u_1, \dots, \vec{a}_k.\, u_k) \equiv \mathrm{F}(t_1, \dots, t_p, \vec{a}_1.\, u_1', \dots, \vec{a}_k.\, u_k')} \tag{7}$$

2.3 Presentation of Predicate Logic

We now describe disjunctive predicate logic as a parameterized algebraic theory. Predicates range over a domain of discourse, which is a signature of parameters. In this paper we only consider two signatures of parameters: (1) the empty one; (2) the signature of natural numbers; but other signatures can be accommodated straightforwardly.

The parameterized algebraic theory is given in Figure 1. It has a constant symbol $\bot$ and a binary function symbol $\vee$. There is a unary function symbol (=:=) which takes two parameters, and a function symbol $\exists$ which binds a parameter. We use an infix notation for $\vee$ and =:=. Term formation (5) yields

$$\frac{}{\Gamma|\Delta \vdash \bot} \qquad \frac{\Gamma|\Delta \vdash t \quad \Gamma|\Delta \vdash u}{\Gamma|\Delta \vdash t \vee u} \qquad \frac{\Gamma \vdash t_1 \quad \Gamma \vdash t_2 \quad \Gamma|\Delta \vdash u}{\Gamma|\Delta \vdash (t_1 =:= t_2)u} \qquad \frac{\Gamma, a \mid \Delta \vdash u}{\Gamma \mid \Delta \vdash \exists a.\, u}$$

The string $(a =:= b)x[]$ can be thought of as the predicate $(a = b) \wedge x[]$, or as the logic program 'unify a and b, and then continue as x'. Note that we do not have arbitrary conjunctions in our algebraic theory. However, if we understand predicate variables as continuation variables, then substitution behaves like conjunction: e.g. given $\vec{a} \mid x : 0 \vdash t$ and $\vec{a} \mid - \vdash u$, the expression $t[^u/_x]$ can be understood as $t \wedge u$. We return to the idea of 'conjunction as sequential composition' in Section 3.

Laws 1–4 are the laws of semilattices. Laws 5–7 are basic axioms for equality. If we write '$t \leqslant u$' for '$t \vee u \equiv u$', then Laws 8, 9, 10 can be written $\exists a.\, x[] \leqslant x[]$, $x[b] \leqslant \exists a.\, x[a]$, $(a =:= b)x[] \leqslant x[]$. Laws 11 and 12 say that $\vee$ commutes over =:= and $\exists$. In fact, all the operations commute over each other. For instance,

$$a, b \mid - \vdash (a =:= b)\bot \overset{4}{\equiv} \bot \vee (a =:= b)\bot \overset{2}{\equiv} (a =:= b)\bot \vee \bot \overset{10}{\equiv} \bot.$$

Laws 13 and 14 are axioms of Peano arithmetic. Law 15 is 'occurs check'.

We can derive $a, b \mid y : 0 \vdash (a =:= b)y[] \equiv (b =:= a)y[]$. First, notice that $a, b \mid y \colon 0 \vdash (a =:= b)(a =:= a)y[] \equiv (a =:= b)(b =:= a)y[]$ is an instance of Law 6, under the substitution $[^{c.((c=:=a)y[])}/_x]$. Now,

$$\begin{aligned} a, b \mid y : 0 \vdash (a =:= b)y &\overset{5}{\equiv} (a =:= b)(a =:= a)y[] \overset{6}{\equiv} (a =:= b)(b =:= a)y[] \\ &\overset{7}{\equiv} (b =:= a)(a =:= b)y[] \overset{6}{\equiv} (b =:= a)(b =:= b)y[] \overset{5}{\equiv} (b =:= a)y[]. \end{aligned}$$

Signature:	$\bot : (0 \mid [])$ $\quad \vee : (0 \mid [0,0])$ $\quad (=:=) : (2 \mid [0])$ $\quad \exists : (0 \mid [1])$
Equations:	1. $- \mid x, y, z : 0 \vdash (x[] \vee y[]) \vee z[] \equiv x[] \vee (y[] \vee z[])$
	2. $- \mid x, y : 0 \vdash x[] \vee y[] \equiv y[] \vee x[]$
	3. $- \mid x : 0 \vdash x[] \vee x[] \equiv x[]$
	4. $- \mid x : 0 \vdash \bot \vee x[] \equiv x[]$
	5. $a \mid x : 0 \vdash (a =:= a)x[] \equiv x[]$
	6. $a, b \mid x : 1 \vdash (a =:= b)x[a] \equiv (a =:= b)x[b]$
	7. $a, b, c, d \mid x : 0 \vdash (a =:= b)(c =:= d)x \equiv (c =:= d)(a =:= b)x$
	8. $- \mid x : 0 \vdash (\exists a.\, x[]) \vee x[] \equiv x[]$
	9. $b \mid x : 1 \vdash x[b] \vee \exists a.\, x[a] \equiv \exists a.\, x[a]$
	10. $a, b \mid x : 0 \vdash ((a =:= b)x[]) \vee x[] \equiv x[]$
	11. $a, b \mid x, y : 0 \vdash (a =:= b)(x[] \vee y[]) \equiv ((a =:= b)x[]) \vee ((a =:= b)y[])$
	12. $- \mid x, y : 1 \vdash \exists a.\, (x[a] \vee y[a]) \equiv \exists a.\, x[a] \vee \exists a.\, y[a]$
Additional equation schemes when the signature of parameters is natural numbers:	
	13. $a \mid x : 0 \vdash (\mathsf{z} =:= \mathsf{s}(a))x[] \equiv \bot$
	14. $a, b \mid x : 0 \vdash (\mathsf{s}(a) =:= \mathsf{s}(b))x[] \equiv (a =:= b)x[]$
	15. $a \mid x : 0 \vdash (a =:= \mathsf{s}^n(a))x[] \equiv \bot \quad \forall n > 0$, where $\mathsf{s}^2(a) = \mathsf{s}(\mathsf{s}(a))$, etc.

Fig. 1. A presentation of the parameterized theory of disjunctive predicate logic

A subtle point is that the context cannot be omitted. When the signature of parameters is empty, the equation $- \mid x : 0 \vdash \exists a.\, x[] \equiv x[]$ is not derivable, although we do have $a \mid x : 0 \vdash \exists a.\, x[] \overset{9}{\equiv} x[] \vee \exists a.\, x[] \overset{2}{\equiv} (\exists a.\, x[]) \vee x[] \overset{8}{\equiv} x[]$. (To instantiate law 9 we applied the substitution $[{}^{a.x[]}/_{x}]$.)

2.4 Other Examples of Parameterized Algebraic Theories

Any classical algebraic theory can be understood as a parameterized one in which the function symbols take no parameters and all the valences are 0. A slightly more elaborate example is the 2-sorted theory of modules over an unspecified ring, which is an algebraic theory parameterized in the signature of rings.

For any signature we have the following theory of computations over a memory cell. There are two function symbols in the parameterized algebraic theory: $\mathsf{w} : (1 \mid [0])$ and $\mathsf{r} : (0 \mid [1])$. The intuition is that $\mathsf{w}(a, x[])$ writes a to memory and continues as x, while $\mathsf{r}(a.\, x[a])$ reads from memory, binds the result to a, and continues as x. The presentation has three equations (c.f. [35]):

$$x[] \equiv \mathsf{r}(a.\, \mathsf{w}(a, x[])) \qquad \mathsf{w}(a, \mathsf{w}(b, x[])) \equiv \mathsf{w}(b, x[]) \qquad \mathsf{w}(a, \mathsf{r}(b.\, x[b])) \equiv \mathsf{w}(a, x[a])$$

The first equation says that if you read a, then write a, then continue as x, then you may as well just run x. The second equation says that if you write to memory twice then it is the second write that counts. The third equation says that if you read b after a write a, then b will be a and the read is unnecessary.

If the parameterizing signature is the signature of natural numbers then there is an expression $\mathsf{r}(a.\,\mathsf{w}(\mathsf{s}(a), x))$, which increments the memory and continues as x.

2.5 Set-Theoretic Models

We briefly discuss models of parameterized algebraic theories. As we will see, the set-theoretic notion of model is not a complete way to understand theories, but it is sound and so we are able to use it to verify consistency.

A set-theoretic structure for a parameterized algebraic theory is given by two sets: a set π of parameters, and a set M which is the carrier. For each n-ary function symbol f in the signature of parameters, a function $\mathrm{f} : \pi^n \to \pi$ must be given. For each function symbol $\mathrm{F} : (p \mid \vec{n})$ in the theory, a function $\mathrm{F} : \pi^p \times \prod_{j=1}^{|\vec{n}|} M^{(\pi^{n_j})} \to M$ must be given. Here $M^{(\pi^{n_j})}$ is the set of functions from (π^{n_j}) to M. It is routine to extend this to all terms, interpreting a term-in-context $\vec{a} \mid \vec{x} : \vec{n} \vdash t$ as a function $[\![t]\!] : \pi^{|\vec{a}|} \times \prod_{j=1}^{|\vec{n}|} M^{(\pi^{n_j})} \to M$. We say that a structure is a *model* when for each equation $\Gamma \mid \Delta \vdash t \equiv u$ in the theory the corresponding functions are equal: $[\![t]\!] = [\![u]\!]$. This interpretation is sound:

Proposition 1. *If $\vec{a} \mid \vec{x} : \vec{n} \vdash t \equiv u$ is derivable in a parameterized algebraic theory, then $[\![t]\!] = [\![u]\!]$ in all models.*

Consider the theory of disjunctive predicate logic over the signature for natural numbers. We can let π be the set $\mathbb{N}$ of natural numbers, and then we must provide a set M together with an element $\bot$ and three functions: $\vee : M \times M \to M$, $(=:=) : \mathbb{N} \times \mathbb{N} \times M \to M$, and $\exists : M^{\mathbb{N}} \to M$. In fact, this forms a model if and only if M is a countable semilattice, with $\bot$ and $\vee$ supplying the finite joins and $\exists$ supplying the countably infinite joins. By the soundness result, the consistency of our theory is witnessed by giving a non-trivial countable-join-semilattice.

There are two things that are unsatisfactory about set-theoretic models of disjunctive predicate logic. First, it is often best not to think of $\exists$ as a countable join: in logic programming it is better to think of $\exists$ as introducing a free logic variable. Second, the interpretation of $(=:=)$ is necessarily fixed, as we now explain.

2.6 Incompleteness of Set-Theoretic Models

Classical universal algebra is complete for set-theoretic models: if an equation is true in all algebras, then it is derivable. However, set-theoretic models are *not* complete for parameterized algebraic theories: some equations are true in all set-theoretic models but not derivable. In any set-theoretic structure for disjunctive predicate logic, the three equations

$$(a =:= a)x[] \equiv x[] \qquad (a =:= b)\bot \equiv \bot \qquad (a =:= b)x[a] \equiv (a =:= b)x[b] \tag{8}$$

entirely determine (=:=). This is because two elements a, b of π are either equal or not equal. In any structure satisfying the three equations we *must* have $(a =:= b)(x) = x$ when $a = b$ and $(a =:= b)(x) = \bot$ when $a \neq b$. The first case, when $a = b$, is the first equation in (8). To establish the second case, when $a \neq b$, we define a function $\delta^a_x : \pi \to M$ by setting $\delta^a_x(c) \stackrel{\text{def}}{=} x$ if $c = a$ and $\delta^a_x(c) \stackrel{\text{def}}{=} \bot$ if $c \neq a$, so that $(a =:= b)(x) = (a =:= b)\delta^a_x(a) = (a =:= b)\delta^a_x(b) = (a =:= b)\bot = \bot$.

Thus any set-theoretic structure satisfying the laws in (8) will also satisfy law 7 in Figure 1: $(a =:= b)(c =:= d)x[] \equiv (c =:= d)(a =:= b)x[]$. But that is not derivable from (8). To resolve this incompleteness we must move to a constructive set theory in which equality is not two-valued, which is the essence of Section 5.

2.7 Other Equational Approaches to Logic

The syntax for parameterized algebraic theories is reminiscent of logical frameworks such as Type Framework [4] although it is much simpler.

Our syntax is also similar to Aczel's syntax [1] which forms the basis of the second-order algebraic theories of Fiore et al. [18,19]. The key difference is that we do not allow second-order variables to take second-order terms as parameters, e.g. $x[y[] \vee z[]]$. This restriction allows us to make a connection with programming languages (§3) and a simple categorical model theory (§5).

By far the most studied equational approach to logic is Tarski's cylindric algebra. Cylindric algebra encodes the binding structures of predicate logic into classical universal algebra. Although cylindric algebra can provide a foundation for concurrent constraint programming [39], it does not extend easily to higher typed programming languages. Our parameterized algebraic theory does (§3).

There have been several proposals for 'nominal algebra' [13,20]. Gabbay and Matthijssen used this to describe first-order logic [20] and the author has earlier developed a nominal-style presentation of semilattices and equality [41, §6]. However, it is unclear how to combine a theory of nominal algebra with programming language primitives in a canonical way. Although the free model construction [13] yields a monad on the category of nominal sets, it does not yield a strength for the monad, and so Moggi's framework [32] does not apply to nominal algebraic theories. Moggi's framework *does* apply to parameterized algebraic theories (§5).

Kurz and Petrişan [28] have shown how to understand cylindric algebra and nominal algebra within the framework of presheaf categories. In Section 5 we will demonstrate that parameterized algebraic theories can be understood as algebraic theories enriched in a presheaf category.

Bronsard and Reddy [12] axiomatized a theory of disjunction, conjunction, existentials and if-then-else, and they gave a completeness result for domain theoretic models, providing a more axiomatic account than earlier domain theoretic models of logic programming (e.g. [21,33,38]). Our work strengthens that early development by moving away from concrete models, which are not a complete way to study algebraic theories with binding (see §2.6). This allows us to give a canonical status to our algebraic theory (Theorem 1).

3 Extending the Algebraic Theory to a Programming Language

Plotkin and Power have proposed that algebraic theories determine notions of computational effect [35]. One can build a higher-order functional programming language in which each type is a model of the algebraic theory, so that the algebraic structure provides the impurities in the functional programming language. We demonstrate this by picking out a fragment of Standard ML, eliciting the algebraic structure of each type by identifying suitable generic effects [36]. For instance, the generic effect for disjunction, $\vee$, is an impure function <u>choose</u>: `unit → bool`, to be thought of as returning an undetermined boolean. Our implementation is thus a structure for the following ML signature.

```
infix 3 =:=
sig
  (* SIGNATURE OF PARAMETERS *)          (* MAIN SIGNATURE *)
  (* param is the domain                 (* Presented using
             of discourse *)                generic effects *)
  type param                             val choose : unit → bool
  val succ : param → param               val fail : unit → 'a
  val zero : param                       val =:= : param * param → unit
                                         val free : unit → param end
```

The algebraic operations at each type can be recovered from the generic effects:

```
      t ∨ u ≝ if choose() then t else u     ⊥ ≝ fail()
(a =:= b)(t) ≝ a=:=b ; t                 ∃a. t ≝ let val a=free() in t end
```

In this ML signature, there are two ways to define addition, firstly as a function:

```
- fun add(a,b) = if choose () then a =:= zero ; b
                 else let val a' = free()
                      in a =:= succ a' ; add(a',succ(b)) end
val add = fn : param * param → param
```

and secondly as a predicate:

```
- fun add'(a,b,c) = if choose () then a =:= zero ; b =:= c
                    else let val a' = free() val c' = free() in
                         a =:= succ a' ; c =:= succ c' ; add'(a',b,c') end
val add' = fn : param * param * param → unit
```

This demonstrates a type isomorphism:

```
- fun iso f a = let val b = free() in (b,f(a,b)) end
val iso = fn : ('a * param → 'c) → ('a → param * 'c)
- fun inverse g a = let val (b',c) = g(a) in b =:= b' ; c end
val inverse = fn : ('a → param * 'c) → ('a * param → 'c)
```

Notice that sequencing (semicolon) is like conjunction.

Our implementation of the ML signature (see online appendix) uses references and `callcc`. One way to view the laws in Figure 1 is as an axiomatic account of which optimizations are allowed [24]. Our implementation certainly doesn't implement Law 2 (commutativity), which would need parallel execution. Does our implementation capture the other laws? A proper answer to this question would need a theory of observational equivalence for this fragment of ML, for instance extending [22] to parameterized algebraic theories, which we leave for future work.

Our lightweight approach to implementation is partly inspired by Eff [9], a new language for algebraic effects. A related approach is to use monads in Haskell, following [11,40], since in our language the type construction `unit → (-)` is equipped with the structure of a monad.

4 Representation of Terms

In Figure 1 we have presented an algebraic theory of disjunctive predicate logic. In Section 3 we have seen that the theory provides a reasonable account of the basic phenomena in logic programming. However, the choice of the presentation, which operations and which equations, is somewhat arbitrary. We now justify the theory by giving a canonical representation of the terms modulo the equations.

As a first step, we consider the theory of semilattices, the fragment of disjunctive predicate logic restricted to $\vee$ and $\bot$. A term built from $\vee$ and $\bot$ is determined by the variables that appear in the term. For instance, the terms $- \mid v, x, y, z : 0 \vdash (x[] \vee z[]) \vee v[]$ and $- \mid v, x, y, z : 0 \vdash v[] \vee ((z[] \vee x[]) \vee \bot)$ both contain the same variables $\{v, x, z\}$, and they are equal. We are able to get a similar result for full disjunctive predicate logic, but it is more complicated. For instance, consider the following term.

$$- \mid x : 2 \vdash \exists a.\, x[a, a] \tag{9}$$

To understand this term as a 'subset' of $\{x\}$, we make the following observation. The subset $\{v, x, z\}$ of $\{v, x, y, z\}$ provides a weakening principle: for any algebraic theory, any term in context $\{v, x, z\}$ can also be understood as a term in context $\{v, x, y, z\}$. The term (9) also describes a weakening principle: in any parameterized algebraic theory, a term in context $(x : 1)$ can be understood as a term in context $(x : 2)$, by substituting every occurrence of $x[t]$ by $x[t, t]$.

This motivates us to define a category whose objects are contexts and whose morphisms are substitutions. We investigate sieves, which are sets of substitutions subject to a closure condition. Our representation theorem provides a correspondence between terms of disjunctive predicate logic and sieves.

Subsets and Sieves. The concept of 'subset' is not a priori a category-theoretic notion because it is defined in terms of elements of sets rather than morphisms. One category-theoretic notion of 'subset' is the notion of sieve.

Definition 2. *Let $\mathcal{C}$ be a category, and let X be an object in $\mathcal{C}$. A* sieve *S on X is a class of morphisms with codomain X which is closed under precomposition: $f \in S \implies fg \in S$. Every morphism $f \colon Y \to X$ generates a sieve on X as follows: $[f] \stackrel{\text{def}}{=} \{g \colon Z \to X \mid \exists h : Z \to Y.\, g = fh\}$. A sieve S on X is* singleton-generated *if it is of this form.*

For two morphisms $f\colon Y \to X$ and $f'\colon Y' \to X$, the following are equivalent: (i) they generate the same sieve ($[f] = [f']$); (ii) $f \in [f']$ and $f' \in [f]$; (iii) there are morphisms $g\colon Y \to Y'$ and $g'\colon Y' \to Y$ such that $f = f'g$ and $f' = fg'$.

Singleton generated sieves can be understood as a category-theoretic version of subset. For instance, in the category of sets, a function $f\colon X \to Y$ generates the same sieve of Y as its image $f(X) \rightarrowtail Y$ (assuming choice). More generally, a singleton-generated sieve on an object X of a category $\mathcal{C}$ is a subobject of X in the regularization of $\mathcal{C}$ [23, A1.3.10(d)]. The importance of sieves and presheaves to logic programming has been observed earlier [16,26,27].

A Category of Contexts. We will describe a correspondence between terms of disjunctive predicate logic and sieves in a category whose objects are contexts and whose morphisms are substitutions.

In a parameterized algebraic theory, the context has two components $(\Gamma|\Delta)$: variables in Γ ranging over parameters and variables in Δ ranging over terms. The objects of our category focus on the second component Δ. Since the names of variables are irrelevant, we represent a context by a list of numbers.

The morphisms of our category are simultaneous substitutions. To motivate, consider the following derived typing rule for substituting variables for variables.

$$\frac{\vec{a} \vdash t_1 \ \dots\ \vec{a} \vdash t_n \qquad - \mid \vec{x} : \vec{m}, y : |\vec{a}| \vdash u}{- \mid \vec{x} : \vec{m}, z : n \vdash u[{}^{\vec{a}.z[t_1,\dots,t_n]}/_{y}]} \tag{10}$$

Definition 3. *Let $\mathbb{S}$ be a signature (of parameters, as in §2.1). The objects of the category* $\mathbf{Ctx}(\mathbb{S})$ *are lists of natural numbers. A morphism $\vec{m} \to \vec{n}$ comprises a function $f\colon |\vec{m}| \to |\vec{n}|$ together with, for $1 \leqslant i \leqslant |\vec{m}|$ and $1 \leqslant j \leqslant n_{f(i)}$, a term $a_1, \dots a_{m_i} \vdash t_{i,j}$ in the signature of parameters. Morphisms compose by composing functions and substituting terms for variables. The identity morphism is built from variables.*

The category theorist will recognize $\mathbf{Ctx}(\mathbb{S})$ as the free finite coproduct completion of the Lawvere theory for the signature $\mathbb{S}$. Lawvere theories are widely regarded as important to the foundations of logic programming (e.g. [8,26,6,27]).

For any term $- \mid \vec{x}\colon \vec{m} \vdash u$ in any parameterized algebraic theory, and any morphism $(f, \vec{t}) : \vec{m} \to \vec{n}$ in $\mathbf{Ctx}(\mathbb{S})$, notice that we can build a term by substitution, following (10),

$$- \mid \vec{y}\colon \vec{n} \vdash (f,\vec{t}) \bullet u \stackrel{\text{def}}{=} u[{}^{a_1\dots a_{m_1}.y_{f(1)}[\vec{t}_1]}/_{x_1}]\dots[{}^{a_1\dots a_{m_{|\vec{m}|}}.y_{f(|\vec{m}|)}[\vec{t}_{|\vec{m}|}]}/_{x_{|\vec{m}|}}]$$

so that substitution respects composition: $(g,\vec{v}) \bullet ((f,\vec{u}) \bullet t) = ((g,\vec{v}) \cdot (f,\vec{u})) \bullet t$.

Representation Theorem. We now state our representation theorem for the theory of disjunctive predicate logic. We focus on terms with no free parameters, returning to this point later.

Given a morphism $(f, \{b_1, \dots, b_{m_i} \vdash t_{i,j}\}_{i \leqslant |\vec{m}|, j \leqslant n_{f(i)}}) : \vec{m} \to \vec{n}$ in $\mathbf{Ctx}(\mathbb{S})$ we define the following term:

$$\wr f, \vec{t} \,\wr \stackrel{\text{def}}{=} - \mid \vec{x} : \vec{n} \vdash \bigvee_{i=1}^{|\vec{m}|} \exists b_1. \dots \exists b_{m_i}. x_{f(i)}[t_{i,1} \dots t_{i,n_{f(i)}}] \tag{11}$$

Theorem 1. *Let $\mathbb{S}$ be either the empty signature or the signature for natural numbers. Let $\vec{n}$ be a list of numbers. The construction $\wr - \int$ induces a bijective correspondence between:*

- *terms in context, $- \mid \vec{x} : \vec{n} \vdash t$, modulo the equivalence relation in Figure 1;*
- *singleton-generated sieves on $\vec{n}$ in the category* $\mathbf{Ctx}(\mathbb{S})$.

(The Theorem can be established for a different signature of parameters by finding appropriate analogues of Laws 13–15.)

Outline Proof of Theorem 1. To prove the representation theorem, we first characterize morphisms into $\vec{n}$ as terms modulo a fragment of the theory. We then show that two morphisms generate the same sieve if and only if the corresponding terms are equal in the full theory.

For the first step, we show that the construction $\wr - \int$ determines a bijective correspondence between morphisms into $\vec{n}$ and terms-in-context modulo a fragment of the theory in Figure 1. The fragment is given by laws 1, 4, 5, 6, 7, 11 and 12 from Figure 1, and schemes 13–15 where relevant, together with the following five laws, which are derivable from laws 2, 3 and 8–10 in Figure 1 but not from the other laws.

$$- \mid x\colon 0 \vdash \bot \vee x \equiv x \qquad a, b \mid - \vdash (a =:= b)\bot \equiv \bot$$

$$- \mid - \vdash \exists a.\, \bot \equiv \bot \qquad b, c \mid x\colon 1 \vdash (b =:= c)\exists a.\, x[a] \equiv \exists a.\, (b =:= c)x[a]$$

$$b \mid x\colon 1 + n \vdash \exists c_1.\, \ldots \exists c_n.\, x[b, \vec{c}] \equiv \exists a.\, \exists c_1.\, \ldots \exists c_n.\, (a =:= b)x[a, \vec{c}]$$

The first four laws are commutativity conditions; the last one is roughly introduction and elimination for $\exists$.

Note that every term can be rewritten to the form in (11) using the laws in this fragment of the theory. We first pull the disjunctions to the front, then the existentials, and then we use the remaining axioms to rearrange and eliminate the equality tests. We thus have a bijective correspondence between terms in context $\vec{n}$ modulo this fragment of the theory, and morphisms into $\vec{n}$ in $\mathbf{Ctx}(\mathbb{S})$.

The second step is to show that that two morphisms in $\mathbf{Ctx}(\mathbb{S})$ determine the same sieve if and only if the corresponding terms (11) can be proven equal using the laws in Figure 1. We show that $[f, \vec{t}] \subseteq [g, \vec{u}]$ if and only if $\wr f, \vec{t} \int \leqslant \wr g, \vec{u} \int$.

Terms with Free Parameters. Let $T^{\exists\vee}(p|\vec{n})$ be the set of terms in the context $(a_1, \ldots, a_p|\vec{x} : \vec{n})$, modulo the equivalence relation. If we write $\mathrm{Sieves}_1(\vec{n})$ for the set of singleton-generated sieves on $\vec{n}$ in $\mathbf{Ctx}(\mathbb{S})$, then Theorem 1 provides a natural bijection $T^{\exists\vee}(0|\vec{n}) \cong \mathrm{Sieves}_1(\vec{n})$.

We now briefly consider the situation where the parameter context p is non-empty, by exhibiting a bijection $T^{\exists\vee}(p|[n_1, \ldots, n_k]) \cong T^{\exists\vee}(0|[p + n_1, \ldots, p + n_k])$.

To go from left to right we substitute $\vec{a}, \vec{b} \mid x_i\colon p + n_i \vdash x_i[b_1 \ldots b_p, a_1 \ldots a_{n_i}]$ for each variable $x_i[a_1 \ldots a_{n_i}]$ $(i \leqslant k)$, and then existentially quantify all the free variables $\vec{b}$. From right to left we substitute

$$a_1 \,..\, a_{n_i} \,..\, a_{p+n_i+n}, b_1 \,..\, b_p \mid x_i : m_i \vdash (a_{n_i+1} =:= b_1) \dots (a_{n_i+p} =:= b_p) x_i[a_1 \dots a_{n_i}]$$

for each variable x_i, yielding a term with free variables $\vec{b}$ (c.f. `iso` in §3).

5 Enriched Clones

We conclude by giving our general syntactic framework of parameterized algebraic theories a canonical status by reference to enriched category theory. The importance of enriched monads for programming language semantics has long been recognized [32]. We show that parameterized algebraic theories characterize a class of enriched monads.

In the previous section we described a bijection between the set of terms of disjunctive predicate logic and the set of sieves in a category of contexts. However, the set of terms is not a mere set: it also has a substitution structure. We characterize this abstractly by introducing enriched clones.

Definition 4. *Let $(\mathcal{V}, \otimes, I)$ be a symmetric monoidal category. Let $\mathcal{C}$ be a $\mathcal{V}$-enriched category, and $J \colon \mathcal{A} \subseteq \mathcal{C}$ be a full sub-$\mathcal{V}$-category. An* enriched clone *is given by*
1. For each $A \in \mathcal{A}$, an object TA in $\mathcal{C}$;
2. A morphism $\eta_A : I \to \mathcal{C}(JA, TA)$ in $\mathcal{V}$ for all $A \in \mathcal{A}$;
3. A morphism $*_{A,B} : \mathcal{C}(JA, TB) \to \mathcal{C}(TA, TB)$ in $\mathcal{V}$ for all $A, B \in \mathcal{A}$
such that the following diagrams commute:

$$\begin{array}{ccc} I \otimes \mathcal{C}(JA,TB) & \xrightarrow{\eta \otimes *} & \mathcal{C}(JA,TA) \otimes \mathcal{C}(TA,TB) \\ & \searrow^{\lambda} & \downarrow{\scriptstyle composition} \\ & & \mathcal{C}(JA,TB) \end{array} \qquad \begin{array}{ccc} I & \xrightarrow{\eta_A} & \mathcal{C}(JA,TA) \\ & \searrow^{\mathrm{id}_{TA}} & \downarrow{\scriptstyle *} \\ & & \mathcal{C}(TA,TA) \end{array}$$

$$\begin{array}{ccccc} \mathcal{C}(JA,TB) \otimes \mathcal{C}(JB,TC) & \xrightarrow{*} & \mathcal{C}(JA,TB) \otimes \mathcal{C}(TB,TC) & \xrightarrow{comp} & \mathcal{C}(JA,TC) \\ {\scriptstyle *\otimes*}\downarrow & & & & \downarrow{\scriptstyle *} \\ \mathcal{C}(TA,TB) \otimes \mathcal{C}(TB,TC) & & \xrightarrow[composition]{\hspace{6em}} & & \mathcal{C}(TA,TC) \end{array}$$

The original notion of abstract clone (e.g. [14, Ch. III]) appears when $\mathcal{V} = \mathcal{C} = \mathcal{S}et$ and $\mathcal{A}$ comprises natural numbers considered as sets. When $\mathcal{V} = \mathcal{S}et$ then enriched clones have been called 'Kleisli structures' (e.g. [15, §7]) and 'relative monads' [5].

We now turn to parameterized algebraic theories. The signature of parameters induces a Lawvere theory $\mathbb{S}$ which is a category whose objects are natural numbers and where a morphism $m \to n$ is a family of n terms over m parameter variables. We are interested in the category $\hat{\mathbb{S}}$ of presheaves on the Lawvere theory $\mathbb{S}$, that is, the category of contravariant functors $\mathbb{S}^{\mathrm{op}} \to \mathcal{S}et$ and natural transformations between them. As we will see shortly, a presheaf can be understood as a set with substitution structure (see also [26,17]). Notice that $\mathbf{Ctx}(\mathbb{S})$ (Def. 3) can be understood as a full subcategory of $\hat{\mathbb{S}}$ once we understand a context $[n_1, \ldots, n_k]$ as the presheaf $\coprod_{i=1}^{k} \mathbb{S}(-, n_i)$. We let $J \colon \mathbf{Ctx}(\mathbb{S}) \subseteq \hat{\mathbb{S}}$ be the embedding. To put it another way, $\mathbf{Ctx}(\mathbb{S})$ is the completion of $\mathbb{S}$ under coproducts, and $\hat{\mathbb{S}}$ is the completion under all colimits. We consider $\hat{\mathbb{S}}$ as a cartesian closed category, i.e. self-enriched. The function space $\hat{\mathbb{S}}(J\vec{n}, F)$ is given by context extension: $\hat{\mathbb{S}}(J\vec{n}, F)(p) = \prod_{i=1}^{|\vec{n}|} F(p + n_i)$.

Every presentation of a parameterized algebraic theory gives rise to an enriched clone for $\mathcal{V} = \mathcal{C} = \hat{\mathbb{S}}$ and $\mathcal{A} = \mathbf{Ctx}(\mathbb{S})$.

1. The data $T(\vec{n})$ assigns a presheaf to each context $\vec{n}$. The set $(T(\vec{n}))(p)$ is the set of terms in context $(p|\vec{n})$ modulo the equations. The functorial action of $T(\vec{n})$ corresponds to the left-hand substitution structure in (6).
2. The data $\eta_{\vec{n}}$ provides an element of the set $\hat{\mathbb{S}}(\vec{n}, P)(0)$, which identifies the variables among the terms.
3. The right-hand substitution structure in (6) gives natural transformations $T\vec{m} \times \hat{\mathbb{S}}(J\vec{m}, T\vec{n}) \to T\vec{n}$. By currying, this supplies the data $*_{\vec{m},\vec{n}}$.

The three commuting diagrams are easy substitution lemmas.

Theorem 2. *Every enriched clone for $\mathcal{V} = \mathcal{C} = \hat{\mathbb{S}}$ and $\mathcal{A} = \mathbf{Ctx}(\mathbb{S})$ arises from a presentation of a parameterized algebraic theory.*

In general, when J is dense, enriched clones can be understood as monoids in the multicategory whose objects are $\mathcal{V}$-functors $\mathcal{A} \to \mathcal{C}$, and where an n-ary morphism $F_1, \ldots, F_n \to G$ between $\mathcal{V}$-functors is an extra-natural family of morphisms $\mathcal{C}(JA_0, F_1A_1) \otimes \cdots \otimes \mathcal{C}(JA_{n-1}, F_nA_n) \to \mathcal{C}(JA_0, GA_n)$ in $\mathcal{V}$. When this multicategory has tensors then we arrive at the situation considered by Kelly and Power [25, §5]. They focus on the situation where $\mathcal{A}$ comprises the finitely presentable objects of $\mathcal{C}$, but it seems reasonable to replace 'finitely presentable' with another well-behaved notion of finiteness [29,2,10,42]. We consider sifted colimits [2,3,29], i.e. colimits that commute with products in the category of sets, which leads us to the notions of strongly finitely presentable object and strongly accessible category [2] (aka generalized variety [3]).

Proposition 2 (c.f. [25], §5). *Let $\mathcal{V}$ be strongly finitely accessible as a closed category. Let $J \colon \mathcal{A} \subseteq \mathcal{V}$ comprise the strongly finitely presentable objects. Let $T \colon \mathcal{A} \to \mathcal{V}$ be a $\mathcal{V}$-functor. To equip T with the structure of an enriched clone is to equip the left Kan extension of T along J with the structure of an enriched monad.*

Parameterized algebraic theories fit the premises of this proposition. The presheaf category $\hat{\mathbb{S}}$ is strongly finitely accessible as a closed category, and $\mathbf{Ctx}(\mathbb{S})$ comprises the strongly finitely presentable objects (up to splitting idempotents).

Corollary 1. *To give a parameterized algebraic theory is to give a sifted-colimit-preserving enriched monad on $\hat{\mathbb{S}}$.*

Summary. We have shown that our framework for parameterized algebraic theories (§2) is a syntactic formalism for enriched clones (§5). For our theory of disjunctive predicate logic, which has applications to logic programming (§3), the clones can be represented abstractly as sieves (§4).

Acknowledgements. Thanks to reviewers for helpful feedback, and to M Fiore, M Hyland, O Kammar, A Kurz, P Levy, P-A Melliès, R Møgelberg, G Plotkin, J Power and T Uustalu for discussions.

References

1. Aczel, P.: A general Church-Rosser theorem (1978)
2. Adámek, J., Borceux, F., Lack, S., Rosický, J.: A classification of accessible categories. J. Pure Appl. Algebra 175(1-3), 7–30 (2002)
3. Adámek, J., Rosický, J.: On sifted colimits and generalized varieties. Theory Appl. Categ. 8(3), 33–53 (2001)
4. Adams, R.: Lambda-free logical frameworks. Ann. Pure Appl. Logic (to appear)
5. Altenkirch, T., Chapman, J., Uustalu, T.: Monads Need Not Be Endofunctors. In: Ong, L. (ed.) FOSSACS 2010. LNCS, vol. 6014, pp. 297–311. Springer, Heidelberg (2010)
6. Amato, G., Lipton, J., McGrail, R.: On the algebraic structure of declarative programming languages. Theor. Comput. Sci. 410(46), 4626–4671 (2009)
7. Antoy, S., Hanus, M.: Functional logic programming. C. ACM 53(4), 74–85 (2010)
8. Asperti, A., Martini, S.: Projections instead of variables: A category theoretic interpretation of logic programs. In: Proc. ICLP 1989 (1989)
9. Bauer, A., Pretnar, M.: Programming with algebraic effects and handlers. arXiv:1203.1539v1
10. Berger, C., Melliès, P.-A., Weber, M.: Monads with arities and their associated theories. J. Pure Appl. Algebra 216(8-9), 2029–2048 (2012)
11. Braßel, B., Fischer, S., Hanus, M., Reck, F.: Transforming Functional Logic Programs into Monadic Functional Programs. In: Mariño, J. (ed.) WFLP 2010. LNCS, vol. 6559, pp. 30–47. Springer, Heidelberg (2011)
12. Bronsard, F., Reddy, U.S.: Axiomatization of a Functional Logic Language. In: Kirchner, H., Wechler, W. (eds.) ALP 1990. LNCS, vol. 463, pp. 101–116. Springer, Heidelberg (1990)
13. Clouston, R.A., Pitts, A.M.: Nominal equational logic. In: Computation, Meaning, and Logic. Elsevier (2007)
14. Cohn, P.M.: Universal algebra, 2nd edn. D Reidel (1981)
15. Curien, P.-L.: Operads, clones and distributive laws. In: Operads and Universal Algebra. World Scientific (2012)
16. Finkelstein, S.E., Freyd, P.J., Lipton, J.: Logic Programming in Tau Categories. In: Pacholski, L., Tiuryn, J. (eds.) CSL 1994. LNCS, vol. 933, pp. 249–263. Springer, Heidelberg (1995)
17. Fiore, M., Plotkin, G., Turi, D.: Abstract syntax and variable binding. In: Proc. LICS 1999 (1999)
18. Fiore, M., Hur, C.-K.: Second-Order Equational Logic (Extended Abstract). In: Dawar, A., Veith, H. (eds.) CSL 2010. LNCS, vol. 6247, pp. 320–335. Springer, Heidelberg (2010)
19. Fiore, M., Mahmoud, O.: Second-Order Algebraic Theories. In: Hliněný, P., Kučera, A. (eds.) MFCS 2010. LNCS, vol. 6281, pp. 368–380. Springer, Heidelberg (2010)
20. Gabbay, M.J., Mathijssen, A.: One and a halfth order logic. J. Logic Comput. 18 (2008)

21. Jagadeesan, R., Panangaden, P., Pingali, K.: A fully abstract semantics for a functional language with logic variables. In: LICS 1989 (1989)
22. Johann, P., Simpson, A., Voigtländer, J.: A generic operational metatheory for algebraic effects. In: LICS 2010 (2010)
23. Johnstone, P.T.: Sketches of an Elephant. OUP (2002)
24. Kammar, O., Plotkin, G.D.: Algebraic foundations for effect-dependent optimisations. In: Proc. POPL 2012 (2012)
25. Kelly, G.M., Power, A.J.: Adjunctions whose counits are coequalisers. J. Pure Appl. Algebra 89, 163–179 (1993)
26. Kinoshita, Y., Power, A.J.: A fibrational Semantics for Logic Programs. In: Herre, H., Dyckhoff, R., Schroeder-Heister, P. (eds.) ELP 1996. LNCS, vol. 1050, pp. 177–191. Springer, Heidelberg (1996)
27. Komendantskaya, E., Power, J.: Coalgebraic Semantics for Derivations in Logic Programming. In: Corradini, A., Klin, B., Cîrstea, C. (eds.) CALCO 2011. LNCS, vol. 6859, pp. 268–282. Springer, Heidelberg (2011)
28. Kurz, A., Petrişan, D.: Presenting functors on many-sorted varieties and applications. Inform. Comput. 208(12), 1421–1446 (2010)
29. Lack, S., Rosický, J.: Notions of Lawvere theory. Appl. Categ. Structures 19(1) (2011)
30. Melliès, P.-A.: Segal condition meets computational effects. In: Proc. LICS 2010 (2010)
31. Møgelberg, R.E., Staton, S.: Linearly-Used State in Models of Call-by-Value. In: Corradini, A., Klin, B., Cîrstea, C. (eds.) CALCO 2011. LNCS, vol. 6859, pp. 298–313. Springer, Heidelberg (2011)
32. Moggi, E.: Notions of computation and monads. Inform. Comput. 93(1) (1991)
33. Moreno-Navarro, J.J., Rodríguez-Artalejo, M.: Logic programming with functions and predicates. J. Log. Program 12(3&4), 191–223 (1992)
34. Plotkin, G.: Some Varieties of Equational Logic. In: Futatsugi, K., Jouannaud, J.-P., Meseguer, J. (eds.) Goguen Festschrift. LNCS, vol. 4060, pp. 150–156. Springer, Heidelberg (2006)
35. Plotkin, G., Power, J.: Notions of Computation Determine Monads. In: Nielsen, M., Engberg, U. (eds.) FOSSACS 2002. LNCS, vol. 2303, pp. 342–356. Springer, Heidelberg (2002)
36. Plotkin, G.D., Power, J.: Algebraic operations and generic effects. Appl. Categ. Structures 11(1), 69–94 (2003)
37. Plotkin, G., Pretnar, M.: Handlers of Algebraic Effects. In: Castagna, G. (ed.) ESOP 2009. LNCS, vol. 5502, pp. 80–94. Springer, Heidelberg (2009)
38. Reddy, U.S.: Functional Logic Languages, Part I. In: Fasel, J.H., Keller, R.M. (eds.) Graph Reduction 1986. LNCS, vol. 279, pp. 401–425. Springer, Heidelberg (1987)
39. Saraswat, V.A., Rinard, M.C., Panangaden, P.: Semantic foundations of concurrent constraint programming. In: Proc. POPL 1991, pp. 333–352 (1991)
40. Schrijvers, T., Stuckey, P.J., Wadler, P.: Monadic constraint programming. J. Funct. Program. 19(6) (2009)
41. Staton, S.: Relating Coalgebraic Notions of Bisimulation. In: Kurz, A., Lenisa, M., Tarlecki, A. (eds.) CALCO 2009. LNCS, vol. 5728, pp. 191–205. Springer, Heidelberg (2009)
42. Velebil, J., Kurz, A.: Equational presentations of functors and monads. Math. Struct. in Comp. Science 21 (2011)

Strategies as Profunctors

Glynn Winskel

University of Cambridge Computer Laboratory, England

Abstract. A new characterization of nondeterministic concurrent strategies exhibits strategies as certain discrete fibrations—or equivalently presheaves—over configurations of the game. This leads to a lax functor from the bicategory of strategies to the bicategory of profunctors. The lax functor expresses how composition of strategies is obtained from that of profunctors by restricting to 'reachable' elements, which gives an alternative formulation of the composition of strategies. It provides a fundamental connection—and helps explain the mismatches—between two generalizations of domain theory to forms of intensional domain theories, one based on games and strategies, and the other on presheaf categories and profunctors. In particular cases, on the sub-bicategory of rigid strategies which includes 'simple games' (underlying AJM and HO games), and stable spans (specializing to Berry's stable functions, in the deterministic case), the lax functor becomes a pseudo functor. More generally, the laxness of the functor suggests what structure is missing from categories and profunctors in order that they can be made to support the operations of games and strategies. By equipping categories with the structure of a 'rooted' factorization system and ensuring all elements of profunctors are 'reachable,' we obtain a pseudo functor embedding the bicategory of strategies in the bicategory of reachable profunctors. This shift illuminates early work on bistructures and bidomains, where the Scott order and Berry's stable order are part of a factorization system, giving a sense in which bidomains are games.

1 Introduction

A very general definition of nondeterministic concurrent strategy between games represented by event structures, has recently been given—see [1] for further background and examples. Building on this work and a new characterization of strategies (Lemma 1) we exhibit a strategy in a game as a presheaf, and a strategy between games as a profunctor. This exposes a lax functor from a bicategory of games and strategies to the bicategory of profunctors. In several well-known sub-bicategories of games the lax functor becomes a pseudo functor.

This somewhat technical result is significant because both strategies and profunctors have been central to generalizations of domain theory to forms of intensional domain theories.[1] Game semantics has been strikingly successful in

[1] A discussion of the limits of traditional domain theory, and the form generalizations should take can be found in [2,3].

F. Pfenning (Ed.): FOSSACS 2013, LNCS 7794, pp. 418–433, 2013.

providing denotational semantics in close agreement with operational semantics, following on from seminal work on game semantics and PCF [4,5]. However, being rooted in sequential games, traditional game semantics does not integrate smoothly with concurrent computation. Profunctors themselves provide a rich framework in which to generalize domain theory in a way that is arguably closer to that initiated by Dana Scott than game semantics; we refer the reader to Hyland's case for such a generalization [2] and to the relevance of presheaf categories and profunctors to concurrent computation [6]. But the mathematical abstraction comes at a price: it can be hard to give an operational reading to denotations as profunctors. The lax functor from strategies to profunctors provides a fundamental connection between the two approaches. Indeed it exhibits composition of strategies as essentially composition of profunctors but restricted to those elements which are 'reachable'; roughly, they are 'reachable' in the sense of satisfying the causal-dependency constraints of both components of the composition, whereas profunctor composition allows elements merely when input matches output. The lax functor helps explain the mismatches between the two approaches and how we might marry them to obtain the benefits of both.

Arguably the concept of strategy is potentially as fundamental as that of relation. But for this potential to be seen and realized the concept needs to be developed in sufficient generality. This is one motivation for grounding strategies in a general model for concurrent computation. Doing so has exposed unexpected characteristics of strategies, which carry the concept of strategy into new terrain.

A surprise in developing this work has been the central role taken by the *reversal*, or undoing, of (compound) moves of Opponent. The idea first appears rather formally in Lemma 1 where, in a strategy, reversals of Opponent moves satisfy the same property as moves of Player. It then takes on a key role in characterizing a concurrent strategy in a game as a discrete fibration which preserves Player moves and reversals of Opponent moves (Theorem 1). Pushing the idea to completion we are led to a view of games as factorization systems in which 'left' maps stand for the reversal of compound Opponent moves while 'right' maps stand for compound Player moves. Section 9 gives a sketch of how concurrent games and strategies essentially form a sub-bicategory within a bicategory of strategies between games as rooted factorization systems.

2 Event Structures and Stable Families

An *event structure* comprises $(E, \mathrm{Con}, \leq)$, consisting of a set E, of *events* which are partially ordered by $\leq$, the *causal dependency relation*, and a nonempty *consistency relation* Con consisting of finite subsets of E, which satisfy

$$\{e' \mid e' \leq e\} \text{ is finite for all } e \in E,$$
$$\{e\} \in \mathrm{Con} \text{ for all } e \in E,$$
$$Y \subseteq X \in \mathrm{Con} \implies Y \in \mathrm{Con}, \quad \text{and}$$
$$X \in \mathrm{Con} \;\&\; e \leq e' \in X \implies X \cup \{e\} \in \mathrm{Con}.$$

The *configurations*, $\mathcal{C}^\infty(E)$, of an event structure E consist of those subsets $x \subseteq E$ which are

Consistent: $\forall X \subseteq x.\ X$ is finite $\Rightarrow X \in \text{Con}$, and
Down-closed: $\forall e, e'.\ e' \leq e \in x \implies e' \in x$.

Often we shall be concerned with just the finite configurations of an event structure. We write $\mathcal{C}(E)$ for the set of *finite* configurations.

We say that events e, e' are *concurrent*, and write $e\ co\ e'$ if $\{e, e'\} \in \text{Con}\ \&\ e \not\leq e'\ \&\ e' \not\leq e$. In games the relation of *immediate* dependency $e \rightarrow e'$, meaning e and e' are distinct with $e \leq e'$ and no event in between, will play a very important role. For $X \subseteq E$ we write $[X]$ for $\{e \in E \mid \exists e' \in X.\ e \leq e'\}$, the down-closure of X; note if $X \in \text{Con}$, then $[X] \in \text{Con}$.

Operations such as synchronized parallel composition are awkward to define directly on the simple event structures above. It is useful to broaden event structures to stable families, where operations are often carried out more easily, and then turned into event structures by the operation Pr below.

A *stable family* comprises $\mathcal{F}$, a nonempty family of finite subsets, called *configurations*, which satisfy:
Completeness: $\forall Z \subseteq \mathcal{F}.\ Z\uparrow \implies \bigcup Z \in \mathcal{F}$;
Coincidence-freeness: For all $x \in \mathcal{F}$, $e, e' \in x$ with $e \neq e'$,

$$\exists y \in \mathcal{F}.\ y \subseteq x\ \&\ (e \in y \iff e' \notin y);$$

Stability: $\forall x, y \in \mathcal{F}.\ x \uparrow y \implies x \cap y \in \mathcal{F}$.
Above, $Z\uparrow$ means $\exists x \in \mathcal{F} \forall z \in Z.\ z \subseteq x$, and expresses the compatibility of Z in $\mathcal{F}$; we use $x \uparrow y$ for $\{x, y\}\uparrow$. We call elements of $\bigcup \mathcal{F}$ *events* of $\mathcal{F}$.

Proposition 1. *Let x be a configuration of a stable family $\mathcal{F}$. For $e, e' \in x$ define*

$$e' \leq_x e \text{ iff } \forall y \in \mathcal{F}.\ y \subseteq x\ \&\ e \in y \implies e' \in y.$$

When $e \in x$ define the prime configuration

$$[e]_x = \bigcap \{y \in \mathcal{F} \mid y \subseteq x\ \&\ e \in y\}.$$

Then $\leq_x$ is a partial order and $[e]_x$ is a configuration such that

$$[e]_x = \{e' \in x \mid e' \leq_x e\}.$$

Moreover the configurations $y \subseteq x$ are exactly the down-closed subsets of $\leq_x$.

Proposition 2. *Let $\mathcal{F}$ be a stable family. Then,* $\text{Pr}(\mathcal{F}) =_{\text{def}} (P, \text{Con}, \leq)$ *is an event structure where:*

$$P = \{[e]_x \mid e \in x\ \&\ x \in \mathcal{F}\},$$
$$Z \in \text{Con} \text{ iff } Z \subseteq P\ \&\ \bigcup Z \in \mathcal{F} \text{ and,}$$
$$p \leq p' \text{ iff } p, p' \in P\ \&\ p \subseteq p'.$$

A (partial) map of stable families $f : \mathcal{F} \to \mathcal{G}$ is a partial function f from the events of $\mathcal{F}$ to the events of $\mathcal{G}$ such that for all configurations $x \in \mathcal{F}$,

$$fx \in \mathcal{G} \;\&\; (\forall e_1, e_2 \in x.\; f(e_1) = f(e_2) \implies e_1 = e_2)\,.$$

Maps of event structures are maps of their stable families of configurations. Maps compose as functions. We say a map is *total* when it is total as a function.

Pr is the right adjoint of the "inclusion" functor, taking an event structure E to the stable family $\mathcal{C}(E)$. The unit of the adjunction $E \to \mathrm{Pr}(\mathcal{C}(E))$ takes an event e to the prime configuration $[e] =_{\mathrm{def}} \{e' \in E \mid e' \le e\}$. The counit $max : \mathcal{C}(\mathrm{Pr}(\mathcal{F})) \to \mathcal{F}$ takes prime configuration $[e]_x$ to its maximum event e; the image of a configuration $x \in \mathcal{C}(\mathrm{Pr}(\mathcal{F}))$ under the map max is $\bigcup x \in \mathcal{F}$.

Definition 1. Let $\mathcal{F}$ be a stable family. We use $x {-\!\!\subset} y$ to mean y covers x in $\mathcal{F}$, *i.e.* $x \subsetneq y$ in $\mathcal{F}$ with nothing in between, and $x \overset{e}{-\!\!\subset} y$ to mean $x \cup \{e\} = y$ for $x, y \in \mathcal{F}$ and event $e \notin x$. We sometimes use $x \overset{e}{-\!\!\subset}$, expressing that event e is enabled at configuration x, when $x \overset{e}{-\!\!\subset} y$ for some y. W.r.t. $x \in \mathcal{F}$, write $[e)_x =_{\mathrm{def}} \{e' \in E \mid e' \le_x e \;\&\; e' \ne e\}$, so, for example, $[e)_x \overset{e}{-\!\!\subset} [e]_x$. The relation of *immediate* dependence of event structures generalizes: with respect to $x \in \mathcal{F}$, the relation $e \rightarrowtriangle_x e'$ means $e \le_x e'$ with $e \ne e'$ and no event in between.

3 Process Operations

Products. Let $\mathcal{A}$ and $\mathcal{B}$ be stable families with events A and B, respectively. Their product, the stable family $\mathcal{A} \times \mathcal{B}$, has events comprising pairs in $A \times_* B =_{\mathrm{def}} \{(a, *) \mid a \in A\} \cup \{(a, b) \mid a \in A \;\&\; b \in B\} \cup \{(*, b) \mid b \in B\}$, the product of sets with partial functions, with (partial) projections π_1 and π_2—treating $*$ as 'undefined'—with configurations $x \in \mathcal{A} \times \mathcal{B}$ iff

$$\begin{aligned}
&x \text{ is a finite subset of } A \times_* B \text{ s.t. } \pi_1 x \in \mathcal{A} \;\&\; \pi_2 x \in \mathcal{B},\\
&\forall e, e' \in x.\; \pi_1(e) = \pi_1(e') \text{ or } \pi_2(e) = \pi_2(e') \Rightarrow e = e', \&\\
&\forall e, e' \in x.\; e \ne e' \Rightarrow \exists y \subseteq x.\; \pi_1 y \in \mathcal{A} \;\&\; \pi_2 y \in \mathcal{B} \;\&\; (e \in y \iff e' \notin y)\,.
\end{aligned}$$

Right adjoints preserve products. Consequently we obtain a product of event structures A and B by first regarding them as stable families $\mathcal{C}(A)$ and $\mathcal{C}(B)$, forming their product $\mathcal{C}(A) \times \mathcal{C}(B), \pi_1, \pi_2$, and then constructing the event structure

$$A \times B =_{\mathrm{def}} \mathrm{Pr}(\mathcal{C}(A) \times \mathcal{C}(B))$$

and its projections as $\Pi_1 =_{\mathrm{def}} \pi_1 max$ and $\Pi_2 =_{\mathrm{def}} \pi_2 max$.

Restriction. The *restriction* of $\mathcal{F}$ to a subset of events R is the stable family $\mathcal{F} \upharpoonright R =_{\mathrm{def}} \{x \in \mathcal{F} \mid x \subseteq R\}$. Defining $E \upharpoonright R$, the restriction of an event structure E to a subset of events R, to have events $E' = \{e \in E \mid [e] \subseteq R\}$ with causal dependency and consistency induced by E, we obtain $\mathcal{C}(E \upharpoonright R) = \mathcal{C}(E) \upharpoonright R$.

Proposition 3. *Let $\mathcal{F}$ be a stable family and R a subset of its events. Then,* $\mathrm{Pr}(\mathcal{F} \upharpoonright R) = \mathrm{Pr}(\mathcal{F}) \upharpoonright max^{-1} R$.

Synchronized Compositions. Synchronized parallel compositions are obtained as restrictions of products to those events which are allowed to synchronize or occur asynchronously according to the specific synchronized composition. For example, the synchronized composition of Milner's CCS on stable families $\mathcal{A}$ and $\mathcal{B}$ (with labelled events) is defined as $\mathcal{A} \times \mathcal{B} \upharpoonright R$ where R comprises events which are pairs $(a, *)$, $(*, b)$ and (a, b), where in the latter case the events a of $\mathcal{A}$ and b of $\mathcal{B}$ carry complementary labels. Similarly, synchronized compositions of event structures A and B are obtained as restrictions $A \times B \upharpoonright R$. By Proposition 3, we can equivalently form a synchronized composition of event structures by forming the synchronized composition of their stable families of configurations, and then obtaining the resulting event structure—this has the advantage of eliminating superfluous events earlier.

Projection. Event structures support a simple form of hiding. Let $(E, \leq, \text{Con})$ be an event structure. Let $V \subseteq E$ be a subset of 'visible' events. Define the *projection* of E on V, to be $E{\downarrow}V =_{\text{def}} (V, \leq_V, \text{Con}_V)$, where $v \leq_V v'$ iff $v \leq v'$ & $v, v' \in V$ and $X \in \text{Con}_V$ iff $X \in \text{Con}$ & $X \subseteq V$.

4 Event Structures with Polarities

Both a game and a strategy in a game are to be represented by an event structure with polarity, which comprises (E, pol) where E is an event structure with a polarity function $pol : E \to \{+, -\}$ ascribing a polarity + (Player) or − (Opponent) to its events. The events correspond to (occurrences of) moves. Maps of event structures with polarity are maps of event structures which preserve polarity.

Dual and Parallel Composition of Games. The *dual*, $E^\perp$, of an event structure with polarity E comprises a copy of the event structure E but with a reversal of polarities. We write $\overline{e} \in E^\perp$ for the event complementary to $e \in E$ and *vice versa.* The operation $A \| B$—a simple parallel composition of games—simply forms the disjoint juxtaposition of A, B, two event structures with polarity; a finite subset of events is consistent if its intersection with each component is consistent.

5 Pre-strategies

Let A be an event structure with polarity, thought of as a game. A *pre-strategy in A* represents a nondeterministic play of the game and is defined to be a total map $\sigma : S \to A$ from an event structure with polarity S. Two pre-strategies $\sigma : S \to A$ and $\tau : T \to A$ in A will be essentially the same when they are isomorphic, *i.e.* there is an isomorphism $\theta : S \cong T$ such that $\sigma = \tau\theta$; then we write $\sigma \cong \tau$.

Let A and B be event structures with polarity. Following Joyal [7], a pre-strategy *from A to B* is a pre-strategy in $A^\perp \| B$, so a total map $\sigma : S \to A^\perp \| B$. It thus determines a span

$$A^\perp \xleftarrow{\sigma_1} S \xrightarrow{\sigma_2} B\,,$$

of event structures with polarity where σ_1, σ_2 are *partial* maps and for all $s \in S$ either, but not both, $\sigma_1(s)$ or $\sigma_2(s)$ is defined. We write $\sigma : A \twoheadrightarrow B$ to express that σ is a pre-strategy from A to B. Note a pre-strategy σ in a game A coincides with a pre-strategy from the empty game $\sigma : \varnothing \twoheadrightarrow A$.

5.1 Composing Pre-strategies

Consider two pre-strategies $\sigma : A \twoheadrightarrow B$ and $\tau : B \twoheadrightarrow C$ as spans:

$$A^\perp \xleftarrow{\sigma_1} S \xrightarrow{\sigma_2} B \qquad B^\perp \xleftarrow{\tau_1} T \xrightarrow{\tau_2} C .$$

Their composition $\tau \odot \sigma : A \twoheadrightarrow C$ is defined as a synchronized composition, followed by projection to hide internal synchronization events. It is convenient to build the synchronized composition from the product of stable families $\mathcal{C}(S) \times \mathcal{C}(T)$, with projections π_1 and π_2, as

$$\mathcal{C}(T) \odot \mathcal{C}(T) =_{\text{def}} \mathcal{C}(S) \times \mathcal{C}(T) \restriction R , \text{ where}$$

$$\begin{aligned} R = \; & \{(s,*) \mid s \in S \;\&\; \sigma_1(s) \text{ is defined}\} \cup \{(*,t) \mid t \in T \;\&\; \tau_2(t) \text{ is defined}\} \cup \\ & \{(s,t) \mid s \in S \;\&\; t \in T \;\&\; \sigma_2(s) = \overline{\tau_1(t)} \text{ with both defined}\} . \end{aligned}$$

Define $T \odot S =_{\text{def}} \Pr(\mathcal{C}(T) \odot \mathcal{C}(S)) \downarrow V$, where

$$\begin{aligned} V = \; & \{p \in \Pr(\mathcal{C}(T) \odot \mathcal{C}(S)) \mid \exists s \in S.\; max(p) = (s,*)\} \cup \\ & \{p \in \Pr(\mathcal{C}(T) \odot \mathcal{C}(S)) \mid \exists t \in T.\; max(p) = (*,t)\} . \end{aligned}$$

The span $\tau \odot \sigma$ comprises maps $\upsilon_1 : T \odot S \to A^\perp$ and $\upsilon_2 : T \odot S \to C$, which on events p of $T \odot S$ act so $\upsilon_1(p) = \sigma_1(s)$ when $max(p) = (s,*)$ and $\upsilon_2(p) = \tau_2(t)$ when $max(p) = (*,t)$, and are undefined elsewhere.

5.2 Concurrent Copy-cat

Let A be an event structure with polarity. The copy-cat strategy from A to A is an instance of a pre-strategy, so a total map $\gamma_A : \mathbb{C}_A \to A^\perp \| A$, based on the idea that Player moves, of +ve polarity, always copy previous corresponding moves of Opponent, of −ve polarity. For $c \in A^\perp \| A$ we use $\overline{c}$ to mean the corresponding copy of c, of opposite polarity, in the alternative component. Define $\mathbb{C}_A$ to comprise the event structure with polarity $A^\perp \| A$ together with extra causal dependencies $\overline{c} \leq_{\mathbb{C}_A} c$ for all events c with $pol_{A^\perp \| A}(c) = +$.

Proposition 4. *Let A be an event structure with polarity. Then $\mathbb{C}_A$ is an event structure with polarity. Moreover,*
$x \in \mathcal{C}(\mathbb{C}_A)$ *iff* $x \in \mathcal{C}(A^\perp \| A)$ & $\forall c \in x.\; pol_{A^\perp \| A}(c) = + \implies \overline{c} \in x .$

The *copy-cat* pre-strategy $\gamma_A : A \twoheadrightarrow A$ is defined to be the map $\gamma_A : \mathbb{C}_A \to A^\perp \| A$ where γ_A is the identity on the common set of events.

6 Strategies

The main result of [1] is that two conditions on pre-strategies, *receptivity* and *innocence*, are necessary and sufficient for copy-cat to behave as identity w.r.t. the composition of pre-strategies. Receptivity ensures an openness to all possible moves of Opponent. A pre-strategy σ is *receptive* iff

$$\sigma x \xrightarrow{a}\!\subset \ \&\ pol_A(a) = - \text{ implies } \exists! s \in S.\ x \xrightarrow{s}\!\subset \ \&\ \sigma(s) = a\,.$$

Innocence restricts the behaviour of Player; Player may only introduce new relations of immediate causality of the form $\ominus \rightarrowtriangle \oplus$ beyond those imposed by the game. A pre-strategy σ is *innocent* when

$$s \rightarrowtriangle s' \text{ and } pol(s) = + \text{ or } pol(s') = - \text{ implies } \sigma(s) \rightarrowtriangle \sigma(s').$$

Copy-cat behaves as identity w.r.t. composition, *i.e.* $\sigma \circ \gamma_A \cong \sigma$ and $\gamma_B \circ \sigma \cong \sigma$, for a pre-strategy $\sigma : A \twoheadrightarrow B$, iff σ is receptive and innocent; copy-cat pre-stategies $\gamma_A : A \twoheadrightarrow A$ are receptive and innocent [1].

This result motivates the definition of a *strategy* as a pre-strategy which is receptive and innocent. We obtain a bicategory, *Strat*, in which the objects are event structures with polarity—the games, the arrows from A to B are strategies $\sigma : A \twoheadrightarrow B$ and the 2-cells are maps of spans. The vertical composition of 2-cells is the usual composition of maps of spans. Horizontal composition is given by the composition of strategies $\odot$ (which extends to a functor on 2-cells via the functoriality of synchronized composition).

6.1 A New Characterization of Concurrent Strategies

Let x and x' be configurations of an event structure with polarity. Write $x \subseteq^- x'$ to mean $x \subseteq x'$ and $pol(x' \setminus x) \subseteq \{-\}$, *i.e.* the configuration x' extends the configuration x solely by events of –ve polarity. Similarly, write $x \subseteq^+ x'$ to mean $x \subseteq x'$ and $pol(x' \setminus x) \subseteq \{+\}$. With this notation in place we can give an attractive characterization of concurrent strategies, key to this paper.

Lemma 1. *A strategy in a game A comprises $\sigma : S \to A$, a total map of event structures with polarity, such that*
(i) whenever $y \subseteq^+ \sigma x$ in $\mathcal{C}(A)$ there is a (necessarily unique) $x' \in \mathcal{C}(S)$ so that $x' \subseteq x$ & $\sigma x' = y$, i.e.

$$\begin{array}{ccc} x' & \cdots\subseteq\cdots & x \\ {\scriptstyle\sigma}\,\vdots & & \downarrow{\scriptstyle\sigma} \\ y & \subseteq^+ & \sigma x\,, \end{array}$$

and
(ii) whenever $\sigma x \subseteq^- y$ in $\mathcal{C}(A)$ there is a unique $x' \in \mathcal{C}(S)$ so that $x \subseteq x'$ & $\sigma x' = y$, i.e.

$$\begin{array}{ccc} x & \cdots\subseteq\cdots & x' \\ {\scriptstyle\sigma}\downarrow & & \vdots\,{\scriptstyle\sigma} \\ \sigma x & \subseteq^- & y\,. \end{array}$$

7 Strategies as Discrete Fibrations

Condition (i) of Lemma 1, concerning the order $\subseteq^+$, is familiar from discrete fibrations (*cf.* Definition 2 below) while condition (ii), concerning $\subseteq^-$, is dual—the order $\subseteq^-$ simply points in the wrong direction. This suggests building a new relation $\sqsubseteq$ associated with an event structure with polarity out of compositions of $\subseteq^+$ with the *reversed* order $\supseteq^-$. In fact $\sqsubseteq$, the relation so obtained, is a partial order and instances of $\sqsubseteq$ always factor uniquely as an instance of $\supseteq^-$, associated with the reversal or undoing of Opponent moves, followed by $\subseteq^+$, the performance of Player moves (Section 7.1). We call the order $\sqsubseteq$ the *Scott* order because increasing w.r.t. $\sqsubseteq$ is associated with more +ve events (think more output) and less −ve events (think less input)—reminiscent of the pointwise order on functions in domain theory.

The seemingly formal Scott order will be the key to a new understanding of strategies as discrete fibrations (Theorem 1). Discrete fibrations are a reformulation of presheaves so reveal strategies $\sigma : S \to A$ in a game A as certain presheaves over $(\mathcal{C}(A), \sqsubseteq_A)$ (Section 7.2). Through the fortuitous way in which the Scott order interacts with the dual and parallel operations on games a strategy between games $\sigma : A \rightarrow\!\!\!\!\!\rightarrow B$ turns into a presheaf over $(\mathcal{C}(A), \sqsubseteq_A)^{\mathrm{op}} \times (\mathcal{C}(B), \sqsubseteq_B)$, *i.e.* a profunctor from $(\mathcal{C}(A), \sqsubseteq_A)$ to $(\mathcal{C}(B), \sqsubseteq_B)$ (Section 7.3).

7.1 The Scott Order in Games

Let A be an event structure with polarity. The $\subseteq$-order on its configurations decomposes into two more fundamental orders $\subseteq^-$ and $\subseteq^+$. Define the *Scott* order, between configurations $x, y \in \mathcal{C}(A)$, by

$$x \sqsubseteq_A y \iff x \supseteq^- x \cap y \subseteq^+ y\,.$$

We use $\supseteq^-$ as the converse order to $\subseteq^-$. The properties of the Scott order are summarised in the next proposition. In particular,

$$x \sqsubseteq_A y \text{ iff } x \supseteq^- \cdot \subseteq^+ \cdot \supseteq^- \cdots \supseteq^- \cdot \subseteq^+ y\,.$$

Proposition 5. *Let A be an event structure with polarity.*
(i) If $x \subseteq^+ w \supseteq^- y$ *in* $\mathcal{C}(A)$*, then* $x \supseteq^- x \cap y \subseteq^+ y$ *in* $\mathcal{C}(A)$*.*
(ii) The relation $\sqsubseteq_A$ *is the transitive closure of the relation* $\supseteq^- \cup \subseteq^+$*.*
(iii) $(\mathcal{C}(A), \sqsubseteq_A)$ *is a partial order for which whenever* $x \sqsubseteq_A y$ *there is a unique* z*, viz..* $x \cap y$*, for which* $x \supseteq^- z \subseteq^+ y$*.*

Proof. (i) Assume $x \subseteq^+ w \supseteq^- y$ in $\mathcal{C}(A)$. Clearly $x \supseteq x \cap y$. Suppose $a \in x$ and $pol_A(a) = +$. Then $a \in w$, and because only −ve events are lost from w in $w \supseteq^- y$ we obtain $a \in y$, so $a \in x \cap y$. It follows that $x \supseteq^- x \cap y$, as required. Similarly, $x \cap y \subseteq^+ y$. (ii) Directly from (i). (iii) Clearly $\sqsubseteq$ is reflexive. Supposing $x \sqsubseteq y$, *i.e.* $x \supseteq^- x \cap y \subseteq^+ y$ in $\mathcal{C}(A)$ we see that the +ve events of x are included in y, and the −ve events of y are included in x. Hence if $x \sqsubseteq y$ and $y \sqsubseteq x$ in $\mathcal{C}(A)$ then x and y have the same +ve and −ve events and so are equal. Transitivity follows by (ii). Unique-factorization follows from the fact that when $x \supseteq^- z \subseteq^+ y$ necessarily $z = x \cap y$, as is easy to show. □

7.2 Strategies in Games as Presheaves

Let A be an event structure with polarity. We shall show how strategies in A correspond to cerain fibrations, so presheaves, over the order $(\mathcal{C}(A), \sqsubseteq_A)$. We concentrate on discrete fibrations over partial orders.

Definition 2. A *discrete fibration* over a partial order $(Y, \sqsubseteq_Y)$ is a partial order $(X, \sqsubseteq_X)$ and an order-preserving function $f : X \to Y$ such that

$$\forall x \in X, y' \in Y.\ y' \sqsubseteq_Y f(x) \implies \exists! x' \sqsubseteq_X x.\ f(x') = y' .$$

Via the Scott order we can recast strategies $\sigma : S \to A$ as those discrete fibrations $F : (\mathcal{C}(S), \sqsubseteq_S) \to (\mathcal{C}(A), \sqsubseteq_A)$ which preserve $\varnothing$, $\supseteq^-$ and $\subseteq^+$ in the sense that $F(\varnothing) = \varnothing$ while $x \supseteq^- y$ implies $F(x) \supseteq^- F(y)$, and $x \subseteq^+ y$ implies $F(x) \subseteq^+ F(y)$, for $x, y \in \mathcal{C}(S)$:

Theorem 1. *(i) Let $\sigma : S \to A$ be a strategy in game A. The map σ“ taking a finite configuration $x \in \mathcal{C}(S)$ to $\sigma x \in \mathcal{C}(A)$ is a discrete fibration from $(\mathcal{C}(S), \sqsubseteq_S)$ to $(\mathcal{C}(A), \sqsubseteq_A)$ which preserves $\varnothing$, $\supseteq^-$ and $\subseteq^+$.*
(ii) Suppose $F : (\mathcal{C}(S), \sqsubseteq_S) \to (\mathcal{C}(A), \sqsubseteq_A)$ is a discrete fibration which preserves $\varnothing$, $\supseteq^-$ and $\subseteq^+$. There is a unique strategy $\sigma : S \to A$ such that $F = \sigma$“.

Proof. (i) That σ“ forms a discrete fibration is a direct corollary of Lemma 1. As a map of event structures with polarity, σ“ automatically preserves $\varnothing$, $\supseteq^-$ and $\subseteq^+$. (ii) Assume F is a discrete fibration preserving $\varnothing$, $\supseteq^-$ and $\subseteq^+$. First observe a consequence, that if $x \subseteq^+ x'$ in $\mathcal{C}(S)$ and $F(x) \subseteq^+ y'' \subseteq F(x')$ in $\mathcal{C}(A)$, then there is a unique $x'' \in \mathcal{C}(S)$ such that $x \subseteq^+ x'' \subseteq x'$ and $F(x'') = y''$. (An analogous observation holds with + replaced by −.) Suppose now $x \overset{+}{\longrightarrow\!\!\!\!\subset} x'$ in $\mathcal{C}(S)$—where we write $x \overset{+}{\longrightarrow\!\!\!\!\subset} x'$ to abbreviate $x \overset{s}{\longrightarrow\!\!\!\!\subset} x'$ for some +ve $s \in S$. As F preserves $\subseteq^+$, $F(x) \subseteq^+ F(x')$. The observation implies $F(x) \overset{+}{\longrightarrow\!\!\!\!\subset} F(x')$ in $\mathcal{C}(A)$. Similarly, $x \overset{-}{\longrightarrow\!\!\!\!\subset} x'$ implies $F(x) \overset{-}{\longrightarrow\!\!\!\!\subset} F(x')$.

Define the relation $\approx$ between prime intervals $[x, x']$, where $x \longrightarrow\!\!\!\!\subset x'$, as the least equivalence relation such that $[x, x'] \approx [y, y']$ if $x \longrightarrow\!\!\!\!\subset y$ and $x' \longrightarrow\!\!\!\!\subset y'$ with $y \neq x'$. For configurations of an event structure, $[x, x'] \approx [y, y']$ iff $x \overset{e}{\longrightarrow\!\!\!\!\subset} x'$ and $y \overset{e}{\longrightarrow\!\!\!\!\subset} y'$ for some common event e. As F preserves coverings it preserves $\approx$. Consequently we obtain a well-defined function $\sigma : S \to A$ by taking s to a if an instance $x \overset{s}{\longrightarrow\!\!\!\!\subset} x'$ is sent to $F(x) \overset{a}{\longrightarrow\!\!\!\!\subset} F(x')$. Clearly σ preserves polarities.

By induction on the length of covering chains $\varnothing \overset{s_1}{\longrightarrow\!\!\!\!\subset} x_1 \overset{s_2}{\longrightarrow\!\!\!\!\subset} \cdots \overset{s_n}{\longrightarrow\!\!\!\!\subset} x_n = x$ and the fact that F preserves $\varnothing$ and coverings, $\varnothing = F(\varnothing) \overset{\sigma(s_1)}{\longrightarrow\!\!\!\!\subset} F(x_1) \overset{\sigma(s_2)}{\longrightarrow\!\!\!\!\subset} \cdots \overset{\sigma(s_n)}{\longrightarrow\!\!\!\!\subset} F(x_n) = F(x)$ with $\sigma x = F(x) \in \mathcal{C}(A)$. Moreover we cannot have $\sigma(s_i) = \sigma(s_j)$ for distinct i, j without contradicting F preserving coverings. This establishes $\sigma : S \to A$ as a total map of event structures with polarity. The assumed properties of F directly ensure that σ satisfies the two conditions of Lemma 1 required of strategy. □

As discrete fibrations correspond to presheaves, Theorem 1 entails that strategies $\sigma : S \to A$ correspond to (certain) presheaves over $(\mathcal{C}(A), \sqsubseteq_A)$—the presheaf for σ is a functor $(\mathcal{C}(A), \sqsubseteq_A)^{\mathrm{op}} \to \mathbf{Set}$ sending y to the fibre $\{x \in \mathcal{C}(S) \mid \sigma x = y\}$.

7.3 Strategies between Games as Profunctors

A strategy $\sigma : A \nrightarrow B$ determines a discrete fibration σ" over $(\mathcal{C}(A^\perp \| B), \sqsubseteq_{A^\perp \| B})$. But

$$(\mathcal{C}(A^\perp \| B), \sqsubseteq_{A^\perp \| B}) \cong (\mathcal{C}(A^\perp), \sqsubseteq_{A^\perp}) \times (\mathcal{C}(B), \sqsubseteq_B) \quad (1)$$
$$\cong (\mathcal{C}(A), \sqsubseteq_A)^{\mathrm{op}} \times (\mathcal{C}(B), \sqsubseteq_B). \quad (2)$$

The first step (1) relies on the correspondence between a configuration of $A^\perp \| B$ and a pair, with left component a configuration of $A^\perp$ and right component a configuration of B. In the last step (2) we are using the correspondence between configurations of $A^\perp$ and A induced by the correspondence $a \leftrightarrow \overline{a}$ between their events: a configuration x of $A^\perp$ corresponds to a configuration $\overline{x} =_{\mathrm{def}} \{\overline{a} \mid a \in x\}$ of A. Because $A^\perp$ reverses the roles of $+$ and $-$ in A, the order $x \sqsubseteq_{A^\perp} y$, *i.e.* $x \supseteq^- x \cap y \subseteq^+ y$ in $\mathcal{C}(A^\perp)$, corresponds to the order $\overline{y} \sqsubseteq_A \overline{x}$, *i.e.* $\overline{y} \supseteq^- \overline{x} \cap \overline{y} \subseteq^+ \overline{x}$ in $\mathcal{C}(A)$, so $\overline{x} \sqsubseteq_A^{\mathrm{op}} \overline{y}$.

It follows that a strategy $\sigma : S \to A^\perp \| B$ determines a discrete fibration

$$\sigma\text{"} : (\mathcal{C}(S), \sqsubseteq_S) \to (\mathcal{C}(A), \sqsubseteq_A)^{\mathrm{op}} \times (\mathcal{C}(B), \sqsubseteq_B)$$

where $\sigma\text{"}(x) = (\overline{\sigma_1 x}, \sigma_2 x)$, for $x \in \mathcal{C}(S)$. One way to define a *profunctor* from $(\mathcal{C}(A), \sqsubseteq_A)$ to $(\mathcal{C}(B), \sqsubseteq_B)$ is as a discrete fibration over $(\mathcal{C}(A), \sqsubseteq_A)^{\mathrm{op}} \times (\mathcal{C}(B), \sqsubseteq_B)$. Hence the strategy σ determines a profunctor[2] $\sigma\text{"} : (\mathcal{C}(A), \sqsubseteq_A) \nrightarrow (\mathcal{C}(B), \sqsubseteq_B)$.

8 A Lax Functor from Strategies to Profunctors

We now study how the operation from strategies σ to profunctors σ" preserves identities and composition.

8.1 Identity

The operation $(-)$" preserves identities:

Lemma 2. *Let A be an event structure with polarity. For $x \in \mathcal{C}(A^\perp \| A)$,*

$$x \in \mathcal{C}(\mathbb{C}_A) \quad \textit{iff} \quad x_2 \sqsubseteq_A \overline{x}_1,$$

where $x_1 \in \mathcal{C}(A^\perp)$ and $x_2 \in \mathcal{C}(A)$ are the projections of x to its components.

Proof. From Proposition 4, we deduce: $x \in \mathcal{C}(CC_A)$ iff (i) $\overline{x}_1^+ \supseteq x_2^+$ and (ii) $\overline{x}_1^- \subseteq x_2^-$, where $z^+ = \{a \in z \mid pol_A(a) = +\}$ and $z^- = \{a \in z \mid pol_A(a) = -\}$ for $z \in \mathcal{C}(A)$. It remains to argue that (i) and (ii) iff $x_2 \supseteq^- \overline{x}_1 \cap x_2 \subseteq^+ \overline{x}_1$. □

Corollary 1. *Let A be an event structure with polarity. The profunctor γ_A" of the copy-cat strategy γ_A is an identity profunctor on $(\mathcal{C}(A), \sqsubseteq_A)$.*

Proof. The profunctor $\gamma_A\text{"} : (\mathcal{C}(A), \sqsubseteq_A) \nrightarrow (\mathcal{C}(A), \sqsubseteq_A)$ sends $x \in \mathcal{C}(\mathbb{C}_A)$ to $(\overline{x}_1, x_2) \in (\mathcal{C}(A), \sqsubseteq_A)^{\mathrm{op}} \times (\mathcal{C}(A), \sqsubseteq_A)$ precisely when $x_2 \sqsubseteq_A \overline{x}_1$. It is thus an identity on $(\mathcal{C}(A), \sqsubseteq_A)$. □

[2] Most often a profunctor from $(\mathcal{C}(A), \sqsubseteq_A)$ to $(\mathcal{C}(B), \sqsubseteq_B)$ is defined as a functor $(\mathcal{C}(A), \sqsubseteq_A) \times (\mathcal{C}(B), \sqsubseteq_B)^{\mathrm{op}} \to \mathbf{Set}$, *i.e.*, as a presheaf over $(\mathcal{C}(A), \sqsubseteq_A)^{\mathrm{op}} \times (\mathcal{C}(B), \sqsubseteq_B)$, and as such corresponds to a discrete fibration.

8.2 Composition

We need to relate the composition of strategies to the standard composition of profunctors. Let $\sigma : S \to A^{\perp} \| B$ and $\tau : T \to B^{\perp} \| C$ be strategies, so $\sigma : A \mathrel{\rlap{\hskip.5em+}{\longrightarrow}} B$ and $\tau : B \mathrel{\rlap{\hskip.5em+}{\longrightarrow}} C$. Abbreviating, for instance, $(\mathcal{C}(A), \sqsubseteq_A)$ to $\mathcal{C}(A)$, strategies σ and τ give rise to profunctors $\sigma\text{``} : \mathcal{C}(A) \mathrel{\rlap{\hskip.5em+}{\longrightarrow}} \mathcal{C}(B)$ and $\tau\text{``} : \mathcal{C}(B) \mathrel{\rlap{\hskip.5em+}{\longrightarrow}} \mathcal{C}(C)$. Their composition is the profunctor $\tau\text{``} \circ \sigma\text{``} : \mathcal{C}(A) \mathrel{\rlap{\hskip.5em+}{\longrightarrow}} \mathcal{C}(C)$ built, as now described, as a discrete fibration from the discrete fibrations $\sigma\text{``} : \mathcal{C}(S) \to \mathcal{C}(A)^{\mathrm{op}} \times \mathcal{C}(B)$ and $\tau\text{``} : \mathcal{C}(T) \to \mathcal{C}(B)^{\mathrm{op}} \times \mathcal{C}(C)$.

First, we define the set of *matching pairs*,

$$M =_{\mathrm{def}} \{(x, y) \in \mathcal{C}(S) \times \mathcal{C}(T) \mid \sigma_2 x = \overline{\tau_1 y}\},$$

on which we define $\sim$ as the least equivalence relation for which

$$(x, y) \sim (x', y') \quad \text{if } x \sqsubseteq_S x' \ \&\ y' \sqsubseteq_T y \ \&\ \sigma_1 x = \sigma_1 x' \ \&\ \tau_2 y' = \tau_2 y\,.$$

Define an order on equivalence classes $M/\sim$ by:

$$\begin{aligned} m \sqsubseteq m' \text{ iff } & m = \{(x, y)\}_{\sim} \ \&\ m' = \{(x', y')\}_{\sim} \ \&\ x \sqsubseteq_S x' \ \&\ y \sqsubseteq_T y' \ \& \\ & \sigma_2 x = \sigma_2 x' \ \&\ \tau_1 y = \tau_1 y', \end{aligned}$$

for some matching pairs $(x, y), (x', y')$—so then $\sigma_2 x = \sigma_2 x' = \overline{\tau_1 y} = \overline{\tau_1 y'}$. The relation $\sqsubseteq$ above is easily seen to be a partial order on $M/\sim$. The profunctor composition $\tau\text{``} \circ \sigma\text{``}$ is given as

$$\tau\text{``} \circ \sigma\text{``} : M/\sim \ \to \ \mathcal{C}(A)^{\mathrm{op}} \times \mathcal{C}(C)\,, \text{ acting so } \{(x, y)\}_{\sim} \mapsto (\overline{\sigma_1 x}, \tau_2 y)$$

—it inherits from $\sigma\text{``}$ and $\tau\text{``}$ the property of being a discrete fibration.

It is *not* the case that $(\tau \odot \sigma)\text{``}$ and $\tau\text{``} \circ \sigma\text{``}$ coincide up to isomorphism. The profunctor composition $\tau\text{``} \circ \sigma\text{``}$ will generally contain extra equivalence classes $\{(x, y)\}_{\sim}$ for matching pairs (x, y) which are "unreachable." Although $\sigma_2 x = \overline{\tau_1 y}$, equals z say, automatically for a matching pair (x, y), the configurations x and y may impose incompatible causal dependencies on their 'interface' z so never be realized as a configuration in the synchronized composition $\mathcal{C}(T) \odot \mathcal{C}(S)$ used in building the composition of strategies $\tau \odot \sigma$.

Example 1. Let A and C both be the empty event structure $\varnothing$. Let B be the event structure consisting of the two concurrent events b_1, assumed $-$ve, and b_2, assumed $+$ve in B. Let the strategy $\sigma : \varnothing \mathrel{\rlap{\hskip.5em+}{\longrightarrow}} B$ comprise the event structure $s_1 \rightarrowtriangle s_2$ with s_1 $-$ve and s_2 $+$ve, $\sigma(s_1) = b_1$ and $\sigma(s_2) = b_2$. In $B^{\perp}$ the polarities are reversed so there is a strategy $\tau : B \mathrel{\rlap{\hskip.5em+}{\longrightarrow}} \varnothing$ comprising the event structure $t_2 \rightarrowtriangle t_1$ with t_2 $-$ve and t_1 $+$ve yet with $\tau(t_1) = \overline{b}_1$ and $\tau(t_2) = \overline{b}_2$. The equivalence class $\{(x, y)\}_{\sim}$, where $x = \{s_1, s_2\}$ and $y = \{t_1, t_2\}$, would be present in the profunctor composition $\tau\text{``} \circ \sigma\text{``}$, in addition to $\{(\varnothing, \varnothing)\}_{\sim}$, whereas $\tau \odot \sigma$ would be the empty strategy and accordingly the profunctor $(\tau \odot \sigma)\text{``}$ only has a single element, $\varnothing$. □

8.3 Laxness

This section establishes the exact relation between the two compositions $(\tau \odot \sigma)``$ and $\tau`` \circ \sigma``$. The proofs use that the equivalence relation $\sim$ between matching pairs is generated by a single-step relation:

Lemma 3. *On matching pairs, define*

$$(x,y) \rightsquigarrow_1 (x',y') \;\; \text{iff} \;\; \exists s \in S, t \in T.\; x \overset{s}{\longrightarrow\!\!\subset} x' \;\&\; y \overset{t}{\longrightarrow\!\!\subset} y' \;\&\; \sigma_2(s) = \overline{\tau_1(t)}\,.$$

The smallest equivalence relation including $\rightsquigarrow_1$ coincides with the relation $\sim$.

Now we make precise what it means for a matching pair to be reachable.

Definition 3. For (x,y) a matching pair, define

$$x \cdot y =_{\mathrm{def}} \{(s,*) \mid s \in x \;\&\; \sigma_1(s) \text{ is defined}\} \cup \{(*,t) \mid t \in y \;\&\; \tau_2(t) \text{ is defined}\} \cup$$
$$\{(s,t) \mid s \in x \;\&\; t \in y \;\&\; \sigma_2(s) = \overline{\tau_1(t)}\}\,.$$

Say (x,y) is *reachable* if $x \cdot y \in \mathcal{C}(T) \odot \mathcal{C}(S)$, and *unreachable* otherwise.

For $z \in \mathcal{C}(T) \odot \mathcal{C}(S)$ say a *visible prime* of z is a prime of the form $[(s,*)]_z$, for $(s,*) \in z$, or $[(*,t)]_z$, for $(*,t) \in z$.

We can specify when a matching pair is reachable without invoking the composition of strategies, important for the generalization in Section 9:

Proposition 6. *A matching pair (x,y) is reachable iff there is a sequcnce of matching pairs $(\emptyset,\emptyset) = (x_0,y_0), \cdots, (x_i,y_i), (x_{i+1},y_{i+1}), \cdots, (x_n,y_n) = (x,y)$ such that for all i, either $(x_i,y_i) \rightsquigarrow_1 (x_{i+1},y_{i+1})$*

or $\exists s \in S.\; x_i \overset{s}{\longrightarrow\!\!\subset} x_{i+1} \;\&\; y_i = y_{i+1} \;\&\; \sigma_1(s)$ is defined

or $\exists t \in T.\; y_i \overset{t}{\longrightarrow\!\!\subset} y_{i+1} \;\&\; x_i = x_{i+1} \;\&\; \tau_2(t)$ is defined.

(The relation $\rightsquigarrow_1$ is that introduced in Lemma 3.)

Theorem 2 below provides the precise relation between $(\tau \odot \sigma)``$ and $\tau`` \circ \sigma``$. Its proof requires that reachable matching pairs are $\sim$-equivalent iff they are associated with the same configuration in $T \odot S$, the import of (ii) in the next lemma.

Lemma 4. *(i) If (x,y) is a reachable matching pair and $(x,y) \sim (x',y')$, then (x',y') is a reachable matching pair. (ii) Whenever (x,y), (x',y') are reachable matching pairs, $(x,y) \sim (x',y')$ iff $x \cdot y$ and $x' \cdot y'$ have the same visible primes.*

Proof. We use Lemma 3 characterizing $\sim$ in terms of $\rightsquigarrow_1$.
(i) Suppose $(x,y) \rightsquigarrow_1 (x',y')$ or $(x',y') \rightsquigarrow_1 (x,y)$. By inspection of the construction of the product of stable families in Section 3, if $x \cdot y \in \mathcal{C}(T) \odot \mathcal{C}(S)$ then $x' \cdot y' \in \mathcal{C}(T) \odot \mathcal{C}(S)$.

(ii) *"If"*: Suppose $x \cdot y$ and $x' \cdot y'$ have the same visible primes, forming the set Q. Then $z =_{\rm def} \bigcup Q \in \mathcal{C}(T)\odot\mathcal{C}(S)$, being the union of a compatible set of configurations in $\mathcal{C}(T)\odot\mathcal{C}(S)$. Moreover, $z \subseteq x \cdot y, x' \cdot y'$. Take a covering chain

$$z \overset{e_1}{-\!\!\subset} \cdots z_i \overset{e_i}{-\!\!\subset} z_{i+1} \cdots \overset{e_n}{-\!\!\subset} x \cdot y$$

in $\mathcal{C}(T)\odot\mathcal{C}(S)$. Each $(\pi_1 z_i, \pi_2 z_i)$ is a matching pair. Necessarily, $e_i = (s_i, t_i)$ for some $s_i \in S$, $t_i \in T$, with $\sigma_2(s_i) = \overline{\tau_1(t_i)}$, again by the definition of $\mathcal{C}(T)\odot\mathcal{C}(S)$. Thus

$$(\pi_1 z_i, \pi_2 z_i) \rightsquigarrow_1 (\pi_1 z_{i+1}, \pi_2 z_{i+1}) .$$

Hence $(\pi_1 z, \pi_2 z) \sim (x, y)$, and similarly $(\pi_1 z, \pi_2 z) \sim (x', y')$, so $(x, y) \sim (x', y')$. *"Only if"*: It suffices to observe that if $(x, y) \rightsquigarrow_1 (x', y')$, then $x \cdot y$ and $x' \cdot y'$ have the same visible primes. But if $(x, y) \rightsquigarrow_1 (x', y')$ then $x \cdot y \overset{(s,t)}{-\!\!\subset} x' \cdot y'$, for some $s \in S, t \in T$, and no visible prime of $x' \cdot y'$ contains (s, t). □

Theorem 2. *Let $\sigma : A \twoheadrightarrow B$ and $\tau : B \twoheadrightarrow C$ be strategies. Defining*

$$\varphi_{\sigma,\tau} : \mathcal{C}(T\odot S) \to M/\sim \quad by \quad \varphi_{\sigma,\tau}(z) = \{(\Pi_1 z, \Pi_2 z)\}_\sim ,$$

where $\Pi_1 z = \pi_1 \bigcup z$ and $\Pi_2 z = \pi_2 \bigcup z$, yields an injective, order-preserving function from $(\mathcal{C}(T\odot S), \sqsubseteq_{T\odot S})$ to $(M/\sim, \sqsubseteq)$—its range is precisely the equivalence classes $\{(x, y)\}_\sim$ for reachable matching pairs (x, y). The diagram

$$\begin{array}{ccc} (\mathcal{C}(T\odot S), \sqsubseteq_{T\odot S}) & \xrightarrow{\varphi_{\sigma,\tau}} & (M/\sim, \sqsubseteq) \\ {\scriptstyle (\tau\odot\sigma)``}\big\downarrow & \swarrow{\scriptstyle \tau``\circ\sigma``} & \\ (\mathcal{C}(A), \sqsubseteq_A)^{\rm op} \times (\mathcal{C}(C), \sqsubseteq_C) & & \end{array}$$

commutes.

Proof. For $z \in \mathcal{C}(T\odot S)$, we obtain that $\varphi_{\sigma,\tau}(z) = (\Pi_1 z, \Pi_2 z) = (\pi_1 \bigcup z, \pi_2 \bigcup z)$ is a matching pair, from the definition of $\mathcal{C}(T)\odot\mathcal{C}(S)$; it is clearly reachable as $\pi_1 \bigcup z \cdot \pi_2 \bigcup z = \bigcup z \in \mathcal{C}(T)\odot\mathcal{C}(S)$. For any reachable matching pair (x, y) let z be the set of visible primes of $x \cdot y$. Then, $z \in \mathcal{C}(T\odot S)$ and, by Lemma 4(ii), $(\Pi_1 z, \Pi_2 z) \sim (x, y)$ so $\varphi_{\sigma,\tau}(z) = \{(x, y)\}_\sim$. Injectivity of $\varphi_{\sigma,\tau}$ follows directly from Lemma 4(ii).

To show that $\varphi_{\sigma,\tau}$ is order-preserving it suffices to show if $z -\!\!\sqsubset z'$ in $(\mathcal{C}(T\odot S), \sqsubseteq)$ then $\varphi_{\sigma,\tau}(z) \sqsubseteq \varphi_{\sigma,\tau}(z')$ in $(M/\sim, \sqsubseteq)$. (The covering relation $-\!\!\sqsubset$ is w.r.t. $\sqsubseteq$.) If $z -\!\!\sqsubset z'$ then either $z \overset{p}{-\!\!\subset} z'$, with p +ve, or $z' \overset{p}{-\!\!\subset} z$, with p −ve, for p a visible prime of $\mathcal{C}(T)\odot\mathcal{C}(S)$, *i.e.* with $max(p)$ of the form $(s, *)$ or $(*, t)$. We concentrate on the case where p is +ve (the proof when p is −ve is similar). In the case where p is +ve,

$$\Pi_1 z \cdot \Pi_2 z = \bigcup z \subseteq \bigcup z' = \Pi_1 z' \cdot \Pi_2 z'$$

in $\mathcal{C}(T)\odot\mathcal{C}(S)$ and there is a covering chain

$$\bigcup z = w_0 \overset{(s_1,t_1}{-\!\!\subset} w_1 \cdots \overset{(s_n,t_n)}{-\!\!\subset} w_n \overset{max(p)}{-\!\!\subset} \bigcup z'$$

in $\mathcal{C}(T)\odot\mathcal{C}(S)$. Each w_i, for $0 \le i \le m$, is associated with a reachable matching pair $(\pi_1 w_i, \pi_2 w_i)$ where $\pi_1 w_i \cdot \pi_2 w_i = w_i$. Also $(\pi_1 w_i, \pi_2 w_i) \leadsto_1 (\pi_1 w_{i+1}, \pi_2 w_{i+1})$, for $0 \le i < m$. Hence $(\Pi_1 z, \Pi_2 z) \sim (\pi_1 w_n, \pi_2 w_n)$, by Lemma 3. If $max(p) = (s, *)$ then $\pi_1 w_n \stackrel{s}{\longrightarrow\!\!\subset} \Pi_1 z'$, with s +ve, and $\pi_2 w_n = \Pi_2 z'$. If $max(p) = (*, t)$ then $\pi_1 w_n = \Pi_1 z'$ and $\pi_2 w_n \stackrel{t}{\longrightarrow\!\!\subset} \Pi_2 z'$, with t +ve. In either case $\pi_1 w_n \sqsubseteq_S \Pi_1 z'$ and $\pi_2 w_n \sqsubseteq_T \Pi_2 z'$ with $\sigma_2 \pi_1 w_n = \sigma_2 \Pi_1 z'$ and $\tau_1 \pi_2 w_n = \tau_1 \Pi_2 z'$. Hence, from the definition of $\sqsubseteq$ on $M/\sim$,

$$\varphi_{\sigma,\tau}(z) = \{(\Pi_1 z, \Pi_2 z)\}_\sim = \{(\pi_1 w_n, \pi_2 w_n)\}_\sim \sqsubseteq \{(\Pi_1 z', \Pi_2 z')\}_\sim = \varphi_{\sigma,\tau}(z')\,.$$

It remains to show commutativity of the diagram. Let $z \in \mathcal{C}(T\odot S)$. Then,

$$(\tau``\circ\sigma``)(\varphi_{\sigma,\tau}(z)) = (\tau``\circ\sigma``)(\{(\Pi_1 z, \Pi_2 z)\}_\sim) = (\overline{\sigma_1 \Pi_1 z},\ \tau_2 \Pi_2 z) = (\tau\odot\sigma)``(z)\,,$$

via the definition of $\tau\odot\sigma$—as required. □

Because $(-)``$ does not preserve composition up to isomorphism but only up to the transformation φ of Theorem 2:

Corollary 2. *The operation $(-)``$ forms a* lax *functor from Strat, the bicategory of strategies, to $\mathcal{P}rof$, that of profunctors; identities are preserved up to the isomorphism of Corollary 1 while composition is preserved up to φ of Theorem 2.*

Despite laxness, the relation between strategy composition and profunctor composition is surprisingly straightforward: *the composition of strategies, viewed as a profunctor, is given by restricting the composition of profunctors to reachable matching pairs.*

In special cases composition is preserved up to isomorphism because all the relevant matching pairs are reachable. Say a strategy σ is *rigid* when the components σ_1, σ_2 preserve causal dependency when defined. In fact, rigid strategies form a sub-bicategory of *Strat*. For composable rigid strategies σ and τ we do have $(\tau\odot\sigma)`` \cong \tau``\circ\sigma``$. Stable spans (including Berry's stable functions), those strategies between games where all moves are +ve [1], and simple games [8,9] lie within the bicategory of rigid strategies.

9 Games as Factorization Systems

The results of Section 7.1 show an event structure with polarity determines a factorization system [10]; the 'left' maps are given by $\supseteq^-$ and the 'right' maps by $\subseteq^+$. More specifically they form an instance of a *rooted* factorization system $(\mathbb{X}, \to_L, \to_R, 0)$ where maps $f : x \to_L x'$ are the 'left' maps and $g : x \to_R x'$ the 'right' maps of a factorization system on a small category $\mathbb{X}$, with distinguished object 0, such that any object x of $\mathbb{X}$ is reachable by a chain of maps:

$$0 \leftarrow_L \cdot \to_R \cdots \leftarrow_L \cdot \to_R x\,;$$

and two 'confluence' conditions hold:

$$x_1 \to_R x \;\&\; x_2 \to_R x \implies \exists x_0.\ x_0 \to_R x_1 \;\&\; x_0 \to_R x_2\,, \quad \text{and its dual}$$
$$x \to_L x_1 \;\&\; x \to_L x_2 \implies \exists x_0.\ x_1 \to_L x_0 \;\&\; x_2 \to_R x_0\,.$$

Think of objects of $\mathbb{X}$ as configurations, the R-maps as standing for (compound) Player moves and L-maps for the reverse, or undoing, of (compound) Opponent moves in a game.

The characterization of strategy, Lemma 1, exhibits a strategy as a discrete fibration w.r.t. $\sqsubseteq$ whose functor preserves $\varnothing$, $\supseteq^-$ and $\subseteq^+$. This generalizes. Define a strategy in a rooted factorization system to be a functor from another rooted factorization system preserving 0, L-maps, R-maps and forming a discrete fibration. To obtain strategies *between* rooted factorization systems we again follow the methodology of Joyal [7], and take a strategy from $\mathbb{X}$ to $\mathbb{Y}$ to be a strategy in the dual of $\mathbb{X}$ in parallel composition with $\mathbb{Y}$. Now the dual operation becomes the opposite construction on a factorization system, reversing the roles and directions of the 'left' and 'right' maps. The parallel composition of factorization systems is given by their product. Composition of strategies is given essentially as that of profunctors, but restricting to reachable elements—the definition of reachable element is a direct generalization of Proposition 6. The bicategory of concurrent strategies is equivalent to the sub-bicategory in which the objects and strategies are on rooted factorization systems of the form of $((\mathcal{C}(A), \sqsubseteq_A), \supseteq^-, \subseteq^+, \varnothing)$ for an event structure with polarity A.

One pay-off of the increased generality is that bistructures, a way to present Berry's bidomains as factorization systems [11], inherit a reading as games. The new view also allows us to formalize strategies in some games based on moves as vectors, as in some games of chase, in which moves of Player (as hunter) and Opponent (as prey) may be translations in space or changes in velocity. Details will appear elsewhere.

Acknowledgments. Thanks to Pierre Clairambault, Marcelo Fiore, Julian Gutierrez, Thomas Hildebrandt, Martin Hyland, Alex Katovsky, Samuel Mimram, Gordon Plotkin, Silvain Rideau and Sam Staton for helpful remarks. The support of Advanced Grant ECSYM of the ERC is acknowledged with gratitude.

References

1. Rideau, S., Winskel, G.: Concurrent strategies. In: LICS 2011. IEEE Computer Society (2011)
2. Hyland, M.: Some reasons for generalising domain theory. Mathematical Structures in Computer Science 20(2), 239–265 (2010)
3. Winskel, G.: Events, causality and symmetry. Comput. J. 54(1), 42–57 (2011)
4. Abramsky, S., Jagadeesan, R., Malacaria, P.: Full abstraction for PCF. Inf. Comput. 163(2), 409–470 (2000)
5. Hyland, J.M.E., Ong, C.H.L.: On full abstraction for PCF: I, II, and III. Inf. Comput. 163(2), 285–408 (2000)
6. Cattani, G.L., Winskel, G.: Profunctors, open maps and bisimulation. Mathematical Structures in Computer Science 15(3), 553–614 (2005)
7. Joyal, A.: Remarques sur la théorie des jeux à deux personnes. Gazette des sciences mathématiques du Québec 1(4) (1997)

8. Hyland, M.: Game semantics. In: Pitts, A., Dybjer, P. (eds.) Semantics and Logics of Computation. Publications of the Newton Institute (1997)
9. Harmer, R., Hyland, M., Melliès, P.A.: Categorical combinatorics for innocent strategies. In: LICS 2007. IEEE Computer Society (2007)
10. Joyal, A.: Factorization systems. Joyal's CatLab (2012), `http://ncatlab.org/joyalscatlab/`
11. Curien, P.L., Plotkin, G.D., Winskel, G.: Bistructures, bidomains, and linear logic. In: Proof, Language, and Interaction, Essays in Honour of Robin Milner, pp. 21–54. MIT Press (2000)

Generalised Name Abstraction for Nominal Sets

Ranald Clouston*

Logic and Computation Group, Research School of Computer Science,
The Australian National University, Canberra, ACT 0200, Australia
ranald.clouston@anu.edu.au

Abstract. The Gabbay-Pitts nominal sets model provides a framework for reasoning with names in abstract syntax. It has appealing semantics for name binding, via a functor mapping each nominal set to the 'atom-abstractions' of its elements. We wish to generalise this construction for applications where sets, lists, or other patterns of names are bound simultaneously. The atom-abstraction functor has left and right adjoint functors that can themselves be generalised, and their generalisations remain adjoints, but the atom-abstraction functor in the middle comes apart to leave us with two notions of generalised abstraction for nominal sets. We give new descriptions of both notions of abstraction that are simpler than those previously published. We discuss applications of the two notions, and give conditions for when they coincide.

Keywords: name binding, nominal sets, adjoint functors.

1 Introduction

Programming languages and formal calculi frequently feature syntactic constructs called *names*, or *object-level variables*. When combined with *binding* constructs, rendering names anonymous within their scope, technical and conceptual issues arise that have attracted much research in recent years. Some such constructs bind not just a single name at a time, but a set, list, or *pattern* of names; as yet there is no convincing general theory for such binders [1].

One of the most prominent approaches to names and binding is the *nominal sets* model [10,16]. Nominal sets provide a mathematical model in which names exist as first-class citizens, manipulated by permutations. This model supports reasoning techniques that closely match pen-and-paper practice, in which we often explicitly manipulate bound names. Nominal sets have inspired a literature too extensive to summarise here, with nominal variants offered of everything from equational logic [4], to interactive theorem proving [24], to game semantics [23].

Binding in nominal sets is elegantly captured by a construction called *atom-abstraction*; for each nominal set X we define a new nominal set, written $[\mathbb{A}]X$, which can be thought of in two ways: as a set of pairs of names and X-elements

* The author gratefully acknowledges his discussions with Andrew Pitts, Michael Norrish, and Barry Jay, and the comments of the anonymous reviewers.

F. Pfenning (Ed.): FOSSACS 2013, LNCS 7794, pp. 434–449, 2013.

quotiented by legal renamings, and as a function mapping free names to X-elements. For example, $[\mathbb{A}]\mathbb{A}$ (names abstracted over names) contains pairs such as (a, a), equivalent to (b, b) but not equivalent to (b, a), or alternatively contains functions such as $\lambda x.x$ and $\lambda x.a$.

It is worth dwelling on this dual description of $[\mathbb{A}]X$ as (quotiented) product and (partial) function space. We can use the language of category theory to capture this intuition more firmly. The map $X \mapsto [\mathbb{A}]X$ extends to an endofunctor $[\mathbb{A}]$- on the category of nominal sets with both left and right adjoints:

$$\text{-} \otimes \mathbb{A} \dashv [\mathbb{A}]\text{-} \dashv \text{-}_{[\mathbb{A}]} \,. \tag{1}$$

Attention was first drawn to this chain of adjunctions by [15]. It immediately tells us that atom-abstraction has nice properties, preserving all limits and colimits. More intriguingly, the leftmost functor comes via a sort of product, and the rightmost functor via a sort of function space. This can be compared to the cartesian closure adjunction $\text{-} \times X \dashv X \to \text{-}$, which gives the standard relationship between products and function spaces. Hence $[\mathbb{A}]X$ is somehow amphibious, both a product and a function space. As [6] puts it, atom-abstractions are "constructed like a pair... but destructed like a partial function".

Atom-abstraction nominal sets as a semantics for binding one name at a time are now well established; what of applications involving more complex binding? Such applications, such as the functional programming operator *letrec* that binds lists of names simultaneously, are known to be imperfectly modelled by one-at-a-time binding - see [1,21,24] for varied discussions on this point. This paper explores and critiques the fundamental mathematics that could provide a nominal sets semantics for such general binding. Concretely, we will try to generalise the atoms-abstraction construction $[\mathbb{A}]Y$ by replacing the set of names $\mathbb{A}$ with other nominal sets X, such as the nominal set of lists of names.

The adjunctions of (1) offer one approach to this generalisation, as the functors $\text{-}\otimes\mathbb{A}$ and $\text{-}_{[\mathbb{A}]}$ easily generalise to functors $\text{-}\otimes X$ and $\text{-}_{[X]}$. These generalisations remain left and right adjoints respectively, but an interesting thing happens: the right adjoint of $\text{-} \otimes X$ and left adjoint of $\text{-}_{[X]}$ do not in general coincide. Thus we have two notions of generalised abstraction, one of which is constructed like a pair, and the other of which is destructed like a partial function. Summarising our situation, we have

$$\text{-} \otimes X \dashv X \twoheadrightarrow\!\!\ast\, \text{-} \quad \text{and} \quad [X]\text{-} \dashv \text{-}_{[X]} \tag{2}$$

where $X \twoheadrightarrow\!\!\ast\, \text{-}$ and $[X]$- coincide in the case that $X = \mathbb{A}$, but not in general.

The left hand adjunction of (2) is known to exist for general category theoretic reasons, because $\otimes$, called here the *separated product*, is related to the 'Day convolutions' of [5]. We call its closure the *separating function space*, and use the 'magic wand' notation from the logic of Bunched Implications (BI) [18]. However this category theoretic view does not yield an accessible nominal set theoretic construction, and such concrete constructions have been key to the succesful applications of nominal techniques. The closest we have to such a construction is [19, Sec. 3.3], which involves a tricky quotient on partial functions and selection of canonical members from each equivalence class. In Sec. 3 we will give a considerably simpler and more intuitively appealing construction.

Conversely, the construction $[X]$- was established in [9, Sec. 6] and implemented in the language FreshML [22]. It has been given the name *generalised abstraction* and notation matching that name; we will harmonise with the literature on this, at the risk of implying it is the only generalisation of atom-abstraction worth considering. To our knowledge, the right hand adjunction of (2) was first explicitly observed in the unpublished [17], although it is a corollary of the earlier [19, Prop. 3.3.32]; in Sec. 4 we will sketch this adjunction, along with a novel treatment of generalised abstractions as equalisers.

Sec. 5 will give necessary and sufficient conditions under which these notions of abstraction coincide, as they do for $\mathbb{A}$. The 'sufficient' direction is due to [19, Sec. 10.3]; here we restate that proof in terms of our simple notion of separating functions, and give the 'necessary' direction also. Finally, Sec. 6 will look at applications and limitations of these mathematical developments.

2 Nominal Sets

This section gives us a brief overview of nominal sets; [16] or [3, Cha. 2] provide more leisurely introductions to the area.

Definition 2.1. *Fix a countably infinite set* $\mathbb{A}$ *of* atoms*. The set* Perm *of (finite)* permutations *consists of all bijections* $\pi : \mathbb{A} \to \mathbb{A}$ *whose non-trivial domain*

$$supp(\pi) \triangleq \{a \mid \pi(a) \neq a\} \tag{3}$$

is finite.

Perm *is a group, with multiplication as permutation composition,* $\pi'\pi(a) = \pi'(\pi(a))$*, and identity as the permutation* ι *leaving all atoms unchanged.*

Example 2.2. The *transpositions* $(a\ b)$ map $a \mapsto b$, $b \mapsto a$ and leave all other atoms unchanged. Now let

$$\mathbb{A}^{(n)} \triangleq \{(a_1, \ldots, a_n) \in \mathbb{A}^n \mid a_i \neq a_j \text{ for } 1 \leq i < j \leq n\} \ . \tag{4}$$

All the tuples of atoms we use in this paper will be so disjoint. Take $\vec{a} = (a_1, \ldots, a_n), \vec{a}' = (a'_1, \ldots, a'_n) \in \mathbb{A}^{(n)}$ with mutually disjoint underlying sets. Then their *generalised transposition* is

$$(\vec{a}\,\vec{a}') \triangleq (a_1\,a'_1) \cdots (a_n\,a'_n) \ .$$

Definition 2.3. *A* Perm-set *is a set* X *equipped with a function, or* Perm-action*,* $(\pi, x) \mapsto \pi \cdot x$ *from* Perm $\times X$ *to* X *such that* $\iota \cdot x = x$ *and* $\pi \cdot (\pi' \cdot x) = \pi\pi' \cdot x$*.*

Given such a Perm*-set* X *we say that a set of atoms* $\overline{a} \subseteq \mathbb{A}$ supports $x \in X$ *if for all* $\pi \in$ Perm*,* $supp(\pi) \cap \overline{a} = \emptyset$ *implies that* $\pi \cdot x = x$*.*

Definition 2.4. *A* nominal set *is a* Perm*-set* X *with the* finite support property*: for each* $x \in X$ *there exists some finite* $\overline{a} \subseteq \mathbb{A}$ *supporting* x*.*

If an element x *is finitely supported then there is a unique least such support set [10, Prop. 3.4], which we write* $supp(x)$ *and call* the support of x*.*

Given nominal set elements $x \in X, y \in Y$*, if* $supp(x) \cap supp(y) = \emptyset$ *then we write* $x \# y$*, and say that* x *is* fresh *for* y*, or equivalently that* y *is fresh for* x*.*

Remark 2.5. The atoms $\mathbb{A}$ can be seen as a set of *names*, and the support of a nominal set element as its set of *free names*. The finite support condition reflects that for most notions of syntax, terms may have only finitely many free names. It is a useful condition because it allows us to uniquely define *the* support of an element, and hence always find names that are fresh for, or *not free in*, that element.

Example 2.6. (i) Any set is a nominal set under the trivial Perm-action $\pi \cdot x = x$; then $supp(x) = \emptyset$.

(ii) $\mathbb{A}$ is a nominal set given $\pi \cdot a = \pi(a)$; $supp(a) = \{a\}$.

(iii) Perm is a nominal set given the conjugation action $\pi \cdot \pi' = \pi\pi'\pi^{-1}$; support is as (3).

(iv) $\mathcal{P}_{fin}(\mathbb{A})$, the set of finite sets of atoms, is nominal given the element-wise Perm-action; $supp(\overline{a}) = \overline{a}$.

(v) Given nominal sets X, Y, the usual product $X \times Y$ is nominal given the element-wise Perm-action; $supp((x, y)) = supp(x) \cup supp(y)$. We write the n-fold product of X as X^n.

(vi) Define a subset of $X \times Y$ by

$$X \otimes Y \triangleq \{(x, y) \in X \times Y \mid x \# y\} .$$

This is nominal with the same element-wise action and supports as $X \times Y$, and is called the *separated product* of X and Y. The n-fold separated product of X is written $X^{(n)}$, as with (4).

(vii) The usual disjoint union $X + Y$, with typical members $(x, 1)$ or $(y, 2)$, is nominal given the Perm-actions and supports inherited from X, Y. Where there is no confusion we will omit the indices.

Definition 2.7. *We can define a* Perm*-action on the functions* $f : X \to Y$ *by applying the evident action to their graphs, i.e.*

$$(\pi \cdot f)(x) \triangleq \pi \cdot (f(\pi^{-1} \ x)) \tag{5}$$

Functions between nominal sets are not necessarily finitely supported under this action; we call the functions that are so the finitely supported functions.

A function that has empty support under this action is called an equivariant function*; this property has the equivalent formulation*

$$\pi \cdot (f(x)) = f(\pi \cdot x) .$$

The category of nominal sets*, written* $\mathcal{N}om$*, has as objects, nominal sets, and as arrows, equivariant functions between them.*

Example 2.8. (i) The permutations of Perm are all finitely supported functions $\mathbb{A} \to \mathbb{A}$: compare Ex. 2.6(iii) and (5).

(ii) For each nominal set X, the map $x \mapsto supp(x)$ is an equivariant function $X \to \mathcal{P}_{fin}(\mathbb{A})$.

(iii) If X is a subset of the nominal set Y, and X is closed under Y's Perm-action, then we call X a *nominal subset* of Y; the inclusion function is obviously equivariant. For example, $X \otimes Y$ is a nominal subset of $X \times Y$.

Remark 2.9. $\mathcal{N}om$ has much categorial structure; it is a Grothendiek topos. We in particular note, without proof, that it has initial object $\emptyset$ and terminal object $1 = \{\bullet\}$ under the trivial Perm-actions (Ex. 2.6(i)), that Ex. 2.6(v) and (vii) define its binary product and coproducts, and that its exponential from X to Y is the nominal set of functions $X \to Y$ finitely supported under (5).

Lemma 2.10. *The following are useful basic facts about nominal sets. We will use them often in this paper, usually without specific reference:*

(i) Given an atom $a \in \mathbb{A}$, $a \# x$ if and only if for **some** *atom $a' \# (a, x)$, we have $(a\, a') \cdot x = x$, if and only if for* **any** *atom $a' \# (a, x)$, we have $(a\, a') \cdot x = x$*

(ii) If f is equivariant then $supp(f(x)) \subseteq supp(x)$.

(iii) If the permutations π, π' coincide on their restrictions to $supp(x)$, then $\pi \cdot x = \pi' \cdot x$.

We now present the standard atom-abstraction construction, which gives semantics for name binding operations such as λ-abstraction:

Definition 2.11. *Given a nominal set X we define a relation on $\mathbb{A} \times X$ by*

$$(a, x) \sim (a', x') \Leftrightarrow (a\, b) \cdot x = (a'\, b) \cdot x'$$

for some atom $b \# (a, a', x, x')$. This defines an equivalence relation; write the class containing (a, x) as $\langle a \rangle x$ and call such a class the atom-abstraction *of a on x. This relation is equivariant, i.e. $(a, x) \sim (a', x')$ implies $(\pi(a), \pi \cdot x) \sim (\pi(a'), \pi \cdot x')$.*

The set of atom-abstractions of a on x as a ranges over $\mathbb{A}$ and x over X hence forms a nominal set under the action $\pi \cdot \langle a \rangle x \triangleq \langle \pi(a) \rangle (\pi \cdot x)$. Write this nominal set $[\mathbb{A}]X$, and call its members the atom-abstractions on *X.*

This construction extends to a functor $[\mathbb{A}]\text{-} : \mathcal{N}om \to \mathcal{N}om$ by, given equivariant $f : X \to Y$ and atom-abstraction $\langle a \rangle x \in [\mathbb{A}]X$, the map $([\mathbb{A}]f)(\langle a \rangle x) = \langle a \rangle (f(x))$.

Definition 2.12. *The separated product construction $X \otimes Y$ of Ex. 2.6(vi) extends to a monoidal operation on $\mathcal{N}om$ under the evident action on equivariant functions. This gives rise to a functor $\text{-} \otimes X : \mathcal{N}om \to \mathcal{N}om$ for any X.*

Definition 2.13. *For any nominal sets X, Y define the nominal set of* freshening functions *from X to Y by*

$$Y_{[X]} \triangleq \{f : X \to Y \mid f \text{ is finitely supported, and } \forall x \in X.\, x \# f(x)\}\ .$$

Equivalently, $Y_{[X]}$ is the set of functions whose graphs draw elements from $X \otimes Y$.

This extends to a functor $\text{-}_{[X]} : \mathcal{N}om \to \mathcal{N}om$ in the evident manner: given equivariant $g : Y \to Z$ and freshening function $f \in Y_{[X]}$, we have $g_{[X]}(f) = g \circ f$.

Theorem 2.14. *We have the adjunctions*

$$\text{-} \otimes \mathbb{A} \dashv [\mathbb{A}]\text{-} \dashv \text{-}_{[\mathbb{A}]}$$

Proof. By Thms. 3.6, 4.5, and 5.5 below. We note here only that the co-unit of the left hand adjunction, with components $\varepsilon_X : [\mathbb{A}]X \otimes \mathbb{A} \to X$, is called *concretion*, and is defined by

$$\varepsilon_X(\langle a \rangle x, b) = (a\ b) \cdot x\ .$$

It is in this sense that atom-abstraction is destructed like a function space.

3 Separating Functions

In this section we will define the *closure* of the separated product $\otimes$: a binary connective $-\!\!*$ with the adjoint property $\text{-} \otimes X \dashv X -\!\!* \text{-}$ for any nominal set X. Such a closure condition can be compared to the standard cartesian closure relating products and function spaces. As $\otimes$ is a sort of product, we will not be surprised to find that $-\!\!*$ forms a sort of function space.

Definition 3.1. *Given sets X, Y, the* partial functions $f : X \rightharpoonup Y$ *are the functions $f : dom(f) \to Y$ for $dom(f) \subseteq X$. We say $f(x) \downarrow$, and that $f(x)$* converges, *if $x \in dom(f)$. We say $f(x) \uparrow$, and that $f(x)$* diverges, *if $x \in X - dom(f)$.*

Given nominal sets X, Y and a partial function $f : X \rightharpoonup Y$, we can define a Perm*-action on f by the* Perm*-action on its graph, i.e.*

$$(\pi \cdot f)(x) = \begin{cases} \pi \cdot f(\pi^{-1} \cdot x) & \text{if } f(\pi^{-1} \cdot x) \downarrow \\ \uparrow & \text{if } f(\pi^{-1} \cdot x) \uparrow \end{cases} \tag{6}$$

The nominal set $X \rightharpoonup_{fs} Y$ is the set of partial functions $X \rightharpoonup Y$ finitely supported under (6).

Definition 3.2. *Given nominal sets X, Y a* separating function *f from X to Y is a finitely supported partial function satisfying*

(i) $f(x) \downarrow$ if and only if $f \# x$;
(ii) $supp(f) = \bigcup_{x \in dom(f)} supp(f(x)) - supp(x)$;

This defines a nominal subset of $X \rightharpoonup_{fs} Y$, which we write $X -\!\!* Y$ and call the separating function space *from X to Y.*

Lemma 3.3. *For any finitely supported partial function $f : X \rightharpoonup Y$, Def. 3.2(ii) is equivalent to*

$$supp(f) \subseteq \bigcup_{x \in dom(f)} supp(f(x)) - supp(x) \ . \tag{7}$$

Proof. The converse holds for any finitely supported f: Say $a \in supp(f(x)) - supp(x)$ for some $x \in dom(f)$. Take $a' \# (a, x, f)$, so $(a\ a') \cdot f(x) \neq f(x)$. To see that $a \in supp(f)$ we will show that $f, (a\ a') \cdot f$ disagree on x.

$a, a' \# x$ implies that $(a\ a') \cdot x = x$, so $f(x) \downarrow$ implies that $((a\ a') \cdot f)(x)$ converges to $(a\ a') \cdot f(x)$ by (6). This is not equal to $f(x)$, as required.

Remark 3.4. How is Def. 3.2 motivated? The proposed adjunction requires that we evaluate $f(x)$ only in the case that $(f, x) \in (X \twoheadrightarrow Y) \otimes X$; that is, where $f \# x$. It therefore must be that f is partial and that its domain be entirely determined by its support; otherwise we would not have total evaluation, or would have non-identical functions that evaluate identically. This restriction (here, Def. 3.2(i)) is also found in [19, Sec. 3.3].

A counter-example helps to motivate Def. 3.2(ii). Following the 'sharing interpretation' of the logic BI [18, Cha. 9], if we interpret names as resources, and supports as the resources claimed by each element, then $\otimes$ is a 'non-sharing' product, and we would expect its closure to consist of functions that cannot access their arguments' supports. But consider the partial function $\uparrow_a : \mathbb{A} \rightharpoonup 1$ that diverges on a and converges elsewhere. This obeys Def. 3.2(i), as $supp(\uparrow_a) = \{a\}$, yet it needs to access its argument's support to determine divergence. However the right hand side of (7) is empty, so $\uparrow_a \notin \mathbb{A} \twoheadrightarrow 1$, so Def. 3.2(ii) fails.

Example 3.5. (i) If all elements of Y have empty support, then $X \twoheadrightarrow Y$ contains exactly the equivariant total functions $X \to Y$. In particular, $X \twoheadrightarrow 1 \cong 1$.

(ii) The separating functions $\mathbb{A} \twoheadrightarrow \mathbb{A}$ are the identity and, for each $a \in \mathbb{A}$, the partial functions f_a defined by $f_a(b) = a$ if $a \neq b$, and diverging on a.

(iii) The separating functions $(1 + \mathbb{A}) \twoheadrightarrow \mathbb{A}$ are defined, for each $a \in \mathbb{A}$, by mapping $\bullet \mapsto a$ and then (1) as the identity on $\mathbb{A}$ except diverging on a, (2) sending all atoms to a except diverging on a, or (3) for any $b \neq a$, sending all atoms to b except diverging on $\{a, b\}$.

Theorem 3.6. *The definition of separating function spaces extends to a functor $X \twoheadrightarrow \text{-} : \mathcal{N}om \to \mathcal{N}om$ for any nominal set X, with the adjoint property*

$$\text{-} \otimes X \dashv X \twoheadrightarrow \text{-}$$

Proof. The bijection $\mathcal{N}om(Z \otimes X, Y) \cong \mathcal{N}om(Z, X \twoheadrightarrow Y)$ is given via the co-unit, whose components $\varepsilon_Y : (X \twoheadrightarrow Y) \otimes X \to Y$ are the usual evaluation functions:

$$\varepsilon_Y(f, x) = f(x)$$

Each ε_Y is straightforwardly total and equivariant. We must show that for any $f : Z \otimes X \to Y$ there is a unique $\hat{f} : Z \to (X \twoheadrightarrow Y)$ such that

$$\begin{array}{ccccc} Z & & Z \otimes X & & \\ \downarrow{\scriptstyle \hat{f}} & & \downarrow{\scriptstyle \hat{f}\otimes X} & \searrow{\scriptstyle f} & \\ X \twoheadrightarrow Y & & (X \twoheadrightarrow Y) \otimes X & \xrightarrow[\varepsilon_Y]{} & Y \end{array} \tag{8}$$

commutes. For each $z \in Z$ let

$$\overline{a}_z \triangleq \bigcup_{(x \in X) \# z} supp(f(z,x)) - supp(x) \ . \tag{9}$$

The map $z \mapsto \overline{a}_z$ is straightforwardly equivariant, so $\overline{a}_z \subseteq supp(z)$ by Lem. 2.10(ii). Now let $\hat{f} : Z \to (X \twoheadrightarrow Y)$ be

$$\hat{f}(z)(x) \triangleq \begin{cases} f((\vec{a}\,\vec{a}') \cdot z, x) & \text{if } \overline{a}_z \# x \\ \uparrow & \text{otherwise.} \end{cases} \tag{10}$$

where $\vec{a}$ is an ordering of $supp(z) \cap supp(x)$ and $\vec{a}'$ is a tuple of the same size fresh for (z, x). Then where $\hat{f}(z)(x)\downarrow$ we may picture the elements' supports as

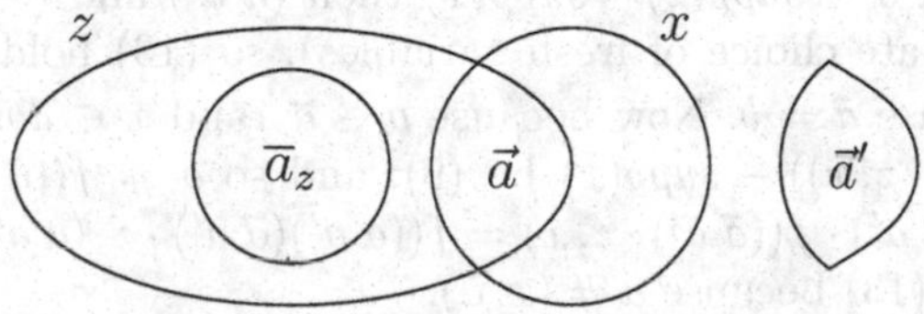

Evidently $(\vec{a}\,\vec{a}') \cdot z \# x$, so $f((\vec{a}\,\vec{a}') \cdot z, x)$ is well-defined. To see that $\hat{f}$ is **well-defined** we must check that it does not depend on our choice of fresh $\vec{a}'$. To do this we first prove

$$\vec{a} \# f(z, (\vec{a}\,\vec{a}') \cdot x) \ . \tag{11}$$

Suppose b is in the support of both sides of (11). Now $(b, z) \# (\vec{a}\,\vec{a}') \cdot x$, so $b \in \overline{a}_z$ by (9). But $\overline{a}_z \# x$ and $b \in supp(x)$ by definition, so by contradiction (11) holds. Because f is equivariant we can then apply $(\vec{a}\,\vec{a}')$ to both sides of (11) to get $\vec{a}' \# f((\vec{a}\,\vec{a}') \cdot z, x)$, which is sufficient to conclude that our choice of fresh $\vec{a}'$ is arbitrary. $\hat{f}$ is hence a function; equivariance is straightforward.

The diagram (8) **commutes** because if we start at $(z, x) \in Z \otimes X$ then $\overline{a}_z \# x$, and $\vec{a}$ in (10) is empty, so $\hat{f}(z)(x)$ converges to $f(z, x)$.

We next confirm that $\hat{f}(z)$ is a **separating function**. $\hat{f}(z)(x)\downarrow$ iff $\overline{a}_z \# x$ by (10), so Def. 3.2(i) holds if $supp(\hat{f}(z)) = \overline{a}_z$. Def. 3.2(ii) asks further that $supp(\hat{f}(z))$ equals

$$\bigcup_{x \in dom(\hat{f}(z))} supp(\hat{f}(z)(x)) - supp(x) \ . \tag{12}$$

We start by showing that $\overline{a}_z$ equals (12). Taking $a \in \overline{a}_z$, there exists $x \in X$ such that $x \# z$ and $a \in supp(f(z,x)) - supp(x)$. Then $\hat{f}(z)(x) = f(z,x)$, so a is in (12). Conversely, take a in (12). There exists $x \in dom(\hat{f}(z))$ such that $a \in supp(\hat{f}(z)(x)) - supp(x)$. Now $\hat{f}(z)(x) = f((\vec{a}\,\vec{a}') \cdot z, x)$, and $a \# x$ implies $a \# \vec{a}$, and f is equivariant, so we can apply $(\vec{a}\,\vec{a}')$ to both sides of $a \in supp(\hat{f}(z)(x)) - supp(x)$ to yield $a \in supp(f(z,(\vec{a}\,\vec{a}') \cdot x)) - supp((\vec{a}\,\vec{a}') \cdot x)$. But $(\vec{a}\,\vec{a}') \cdot x \# z$, so $a \in \overline{a}_z$ by (9).

By Lem 3.3 we need only now show (by contrapositive) that $supp(\hat{f}(z)) \subseteq \overline{a}_z$. The equivariance of $\hat{f}$ and Lem. 2.10(ii) tell us that $a \# z$ implies $a \# \hat{f}(z)$, so we need only consider $a \in supp(z) - \overline{a}_z$. We will show that, given fresh atom a',

$$\hat{f}(z) = (a\,a') \cdot \hat{f}(z)\,.$$

First, convergence: for any $x \in X$, $\hat{f}(z)(x)\downarrow$ iff $\overline{a}_z \# x$ by (10), iff $\overline{a}_z \# (a\,a') \cdot x$ because $a, a' \# \overline{a}_z$, iff $\hat{f}(z)((a\,a') \cdot x)\downarrow$, iff $((a\,a') \cdot \hat{f}(z))(x)\downarrow$ by (6).

Now, taking any x on which they converge, we will show that $\hat{f}(z)(x) = ((a\,a') \cdot \hat{f}(z))(x)$. By $\hat{f}$'s equivariance and (10), this asks that

$$f((\vec{a}\,\vec{a}') \cdot z, x) = f((\vec{b}\,\vec{b}')(a\,a') \cdot z, x) \tag{13}$$

where $\vec{a}$ orders $supp(z) \cap supp(x)$, $\vec{b}$ orders $supp((a\,a') \cdot z) \cap supp(x)$, and $\vec{a}', \vec{b}'$ are chosen fresh. If $a \in supp(z) \cap supp(x)$ then $(\vec{a}\,\vec{a}')$ and $(\vec{b}\,\vec{b}')(a\,a')$ coincide (given an appropriate choice of fresh variables), so (13) holds. Otherwise, say $a \# x$, in which case $\vec{a} = \vec{b}$. Now because $a \notin \overline{a}_z$ and $x \in dom(\hat{f}(z))$, we have $a \notin supp(f((\vec{a}\,\vec{a}') \cdot z, x)) - supp(x)$ by (9), and so $a \# f((\vec{a}\,\vec{a}') \cdot z, x)$. Hence $f((\vec{a}\,\vec{a}') \cdot z, x) = (a\,a') \cdot f((\vec{a}\,\vec{a}') \cdot z, x) = f((a\,a')(\vec{a}\,\vec{a}') \cdot z, (a\,a') \cdot x)$. This is the right hand side of (13) because $a \# (x, \vec{a})$.

Finally, confirmation that $\hat{f}$ is **uniquely determined** is routine.

Corollary 3.7. *$X \twoheadrightarrow\!\!*\,$- extends to a bifunctor -$\twoheadrightarrow\!\!*$- : $\mathcal{N}om^{op} \times \mathcal{N}om \to \mathcal{N}om$.*

Proof. Given equivariant $g : X' \to X, h : Y \to Y'$, apply the adjunction to

$$(X \twoheadrightarrow\!\!* Y) \otimes X' \xrightarrow{id \otimes g} (X \twoheadrightarrow\!\!* Y) \otimes X \xrightarrow{\varepsilon_Y} Y \xrightarrow{h} Y'$$

Theorem 3.8. *$X \twoheadrightarrow\!\!*$ - has no right adjoint in general.*

Proof. It suffices to find some nominal set X, and some colimit in $\mathcal{N}om$, such that this colimit is not preserved by $X \twoheadrightarrow\!\!*$-. Now $\mathbb{A}^2 \twoheadrightarrow\!\!* 1 \cong 1$, so $(\mathbb{A}^2 \twoheadrightarrow\!\!* 1) + (\mathbb{A}^2 \twoheadrightarrow\!\!* 1)$ has two elements. $\mathbb{A}^2 \twoheadrightarrow\!\!* (1 + 1)$, on the other hand, contains four elements: the maps $(a, b) \mapsto (\bullet, i)$ for $i = 1$ or 2, the map

$$(a, b) \mapsto \begin{cases} (\bullet, 1) & \text{if } a = b \\ (\bullet, 2) & \text{otherwise} \end{cases}$$

and its converse.

4 Generalised Abstraction

Lemma 4.1. *Given nominal sets X, Y, we can define an equivalence relation $\sim$ on $X \times Y$ by setting, for all permutations $\pi \# (supp(y) - supp(x))$,*

$$(x, y) \sim \pi \cdot (x, y)$$

recalling that $\pi \cdot (x, y) = (\pi \cdot x, \pi \cdot y)$.

Proof. For reflexivity, set $\pi = \iota$. For symmetry, $(\pi \cdot x, \pi \cdot y) \sim \pi^{-1} \cdot (\pi \cdot x, \pi \cdot y)$ by the equivariance of *supp* and set minus. Transitivity is similarly straightforward.

Definition 4.2. *Write $X \times Y$ modulo $\sim$ as $[X]Y$. Call its members the X-*abstractions on Y. *Write the equivalence class containing (x, y) as $\langle x \rangle y$. Call the construction, as X ranges across all nominal sets,* generalised abstraction.

Lemma 4.3. *$[X]Y$ is a nominal set under the action $\pi \cdot \langle x \rangle y = \langle \pi \cdot x \rangle (\pi \cdot y)$, and*

$$supp(\langle x \rangle y) = supp(y) - supp(x)$$

Proof. Standard nominal techniques.

Example 4.4. (i) $[\mathbb{A}]X$ is the familiar notion of atom-abstraction from Def. 2.11. For example, $[\mathbb{A}]\mathbb{A}$ contains the emptily supported element $\langle a \rangle a = \langle b \rangle b = \cdots$ and, for each atom a, the element $\langle b \rangle a = \langle c \rangle a = \cdots$ for $b, c, \ldots \neq a$, supported by $\{a\}$. Compare with Ex. 3.5(ii).

(ii) $[\mathbb{A} + 1]\mathbb{A}$ contains all the elements of $[\mathbb{A}]\mathbb{A}$ plus, for each $a \in \mathbb{A}$, the element $\langle \bullet \rangle a$ supported by $\{a\}$. Contrast with Ex. 3.5(iii), which had no emptily supported element.

Theorem 4.5. *Generalised abstraction extends to a functor $[X]\text{-} : \mathcal{N}om \to \mathcal{N}om$ for any nominal set X, with the adjoint property*

$$[X]\text{-} \dashv \text{-}_{[X]}$$

where $\text{-}_{[X]}$ is the freshening function *functor of Def. 2.13*

Proof. In the case $X = \mathbb{A}$, the proof is [8, Thm. 9.6.6], although the language of adjunctions is not explicitly used there. The general situation [17] holds similarly: the bijection $\mathcal{N}om(Y, Z_{[X]}) \cong \mathcal{N}om([X]Y, X)$ is defined by mapping equivariant $f : Y \to Z_{[X]}$ to $g : [X]Y \to Z$, where $g(\langle x \rangle y) = f(y)(x)$. The unit of the adjunction has components $\eta_Y : Y \to ([X]Y)_{[X]}$ defined by $\eta_Y(y)(x) = \langle x \rangle y$.

Theorem 4.6. *$[X]-$ has no left adjoint in general.*

Proof. It suffices to find some limit in $\mathcal{N}om$ not preserved by $[X]$- for some X. $[\mathbb{A}^2]1$ has two distinct objects:

(i) $\langle (a, a) \rangle \bullet$;
(ii) $\langle (a, b) \rangle \bullet$ (where $a \neq b$).

They are not equal because there is no permutation π such that $\pi \cdot ((a,a),\bullet) = ((a,b),\bullet)$. $[\mathbb{A}^2]1$ is therefore not terminal.

The next theorem gives a novel description of generalised abstractions as (isomorphic to) certain nominal subsets of finite atom-set abstractions.

Theorem 4.7. *Given nominal sets X, Y, $[X]Y$ is the equaliser of functions $\top, f$ from $[\mathcal{P}_{fin}(\mathbb{A})](X \times Y)$ to $2 = \{\top, \bot\}$, defined as:*

(i) The constant function $\top$;

(ii) $f(\langle \overline{a} \rangle (x,y)) = \begin{cases} \top & \textit{if } supp(x) = \overline{a} \\ \bot & \textit{otherwise.} \end{cases}$

Proof. The forgetful functor $\mathcal{N}om \to \mathcal{S}et$ reflects equalisers, so we need only observe that $[X]Y$ is isomorphic to the usual equaliser via the bijective map $\langle x \rangle y \mapsto \langle supp(x) \rangle (x,y)$.

5 When Do Separating Functions and Generalised Abstraction Coincide?

From Exs. 3.5(iii) and 4.4(ii) and Thms. 3.8 and 4.6, we see that $X \twoheadrightarrow\!\!\ast\, Y$ and $[X]Y$ are not always isomorphic. Thm. 5.5 will specify when they do coincide.

Definition 5.1. *A nominal set X is* transitive *if for all $x, y \in X$, there exists $\pi \in \mathsf{Perm}$ such that $\pi \cdot x = y$.*
X is strong[1] *if, for all $\pi \in \mathsf{Perm}$, $\pi \cdot x = x$ implies $\pi \# x$.*

Example 5.2. (i) $\mathbb{A}$ is transitive and strong.
(ii) $\mathcal{P}_2(\mathbb{A})$, the nominal set of unordered pairs of atoms with the element-wise Perm-action, is transitive but not strong: $(a\ b) \cdot \{a,b\} = \{a,b\}$, but it is not the case that $(a\ b) \# \{a,b\}$.
(iii) $\mathbb{A}^2$ is strong but not transitive: (a,a) and (a,b), where $a \neq b$, occupy different orbits.
(iv) $\mathcal{P}_{fin}(\mathbb{A})$ is neither transitive nor strong.

Lemma 5.3. *Say $X \twoheadrightarrow\!\!\ast\, Y \cong [X]Y$ for all Y. Then*

(i) X is non-empty;
(ii) X is transitive;
(iii) X is strong.

Proof. (i) $\emptyset \twoheadrightarrow\!\!\ast\, Y$ contains the empty function, while $[\emptyset]Y$ is empty always.
(ii) Say we have $x, x' \in X$ with no permutation between them. $\langle x \rangle \bullet$ and $\langle x' \rangle \bullet$ are therefore distinct elements of $[X]1$, which then differs from $X \twoheadrightarrow\!\!\ast\, 1 \cong 1$.

[1] This property was introduced by [19] and called *essentially simple*; we use the more widely used *strong* from [23].

(iii) Say X is not strong, so we have $\pi \cdot x = x$ with $supp(\pi) \cap supp(x) \neq \emptyset$. Say $supp(x) - supp(\pi)$ has n elements, and suppose for contradiction that there existed a total equivariant function $f : X \to \mathbb{A}^{(n+1)}$. $supp(f(x)) \subseteq supp(x)$ by Lem. 2.10(ii), and because $supp(f(x))$ has $n+1$ elements it is too big to contain just $supp(x) - supp(\pi)$; it must contain some $a \in supp(x) \cap supp(\pi)$. Now $\pi \cdot f(x) = f(\pi \cdot x) = f(x)$, but this Perm-action is just the element-wise action on lists, so $\pi(a) = a$. This contradicts $a \in supp(\pi)$, so there are no equivariant functions $X \to \mathbb{A}^{(n+1)}$. But the total equivariant functions are the only emptily supported elements of separating function spaces, so $X \twoheadrightarrow\!\!* \,\mathbb{A}^{(n+1)}$ has no emptily supported elements.

Conversely, let $\vec{a}$ be an ordering of $(supp(x) - supp(\pi)) \cup \{a\}$, where $a \in supp(\pi) \cap supp(x)$. Then $\langle x \rangle \vec{a}$ is an emptily supported element of $[X](\mathbb{A}^{(n+1)})$ by Lem. 4.3.

Lemma 5.4. *Say X is transitive and strong. Then we can define an equivariant function, called* generalised concretion, $[X]Y \otimes X \to Y$ *sending* $(\langle x \rangle y, \pi \cdot x) \mapsto \pi \cdot y$, *where* $\pi \# \langle x \rangle y$.

Proof. We first check that for any $(\langle x \rangle y, x') \in [X]Y \otimes X$ there exists such a permutation π. Because X is transitive there exists a permutation π' such that $\pi' \cdot x = x'$. Let $\vec{a}$ order $supp(\pi') \cap supp(\langle x \rangle y)$, so $\vec{a} \# (x, x')$, and let $\vec{a}'$ be a fresh copy. Then $(\vec{a}\ \vec{a}')\pi'(\vec{a}\ \vec{a}') \cdot x = (\vec{a}\ \vec{a}')\pi' \cdot x = (\vec{a}\ \vec{a}') \cdot x' = x'$. For any $a \in supp(\langle x \rangle y)$ either $a \# \pi'$ or $a \in supp(\vec{a})$; either way $a \# (\vec{a}\ \vec{a}')\pi'(\vec{a}\ \vec{a}')$.

Concretion does not depend on choice of permutation: say $(\pi, \pi') \# \langle x \rangle y$ and $\pi \cdot x = \pi' \cdot x$. Then $\pi^{-1}\pi' \cdot x = x$, and so, because X is strong, $\pi^{-1}\pi' \# x$. But $supp(\pi^{-1}\pi') \subseteq supp(\pi, \pi')$, and this is fresh for $\langle x \rangle y$, so $\pi^{-1}\pi' \# y$, so $\pi \cdot y = \pi' \cdot y$. Concretion also does not depend on choice of representative: say $(x, y) \sim \pi \cdot (x, y)$, and $(\langle x \rangle y, \pi' \cdot x) \mapsto \pi' \cdot y$. Applying π to both sides of $\pi^{-1}\pi' \# \langle x \rangle y$ yields $\pi'\pi^{-1} \# \pi \cdot \langle x \rangle y$, and so concretion maps $(\pi \cdot \langle x \rangle y, \pi'\pi^{-1} \cdot (\pi \cdot x)) \mapsto \pi'\pi^{-1} \cdot (\pi \cdot y) = \pi' \cdot y$. Concretion is then well-defined; equivariance is straightforward.

Theorem 5.5. $X \twoheadrightarrow\!\!* \,Y \cong [X]Y$ *iff X is non-empty, transitive and strong.*

Proof. The left-to-right direction is Lem. 5.3. Conversely, the map $g : [X]Y \to (X \twoheadrightarrow\!\!* \,Y)$ is got by applying the adjunction of Thm. 3.6 to the generalised concretion of Lem. 5.4. Unpacking the adjunction, $\overline{a}_{\langle x \rangle y} = supp(\langle x \rangle y)$, and so

$$g(\langle x \rangle y)(\pi \cdot x) = \begin{cases} \pi \cdot y & \text{if } \pi \cdot x \# \langle x \rangle y \\ \uparrow & \text{otherwise.} \end{cases} \qquad (14)$$

The converse map $h : (X \twoheadrightarrow\!\!* \,Y) \to [X]Y$ is $h(f) \triangleq \langle x \rangle (f(x))$ for $x \# f$. Such an x exists because X is non-empty. Equivariance of h is easy, but we must confirm that h does not depend on our choice of x: given $x, x' \# f$ there exists π such that $\pi \# f$ and $\pi \cdot x = x'$. Then $\langle x' \rangle (f(x')) = \pi \cdot \langle x \rangle (f(x))$, which equals $\langle x \rangle f(x)$ because $\pi \# f$ implies $\pi \# supp(f(x)) - supp(x)$ by Lem. 3.3.

$h \circ g(\langle x \rangle y) = \langle \pi \cdot x \rangle (g(\langle x \rangle y)(\pi \cdot x))$ for $\pi \cdot x \# \langle x \rangle y$ and, without loss of generality, $\pi \# \langle x \rangle y$. This yields $\pi \cdot (\langle x \rangle y) = \langle x \rangle y$. Conversely,

$(g \circ h(f))(x) = g(\langle \pi \cdot x \rangle f(\pi \cdot x))(x)$ for $\pi \cdot x \# f$. Say $x \# f$; then by Def. 3.2, $x \# supp(f(\pi \cdot x)) - supp(\pi \cdot x)$, so $x \# \langle \pi \cdot x \rangle f(\pi \cdot x)$. Hence by (14), $(g \circ h(f))(x) = \pi^{-1} \cdot f(\pi \cdot x)$. But without loss of generality $\pi \# f$, so $\pi^{-1} \cdot f(\pi \cdot x) = f(x)$. If x is not fresh for f then $(g \circ h(f))(x) \uparrow$ as required.

Remark 5.6. The proof given under Thm. 5.5 is presented in more abstract form by [19, Prop. 10.3.7], which shows that non-empty, transitive, strong nominal sets - called there *name-like objects* - produce a 'category with binding structure'. The converse (here, Lem. 5.3) is, however, new to this paper. We also note that [19, Lem. 10.3.6] gives a succinct description of name-like objects in $\mathcal{N}om$: a nominal set is non-empty, transitive and strong iff it is isomorphic to $\mathbb{A}^{(n)}$ for $n \geq 0$. This gives an easy criterion for which our generalisations of name binding coincide, so that they may be constructed like a pair and destructed like a partial function.

6 Applications and Further Work

The logic of Bunched Implications (BI). The categorial structure specified in Sec. 2, and by Thm. 3.6, makes $\mathcal{N}om$ a bi-cartesian doubly closed category, and hence a model of BI [18]. BI is a logic where additive intuitionistic logic sits alongside multiplicative (substructural) intuitionistic logic. The additive logic is interpreted by cartesian closure and finite coproducts, while the multiplicative logic is interpreted by $\otimes$, $-\!\!*$, and $\otimes$'s identity. Now $\otimes$ has as identity the terminal object 1, which makes $\mathcal{N}om$ a model of *affine* BI, where we have *weakening* via canonical 'projection' functions $X \leftarrow X \otimes Y \rightarrow Y$, but do not have *contraction*, as for example there is no equivariant function $\mathbb{A} \rightarrow \mathbb{A} \otimes \mathbb{A}$.

We have thus described an appealingly concrete model for (affine) BI. It is, in fact, very close to the functor category $\mathcal{S}et^{\mathcal{I}}$ discussed in [18, Sec. 9.3], where $\mathcal{I}$ is the category of finite sets (without loss of generality, finite sets of atoms), and injections between them. We can think of such a functor as mapping each finite $\overline{a} \subseteq \mathbb{A}$ to the subset of elements it supports. $\mathcal{N}om$ is known to be equivalent to the category of *pullback-preserving* functors $\mathcal{I} \rightarrow \mathcal{S}et$, called the *Schanuel topos*. Pullback preservation means that elements' supports are closed under intersection, necessary for the definition of freshness. The inclusion functor *cell* : $\mathcal{I} \rightarrow \mathcal{S}et$ of [18, Sec. 9.3] then corresponds to the nominal set of atoms $\mathbb{A}$. We hope that the concrete constructions of this paper will facilitate the application of BI to reasoning about names as resources, continuing the work of [20].

$\mathcal{N}om$-enriched Categories. $\mathcal{N}om$ has two monoidal products $\times, \otimes$ with which it might make sense to explore *enriched category theory* [14]. For example, it is $\otimes$ that is used in [7, Sec. 5] to define their 'freshness environments'. As both these monoidal products are closed, we have two different notions of *internal hom* induced, for which we now have two concrete descriptions.

Nominal Isabelle. The Nominal Isabelle package for interactive theorem proving over abstract syntax now supports various notions of generalised binding [24]. Perhaps surprisingly, this work is not explicitly based on the established notion of generalised abstraction. Nonetheless it is clear that [24]'s set-binding and list-binding are the abstractions of $\mathcal{P}_{fin}(\mathbb{A})$, and the nominal set $\mathbb{A}^*$ of finite lists of atoms, respectively. It is hoped this paper will bring the concept of generalised abstraction back into view, and that our new Thm. 4.7 will help unify this notion. Nominal Isabelle's notion of 'set+-binding' goes further than $\mathcal{P}_{fin}(\mathbb{A})$-abstraction by ignoring 'vacuous binders', so for example $\langle\{a\}\rangle b \approx_\alpha \langle\emptyset\rangle b$ for $a \neq b$. This is a quotient of the $\mathcal{P}_{fin}(\mathbb{A})$-abstraction.

The Pure Pattern Calculus and Other Pattern Binding. [12] presents a formal calculus for pattern matching[2], with term syntax

$$t ::= a \mid tt \mid t \rightarrow_{\overline{a}} t$$

where $a \in \mathbb{A}$ and $\overline{a} \in \mathcal{P}_{fin}(\mathbb{A})$. We call a term $p \rightarrow_{\overline{a}} t$ a *case*, linking the *pattern* p to the *body* t via the *binding variables* $\overline{a}$. Free atoms are defined by

$$fa(a) \triangleq a, \quad fa(t\,u) \triangleq fa(t) \cup fa(u), \quad fa(p \rightarrow_{\overline{a}} t) \triangleq (fa(p) \cup fa(t)) - \overline{a} \ . \tag{15}$$

α-conversion is the congruence generated by

$$p \rightarrow_{\overline{a}} t \approx_\alpha \{a \leftarrow b\}p \rightarrow_{\{a \leftarrow b\}\overline{a}} \{a \leftarrow b\}t$$

where $b \notin fa(p) \cup fa(t) \cup \overline{a}$ and where $\{a \leftarrow b\}$ is the substitution of b for a. As b is chosen fresh we can use permutations $(a\ b)$ instead of substitutions, and hence prove that the nominal set of pure pattern calculus terms quotiented by α-equivalence is the initial algebra for this endofunctor on $\mathcal{N}om$:

$$F \triangleq \mathbb{A} + (\text{-} \times \text{-}) + [\mathcal{P}_{fin}(\mathbb{A})](\text{-} \times \text{-}) \ .$$

Other inductive definitions in [12], such as the free atom function (15) above, are also F-algebras, with their actions on terms defined in the usual way as the unique homomorphism from the initial algebra. The final coalgebra for F gives, as we would expect, a sensible notion of infinitary terms (following e.g. [11]).

This calculus also offers a convenient piece of syntactic sugar:

$$p \rightarrow t \triangleq p \rightarrow_{fa(p)} t \ .$$

We could ask what would happen if we took this sort of construction as basic, rather than sugar, as is quite common - see for example the ρ-calculus [2]. As [1] says, "binding all of the distinct names of a term in another term seems to be a common enough case to deserve special attention and notation". Generalised abstraction provides exactly this notion. However induction over this construction becomes problematic. We would like to have a functor that acts on nominal sets by

$$GX = [X]X \ .$$

[2] A more recent formulation of pattern calculus [13] makes a distinction between *variables* and *matchables* which adds complexity we do not attempt to discuss here.

However it is not clear to us that such a bifunctor [-]- can be defined on the arrows of $\mathcal{N}om$; we therefore do not have a generalised abstraction variant of Cor. 3.7. The naive definition $G\,f(\langle x\rangle x') = \langle f(x)\rangle f(x')$ fails because equivariant functions can shrink supports of the abstracted elements. For example, given $\bullet{+}id : \mathbb{A}{+}\mathbb{A} \to 1{+}\mathbb{A}$, we would have $G(\bullet{+}id)$ mapping $\langle(a,1)\rangle(a,2) \mapsto \langle\bullet\rangle a$, and this sends an emptily supported element to an non-emptily supported element, violating Lem. 2.10(ii). Hence even though standard category theory gives us a notion of induction on $\mathcal{P}_{fin}(\mathbb{A})(Tm \times Tm)$, where Tm is the set of terms modulo α-equivalence, we do not yet have such a notion for its nominal subset $[Tm]Tm$. Finding a robust induction principle for such constructions is a topic for future research.

References

1. Cheney, J.: Towards a general theory of names, binding and scope. In: MERLIN, pp. 33–40. ACM (2005)
2. Cirstea, H., Kirchner, C.: The rewriting calculus - part I. Log. J. IGPL 9(3), 339–375 (2001)
3. Clouston, R.: Equational Logic for Names and Binders. Ph.D. thesis, University of Cambridge (2009)
4. Clouston, R., Pitts, A.M.: Nominal equational logic. ENTCS 172, 223–257 (2007)
5. Day, B.: On closed categories of functors. Lecture Notes in Math. 137, 1–38 (1970)
6. Dowek, G., Gabbay, M.J.: From nominal sets binding to functions and λ-abstraction: connecting the logic of permutation models with the logic of functions, arXiv (2011)
7. Fiore, M., Hur, C.K.: Term equational systems and logics. In: MFPS. ENTCS, vol. 218, pp. 171–192 (2008)
8. Gabbay, M.J.: A Theory of Inductive Definitions with Alpha-Equivalence. Ph.D. thesis, Cambridge University (2001)
9. Gabbay, M.J.: FM-HOL, a higher-order theory of names. In: 35 Years of Automath (2002)
10. Gabbay, M.J., Pitts, A.M.: A new approach to abstract syntax with variable binding. Formal Aspects Comput. 13, 341–363 (2002)
11. Jacobs, B., Rutten, J.: A tutorial on (co)algebras and (co)induction. BEATCS 62, 222–259 (1997)
12. Jay, B., Kesner, D.: Pure Pattern Calculus. In: Sestoft, P. (ed.) ESOP 2006. LNCS, vol. 3924, pp. 100–114. Springer, Heidelberg (2006)
13. Jay, B., Kesner, D.: First-class patterns. J. Funct. Program. 19, 191–225 (2009)
14. Kelly, G.M.: Basic concepts of enriched category theory, LMS Lecture Note Series, vol. 64. Cambridge University Press (1982)
15. Menni, M.: About И-quantifiers. Appl. Categor. Struct. 11(5), 421–445 (2003)
16. Pitts, A.M.: Nominal Logic: A First Order Theory of Names and Binding. In: Kobayashi, N., Babu, C. S. (eds.) TACS 2001. LNCS, vol. 2215, pp. 219–242. Springer, Heidelberg (2001)
17. Pitts, A.M.: Nominal sets, the metamathematics of names (2011) (unpublished manuscript)
18. Pym, D.J.: The Semantics and Proof Theory of the Logic of Buunched Implications. Applied Logic Series, vol. 26. Kluwer Academic Publishers (2002)

19. Schöpp, U.: Names and Binding in Type Theory. Ph.D. thesis, University of Edinburgh (2006)
20. Schöpp, U., Stark, I.: A Dependent Type Theory with Names and Binding. In: Marcinkowski, J., Tarlecki, A. (eds.) CSL 2004. LNCS, vol. 3210, pp. 235–249. Springer, Heidelberg (2004)
21. Sewell, P., Nardelli, F.Z., Owens, S., Peskine, G., Ridge, T., Sarkar, S., Strnisa, R.: Ott: Effective tool support for the working semanticist. J. Funct. Program. 20(1), 71–122 (2010)
22. Shinwell, M.R., Pitts, A.M., Gabbay, M.J.: FreshML: Programming with binders made simple. In: ICFP. SIGPLAN Notices, vol. 38, pp. 263–274 (2003)
23. Tzevelekos, N.: Nominal Game Semantics. Ph.D. thesis, University of Oxford (2008)
24. Urban, C., Kaliszyk, C.: General Bindings and Alpha-Equivalence in Nominal Isabelle. In: Barthe, G. (ed.) ESOP 2011. LNCS, vol. 6602, pp. 480–500. Springer, Heidelberg (2011)

Author Index